17th International Conference of the International Building Performance Simulation Association (BS 2021)

Bruges, Belgium
1 – 3 September 2021

ISBN: 978-1-7138-4275-0

Printed from e-media with permission by:

Curran Associates, Inc.
57 Morehouse Lane
Red Hook, NY 12571

Some format issues inherent in the e-media version may also appear in this print version.

This is a preliminary printing of the Building Simulation 2017 Proceedings and is produced by kind permission of the International Building Performance Simulation Association (IBPSA)

Printed with permission by Curran Associates, Inc. (2022)

For permission requests, please contact International Building Performance Simulation Association (IBPSA) at the address below.

International Building Performance Simulation Association (IBPSA)
c/o Malcolm Cook
Loughborough University-Civil Building
Leics LE11 3TU
United Kingdom

malcolm.cook@lboro.ac.uk

Additional copies of this publication are available from:

Curran Associates, Inc.
57 Morehouse Lane
Red Hook, NY 12571 USA
Phone: 845-758-0400
Fax: 845-758-2633
Email: curran@proceedings.com
Web: www.proceedings.com

17TH INTERNATIONAL IBPSA BUILDING SIMULATION CONFERENCE

Book of Abstracts

KU LEUVEN GHENT UNIVERSITY IBPSA NVL boydens engineering part of Sweco daidalos peutz

Program schedule

	Tuesday 31/8 - Belfry -	IMPACT DAY Wednesday 1/9 - Concert Hall-	Thursday 2/9 - Belfry & Crowne Plaza-	Friday 3/9 - Belfry -
8:00				
8:30		Opening Ceremony	Parallel Sessions	Parallel Sessions
9:00	Workshops	Keynotes		
9:30				
10:00		Break	Break	Break
10:30		Parallel Sessions	Parallel Sessions	Parallel Sessions
11:00				
11:30				
12:00		Lunch	Lunch	Lunch
12:30				
13:00		Parallel Sessions		
13:30			Parallel Sessions	Parallel Sessions
14:00				
14:30		Parallel Sessions		
15:00			Parallel Sessions	Parallel Sessions
15:30				
16:00		Break		
16:30		IBPSA Fellows and Awards	Break	Closing Session
17:00		Keynote Student Competition Award	Parallel Sessions	
17:30				A Last Toast
18:00		Thoughts of Industry		
18:30		Break		
19:00		Keynotes		
19:30	Keynote		Conference Dinner *@ De Halve Maan*	
20:00	Music Competition Award	Closing Food for Thoughts		
20:30		Food for Body *Networking Dinner*		
21:00	Welcome Reception			
21:30		Food for Soul *Concert*		
22:00				
22:30				

Detailed Program

Tuesday, 31/Aug/2021

8:00 - 9:00
Opening registration desk
Location: **Cityhall (Belfry)**

The registration desk will open at 08:00, and will remain open until 19:30.

9:00 - 10:30
Workshop 2: Urban Energy Lab 4.0 - An integrated step towards a sustainable future of city districts
Location: **Cityhall (Belfry) - Room 2**
Chair: **Christian Vering**, RWTH Aachen University
Chair : **Dirk Müller**, RWTH Aachen

10:30 - 12:30
Workshop 4: Introduction to statistical learning - application to building energy data analysis
Location: **Cityhall (Belfry) - Room 3**
Chair: **Simon Rouchier**, Université Savoie Mont Blanc

11:00 - 12:30
Workshop 3: SIM-VICUS - A new and open source approach for dynamic Building Energy Performance Simulation
Location: **Cityhall (Belfry) - Room 2**
Chair: **Stephan Hirth**, TU Dresden

13:30 - 17:00
Workshop 1: Introduction to the BOPTEST framework for simulation-based benchmarking of advanced controllers
Location: **Cityhall (Belfry) - Room 1**
Chair: **Javier Arroyo**, KU Leuven

14:00 - 17:00
Technical tour: Lam Gods: Technical tour: Lam Gods
Location: **Sint-Baafs Cathedral**

19:30 - 20:15
Keynote Ardeshir Mahdavi - The Sound of Space: Reflections on Architecture, Acoustics, and Music
Location: **Cityhall (Belfry)**
Chair: **Dirk Saelens**, KU Leuven / EnergyVille

20:15 - 20:45
Awarding Ceremony for the winner of the Music Competition
Location: **Cityhall (Belfry)**
Chair: **Dirk Saelens**, KU Leuven / EnergyVille

Frank Deleu will talk about the Carillon at the Belfry.
Awarding ceremony for the winner of the music competition.
Wim Berteloot will play the winning composition on the Carillon.

20:45 - 22:00
Welcome Reception
Location: **Cityhall (Belfry)**

Wednesday, 01/Sept/2021

7:00 - 8:30 **Opening registration desk**
Location: **Concert Hall**

The registration desk will open at 07:00, and will remain open until 20:00.

8:30 - 9:00 **Plenary Opening Ceremony**
Location: **Concert Hall - Concertzaal**

Chairs: **Lieve Helsen,** KU Leuven / Energyville and **Wim Boydens**, boydens engineering, part of Sweco / Ghent University
Conference Opening - Food for thoughts
IBPSA and IBPSA NVL Welcome words (Lori McElroy and Dirk Saelens)
Opening Innovation Speech by Minister of Innovation Hilde Crevits

9:00 - 9:15 **Keynote: Thoughts of leading Industry by Martin Dieryckx, General Manager Environmental Research Center Daikin Europe**
Location: **Concert Hall - Concertzaal**

9:15 - 10:00 **Keynote: Climate Change, The European Green Deal and the Building sector by Jos Delbeke - former Director-General of the European Commission's DG Climate Action**
Location: **Concert Hall - Concertzaal**

10:00 - 10:30 **Coffee Break**
Location: **Concert Hall - Foyers**

10:30 - 12:00

Session	Location	Chair(s)
Session W1.1: Practice and industry related case studies	**Concert Hall - Forum 6**	**Mathias Bouquerel**; **Thomas Bockelandt**
Session W1.2: Student Modelling Competition: Can we build a building without HVAC and achieve good comfort in Belgium (like 2226 in Austria)?	**Concert Hall - Artiestenfoyer**	**Elisa Van Kenhove**
Session W1.3: Buildings paving the way for the energy transition	**Concert Hall - Studio 1**	**Wilfried GJHM van Sark**; **Eric Deen**
Session W1.4: Climate change and bioclimatic design	**Concert Hall - Studio 2**	**Georgios Kyriakodis**; **Jérôme Le Dréau**
Session W1.5: The role of occupants	**Concert Hall - Kamermuziekzaal**	**Fabian Ochs**
Session W1.6: Improving indoor environmental quality	**Concert Hall - Concertzaal**	**Hilde Breesch**; **Dimitrov Bolashikov**
Session W1.7 (Online Track): Ensuring high quality building simulations	**Virtual Meeting Room 1**	
Session W1.8 (Online Track): Buildings paving the way for the energy transition	**Virtual Meeting Room 2**	
Session W1.9 (Online Track): Ensuring high quality building simulations/BIM	**Virtual Meeting Room 3**	

12:00 - 13:00 **Lunch**
Location: **Foyers (Concert Hall) and 'het Zand' square outside (front of building)**

13:00 - 14:30

Session	Location	Chair(s)
Session W2.1: Practice and industry related case studies	**Concert Hall - Forum 6**	**Dorota Brzezińska**; **Friedl Decock**
Session W2.2: Ensuring high quality building simulations	**Concert Hall - Artiestenfoyer**	**Laura Carnieletto**; **Frederik Maertens**
Session W2.3: Buildings paving the way for the energy transition	**Concert Hall - Studio 1**	**Alessia Arteconi**; **Alain Heeren**
Session W2.4: Climate change and bioclimatic design	**Concert Hall - Studio 2**	**Shady Attia**; **Yasuyuki Shiraishi**
Session W2.6: The role of occupants	**Concert Hall - Kamermuziekzaal**	**Andrea Gasparella**; **Damien Picard**
Session W2.7 (Online Track): Ensuring high quality building simulations	**Virtual Meeting Room 1**	
Session W2.8 (Online Track): Buildings paving the way for the energy transition	**Virtual Meeting Room 2**	
Session W2.9 (Online Track): Improving indoor environmental quality	**Virtual Meeting Room 3**	

Time							
14:40 - 16:10	**Session W3.1: Practice and industry related case studies** Location: **Concert Hall - Forum 6** Chair: **Jeffrey Spitler** Chair : **Anna Bortkiewicz**	**Session W3.2: Ensuring high quality building simulations** Location: **Concert Hall - Artiestenfoyer** Chair: **Steffen Petersen** Chair : **Kristof Vlieghe**	**Session W3.3: Buildings paving the way for the energy transition** Location: **Concert Hall - Studio 1** Chair: **Dariusz Heim** Chair : **Sorin Comsa**	**Session W3.4: Historical and heritage buildings** Location: **Concert Hall - Studio 2** Chair: **Staf Roels** Chair : **Wouter Van de Walle**	**Session W3.5: Improving indoor environmental quality** Location: **Concert Hall - Kamermuziekzaal** Chair: **Natalia Giraldo Vasquez** Chair : **Jean-Baptiste BOUVENOT**	**Session W3.7 (Online Track): Ensuring high quality building simulations** Location: **Virtual Meeting Room 1**	**Session W3.8 (Online Track): The role of occupants** Location: **Virtual Meeting Room 2**	**Session W3.9 (Online Track): Buildings paving the way for the energy transition** Location: **Virtual Meeting Room 3**

16:10 - 16:40 **Short break**
Location: **Concert Hall - Foyers**

16:40 - 17:10 **IBPSA Fellows & Awards**
Location: **Concert Hall - Concertzaal**
Chair: **Michael Kummert**

17:10 - 18:00 **Keynote: Are "Low-Tech" buildings like the '2226' a solution towards climate neutrality? by Lars Junghans - Associate Professor at the University of Michigan's Taubman College of Architecture and Urban Planning**
Location: **Concert Hall - Concertzaal**

This keynote lecture will be followed by the **Student Modelling Competition Award**.

18:00 - 18:30 **Keynote: Thoughts of leading Industry in the heart of the energy transition, by Leonie Assheuer, EU Affairs Manager at Viessmann**
Location: **Concert Hall - Concertzaal**

18:30 - 19:00 **Short Break**

19:00 - 19:10 **Opening evening session by Dirk De Fauw, Mayor of Bruges**
Location: **Concert Hall - Concertzaal**

19:10 - 19:45 **Keynote: Sustainable architecture is a conscious and sound approach, by Nguyen Hoang Manh - founder of MIA Design Studio, Vietnam**
Location: **Concert Hall - Concertzaal**

19:45 - 20:10 **Keynote: Vision of a leading global engineering company, towards transforming society together, by Andreas Gyllenhammar CSO 'Chief Sustainability Officer' of SWECO group (to be confirmed)**
Location: **Concert Hall - Concertzaal**

20:10 - 21:30 **Food for body - Networking dinner**
Location: **Concert Hall - Foyers**

21:30 - 23:00 **Food for soul - concert: young talents ... promising a bright future ahead**
Location: **Concert Hall**

performed by **Emmy d'Arc, Sam De Nef and band**

Thursday, 02/Sept/2021

8:30 - 10:00	**Session T1.1: Practice and industry related case studies** Location: **Cityhall (Belfry) - Room 1** Chair: **Mathias Bouquerel** Chair : **Jeroen Van der Veken**	**Session T1.2: Ensuring high quality building simulations** Location: **Cityhall (Belfry) - Room 2** Chair: **Thierry Duforestel**	**Session T1.3: Buildings paving the way for the energy transition** Location: **Cityhall (Belfry) - Room 3** Chair: **Wilfried GJHM van Sark**	**Session T1.4: Improving indoor environmental quality** Location: **Cityhall (Belfry) - Room 4** Chair: **Dariusz Heim** Chair : **Hayder Alsaad**	**Session T1.5: Climate change and bioclimatic design** Location: **Crowne Plaza - De Burgh Room** Chair: **Laura Carnieletto**	**Session T1.6: Ensuring high quality building simulations** Location: **Crowne Plaza - Arnulf Room** Chair: **Jérôme Le Dréau**			
10:00 - 10:30	**Coffee Break** Location: **Belfry**								
10:30 - 12:00	**Session T2.1: Practice and industry related case studies** Location: **Cityhall (Belfry) - Room 1** Chair: **Eline Himpe**	**Session T2.2: Ensuring high quality building simulations** Location: **Cityhall (Belfry) - Room 2** Chair: **Frank De Troyer** Chair : **Nils Artiges**	**Session T2.3: Buildings paving the way for the energy transition** Location: **Cityhall (Belfry) - Room 3** Chair: **Alessia Arteconi**	**Session T2.4: Improving indoor environmental quality** Location: **Cityhall (Belfry) - Room 4** Chair: **Catherine Gorle** Chair : **Massimo Palme**	**Session T2.5: Climate change and bioclimatic design** Location: **Crowne Plaza - De Burgh Room** Chair: **Francesco De Luca** Chair : **Georgios Kyriakodis**	**Session T2.6: Ensuring high quality building simulations** Location: **Crowne Plaza - Arnulf Room** Chair: **Jeffrey Spitler**	**Session T2.7 (Online Track): Ensuring high quality building simulations** Location: **Virtual Meeting Room 1**	**Session T2.8 (Online Track): Buildings paving the way for the energy transition** Location: **Virtual Meeting Room 2**	**Session T2.9 (Online Track): Practice and industry related case studies** Location: **Virtual Meeting Room 3**
12:00 - 13:30	**Lunch** Location: **Belfry**								
13:30 - 15:00	**Session T3.1: Practice and industry related case studies** Location: **Cityhall (Belfry) - Room 1** Chair: **Joel Neymark** Chair : **Jacopo Vivian**	**Session T3.2: Ensuring high quality building simulations** Location: **Cityhall (Belfry) - Room 2** Chair: **Dorota Brzezińska**	**Session T3.3: Buildings paving the way for the energy transition** Location: **Cityhall (Belfry) - Room 3** Chair: **James O'Donnell**	**Session T3.4: Improving indoor environmental quality** Location: **Cityhall (Belfry) - Room 4** Chair: **Hilde Breesch** Chair : **Ricardo Forgiarini Rupp**	**Session T3.5: Ensuring high quality building simulations** Location: **Crowne Plaza - De Burgh Room** Chair: **Jelle Laverge** Chair : **Benjamin Kadoch**	**Session T3.6: The role of occupants** Location: **Crowne Plaza - Arnulf Room** Chair: **Romana Markovic** Chair : **Mathieu Schumann**	**Session T3.7 (Online Track): Ensuring high quality building simulations** Location: **Virtual Meeting Room 1**	**Session T3.8 (Online Track): Climate change and bioclimatic design** Location: **Virtual Meeting Room 2**	**Session T3.9 (Online Track): Improving indoor environmental quality** Location: **Virtual Meeting Room 3**
15:00 - 16:30	**Session T4.1: Ensuring high quality**	**Session T4.2: Ensuring high quality**	**Session T4.3: Buildings paving the way**	**Session T4.4: Improving indoor environmental quality**	**Session T4.5: Meet the peers** Location: **Crowne Plaza**	**Session T4.6: The role of occupants** Location:	**Session T4.7 (Online Track): Practice**	**Session T4.8 (Online Track): Climate**	**Session T4.9 (Online Track): Ensuring high quality**

	building simulations Location: **Cityhall (Belfry) - Room 1** Chair: **Agnese Salvati** Chair : **Hicham Johra**	**building simulations** Location: **Cityhall (Belfry) - Room 2** Chair: **Alessandro Dama** Chair : **Jérôme Le Dréau**	**for the energy transition** Location: **Cityhall (Belfry) - Room 3** Chair: **Elli Nikolaidou** Chair : **Vojtech Zavrel**	Location: **Cityhall (Belfry) - Room 4** Chair: **J. Alstan Jakubiec** Chair : **Hayder Alsaad**	**- De Burgh Room** Chair: **Elisa Van Kenhove** **Presenters:** Bjarne Olesen, Lieve Helsen, Catherine Gorle and Jeffrey Spitler	**Crowne Plaza - Arnulf Room** Chair: **Jérôme Le Dréau** Chair : **Chantal Basurto**	**and industry related case studies** Location: **Virtual Meeting Room 1**	**change and bioclimatic design** Location: **Virtual Meeting Room 2**	**building simulations** Location: **Virtual Meeting Room 3**
16:30 - 17:00	**Coffee Break** Location: **Belfry**								
17:00 - 18:30	**Session T5.1: Buildings paving the way for the energy transition** Location: **Cityhall (Belfry) - Room 1** Chair: **James O'Donnell**	**Session T5.2: Ensuring high quality building simulations** Location: **Cityhall (Belfry) - Room 2** Chair: **Andreas Nicolai**	**Session T5.3: Buildings paving the way for the energy transition** Location: **Cityhall (Belfry) - Room 3** Chair: **Christoph Nytsch-Geusen**	**Session T5.4: Improving indoor environmental quality** Location: **Cityhall (Belfry) - Room 4** Chair: **Shady Attia** Chair : **Massimo Palme**	**Session T5.5: Ensuring high quality building simulations** Location: **Crowne Plaza - De Burgh Room** Chair: **Fabian Ochs**	**Session T5.6: Buildings paving the way for the energy transition** Location: **Crowne Plaza - Arnulf Room** Chair: **Catherine Gorle**	**Session T5.7 (Online Track): Ensuring high quality building simulations** Location: **Virtual Meeting Room 1**	**Session T5.8 (Online Track): The role of occupants** Location: **Virtual Meeting Room 2**	**Session T5.9 (Online Track): Ensuring high quality building simulations** Location: **Virtual Meeting Room 3**
18:30 - 19:30	**Free time**								
19:30 - 23:00	**Conference Dinner** Location: **Brewery Halve Maan**								

Friday, 03/Sept/2021

8:30 - 10:00	**Session F1.1: Ensuring high quality building simulations** Location: **Cityhall (Belfry) - Room 1** Chair: **Thierry Duforestel**, EDF R&D	**Session F1.2: Ensuring high quality building simulations** Location: **Cityhall (Belfry) - Room 2** Chair: **Andrea Gasparella**, Free University of Bozen - Bolzano Chair : **Aurelien Bres**, AIT Austrian Institute of Technology	**Session F1.3: Buildings paving the way for the energy transition** Location: **Cityhall (Belfry) - Room 3** Chair: **Eline Himpe**, Ghent University	**Session F1.4: Practice and industry related case studies** Location: **Cityhall (Belfry) - Room 4** Chair: **Romana Markovic**, KIT			
10:00 - 10:30	**Coffee Break** Location: **Cityhall (Belfry)**						
10:30 - 12:00	**Session F2.1: Ensuring high quality building simulations** Location: **Cityhall (Belfry) - Room 1** Chair: **Jacopo Vivian**, University of Padua Chair : **Jelle Laverge**, Ghent University	**Session F2.2: Ensuring high quality building simulations** Location: **Cityhall (Belfry) - Room 2** Chair: **Joel Neymark**, J. Neymark & Associates Chair : **Ádám Bognár**, Eindhoven University of Technology	**Session F2.3: Buildings paving the way for the energy transition** Location: **Cityhall (Belfry) - Room 3** Chair: **Tarek Rakha**, Georgia Institute of Technology Chair : **Yasuyuki Shiraishi**, The University of Kitakyushu	**Session F2.4: The role of occupants** Location: **Cityhall (Belfry) - Room 4** Chair: **Natalia Giraldo Vasquez**, Federal University of Santa Catarina Chair : **Mathieu Schumann**, EDF	**Session F2.5 (Online Track): Practice and industry related case studies** Location: **Virtual Meeting Room 1**	**Session F2.6 (Online Track): Improving indoor environmental quality** Location: **Virtual Meeting Room 2**	**Session F2.7 (Online Track): Buildings paving the way for the energy transition / optimisation /control** Location: **Virtual Meeting Room 3**
12:00 - 13:30	**Lunch** Location: **Belfry**						
13:30 - 15:00	**Session F3.1: Improving indoor environmental quality** Location: **Cityhall (Belfry) - Room 1** Chair: **J. Alstan Jakubiec**, University of Toronto Chair : **Ricardo Forgiarini Rupp**, Technical University of Denmark	**Session F3.2: Ensuring high quality building simulations** Location: **Cityhall (Belfry) - Room 2** Chair: **Andreas Nicolai**, TU Dresden	**Session F3.3: Buildings paving the way for the energy transition** Location: **Cityhall (Belfry) - Room 3** Chair: **Elli Nikolaidou**, University of Bath	**Session F3.4: Climate change and bioclimatic design** Location: **Cityhall (Belfry) - Room 4** Chair: **Vojtech Zavrel**, CTU in Prague, Faculty of Mechanical Engineering Chair : **Sammy Rogmans**, Sweco	**Session F3.5 (Online Track): Practice and industry related case studies** Location: **Virtual Meeting Room 1**	**Session F3.6 (Online Track): Ensuring high quality building simulations** Location: **Virtual Meeting Room 2**	**Session F3.7 (Online Track): The role of occupants** Location: **Virtual Meeting Room 3**
15:00 - 16:30	**Session F4.1: Improving indoor environmental**	**Session F4.2: Ensuring high quality building simulations**	**Session F4.3: Buildings paving the way for the energy**	**Session F4.4: Climate change and bioclimatic design**	**Session F4.5 (Online Track): Practice and industry related**	**Session F4.6 (Online Track): Ensuring high quality building**	**Session F4.7 (Online Track): Improving**

	quality Location: **Cityhall (Belfry) - Room 1** Chair: **Alessandro Dama**, Politecnico di Milano Chair : **Ádám Bognár**, Eindhoven University of Technology	Location: **Cityhall (Belfry) - Room 2** Chair: **Nils Artiges**, Univ. Grenoble Alpes, CNRS, Grenoble INP, G2Elab, F-38000 Grenoble, France	**transition** Location: **Cityhall (Belfry) - Room 3** Chair: **Steffen Petersen**, Aarhus University Chair : **Jeroen Van der Veken**, BBRI	Location: **Cityhall (Belfry) - Room 4** Chair: **Tarek Rakha**, Georgia Institute of Technology Chair : **Chantal Basurto**, Idiap Research Institute	**case studies** Location: **Virtual Meeting Room 1**	**simulations** Location: **Virtual Meeting Room 2**	**indoor environmental quality** Location: **Virtual Meeting Room 1**

16:30 - 17:30

Closing Session with Award Ceremony
Location: **Cityhall (Belfry)**
Chair: **Lieve Helsen**, KU Leuven
Chair : **Wim Boydens**, boydens engineering / UGent

BS2021 Awards (Michael Kummert, Jelle Laverge, Friedl Decock)
BS2021 thanks and final remarks (Paul Strachan, Dirk Saelens, Lieve Helsen, Wim Boydens)
IBPSA Announcements (Paul Strachan)
Adjourn (Lori McElroy)

17:30 - 18:00

Last toast towards a long lasting taste of Bruges
Location: **Cityhall (Belfry)**

Abstracts

Workshops

List of abstracts
(in order by program schedule)

Workshop 2

Paper ID 115: Urban Energy Lab 4.0 - An integrated step towards a sustainable future of city districts

Christian Vering*, Rita Streblow, Dirk Müller

RWTH Aachen University, Germany

The project brings together scientists from E.ON Energy Research Center (E.ON ERC), Centers for Wind Power Drives (CWD) and the Chair of Energy Efficient Building (E3D). The project, coordinated by Prof. Rita Streblow and Prof. Dirk Müller, creates new experimental facilities at the RWTH Aachen. The new infrastructure expands the possibilities to solve current research questions of the energy transition in future cities

Together, the participating scientists want to create a unique and highly networked infrastructure for the conception and analysis of new urban energy supply concepts. Future urban energy systems are characterised by decentralised generation, sector coupling and the new possibilities of digitisation. The planning, construction and operation of components of these energy systems must therefore meet a wide range of requirements that can only be met through interdisciplinary and synergetic interaction between construction technology, technical building equipment, network planning and automation.

The new research infrastructure is a flexible experimental field for controllable experiments from the supply of a room to the energetic observation of a city quarter. New technologies can be tested and sustainable solution concepts for a safe and environmentally friendly urban energy supply can be developed. The necessary scaling from real scale to laboratory scale is done using a hardware-in-the-loop platform. A power supply system with the size of an entire city is replaced in some areas by complex simulations so that real building blocks of the energy system can be operated in the laboratory under realistic boundary conditions. This hardware-in-the-loop approach is unique in this scale and with the planned flexible expandability and its interdisciplinary structure.

The following investments will be made over the next three years for this networked laboratory of a city quarter:

- Test environment for investigating user comfort and user behaviour (Indoor Climate Laboratory subproject)

- Test benches for the inspection of building components (FlexFass subproject),

- Refrigerant laboratory for the electrification of heat supply with heat pumps (subproject HELENA, by Christian Vering)

- High-performance systems for real-time mapping of energy flows in a city quarter (subproject RTLab).

All elements are networked with the help of intelligent automation and supplemented by simulations (subproject InFIS).

Keywords: Sustainable Cities, City in the Lab, Infrastructure design

Workshop 4

Paper ID 277: Introduction to statistical learning - application to building energy data analysis

Simon Rouchier*

Université Savoie Mont Blanc, France

With the increasing availability of data from sensor networks, smart meters and building management systems, and the increasing popularity of data science, building energy researchers now have promising opportunities to monitor and understand the performance of buildings in operation. Applications of data analysis are manifold: predicting energy use; assessing the effectiveness of energy conservation measures; estimating the thermal properties of the envelope or the efficiency of HVAC systems; identifying faults during operation; etc.

Many engineers and researchers however lack the tools for a critical analysis of their results.

This caution is particularly important as the dimensionality of a problem (i.e. the number of

unknown parameters) increases.

This workshops aims at providing a gentle introduction to statistical learning applied to building energy data. After a general description of statistical modelling and the process of data analysis, the participants will be free to browse the individual tutorials provided by the author. These tutorials are notebooks, written in the Python programming language, and made available online. They will be organised in independent categories:

- Prediction of energy use with simple statistical models;
- Bayesian inference with the Stan platform;
- State-space modelling with the PySIP library;
- Introduction to machine learning with the scikit-learn library.

The workshop does not require the purchase of any commercial software. The primary target audience are students and researchers new to the field, but more experienced researchers are welcome as well.

Resources:

The author's page containing the tutorials: https://buildingenergygeeks.org/

The PySIP library for state-space inference and prediction in Python: https://github.com/locie/pySIP

The Stan platform for Bayesian inference: https://mc-stan.org/

The scikit-learn library for machine learning: https://scikit-learn.org/stable/

Link to international framework:

The workshop will show examples of techniques that were used in the frame of the IEA- EBC Annex 71: Building energy performance assessment based on in situ measurements.

Keywords: statistical learning, data, inference, prediction

Workshop 3

Paper ID 11116: The novel dynamic building energy performance simulation tool SIM-VICUS

Stephan Hirth*, Andreas Nicolai

TU Dresden, Germany

(1) scienticts / practitioners / students

(2) In this workshop we introduce the new dynamic Building Energy Performance Simulation tool SIM-VICUS. Therefore two tutorials will be introduced with a complex and simple building geomtry in order to illustrate an efficient workflow.

Keywords: simulation, energy, building, performance, open-source

Workshop 1

Paper ID 11134: Introduction to the BOPTEST framework for simulation-based benchmarking of advanced controllers

Javier Arroyo* (1,2,3), David Blum (4)

1: Department of Mechanical Engineering, KU Leuven, Heverlee, Belgium; 2: EnergyVille, Thor Park, Waterschei, Belgium; 3: Flemish Institute for Technological Research (VITO), Mol, Belgium; 4: Lawrence Berkeley National Laboratory, Berkeley, USA

The Building Optimization Performance Test (BOPTEST) framework is used to evaluate the performance of building HVAC control algorithms by enabling co-simulation between detailed building models and external controllers written by users in separate software. The framework provides a standard simulation environment and is organized into specific building test cases. The detailed building models are implemented in Modelica using libraries developed within the IBPSA Project 1 and packaged as Functional Mockup Units for simulation within the standardized simulation environment. An API for this environment allows users to advance the simulation of the detailed building model with input from their control algorithms, receive data back from the detailed building model, and report key performance indicators, such as total energy use, operating cost, and thermal discomfort, after co-simulation is complete.

The focus of the workshop is to provide a hands-on introduction to the BOPTEST framework and software, which aims to enable benchmarking and facilitate the development of advanced control strategies for buildings. The key objectives are to provide participants with:

- an overview of the BOPTEST project goals and approach

- a guided, hands-on experience with the software

- an opportunity to use the software for their own research and development

- an opportunity to provide feedback for project development

The central parts of the workshop are two hands-on exercises that allow users to interact with BOPTEST software by working in teams. The first exercise will be a guided, introductory tutorial to the software and the second exercise will be an extension that encourages participants to use the software to answer a challenge. Throughout the exercises and associated presentations, there will be ample time for questions and discussion.

The target audience is building simulation users for the purpose of advanced HVAC control algorithm development and testing. The public may be broad, from beginners to advanced control experts.

BOPTEST is being developed within work package 1.2. of IBPSA Project 1. See: https://ibpsa.github.io/project1/ for more information, and https://github.com/ibpsa/project1-boptest for the repository hosting this project.

Keywords: BOPTEST, building simulation, HVAC control, optimal control, benchmarking

Traditional and Interactive Presentations

List of abstracts
(in order by program schedule)

Session W1.1: Practice and industry related case studies

Paper ID 30355: A BIM-based Business Process Model to support LEED® certification in retrofit projects

Letizia D'Angelo* (1,2,3), Magdalena Hajdukiewicz (1,2,3), Federico Seri (1,2,3), Marcus Keane (1,2,3)

1: School of Engineering, College of Science and Engineering, National University of Ireland, Galway; 2: Informatics Research Unit for Sustainable Engineering (IRUSE), National University of Ireland Galway; 3: Ryan Institute, National University of Ireland Galway

AIM AND APPROACH

Energy efficiency plays a vital role in reducing the impact of the buildings' energy costs and avoid greenhouse gas emissions, but the global rate of progress is slowing.

The aim of the EU's Energy Efficiency Directive (EED) is to help citizens and public authorities to better manage their energy consumption. If successful, this will bridge the gap between existing framework directives and national energy efficiency measures.

Within this objective, the LEED certifications may improve the achievement of these results. In fact, LEED certified buildings not only increase the building efficiency, they also add value to the properties and represent an ethical system for sustainability. Furthermore, those certified buildings can qualify for discounted insurance, tax breaks and other incentives.

In this paper, the LEED process will be analysed utilising a methodological approach aimed at defining an innovative Business Process Model (BPM) using the benefits of the BIM.

SCIENTIFIC INNOVATION AND RELEVANCE

The advantages of using a detailed BIM Project Execution Plan (BEP) are well recognised in the marketing sector for achieving a high-quality sustainability rating in a short period of time, but both a planning guide and the benefits have only been demonstrated in new construction projects.

The aim of this research is to create a business process model (BPM) that clearly underline the workflow of retrofitting existing buildings using the benefit of BIM.

In a renovation projects there are many use cases of BIM that have to be considered in the process such as the existing conditions modeling (enhance the accuracy of existing conditions documentation), cost estimation (cost effects of additions and modifications), energy analysis (increase efficiency of structural, electrical and mechanical system) and so on.

This paper will focus on increasing the effectiveness of sustainability goals, the LEED evaluation, so the detailed process map (BPM) of this BIM use has been carried out using the Business Process Modeling and Notations (BPMN) v2.0, an established modeling language able to graphically represent the overall process.

PRELIMINARY RESULTS AND CONCLUSIONS

This research presents an innovative methodological approach aimed at defining a novel BIM based BPM for the acquisition of the credit "Optimize Energy Performance" in the process of a LEED certification for an existing building.

A thermal model of the building "as is" (baseline) has been created using the software IES-VE and has been compared with its retrofitted scenario (proposed). After the insulation of the envelope and the combined use of a more efficient heating/cooling system and renewable sources such as a new photovoltaic system and solar panels the percent savings of the energy use has been the 64%, reaching 18 points out of 18.

MAIN REFERENCES

(1) "BIM Project Execution Planning Guide – Version 2.0." April 16, 2010 The Pennsylvania State University, University Park, PA, USA

(2) ASHRAE (2019) ASHRAE Standard 90.1 - Energy Standard for Buildings Except Low-Rise Residential Buildings.

(3) Building Certification: What Does LEED Really Mean? Available at: https://dfw.cbslocal.com/2013/04/10/building-certification-what-does-leed-really-mean/.

(4) D'Angelo, L. et al. (2019) 'BIM-based Business Process Model to support systematic deep renovation of buildings" doi:10.26868/25222708.2019.210481

(5) LEED, Reference guide for Buidling Design and Construction (2014), U.S. Green Building Council

Keywords: BIM, existing buildings, LEED, business process model

Paper ID 30154: Ecological and economic analysis of low-temperature district heating in typical residential areas

Jan Stock* (1), Dominik Hering (1), André Xhonneux (1), Dirk Müller (1,2)

1: Forschungszentrum Jülich GmbH, Institute of Energy and Climate Research, Energy Systems Engineering (IEK-10), Jülich, Germany; 2: RWTH Aachen University, E.ON Energy Research Center, Institute for Energy Efficient Buildings and Indoor Climate, Aachen, Germany

AIM AND APPROACH

District heating is a key heating supply technology for the building sector. Decreasing the operation temperatures of district heating allows to increase efficiency and to use new sources of heat, such as low temperature waste heat or sustainable energy sources. Low-temperatures district heating networks, also known as district heating networks of the fourth generation, have operation temperatures up to 60°C. [1]

The aim of this investigation is to compare low-temperature district heating for typical residential areas.

We use typical building types and characteristics of typical residential areas in Germany to create dynamic simulation models in MODELICA [2,3]. Based on the simulation results, we evaluate installation and operating costs relative to supplied heat.

We conclude the ecologic and economic investigations with a classification of the most efficient low-temperature district heating integrations in the investigated residential areas.

SCIENTIFIC INNOVATION AND RELEVANCE

Successful heat supply with low-temperature district heating already has been shown for old and modern building types [4,5]. An efficient integration of low-temperature district heating depends on the supplied building types and their specific heat demands and temperature requirements [6].

We present an investigation of typical residential areas with varying building size and age. We use typical building types of the German building stock and combine them to typical residential areas. We use open-source Python tools for automated model generation, which allows us to create multiple dynamic simulation models [7]. For this evaluation, we investigate nine typical residential areas with varying characteristics.

With the results of this investigation, we show the most efficient and beneficial use of a low temperature heating source for building heating supply.

PRELIMINARY RESULTS AND CONCLUSIONS

The results show that an efficient heat supply depends on different building specifications. A network with a 40°C heat source can efficiently supply residential areas with many modern single-family houses. Especially a high amount of direct heat by the network and a high space heating share leads to an efficient use of supplied heat. With a 60°C low-temperature district heating network housing areas with a high number of old buildings can be supplied efficiently. In contrast to that, the supply of a residential area with many modern multi-family houses needs a big amount of additional electricity for heat pump operation to support high temperature requirements. Reference scenarios with air source heat pumps show a higher primary energy consumption than the supply with low-temperature district heating. The cost evaluation of the low-temperature district heating scenarios shows that residential areas with a high heat density have the smallest specific investment costs refer to possible primary energy saving.

MAIN REFERENCES

[1] Lund et al.: 4th Generation District Heating (4GDH). In: Energy 68 (2014), S. 1–11

[2] Loga et al.: Deutsche Wohngebäudetypologie: Beispielhafte Maßnahmen zur Verbesserung der Energieeffizienz von typischen Wohngebäuden ; erarbeitet im Rahmen der EU-Projekte TABULA - "Typology approach for building stock energy assessment". 2., erw. Aufl. Darmstadt : IWU, 2015 http://www.building-typology.eu/downloads/public/docs/brochure/DE_TABULA_TypologyBrochure_IWU.pdf.

– ISBN 9783941140479

[3] Remmen et al.: TEASER: an open tool for urban energy modelling of building

stocks. In: Journal of Building Performance Simulation 11 (2018), Nr. 1, S. 84–98

[4] Brand et al.: Renewable-based low-temperature district heating for existing buildings in various stages of refurbishment. In: Energy 62 (2013), S. 311–319

[5] Schmidtet al.: Development of an Innovative Low Temperature Heat Supply Concept for a New Housing Area. In: Energy Procedia 116 (2017), S. 39–47

[6] Köfinger et al.: Low temperature district heating in Austria: Energetic, ecologic and

economic comparison of four case studies. In: Energy 110 (2016), S. 95–104

[7] Fuchs et al.: Workflow automation for combined modeling of buildings and

district energy systems. In: Energy 117 (2016), S. 478–484

Keywords: low-temperature district heating (LTDH), district heating simulation, decentral heat pumps, residential area, waste heat

Paper ID 31148: Multidisciplinary BIM-model for an integrative design approach: study case of a car dealership building in Brussels

Nassim Jamali*, Pedro Marques, Jolien De Clerck, Wim Boydens

Boydens Engineering, Belgium

AIM AND APPROACH

Elaboration of BIM model permitting a holistic approach of the MEP design.

The automation of workflow enables the increase of iteration of simulations. This leads to high quality optimization of the design.

SCIENTIFIC INNOVATION AND RELEVANCE

• Multidisciplinary approach on MEP design

• Translation from BIM to BEM and other models

• Integrative design using simulations from early-stage

PRELIMINARY RESULTS AND CONCLUSIONS

The iterations of the design brought the project to be Net Zero Energy Building and Carbon Neutral.

MAIN REFERENCES

Boydens Standards

Keywords: BIM, BEM, Multidisciplinary

Paper ID 30608: Facility location problem in thermal network design

Tristan André Rey*, Jessen Page

HES-SO, Switzerland

AIM AND APPROACH

Energy transition requires efficient energy production and distribution systems. For this purpose, a methodology is developed to extract the energy of ground water and to distribute it optimally in a neighborhood.The first step of the methodology consists in the geographical clustering of the buildings, aiming to reduce the complexity of the problem. Then, knowing the energy demand of the neighborhood and potential locations for ground water wells, the resolution of an assignment problem (capacitated facility location splittable problem - CFLS problem) is performed. Each ground water well can be drilled or not and has a specific amount of power available. The goal of the resolution is then to determine which drillings need be executed and which buildings they should supply, while minimizing the total cost of the operation. To do so, this assignment problem is solved using a bespoke solving method and the results are compared to those obtained with a mixed-integer linear programming (MILP) based solving method. The methodology ends with the design of the topology of a thermal network. This design process uses street network data and is based on the resolution of a minimum Steiner tree.

SCIENTIFIC INNOVATION AND RELEVANCE

The temperature increase associated to climate change implies an increasing demand for cooling. This demand needs to be fulfilled in an efficient way and the use of ground water for this purpose becomes more frequent. Ground water can be used for heating or cooling and is thus of prime interest for district networks. However, the introduction of multiple sources complexifies the design process. The resolution of the CFLS problem is well-known by the optimization and operational research community and is applied here to the design of thermal networks, as it meets the expectations required for these designs.

PRELIMINARY RESULTS AND CONCLUSIONS

The applicability of the methodology is validated with a case study. The bespoke solving method generates multiple solutions using a heuristic algorithm while the MILP based solving method generates multiple solutions by the way of integer cuts. Key performance indicators (KPIs) are implemented and an a priori optimal solution is chosen using parallel coordinates for a multi-criterion based decision.

MAIN REFERENCES

- Madhukar R. Korupolu, C. Greg Plaxton, and Rajmohan Rajaraman. “Analysis of a Local Search Heuristic for Facility Location Problems”. In: Journal of Algorithms 37 (2000), pp. 146-188

- Stéphane Joos, François Maréchal, Jérémy Unternährer, Stefano Moret. “Spatial clustering for district heating integration in urban energy systems: application to geothermal energy”. In: Applied energy (2016)

- Luc Girardin, “A GIS-based Methodology for the Evaluation of Integrated Energy Systems in Urban Area”. EPFL, 2012

Keywords: Thermal network design, Capacitated facility location splittable problem, mixed-integer linear programming, minimum Steiner tree

Paper ID 30901: MILP based approach for the preliminary investigation of thermal networks in urban areas

Francesca Belfiore (1,2), Tristan Rey (1), Jessen Page* (1), François Maréchal (2)

1: HES-SO Valais-Wallis, Switzerland; 2: École Polytechnique Fédérale de Lausanne (EPFL), Lausanne, Switzerland

AIM AND APPROACH

Distric Energy Systems (DES) have been widely recognized as crucial actors during the ongoing Energy Transition, demanded by the challenging mid-century emission reduction target. While district heating still represents the most common type of DES, the so-called “anergy network” allows to exploit synergies between heating and cooling sectors representing a promising opportunity to improve energy efficiency in urban areas. In this framework, this study aims at proposing cost optimal solutions for the design and operation of a DES in a neighborhood characterized by a variegate building stock, per year of construction and dwelling affectation. In particular different network types (heating, cooling and anergy) are investigated together with the resulting degree of centralization and respectively decentralization of heat and cold production. The problem is formulated as a Mixed Integer Linear Programming (MILP), with binary variables linked to the selection and scheduling of technologies as well as the existence and type of pipelines, while continuous variables are associated to sizes and operating loads. A set of potential solutions is generated by the means of integer cut constraints on the network design. The impact of the users temperature on the system design is also investigated.

SCIENTIFIC INNOVATION AND RELEVANCE

Distric Energy System (DES) represents a promising opportunity to boost the decarbonisation of the urban sector, promoting system electrification and the integration of more sustainable technologies. To fully exploit the benefits of DES all the components involved in the district must be optimized, from the heat source to the final users, including the distribution network. However, the optimization of such complex systems is a challenging tasks for several reasons: it includes both spatial and temporal aspects, consumption profiles may vary in stochastic manner and it involves a large number of decision variables, both continuous and binaries. This works proposes a MILP framework to investigate different types of networks within the district and the interaction with the energy system at the building level. Both temporal and spatial clustering are employed to reduce the problem size. The network temperature, often assumed constant during a summer/winter operation is also subject to optimization, with either a constant profile or a linear trend with the ambient temperature. Two different approaches to optimize the network temperature levels are compared: the use of a heuristic master level optimization and an integer cut-constraint applied to the MILP.

PRELIMINARY RESULTS AND CONCLUSIONS

Preliminary results show the crucial role of optimization in defining optimum solutions among the many available options: an anergy network coupled with decentralized heat pumps and electrically driven vapour compression chillers, a heating network whose temperature level implies a certain degree of decentralization of heat production and eventually a cooling network. The distribution of the temperature demand among the users, resulting from the diversity of the building stock also affects the set of optimal solutions, suggesting the importance of considering building refurbishment options among the decision variables. Due to the large capital investment of network pipeline, operational constraints such as the maximum allowed velocity play also a role in identifying the cost optimal solutions. In the case of a partially decentralized heating network the overall system efficiency is linked to the coefficient of performance of the central and decentralized heat pumps and the homogeneity of the temperature demanded by the different buildings.

MAIN REFERENCES

REN21. 2020. Renewables 2020 Global Status Report (Paris: REN21 Secretariat). ISBN 978-3-948393-00-7.

IEA (2019). The Critical Role of Buildings. (Paris IEA).

5th generation district heating and cooling systems: A review of existing cases in Europe. Buffa et al. Renewable and Sustainable Energy Reviews, April 2019.

Distributed or centralized? Designing district-level urban energy systems by a hierarchical approach considering demand uncertainties. R. Jing at al. Applied Energy, vol 252, October 2019.

4th Generation District Heating (4GDH): Integrating smart thermal grids into future sustainable energy systems. H. Lund et al. Energy, vol. 68, no. Supplement C, pp. 1–11, Apr. 2014.

Multi-objective, multi-period optimization of district energy systems: Networks design. S. Fazlollahi et al. Computer Aided Chemical Engineering, ser. 23 European Symposium on Computer Aided Process Engineering.

MIP approach for designing heating systems in residential buildings and neighbourhoods. H. Harb at al. Journal of Building Performance Simulation, vol. 9, no. 3, pp. 316–330, May 2016. Available: https://doi.org/10.1080/19401493.2015.1051113

Performance and profitability perspectives of a CO2 based district energy network In Geneva's City Centre. S. Henchoz at al. Energy, vol 85, June 2015.

Keywords: District energy system, Heat pumping integration, District heating and cooling, Mixed integer linear programming

Paper ID 30978: MUSEGRIDS tool: energy performance calculation based on EPBD standards for EU and abroad

Gema Hernández Moral*, Victor Iván Serna González, Francisco Javier Miguel, César Valmaseda Tranque

Fundacion CARTIF, Spain

AIM AND APPROACH

The implementation of energy policies with a view to reduce CO2 emissions poses specific challenges, especially to public authorities, who should define specific objectives, and evaluate the adequacy and impact of energy actions proposed. Nevertheless, an appropriate analysis is highly time-consuming due to the lack of tools. In this process, the first step is to establish the baseline energy status of the area of study. Only with this knowledge, in particular of the residential sector (main CO2 emissions contributor in cities), is it possible to plan for a low carbon economy. In this context, the MUSEGRIDS tool will support energy planners by matching energy needs to energy supply of the building stock at local scale. In this paper, the first step of the approach is explained: the attainment of the energy demand. This is calculated at building level, but then aggregated at other scales. To do so, publicly available data sources, building typologies (such as those coming from TABULA or the Building Stock Observatory), and reference data coming from Open Street Maps and INSPIRE-based data are deployed. This allows to automatically calculate hourly energy demand and consumption values based on the ISO 52000 standards family.

SCIENTIFIC INNOVATION AND RELEVANCE

By basing the proposed tool on Energy Performance Certificates (EPC) calculation standards specified in the Energy Performance Directive of Buildings (EPBD 2018/844/EU), in particular ISO 52016; the MUSEGRIDS tool achieves two main goals: it aids in the implementation of energy directives and energy actions by offering an easy to use tool to identify areas in need and, secondly, it promotes the EPBD, by making use of one of its main instruments to measure energy performance. The energy performance of buildings is calculated at a building scale, but then aggregated at coarser levels of detail, in order to enable different actors to complement their decision-making process in the most appropriate way. This is performed by exploiting publicly available data (from building typologies, open street maps or based on the INSPIRE Directive). Thus, based on this accessible-to-all approach it is argued how it could be replicated in a European context, as well as in a global context, and how the input data can be more or less accurate depending on this scope.

PRELIMINARY RESULTS AND CONCLUSIONS

The developed tool provides three main outputs: (1) GIS file with geo-located information for each building of a municipality in terms of energy needs (cooling and heating demand), DHW, energy per type of fuel, CO2 emissions per year; (2) the same information aggregated for all the building but disaggregated by time (per hour); (3) raster files with the same information mapped in a 100x100m2 grid.

The tool has been evaluated and validated in different ways, and tested in three locations: Osimo (IT), Oud-Heverlee (BE) and Aranda de Duero (ES). In this process, failures in the combination of data sources, mapping, calculation process have been detected and also the accuracy of the results tested based on the comparison with real data.

All in all, even when the tool has the ambition to be applied worldwide, the main barrier is the lack of completeness in the source data: both in terms of geometric and buildings identification data (from OSM and INSPIRE-based) or building characterisation data (e.g in Tabula-episcope for some countries or regions). The systematic and harmonised characterisation of buildings and their typologies would highly benefit this process and contribute to more robust decision-making in the field of energy performance in buildings.

MAIN REFERENCES

Directive 2018/844/EU. European Parliament and the Council of 30 May 2018 "On the Energy Performance of Buildings":https://eur-lex.europa.eu/legal-content/EN/TXT/?uri=uriserv%3AOJ.L_.2018.156.01.0075.01.ENG[retrieved: Jul, 2020]

Howard B, Parshall L, Thompson J, Hammer S, Dickinson J and Modi V 2012 Spatial distribution of urban building energy consumption by end useEnergy Build.45 141–51

Kavgic M, Mavrogianni A, Mumovic D, Summerfield A, Stevanovic Z and Djurovic-PetrovicM 2010 A review of bottom-up building stock models for energy consumption in the residential sectorBuild. Environ.45 1683–97

Directive (2007/2/EC) establishing an Infrastructure for Spatial Information in the European Community (INSPIRE): https://eur-lex.europa.eu/legal-content/EN/ALL/?uri=CELEX%3A32007L0002 [retrieved: Jul, 2020]

Miranda Pereira I and Sad de Assis E 2013 Urban energy consumption mapping for energy management Energy Policy59257–69

Keywords: EPBD, ISO standards, public data sources, building typologies, Open Street Maps

Paper ID 30971: Optimal operation of distributed energy systems - a game theory based approach

Sarah Henn*, Sarah Welter, Tanja Osterhage, Dirk Müller

RWTH Aachen University, E.ON Energy Research Center, Institute for Energy Efficient Buildings and Indoor Climate, Germany

AIM AND APPROACH

Within the ongoing transformation of the energy system, we observe a number of key drivers that require a conversion to localized energy system operation solutions [1]. On the one hand, there is an increase of distributed energy resources and a higher demand of electricity due to electrification of mobility and heating sector. On the other hand, there are limited distribution grid capacities.

These factors lead to a growing need for flexibilization of energy demand, for instance by intelligently operated energy storage systems as well as district or building energy systems. Local energy systems must be managed by means of a whole system approach, which is robust and allows for interconnecting localized solutions [2]. Moreover, it is necessary that this solution is able to deal with competing requirements on limited resources.

This study aims to evolve a solution, fitting all these requirements, by using a game theory based approach. Distributed coordination relying on game theoretic principles allows for an interconnected, autonomous and non-discriminatory operation. Within the study we evolve a game theoretic approach and simulatively apply it to a residential neighbourhood. We compare the consequential building energy system operation to both central and a fully decentral operation optimization.

SCIENTIFIC INNOVATION AND RELEVANCE

Current literature reviews on optimal district operation treat game theory as a promising approach in local energy trading [1] [3]. Reynolds states that "the autonomous, distributed, and heterogeneous nature of the smart grid make game theory well suited to smart grid problems" [2].

However, many studies investigated either load or feed-in situations. As neighbourhoods of the future will certainly have to meet both requirements, we introduce an approach which tackles both. Furthermore, numerous studies compared the game theoretic operation to an operation without usage of flexibilities, e.g. [4], [5] and [6]. A comparison with intelligent alternative approaches of operation was rarely made, e.g. in [7].

In our game-theory approach, the households of a district play a dynamic non-cooperative game in order to find their optimal energy consumption and feed-in schedule, as e.g. in [8]. Therefore, the players act in a continuous strategy space, optimizing their pay-off with limited information of the trading behaviors of their peers. We derive the closed-form expression of the best response by means of the adjusted, above described MILP sub problems. Thus, the Nash-equilibrium of the whole game is calculated iteratively.

PRELIMINARY RESULTS AND CONCLUSIONS

Our results showed that the game theoretic approach offers advantages only in terms of grid supportiveness, as high full-load hours are realized. With regard to costs, energy consumption, emissions and computing effort the compared approaches were preferable. Central coordination schemes were best in terms of self-consumption ratio, autarky ratio and hours of autonomy from the grid.

Even in the neighborhood surveyed, with homogeneous building types and homogeneous BES, there was a high potential for energy trading or sharing among the residents. This led to significantly lower costs for a cost balance regarding the whole neighbourhood instead of a balance by building. Thus, considering neighbourhoods as a whole system holds potential monetary savings for residents depending on the billing scheme as well as possibilities for future business models of energy suppliers and district operators.

Nevertheless, user privacy and autonomy should be compromised as minimally as possible when designing a neighborhood control system. Here, we believe that distributed approaches generally offer a favorable tradeoff between control performance and autonomy.

MAIN REFERENCES

[1] Reynolds, J. (2019). Real-time and semantic energy management across buildings in a district configuration (Doctoral dissertation, Cardiff University).

[2] Zhou, Y., Wu, J., Long, C., & Ming, W. (2020). State-of-the-art analysis and perspectives for peer-to-peer energy trading. Engineering.

[3] Zia, M. F., Elbouchikhi, E., & Benbouzid, M. (2018). Microgrids energy management systems: A critical review on methods, solutions, and prospects. Applied energy, 222, 1033-1055.

[4] Witte, P. and M. Kaltschmitt (2017). Dezentrale Steuerung eines Pools von Wärmepumpen auf Basis spieltheoretischer Methoden. Zeitschrift für Energiewirtschaft 41 (4), 237-259.

[5] Pilz, M., L. Al-Fagih, and E. Pfuegel (2017). Energy Storage Scheduling with an Advanced Battery Model: A Game-Theoretic Approach. Inventions 2 (4), 30.

[6] Nguyen, H. K., J. B. Song, and Z. Han (2015). Distributed Demand Side Management with Energy Storage in Smart Grid. IEEE Transactions on Parallel and Distributed Systems 26 (12), 3346-3357.

[7] Yaagoubi, N. and H. T. Mouftah (2014). User-aware game theoretic approach for demand management. IEEE Transactions on Smart Grid 6 (2), 716-725.

[8] Pilz, M., & Al-Fagih, L. (2019). A dynamic game approach for demand-side management: scheduling energy storage with forecasting errors. Dynamic Games and Applications, 1-33.

Keywords: MILP optimization, game theory, district energy systems, distributed energy scheduling strategy

Paper ID 30516: Impact of 2-zone recirculation on IAQ and energy performance of a demand controlled MEV system

Ivan Pollet* (1), Bavo De Maré (1), Steven Delrue (1), Frederik Losfeld (1), Jelle Laverge (2), Samuel Caillou (3)

1: Research, Renson Ventilation, Waregem, Belgium; 2: Building Physics Research Group, Ghent University, Gent, Belgium; 3: Belgian Building Research Institute (BBRI), Limelette, Belgium

AIM AND APPROACH

Demand controlled mechanical extract ventilation (MEV) systems with natural air inlets and central mechanical ventilation systems with heat recovery (MVHR) are usually used in the Belgian residential sector to control IAQ. In order to balance the supply and extract air flow rates of MVHR systems, air can be transferred mechanically between habitable rooms to reduce the supply capacity of the unit, called a 2-zone recirculation or cascade system. A similar approach can be applied to balance MEV systems by transferring air from the night to the day zone, while reducing natural air supply. CONTAM simulations were performed on several residential models (without or with open kitchen, natural or mechanical indoor air transfers, with or without air supply in the living room) to analyze the impact on IAQ (CO2, RH) and energy efficiency EE (heating losses) of MEV systems.

SCIENTIFIC INNOVATION AND RELEVANCE

Internationally, recirculation MEV systems were limited studied [1;2;3;4]. The performance depends mainly on the applied control strategy as a function of the dwelling structure. Especially in case of demand controlled MEV systems, a smart control of the indoor air transfer is crucial to guarantee the IAQ in all habitable rooms. The pollutants transferred from the night to the day zone have to be evacuated from the day zone, to prevent backflow to the night zone. Cascading also allows to limit the natural air supply in the day zone which affects the cross flows through the dwelling, and in that way the local IAQ and overall energy consumption. A MEV system with direct extraction from all rooms, functional as well as habitable ones, and air flow rates controlled on CO2, VOC or RH, was used as benchmark for IAQ and EE.

PRELIMINARY RESULTS AND CONCLUSIONS

Preliminary results clearly prove that pollutants cannot be stored indoors, contrary to the storage of heat or cold. Pollutants can be transferred within the house, but finally have to be extracted to prevent unwanted backflow. The indoor air transfer can be carried out by means of ducts combined with a central transfer fan or by means of individual room transfer fans connected to the circulations spaces. The impact of both configurations on IAQ and EE is limited.

The cascade principle is promising in case of an open kitchen where the extraction is controlled on CO2 and RH to allow a free transfer between the living room and the kitchen. Due to no air supply vents in the living room/open kitchen, the extraction in the kitchen must be increased to maintain the same reference IAQ. The IAQ in the bedrooms is slightly worse, since door undercuts providing air supplied from the circulation spaces, were not available anymore. Further modelling and simulation research must allow to optimize the performance of a 2-zone recirculation MEV system to a full extraction MEV systems by optimizing the control strategies of the mechanical transfer devices in relation to the extract flows from the functional rooms.

MAIN REFERENCES

[1] Rojas, G. (2015). Optimization Potentials for Mechanical Ventilation of Energy Efficient Housing - Simulation and Evaluation Methods. Thesis, 147 p. DOI: 10.13140/RG.2.1.4959.2404

[2] Rojas, G., Pfluger, R., Feist, W. (2015). Cascade ventilation – air exchange efficiency in living rooms without separate supply air. Energy and Buildings, 100, 27-33.

[3] Van Gaever, R., Laverge, J., Caillou, S. (2016). A comparison of different ventilation strategies for dwellings in terms of airflow rates and airflow paths. 14th Indoor Air Conference, Ghent, 3-8 July 2016, paper 786.

[4] Van Gaever, R., Caillou, S., Pecceu, S. (2019). Minimising the influence of the stack effect and wind on the operation of mechanical exhaust ventilation systems. 40th AIVC Conference, Ghent (Belgium), 15-16 October 2019, 263-271.

Keywords: air recirculation, MEV, demand controlled ventilation

Paper ID 31103: Global sensitivity analysis of a bottom-up building stock energy model

Sebastian Forthuber, Lukas Kranzl, Andreas Müller, Iná Maia*

TU Wien, Austria

AIM AND APPROACH

For various planning and policy issues the estimation of future development of heating and cooling demand is of great importance. However techono-socio-economic building stock energy models used for energy and emission development projections and policy assessment involve considerable uncertainty. In this paper we assess the building stock model Invert/EE-Lab (www.invert.at), using the Elementary Effects Method, a method appropriate for the degree of complexity of the model.

We provide exemplary results for selected parameters for selected countries that have been acquired in the IEA EBC Annex 70 – Building Energy Epidemiology project. Invert/EE-Lab is a dynamic bottom-up model that evaluates the effects of economic and regulatory conditions on future development of total energy demand, energy carrier mix, CO2 emission reduction and costs.The model builds on input data including a disaggregated building stock database on country level, supply technologies, regional climate, energy prices and energy carrier potentials as well as behavioural aspects and investment decision criteria. Within this article we focus on the analysis of the influence of relevant indicators such as interest rates, costs, energy prices, selected technical parameters and behavioural aspects on the final energy demand, respectively related energy carrier shares or installed capacities.

SCIENTIFIC INNOVATION AND RELEVANCE

The Elementary Effects method can be seen as a randomized "One-At-a-Time" design. Elementary effects for each input are computed from different points in the input space, leading to to mean and standard deviation that can be taken as a measure of importance of a specific input variable and its interactions with other inputs. This method is applied to the building stock model Invert/EE-Lab. The key approach of the model Invert/EE-Lab is to describe the building stock, heating, cooling and hot water systems on disaggregated level, calculate related energy needs and delivered energy, determine reinvestment cycles and new investment of building components and technologies. The core of the tool is a myopical, multinominal logit approach, which optimizes objectives of agents.

For the application of the EE-method to Invert/EE-Lab we use the SAFE toolbox to generate the input samples used for the model iterations as well as for analysis and visualization. The great variety of input paramters was broken down to 11 parameters, comprising of interest rates, different investment decision parameters, heating system and renovation costs, energy prices, lifetime of buildings and heatpump coefficient of performance, as well as factors representing user behaviour.

PRELIMINARY RESULTS AND CONCLUSIONS

The analysis was carried out with respect to the output variables installed capacity of heatpumps and share of final energy demand related to heatpumps and natural gas heating systems. The results for installed heatpump capacity show different levels of importance and interconnections for the evaluated input variables. Whereas the parameter economic weight can be considered to be of significant influence and great interconnection with other variables, others like interest rates, costs and behavioural factors can be considered as non influential on the observed output variable. It has to be mentioned tough, that the results are highly dependent on the selected value ranges for the input parameters, as well as on country specific datasets and presets. The results also provide confidence bounds and convergence investigations which lead to differentiated analysis of the parameter impacts. These is explained in more detail in the full paper. The sensitivity analysis carried out through the Elementary effects method provides valuable insights for the influence of various input parameters. The insights gained can be used for improved scenario development as well as deeper result interpretation trough better model understanding.

MAIN REFERENCES

Campolongo, F., S. Tarantola and A. Saltelli. (1999). "Tackling quantitatively large dimensionality problems". Computer Physics Communications. 1999 (1–2): 75–85. Bibcode:1999CoPhC.117...75C. doi:10.1016/S0010-4655(98)00165-9.

Müller, A., 2015. Energy Demand Assessment for Space Conditioning and Domestic Hot Water: A Case Study for the Austrian Building Stock (PhD-Thesis). Technische Universität Wien, Wien.

Kranzl, L., Hummel, M., Müller, A., Steinbach, J., 2013. Renewable heating: Perspectives and the impact of policy instruments. Energy Policy. https://doi.org/10.1016/j.enpol.2013.03.050 Invert/EE-Lab [Model website], URL http://invert.at/ (accessed 4.4.17).

Morris, M.D., 1991. Factorial Sampling Plans for Preliminary Computational Experiments. Technometrics 33, 161–174. https://doi.org/10.1080/00401706.1991.10484804

Pianosi, F., Sarrazin, F., Wagener, T., 2015. A Matlab toolbox for Global Sensitivity Analysis. Environmental Modelling & Software 70, 80–85. https://doi.org/10.1016/j.envsoft.2015.04.009

Keywords: Building stock simulation, uncertainty analysis, modelling up to national level

Paper ID 31137: Heat zoning- and district heating simulation tool

Raf De Herdt, Joris Dedecker, Thomas Koch, Pedro Pattijn*

Ingenium NV, Belgium

AIM AND APPROACH

A lot of ambitious cities in Flanders are looking for a way to replace the fossil fuel heating systems for building heating to decarbonize the cities. A GIS-based simulation tool can help local governments in finding opportunities for local heat exchange and support the choice for a certain heat strategy.

SCIENTIFIC INNOVATION AND RELEVANCE

- GIS-based tool, based on geographical open source data
- Analyzing heat demand on an urban or regional scale
- Analyzing the renovation potential on an urban or regional scale
- Analyzing potential heat sources on an urban or regional scale
- Easy routing and dimensioning of district heating grids
- Generation of environmental and financial cost

PRELIMINARY RESULTS AND CONCLUSIONS

GIS-based tools, based on geographical open data can be useful to acquire insight in the heat flows in the city, detecting key buildings and possible locations for local heat exchange and support the choice of the local government for a heat strategy for each district to replace fossil fuels. It can allow an analysis of the potential of district heating for individual districts.

MAIN REFERENCES

https://episcope.eu/welcome/ (2021)

Keywords: GIS, Urban Scale, Heat strategy

Session W1.4: Climate change and bioclimatic design

Paper ID 30769: The impact of climate change on the reliable optimization for energy and economic refurbishment of a residential building in Italy

Amedeo Pezzi*, Marco Manzan, Paolo Rosato

University of Trieste, Italy

AIM AND APPROACH

This paper investigates the effects of data uncertainty in an optimization process applied to the refurbishment practice of a residential building focusing in particular on the effects produced by climate change. Both heating and cooling performances are considered in the optimization process. The refurbishment measures considered in this work are different levels of insulation of the opaque surfaces, the substitution of existing windows with more efficient ones and interventions on the plant system like the replacement of the generator. In order to consider also the evolution of the economic situation in which the building is supposed to work a stochastic variation of economic indexes and fuels cost is considered. Using modeFRONTIER software for the optimization and EnergyPlus and Python scripts for the numerical energy simulation, this paper searches for robust refurbishment solutions, that are the solutions that can maintain their optimality for a given range of the input parameters.

SCIENTIFIC INNOVATION AND RELEVANCE

This paper couples the benefic effects of the optimization methods with the concept of robustness of the proposed solutions, ensuring that said solutions could maintain their optimality also in case of input data variability. Moreover, the effect of climate change on the refurbishment process is introduced as a stochastic input, calibrating the improvement interventions considering also the future environment in which the refurbished solution will work.

PRELIMINARY RESULTS AND CONCLUSIONS

Preliminary results have been obtained through the imposition of the maximization of the Net Present Value and the minimization of the Primary Energy consumptions as optimization objectives.

The implementation of economic stochastic parameters and of the effects of climate change in the input data pool is expected to confer to the process major reliability about the future performance of the refurbished system in a broader set of different working conditions. Moreover, the climatic data variation is expected to highlight how the different external conditions could influence the importance of choosing between interventions more focused on winter or summer season and how this choice could alter the performances in the other season.

MAIN REFERENCES

Marco Manzan, Giorgio Lupato, Amedeo Pezzi, Paolo Rosato, Alberto Clarich. Reliability-based optimization for energy refurbishment of a social housing building. Energies 2020, 13, 2310.

Giorgio Lupato, Marco Manzan, Amedeo Pezzi, The effect of climatic data on building performance optimization, BSO18 Building Simulation and Optimization, 4th IBPSA-England Conference BSO 2018, Cambridge, 11th-12th September, 2018.

Pierangioli L., Cellai G., (2016). The Impact of Climate change on energy-efficient refurbishment of social housing stock in Italy, 3rd IBPSA-England Conference BSO 2016, Newcastle, 12th-14th September.

Stephen E Belcher, Jacob N. Hacker, D.S. Powell. Constructing design weather data for future climates. Building Service Engineering Research and Technology 2005, 26.

Keywords: Optimization, Climate Change, Refurbishment

Paper ID 30116: Integrated design of refrigerant, heat pump and system components: Process intensification for optimized heat pump systems

Christian Vering*, Sebastian Ostlender, Fabian Wuellhorst, Philipp Mehrfeld, Dirk Müller

RWTH Aachen University, Germany

AIM AND APPROACH

In the future, heat pumps enable a sustainable building sector heat supply by using a power source that is mostly driven by renewable energies. However, high investment and operational costs currently restrain the market penetration of the technology. Conventional designs lead to oversized heat pumps to ensure security of supply. Oversizing might consequently lead to efficiency losses due to part load operation, which re-duces the overall ecologic potential of heat pumps. To fully assess the potential in the design process of heat pump systems, decisions on three levels must be considered simultaneously: Refrigerant, heat pump unit and system level components. In the context of this paper, the latter includes the heat pump, the heat-ing rod, the buffer tank and the domestic hot water tank as well as the control system. All three levels interact with each other. However, current design methods consider these levels separately. A procedure for the integrated design of the three levels does not exist for building energy systems.

SCIENTIFIC INNOVATION AND RELEVANCE

In process engineering however, the method of process intensification offers a systematic approach for an integrated design. This method is applicable for the design of heat pump systems. In this paper, we present systematic process intensification of heat pump systems and select metrics to optimize. The refrigerant cycle of the heat pump is modelled using a performance map approach, which is obtained by stationary simulations for each combination of fluid and process flowsheet. Specifying a set of fluids and process flow-sheets, a superstructure is derived. Subsequently, a multi objective genetic algorithm solves the mixed inte-ger nonlinear optimization problem. Annual simulations ensure that the influence of the operation on the design is considered respectively. To account for varying system constraints, scenarios are specified. A sce-nario includes weather data, building parameters and geographical location.

PRELIMINARY RESULTS AND CONCLUSIONS

The presented method enables the integrated, cost and emission optimal design of fluid, heat pump and system components simultaneously. Compared to conventional design methods, annualised costs are re-duced by 15 % and emissions by 10 % using the same fluid and process flowsheet. Concurrent, the nominal thermal power of the heat pump is reduced by up to 50 %. Furthermore, the best choice of working fluid and process flowsheet is given due to the superstructure-based approach. We solve the optimization prob-lem for different typical buildings of the German building stock. Therefore, we obtain a set of relevant refrigerants, process flowsheets and sizes of system components. This set may guide decisionmakers towards a sustainable and affordable heat supply of the German building sector. All presented results underline the po-tential of systematic process intensification to establish cost-effective and sustainable supply systems in the building sector.

MAIN REFERENCES

Lutze, Philip, Rafiqul Gani, and John M. Woodley. "Process intensification: a perspective on process synthesis." Chemical Engineering and Processing: Process Intensification 49.6 (2010): 547-558.

Vering, Christian; Brinkmann, Katharina; Maier, Laura; Lauster, Moritz; Mueller, Dirk. (2019). Vergleich der Jahresarbeitszahlen normativ ausgelegter Wärmepumpensysteme - Teil 1. HLH. 1. 68-71.

Schilling, Johannes, Christian Horend, and André Bardow. "Integrating superstructure-based design of molecules, processes and flowsheets." AIChE Journal: e16903.

Keywords: Heat pump design, genetic algorithm, cost and emission optimization

Paper ID 30851: Modeling the impact of Climate Change on Future Heating Demand in Different Types of Buildings in the Belgian residential building stock

Essam Elnagar* (1), Sébastien Doutreloup (2), Vincent Lemort (1)

1: Thermodynamics Laboratory, Aerospace and Mechanical Engineering Department, Faculty of Applied Sciences, Université de Liège, Belgium; 2: Laboratory of Climatology, Department of Geography, UR SPHERES, University de Liège, Belgium

AIM AND APPROACH

In recent years, global warming has a large impact on many aspects of the environment and human activities in buildings. Energy consumption for heating and cooling is directly affected by climate change. The residential buildings of Belgium have an average energy consumption 70% higher than the EU average. The behavior of Belgian residential building stock in terms of energy consumption is very valuable as it will help policy makers to formulate and design targeted measures for global warming aimed at improving energy efficiency and establishing current legal standards and benchmarks in the energy sector. The objective of the present study is to upscale the impact of climate change on the residential building and to evaluate its influence on the heating demand at the national level. This paper first presents the entire housing stock in Belgium which is divided in 992 cases for the reference year 2012. A weighting factor to represent their occurrence in the existing Belgian building stock is associated to each building type. The energy simulation are done for 12 case studies representing the different building types.

SCIENTIFIC INNOVATION AND RELEVANCE

A tree structure model defining Belgian housing typology was created, characterizing Belgian residential building stock in terms of various parameters like building age, scale, level of insulation and energy vectors dedicated to domestic hot water production and space heating. The entire model is validated at national level on the basis of the gas and electricity profiles and the profiles generated are compared to available Synthetic Load Profiles (SLP). The simulation model assumes a uniform repartition of residential buildings over Belgium. The model has also been used for accounting of future climate forecasts for the period 2012 to 2100 and consideration of production and emissions systems for the space cooling. The heating and cooling degree day approach is used to assess the effect of climate change on building energy use with potential weather data by calculating new heating and cooling degree days. The multi-zone model calculation is carried out by improving the model to consider the thermal transmission and cross ventilation between different zones.

PRELIMINARY RESULTS AND CONCLUSIONS

A dynamic building simulation model is used to focus on the future evolution of building heating energy demands. This study aims to assess the evolution of the profiles of final energy consumptions at the Belgium level with the predicted evolution of the climate until the end of the current century. 12 case studies are simulated representing 4 different building types to assess the impact of climate change on the heating demand in the different types of buildings considering the different factors of insulation and years of construction. The results show a significant decrease in the future heating demand in the upcoming years especially between the period 2050-2100.

MAIN REFERENCES

1. An analysis on energy efficiency initiatives in the building stock of Liege, Belgium. MK Singh, S Mahapatra, J Teller - Energy policy, 2013 - Elsevier.

2. Climate change impacts on building heating and cooling energy demand in Switzerland. T Frank - Energy and buildings, 2005 - Elsevier.

3.The characteristics and the energy behaviour of the residential building stock of Cyprus in view of Directive 2002/91/EC. GP Panayiotou, SA Kalogirou, GA Florides... - Energy and ..., 2010 - Elsevier.

4.Impact of climate change heating and cooling energy use in buildings in the United States. H Wang, Q Chen - Energy and Buildings, 2014 - Elsevier.

Keywords: climate change, building energy use, global warming, energy demand

Paper ID 30125: Climate change sensitive overheating assessment in dwellings: a case study in Belgium

Ramin Rahif* (1), Abdulrahman Fani (1), Piotr Kosinski (2), Shady Attia (1)

1: Sustainable Building Design Lab, Dept. UEE, Faculty of Applied Sciences, Université de Liege, Belgium; 2: Faculty of Geoengineering, University of Warmia and Mazury in Olsztyn, Heweliusza Street 10, 10-724 Olsztyn, Poland

AIM AND APPROACH

Due to the current rate of global warming, overheating in buildings are expected to be more frequent and intense in future climates. High indoor temperatures affect occupants' comfort, productivity, and health. As a result, our research is intended to investigate the vulnerability of dwellings to overheating risk in relation to the climate change. More specifically, this paper is aimed at quantifying the impact of high outdoor temperatures on building thermal conditions and long-term annual overheating risk.

By exploring a large body of the literature and standards, we decided to select the overheating assessment method suggested by (Hamdy et al., 2017). Three metrics are used namely, Indoor Overheating Degree (IOD), Ambient Warmness Degree (AWD), and Building Climate Vulnerability Factor (BCVF). The overheating risk is assessed under four climate scenarios representing historical and future scenarios. This method accounts for both severity and frequency of overheating taking into account zonal occupancy profiles and thermal comfort models. Finally, the potential of ventilative cooling is assessed in mitigation of overheating risk. The abovementioned method is applied to a lightweight single-family Passive House with total surface area of 174 m2 located in Eupen, Belgium.

SCIENTIFIC INNOVATION AND RELEVANCE

This study applies state-of-the-art method for climate change sensitive overheating assessment in a new building in Belgium. The new buildings are more likely to become overheated due to applied measures for energy-saving purposes during the winter such as high insulation and airtightness levels. Thus, the current study assigns a lower outdoor base temperature in calculation of AWD. Also, our focus is on long-term annual overheating risk rather than short-term heatwave events or seasonal overheating phenomenon (Hamdy et al., 2017). Three fit-to-purpose metrics are introduced by modifying the calculation period of IOD, AWD, and BCVF metrics. This approach allows us to distinguish between short- and long-term overheating assessments.

Our research contributes to overheating evaluation methods in residential buildings. The research outcomes can be used to assess the cooling systems' performance and effectiveness of adaptation measures in changing climate. Overall, this paper provides essential basis to improve indoor thermal conditions and climate change resilient design of buildings which are within the scope of IBPSA 2021 conference.

PRELIMINARY RESULTS AND CONCLUSIONS

The results of the current study are as following: (a) distribution of annual running mean outdoor temperature under four climate scenarios, (b) ranges of annual mean, maximum, and minimum indoor operative temperature for base case and for minimum and maximum ventilation rates under four climate scenarios, (c) yearly IOD and AWD values for base case and for minimum and maximum ventilation rates under four climate scenarios, (d) climate change vulnerability assessment for base case and for minimum and maximum ventilation rates using BCVF metric, (e) evaluation of ventilative cooling potential to reduce the overheating risk under four climate scenarios.

To conclude, this research gives an insight on whether the new buildings that are in compliance to the Passive House standard are able to suppress the increase in average outdoor temperature in future climates. It also reveals the potential of ventilative cooling as a mitigation strategy and predicts to what extend its potential will decrease as global warming increases.

MAIN REFERENCES

ASHRAE, A. (2017). Standard 55-2017. Thermal environmental conditions for human occupancy.

Butcher, K., and Craig, B. "Chartered Institution of Building Services Engineers," Environ. Des. CIBSE Guide Eighth Ed. Ed CIBSE Guide Chart. Inst. Build. Serv. Eng. Lond., 2015.

Carlucci, S., & Pagliano, L. (2012). A review of indices for the long-term evaluation of the general thermal comfort conditions in buildings. Energy and Buildings, 53, 194-205.

Carlucci, S., Bai, L., de Dear, R., & Yang, L. (2018). Review of adaptive thermal comfort models in built environmental regulatory documents. Building and Environment, 137, 73-89.

CEN, EN 16798-1: Energy performance of buildings–ventilation for buildings–Part 1: Indoor environmental input parameters for design and assessment of energy performance of buildings addressing indoor air quality, thermal environment, lighting and acoustics. European Committee for Standardization Brussels, 2019.

Hamdy, M., Carlucci, S., Hoes, P. J., & Hensen, J. L. (2017). The impact of climate change on the overheating risk in dwellings—A Dutch case study. Building and Environment, 122, 307-323.

Khovalyg, D., Kazanci, O. B., Halvorsen, H., Gundlach, I., Bahnfleth, W. P., Toftum, J., & Olesen, B. W. (2020). Critical review of standards for indoor thermal environment and air quality. Energy and Buildings, 109819.

Keywords: thermal discomfort, global warming, long-term thermal discomfort, thermal comfort

Session W1.5: The role of occupants

Paper ID 30700: Influence of the hydronic loop configuration on the energy performance of a CO2 heat pump for domestic hot water production in a multi-family building

Matteo Dongellini*, Claudia Naldi, Gian Luca Morini

Department of Industrial Engineering, University of Bologna, Italy

AIM AND APPROACH

Nowadays the impact of domestic hot water (DHW) production on the overall energy consumption of a building is significantly increasing. In fact, a strong effort has been made to improve the building envelope insulation properties and the efficiency of HVAC systems for space heating and cooling, while, on the contrary, a lower attention has been focused on the reduction of the energy need related to DHW preparation. Although it accounts for about the 19% of the total energy demand of European Union residential sector [1], this percentage is expected to increase up to 50% in the next years [2]. CO2 heat pumps are promising solutions to achieve significant energy savings for DHW production; furthermore, CO2 is characterized by a low GWP value, is economic and environmentally sustainable [3]. In this paper the annual energy performance of a centralized plant for DHW production in a multi-family building located in Bologna (Italy) and based on an air-to-water CO2 transcritical heat pump has been assessed by means of TRNSYS and compared with the results of a monitoring campaign performed for three months during the winter season of 2017-2018.

SCIENTIFIC INNOVATION AND RELEVANCE

The multi-family residential building considered in this work is composed by 7 stories and 27 flats, for a total useful surface of about 3850 square meters. A particular effort has been made to determine the hot water tap profile of the whole building: the method presented in Reference [4] has been used and adapted to the selected application. A hourly draw-off profile which takes into account the contemporaneity factor of the hot water request among all the apartments has been defined for both workdays and weekends. Moreover, the dynamic model of the CO2 heat pump has been developed with the cooperation of the heat pump manufacturer, who provided the performance data of the unit and its control logic. In order to decrease the temperature of the fresh water entering the heat pump, two water storages are connected in series with the unit. This work allows to evaluate the influence of the hydronic loop configuration on the energy performance of a CO2 heat pump: dynamic simulations evidenced that the configuration initially adopted for the DHW distribution was not able to exploit the maximum energy saving potential of this kind of heat pump units.

PRELIMINARY RESULTS AND CONCLUSIONS

The comparison between experimental and numerical results showed that the CO2 heat pump effective energy performance were much lower than that expected if the heat pump had been conducted in an optimal way. More in detail, the measured average heat pump performance factor (SPF) during the monitored period was 1.66, while, on the other hand, the unit energy performance calculated with dynamic simulations for the same interval could reach values close to 4.70. The analysis pointed out that this strong reduction of the heat pump performance was caused by the configuration of the DHW loop: with the adopted solution, thermal stratification within the storage was not obtained and the temperature of the water at the inlet of the heat pump was very high. For this reason, the unit efficiency dramatically decreases. Experimental data confirms this hypothesis: the temperature of the water stream entering the heat pump was around 40°C, almost 25 K higher than fresh water temperature introduced in the thermal storage vessel from the aqueduct. The results obtained in this work highlight how the energy performance of CO2 heat pumps is significantly influenced by the configuration of the DHW loop and especially by the layout of the thermal storages.

MAIN REFERENCES

[1] T. Kitzberger, D. Kilian, J Cotik, T. Proll, Comprehensive analysis of the performance and intrinsic energy losses of centralized Domestic hot Water (DHW) systems in commercial (educational) buildings, Energy and Buildings 195 (2019), 126-138.

[2] A. Bertrand, A. Mastrucci, N. Schuler, R. Aggoune, F. Maréchal. Characterisation of domestic hot water end-uses for integrated urban thermal energy assessment and optimization, Applied Energy 186 (2017), 152-166.

[3] k. Visser, Transcritical CO2 refrigeration systems for building cooling and heating reduce energy and cooling water consumption, emissions and the legionella danger, Proceedings of the 8th International Conference on Advances in Applied Science and Environmental Engineering (ASEE 2018), 3-4 February 2018, Kuala Lumpur, Malaysia.

[4] K. Ahmed, P. Pylsy, J. Kurnitski, Hourly Consumption profiles of domestic hot water for different occupant groups in dwellings, Solar Energy 137 (2016), 516-530.

Keywords: CO2 heat pump, dynamic simulation, DHW production, monitoring campaign, model calibration

Paper ID 30968: Methods for determining occupant behavioural models for energy-efficient retrofitting of 20th-century buildings

Antonella Mastrorilli (1), Roberta Zarcone (2), Chenafi Sabrina (1), Colonneau Téva* (1)

1: Laboratoire LACTH, Ecole Nationale Supérieure d'Architecture et de Paysage de Lille, 2 Rue Verte, 59650 Villeneuve-d'Ascq, France; 2: Laboratoire GSA, Ecole Nationale Supérieure d'Architecture Paris-Malaquais, 14 rue Bonaparte, 75006 Paris

AIM AND APPROACH

A study carried out within the project "Rethinking innovation. Know and manage the legacies of experiment and innovative social housing from the decade 1968-78" funded by the French Ministry of Cultural, is presented in this paper. The objective of this research is to highlight the influence of the variable occupancy in an energy renovation scenario.

We present the methodology developed for the construction of a "detailed" inhabitant profile on a case study of the social housing "Residence Salamandre" in Villeneuve d'Ascq.

By using an interoperable work, we produced a digital model informed by BIM methods, combining the sharing of information tasks of existing construction and site conditions.

From the typo-morphological analysis of the housing modules, a phase of data collection on the lifestyles of different occupants of the residence was developed. The balance sheet of energy consumed by year (provide by inhabitants) and anonymous surveys made it possible to link the aspect of daily consumption to the question of lifestyles.

In this paper, we present the impact of internal contributions on energy performance by comparing the results with those resulting from the application of usage scenarios in accordance with the French RT2012 standard.

SCIENTIFIC INNOVATION AND RELEVANCE

Today, the influence of occupants' behaviour is oversimplified during the analysis phases prior to energy renovation operations.

Amongst other variables, the lifestyles of the occupants remains one of the most difficult to control. However, it seems to play a decisive role because it represents one of the main factors of discrepancy between the phases of energy renovation and the actual functioning of a building.

Assessing energy needs and performance of existing buildings therefore, requires calculation tools able to produce results that are closest to reality. However, the different energy simulation softwares show many disparities which often leads to different results. In addition, the application of thermal regulations requires the use of referenced conventional calculation methods (DPE, TH BCE Method), certifying a level of energy performance to reach, by combining the analysis of "real" data with those of predetermined data, from hypothetical use scenarios.

This research attempted to develop dynamic energy simulation methods, offering the most realistic representation of energy needs in use conditions. We have defined a "fine" analysis methodology which allowed to take into account the variable of occupancy in housing. This will make us quantify its impact on energy needs and prefigure the most appropriate energy intervention.

PRELIMINARY RESULTS AND CONCLUSIONS

According to the simulations carried out through numerical modelling and the characterisation of a "detailed" user profile - for a family of three people - the first results obtained have demonstrated the impact of Inhabitants users in the calculation of consumption needs.

In fact, in the case of the "detailed" user profile, the internal contributions cover a total of 40% of the home's heating needs, i.e. 10% more than the results obtained considering the conventional profile from the French standard RT2012.

The results then highlighted a striking observation: Through dynamic energy building simulation, the occupancy variable in "detailed" user profile shows more efficient alternative than the intervention in external over-insulation of the Salamandre residence, with a strong aesthetic and constructive impact as a whole. Resorting to a "fine" analysis on a case-by-case basis - provide by actual measurements of occupancy variables in the dwelling - might then be more appropriate, in order to preserve the original aesthic and material qualities of building heritage objects.

Therefore, the conceptual and technical results obtained in this research aim to generalize this analysis methodology prior to each energy renovation project for existing buildings.

MAIN REFERENCES

Fabi V, Andersen RV, Corgnati SP, Olesen BW. A methodology for modelling energy-related human behaviour: application to window opening behaviour in residential buildings. Build Simul 2013 6:415

Andersen RV, Toftum J, Andersen KK, Olesen BW (2009), "Survey of occupant behaviour and control of indoor environment", Danish dwellings. Energy and Buildings, 41: 11–16

Andersen RV (2012), "The influence of occupants' behaviour on energy consumption investigated in 290 identical dwellings and in 35 apartments", Proceedings of Healthy Buildings 2012, Brisbane, Australia.

Brundrett GW (1997), "Ventilation: A behavioural approach", International Journal of Energy Research, 1: 289–298. Emery AF, Kippenhan CJ (2006), "A long term of residential home heating consumption and the effect of occupant behavior on homes in the Pacific Northwest constructed according to improved thermal standards", Energy, 31: 677-693.

Geslin F., Le bâti ancien appelle des solutions non standardisées, Les cahiers techniques du bâtiment, (https://www.cahiers-techniques-batiment. fr/article/le-bati-ancien-appelle-des-solutions-non standardisees.32324, consulté le 28/03/2017).

Nicol JF, Humphreys MA (2004), "A stochastic approach to thermal comfort-occupant behaviour and energy use in buildings", ASHRAE Transactions, 110(2): 554–68.

Keywords: occupants' behaviour, energy renovation scenario, Dynamic simulations

Paper ID 31106: Human in the Loop: perceived based control as the key to enhance buildings' performance

Davide Calì, Christian Ankerstjerne Thilker, Sebastian Arcos Specht, Jaume Palmer Real, Henrik Madsen, Bjarne W. Olesen*

DTU, Denmark

AIM AND APPROACH

The performance gap of existing and new buildings [1], both in terms of energy and occupants' comfort, jeopardizes the effort to reach deep decarbonization target. One of the main issues causing high CO2 emissions of buildings is related to VOLATILITY. Buildings are mostly planned and controlled based on assumptions and fixed schedules which might were valid in the Sixties. However, our society evolved: For example, residential buildings where family lives are often empty during the day while both parents go to work and children stay until afternoon at schools; in parallel, work-from-home became reality, also several times a week. Moreover, not only the demand for comfort is volatile: to minimize buildings' impact on climate change, we have to maximize the use of renewable energy sources. As a consequence, the production of energy is non-projectable. Matching the volatile usage of buildings with intermittent energy production can help both enhancing personal comfort and reducing CO2 emissions caused by the existing building stock. In this work, we propose a Human-in-the-Loop approach, where occupants are traced within a building, and, when they desire it, can provide feedback about their perceived comfort in specific rooms: this feedback is then used to control the building.

SCIENTIFIC INNOVATION AND RELEVANCE

The provision of flexibility services to the energy grids became more popular in the last decade. However, too often flexibility projects have a strong focus on the quality of the services provided to the grid, and do not actively consider real occupants needs. Eventually, flexibility services are connected to a deficit of indoor comfort (e.g. accepting lower indoor temperature or lower air change per hour). Through the use of our tracing and feedback app "FEEDME" and a network of IoT sensors connected to our vendor-neutral monitoring platform CLIMIFY [2], we gather information regarding the number of occupants within a room, their live feedback on the indoor climate, and their past preferences at given indoor conditions. This information can be used to:

1. Simplify the way buildings are controlled: instead of asking the occupants to choose set points, we ask how she/he feels, and control the building accordingly

2. Take into account the diversity of occupants' needs, and also the way those needs change during the day in the control of buildings - Mediate among different needs in a democratic way;

3. Optimize energy use, CO2 emissions and indoor climate in buildings through model predictive control and AI.

PRELIMINARY RESULTS AND CONCLUSIONS

FEEDME and CLIMIFY are currently being tested in a school with 13 classrooms, located in Denmark (a second demonstration office building is under arrangement). In the school, we gathered the feedback of occupants on their perceived thermal comfort (using a 5-steps scale) for a period of 6 weeks. In a first 3-week period, the set-point of smart thermostats was for all classrooms set to 22°C: 51% of the occupants were fully satisfied with the indoor climate, while over 10% were completely dissatisfied. In the second 3-week period, the set points of each classroom were manually adjusted accordingly to the received feedback of the previous 3 weeks. As a result, over 63% of occupants were fully satisfied, only 6% was fully unsatisfied [5]. In this first experiment, we only adjusted the set-point once, and accordingly to the location were the feedback was given only. A real time optimization, considering the exact occupants in each classroom and adjusting also other set-points, such as ventilation and blinds, could further enhance occupants experience in buildings. FEEDME keeps the occupants in the middle of the control loop but minimizes human errors. Moreover, the indoor climate enhancement is reflected into higher productivity [4, 5].

MAIN REFERENCES

[1] D Calì, T Osterhage, R Streblow, D Müller, Energy performance gap in refurbished German dwellings: Lesson learned from a field test - Energy and buildings, 127 (2016), 1146-1158. https://doi.org/10.1016/j.enbuild.2016.05.020

[2] Calì, D., Kindler, E., Ebrahimy, R., Bacher, P., Hu, K. S., Østrup, M. L., Bachalarz, M., & Madsen, H. (2019). climify.org: an online solution for easy control and monitoring of the indoor environment. E3S Web of Conferences, 111. https://doi.org/10.1051/e3sconf/201911105006

[3] D. Calì, Results of data analysis and optimization algorithms - Technical Report, DTU, 2020.

[4] P. Wargocki, J.A. Porras-Salazar, S. Contreras-Espinoza, The relationship between classroom temperature and children's performance in school, Build. Environ. 157 (2019) 197–204. https://doi:10.1016/j.buildenv.2019.04.046

[5] P. Wargocki, J.A. Porras-Salazar, S. Contreras-Espinoza, W. Bahnfleth, The relationships between classroom air quality and children's performance in school, Build. Environ. 173 (2020). https://doi:10.1016/j.buildenv.2020.106749

Keywords: Human in the Loop, Control, Indoor Climate, Demand Side Management, Comfort

Paper ID 30940: A simulation workflow for exposure characterisation of daylit spaces based on occupant gaze orientation

Mandana Sarey Khanie (1), Mikkel Kofod Pedersen (1), Trine Illum* (1), Rasmus Nielsen (1), Thorbjøn Asmussen (2)

1: Technical University of Denamrk, Denmark; 2: VELUX A/S, Hoersholm, Denmark

AIM AND APPROACH

This paper represents a simulation workflow for characterization of spectral exposure depending on occupants' position and gaze behavior in buildings. The project uses existing gaze movement database as well as occupant-tracking data obtained in a pilot study for its development. The existing database has been obtained at a daylight lab at Freiburg, Germany, in a user-assessment study where eye-tracking systems were used to record visual responses to the luminous environment [1]. The pilot study was done over a period of 2 months where occupants' orientations were tracked using an image-based sensors recoding dwells and movements of occupant. The dwell and track data from the pilot study were used to define exposure ranges to spectral effectiveness of the space. Using the two databases , a Grasshopper3D tool was developed to demonstrate gaze behavior[1], [2], exposure to illumination levels, and the exposure to spectral lighting, thus allows for exposure characterization of the space at the eye level at each given position. The latter was processed using Lark Spectral Lighting tool [3] to account for photopic, Rea[4] and Lucas [5]circadian illuminance. Using the tool, a simulation study was done and the results as well as the work flow are presented here.

SCIENTIFIC INNOVATION AND RELEVANCE

Despite different existing wavelength-dependent models to predict spectral-effectiveness of light[6], [7], these methods can only predict the health potentials in a space with assumption of static building occupants on fixed pre-defined points[8]. With great benefits on our well-being [9], [10], an exposure characterization of space for daylight, based on dynamic occupant behavior is a step forward. This step allows for better understanding occupant well-being indoors based on actual occupant's position and orientation. The dynamic human behavior to light exposure has been addressed in fewer studies where photometric measurements and eye-tracking methods were coupled for observations of gaze or eye responses to light [1], [11]. Building up on these exiting methodologies, here the occupant light-driven behavior is used to predict the exposure to spectral lighting and illumination levels in space. In addition, data gathered by the sensors are processed to show the actual exposure levels in real time. The developed tool and processing method can be used in design phases to introduce interventions, e.g. change of interior layout, for optimal lighting solutions.

PRELIMINARY RESULTS AND CONCLUSIONS

The simulation study provided a clear demonstration of visual patterns and exposure patterns at different points in space based on the dynamic occupant behavior. The health potential of the selected dominant gaze orientation was evaluated at each position. From the Lark tool [6], the values had to be converted as the selected threshold values used the units Circadian Stimulus (CS) and Equivalent Melanopic Lux (EML). The threshold value CS explained the optimal stimulus throughout the working day, where specific values are given for each hour, while the threshold value EML gave a fixed value over all hours. While most positions in the models showed satisfactory during summer time and under sunny conditions, only fewer positions would reach the thresholds under overcasts skies. Hence, optimized use of such areas in certain climatic regions proves to be essential. Moreover, as in north façade and lack of visual discomfort allows for orienting towards windows, higher health potentials can be achieved. It can therefore be concluded that the orientation in space is crucial for the vertical illuminance measured at the eye, but it is still important that the visual comfort is maintained.

MAIN REFERENCES

1. Sarey Khanie, M., et.al. Gaze and discomfort glare, Part 1: Development of a gaze-driven photometry. Light. Res. Technol. (2016)

2.Sarey Khanie, M. et al. A Gaze Visualizer tool for Grasshopper3d. in SimBuild – USA, (2018)

3. Inanici, M., et.al. Spectral daylighting simulations, Department of Architecture, Seattle, USA

4. Rea, M. The lumen seen in a new light, Light. Res. Technol. 47, (2015)

5.Lucas, R. et al. Measuring and using light in the melanopsin age. Trends in Neurosciences (2014).

6.Amundadottir,M. et.al. Unified framework to evaluate non-visual spectral effectiveness of light for human health. Light. Res. Technol. 49, (2017)

7. Rea, M., et.al. A new approach to understanding the impact of circadian disruption on human health. J Circadian Rhythm. 6, (2008)

8.Amundadottir, M. et.al. A human-centric approach to assess daylight in buildings for non-visual health potential, visual interest and gaze behavior. Build. Environ. 113, (2017)

9. Lockley, S. Circadian Rhythms: Influence of Light in Humans. Encycl. Neurosci. 2, (2009)

10.Birchler-Pedross, A. et al. Subjective Well-Being Is Modulated by Circadian Phase, Sleep Pressure, Age, and Gender. J. Biol. Rhythms 24, (2009)

11.Lin, Y. et al. Eye movement and pupil size constriction under discomfort glare. Invest. Ophthalmol. Vis. Sci. 56, (2015)

Keywords: spectral lighting, daylighting, spectral exposure, gaze behaviour

Paper ID 30868: Quantifying household specific self-consumption of photovoltaic-based power generation in energy efficient buildings – a comprehensive parametric study to increase the reliability of energy consulting

André Müller* (1,2), Johannes Koert (2), Patrick Wörner (2)

1: Institute for Housing and Environment, Germany; 2: Institute of Concrete and Masonry Structures, Technische Universität Darmstadt, Germany

AIM AND APPROACH

The energy consumption from private households is responsible for a substantial share of the total greenhouse gas emissions in Germany. For this reason, German legislation promotes the climate-neutral operation of buildings until 2050. Consequently, building related power generation from renewable energies can be considered in the calculation of energy performance according to the German Energy Saving Ordinance (EnEV). However, the EnEV calculation rules allow neither for a proper estimation of the amount of renewable energy generated locally from PV nor for the estimation of private electricity consumption. Thus, the assessments of PV systems to be installed on buildings lack reliability and highly depend on the modelling skills of the energy consultant. To overcome this barrier for the implementation of building-related PV an easy applicable matrix of energy generation and private consumption for different locations as well as types of buildings and households is developed. This is achieved by feeding IDA ICE building simulation models with profiles generated by the user behaviour model 'PeakTime'. Typical user behaviour of certain household types is translated to annual power consumption profiles, added to the load profiles of building archetypes and compared to the electrical power generation profiles from a PV system.

SCIENTIFIC INNOVATION AND RELEVANCE

A variety of studies already exist on the self-consumption of power production from building related PV systems. However, most of these narrowed research down to a single or respectively a few real or typical households' electricity consumption. Thus, the results allow for a plausible estimation of self-consumption, but not for a detailed prognosis and quantification as the source of reliable energy consulting. The basis of the presented investigation on household-type specific PV self-consumption is a recently developed stochastic user behaviour model, which allows the generation of load profiles for power consumption of various household configurations. Thereby, household specific circumstances, e.g. the co-use of electrical appliances by two or more household members, and their effect on households' load profiles are reflected. The relevance arises from the fact that, besides calculated standard load profiles for household types, a distribution of self-consumption from PV systems emerges. To allow for a proper comparison of available technologies and energy performance levels, the household load profiles are processed in a parametric study within the building simulation software IDA ICE. The results of these simulations are categorized and summarized to achieve applicability in the context of energy consulting and allow for more profound decisions of building owners.

PRELIMINARY RESULTS AND CONCLUSIONS

The preliminary results illustrate the fact, that power consumption of household types differ dependent on the status of employment household members and the presence of kids, respectively. At the same time, the amount of self-consumption from PV generated power is highly correlated to the user behaviours underlying the household types. While energy performance calculations according to German EnEV calculation rules cannot reflect these variations, the performed parametric study gives reliable values for self-consumption in general as well as ratios of solar coverage of the use cases heating and cooling, hot water demand as well as other electrical appliances of a household. Thereby, the preliminary results build on a reduced number of available household types and represent an intermediate step on the way of making available realistic user behaviour data for energy performance calculations as well as dynamic building simulations.

MAIN REFERENCES

Yan, Da ; Hong, Tianzhen; Dong, Bing; Mahdavid, Ardeshir; D'Oca, Simona; Gaetanie, Isabella; Fenga, Xiaohang (2017): IEA EBC Annex 66: Definition and simulation of occupant behavior in buildings. Energy and Buildings, Volume 156, Pages 258-270.

Wörner, Patrick (in Press): Einfluss des Nutzerverhaltens auf den Stromverbrauch in Wohngebäuden - Entwicklung eines komplexen Simulationsmodells für energetische Analysen. Dissertation. Institute of Concrete and Masonry Structures, Technische Universität Darmstadt. Graubner, Carl-Alexander (Ed.)

Loga, Tobias; Frank, Milena (2016): Photovoltaic power generation to cover domestic power demand in Passive House: A parameter study. 20th International Passive House Conference 2016: 22th – 23th April 2016, Darmstadt: Proceedings / Passive House Institute. Feist, Wolfgang (Ed.)

https://www.iwu.de/fileadmin/user_upload/dateien/energie/neh_ph/2016_passivhaustagung_LogaFrank_PVEigendeckungImPassivhaus.pdf (German version; English version is available to the author as print version only; last access: 31.07.2020)

Gaetani, Isabella; Hoes, Pieter-Jan; Hensen, Jan L.M. (2018): Estimating the influence of occupant behavior on building heating and cooling energy in one simulation run. Applied Energy, Volume 223, Pages 159-171.

BMUB (2016): Climate Action Plan 2050 – Principles and goals of the German government's climate policy. Federal Ministry for the Environment, Nature Conservation, Building and Nuclear Safety (Ed.).

https://www.bmu.de/en/publication/climate-action-plan-2050/ (Last access: 31.07.2020)

Keywords: user behaviour, dynamic building simulation, residential buildings, self-consumption, photovoltaic

Session W1.6: Improving indoor environmental quality

Paper ID 30182: Computational assessment of occupant-centric radiant cooling solutions

Helene Teufl*, Ardeshir Mahdavi

Department of Building Physics and Building Ecology, TU Wien, Vienna, Austria

AIM AND APPROACH

The aim of this contribution is to computationally model and evaluate a number of previously proposed alternative radiant cooling designs [1]. The development of these solutions is guided by three main concepts. The first concept aims at positioning cooling elements in close proximity to occupants. This approach is suggested to enhance both energy efficiency and personal control. The second concept concerns the surface temperature of the radiant element. In comparison to classical radiant cooling solutions, the alternative strategies are meant to allow for and accommodate surface condensation via integrated drainage elements. Thus, lower panel surface temperatures are possible. The third concept aims at incorporating an appropriately selected vegetation layer into the designs [1, 2]. Such a layer could contribute to the appearance of the radiant panels. Moreover, placed below the vertical radiant panel, the container for the layer's substrate can act as the collector of condensed water. The paper entails a detailed computational examination of the effectiveness of the proposed solutions.

SCIENTIFIC INNOVATION AND RELEVANCE

Buildings' increasing cooling energy use has been attributed to phenomena such as global warming and urban heat islands [3]. This underlines the need for innovative cooling solutions. In this context, radiant cooling technologies have been promoted because they have the potential to improve the energy efficiency as well as occupants' thermal comfort [4-7]. Nonetheless, some important aspects must be considered when designing and implementing radiant cooling systems (e.g., buildings' climatic context, position of the radiant elements relative to occupants, water vapor condensation risk). The previously mentioned alternative radiant cooling solutions are intended to address these aspects [1]. The contribution highlights the potential of these solutions in view of energy efficiency and thermal comfort.

PRELIMINARY RESULTS AND CONCLUSIONS

The paper presents the result of a virtual examination of prototype designs of occupant-centric radiant cooling systems in an office setting. A software solution is developed to probe multiple design variations including panels positioned on one or two sides of typical workstations. The effectiveness of the solutions is explored for different climatic regions (from hot dry to hot humid) and multiple surface temperature regimes (involving both temperatures above and below the dew-point temperature) are considered. The outcome of the case study shows that occupant-centric radiant cooling can improve both energy efficiency and users' thermal comfort. Moreover, limitations of the proposed configurations – particularly in extremely hot and humid conditions – are illustrated, and potential complementary measures (i.e., additional convective cooling via task elements) are examined. The developed software solution can offer a preliminary virtual assessment of a variety of technical solutions in a variety of climatic boundary conditions.

MAIN REFERENCES

1. Mahdavi, A., Teufl, H., 2020. Occupant-centric radiant cooling solutions. Requirements, designs, assessment. In PLEA 2020 (Ed.): Proceedings of the PLEA 2020 Conference – to appear.

2. Gau, U., 2005. Eine mobile Kühlwand für Büroräume. Master Thesis: TU Wien.

3. IEA, 2018. The Future of Cooling - Opportunities for energy-efficient air conditioning. https://www.iea.org/reports/the-future-of-cooling

4. Rhee, K.N., Kim, K.W., 2015. A 50 year review of basic and applied research in radiant heating and cooling systems for the built environment. Building and Environment 91, pp. 166–190. DOI: 10.1016/j.buildenv.2015.03.040.

5. Rhee, K.N., Olesen, B.W., Kim, K.W., 2017. Ten questions about radiant heating and cooling systems. Building and Environment 112, pp. 367–381. DOI: 10.1016/j.buildenv.2016.11.030.

6. Tian, Z., Love, J.A., 2009. Energy performance optimization of radiant slab cooling using building simulation and field measurements. Energy and Buildings 41 (3), pp. 320–330. DOI: 10.1016/j.enbuild.2008.10.002.

7. Memon, R.A., Chirarattananon, S., Vangtook, P., 2008. Thermal comfort assessment and application of radiant cooling: A case study. Building and Environment 43 (7), pp. 1185–1196. DOI: 10.1016/j.buildenv.2006.04.025.

Keywords: Climate Change, Cooling Demand, Radiant Cooling, Condensation Risk

Paper ID 30754: Occupant-centric control of transparent dynamic façades through an integrated co-simulation framework

Luigi Giovannini*, Manuela Baracani, Fabio Favoino, Valentina Serra

Polytechnic University of Turin, Italy

AIM AND APPROACH

The use of adaptive transparent envelope technologies (adaptive glazings, dynamic shading devices, etc..) in buildings could yield significant performance improvements compared to static solutions, due to their ability of modulating the incoming solar radiation according to external inputs. As a consequence, these components are difficult to operate, as their behaviour simultaneously affects different physical domains, interdependent and often conflicting. Current research trends have mainly focused on controlling these technologies to minimise building energy use, while comfort aspects (thermal and visual comfort) related to their operation are generally overlooked. Nevertheless, understanding the relationship between different performance requirements in operating such dynamic systems is of outmost importance.

In this framework, this paper presents a simulation framework to evaluate the performance of adaptive transparent façades in a comprehensive way, and this is demonstrated by evaluating the performance of different mono-objective control strategies, aimed at the optimisation of either energy or comfort aspects. This is done for an office case study located in Rome and equipped with an active adaptive façade technology modulating the entering solar radiation. This enabled investigating the strengths and drawbacks of the control strategies considered, as well as their impact over physical domains these were not conceived to optimise.

SCIENTIFIC INNOVATION AND RELEVANCE

Current Building Performance Simulation tools have limited capability in evaluating the performance of transparent adaptive technologies [1], including the influence of the control strategy, due to their inability to: (a) flexibly vary the thermo-optical properties of the materials; (b) evaluate in an integrated way the mutual effect of an adaptive envelope technology in different physical domains. As a result, most of the studies evaluating the performance of adaptive envelope components focus only on their effect on energy related aspects, visual or thermal comfort aspects, not taking into account the mutual influence between these domains. This paper presents an integrated simulation framework enabling the evaluation of the performance of active adaptive envelope technologies according to energy, thermal and daylight related aspects in a coupled way. This was done by managing together BPS tools aimed at evaluating the different aspects influenced by the adaptive component behaviour. Specifically, the use of Rhinoceros parametric plugin Grasshopper and its add-on Ladybug/Honeybee allowed different BPS tools to be managed through the same interface. EnergyPlus was used for the thermal and energy analysis while DAYSIM and Radiance were used for the daylight analysis. Finally, the data integration between these tools was achieved through purposely built scripts.

PRELIMINARY RESULTS AND CONCLUSIONS

The results obtained for the office case study considered show that the control strategies analysed, when aimed at optimising a single objective (i.e. energy performance, visual or thermal comfort, etc..), have, at different extents, drawbacks on the other domains they influence. To prevent or reduce these drawbacks, the control strategies should be devised as most suitable trade-off between energy and comfort related aspects. Moreover, the specific features of the space considered, as well as the climate of the site, show to have a significant influence on the effectiveness of a control strategy. Based on the results obtained, in order to successfully meet the desired goals, a specific control strategy should be purposely conceived or optimised for each specific space, taking into account its characteristics and the local climate as well. The integrated simulation approach proposed in the present work can play a key role in the conception and/or optimisation of such control strategies: its ability to overcome the main gaps in the currently available simulation strategies allows in fact a simultaneous and accurate evaluation of the effects of the behaviour of an active adaptive envelope component on energy, visual comfort and thermal comfort related aspects.

MAIN REFERENCES

[1] R.C.G.M. Loonen, F. Favoino, J. Hensen, M. Overend. Review of current status, requirements and opportunities for building performance simulation of adaptive facades, Journal of Building Performance Simulation 10(2) (2017) 205-223.

[2] L. Giovannini, F. Favoino, A. Pellegrino, V.R.M. Lo Verso, V. Serra, M. Zinzi. Thermochromic glazing performance: From component experimental characterisation to whole building performance evaluation. Applied Energy 251 (2019) 11335.

[3] L. Giovannini, F. Favoino, V.R.M. Lo Verso, A. Pellegrino, V. Serra. A Simplified Approach for the Annual and Spatial Evaluation of the Comfort Classes of Daylight Glare Using Vertical Illuminances. Buildings 8 (2018), 171.

[4] E. Arens, T. Hoyt, X. Zhou, L. Huang, H. Zhang, S. Schiavon. Modeling thecomfort effects of short-wave solar radiation indoors. Building and Environment 88 (2015), 3-9.

[5] F. Favoino, F. Fiorito, A. Cannavale, G. Ranzi, M. Overend. Optimal control and performance of photovoltachromic switchable glazing for building integration in temperate climates. Applied Energy 178 (2016), 943-961.

[6] J.M. Dussault, L. Gosselin. Office buildings with electrochromic windows: A sensitivity analysis of design parameters on energy performance, and thermal and visual comfort. Energy and Buildings, 153 (2017), 50-62.

Keywords: Active Adaptive façades, Indoor environmental Quality, Integrated Simulation, Energy Performance; Visual Comfort

Paper ID 30255: Optimization workflow for the design of efficient shading control strategies

Abel Sepúlveda*, Francesco De Luca, Jarek Kurnitski

Tallinn University of Technology, Estonia

AIM AND APPROACH

The main aim of this investigation is to study optimal shading strategies to improve the indoor visual comfort and energy performance at educational buildings in Estonia.

The goals are the following:

1. Optimization of control algorithms for internal roller fabric systems to achieve a suitable daylight provision and glare performance.

2. To prove the viability of these different shading algorithms to contribute achieving the energy consumption of nearly energy zero buildings (nZEB) according to the Estonian regulation [1, 2] and minimum requirements in terms of daylight provision and glare protection defined by the novel European standard EN 17037:2018 [3].

We used a simulation-based methodology and single zone approach (room level) [4]. The software used are Radiance and EnergyPlus for daylight and energy calculations, respectively. The cases study consists in three existing auditoriums with different orientations (east, south and west) located in Tallinn University of Technology campus. We considered interior roller fabrics as main shading device. Four different complex fenestration systems are analyzed and shading controls based on vertical illuminance, direct normal irradiance and daylight glare probability (DGP) are tested.

SCIENTIFIC INNOVATION AND RELEVANCE

Visual comfort has high impact on students' academic performance and health. Estonian energy consumption requirements aim to reach the nearly Energy Zero Building (nZEB) category for new and renovated buildings. In addition, the fulfillment of the actual Estonian daylight requirements based on daylight factor have been proved insufficient to ensure a realistic daylight provision in residential, office and educational buildings [5]. There is a lack of consideration of daylight glare phenomenon during early stages and refurbishment plans in Estonia despite of the low sun altitude [6, 7]. Furthermore, the fulfillment of requirements in terms of daylight provision and glare protection defined by the European standard EN 17037:2018 in combination with nZEB energy requirements defined by the Estonian regulations must be studied.

PRELIMINARY RESULTS AND CONCLUSIONS

It is possible to achieve suitable trade-off between daylight provision, glare protection and energy performance in educational buildings in Estonia. This achievement can be reached combining interior roller fabric and shading control strategies based on variables such as vertical illuminance at eye level, direct normal solar radiation and DGP are crucial to achieve an adequate overall performance. A suitable choice of the fabric and shading control algorithm thresholds helps to optimize the overall performance depending on the room orientation, surrounding buildings and weather conditions. Moreover, the use of the interior roller fabric is key to improve glare protection without increasing lighting consumption. This study would help architects and practitioners to choose adequate CFSs and their control at both, in early stages and refurbishment plans of educational buildings in Estonia.

MAIN REFERENCES

[1] Estonian Government, Minimum requirements for energy performance. Annex 68, RT I, 24.01.2014, 3, (2012). https://www.riigiteataja.ee/en/eli/520102014001.

[2] Estonian Government, Ordinance N° 58. Methodology for calculating the energy performance of buildings. RTI,09.06.2015, 21, (2015).

[3] European comission, BS EN 17037:2018: Daylight in buildings, (2018). https://www.en-standard.eu/bs-en-17037-2018-daylight-in-buildings/.

[4] B. Bueno, A. Sepúlveda, A Specific Building Simulation Tool for the Design and Evaluation of Innovative Fenestration Systems and their Control, in: Nternational Build. Perform. Simul. Assoc. -IBPSA- Build. Simul. 2019. 16th Conf. IBPSA. Proc. Rome, Italy, 2-4 Sept. 2019, 2019. http://publica.fraunhofer.de/dokumente/N-565205.html.

[5] F. De Luca, M. Kiil, R. Simson, J. Kurnitski, R. Murula, Evaluating Daylight Factor Standard through Climate Based Daylight Simulations and Overheating Regulations in Estonia, in: Proc. Build. Simul. 2019 16th Conf. Int. Build. Perform. Simul. Assoc., 2019.

[6] Sepúlveda, A., De Luca, F., Thalfeldt, M., & Kurnitski, J. (2020). Analyzing the fulfillment of daylight and overheating requirements in residential and office buildings in Estonia. Building and Environment, 107036.

[7] F. De Luca, T. Dogan, J. Kurnitski, Methodology for determining fenestration ranges for daylight andenergy efficiency in Estonia, in: Simul. Ser., 2018. https://doi.org/10.22360/simaud.2018.simaud.007.

Keywords: daylighting, nzEB, cold climate, glare, shading control algorithm

Paper ID 30139: Tool Development for Automatic Simulation of central and decentral Heat Supply Scenarios and Application to a district in the City of Mainz, Germany (SimStadt 2.0 project)

Verena Weiler (1), Eric Duminil* (1), Bodo Balbach (2), Bastian Schröter (1)

1: University of Applied Sciences Stuttgart, Germany; 2: Mainzer Stadtwerke AG

AIM AND APPROACH

Often enough, when new constructions or changes in existing built-up areas are planned, energetic assessments such as the choice among heat supply options, are done at a late stage in the process, with key decisions on the area already made. Our tool, called SimStadt, enables local decision makers to perform an early-stage analysis of possible planning scenarios, with limited data requirements on the technical backgrounds and exact design of potential future scenarios.

SimStadt is a scientific workflow management platform, which can be coupled to a number of external tools and libraries. The existing functionalities have been described in various publications ([1]–[4]) and shall not be the focus here. The new developments in the (ongoing) SimStadt 2.0 project are the connection of heat demand calculations to heat supply models. They are modelled in INSEL (www.insel.eu), with the district heating network dimensioned in more detail in STANET (www.stafu.de/en). Additionally, an energy components library contains relevant parameters for various heat supply models. The new development was tested with a district of 65 buildings in Mainz, Germany, where options for a central network system were compared against a decentral air-water heat pump system along technical and economic indicators.

SCIENTIFIC INNOVATION AND RELEVANCE

Most urban building energy simulation tools (UBEM) need a large amount of input data and/or proficient users ([5], [6]). One innovation in the SimStadt 2.0 project and tool development lies in the fact that the simulation environment can be used by non-experts while maintaining sufficient results accuracy for strategic decision-making and comparing various options in a scenario-based approach. This semi-automatic process is facilitated by the architectural design of the SimStadt platform: a CityGML model used as key input contains most of the required information on building geometry; second, a pre-defined set of heating system configurations with underlying, INSEL-based simulation models as well as a library of component specifications from the most important manufacturers enables the user to simply choose one or more heat supply options.

Simulation results are given as tables containing important technical and economic parameters such as total investment and operating cost and GHG emissions of a given scenario, and can be visualized via 3D maps. Thus, multiple scenarios can be easily compared against each other.

PRELIMINARY RESULTS AND CONCLUSIONS

Our approach was validated by comparing the modelling results for CHP plus centralized heating network with measured data from 2017. While actual data give a total annual heat demand of 2,005MWh, our model gives 2,144MWh, i.e. a deviation of 7%, with the simulation mainly over-estimating space heating demand for office buildings. This can partly be attributed to refurbishments that are not yet accounted for in the CityGML building data that serves as input to the simulation. Comparing the values on the supply side, measured data give 2,924MWh, which we underestimate by only 4%. When looking at the monthly comparison, differences are higher especially in March and November, when a typical heating season in Germany starts or ends (-34% and +65%, respectively). Comparing hourly values, the mean bias error is less than 10%.

Results from the model that applies decentral heat pumps in all buildings cannot be validated with measured data, since it is theoretical. However, the model was validated by comparing it to another heat pump model that itself had already been validated with measured data from a different project.

Future analysis will be aimed at identifying the reasons for the differences between measurement and simulation and consequently reducing them.

MAIN REFERENCES

[1] R. Nouvel et al., "SIMSTADT , A NEW WORKFLOW-DRIVEN URBAN ENERGY SIMULATION PLATFORM FOR CITYGML CITY MODELS," in CISBAT 2015, 2015, pp. 889–894.

[2] U. Eicker, D. Monien, É. Duminil, and R. Nouvel, "Energy performance assessment in urban planning competitions," Appl. Energy, vol. 155, pp. 323–333, 2015.

[3] M. Zirak, V. Weiler, M. Hein, and U. Eicker, "Urban models enrichment for energy applications: Challenges in energy simulation using different data sources for building age information," Energy, vol. 190, p. 116292, 2019.

[4] V. Weiler and U. Eicker, "Individual Domestic Hot Water Profiles for Building Simulation at Urban Scale," in IBPSA Building Simulation Conference, 2019.

[5] J. Allegrini, K. Orehounig, G. Mavromatidis, F. Ruesch, V. Dorer, and R. Evins, "A review of modelling approaches and tools for the simulation of district-scale energy systems," Renew. Sustain. Energy Rev., vol. 52, pp. 1391–1404, 2015.

[6] W. Li, Y. Zhou, K. Cetin, J. Eom, Y. Wang, and G. Chen, "Modeling urban building energy use: A review of modeling approaches and procedures," Energy, vol. 141, pp. 2445–2457, 2017.

Keywords: UBEM, energy supply modelling, case study

Session W1.7 (Online Track): Ensuring high quality building simulations

Paper ID 30352: Design optioneering for the definition of technological solution of envelope using BIM

Frida Bazzocchi (1), Neri Banti* (1), Carlo Biagini (2), Cecilia Ciacci (1), Vincenzo Di Naso (1)

1: University of Florence, DICEA (Department of Civil and Environmental Engineering), Italy; 2: University of Florence, DiDA (Department of Architecture), Italy

AIM AND APPROACH

The construction sector has recently undergone an evolution both in terms of performance requirements for buildings and in relation to the management techniques of design processes to make them easier to satisfy. This is the case of BIM Information Modelling (BIM), able to ensure digital and multidisciplinary design management. Combining the parametric nature of this modelling and its possible integration with computational design techniques, the research conducted has developed a simulation tool able to guide the designer in choosing between multiple alternative technological solutions. By writing a visual programming code (VPL), a design optioneering algorithm has been set up in order to perform technological choices. The method was applied to a horizontal partition towards an unheated space and to a vertical perimeter wall. The instruments used belong to the Autodesk Revit suite: Dynamo and Project Refinery, used for code implementation and iteration, respectively. The global outputs considered were the thickness, surface mass, thermal transmittance, cost and global warming potential (GWP) of the proposed stratigraphies. Part of the script was dedicated to verifying intertial condensation in relation to the local climate conditions.

SCIENTIFIC INNOVATION AND RELEVANCE

The construction of sustainable buildings forces the designer to make choices that generally lead to compromise solutions between structure, energy and economy issues. The application of computer and digital approaches is particularly suitable for such multifactorial optimization problems. The availability of new software easily manageable even by non-experts contributed significantly to the implementation of optimization techniques in the building industry. However, most of the studies in literature focus on the overall energy analysis of the buildings under consideration. Instead, the research proposes a methodology that can considerably simplify the decision-making process through generative design techniques. The latter, although not yet fully experimented in engineering, allow to exploit the computational potential of modern computers for comparative analysis between several alternative solutions otherwise not confrontable. The VPL script was designed and written entirely autonomously and therefore no specific reference can be found in literature. The possibility to filter the generated solutions allows to evaluate different and often conflicting aspects in order to better respond to different needs, especially in the preliminary design stage.

PRELIMINARY RESULTS AND CONCLUSIONS

For the floor, made with a lightened reinforced concrete slab, 4 variants of thermal insulation (EPS, XPS, wood fiber, rock wool) and 5 different thicknesses (4-6-8-10-12 cm) were considered, leading to the elaboration of 400 alternatives. After discarding solutions with high GWP index values an additional filter was applied to exclude configurations with too high thicknesses and reduced thermal trasmittance values. The most suitable stratigraphy was the one using a double layer of wood fiber for insulation, although 25% more expensive than tradional solutions. For the perimeter wall, a stratigraphy consisting of an external advanced screen façade with stone covering, thermal insulation, reinforced concrete partition and gypsum-fibre false-wall has been hypothesized. For the external insulation, the use of aerogel was tested in addition to EPS, rock wool and glass wool but then excluded due to the high costs. On the inner side, the comparison involved the construction of a thermo-acoustic insulation layer (rock wool or glass wool) or the maintenance of a gap in the false-wall. Among the 360 different stratigraphies proposed, the solution with rock wool on both sides is the best compromise.

MAIN REFERENCES

• Cecconi, Fulvio Re, Lavinia C. Tagliabue, Sebastiano Maltese, and Martina Zuccaro, 'A Multi-Criteria Framework for Decision Process in Retrofit Optioneering through Interactive Data Flow', Procedia Engineering, 180 (2017), 859–69 <https://doi.org/10.1016/j.proeng.2017.04.247>

• Evins, Ralph, 'A Review of Computational Optimisation Methods Applied to Sustainable Building Design', Renewable and Sustainable Energy Reviews, 22 (2013), 230–45 <https://doi.org/10.1016/j.rser.2013.02.004>

• Marzouk, Mohamed, Shimaa Azab, and Mahmoud Metawie, 'BIM-Based Approach for Optimizing Life Cycle Costs of Sustainable Buildings', Journal of Cleaner Production, 188 (2018), 217–26 <https://doi.org/10.1016/j.jclepro.2018.03.280>

• Pavan, A, C Mirarchi, and M Giani, BIM: Metodi e Strumenti. Progettare, Costruire e Gestire Nell'era Digitale, Costruzioni, Architettura e Design (Tecniche Nuove, 2017) <https://books.google.it/books?id=u0GJswEACAAJ>

• Tuhus-Dubrow, Daniel, and Moncef Krarti, 'Genetic-Algorithm Based Approach to Optimize Building Envelope Design for Residential Buildings', Building and Environment, 45.7 (2010), 1574–81 <https://doi.org/10.1016/j.buildenv.2010.01.005>

• Vanossi, A, and M Imperadori, 'BIM e Optioneering in Edifici S/R Di Piccola Scala BIM and Optioneering in Dry Technology Small-Scale Building', 2015 <https://www.researchgate.net/profile/Andrea_Vanossi/publication/276410657_BIM_e_option

Keywords: Design optioneering, BIM, Envelope, Multicriteria simulation

Paper ID 30246: Uncertainty analysis of building energy analysis based on replicated Latin Hypercube sampling

Wei Tian (1), Miaorou Jin (1), Pieter de Wilde* (2), Xing Fu (3), Jiaxin Shi (1)

1: Tianjin University of Science and Technology, China; 2: Plymouth University, UK; 3: Tianjin Architectural Design Institute, China

AIM AND APPROACH

Uncertainty analysis has been used in building energy analysis to provide robust predictions of energy performance of buildings due to inherent uncertain variables. Latin Hypercube sampling (LHS) is commonly used to obtain reliable results in order to reduce computational cost by using stratified sampling techniques. Most of previous studies apply LHS only one single time to obtain variations of energy performance of buildings. This research is focused on the comparison of one time LHS and replicated LHS to obtain more stable energy predictions for buildings. The replicated LHS generates LHS data of a sample size n with different sequences of random numbers. The results from these independent LHS data are analyzed to determine whether there are consistent uncertainty analysis results of building energy performance. A 3-storey office building is used as a case study to demonstrate the application of the replicated LHS technique. The energy simulation is carried out with the EnergyPlus program and the statistical analysis is carried out with the R computational environment.

SCIENTIFIC INNOVATION AND RELEVANCE

The replicated LHS technique can provide more robust results in comparison with one time LHS in building energy analysis by providing proper comparison results from these different LHS data. The visualization of uncertainty analysis results can be plotted from the CDF (cumulative distributions functions) of heating and cooling energy data to evaluate the similarity or difference from these various LHS samplings. Formal statistical results of confidence intervals for building energy can be also derived by assuming a t-distribution for different independent LHS results to assess the convergence of uncertainty analysis. Moreover, additional sampling results can easily be added in uncertain analysis of building energy using this replicated LHS methods since the LHS data are independent each other. This replicated sampling technique can be also used for sensitivity analysis to properly determine the convergence of a ranking importance of factors influencing building energy performance.

PRELIMINARY RESULTS AND CONCLUSIONS

The replicated LHS can provide reliable estimation of the uncertainty prediction performance of buildings by running several independent samplings. In contrast, the energy results from one time LHS may be unstable for a small sampling number. The sample size and the replicated number for using the replicated LHS method in building energy analysis can be determined by several factors, including number of input variables, relationships between inputs and outputs, and requirement for sampling convergence. A small number of LHS sampling combined with several independent LHS data is recommended to provide convergence information of uncertain energy performance analysis of buildings. Final analysis can be conducted by combining all the LHS sampling data to provide better results. Further analysis is required to understand the combining effects of various LHS methods, such as optimal and augmented sampling.

MAIN REFERENCES

[1] Hansen C W, Helton J C, Sallaberry C J. Use of replicated Latin hypercube sampling to estimate sampling variance in uncertainty and sensitivity analysis results for the geologic disposal of radioactive waste. Reliability Engineering & System Safety, 2012, 107: 139-148.

[2] Tian W, Heo Y, de Wilde P, et al. A review of uncertainty analysis in building energy assessment. Renewable and Sustainable Energy Reviews, 2018, 93: 285-301.

[3] Li H, Wang S, Coordinated robust optimal design of building envelope and energy systems for zero/low energy buildings considering uncertainties, Applied Energy, 2020, 265:114779.

[4] Ina De Jaeger, Glenn Reynders, Dirk Saelens, Quantifying Uncertainty Propagation For The District Energy Demand Using Realistic Variations On Input Data, IBPSA 2019, Rome, Italy, September 2-4, 2019.

[5] Adrian Chong, Song Chao, A Framework For The Continuous Calibration Of Building Energy Models With Uncertainty, IBPSA 2019, Rome, Italy, September 2-4, 2019.

[6] Pang Z, O'Neill Z, Li Y, et al. The role of sensitivity analysis in the building performance analysis: A critical review. Energy and Buildings, 2020, 209: 109659.

Keywords: Building energy analysis, uncertainty analysis, Latin hypercube sampling, confidence interval

Paper ID 30501: Quality check of openBIM input data for thermal building simulation based on mvdXML

Andreas Geiger*, Karl-Heinz Häfele, Veit Hagenmeyer

Karlsruhe Institute of Technology, Germany

AIM AND APPROACH

Today, Building Information Modelling (BIM) covers many domains of the building life cycle. More and more software tools in the area of building performance simulation support BIM by importing openBIM data according to the buildingSMART standard Industry Foundation Classes (IFC).

However, success and quality of building performance simulations depends strongly on quality, completeness, and consistency of BIM data. Therefore, it is necessary to check in this sense the content of openBIM data before transferring it into a respective simulation tool.

To describe data exchange scenarios in general, Model View Definitions (MVD) allow the definition of subsets of the corresponding IFC data model and enforce certain IFC attributes or types to be mandatory. With Model View Definition XML (mvdXML) – also provided by buildingSMART – MVDs can be formalized and therefore are testable by software tools.

The present paper introduces a first approach for MVD rulesets to cover exchange requirements for thermal building simulations, including formalizing the MVD ruleset as mvdXML. To verify and apply these rules, an additional module for the in-house software tool IFCExplorer is developed. This module can load, edit, save and check mvdXML rules. One key focus of the development is the representation of the checking results.

SCIENTIFIC INNOVATION AND RELEVANCE

Parallel to the development of IFC4, buildingSMART has developed mvdXML as a neutral and machine-readable format for the documentation of model view definitions (MVD). An MVD represents a subset of an IFC scheme for a specific use case, in order to ensure a reliable exchange of information for specific requirements. With ReferenceView and DesignTransferView, buildingSMART provides two official MVDs for IFC4. Since 2013, the mvdXML format additionally supports validation functionality. However, it turns out that the functional scope of mvdXML is not sufficient for some practical requirements.

For instance, the official buildingSMART MVDs are not designed for the requirements of thermal building simulations. They are missing most of the relevant rules to ensure that a simulation can be performed. Even applying all rules supported by mvdXML, there are certain limitations in checking the IFC models.

The present paper analyzes the requirements that must be fulfilled to use an architecture model for a thermal building simulation. The current version of mvdXML is evaluated on whether these requirements can be met using the existing rule syntax. Limitations of the this mvdXML version are shown. Corresponding extensions of the mvdXML standard are proposed and verified by a reference implementation in the tool IFCExplorer.

PRELIMINARY RESULTS AND CONCLUSIONS

Target of the current research is to set up an integrated workflow for thermal building simulations based on openBIM architectural models. One important module of this workflow is the quality check of openBIM architectural input data. By developing this module, the following results are achieved and are discussed in the present paper:

- Concept of a Model View Definition and an initial set of requirements,
- Manual generation of the corresponding mvdXML,
- Implementation of the mvdXML rule checking engine in the software tool IFCExplorer,
- Presentation of the checking results in an interactive way.

All results are in a prototypical stage. The implementation shows that the current mvdXML specification is not covering all requirements for a comprehensive model check. Proposals for extensions of mvdXML are collected and will be introduced to the mvdXML developer group. The checking engine can process all major rule types. Particular attention is paid to the reporting of the test results. To interpret and evaluate the results, no specific knowledge of MVD and mvdXML should be necessary.

The functionalities of the module are demonstrated with a selection of openBIM models of existing buildings.

MAIN REFERENCES

[1] - Chipman, T., Liebich, T., Weise, M., 2016, „Specification of a standardized format to define and exchange", buildingMSART International Ltd., 15.02.2016

[2] - Liebich, T., Geiger, A., Katranuschkov, P., Linhard, K., Steinmann, R., Weise, M., 2011, „mvdXML specification, mvdXML Schema", buildingSMART International

[3] - Weise, M., Nisbet, N., Liebich, T., Benghi, C., 2016, „IFC model checking based on mvdXML 1.1"

[4] - Hietanen, J., 2006, „IFC model view definition format", buildingSMART International

[5] - ISO 16739. 2018. ISO 16739:2018 Industry Foundation Classes (IFC) for data sharing in the construction and facility management industries. 2018

Keywords: openBIM, IFC, mvdXML, rule checking, building simulation

Paper ID 30586: Cross-over analysis of building-stock modelling approaches for bottom-up engineering model

Usama Bin Perwez*, Yohei Yamaguchi, Yoshiyuki Shimoda

Osaka University, Japan

AIM AND APPROACH

In building stock modelling (BSM), several methodologies have been developed to quantitatively improve the robustness and the consideration of heterogeneity of building stocks but most of these studies have not focused on identifying the impact of these methodologies on the performance of urban building energy models (UBEMs). To further understand the limitations of various building stock modelling methodologies, this paper presents the evaluation of three building stock modelling methodologies, geo-referenced [1], sample-based and sample-free models [2][3][4], to assess the accuracy and added-value of these approaches for use in bottom-up engineering model. The first method uses geographical information systems (GIS) to incorporate geo-referenced dataset within the model, while other two methods follow a synthetic approach by using iterative proportional fitting (IPF) [5] and Monte Carlo random sampling [6] to generate sample-based and sample-free building stocks respectively. These methods have been implemented on Tokyo commercial building stock as a case study, to further explore the strategies to overcome the limitation of uncertainty and availability of data in the implementation of UBEM.

SCIENTIFIC INNOVATION AND RELEVANCE

To the best of the author's knowledge, this study presents a comparison of three major building stock approaches for the first time. The study points out the advantages and disadvantages of synthetic building stock approaches in comparison to a GIS-based building stock approach by using detailed dynamic building simulation tool. The study aims to contribute to the literature development by:

1. To quantify the accuracy and added-value of these building stock modelling approaches for use in bottom-up engineering model.

2. To provide a granular level framework in choosing the right building stock modelling approach by assessing the availability (or uncertainty) of data at either aggregated or disaggregated level.

3. To demonstrate the use of these approaches in predicting the energy demand and GHG emissions of the commercial building stock of Tokyo, Japan.

PRELIMINARY RESULTS AND CONCLUSIONS

This study has focused to determine the accuracy and added value of the building stock modelling approaches, in terms of heterogeneity, data dimensionality, integration and non-linear interactions within stock, for use in bottom-up engineering model. The building stock modelling methodologies are implemented for the commercial building stocks (office, retail, school, hotel and hospital segments) of Tokyo, to evaluate the cross-sectional and longitudinal enrichment of building attributes and heating, ventilation, and air-conditioning (HVAC) stock. The preliminary results show that sample-free synthetic approach can incorporate multiple input distributions (building age and material stock) using building census or survey data, while geo-referenced approach provides additional key determinants such as building typology (shape coefficient and indices) and urban morphological attributes. This shows that synthetic approach can be extended to commercial building stock, which mostly has a poorer data availability than residential building stock, that further allows to encompass modelling of a typical mixed-use urban environment. Moreover, this cross-over analysis will provide a granular level framework to assist the city level planners and policy makers in choosing the right building stock modelling approach for predicting the energy demand and GHG emissions of the commercial building stock.

MAIN REFERENCES

1) M. Österbring, É. Mata, L. Thuvander, M. Mangold, F. Johnsson, H. Wallbaum: A differentiated description of building-stocks for a geo referenced urban bottom-up building-stock model, Energy and Buildings, Vol. 120, pp. 78-84, 2016.

2) K. Hermes, M. Poulsen: A review of current methods to generate synthetic spatial microdata using reweighting and future directions, Computers, Environment and Urban Systems, Vol. 36, Issue 4, pp. 281-290, 2012.

3) M. Lenormand, G. Deffuant: Generating a synthetic population of individuals in households: sample-free vs sample-based methods, J. Artif. Soc. Soc. Simul., Vol. 16, pp. 1–10, 2013.

4) C. Nägeli, C. Camarasa, M. Jakob, G. Catenazzi, Y. Ostermeyer: Synthetic building stocks as a way to assess the energy demand and greenhouse gas emissions of national building stocks, Energy and Buildings, Vol. 173, Pages 443-460, 2018.

5) R. Beckman, K. Baggerly, M. McKay: Creating synthetic baseline populations, Transportation Research Part A: Policy and Practice, Vol. 30, Issue 6, pp. 415-429,1996

6) F. Gargiulo, S. Ternes, S. Huet, G. Deffuant: An Iterative Approach for Generating Statistically Realistic Populations of Households, PLOS ONE 5(1): e8828, 2010.

Keywords: Building stock modelling (BSM), Bottom-up engineering model, Urban building energy modelling (UBEM), Commercial buildings

Paper ID 30888: Automatic IFC data enrichment with space geometries for Building Energy Performance Simulations

Georgios Nektarios Lilis*, Kyriakos Katsigarakis, Dimitrios Rovas

University College London, London, UK

AIM AND APPROACH

Recently, there is an increased attention to methods for automatic building energy performance simulation (BEPS) model generation from building information models (BIMs) for evaluation of building retrofitting scenarios [1]. These BEPS models require information on the geometry of the building space volumes for correct second-level space boundary topology generation [2], [6]. Sometimes, this space volume geometrical information is incorrectly defined or not defined at all in the building information models. The present work introduces an algorithm for the generation of the geometric representation of the building space volumes from their surrounding architectural building elements defined in the BIMs. The algorithm has three stages. Initially, the boundary representations of all architectural building elements (walls, slabs, ...) are extracted from their geometric representations. In the second stage, the common boundary surfaces of these representations are calculated. Finally, by subtracting the common boundary surfaces, from the boundary representation surfaces, the first-level space boundary surfaces [6] are obtained. These first-level space boundary surfaces are merged to form the internal building space geometrical representations.

SCIENTIFIC INNOVATION AND RELEVANCE

Attempts for building space volume extraction from 3D building models using graph methods have been documented [3]. Here a novel approach based on BIMs conforming to the International Foundation Class 4 (IFC4) standard [4] is introduced. The deserialization of the IFC4 models are performed using a purpose-built library; in addition, a geometric library has been developed for boundary representation generation and dedicated geometric operations. The space definitions can be quite useful for a number of cases, including the process of automating the energy model generation. The algorithm generates the internal building space volumes using the physical obstructions (walls, slabs, ...) as space separation boundaries and supports linear as well curved surrounding constructions. Possible user or IFC exporter errors using manual building space definitions can be also avoided by the use of the introduced process. The remaining surfaces of the boundary representations of the building constructions after their common boundaries and first-level space boundary surfaces are removed, form an outer building shell which is essentially the building facade. Finally, the obtained first-level space boundary surfaces [6] are used together with the BIM library to enrich the input IFC4 file with the building space geometrical representations.

PRELIMINARY RESULTS AND CONCLUSIONS

To guarantee accurate results, an important prerequisite of the introduced algorithm is that the input BIM IFC4 file is geometric error-free i.e. all boundary representations of the architectural elements have no missing surfaces and they do not intersect with each other [5]. If intersections (clashes) among building conclusions exist, the common edges of the first-level space boundary surface near the area of the clash, cannot be defined and therefore the boundary representations of the respective building spaces cannot be reproduced. Finally, a missing surface in the boundary representation of an architectural element, will result in a missing surface in the first-level space boundaries and the respective building space volumes and therefore should be detected and corrected. To illustrate the broad applicability, the method is tested on real complex building cases. Apart from the internal building space volumes the technique also identifies the external facade surfaces of the investigated buildings. The development of the introduced algorithm is supported by the H2020 research project BIMERR.

MAIN REFERENCES

[1] Giannakis, G. I., Katsigrakis, K. I., Lilis, G. N., & Rovas, D. V. (2019) "A Workflow for Automated Building Energy Performance Model Generation Using BIM Data", Building Simulation conference of IBPSA.

[2] Lilis, Georgios N., Georgios I. Giannakis, and Dimitrios V. Rovas. (2017) "Automatic generation of second-level space boundary topology from IFC geometry inputs."Automation in Construction 76: 108-124.

[3] van Treeck, Christoph, and Ernst Rank. (2017) "Dimensional reduction of 3D building models using graph theory and its application in building energy simulation."Engineering with Computers 23.2:109-122.

[4] Liebich, Thomas. (2013) "IFC4—The new buildingSMART standard." IC Meeting. Helsinki, Finland: bSI Publications.

[5] Lilis, G. N., Giannakis, G., Katsigarakis, K., & Rovas, D. V. (2018). A tool for IFC building energy performance simulation suitability checking. In Proc. 12th European Conference on Product andProcess Modelling (ECPPM) (pp. 57-64).

[6] Bazjanac, V. (2010). Space boundary requirements for modeling of building geometry for energy and other performance simulation. In CIB W78: 27th International Conference.

Keywords: BIM, IFC, Simulation, Geometry, Spaces

Session W1.8 (Online Track): Buildings paving the way for the energy transition

Paper ID 30462: Projecting impacts of uncertain climate change on future energy demand

Sandhya Patidar*, David Jenkins, Andrew Peacock

Heriot-Watt University, United Kingdom

AIM AND APPROACH

Energy demand in future is very likely to be impacted by various uncertain changes occurring in our society, such as technological growth, user behaviour, government policy, industrial strategies and climate change [1]. Climate change is one of the key factors effecting the future energy demand, e.g., warming climate could influence winter heating demand and/or summer air-conditioning demand [2]. This paper aims to develop a hybrid system of novel data-driven approaches for conducting an in-depth analysis (evaluating the magnitude) of the impacts of climate change and associated uncertainty on energy demand at both residential and community-level. The new approach is calibrated at a level of individual building demand profile to simulate future-morphed synthetic profiles that can be scaled up to realise, and visualised, the impact in aggregated profiles of communities in a manner that could reflect how individual dwelling profiles might respond to various future scenarios, and how this might be quantified at a regional, community level.

SCIENTIFIC INNOVATION AND RELEVANCE

To attain optimum efficiency, the underpinning modelling approaches will adopt a hybrid structure that will integrate carefully selected cutting-edge statistical, mathematical and machine learning based data processing/modelling approaches. The innovative design of hybrid system involves statistical time-series decomposition technique (STL: A Seasonal –Trend Decomposition procedure based on Loess processes [3]) for extracting trend, seasonal and random components from energy demand and corresponding climatic (e.g. temperature) time-series. The seasonal components of demand and climatic series are processed using wavelet power spectrum analysis. Application of wavelet analysis is strategically intended for identifying and extracting highly significant periodic signals (features) from the seasonal components. The periodic features drawn from the climatic and demand series are used to calibrate a 'Climate module' based on widely applied machine learning algorithm Support Vector Regression (SVR) [4]. The climate module within the hybrid system is designed for projecting the impacts of seasonal/periodic climatic feature on the corresponding seasonal features in demand series. Finally, a demand synthesis module, referred to as HMM-GP [5], that involves a Hidden Markov Model (HMM) coupled with an extreme-value distribution, Generalised Pareto (GP), is integrated within the hybrid system for generating future morphed synthetic demand profiles.

PRELIMINARY RESULTS AND CONCLUSIONS

To demonstrate the potentials of the proposed hybrid system, a small community village, Fintry based in Stirlingshire (Central Scotland), is selected [6]. For the present study electricity demand data available across 115 participating households in the community of (~350 households) at 15- and 30-minute resolution are used. The hybrid system is applied to a small selection of sample buildings from the case-study community to generate several statistically synchronous future-morphed synthetic demand series. These future-morphed simulated series are aggregated to generate community-level demand profiles for future climate change scenarios. A k-mean clustering approach (using statistical mean/median of total demand as key features) is applied to group the buildings in the community. Selection of the sample is weighted proportionally to the size of the clusters, for both sampling and aggregation purpose, to ensure the outputs represent adequate diversity across the community. Climate projection from the UKCP18 database [7] is utilised within the climatic module to generate climate-informed demand series and to explore climatic uncertainty in future energy demand at community-level. The results are thoroughly analysed and validated using appropriate statistical measures, such as box plots, probability density distribution, percentile distribution and autocorrelation function.

MAIN REFERENCES

[1] "Updated energy and emissions projections," HM Government, 2017.

[2] T. Rogers-Hyden, F. Hatton and I. Lorenzoni, "'Energy security' and 'climate change': Constructing UK energy discursive realities," Global Environmental Change, vol. 21, no. 1, pp. 134-142, 2011.

[3] R. B. Cleveland, W. S. Cleveland, J. E. McRae and I. Terpenning, "STL: A Seasonal-Trend Decomposition Procedure Based on Loess," Journal of Official Statistics, vol. 6, no. 1, pp. 3-73, 1990.

[4] M. Awad and R. Khanna, "Support Vector Regression," in Efficient Learning Machines, Berkeley, CA, Springer, Apress, 2015, pp. 67-80.

[5] S. Patidar, D. P. Jenkins, A. Peacock and P. Mccallum, "Time Series Decomposition Approach for Simulating Electricity Demand Profile," in Building Simulation 2019, 16th IBPSA International International Conference, Rome, Italy, 2019.

[6] J. Smith, "http://smartfintry.org.uk/about-smart-fintry/resources/," Smart Fintry, 13 04 2018. [Online]. Available: http://smartfintry.org.uk/wp-content/uploads/2018/04/Smart-Fintry-Innovation-Report-final.pdf. [Accessed 07 2020].

[7] "UK Climate Projections (UKCP18)," Met Office Hadley Centre Climate Programme, UK, 2018.

Keywords: Climate change, energy demand simulation, prediction and forecasting

Paper ID 31042: Analysis of the energy performance of a new ventilated brick wall: behaviour of real scale prototype under different ventilation configuration and weather conditions

Costanza Vittoria Fiorini* (1), Domenico Palladino (2), Cinzia Buratti (3)

1: CIRIAF, University of Perugia, Via G. Duranti 63, 06125 Perugia, Italy; 2: ENEA, Italian National Agency for New Technologies, Energy and Sustainable Economic Development of Engineering, Via Anguillarese, 301, 00123 Rome, Italy; 3: Department of Engineering, University of Perugia, Via G.Duranti 67, 06125 Perugia, Italy

AIM AND APPROACH

Ventilated façade allows to reduce thermal loads during the cooling period.

This work aims to assess the thermal behavior of a ventilated wall with high thermal inertia, made of local construction materials.

According to results obtained at laboratory scale, a real prototype of the ventilated façade was built as a perimeter wall of a gym.

Openings, 250x25 mm2/m of horizontal length, are placed at the bottom and at the top of the wall. Its height, 8.50m, allows to increase the flow rate within the air gap, thanks to the stack-effect.

The southern façade is equipped to monitor air and surface temperature, air velocity, heat flux.

Three experimental campaigns were carried out in winter and summer period with close, partially close, and open holes, with the following aims: comparing non-ventilated, partially ventilated, and ventilated configuration; evaluating the height influence on the temperatures within the ventilation gap and highlighting the potential phase displacement between those temperatures and the outdoor air temperature; monitoring the temperature and air velocity fields in the air gap and inside/outside, in order to evaluate the energy performance of the ventilated wall during two years of monitoring; using experimental results to implement and validate a 3D-CFD model.

SCIENTIFIC INNOVATION AND RELEVANCE

Given the impossibility of correctly assessing the thermal transmittance of this type of wall through the methodology provided by the current reference legislation, the research is aimed at evaluating the in-situ behavior of a real scale prototype, taking into account how the number and position of the openings affect its performance. Openings configuration was first optimized on a prototype of masonry ventilated wall built at laboratory scale and the experimental data monitored on it allowed the validation of a 3D CFD simulation model.Implementation and validation of a new 3D CFD model with experimental results at real scale (indoor/ outdoor air temperature, surface and air-gap temperature) will allow to predict the behaviour of the wall in different periods of the year and in different climate conditions, in order to establish the real benefit of the ventilation. Results related to conventional wall and innovative brick ventilated wall, in terms of energy demand, will be compared both in different cities and climatic conditions. Also in suggesting a ventilated wall made by traditional and local materials lies the peculiarity of this purpose: such a solution guarantees thermal comfort maintaining the constructive tradition (of Umbria region), both for new buildings and for refurbishment interventions.

PRELIMINARY RESULTS AND CONCLUSIONS

Non-ventilated, partially-ventilated and ventilated wall in winter case inside the cavity have surface temperatures and air temperature curves overlapping.

For partially-ventilated configuration cavity temperatures never reach the external peaks, departing from them of about 3°C, allowing heat losses reduction during the heating hours.

Ventilation configurations affect air temperature profiles in the cavity, with an almost linear trend as the height increases, according to stack-effect. For the ventilated configuration in the coldest hours the temperature is higher in the upper part of the wall, in the warmest at 3.55m and 6.5m. At the bottom (2.55m) the lowest temperatures are generally recorded.

In summer conditions inside the cavity the air temperature deviates from the surface ones up to 5°C, following the course of the external temperature in a smaller range (15°C and 20°C respectively), especially at lower temperatures.

Outdoor/indoor air temperatures phase displacement is 1.30 h in the night period and 3 h in the day period. During the daytime the counter-wall maintain low surface temperatures on the inner wall of the cavity: up to 4.5ºC lower than the external.

Summer difference between average flow values for ventilated and partially-ventilated wall is 8 W/m2: the first one guarantees optimal disposal.

MAIN REFERENCES

[1] Buratti, C., Palladino, D., Moretti, E., Di Palma, R., 2018. Development and optimization of a new ventilated brick wall: CFD analysis and experimental validation. Energy Build. 168, 284–297.

[2] Gagliano, A., Aneli, S., 2020. Analysis of the energy performance of an Opaque Ventilated Façade under winter and summer weather conditions. Solar Energy 205, 531-544.

[3] Liu, L., Yu, Z., Zhang, H., 2017. Simulation study of an innovative ventilated facade utilizing indoor exhaust air. In: International Conference on Improving Residential Energy Efficiency, IREE 2017.

[4] C. Marinoscia , G. Semprinia , G.L. Morini , Experimental analysis of the summer thermal performances of a naturally ventilated rainscreen façade building, En- ergy Build. 72 (2014) 280–287 .

[5] S. Saadon , L. Gaillard , S. Giroux-Julien , C. Ménézo , Simulation study of a naturally-ventilated building integrated photovoltaic/thermal (BIPV/T) envelope, Renew. Energy 87 (2016) 517–531 .

[6] C. Buratti , D. Palladino , E. Moretti , Prediction of indoor conditions and thermal comfort using CFD simulations: a case study based on experimental data, Energy Procedia 126 (2017) 115–122 .

Keywords: ventilated wall, CFD, real scale prototype, experimental performance, ventilation openings configuration

Paper ID 30475: Development of an urban sewage state prediction model and case studies for the evaluation of sewage heat utilization potential

Weian Chen* (1), Shohei Miyata (2), Yasunori Akashi (2)

1: National Kaohsiung Normal University, Kaohsiung, Taiwan; 2: University of Tokyo, Japan

AIM AND APPROACH

Sewage heat is a stable heat source for heating and hot water supply through the heat pump due to its small temperature fluctuation affected by season [1]. The utilization of sewage heat is a novel issue within the field of the development of renewable energy, especially the technique of recovering the heat through pipelines [2–5] in a regional scale to maximize heat reutilization by buildings [2,6]. However, the features and utilization potential of sewage heat have not been completely explored so far [7,8]. Therefore, there is still the possibility to draw up better sewage heat utilization plans and strategies to efficiently exploit the heat in urban areas, owing to its high penetration rate of the sewer system, and eventually reduce the environmental load.

This study aims at suggesting an estimation methodology and judging criteria based on an urban sewage state prediction model to evaluate the sewage heat utilization potential in an urban area and recommended utilization objects. In this paper, the evaluation methodology is introduced through case studies that applied the model to an actual area for evaluating the regional sewage heat utilization potential and clarifying which building is proper to utilize the sewage heat.

SCIENTIFIC INNOVATION AND RELEVANCE

An urban sewage state prediction model is proposed in the study, which can predict the sewage state and conduct the regional simulation for evaluating the sewage heat utilization potential to make the regional optimal utilization of sewage heat. Specifically, because the model can be applied to the areas as long as certain statistical databases (original water consumption unit, hot and tap water consumption unit, etc.) and GIS data (building and pipe information) are prepared, the measurement data of sewage flow rate and temperature are not essential in this model; therefore, the estimation method can be adapted to not only existed areas but new areas which are still under planning and drawing up the proper regional energy utilization plan.

This model can simulate the entire circulation from the energy supply side to the demand side, which including sewage physical model and sewage heat utilization system model. The two models were both established from the perspective of the features of heat rejection and heat demand of different building types. Comprehensively, it is expected to apply the model to discuss sewage heat utilization potential in an urban scale; moreover, the relationship between building spatial distribution and their energy consumption can also be clarified.

PRELIMINARY RESULTS AND CONCLUSIONS

The application of this model is first applied to an actual urban area to simulate the comprehensive utilization potential under different utilization rates. Specifically, without considering the particular strategies, the model is attempted to apply to an entire area for confirming its feasibility at the initial step, and clarified the approximate regional sewage heat utilization potential. Regarding the result, while the penetration rate of sewage heat utilization system is 80 %, the regional sewage heat utilization potential can achieve the greatest effect for about 10.69%. It shows that there is an abundant amount of sewage heat existed in the urban area that can be broadly utilized by more buildings. However, there is still a limitation that excessive utilization may lead to the worse effect; For instance, when the penetration rate is 100%, the utilization potential is 10.38%, which is lower than the result of 80%.

Secondly, from the practical view, we focus on specific buildings to compare which objective building is recommended to utilize the sewage heat in priority according to its better energy-saving effect and higher performance of sewage heat utilization system. Briefly, the relationship between different building location and their energy consumption, heat pump COP are analyzed.

MAIN REFERENCES

[1] Ministry of Land, Infrastructure, Transport W management department SD. Sewage heat utilization manual. 2015.

[2] Ichinose T. et al., Regional feasibility study on district sewage heat supply in Tokyo with geographic information system. Sustain Cities Soc 2017;32:235–46.

[3] Culha O. et al., Heat exchanger applications in wastewater source heat pumps for buildings: A key review. Energy Build 2015;104:215–32.

[4] Dürrenmatt DJ, Wanner O. Simulation of the wastewater temperature in sewers with TEMPEST. Water Sci Technol 2008;57:1809–15.

[5] Cui L. et al., Study of sewage heat utilization and heat interchange system that utilizes a network of sewer lines in urban areas (part 14 study of sewage heat amount available). vol. 10, 2014.

[6] Zhang C. et al., Design generalization of urban sewage source heat pump heating and air conditioning engineering. Proc - 3rd Int Conf Meas Technol Mechatronics Autom ICMTMA 2011 2011;1:926–9.

[7] MIKE M. et al., The Evaluation of Sewage Temperature and Flow Rate for Estimating Sewage Temperature and Flow Rate in Sewer Line. Trans Soc Heating, Air-Conditioning Sanit Eng Japan 2014;39:11–21.

[8] Cipolla SS. et al., Heat recovery from urban wastewater: Analysis of the variability of flow rate and temperature. Energy Build 2014;69:122–30.

Keywords: Urban energy supply and demand, Regional energy plan, Sewage heat utilization potential, Renewable energy, Utilization rate

Paper ID 30519: Investigating the control strategies for Breathing Walls during summer: a dynamic simulation study

Andrea Alongi*, Adriana Angelotti, Livio Mazzarella

Politecnico di Milano, Italy

AIM AND APPROACH

The paper aims to study different strategies for the optimal operation of Breathing Walls during summer. The investigation is carried out by performing dynamic simulations on a case study. To this purpose, a transient Finite-Difference numerical model for Breathing Wall components, previously developed in Matlab and validated by the Authors in [1], is coupled with TRNSYS. The case study consists in an office room located in Milan, Italy, provided with Air Conditioning and Mechanical Ventilation and with an air permeable wall.

SCIENTIFIC INNOVATION AND RELEVANCE

The best operation strategy for Breathing Walls in heating dominated climates is almost assessed: it consists in forcing the ventilation air across the permeable walls and roof in order to pre-heat it. Conversely, the optimal use of Breathing Walls in cooling conditions still needs to be explored: some Authors suggested the so-called Exhaust Air Insulation mode [2], namely forcing exhaust air from the indoor environment across the Breathing Walls to reduce envelope heat gains, some others proposed coupling with night free cooling [3]. In addition, pre-cooling of the ventilation air by flowing across the Breathing Walls may be considered. Therefore, the scientific relevance of the paper consists in clarifying the best summer use of this technology in the given context and proposing a control strategy for the airflow direction.

Moreover, the study features the integration of a model specifically developed for Breathing Wall components with an existing BES tool (TRNSYS), which has never been done before. This allows to perform a comprehensive building simulation, where the Breathing Wall can properly be simulated at the same time as a building envelope component and as a part of the mechanical ventilation system.

PRELIMINARY RESULTS AND CONCLUSIONS

Preliminary results of the simulations show that the best operation strategy depends on the characteristics of the room cooling load, namely the relative importance of ventilation load compared to transmission load. Moreover, a key parameter that should be firstly evaluated is the heat recovery efficiency of the Breathing Wall, to be compared with the efficiency of the heat recovery exchanger in case present.

MAIN REFERENCES

[1] Alongi A, Angelotti A, Mazzarella L. A numerical model to simulate the dynamic performance of Breathing Walls, submitted to: Journal of Building Performance Simulation.

[2] Wang J., Du Q., Zhang C., Xu X., Gang W. (2018). Mechanism and preliminary performance analysis of exhaust air insulation for building envelope wall. Energy & Buildings (173) 516-529.

[3] Ascione F, Bianco N and De Stasio C. Dynamic insulation of the building envelope: numerical modelling under transient conditions and coupling with nocturnal free cooling. Appl Therm Eng 2015; 84: 1–14.

Keywords: breathing wall, dynamic insulation, numerical model, building energy simulation, control strategy

Paper ID 30880: An evaluation of the proposed framework to introduce a smart readiness indicator for buildings

Vasiliki Varsami*, Esfand Burman

Institute for Environmental Design and Engineering, University College London (UCL), United Kingdom

AIM AND APPROACH

The Energy Performance of Buildings Directive was amended in 2018. A key objective was to promote the development of smart buildings since they were considered key enablers for future energy grids and systems. In this revision, the Directive called for the development of a new Smart Readiness Indicator (SRI) for buildings. The indicator is expected to provide a common framework across Europe, which evaluates the capacity of a building to use information and communication technologies in order to adapt to the needs of the occupants and the grid.

Although the scheme is not finalised, researchers and industry practitioners have already raised many questions about the methodology and the fair application of the scheme across all European Member States.

This paper reviews the proposed smart-readiness framework and assessment methods. Additionally, data from two existing non-residential buildings have been used to investigate the impact of the simplified and detailed methodologies defined for SRI on the final rating. Through this research and application of the scheme, the study aims to evaluate the strengths and improvement opportunities of the SRI framework and its role on enhancing building performance.

SCIENTIFIC INNOVATION AND RELEVANCE

An ongoing second technical study is finalising the newly proposed framework that is expected to measure the capability of buildings to respond efficiently to the external environment and the demand of the occupants. The scheme has the potential to offer multiple benefits, from improving occupants' well-being to promoting interconnected building communities and smart grids. Moreover, smart buildings can be key enablers for Europe to meet its Paris Agreement goal and keep in line with UN Sustainable Development Goals.

However, an indicator that measures capability can fail to translate into real performance. This could send a wrong message to end users reminiscent of the problem of energy performance gap and the credibility issue facing energy performance certificates. The two separate methodologies that exist may also lead to inconsistent certifications.

Therefore, the study aims to identify the potential issues that may rise up during the introduction of the indicator to the industry and can hinder the success of the scheme as well as improvement opportunities. By doing so, the paper can inform the process of continuous improvement of the scheme and demonstrate how this framework could be used to improve energy efficiency, energy flexibility and building user comfort.

PRELIMINARY RESULTS AND CONCLUSIONS

Two large-scale, public buildings have been chosen as case studies. Both buildings have been designed with innovative, energy-efficient systems but have different environmental approaches and energy needs.

After calculating their smart readiness indicator score using both the simplified and detailed proposed methodology, the results have shown an approximate 10% difference between the two methods. Regarding energy performance, no correlation between the smart readiness results and energy performance of the case studies has been identified. Subsequently, the weighted factors where adjusted according to simulation results to demonstrate how the assessment can account for the building type and climate.

The original scores were compared with results from the literature and the beta testing process of SRI. Although both buildings have been constructed in the past five years, their SRI scores are close to the mean value for non-residential buildings. In particular, of the three key functionalities that smart readiness assesses, namely, energy efficiency, energy flexibility, and response to user needs, both cases scored very low (around 18%) on energy/grid flexibility. By having such a strong focus on demand control and grid integration, the scheme has the potential to promote grid flexibility in buildings and pave the way for future electricity markets.

MAIN REFERENCES

European Commission, (2020). Discussion document – preparation of the delegated act of the smart readiness indicator. In: Meeting of the Expert Group on the Energy Performance of Buildings Directive. Brussels.

Janhunen, E., Pulkka, L., Säynäjoki, A. and Junnila, S., 2019. Applicability of the Smart Readiness Indicator for Cold Climate Countries. Buildings, 9(4), p.102.

Kurnitski, J. & Hogeling, J. (2018) Smart Readiness Indicator (SRI) for buildings not so smart as expected. REHVA Journal. (August 2018), 4.

Märzinger, T. and Österreicher, D., 2019. Supporting the Smart Readiness Indicator—A Methodology to Integrate A Quantitative Assessment of the Load Shifting Potential of Smart Buildings. Energies, 12(10).

Smartreadinessindicator.eu. (2020). Smart Readiness Indicator for Buildings. [online]

Vigna, I., Pernetti, R., Pernigotto, G. and Gasparella, A., 2020. Analysis of the Building Smart Readiness Indicator Calculation: A Comparative Case-Study with Two Panels of Experts. Energies, 13(11), p.2796.

Verbeke, S., Aerts, D., Rynders, G., Ma, Y. and Waide, P., (2020). 3rd Interim Report of the 2nd Technical Support Study on the Smart Readiness Indicator for Buildings. Brussels: VITO NV.

Keywords: smart readiness indicator, framework, evaluation, smart grids, energy flexibility

Session W1.9 (Online Track): Ensuring high quality building simulations/BIM

Paper ID 30156: Automatic generation of second level space boundary geometry from IFC models

Eric Fichter*, Veronika Richter, Jérôme Frisch, Christoph van Treeck

Institute of Energy Efficiency and Sustainable Building (E3D), RWTH Aachen University, Germany

AIM AND APPROACH

Building Information Modeling (BIM) describes a method of networked planning, execution and management of buildings using software. All relevant building data are digitally modelled, combined and recorded. To ensure a holistic planning process, it is advisable to use building models for process planning simulations. In case of an OpenBIM approach, the models can be transferred as IFC files [1]. In simulation practices, however, the models are often of poor quality. This is due to immature export functions and incomplete enrichment. In context of thermal simulations, this includes the lack of second level space boundaries within IFC models, which are necessary for simulation in EnergyPlus [2]. The consequence is a time-consuming manual rework for the simulation expert. An automated enrichment using software that is robust against geometric modelling or export errors and that does not rely on perfect meta-information would be beneficial. Therefore, the goal of this paper is to present a tool that enriches IFC files with correct second level space boundaries. The algorithm will be explained and then applied to an example building. For verification, the enriched IFC model will be simulated using EnergyPlus. The results will be compared with a manually generated simulation model.

SCIENTIFIC INNOVATION AND RELEVANCE

In the last two decades, multiple algorithms and programs for the generation of second level space boundaries were presented and documented in literature [3, 4, 5, 6]. However, most of them are not actively supported anymore, are based on older IFC versions (for example IFC 2x3) or are not publicly accessible. Moreover, all the mentioned algorithms used closed-source libraries. In contrast, the tool presented in this paper is based on widely used and well maintained open-source libraries. Furthermore, no meta information, e.g. the definition of spaces and their relation to building elements, is needed. Nevertheless, if meta information is provided, it could also be used. By avoiding Boolean operations between volumes, the algorithm is robust against modeling errors, such as gaps and non-physical superposition of volumes. Beside the generation of the space boundaries, a watertight and simplified surface geometry is created, in which the faces of the building elements are split according to the boundaries. Moreover, the spaces are generated as well.

PRELIMINARY RESULTS AND CONCLUSIONS

The generated space boundaries are exported to IFC. If necessary, the simplified and corrected geometries can also be exported as boundary representation in a geometry file format. The proposed method significantly reduces the manual effort for model generation and contributes to the derivation of simulation models from BIM data. Therefore, it improves the OpenBIM work process along with the design and optimization of building energy systems. In addition to the use in building energy performance simulation, an application in the pre-processing of geometry for Computational Fluid Dynamics (CFD) simulations is conceivable. The algorithm and the applications will be presented in the paper.

MAIN REFERENCES

[1] ISO 16739-1:2018, Industry Foundation Classes (IFC) for data sharing in the construction and facility management industries — Part 1: Data schema

[2] EnergyPlus Development Team, 2020. EnergyPlus engineering reference: The reference to EnergyPlus calculations. EnergyPlus Version 9.0. US Department of Energy.

[3] van Treeck, C., Rank, E. Dimensional reduction of 3D building models using graph theory and its application in building energy simulation. Engineering with Computers 23, 109–122 (2007).

[4] Rose, C.M., Bazjanac, V. An algorithm to generate space boundaries for building energy simulation. Engineering with Computers 31, 271–280 (2015).

[5] Lilis, Georgios & Giannakis, Georgios & Rovas, Dimitrios. (2016). Automatic generation of second-level space boundary topology from IFC geometry inputs. Automation in Construction. 76.

[6] Jones, Nathaniel & McCrone, C.J. & Walter, B.J. & Pratt, K.B. & Greenberg, D.P. (2013). Automated translation and thermal zoning of digital building models for energy analysis. Proceedings of BS 2013: 13th Conference of the International Building Performance Simulation Association. 202-209.

Keywords: IFC, BIM, Space Boundaries, BEPS, EnergyPlus

Paper ID 30169: Geometrical interoperability of 3D CityGML building models for urban energy use cases

Avichal Malhotra* (1), Yue Pan (2), Jérôme Frisch (1), Christoph van Treeck (1)

1: Institute of Energy Efficiency and Sustainable Building (E3D), RWTH Aachen University, Germany; 2: Nesseler Grünzig Bau Gmbh, Germany

AIM AND APPROACH

Geographical Information Systems (GIS) are often used as the foundation for urban scale thermal simulations. Virtual 3D city models, therefore, do serve as an important entity for analysing the thermal behaviour of buildings at an urban scale. However, the geometrical and semantic information for multiple buildings are openly available only for a relatively smaller number of countries, cities and districts. CityGML [1], a XML based modelling standard, is gaining popularity for city level information modelling and simulations. State- and city-wide 3D CityGML models are also available for countries like Germany, Austria and Switzerland [2]. However, the question about the geometrical interoperability of these models is still unanswered. Within the scope of this paper, the authors will present a CityGML Geometrical Transformation and Validation tool (CityGTV). This tool does facilitate the interoperability of 3D CityGML building models for countries where no or low quality data exists. Furthermore, the CityGTV will allow simulation scientists and urban planners to efficiently transform the building coordinates, validate the geometrical aspects of the buildings and further compute the district level energy performance of the buildings using environments such as Modelica, EnergyPlus, etc. With the tool, the altitude and orientations of individual buildings can also be transform.

SCIENTIFIC INNOVATION AND RELEVANCE

For many of the developed and developing nations, the scarcity of CityGML datasets is quite high. These countries on the other hand have a large potential for energy conservation and green energy production. Facilitating coordinate transformations of the 3D models within different reference systems, the transformed CityGML datasets can be used for multiple use cases and scenarios. Moreover, as the semantic, geometric and sometimes monitoring data availability of the existing built-up buildings is higher, transformed virtual 3D models of these buildings can be used as references for future planning and development of an urban area. One possible use case would be to analyse the solar potential of a city quarter in equatorial regions using the transformed 3D models of buildings from temperate or polar areas. Though CityGML data models only contain the geometrical and semantic information, these could be extended using the CityGML Application Domain Extensions (ADEs) [3]. For the scenario-specific analysis, the building models can be enriched using tools such as CityGML Toolbox [4] and therefore, be simulated for the individual use cases. This tool will also help urban planners and the simulation community to increase the operability of CityGML data models for multiple applications and domains.

PRELIMINARY RESULTS AND CONCLUSIONS

The CityGTV tool is currently being developed using python programming and PyQT [5] schema bindings. The user-friendly architecture and interface of the tool will allow users of different expertise levels to easily transform the building model(s) to their required reference systems. Within the tool the functionalities of geometrical validation and visualization are also foreseen by the authors. In future, the CityGTV tool will undergo an open source development process. The object oriented development methodology adapted by the authors will also facilitate the integration of CityGTV into different tool chains and simulation environments. The current implementation of the tool is being rigorously tested and further developed using open source datasets of large city quarters from cities such as Hamburg, Vienna, Berlin, etc. The transformed building models can be exported as new CityGML datasets which are in the user defined coordinate reference system.

MAIN REFERENCES

[1] G. Gröger, T. Kolbe, C. Nagel und K. Häfele, „OGC City Geography Markup Language (CityGML) Encoding Standard," OGC, 2012.

[2] A. Malhotra, J. Bischof, J. Allan, J. O'Donnel, T. Schwengler, J. Benner und G. Schweiger, „A review on country specific data availability and acquisition techniques for city quarter information modelling for building energy analysis," in Forthcoming IBPSA BauSIM 2020, Graz, 2020.

[3] F. Biljecki, K. Kumar und C. Nagel, „CityGML Application Domain Extension (ADE): overview of developments," Open Geospatial Data, Software and Standards, 27 August 2018.

[4] J. Hütter, „KIT - IAI Homepage," 2018. [Online]. [Zugriff am 09 02 2020].

[5] PyQt, „PyQt Referencing Guide," 2012.

Keywords: CityGML, Interoperabiltity, GIS, Coordinate Reference Systems

Paper ID 30849: Interoperability between BIM and building energy modelling – a case study

Francesco Asdrubali (1), Manuel Manzo (2), Gianluca Grazieschi* (3)

1: University of Roma Tre, Italy; 2: University of Roma Tre, Italy; 3: University of Roma Tre, Italy

AIM AND APPROACH

The paper aims to assess some interoperability issues between Building Energy Modelling (BIM) and dynamic Building Energy Modelling (BEM). The investigation is performed considering as a case study the design of a new residential complex composed of two terraced buildings located in the eastern belt of Rome. The two buildings are very similar and are designed following the most updated regulations about energy efficiency in Italy: they aim at the nearly zero energy standard. During their early design stage, the chosen case studies were modelled using Revit as a BIM platform. The BIM model was exported in Design Builder in order to perform a dynamic simulation of their energy consumptions in an optimization perspective.

SCIENTIFIC INNOVATION AND RELEVANCE

Building Information Modelling is spreading in the recent years due to the legislations that impose it. Some energy simulation techniques have already been implemented in the BIM models since they can contain information about the thermo-physical properties of opaque and transparent surfaces. Moreover, some plug-ins can be added in BIM tools in order to perform energy analysis. However, in most cases, the energy modelling is simplified and stationary. In order to perform a dynamic energy simulation, it is appropriate to export the BIM model into a dynamic energy simulation software. This means that BIM and BEM tools should be interoperable. The interface between the two models still presents some limitations. Some research is needed to improve it and some potential software features for a better BIM to BEM interoperability are suggested.

PRELIMINARY RESULTS AND CONCLUSIONS

The main difficulties and limitations in the interface between the two softwares are evaluated after the application to the case study. The creation of an accurate BIM-borne BEM model is quite time-consuming, laborious and subject to human made errors: the users are required to check the interoperability issues and, in some cases, to fix them manually.

After the description of the limitations found during the application, some suggestions are proposed in order to improve the interoperability between BIM and BEM models.

MAIN REFERENCES

Léa Sattler, Samir Lamouri, Robert Pellerin, Thomas Maigne. Interoperability aims in Building Information Modeling exchanges: a literature review. IFAC-PapersOnLine, Volume 52, Issue 13, 2019, https://doi.org/10.1016/j.ifacol.2019.11.180.

Ehsan Kamel, Ali M. Memari. Review of BIM's application in energy simulation: Tools, issues, and solutions. Automation in Construction, Volume 97, 2019, Pages 164-180, https://doi.org/10.1016/j.autcon.2018.11.008.

Georgios Gourlis, Iva Kovacic, Building Information Modelling for analysis of energy efficient industrial buildings – A case study. Renewable and Sustainable Energy Reviews, Volume 68, Part 2, 2017, Pages 953-963, https://doi.org/10.1016/j.rser.2016.02.009.

Haily Fernald, Seungyeon Hong, Scott Bucking, William O'Brien. BIM to BEM translation workflows and their challenges: a case study using a detailed BIM model. Proceedings of eSim 2018, the 10th conference of IBPSA-Canada Montréal, QC, Canada, May 9-10, 2018.

Keywords: BIM, BEM, energy, modelling, interoperability

Paper ID 30541: Texture mapping and IFC material retrievement for virtual reality applications

Sebastian Duque Mahecha*, Sergi Aguacil Moreno, Alexandre Dennis Stoll, Laurent Deschamps

Building2050 group, Ecole Polytechnique Fédérale de Lausanne (EPFL), Fribourg, Switzerland

AIM AND APPROACH

Successful experiments have been carried out in order to enable interactive IoT device management through immersive environments supported by Virtual-Reality(VR) and Augmented Reality (AR)[1–7]. In these experiments, material texturing plays a key role in making the experience vivid and facilitating spatial orientation and object sight recognition. In parallel, Architecture, Engineering and Construction(AEC) professionals increasingly integrate into their practice Building Information Modelling(BIM) methodologies, notably relying on the IFC data model[8] as the open international standard [9,10]. According to these methodologies, digital models synthesising informed-3D elements function as central source of building data along the building lifecycle. In spite of the evident convenience of using such models for IoT device management, a review of current industry practice and scientific literature in achieving IFC-IoT integration shows that there is still a major difficulty in keeping together IFC parametric information and material textures. Therefore, this article analyses different workflows to manage IoT interactive environments capable of preserving, from the original model, the exported IFC geometric and parametric information, as well as the material-textures affected to the model's elements. More specifically, we describe an automated workflow IFC file > IFC verification definition correcting textures>Game Engine (Unity[11]), allowing the integration of material-textures to an IFC2.3 export.

SCIENTIFIC INNOVATION AND RELEVANCE

To date, workflows available for immersive enabled IoT interactive environments allow retrieving either the material-texture information or the IFC parametric information. In general, using any game-engine software facilitates the transfer of texture information, but makes difficult retrieving the necessary IFC data to assure the IoT system connectivity. On the other hand, using IFC-compatible game-engine applications like Unity, which efficiently handle IFC data, pushes one to let go of texture information in spite the fact that, paradoxically, Unity's engine is particularly performant in terms of texture rendering. The reason for this, is that the current version IFC 4.2 does not suit a robust textures support, which may only be suited in the next major version, that is the IFC 5 [12]. In turn, most modelling software have hardly made it to IFC 2.3 certification [13]. Hence, stressing that under the circumstances it may take yet some years before modelling software providers can support texture information within an IFC based BIM methodology, this article describes an automated workflow to allow the updatable integration of material-textures into an IFC 2.3 export. In this way, it will be possible to work with material textures within a fully integrated BIM methodology.

PRELIMINARY RESULTS AND CONCLUSIONS

Results of a first approach show that, after a relatively easy treatment in Blender+BlenderBIM [14,15], IFC parametric data can incorporate both material information and image-based texture definition. However, this procedure proved highly inefficient in relation to changes and updates in the original IFC exported-model. Thus, our second approach was to introduce our own script, in order to automate the procedure: a "transparent" step matching IFC entities and material-textures. Results are satisfactory in relation to the suppression and modification of elements. Currently we work on changes involving element addition. Reflecting on future applications it is important to highlight that textures, are more than a cosmetic artifice to produce impressive renderings. Being able to transfer material and texture information through workflows for different BIM usages is fundamental in achieving comprehensible immersive experiences, or getting results that are more precise in model-based simulations, such as light reflectance and solar irradiance studies. In the future, this method may as well indicate ways in which we will be able to exploit IFC's embedded material properties to retrieve physical material information as well as texture appearance from the model elements.

MAIN REFERENCES

[1] G.White et al., Augmented reality in IoT, International Conference on Service-Oriented Computing. 11434,(2019)149–160.

[2] D.Jo, G.J.Kim, AR enabled IoT for a smart and interactive environment: A survey and future directions, Sensors 19,(2019).

[3] L.Müller et al., GuideMe: A mobile augmented reality system to display user manuals for home appliances, International Conference on Advances in Computer Entertainment Technology, 8253 (2013)152–167.

[4] K.M.Chang et al., An automated IoT visualization BIM platform for decision support in facilities management, Applied Sciences 8,(2018).

[5] SwissLivingChallenge, NeighborHub,(2017).

[6] S.Tang et al., Automation in Construction A review of building information modeling (BIM) and the internet of things (IoT) devices integration: Present status and future trends, Autom. Constr. 101 (2019)127–139.

[7] B.Dalton, M.Parfitt, Immersive Visualization of Building Information Models, Design Innovation Research Centre working paper 6(2013)1–20.

[8] BuildingSMART, Industry Foundation Classes-IFC,(2020).

[9] M.Poljanšek, Building Information Modelling (BIM) standardization,(2017).

[10] A.Kiviniemi et al., Review of the Development and Implementation of IFC compatible BIM Executive Summary, ERA Build(2008)1–2.

[11] UnityTechnologies, Unity platform,(2005). https://unity.com/.

[12] J.Ouellette, IFC-Problem about Elements with Texture, BuildingSMART forum, (2019). https://forums.buildingsmart.org/.

[13] BuildingSMART, IFC certified software,(2020). https://www.buildingsmart.org/.

[14] R.Ton, Blender Foundation,(2018).

[15] BlenderOrg, BlenderBIM - IfcOpenShell software library,(2019). https://blenderbim.org/.

Keywords: Building Information Modelling, Industry Foundation Classes, Internet of Things, Immersive Virtual Reality, digital material-textures

Paper ID 30201: Proposed integration of utilities in the Energy ADE 2.0

Maximilian Schildt*, Christian Behm, Avichal Malhotra, Sebastian Weck-Ponten, Jérôme Frisch, Christoph van Treeck

Institute of Energy Efficiency and Sustainable Building (E3D), RWTH Aachen University, Germany

AIM AND APPROACH

Energy Performance Simulations often require a large amount of data for building and district level simulations. This data highly depends on the quality, quantity as well as granularity of individual models and parameters. However, for efficiently storing, exchanging and reusing the information, a comprehensive data management system is very important. The management system should contain the relevant data from a district level to individual buildings in terms of geometry, HVAC components as well as network information in order to run energy performance simulations and operation optimization. Database management systems, such as 3D CityDB [1], do allow the storage of semantic data for 3D CityGML [2] building and virtual city models. Furthermore, for structuring the energy specific data, the use of the CityGML Energy Application Domain Extension [3] is foreseen. Currently, a comprehensive structuring of the energy systems data is missing in the Energy ADE 2.0 and therefore, the aim of this paper is to propose an approach to fill the gap for the management system. Moreover, the access to specific HVAC components within the data structure facilitates not only the analysis of energy demands but also the selection of components to cover the demand in a cost- or CO2-efficient manner.

SCIENTIFIC INNOVATION AND RELEVANCE

Based on the investigations over the energy systems within a field test of a collaborative research project, a perspective blueprint for the decentralized energy system transformation is developed. Furthermore, within the scope of this paper, a decisive analysis will be carried out to determine the components of plant and building technologies that should be additionally mapped with its respective degrees of detail and their individual key performance indicators. The components to be mapped will be systematically structured and classified based on the Energy ADE schema into related categories such as energy sources, conversion and transmission. This paper will show an in-depth analysis to determine the coherence of the components that can be mapped using the existing ADEs [4]. For the missing components, however, a modular XML schema definition (XSD) will be developed based on the central dependencies between the established categories. This approach allows the simulations of CO2-emissions and footprints in districts that are linked to a central database. Moreover, this will serve as a basis for optimized choice, dimensioning and operation of energy systems and utilities at an urban scale.

PRELIMINARY RESULTS AND CONCLUSIONS

Presently, the XML schema of Energy ADE has been extended using the classes necessary to map energy systems and utilities on district and building levels. The template of the extended classes has been linked to the 3DCityDB schema in an object-oriented database for implementation of the aforementioned research project. The approach is to attain a database that takes the utilities and energy systems fully into account and furthermore enables a tool-chain for energetic simulations and optimization scenarios. Comparing to Industry Foundation Classes (IFC) [5], CityGML and Energy ADE allow reduced order modelling and thereby reduce the amount of data needed to feed large datasets of buildings and districts into a simulation and optimization tool-chains. The current usage of CityGML Energy ADE and 3DCityDB is being investigated with the vision of reducing data redundancy while saving multi-criteria simulation results and optimization scenarios.

MAIN REFERENCES

[1] Z. Yao, C. Nagel, F. Kunde, G. Hudra, P. Willkomm, A. Donaubauer, T. Adolphi and T. H. Kolbe, "3DCityDB – a 3D geodatabase solution for the management, analysis, and visualization of semantic 3D city models based on CityGML," Open Geospatial Data, Software and Standards, 2018.

[2] G. Gröger, T. Kolbe, C. Nagel and K. Häfele, "OGC City Geography Markup Language (CityGML) Encoding Standard," OGC, 2012.

[3] G. Agugiaro, J. Benner, P. Cipriano and R. Nouvel, "The Energy Application Domain Extension for CityGML: enhancing interoperability for urban energy simulations," Open Geospatial Data, Software and Standards, 5 March 2018.

[4] F. Biljecki, K. Kumar and C. Nagel, "CityGML Application Domain Extension (ADE): overview of developments," Open Geospatial Data, Software and Standards, 27 August 2018.

[5] buildingSMART, "IFC Specifications Database," 18 06 2020. [Online]. Available: https://technical.buildingsmart.org/standards/ifc/ifc-schema-specifications/. [Accessed 06 07 2020].

Keywords: CityGML, Energy ADE, Utilities, HVAC, Database

Session W2.1: Practice and industry related case studies

Paper ID 30321: Simulation support for the design of ventilated windows: a case study

Shiva Najaf Khosravi*, Ardeshir Mahdavi

Vienna University of Technology, Austria

AIM AND APPROACH

Efforts to develop advance building envelope components include, among other things, ventilated window constructions. These windows utilize the airflow between glazing panes to preheat outside air before it enters indoors. In the framework of a previous research project, an instance of such a system was implemented in a testbed and was subjected to experimental studies under real conditions. This system comprises a single glazing unit as the outer pane and double glazing as the inner wing. The thermal behavior of the system and its capacity to preheat ventilation air was investigated. In the present contribution, we report on high-resolution 3D steady-state CFD simulation of this component. Model results were compared with measurements. The aim was to explore the fidelity of the CFD model and its potential toward design support. Specifically, variables related to the size of the outlet opening as well as the position of double glazing were examined under.

SCIENTIFIC INNOVATION AND RELEVANCE

In the last few decades, the thermal behavior of ventilated windows, or supply air windows, has received much attention in the literature [1-5]. This type of window, which can be deployed in both new and old buildings, is meant to provide in the cold season preheated ventilated air [6,7]. Nevertheless, predicting the thermal performance of a ventilated window is not a trivial task. Temperature and airflow patterns emerge from numerous dynamic interrelated thermal processes. These processes depend on the geometry and thermophysical properties of ventilated window components, building itself, and weather conditions [2, 5, 7]. The present research aimed to characterize a ventilated window in terms of its flow field (velocity and temperature distribution) while considering the influence of solar energy transmittance on temperature and airflow fields. This was done by numerically modeling the fluid dynamic and heat transfer in a ventilated window and solving the model in a CFD code. Furthermore, the experimental measurement data was applied to evaluate the reliability of the CFD model.

PRELIMINARY RESULTS AND CONCLUSIONS

The CFD model utility was demonstrated in terms of the design optimization of the supply air (ventilated) window. Specifically, we documented the thermal behavior of the ventilated window and evaluated the risk of condensation. A numerical model was created with a CFD commercial software in order to simulate the working of the ventilation outlets vents and the glazing unit's performance. The CFD investigation indicates that ventilated windows can provide preheated fresh air to the buildings' interior spaces. Simulation results suggest, in this case, that smaller inlet opening dimensions may be conducive to obtaining better performance regarding the temperature and volume flow of the incoming air. Furthermore, the application of double-layered glazing on the outside wing can lower the condensation risk during the cold season.

MAIN REFERENCES

1. Gosselin, J.R. and Q. Chen, A computational method for calculating heat transfer and airflow through a dual-airflow window. Energy and Buildings, 2008. 40(4): p. 452-458.

2. Bhamjee, M., A. Nurick, and D.M. Madyira, An experimentally validated mathematical and CFD model of a supply air window: Forced and natural flow. Energy and Buildings, 2013. 57: p. 289-301.

3. Carlos, J.S., Real climate experimental study of two double window systems with preheating of ventilation air. Energy and Buildings, 2010. 42(6): p. 928-934.

4. McEvoy, M.E., R.G. Southall, and P.H. Baker, Test cell evaluation of supply air windows to characterise their optimum performance and its verification by the use of modelling techniques. Energy and Buildings, 2003. 35(10): p. 1009-1020.

5. Laverge, J., Condensation in a closed cavity double skin facade: a model for risk assessment. 2010.

6. Hasse, M. and T. Wigenstad, Condensation Issues in Ventilated Façades, in Materials with Complex Behaviour II. Advanced Structured Materials, A. Öchsner, L.F.M.d. Silva, and H. Altenbach, Editors. 2012, Springer: Berlin.

7. Tanimoto, J. and K. i. Kimura, Simulation study on an air flow window system with an integrated roll screen. Energy and Buildings, 1997. 26(3): p. 317-325.

Keywords: Ventilated window, air flow, optimal spacing, thermal performance

Paper ID 31115: Daylighting simulation using spectral ray tracing

Antony Escudie* (1), Chloé Therreau (1), Tanguy Timmermans (2), Louis Dellieu (2), Grégoire Besse (1)

1: Eclat Digital; 2: AGC Glass Europe

AIM AND APPROACH

Direct sunlight is poor for visual comfort in offices or homes. Areas exposed to direct sunlight are too bright, with illuminances over 10000 lx, making most computer displays unreadable. The contrast between lit zones and dark zones is generally too high, requiring the eye to constantly adapt when looking at different places of the same room : this can be an important cause of eye fatigue.

People will generally react by closing the curtains or masking the sun light by any other mean. Then, the indoor lighting will be too low for comfortable work, and they will switch on artificial lights.

Day-lighting in bright sunlight is a main concern in architecture : using special designs and materials, it is possible to improve significantly the visual comfort. These techniques can be realistically simulated with Ocean, spectral ray tracing software.

SCIENTIFIC INNOVATION AND RELEVANCE

Unlike traditional building light simulation software, no model simplification is necessary : it may work with fully detailed CAD models and several millions of polygons. The same software and model may be used for producing detailed illuminance mappings, as well as realistic renderings for visualization.

PRELIMINARY RESULTS AND CONCLUSIONS

In this simple day-lighting example, we demonstrated the broad possibilities of the Ocean software for lighting design. The exact nature of its calculations allow making quantitative analysis of day-lighting performance. As it is possible to work on detailed CAD models, with complex materials based on measured spectral data, testing virtually any day-lighting system design is possible.

MAIN REFERENCES

Physically Based Rendering : from theory to implementation, M. Pharr et. al, Third Edition, (2004)

Robust Monte Carlo Methods for Light Transport Simulation, E. Veach. Stanford University, (1997)

Keywords: Daylighting, raytracing, spectral redering

Paper ID 30567: Evaluating daylighting performance metrics in LEED v4 for commercial office buildings: What criteria is missing to enhance the visual performance and comfort of occupants?

Riwayat Katia*, Nourhan Elsayed, Tarek Rakha

Georgia Institute of Technology, United States of America

AIM AND APPROACH

The Leadership in Energy and Environmental Design (LEED) is one of the most widely adopted building benchmarks globally, and it encompasses multiple rating criteria for performance. However, such criteria typically focus on reducing energy consumption, and may disregard providing a high-level of occupant satisfaction. Recently, LEED has adopted grid-level Climate-based Daylighting Metrics (CBDM) as one of the possible metrics for daylighting evaluation. This approach focuses on a desk height evaluation of Spatial Daylighting Autonomy (sDA) and Annual Solar Exposure (ASE), which may disregard an occupant's visual comfort at eye level. This paper proposes a more comprehensive approach to daylighting performance evaluation in commercial office space that incorporates vertical-eye level glare evaluation, in addition to desk-level CBDM. Simulations for workspaces in a tall office building in Atlanta, GA, USA at varying orientations and proportions were undergone using Climate Studio. Simulations were conducted with different shading systems to maximize sDA and minimize ASE and Annual Glare. The simulations followed the same thresholds for sDA and ASE as proposed by LEED. However, thresholds for annual glare were set based on proposed recommendations in the literature which were; Likely to be comfortable: $DGP<0.23$, Bounded between comfort and discomfort: $0.23<DGP<0.25$, Likely to be uncomfortable: $DGP>0.25$.

SCIENTIFIC INNOVATION AND RELEVANCE

Overlooking occupants' dynamic behaviours and reducing their needs to a set of limited parameters required by standards (e.g., illuminance thresholds) can cause a decline in satisfaction and consequent energy inefficiency in LEED buildings. The LEED program adopted the use of sDA and ASE to predict daylight sufficiency and probability of glare respectively, (IES LM-83-12, 2012) as one of the three compliance paths for acquiring the 'Daylight & Views' credits in LEED v4 (LEED Reference Guide, 2014). However, LEED did not consider that LM-83-12 also stated that there are other sources of glare besides direct sunlight that are not fully evaluated by the ASE metric. ASE evaluates disability glare when the amount of light is excessive and the occupant can't see, but does not evaluate discomfort glare at the eye level, when there is a range of luminance in a field of vision that causes degradation of visual performance and tiring of the eyes (Carlucci et al., 2015), thus, reducing the level of occupants' satisfaction significantly. This paper's innovation is in proposing a comprehensive approach that includes the addition of Annual Glare evaluations as a metric, with the tested thresholds as the criteria for obtaining the LEED credits.

PRELIMINARY RESULTS AND CONCLUSIONS

Three sidelit private offices in the north, west, and south orientations along with two sidelit open plan offices in the north and east orientations on the 16th floor of 191 Peachtree Tower, Atlanta, GA were investigated. When an overhang and horizontal louvers were placed at 0 degrees in the sidelit private offices in the west and south orientations, the offices acquired all four LEED credits (100% sDA, 0.0% ASE). However, they suffered from 48.4% and 14.8% of disturbing glare percentages respectively, ($0.40<DGP<0.45$) over the year for more than 5% of time. When the same horizontal louvers were tilted at 45 degrees, the annual glare was reduced to 1.6% and 1.7% respectively, with the same values for sDA and ASE. Similarly, when horizontal louvers tilted at 45 degrees were placed with the overhang in the open plan office in the east orientation, the annual glare was reduced from 21.3% to 0.0% with the ASE at 0.0% in both cases. These preliminary results justify further in-depth investigation at different orientations with other shading configurations to evaluate annual glare based on a new recommended threshold. This can guide LEED and other benchmarks to require more stringent and comprehensive daylighting performance metrics.

MAIN REFERENCES

Carlucci et al. (2015). A review of indices for assessing visual comfort with a view to their use in optimization processes to support building integrated design. Renewable and Sustainable Energy Reviews 47, 1016-1033.

IES LM-83-12. (2012). Retrieved April 10, 2017.

Jakubiec, J. A., & Reinhart, C. F. (2012). The 'adaptive zone'- A concept for assessing discomfort glare throughout daylit spaces. Lighting Research and Technology, 44(2), 149–170.

LEED v4 BD+C Reference Guide. (2014). Retrieved April 10, 2017, from U.S. Green Building Council.

Solemma. (2020). Climate Studio. Retrieved March 19, 2020.

Van Den Wymelenberg, K., Inanici, M., & Johnson, P. (2010). The effect of luminance distribution patterns on occupant preference in a daylit office environment. LEUKOS - Journal of Illuminating Engineering Society of North America, 7(2), 103–122.

Van Den Wymelenberg, K., & Inanici, M. (2014). A critical investigation of common lighting design metrics for predicting human visual comfort in offices with daylight. LEUKOS - Journal of Illuminating Engineering Society of North America.

Wilder, R., Mukhopadhyay, J., Femrite, T., & Amende, K. (2019). Evaluating glare in leed certified buildings to inform criteria for daylighting credits. Journal of Green Building, 14(4), 57–76.

Keywords: Daylighting, LEEDv4, Annual Glare, Productivity, Visual Comfort

Paper ID 31146: Designing a Solar Shading Solution parametrically using the direct sun and the view to the outside for a building in Vietnam

Pedro Marques*, Chinh Trieu, Wim Boydens

Studiebureau Boydens, Belgium

AIM AND APPROACH

In hot and humid climates where the sun is a permanent cooling load, façade optimization can be performed deeply with the aid of radiance-based daylight simulations. However, during design phase of a project, the extent of the parametric analysis is generally time-restricted.

For a mixed-use building in Vietnam, a grasshopper script was set up in order to aid the design team in optimizing the depth, width and density of vertical fins on the outer façade. The first instinct was the use of daylight simulations, so that the second skin would allow just the right amount of daylight in - too much leads to glare and overheating, not enough leads to artificial lighting usage.

After the first attempts, the initial setup was unsatisfactory. Simulations were taking too long, and the best results were produced by a densely populated outer facade, where the view to the outside was fractional.

Given the need to answer the time and design constraint of the project, a new approach was developed, using two factors - direct sun penetration and view to the outside. By turning the exercise into a static calculation, one was able to calculate thousands of possibilities for each element of the outer facade.

SCIENTIFIC INNOVATION AND RELEVANCE

The use of static calculations in parametric evaluations can reduce by 50 the amount of time necessarily to run façade studies parametrically. This would mean that, while a radiance-based study can study 100 variations, a static approach could reach 5000. Furthermore, radiance-based simulations are not responsive enough to be used in design team meetings, whereas a static approach is extremely helpful.

From the variety of the final results, the cherry-picked results could be secondly checked more thoroughly through the use of radiance-based simulations.

PRELIMINARY RESULTS AND CONCLUSIONS

The creation of an alternative path to optimization lead to a timely evaluation of the building facade, which worked effectively within the feedback loop of the design team. Thanks to the speedy and personalized approach, the design team changed its design for a better performing second skin.

MAIN REFERENCES

nothing to declare

Keywords: Radiance, daylight, direct sun

Paper ID 30225: Facades, roofs and solar parking yield estimation at Utrecht science park

Tayzer Damasceno de Oliveira* (1,2,3), Luis Fialho (2), Atse Louwen (3,4), Wilfried G.J.H.M. van Sark (3)

1: Accenture, Lisboa, Portugal; 2: University of Évora, Évora, Portugal; 3: Utrecht University, Utrecht, The Netherlands; 4: Eurac Research, Bolzano, Italy

AIM AND APPROACH

The goal of this work is to estimate the yearly potential and estimate the number of Electric vehicles charged per day using solar photovoltaic cells at Utrecht Science Park, located at Utrecht, The Netherlands. The first part is related to use ArcGIS to build the building shapefile with the building's height. The second use starts using the AutoCAD to apply the extrusion to make the building in 3D, Revit with the use of solar analysis tool that helps to find the best places to install the PV system. The third part belongs to the calculus to estimate how many charging stations we can have using the solar parking lots. Three layouts (VC0 [traditional], VC1, and VC2) were used in order to compare the yearly potential. The VC0 (16 767 kW) has the solar panels only on the roofs turned south. VC1 (26 769 kW) has the layout designed for the solar panels on the roofs and solar parking turned to the south direction, and VC2 (37 893 kW) with the cells turned to the east-west direction. And the facades turned south.

SCIENTIFIC INNOVATION AND RELEVANCE

The innovation relates to the attachment of the solar panels inside a building that is already in use, instead of the change the structure to integrate them. The solar parking lot has a great benefit using the parking lots to produce energy and protecting the cars from the weather conditions like hail, snow and etc. In addition, the use of this technology help to make the building a Nearly zero-energy building using the energy inside the building. Simultaneously using the charging station where the employees and students can charge their EV or even some electronic devices. Furthermore, demonstrating to the students the importance of using renewable energies (solar energy on this project) and how it works.

PRELIMINARY RESULTS AND CONCLUSIONS

The VC0 has a production of 13 549 MWh/year, VC1 with 19 708 MWh/year, and VC2 of 24 083 MWh/year, having the performance ratio of 0.71, 0.70, and 0.65. On the side of economic analysis, the NPV is positive for all of them, the Payback has the lowest value for VC0 that is 5.54, VC1 has 6.53, and VC1 with 8.02. The IRR is 14.06% for VC0, 11.83% for VC1, and 9.27% for VC2. The LCOE has the lowest value for VC0, being 0.068 €/kWh for VC0, 0,078 €/kWh for VC1, and 0.093 for VC2. Generally looking, VC0 represents the best option, especially if we take into consideration the economical results, but in the side of innovation, VC1 goes further because the parameters are better than VC2, and for this work, we didn't count the economic returns with the charging of the EV.

MAIN REFERENCES

AHN, Actueel Hoogtebestand Nederland (actual height data of the Netherlands), www.ahn.nl (2021).

Fouad MM, Shihata LA, Mohamed AH. Modeling and analysis of Building Attached Photovoltaic Integrated Shading Systems (BAPVIS) aiming for zero energy buildings in hot regions. Journal of Building Engineering, 21, 18-27, 2019.

Eldin, A. H., Refaey, M., & Farghly, A. (2015). A Review on Photovoltaic Solar Energy Technology and its Efficiency.

Jacobson, M. Z., & Jadhav, V. (2018). World estimates of PV optimal tilt angles and ratios ofsunlight incident upon tilted and tracked PV panels relative to horizontal panels. Solar Energy, 169, 55-66.

Oliveira, T. D. (2020). Facades and solar parking yield estimation at Ultrecht University (Master's thesis, Universidade de Évora).

Reich, N.H., Mueller, B. Armbruster, A., Van Sark, W.G.J.H.M., Kiefer, K., & Reise, Ch. (2012). Performance Ratio revisited: are PR > 90% realistic?. Progress in Photovoltaics, 20, 717-726

Keywords: Solar energy, renewable energy, solar parking lot, building attached photovoltaic

Session W2.2: Ensuring high quality building simulations

Paper ID 30146: Heat and moisture transport through a living wall system designated for greywater treatment

Hayder Alsaad*, Conrad Voelker

Bauhaus-University Weimar, Germany

AIM AND APPROACH

Façade greening systems designated for remediating greywater can help to relieve the water treatment centres while saving on irrigation water (Prodanovic et al. 2017). This study aims to numerically investigate the heat and moisture transport in such systems. As most heat and moisture simulation models cannot simulate the complex impact of vegetation on the simulated parameters, this study was conducted by coupling two simulation tools: ENVI-Met and Delphin. ENVI-Met is a high-resolution meteorological model that can simulate the interaction between urban geometry, vegetation, and the outdoor environment (Bruse and Fleer 1998). Delphin, on the other hand, is a simulation package for coupled heat and moisture transport in porous building materials (Grunewald 2000). In the present study, ENVI-Met was used to calculate the influence of the plants on air temperature, velocity, relative humidity, wind direction, and radiation (long wave and short wave) on the façade. Subsequently, the calculated parameters were then imposed on the façade in Delphin. Thus, ENVI-Met was used to determine the local climate conditions on the façade, which were used to conduct the hygrothermal simulations with Delphin. The hygrothermal simulations had a duration of four years to reach the equilibrium moisture content in the construction.

SCIENTIFIC INNOVATION AND RELEVANCE

With the continuously increasing levels of pollution in cities and rising temperatures due to the urban heat islands, living walls have been growingly investigated because of their promising potential in improving the urban environment. In addition, these systems can have a significant impact on the performance of the walls on which they are mounted. The literature indicates that façade greening can improve the heating demand of the building (Tudiwer and Korjenic 2017). Yet, due to evaporation from the substrate and transpiration from the plants, the relative humidity on the façade can increase (Capener and Sikander 2015). Moreover, as the greening system investigated in this study is meant for greywater treatment, it involves continuous water flow in the substrate of up to 50-75 L/d. An increase in humidity can damage the building material and reduce the energy efficiency of the building by increasing the heat conductivity of the wall layers. As hygrothermal simulations of living walls designated for greywater treatment is not reported in the literature, this study aims to investigate the impact of relatively high exposure to moisture on façades.

PRELIMINARY RESULTS AND CONCLUSIONS

To evaluate the impact of the living wall, two simulation models were created: a façade covered with a living wall and a reference facade with no greening. Both facades had a generic brick structure with a total thickness of 420 mm. The simulations showed that while the living wall was emitting water vapour, it did not increase the humidity content in the structure because of the ventilated air gap between the greening and the façade. In fact, the facade greening protected the wall from wind-driven rain and thus had a 16% less humidity content in the fourth simulation year in comparison to the reference case. The relative humidity of the interior surface of the wall was almost similar in both cases (58.3% with greening and 59.8% without greening). In the summer months (21 June – 21 September), the living wall cooled the façade's surface temperature due to shading, the thermal mass of the substrate, and the passive cooling of the plants. The maximum surface temperature behind the greening was 25°C compared to 40.6°C without greening. In the winter, the greening increased the minimum interior surface temperature by 1.2 K, which indicates an improvement in the thermal resistance of the construction.

MAIN REFERENCES

Bruse, Michael; Fleer, Heribert (1998): Simulating surface–plant–air interactions inside urban environments with a three dimensional numerical model. In Environmental Modelling & Software 13 (3-4), pp. 373–384. DOI: 10.1016/S1364-8152(98)00042-5.

Capener, Carl-Magnus; Sikander, Eva (2015): Green Building Envelopes – Moisture Safety in Ventilated Light-weight Building Envelopes. In Energy Procedia 78, pp. 3458–3464. DOI: 10.1016/j.egypro.2015.11.179.

Grunewald, John (2000): Documentation of the Numerical Simulation Program DIM3.1", Volume 2: User's Guide. Insitute of Building Climatology, Faculty of Architecture, Univesity of Technology Dresden.

Prodanovic, Veljko; Hatt, Belinda; McCarthy, David; Zhang, Kefeng; Deletic, Ana (2017): Green walls for greywater reuse. Understanding the role of media on pollutant removal. In Ecological Engineering 102, pp. 625–635. DOI: 10.1016/j.ecoleng.2017.02.045.

Tudiwer, David; Korjenic, Azra (2017): The effect of living wall systems on the thermal resistance of the façade. In Energy and Buildings 135, pp. 10–19. DOI: 10.1016/j.enbuild.2016.11.023.

Keywords: Hygrothermal simulations, Living wall, Coupling, ENVI-Met, Delphin

Paper ID 30605: A convolutional neural network for the hygrothermal assessment of timber frame walls

Astrid Tijskens, Staf Roels*

KU Leuven, Belgium

AIM AND APPROACH

Timber frame walls typically consist of a wind barrier at the cold exterior side and a vapour barrier at the warm interior side. In cold climates, the vapour barrier must have a higher vapour resistance than the wind barrier, to ensure vapour that entered the construction at the inside can dry out towards outside. However, there are no general guidelines available as to which combinations of wind and vapour barrier are safe in a specific context. Sometimes, a rule of thumb is used, which requires the ratio between the vapour resistances of vapour and wind barrier to be between 5 and 15 or even higher. This rule, however, does not take into account moisture buffering capacity of the structure nor specific climatic aspects, and hence does not guarantee an optimal solution. Because a hygrothermal simulation for every case would be too time-intensive, a metamodel is proposed in the current study, which allows quickly determining adequate combinations of wind and vapour barrier under given conditions. A convolutional neural network for time series is used to replace the hygrothermal simulations, thus allowing flexibility in the desired post-processing.

SCIENTIFIC INNOVATION AND RELEVANCE

The use of neural networks for time series predictions is a fairly novel metamodelling strategy in the field of building physics. When evaluating the hygrothermal performance of a building component in a probabilistic framework, metamodelling strategies have in the past been applied to predict specific and single-valued performance indicators. This approach provides little flexibility and might not provide sufficient information for decision-making. Instead, a metamodel predicting hygrothermal time series, as calculated by the original hygrothermal model, provides more information and allows the user to post-process the output as desired. In [1-2], the authors proved the applicability of convolutional neural networks for hygrothermal calculations of massive brick walls. The current study explores the models to predict the hygrothermal response of timber frame walls.

PRELIMINARY RESULTS AND CONCLUSIONS

First results show that it is possible to replace the time-consuming hygrothermal model with a much faster convolutional neural network, while maintaining high accuracy. Since material properties, such as (humidity dependent) vapour resistance and moisture buffering capacity, play a significant role in the hygrothermal response, the network requires additional input on this, compared to the network from [1-2], resulting in a slightly different network architecture.

MAIN REFERENCES

[1] A. Tijskens, S. Roels, and H. Janssen, "Neural networks for metamodelling the hygrothermal behaviour of building components," Building and Environment, vol. 162, no. June, p. 106282, 2019.

[2] A. Tijskens, H. Janssen, and S. Roels, "Optimising Convolutional Neural Networks to Predict the Hygrothermal Performance of Building Components," Energies, vol. 12, no. 20, p. 3966, 2019.

Keywords: hygrothermal assessment, timber frame walls, metamodelling, convolutional neural network

Paper ID 30726: Besos: a python library that links energyplus with energy hub, optimization and machine learning tools

Theodor Victor Christiaanse (1,2), Paul Westermann (1,2), Will Beckett (1,2), Gaelle Faure (1,2), Ralph Evins* (1,2)

1: Energy in Cities group, Department of Civil Engineering, University of Victoria, BritishColumbia, Canada; 2: Institute for Integrated Energy Systems, University of Victoria, British Columbia, Canada

AIM AND APPROACH

The goal of the BESOS library is to create an easy way for academics start building modelling experiments that involve linking EnergyPlus with optimization, energy system design and machine learning techniques. The Building and Energy Simulation, Optimization and Surrogate-modelling (besos) library provides a Python-based software bridge between EnergyPlus and various other environments. Furthermore, software abstractions are specifically setup such that building energy modellers can quickly create numerous model variations using parametric definitions. These include optimizing the building design and energy systems, and integration of the modeling results with machine learning techniques. Accompanying the Python library, a web platform (BESOS) [1] is freely accessible for academics to use the software, learning through our extensive library of examples and developing new software extensions on top of the existing software paradigms.

SCIENTIFIC INNOVATION AND RELEVANCE

The optimal exploration of building design and operation requires the use of many software tools. Among them, EnergyPlus (E+) is a commonly used physics-based building energy simulation tool. The use of machine learning (ML) is finding adoption among engineers in the field and may have a huge impact on the speed and breath of modelling experiments. ML techniques can perform different tasks that could inform building design and operation such as; (i) capture the dynamics of a physics-based models in a fast and accurate surrogate model [2] reducing the cost of expensive exploration, (ii) large building datasets of E+ or real timeseries data can be analysis on a building-by-building level using ML techniques [3], (iii) retrofit measures may be identified by understanding the timeseries sensor data through black box techniques [4], (iv) forecasting future energy performance of the building stock using ML for forecasting [5].

The potential of these techniques ML techniques is evident. The integration between these new Python-based libraries and physics-based modelling tools used for energy modelling was limited. Our software library gives modellers the ability to create building models that work in E+ and manipulate the inputs and outputs of the E+ model within a Python environment.

PRELIMINARY RESULTS AND CONCLUSIONS

The Python environment allows for the integration with these new novel machine learning libraries such as TensorFlow and Scikit-learn. Furthermore, we have also included links to powerful optimization solvers and energy system design tool Energy Hub. These three tools provide numerous options for energy modelers to combine multivariant data sets and E+ models to create experiments that stack many different techniques into a single interface.

We will discuss the challenges and successes we have had building the software library besos and BESOS platform. A demo of the software capabilities will be shown and demonstration how multi-model ecologies can be built are presented. A list of recent projects and papers that are would not be possible without the platform [2-5]. Finally, we will share how we continue to improve the underlying software for the next possible versions by partnering with computer scientists and making innovative technologies available to the building modelling domain.

MAIN REFERENCES

[1] Faure G, Christiaanse T, Evins R, Baasch GM. BESOS: a Collaborative Building and Energy Simulation Platform. In Proceedings of the 6th ACM International Conference on Systems for Energy-Efficient Buildings, Cities, and Transportation 2019 Nov 13 (pp. 350-351).

[2] Westermann P, Evins R. Surrogate modelling for sustainable building design–A review. Energy and Buildings. 2019 Sep 1;198:170-86.

[3] Baasch G, Wicikowski A, Faure G, Evins R. Comparing gray box methods to derive building properties from smart thermostat data. In Proceedings of the 6th ACM international conference on systems for energy-efficient buildings, cities, and transportation 2019 Nov 13 (pp. 223-232).

[4] Westermann P, Deb C, Schlueter A, Evins R. Unsupervised learning of energy signatures to identify the heating system and building type using smart meter data. Applied Energy. 2020 Apr 15;264:114715.

[5] Westermann P, Braun J, Murphy E, Grieco J, Evins R. Insight Into Predictive Models: On The Joint Use Of Clustering And Classification By Association (CBA) On Building Time Series. In Rome, Italy; [cited 2020 Jul 16]. p. 1564–71. Available from: http://www.ibpsa.org/proceedings/BS2019/BS2019_211236.pdf

Keywords: machine learning, building modelling, energy hub modelling, energyplus

Paper ID 30613: Data-driven black box model of building dynamics

Sophie Bernard* (1,2), Valery Ann Jacobs (1), Bert Belmans (1,3), Arjen Mentens (1), Filip Descamps (1,4), John Lataire (1)

1: Vrije Universiteit Brussel, Belgium; 2: Université libre de Bruxelles, Belgium; 3: Universiteit Antwerpen, Belgium; 4: Daidalos Peutz, Belgium

AIM AND APPROACH

In this research a modelling tool is presented to derive surrogate models of thermal energy transfers in buildings, to support the development and testing of smart control algorithms.

A data-driven approach was used to identify a model able to predict the indoor temperature in a case-study building when an electric heater was turned on. The data about the system was generated by EnergyPlus simulations, which resolve the heat balance equations to simulate the thermal response of a building. The model structure that was selected is a second-order ARMAX transfer function whose parameters were identified with a Least Squares optimization criterion. The model inputs were limited to the heater's power, the global horizontal solar radiation and the outdoor dry-bulb temperature.

SCIENTIFIC INNOVATION AND RELEVANCE

Buildings account for one-third of the global energy consumption in the world, which is more than the industry sector or the transport sector. For environmental and economic reasons, it is thus important to reduce their energy consumption, while preserving a comfortable environment for their occupants. In this context, computer-aided control techniques for building comfort systems can be a valuable asset. Advanced control techniques, such as model predictive control, require supporting system models for development and testing. However, detailed physical models that carefully take system dynamics into account are too computationally intensive for practical applications. This is why surrogate models of low complexity are highly relevant.

The innovative element is that a frequency domain modelling approach is used. This allows to conveniently select the frequency band of interest. Namely, the dynamics are mostly important at low frequencies. Discarding the high frequencies implicitly removes a significant amount of noise.

Also, the use of a data-driven approach implicitly takes into account influences which, in a first principles approach, might have been neglected. That is, the model is validated on the data rather than on physical insight.

PRELIMINARY RESULTS AND CONCLUSIONS

Simulation experiments have been conducted where random binary sequences (a sequence of step inputs) have been applied as the heater's input, and historical weather conditions have been used for the solar radiation and the outside air temperature.

A transfer function model was obtained, describing the relation between the inputs (heater and weather conditions) and the resulting ambient temperature in the room. A data set corresponding to the month of July was used for the identification, and the resulting model was then validated on a data set of the month of December.

It was demonstrated that the model was able to predict fairly accurately the indoor temperature when the building was subject to winter or summer weather conditions. Further improvements and refinements will be carried out, including taking into account the difference in time constants of the heater and the weather conditions.

MAIN REFERENCES

P. Abrahams et al., Method for Building Model Calibration to Assess Overheating Risk in a Passive House in Summer, Proceedings of the 16th IBPSA conference, Rome, Italy, Sept. 2-4, 2019, DOI 10.26868/25222708.2019.210768

S. Mostafavi et al., Model Development for Robust Optimal Control of Building HVAC, Proceedings of the 16th IBPSA conference, Rome, Italy, Sept. 2-4, 2019, DOI 10.26868/25222708.2019.211331

S. Royer, Energy and Buildings, Volume 78, pg. 231-237, A procedure for modeling buildings and their thermal zones using co-simulation and system identification, 2014. DOI 10.1016/j.enbuild.2014.04.013

S. Royer et al., IFAC Proceedings Volumes, Volume 47, Issue 3, Pages 10850-10855, Black-box modeling of buildings thermal behavior using system identification, 2014. DOI 10.3182/20140824-6-ZA-1003.01519

R. Pintelon and J. Schoukens. System Identification: A Frequency Domain Approach. John Wiley, 2nd edition, 2012.

Keywords: Building thermal modelling, EnergyPlus, system identification, black-box model, surrogate modelling

Paper ID 30226: Summer passive strategies assessment based on calibrated building model using on site measurement data

Obaidullah Yaqubi* (1,2,3), Auline Rodler (1,2), Sihem Guernouti (1,2,3), Marjorie Musy (1,2,3)

1: Equipe de recherche BPE, Cerma Ouest, Nantes, France; 2: Institut de Recherche en Sciences et Techniques de la Ville (IRSTV) , Nantes, France; 3: CNRS UMR 6183, GeM, Université de Nantes, France

AIM AND APPROACH

With the changes in worldwide climate conditions, extreme summer heat events will become more frequent and severe rendering buildings uncomfortable. This paper in this context, presents the application of co-simulation on practical design issues of mixed mode ventilated buildings. It is based on a 2-months field study measurement of outdoor and indoor air temperatures and window operation of an existing residential building during the hottest season of the year in Nantes (France) in 2018.

The aim of this study is first to use measured indoor temperature to calibrate a building co-simulation model and second to evaluate how openness ratio of windows and operation of window shutters affect the indoor thermal comfort during summer.

SCIENTIFIC INNOVATION AND RELEVANCE

Two modelling tools, Contam and Trnsys, were coupled to simultaneously simulate airflow and temperature dynamics of the whole building. Since each of the five storeys had similar thermo-physical features, it was decided to consider only the last storey for the purpose of the present study because it is the most sensitive to outdoor conditions. Each piece in the apartment i.e. bedroom, living room, and bathroom was treated as a separate zone. The time step of the simulation has been set to 15 min.The agreement between measured and simulated indoor temperature values at every hour was evaluated with the Coefficient of Variation of the Root Mean Square Error (CV(RMSE)) and Mean Absolute error (MAE).

For assessing the thermal comfort, international adaptive comfort standard methodologies such as EN 16798 and ASHRAE 55 as well as PMV were used to measure and to compare indoor comfort in the apartments.

PRELIMINARY RESULTS AND CONCLUSIONS

After making necessary adjustments to the model, the co-simulation model was calibrated to the lowest (CV(RMSE)) values, between 3 and 5%, on indoor temperatures in the living rooms of 3 apartments and stairwell. Analysing indoor temperature of calibrated simulation building with adaptive thermal comfort indices showed that apartments and stairwell in the last floor of the building experienced higher temperatures than the maximum allowable operative temperature for category I and II of EN 16798 and ASHRAE 55 category of acceptability of 80 and 90% up to 5% of the studied period. Passive strategies such as adjusting the openness ratio of windows at night and day (natural ventilation) and closing window shutters to 0.9 i.e. ratio of no-transparent area during the day proved to reduce overheating risks for category I and II of EN 16798 and ASHRAE 55 acceptability category of 80% but not enough for acceptability category of 90%. The latter may need to use mechanical means to reduce overheating risks.

MAIN REFERENCES

Dols, W.S., and Polidoro, B.J. (2015). CONTAM User Guide and Program Documentation Version 3.2 (Na-tional Institute of Standards and Technology)

ASHRAE-55-2017, «Thermal Environmental Conditions for human occupancy,» Atlanta, 2017.

Bienvenido-Huertas, D., Sánchez-García, D., Rubio-Bellido, C., and Oliveira, M.J. (2020). Influence of adap-tive energy saving techniques on office buildings located in cities of the Iberian Peninsula. Sustain. Cities Soc. 53, 101944

Keywords: simulation, calibration, indoor thermal comfort

Session W2.3: Buildings paving the way for the energy transition

Paper ID 30706: A low-order semi-physical borefield model for optimal control applications

Iago Cupeiro Figueroa* (1), Lieve Helsen (1,2)

1: KU Leuven, Belgium; 2: Energyville, Belgium

AIM AND APPROACH

The following research presents a new borefield model for optimal control applications. The proposed model uses a resistance capacitance (RC) network to model the ground. The first nodes of the ground model, which capture the short-term dynamics, are parametrized based on the physical properties of the ground and the borehole. The other nodes, which capture the long-term dynamics, are parametrized based on a fitting procedure that uses information from the borefield characteristic thermal response (i.e., the g-function [1]).

SCIENTIFIC INNOVATION AND RELEVANCE

In the literature, other approaches can be found for control-oriented modeling of borefields. Data-driven models have the drawback that they require a large amount of data [2]. RC ground models impose a fixed temperature at one of the network ends, capturing only the short-term dynamics and disregarding the borehole interactions [3]. Finally, load-aggregation formulations require a large number of states to capture the load history information [4]. The proposed model captures the long-term behaviour of the borefield relying on its characteristic g-function (i.e., no monitoring data but geometrical and physical information of the borefield is required) and using a low number of states.

PRELIMINARY RESULTS AND CONCLUSIONS

The model fitting procedure is parametrized as a function of the number of states to assess the minimum level of detail needed. Based on a step-response test, it is found that 4 states are enough to capture the long-term dynamics. An excessive number of states can lead to overfitting. The proposed RC model is implemented using Modelica and validated against the IBPSA library borefield model [5] for the short- and the long- term range. The model provides good accuracy with a maximum error under 0.5 °C in the short-term and 0.1 °C in the long-term for a number of states lower than 10 for a step-response and periodic pulse case studies.

MAIN REFERENCES

[1] Eskilson, P. (1987). Thermal analysis of heat extraction boreholes. PhD thesis

[2] Atam, E., Schulte, D. O., Arteconi, A., Sass, I., & Helsen, L. (2018). Control-oriented modeling of geothermal borefield thermal dynamics through Hammerstein-Wiener models. Renewable Energy, 120, 468-477.

[3] Cupeiro Figueroa, I., Picard, D., & Helsen, L. (2020). Short-term modeling of hybrid geothermal systems for Model Predictive Control. Energy and Buildings, 109884.

[4] Laferrière, A., & Cimmino, M. (2018). Model predictive control applied to residential self-assisted ground source heat pumps.

[5] Laferrière, A., Cimmino, M., Picard, D., & Helsen, L. (2020). Development and validation of a full-time-scale semi-analytical model for the short-and long-term simulation of vertical geothermal bore fields. Geothermics, 86, 101788.

Keywords: Control-oriented modeling, borefield modeling, model validation

Paper ID 30380: An Open-AI gym environment for the Building Optimization Testing (BOPTEST) framework

Javier Arroyo* (1,2,3), Carlo Manna (2,3), Fred Spiessens (2,3), Lieve Helsen (1,2)

1: Department of Mechanical Engineering, KU Leuven, Heverlee, Belgium; 2: EnergyVille, Thor Park, Waterschei, Belgium; 3: Flemish Institute for Technological Research (VITO), Mol, Belgium

AIM AND APPROACH

Building HVAC accounts for 15% of the world final energy use [1]. Their primitive controllers for indoor climate control have shown significant room for improvement, and disagree with the superb developments in state-of-the-art technologies like machine learning. One of the main reasons why these technologies are not yet being adopted by the building sector is the lack of rigorous benchmarking to grow in maturity.

This paper describes an OpenAI-Gym environment for the BOPTEST framework to rigorously benchmark different reinforcement learning algorithms among themselves and against controllers of other type by building simulation. The design philosophy of the environment and its different features are introduced. Finally, the environment is demonstrated in one emulator building model to train a reinforcement learning algorithm and compare it against a classical control logic.

SCIENTIFIC INNOVATION AND RELEVANCE

This paper:

- Introduces an OpenAI-Gym environment that enables the interaction with a set of physics-based and highly detailed emulator building models to implement and assess reinforcement learning for the application of building climate control and demand response.

- Demonstrates the functionality of the framework by implementing and evaluating a state-of-the-art reinforcement learning algorithm to one of the building emulator models.

This work fosters a novel interface that bridges the gap between the latest innovations of machine learning and the field of building energy management. The presented environment can be used to assess the performance of reinforcement learning algorithms when implemented in detailed and reliable building emulator models.

PRELIMINARY RESULTS AND CONCLUSIONS

This paper presents an OpenAI-Gym environment for the BOPTEST framework to enable simulation, evaluation and benchmarking of RL algorithms for building energy management. The framework integrates the main elements that are required by every Gym environment, like the "reset" and "step" methods, along with other useful features to conveniently implement and evaluate "state-of-the-art" RL algorithms for building climate control.

The presented environment heavily relies on the already existing BOPTEST functionality and exploits some of its core components like its API and its KPI calculator module. Definition of the reward function, possibly by the user, is especially taken care of.

Finally, a RL algorithm has been tuned, trained, implemented and benchmarked for different hyperparameter settings of the BOPTEST-Gym environment in order to demonstrate the capabilities of the framework for evaluation and benchmarking.

MAIN REFERENCES

[1] International Energy Agency. (2019). Global Status Report for Buildings and Construction: Towards a zero-emissions, efficient and resilient buildings and construction sector

[2] Peirelinck, T., Ruelens, F., & Decnoninck, G. (2018). Using reinforcement learning for optimizing heat pump control in a building model in Modelica. 2018 IEEE International Energy Conference, ENERGYCON 2018, 1–6. https://doi.org/10.1109/ENERGYCON.2018.8398832

[3] Dulac-Arnold, G., Mankowitz, D., & Hester, T. (2019). Challenges of Real-World Reinforcement Learning (2019), RL4RealLife Workshop in the 36th International Conference on Machine Learning (ICML) 2019.

[4] Blum, D., Jorissen, F., Huang, S., Chen, Y., Arroyo, J., Benne, K., and Li. Y. (2019). Prototyping the BOPTEST Framework for Simulation-Based Testing of Advanced Control Strategies in Buildings. In proceedings of the 16th IBPSA Conference, Rome.

Keywords: reinforcement learning, BOPTEST, OpenAI-Gym, building energy management

Paper ID 30721: Open and Reproducible Use Cases for Energy (ORUCE) methodology in systems design and operation: a dwelling photovoltaic self-consumption example.

Sacha Hodencq*, Benoit Delinchant, Frederic Wurtz

Univ. Grenoble Alpes, CNRS, Grenoble INP, G2Elab, 38000 Grenoble, France*

AIM AND APPROACH

The development of low-carbon and decentralized renewable energies in the struggle against climate change, especially in urban areas where more than half of the global primary energy is consumed, raises the following question: how to design an energy project and manage energy flows in order to be optimal from ecological and financial points of view? One way to help stakeholders in this task is to provide open energy modelling environments. In this article, we present an Open and Reproducible Use Case for Energy (ORUCE) based on open data, models and tools for the optimal design and operation of an energy system. The use case is the one of a dwelling with self-consumption of photovoltaic (PV) energy and battery storage use. Two free and open source software will be used: NoLOAD for the design, with a focus on both Life Cycle Assessment (LCA) and energy coverage, and OMEGAlpes for the energy management across a year. Providing an ORUCE makes the study reproducible and available for modifications and derived analysis, in addition to making the models easy to grasp.

SCIENTIFIC INNOVATION AND RELEVANCE

Historical energy system modelling as well as current mainstream approaches are closed and proprietary, even if open energy modelling has a promising emergence (Pfenninger et al. 2018). Open energy modelling actually includes many interests: first, it leads to an improved quality of science through transparency and peer reviewing, avoiding errors, biases or even fraud (Pfenninger et al. 2017). An open and reproducible research reduces parallel efforts and enables to share methods and quantitative work with policy makers, when institutions are increasingly asking for transparency. Finally, openness provides trust and legitimacy for scientific arguments in the public debate over the energy transition (Morrison 2018).

Open energy modelling processes have been described in the literature (Pfenninger et al. 2017; Morrison 2018; Hülk et al. 2018) as going from open data sets including data points and metadata, being processed and used with study assumptions for the open model formulation and solving. The model output and interpretation can be shared in open access scientific publication as well as public communication. We propose an ORUCE definition, method, and an actual example to put this process into practice and make the dwelling photovoltaic self-consumption example reproducible and available for improvements.

PRELIMINARY RESULTS AND CONCLUSIONS

The full article will present the design of a dwelling self-consumption of photovoltaic energy with battery storage according to LCA and energy coverage objectives on the one hand, and its energy management across a year on the other hand.

The open tools, models and scenarios used for the study are directly available in the code repository as source code, but also as Jupyter Notebooks, i.e. documents that contain live code, equations, visualizations and narrative text that can be freely used online. Notebooks include many interests: it capitalizes the results of the study, and the way those results were found as well. Moreover, users can modify them, contribute to the code open development, and distribute their results.

We define Open and Reproducible Use Cases for Energy (ORUCE) as notebooks based on open data and open source tools, clearly presenting the method and assumptions used to obtain and present the results, and ideally linked to open edition publications. These ORUCE can be made available, used and broadcasted in open energy platforms.

ORUCE nourish openness in energy modelling and provide actual case studies, thus leading to better science and cooperation between researchers themselves as well as with citizens and policy makers.

MAIN REFERENCES

Hülk, Ludwig, Berit Müller, Martin Glauer, Elisa Förster, and Birgit Schachler. 2018. 'Transparency, Reproducibility, and Quality of Energy System Analyses – A Process to Improve Scientific Work'. Energy Strategy Reviews 22 (November): 264–69. https://doi.org/10.1016/j.esr.2018.08.014.

Morrison, Robbie. 2018. 'Energy System Modeling: Public Transparency, Scientific Reproducibility, and Open Development'. Energy Strategy Reviews 20 (April): 49–63. https://doi.org/10.1016/j.esr.2017.12.010.

Pfenninger, Stefan, Joseph DeCarolis, Lion Hirth, Sylvain Quoilin, and Iain Staffell. 2017. 'The Importance of Open Data and Software: Is Energy Research Lagging Behind?' Energy Policy 101 (February): 211–15. https://doi.org/10.1016/j.enpol.2016.11.046.

Pfenninger, Stefan, Lion Hirth, Ingmar Schlecht, Eva Schmid, Frauke Wiese, Tom Brown, Chris Davis, et al. 2018. 'Opening the Black Box of Energy Modelling: Strategies and Lessons Learned'. Energy Strategy Reviews 19 (January): 63–71. https://doi.org/10.1016/j.esr.2017.12.002.

Keywords: Open Science, Reproducibility, Optimisation, Energy system, Self-consumption

Paper ID 31133: Is your building automation and control system properly designed and installed?

Muhyiddine Jradi*

University of Southern Denmark, Denmark

AIM AND APPROACH

Considering the urgent need and the large technical potential, an innovative holistic tool, IBACSA, is designed and developed, providing a first-of-its kind instrument for building automation and control systems assessment and smartness evaluation. IBACSA development is carried out as one of the deliverables of the research project BuildCOM - Automated Auditing and Continuous Commissioning of Next Generation Building Management Systems, supported by the Danish Energy Agency through the EUDP program.

Regarding the auditing and evaluation methodology, IBACSA employs a hybrid qualitative-quantitative multi-criteria holistic framework, considering eight major building domains: Heating, Domestic Hot Water, Cooling, Ventilation, Lighting, Dynamic Envelope, Electricity and Monitoring and Control. Each domain includes a set of services, defined based on the Standard EN15232 dealing with building automation and control systems. The total number of services is 60, where each service is characterized by different levels of control functionalities.

In auditing the BACS using IBACSA, the assessor will be evaluating the 60 services, choosing the right control level for each service. IBACSA evaluates the BACS and quantify the impact of the selected control levels against five major impact criteria: (1) Energy efficiency, (2) Maintenance and fault prediction, (3) Energy flexibility, (4) Comfort and (5) Information to occupants.

SCIENTIFIC INNOVATION AND RELEVANCE

With the building sector digitalization and the huge rise in the number of installed sensors, smart meters and IoT devices, Building Automation and Control Systems (BACS) are to play key role in the foreseen future. Thus, considering the key role of the buildings stock in the energy sector along with the large potential of improving the building energy performance and optimizing the systems operation, a well-designed and properly installed and operated BACS is critical to achieving current and future energy efficiency and environmental goals set by the countries worldwide.

The trend in building commissioning is that this process is carried out on the whole building level, evaluating if the building envelope is living up to the standards and that the predicted energy consumption is in line with the building regulations. However, there is an overall lack of information and frameworks aiding in the evaluation of the BACS and ensuring that such systems are properly designed and installed. In addition, there is a need for user-friendly and comprehensive tools and instruments aiding the decision making on the energy effective and optimal design and installation of BACS along with allowing auditing and evaluating various building services and the overall building smartness.

PRELIMINARY RESULTS AND CONCLUSIONS

IBACSA evaluation matrix aims at evaluating each domain by itself allowing quick identification of poorly performing building domains across all impact criteria. So, each of the eight domains is evaluated and scored against each of the five impact criteria in addition to an overall score provided for each domain. In addition, the evaluation matrix also provides an overall assessment of the BACS with one overall score against each of the five impact criteria. Thus, IBACSA provides the auditor with a flexibility to consider different evaluation impacts and it is up to the interest of the auditor to consider what is more important.

To allow building consultants, researchers and end-users to fully benefit from IBACSA implementation in case study buildings, the tool is compiled in a standalone desktop executable application allowing for wide distribution and flexible use. The tool consists of a user-friendly interface along with graphical elements and tabs. In selecting services control levels, a simple drop-down approach is employed. In addition, IBACSA aids the user with the decision-making process in terms of retrofitting the current BACS by allowing very simple evaluation and comparison of the impacts of BACS functionalities selected.

MAIN REFERENCES

Engelsgaard, S., Alexandersen, E.K., Dallaire, J., Jradi, M. (2020). IBACSA: An interactive tool for building automation and control systems auditing and smartness evaluation, Building and Environment 184, 107240.

European Technical Standard EN 15232, Energy Performance of Buildings—Impact of Building Automation, Control, and Building Management (2nd ed.), CEN, Brussels (2012)

Keywords: IBACSA, Smart buildings, Building automation and control systems, Auditing and Commissioning, EN15232 standard

Session W2.4: Climate change and bioclimatic design

Paper ID 30718: Adapting French buildings to future climate: Passive design optimisation

Anaïs Machard* (1,2), Christian Inard (1), Jean-Marie Alessandrini (2), Charles Pelé (2), Jacques Ribéron (3)

1: Laboratoire des Sciences de l'Ingénieur pour l'Environnement (LASIE, UMR CNRMS 7356), La Rochelle Université, 23 Avenue Albert Einstein, 17000 La Rochelle, France; 2: Département Energie et Environnement, Centre Scientifique et Technique du Bâtiment (CSTB), 84 Avenue Jean Jaurès, Champs-sur-Marne, 77447 Marne-la-Vallée CEDEX 2, France; 3: Département Santé et Confort, Centre Scientifique et Technique du Bâtiment (CSTB), 84 Avenue Jean Jaurès, Champs-sur-Marne, 77447 Marne-la-Vallée CEDEX 2, France;

AIM AND APPROACH

In France, buildings have historically been designed to withstand cold winter. In recent years, "bioclimatic" design has become mandatory in the national thermal regulation, to reduce the heating consumption. However, with climate change and more frequent projected heatwaves, there is a new need to design buildings to also be able to support warm summers and not overheat. Future buildings should be able to be resilient to the outdoor changing climate, and the adaptation should start from the design phase. The dangerous predicted increase of air-conditioning penetration and use reinforces the exigency to propose a design consuming as less energy as possible to reduce greenhouse gases emissions. With the development of climate projections and the recent availability to use these future climate data for building thermal simulations, conceiving buildings today to be adapted for the climate of tomorrow is feasible. In this paper, we introduce a new paradigm: Is it possible to design a building that will be adapted both to winter and summer? Are passive cooling solutions enough to maintain acceptable indoor environment comfort level? In our analysis, we propose to optimize a low-energy residential building case-study for future summer conditions, including future heatwaves sequences.

SCIENTIFIC INNOVATION AND RELEVANCE

The algorithm NSGA-II is used to test different building combinations such as thermal inertia, proper use of solar shading, nocturnal ventilation, cool paints, earth-to-air heat exchanger, and indirect adiabatic cooling in different French cities, while the building is modelled with the software EnergyPlus. The objectives are to reduce summer thermal discomfort while reducing the uncertainty related to the future climate projections from CORDEX datasets. The outcomes of the study give indications of best strategies to implement in different locations, while the overall methodology can be re-used to consider risk and uncertainties related to future climate conditions during the design process.

PRELIMINARY RESULTS AND CONCLUSIONS

It is very challenging to propose an optimised passive building design to future heatwaves in France. Traditional summer bioclimatic design, such as the use of solar shading, nocturnal ventilation and cool paints is enough to maintain acceptable indoor level conditions under a future typical climate of an average climate model, but not under higher projections such as future intense heatwaves from high climate models. In order to provide a resilient design to different sorts of future climate conditions, the additional use of semi-active cooling solutions such as earth-to-air heat exchanger, and indirect adiabatic cooling is necessary.

MAIN REFERENCES

Moazami, A.; Carlucci, S.; Nik, V.M.; Geving, S. Towards climate robust buildings: An innovative method for designing buildings with robust energy performance under climate change. Energy Build. 2019, 202, 109378.

Machard, A.; Inard, C.; Alessandrini, J.M.; Pelé, C.; Ribéron, J. A Methodology for assembling future weather files including heatwaves for building thermal simulations from the European Coordinated Regional Downscaling Experiment (EURO-CORDEX) climate data. Energies 2020,13, 3424.

Ascione, F.; Bianco, N.; Francesca De Masi R. ; Maria Mauro, G. ; Vanoli G.P. Resilience of robust cost-optimal energy retrofit of buildings to global warming: A multi-stage, multi-objective approach. Energy and Buildings. 2017, 153, pages 150-167

Chinazzo, G.; Rastogi P.; Andersen, M.; Robustness assessment methodology for the evaluation of building performance with a view to climate uncertainties. 14th International Conference of the International Building Performance Simulation Association, At Hyderabad, India. 2015

Lapisa, R.; Bozonnet, E.; Salagnac, P.; Abadie M.O.; Optimized design of low-rise commercial buildings under various climates – Energy performance and passive cooling strategies. Building and Environment. 2018, 132, pages 83-95

Keywords: Optimization, climate change, summer overheating, building design, passive cooling

Paper ID 30772: Assessing the relevance of updated climate files in the simulation of residential buildings located in semi-arid regions

Emanuela Giancola, María Nuria Sánchez, Helena López*, Jose Antonio Ferrer, María José Jiménez, Silvia Soutullo

Energy efficiency in buildings research unit, CIEMAT, Spain

AIM AND APPROACH

Real building energy consumption is one of the high-priority research and innovation topics in the scientific community [1]. This work quantifies the climate trends registered on the energy performance of residential buildings located in semi-arid regions. The climate of this zone is characterized by hot-dry summers and cold-dry winters with mean annual temperatures below 18ºC and low annual precipitations. This study contributes to improve the knowledge that the global warming has on the building demands under severe climate conditions.

A generic building case placed in the desert of Tabernas (Almería, Spain) has been analyzed. Firstly, one representative residential building has been modelled with the software TRNSYS [2], according to the local characteristics. To evaluate the annual thermal loads, seasonal temperature set-points have been fixed. Secondly, a parametric evaluation has been developed considering different Spanish building regulations, two boundary conditions (4 and 2 external façades) and three climate years. The meteorological conditions have been assessed through the use of one long-term experimental file for Tabernas [3], and two synthetic years downloaded on the Energy plus [4] and the Spanish building code websites [5]. Finally, the obtained results have been compared to quantify the climate trends produced in different building configurations.

SCIENTIFIC INNOVATION AND RELEVANCE

The climate change impact is a major fact affecting many cities, their communities and infrastructures, and especially vulnerable urban populations. This research is focused on the impact that outdated climate files have on the energy simulation of a generic residential building model. The use of this non-realistic datasets produces many uncertainties into the modelling process [6]. To quantify this deviation, one long-term experimental climate file and two representative climate files have been used as inlet information. The comparison between them gives an idea of how the climate change modifies the patterns of the building energy performance. This issue has been discussed in recently published papers ([7], [8]).

This article highlights the necessity of using updated meteorological information to reduce the uncertainties produced by the input data in the simulation process. This can help governments to better adjust its policies and action plans to the current weather conditions.

Although this study has been carried out for a specific case located in the desert of Tabernas (Almería, Spain), generic buildings and normative values for the building envelope have been selected in order to extend the obtained results to buildings with similar construction characteristics located in similar climate zones (BWk Köppen Geiger classification).

PRELIMINARY RESULTS AND CONCLUSIONS

An initial study analyzes the climate trend in a semi-arid region identifying the tendencies over the last decade by an experimental campaign of 10 years carried out by CIEMAT in Tabernas. Long-term experimental database present warmer (+0.3°C) and drier (-16%) summers compared to the Energy plus synthetic year. However, the Spanish building code synthetic summer is significantly colder (-1.6°C) and wetter (+4%). This tendency is strongly supported by the number of tropical nights, around 65 on both experimental and Energy plus years, and clearly higher than 18, the corresponding value for Spanish building code year. However, Energy plus synthetic year presents the warmest annual climate.

Subsequently, the climate impact on the building thermal conditioning is assessed by means of dynamic simulations with TRNSYS. The building configuration with 4 external façades demands more annual thermal loads than the 2 façades configuration. The implementation of more currently constructive normative strongly reduces the annual thermal loads. The climate file provided by the Spanish normative obtains the highest annual loads, followed by the experimental file. The climate file provides by EnergyPlus website reaches the lowest annual loads. These differences are very low as far as stricter constructive conditions are implemented in the building configuration.

MAIN REFERENCES

[1] EBC Executive Committtee. 2019. International Energy Agency. Strategic Plan 2019-2024. Energy in Buildings and Communities Technology Collaboration Programme.

[2] Transient System Simulation Tool (TRNSYS). Available online: https://www.trnsys.com.

[3] Sánchez, M.N., Soutullo, S., Olmedo, R., Bravo, D., Castaño, S., Jiménez, M.J., 2020. An experimental methodology to assess the climate impact on the energy performance of buildings: A ten-year evaluation in temperate and cold desert areas. Applied Energy. 264, 114730.

[4] Energy Plus Weather Database. Available online: https://energyplus.net/weather-region/europe_wmo_region_6/ESP%20%20.

[5] Spanish Building Code. Available online: https://www.codigotecnico.org/index.php/menu-ahorro-energia.html.

[6] W. Tian, Y. Heo, P. de Wilde, Z. Li, D. Yan, C.S. Park, X. Feng, G Augenbroe. 2018. A review of uncertainty analysis in building energy assessment. Renewable and Sustainable Energy Reviews. 93, pp. 285–301.

[7] Soutullo S, Giancola E, Jiménez MJ, Ferrer JA, Sánchez MN. How Climate Trends Impact on the Thermal Performance of a Typical Residential Building in Madrid. Energies 2020;13(1):237.

[8] Wang, L.; Liu, X.; Brown, H. Prediction of the impacts of climate change on energy consumption for a medium-size office building with two climate models. Energy Build. 2017, 157, 218–226.

Keywords: Climate change, updated meteorological file, building energy simulation, generic residential building models

Paper ID 30845: Climate change impact on the future performance of Nearly Zero Energy buildings: A case study base analysis

Mamak Pourabdollahtootkaboni*, Giovanna De Luca, Ilaria Ballarini, Vincenzo Corrado

Politecnico di Torino, Turin, Italy

AIM AND APPROACH

This paper investigates the effects of climate changes on energy performance and overheating risk of a nearly zero energy building (nZEB) in different climate zones in Italy. The study is carried out by analyzing the nZEB requirements such as the annual energy needs for space heating and space cooling (EPH,nd and EPC,nd, respectively), the overall annual total primary energy (EPgl,tot), the mean overall seasonal efficiencies of technical building systems and renewable energy ratio, under different scenarios. "Representative Concentration Pathways (RCPs)"8.5 (business as usual) of emission and concentration scenarios according to the fifth assessment report of the intergovernmental panel on climate change [1], have been applied in this study. Dynamically down-scaled future hourly weather data from the regional climate models (GERICS-REMO 2015) are used in this work to create future typical meteorological year (TMY). To reduce the climate model uncertainties a distribution based, multivariate method for bias-correction of weather data is implemented [2]. Energy simulations are carried out using Energy Plus for the mid-term (from 2056 to 2075) and long term (from 2081 to 2100) period.

SCIENTIFIC INNOVATION AND RELEVANCE

Recent studies reveal that climate change will shift the building's performance pattern in the future. According to the fifth assessment report of the intergovernmental panel on climate change [1], this trend seems to be inexorable. The temperature increase will endure even by assuming the immediate stop of the greenhouse gas emissions due to the already present greenhouse gases in the atmosphere. To counteract this trend, the Energy and Climate Policy Framework for 2030 [3] sets key targets to cut at least 40% in green gas emissions (compared to 1990 levels). Considering building stock large share of Europe's final energy consumption and CO2 emissions, the concept of nearly-zero energy buildings (nZEB) has received considerable attention. However, the performance of nZEBs in the future has not yet been investigated sufficiently. The climate is changing and the compliance with nZEB requirements may not be a guarantee of energy performance and indoor environmental quality. Considering the long-life span of buildings, the performance of nZEBs should be analysed using future weather data, to ensure energy efficiency, sustainability, and climate resilience over time.

PRELIMINARY RESULTS AND CONCLUSIONS

The preliminary results indicate that climate change affects the energy balance and thermal comfort of the nZEB, however, the extent varies among the different climate regions and future scenarios. The simulations show that the increases in cooling energy need overcome the decreases in heating energy need. The electricity use will increase due to more hours out of thermal comfort. Although the energy output of the PV panel system will be increasing in the future, buildings will miss the target of meeting nearly zero energy and the new configuration is needed to keep nZEB goals in the future. These results highlight the significance of considering future weather for energy performance assessment of nZEBs and establishing building adaptation measures for climate change beside nZEB measures, to ensure a holistic approach.

MAIN REFERENCES

[1] Symon, Carolyn. "Climate change: actions, trends and implications for business. The IPCC fifth assessment report, Working Group 1." (2013): 524-82.

[2] Cannon, Alex J. "Multivariate quantile mapping bias correction: An N-dimensional probability density function transform for climate model simulations of multiple variables." Climate dynamics 50.1-2 (2018): 31-49.

[3] European Commission, COM, 15 final, Communication from the Commission to the European Parliament, the Council, the European Economic and Social Committee and the Committee of the Regions, A policy framework for climate and energy in the period from 2020 to 2030, 2014, 2014, https://eur-lex.europa.eu/legal-content/EN/TXT/PDF/?uri=CELEX:52014DC0015&from=EN

Keywords: Climate change, Future weather data, nZEB, Building energy performance

Paper ID 30188: The impact of a high-end climate change scenario on the energy consumption of a Flemish office building

Delphine Ramon*, Karen Allacker

Department of Architecture, Faculty of Engineering Science, KU Leuven, Belgium

AIM AND APPROACH

This paper evaluates the impact of a high-end climate change scenario on the energy consumption for heating and cooling of a Flemish office building. A dynamic building energy simulation is executed with EnergyPlus for a state-of-the-art Flemish office building. Weather data is extracted from a high-resolution climate model for an RCP 8.5 climate change scenario through a typical meteorological year. This model allows to extract weather data on an hourly time resolution and a spatial resolution of 2.8 by 2.8 km. Data is extracted for different locations in Belgium and allows us to analyze broader, regional effects such as distance to the seaside, and orography as well as local effects such as the urban heat island effect. Different locations are selected in Belgium to cover these regional and local effects. The energy consumption is analyzed and validated for the location of the building as well as for other locations in Belgium both for the current and future climate model data. In addition, the change in thermal comfort within the building is analyzed, making use of the Category II limits of the EN 16798-1:2019 standard. The paper will particularly focus on the impact on cooling energy consumption and summer comfort.

SCIENTIFIC INNOVATION AND RELEVANCE

More and more, buildings are analyzed in a future climate context. Also, more and more regional climate model output is used for the weather data of these simulations. Typically, model data with a spatial resolution up to 12.5km is available. The climate model used in this study is a convection-permitting model that typically has a spatial resolution < 4km. At this spatial resolution, the models have a better resolved daily cycle, an explicit representation of the urban heat island effect, and better-resolved land use and orography making them favorable for building energy simulations. The use of climate model data at this spatial resolution in building energy simulations is not known to the authors. Further, evaluations of building performance towards the future in a Belgian context are not elaborated making use of climate model output in dynamic building simulations. These data have an important advantage in comparison to morphing techniques used in weather generator tools, which do not guarantee to keep the physical consistency between weather variables used in the building energy simulation.

PRELIMINARY RESULTS AND CONCLUSIONS

As expected, an increase in cooling load and a decrease in heating load towards the end of this century under a high-end climate change scenario is found. An increase in cooling load up to 20% on monthly basis is found under typical weather conditions for the RCP 8.5 climate change scenario. Under extreme warm weather conditions, an increase of up to 15% is found. Higher increases are found for the office rooms with Southwest orientation. It is expected that due to the system capacity, the increase in the building-specific cooling energy consumption will be smaller but that this will cause an increase in the discomfort. The latter is confirmed in the preliminary results of the full building energy simulation (including the actual HVAC system). Increases of up to 5% were found in cooling energy consumption. However, further verification of the model is needed. Also, the performance will be evaluated across different locations in Belgium. Smaller changes are expected in more rural areas or areas close to the seaside. While bigger changes in discomfort are expected in urban areas.

MAIN REFERENCES

Nik, V. M. (2016). Making energy simulation easier for future climate–Synthesizing typical and extreme weather data sets out of regional climate models (RCMs). Applied Energy, 177, 204-226.

CEN, E. (2007). 15251, Indoor environmental input parameters for design and assessment of energy performance of buildings addressing indoor air quality, thermal environment, lighting and acoustics. European Committee for Standardization, Brussels, Belgium.

Ramon, D., Allacker, K., De Troyer, F., Wouters, H., & van Lipzig, N. P. (2020). Future heating and cooling degree days for Belgium under a high-end climate change scenario. Energy and Buildings, 109935.

Chow, D. H., & Levermore, G. J. (2010). The effects of future climate change on heating and cooling demands in office buildings in the UK. Building Services Engineering Research and Technology, 31(4), 307-323.

Hosseini, M., Tardy, F., & Lee, B. (2018). Cooling and heating energy performance of a building with a variety of roof designs; the effects of future weather data in a cold climate. Journal of Building Engineering, 17, 107-114.

Keywords: Climate change, Operational Energy, Building energy simulation, Climate model, HVAC

Paper ID 31153: Development of a new decision tool for Sustainable COOLing Systems (CORNET SCOOLS)

Jeroen Van der Veken (1), Margot De Pauw* (2)

1: BBRI, Belgium; 2: KCE Thomas More, Belgium

AIM AND APPROACH

Due to climate change and increasing frequency of heat waves, cooling in residential buildings gains importance. The most common way of cooling in buildings is active cooling with the use of 'split units' or central air-conditioning systems, using an electric compressor and standard refrigerant. However, these systems have a relative high energy consumption, and the European parliament has called a phasedown for many of the refrigerants. Sustainable cooling systems can therefore be a superior alternative for conventional systems. These systems are however not often applied due to a lack of real performance data and guidelines for the correct selection and dimensioning. This problem is addressed by the CORNET SCoolS.

SCIENTIFIC INNOVATION AND RELEVANCE

Key Innovations

• Decision support tool for sustainable cooling systems

• Assessment of summer comfort in dwellings

• Cooling loads room per room

Practical Implications

• The uncertainty about climate projections complicate summer comfort evaluations.

• There exist many methods to assess summer comfort, it is always a good idea to store the (hourly) operative temperatures so that other calculations can be done in post-processing

• Be sure to implement a realistic control algorithm when comparing passive and active cooling systems

PRELIMINARY RESULTS AND CONCLUSIONS

A new decision tool has been made to compare different (sustainable) cooling systems and passive cooling strategies in dwellings. To make it future proof, a new (extreme) climatic data file was constructed. However, the selection of the comfort criteria showed to be difficult.

If adaptive temperature boundaries are uses, as is allowed and described in EN16798-1 (2019), the impact of heat waves is somewhat tempered as the temperature criteria go up as well. However, it is not certain if this can be applied straightforward to residential cases. The authors chose to add the absolute comfort criteria of CIBSE Guide A (2015) and the following comfort classes were determined based on the exceeding hours: good (0h), acceptable (<32h), possible (< 64h), uncomfortable.

In general, the simulations show that a lot of sustainable cooling systems, although their lower specific power, can provide good comfort in most of the residential buildings, provided that passive cooling strategies are applied. Without any cooling system this will become difficult.

MAIN REFERENCES

CIBSE Guide A: Environmental design (2015)

IEA EBC. (2018). Annex 62 Ventilative Cooling Design Guide.

IEA EBC (2020). Annex 80 Resilient Cooling

EN16798-1. (2019).

ISO 7730. (2005).

Witkamp M.J., Koornneef W., Wijnja L., Kaspers J., van Weele H., Schouten C. & Plokker W. (2019) Themablad thermisch comfort (https://energielinq. stroomversnelling.nl/kwaliteitskaders/nieuwe-methode-voorkomt-oververhitting-woningen/)

Keywords: Global warming, summer comfort in dwellings, sustainable cooling systems, passive cooling strategies

Session W2.6: The role of occupants

Paper ID 30216: Mapping the gap in user-related building performance simulation models

Ardeshir Mahdavi*, Veselina Bochukova, Christiane Berger

TU Wien, Austria

AIM AND APPROACH

Recently, increasing attention is being paid in the building performance simulation research community to the quality and resolution of representations of building users in simulation models (Mahdavi and Tahmasebi 2019). This is reflected in a host of publications and projects with highly diverse starting points, approaches, and results (Yan et al. 2017, Berger and Mahdavi 2020). Whereas such diversity can be productive and fruitful, it may also involve redundancy and lack of strategic orientation. In this context, it is beneficial to reexamine this subject from two complementary directions. One ontological direction, characterized here as "top-down", pursues the required format and dimensions of a generalized representation of building users that could be distilled down as needed to cater for the informational requirements of specific applications. The second direction, which can be seen as "bottom-up", starts by reverse-engineering the occupant-specific input templates of common performance simulations in view of any existing shared features and structure. We suggest that the pursuit of these two directions reveals a discontinuity or gap, indicating that there is not yet a seamless path leading from occupant-centric ontologies to occupant-related model input requirements of common performance simulation tools.

SCIENTIFIC INNOVATION AND RELEVANCE

The building information modeling research and development community has become increasingly cognizant of the following circumstance: To be truly effective, building information models must go beyond the static representation of buildings' constituent physical components (Mahdavi 2020). Rather, the time-dependent dynamics of processes associated with the design, construction, and operation of buildings must be taken into consideration. A major class of such processes involve the patterns of occupants' presence and behavior in buildings. A comprehensive solution for the respective representational challenges must go beyond ad hoc amendments to existing simulation input routines. Toward this end, a seamless transition from a comprehensive ontology of building users down to specific input schema tailored for individual simulation application would be desirable, but currently hampered to a representational discontinuity. The aforementioned concurrent top-down and bottom-up inquiries can help map this gap and hence suggest approaches to close it.

PRELIMINARY RESULTS AND CONCLUSIONS

The work thus far has led to the definition of a general schema that captures the main dimensions of occupant-related information. These include physical data pertaining to position and movement, physiological data pertaining to state of metabolism and adaptation, cognitive data pertaining to formation of impressions and value attributions, as well as event-based data pertaining to human-building interaction processes. The result suggests that, in order to consistently structure the schema, underlying foundational theories are needed that capture building users' relevant patterns of presence and behavior. Moreover, the bottom-up reverse-engineering of existing simulation applications reveals the potential and challenges toward seamless derivation of locally tailored input information from ontologically structured sources of occupant-related information.

MAIN REFERENCES

Berger, C., and Mahdavi, A. (2020): Review of current trends in agent-based modeling of building occupants for energy and indoor-environmental performance analysis. Building and Environment, 173; 106726.

Mahdavi, A., and Tahmasebi, F. (2019): People in building performance simulation. Building Performance Simulation for Design and Operation - Expanded Second Edition. Hensen, J., Lamberts, R. (Ed.); Routledge, New York, ISBN: 978-1-138-39219-9; pp. 117 - 145.

Mahdavi, A. (2020): Bringing HIM closer to HER. Keynote. Proceedings of SIMAUD: Symposium on Simulation for Architecture and Urban Design. Vienna (Online), 25-26 May 2020. ISBN: 978-1565553712.

Yan, D., Hong, T., Dong, B., Mahdavi, A., D´Oca, S., Gaetani, I., and Feng, X. (2017): IEA EBC Annex 66: Definition and simulation of occupant behavior in buildings. Energy and Buildings, 156; pp. 258 - 270.

Keywords: Performance simulation, building occupants, representational gap

Paper ID 30571: The impact of occupancy prediction accuracy on the performance of model predictive control (MPC) in buildings

Tao Yang*, Fisayo Caleb Sangogboye, Krzysztof Arendt, Konstantin Filonenko, Jonathan Dallaire, Mikkel Baun Kjærgaard, Christian Veje

Center for Energy Informatics, University of Southern Denmark, Denmark

AIM AND APPROACH

This paper aims to investigate the impact of occupancy accuracy on the performance of MPC-based building control. As an experimental setup, a grey-box model representing the heating, ventilation, and air conditioning (HVAC) system in a case study building is developed and calibrated. Subsequently, a number of 3D stereo-vision cameras is deployed to obtain accurate occupancy counts. Based on the obtained measurements, a data-driven model for predicting occupancy count in simulation is developed. In the evaluation, two MPC-based building controllers with different occupancy accuracy are compared based on multiple shooting optimization algorithm. The first of the compared models uses occupancy estimates from the deployed camera, while the second uses occupancy prediction from the data-driven model. The two MPC-based building controllers are further compared with a conventional rule-based controller in terms of energy consumption and indoor thermal discomfort.

SCIENTIFIC INNOVATION AND RELEVANCE

Globally, buildings are responsible for nearly 40% of total energy consumption among other sectors [1]. MPC is a promising and widely investigated strategy employed in buildings to reduce energy consumption while maintaining thermal comfort. While there exists a wide range of parameters for MPC-based building controls, the accurate estimation of occupancy in buildings constitute a major factor for achieving considerable ambient comfort and energy saving within buildings [2-5]. This paper goes beyond simple analysis of the occupant presence and applies a data-driven approach to accurately predict occupant count and subsequently studies the impact of occupancy accuracy on MPC performance, which contributes to better analyze relationship between occupancy information and MPC performance in buildings.

PRELIMINARY RESULTS AND CONCLUSIONS

The developed grey-box model yields good accuracy of capturing thermal dynamics of the system. Using occupancy estimates from the deployed cameras, an MPC-based controller in contrast to a rule-based controller demonstrates better indoor thermal comfort and higher energy consumption due to that MPC prioritizes thermal comfort (hard constraint) over energy. Comparing MPC with two different occupancy accuracy, occupancy predictions with low accuracy can lead to lower energy consumption at the expense of thermal comfort violations. When increasing optimization horizon, MPC-based controller with more accurate occupancy prediction shows larger energy consumption and improved thermal comfort. Besides, the negative influence of prediction error can be partially mitigated by adopting longer optimization horizons.

MAIN REFERENCES

[1] Cao, Xiaodong, Xilei Dai, and Junjie Liu. "Building energy-consumption status worldwide and the state-of-the-art technologies for zero-energy buildings during the past decade." Energy and buildings 128 (2016): 198-213.

[2] Amin Mirakhorli and Bing Dong. Occupancy behavior-based model predictive control for building indoor climate - a critical review. Energy and Buildings, 129: 499–513, 2016.

[3] Justin R Dobbs and Brandon M Hencey. Model predictive hvac control with online occupancy model. Energy and Buildings, 82: 675–684, 2014.

[4] Frauke Oldewurtel, David Sturzenegger, and Manfred Morari. Importance of occupancy information for building climate control. Applied energy, 101: 521–532, 2013.

[5] Siddharth Goyal, Herbert A Ingley, and Prabir Barooah. Occupancy-based zone climate control for energy-efficient buildings: Complexity vs. performance. Applied Energy, 106: 209–221, 2013.

[6] K. Arendt and C. Veje, "MShoot: an Open Source Framework for Multiple Shooting MPC in Buildings," in 16th IBPSA International Conference and Exhibition Building Simulation 2019, Rome, 2-4 September, 2019, 2019.

[7] Antoine Garnier, Julien Eynard, Matthieu Caussanel, and Stéphane Grieu. Predictive control of multizone heating, ventilation and air-conditioning systems in non-residential buildings. Applied Soft Computing, 37: 847–862, 2015.

Keywords: Occupancy prediction, Model predictive control, Optimization, HVAC

Paper ID 30205: Consequences-based graphical model for contextualized occupants' activities estimation in connected buildings

Huynh Phan* (1), Thomas Recht (1), Laurent Mora (1), Stéphane Ploix (2)

1: I2M Bordeaux, University of Bordeaux, CNRS, Arts et Metiers Institute of Technology, Bordeaux INP, F-33400 Talence, France; 2: G-SCOP, Grenoble Institute of Technology, UMR CNRS 5272, 46 Avenue Felix Viallet, 38031 Grenoble Cedex 1, France

AIM AND APPROACH

Occupants' activities are addressed as important factors resulting in the discrepancy between simulated and actual energy consumptions in residential buildings. Many models with statistical data are proposed to better estimate the activities of occupants than conventional approaches. However, these models are hybrid of the statistical data, which contains the information of many dwellings with different habits and characteristics, and not entirely sufficient to represent contextualized activities in a particular household. To solve this problem, data-driven approaches with machine learning techniques are proposed to estimate the activities based on the data measured from on-site sensors. However, they are black-box approaches and difficult to understand. It is preferable to propose an understandable model, which is able to estimate and evaluate the contextualized activities in a particular household. In this contribution, a general approach model taking into account a particular context is proposed to estimate and predict the occupants' activities in a specific household. Specifically, many sensors (CO2, temperature, motions, etc.) are installed to capture the data in the household. Then, combining with the context information, the necessary features are extracted from measured data and are used to build a consequences-based Bayesian Network, which is understandable, and flexible for the contextualized activities estimation.

SCIENTIFIC INNOVATION AND RELEVANCE

Recently, statistical approach is the most popular approach for occupants' activities estimation in buildings. [1,6] used statistical models to estimate the activities using the characteristics of both buildings and households. Though, their models are hybrid of statistical data of different dwellings and not sufficient to estimate and evaluate the contextualized activities in a particular household. To deal with it, [2,7] applied data-based models with machine learning techniques to estimate activities using the measurement data. However, they do not provide understandable outputs and are not adaptable to context's changes. Otherwise, [3,4] proposed agent-based models to simulate the activities of windows/doors. An agent perceives the space's conditions and executes activities to achieve its predefined comfort. However, agent-based models are too complex with many agents. Besides, [5] built Bayesian Network (BN) to estimate the activities of doors based on the measurement data and expert knowledge. BN is an understandable, easy adaptable model for activities estimation in a particular household. However, all studies focus on simple activities (actions of windows/door, presence, etc.) and the structure of the BN is difficult to define. This study proposes a BN with an expert structure for contextualized activities estimation from measurement data, in a specific household.

PRELIMINARY RESULTS AND CONCLUSIONS

A general approach has been proposed in this contribution to estimate the daily activities in a residential building. Dynamic Time-Series Clustering and expert discretization techniques are used to extract features from variables while Information Gain is used for the necessary features selection. Consequences-based Bayesian Network is used as an estimation model based on an expert structure, which is determined by the consequences of the activities. Finally, cross-validation and F1-score are used to validate the proposed model in the testing data. The proposed model was used to estimate some activities such as cooking breakfast, cooking lunch, washing dishes, etc. The testbed is a detached house in France, which includes 5 household members and 11 rooms. The data was collected from numerous installed sensors of ambient, power, motion, windows/doors contacts, etc. The labels of activities were collected for two months by a self-developed mobile application. In this contribution, the activity cooking lunch in the kitchen is presented. Data set covers 45 weekdays from 09/12/2019 to 30/02/2020 (labels collection period). Results show that cooking lunch is mainly linked to the patterns of the usages of frequently involved appliances (microwave, toaster). Cross-validation is used and F1-score is approximately 91%.

MAIN REFERENCES

1. Dorien Aerts. 2015. Occupancy and Activity Modelling for Building Energy Demand Simulations, Comparative Feedback and Residential Electricity Demand Characterisation. PhD Thesis, Vrije Universiteit Brussel.

2. Alaa Alhamoud, Pei Xu, Frank Englert, et al. 2015. Extracting Human Behavior Patterns from Appliance-level Power Consumption Data. Wireless Sensor Networks, Springer International Publishing, 52–67.

3. Yoon Soo Lee and Ali M. Malkawi. 2014. Simulating multiple occupant behaviors in buildings: An agent-based modeling approach. Energy and Buildings 69: 407–416.

4. Khadija Tijani, Ayesha Kashif, Stéphane Ploix, Benjamin Haas, and Julie Dugdale. 2015. Comparison between purely statistical and multi-agent based ap-proaches for occupant behaviour modeling in buildings. arXiv:1510.02225 [cs].

5. Khadija Tijani, Quoc Dung Ngo, Stéphane Ploix, Benjamin Haas, and Julie Dugdale. 2015. Towards a General Framework for an Observation and Knowledge based Model of Occupant Behaviour in Office Buildings. Energy Procedia 78: 609–614.

6. Urs Wilke. 2013. Probabilistic Bottom-up Modelling of Occupancy and Activities to Predict Electricity Demand in Residential Buildings. .

7. Suyang Zhou, Zhi Wu, Jianing Li, and Xiao-ping Zhang. 2014. Real-time Energy Control Approach for Smart Home Energy Management System. Electric Power Components and Systems 42, 3–4: 315–326.

Keywords: contextualized activity estimation, Bayesian Network, human behavior co-simulation

Paper ID 30273: Multi-Agent based simulation of human activity for building and urban scale assessment of residential load curves and energy use

Mathieu Schumann* (1,5), Quentin Reynaud (2), Nicolas Sabouret (3), François Sempé (6), Benoit Charrier (1,5), Jérémy Albouys (1,3,4,5), Yvon Haradji (1), Christian Inard (4,5)

1: EDF R&D, Palaiseau, France; 2: QRCI, Clermont-Ferrand, France; 3: LIMSI-CNRS, Univ. Paris-Sud, Univ. Paris-Saclay, Orsay, France; 4: LaSIE, CNRS, La Rochelle Université, La Rochelle, France; 5: 4evLab, EDF R&D, CNRS, LaSIE, La Rochelle Université, La Rochelle, France; 6: FSCI, Paris, France

AIM AND APPROACH

Because of the crucial impact of human activity on energy consumption (Janda, 2011), many building and urban energy models integrate human activity modeling. An increasing number of research works use Time Use Surveys (Chenu et al, 2006) and stochastic person-based approaches to generate occupancy profiles, window uses, or activity chronograms (Baetens et al, 2016). The goal of this paper is twofold: 1/ to discuss the limits of the existing state-of-the-art approaches on human activity modeling, and 2/ to present our modeling approach: an agent-based, individual-centered simulation of human activity, taking individual decisions and interactions between individuals into account to produce activity diagrams that are consistent at the household scale. Associated with a population generator, and by linking the simulated activity with appliance use, one of the main applications is the calculation of residential households load curves.

Our work puts the simulation of human activity and everyday life decision-making processes at the center of energy consumption, thermal comfort or indoor air quality assessments in households. This approach has its roots in ergonomics studies, which demonstrated the ability of multi-agent systems to simulate a realistic human activity, using innovative validation methodologies such as participatory simulations with real households (Haradji et al, 2012).

SCIENTIFIC INNOVATION AND RELEVANCE

The analyzed approaches appear to be insufficient to account for the dynamics of individual and collective activity and its impact on energy consumption (Happle, 2018), particularly as households and energy communities become more actively integrated in smart grids and involved in flexibility, collective self-consumption or energy exchange schemes. Such configurations involve individual and collective decision-making, as well as the management of new appliances such as electric vehicles.

We suggest that the co-simulation of autonomous agents, whose activities are built upon statistical data (e.g. Time Use Surveys), with building and appliance models (Plessis, 2014) can effectively tackle these new challenges. We propose a fine-grained, 1 minute time step household simulation, able to perform population-wide yearly simulations. Our agent-based model can reach a satisfying level of diversity in human activities, and it supports interactions between autonomous agents and a changing environment (e.g. activity priorities, incentives, energy prices). This model accounts for emerging, reactive, adaptive, and collective behaviors. Moreover, it allows to deal with many human behavioral aspects (e.g. presence, heating adjustment, housing ventilation) within the same decisional process, bringing consistency to the role of occupant in energy assessments.

PRELIMINARY RESULTS AND CONCLUSIONS

We present the agent-based SMACH platform (SiMulation of human Activity and Consumption in Households) (Haradji et al, 2018). We show how the modeling of occupants as intelligent autonomous agents accounts for the individual and collective dynamics of daily life and the related energy consumption, modeled as the result of the interactions between agents and the household's electrical appliances. We demonstrate how the latest developments of the SMACH simulation platform, combining population synthesis, simulation of individual and collective activities as well as electric mobility, answer some of the challenges related to the role of occupants in energy consumption (Happle, 2018). Especially, we show that multi-agent modeling offers a high degree of modularity due to the internal capabilities of the agents to organize themselves, plan their day, or react to events or changes in the environment. This approach is also explicit, and makes it possible to explain why, when, how and with whom the simulated energy consumption is performed. We illustrate the validation approaches and variety of uses of this model in energy questions and engineering applications such as load curve calculation, at an individual household scale as well as at the population scale.

MAIN REFERENCES

Janda, Kathryn B. (2011). Buildings Don't Use Energy: People Do. Architectural Science Review 54 (1): 15–22

Chenu, A. and Lesnard L. (2006) Time Use Surveys: A Review of Their Aims, Methods, and Results. European Journal of Sociology 47 (3): 335–59.

Baetens, R. and Dirk S. (2016). Modelling Uncertainty in District Energy Simulations by Stochastic Residential Occupant Behaviour. Journal of Building Performance Simulation 9 (4)

Haradji Y., Poizat G., Sempé F (2012) Human Activity and social simulation. Proceedings of the 4th Advances in Human Factors and Ergonomics Conference, San Francisco, California, USA

Plessis, G., Edouard A., Haradji, Y. (2014). Coupling Occupant Behaviour with a Building Energy Model - A FMI Application. 10th International Modelica Conference 96:321–26 Lund, Sweden

Haradji Y. et al, (2018) From modeling human activity to modeling for social simulation: between realism and technological innovation. Activités 15-1

Happle G., Fonseca J. A. and Schlueter A. (2018) A review on occupant behavior in urban building energy models. Energy and Buildings, vol. 174. Elsevier Ltd, pp. 276–292

Keywords: Multi-agent simulation, human activity, household energy use, residential load curve

Paper ID 30981: The formulation of a reference load curve to measure energy flexibility

Muhammad Salman Shahid* (1), Benoît Delinchant (1), Béatrice Roussillon (2), Frédéric Wurtz (1), Daniel Llerena (2)

1: G2Elab, 21 Rue des Martyrs, CS 90624, 38031 Grenoble CEDEX 1, France; 2: GAEL, 1241 Rue des Résidences, 38400 Saint-Martin-d'Hères, France

AIM AND APPROACH

Energy consumers have a degree of choice to implement indirect energy flexibility to mitigate network congestion during the intermittence of renewable production. It is essential to measure the impact of each alert for each consumer. This abstract presents a study performed to compare the different methods for formulating a reference load curve for the residential energy consumers. Hypothetically, this reference load curve gives the habitual energy consumption pattern of residential consumer. The aim of creating a reference load curve is to visualize and measure the degree of deviation of consumption load curve from habitual energy consumption load curve of a consumer. The image of reference load curve superposed on consumption load curve is sent to the consumers as part of the feedback, so that they can watch the impact of their efforts as well. For this purpose, certain statistical methods (mean, Kernel Density Distribution) and naïve methods are explored, whereas the advanced methods (RF regression, Neural networks) are in the process of study. The so forth studied methods are analyzed through an indicator that verifies the under-estimation or over-estimation of the reference load curve (to be discussed in detail in the proposed paper).

SCIENTIFIC INNOVATION AND RELEVANCE

Hypothetically, the habitual energy consumption pattern of a residential customer can be visualized in the form of a load curve (hereafter referred as reference load curve) for a standard day. For a particular day, the deviation of consumption load curve from reference load curve gives the measure of the effort made by the residential consumer to implement indirect flexibility. In case of peak shaving, a good effort can be observed if the consumption load curve is under reference load curve. The result should be opposite in the case of load shifting for the period of time to which the load is shifted. For the purpose of experiment, two types of alerts are defined. Orange alert demands the households to implement peak shaving between 6 PM and 8 PM on alert day. Green alert demands the households to implement load shifting from evening to afternoon (between noon and 3 PM). The reference load curve is used to measure the impact of nudge signal on each household for each alert. This in consequence measures the effectiveness of nudge tool to implement indirect energy flexibility in residential sector.

PRELIMINARY RESULTS AND CONCLUSIONS

For half-hourly sampled consumption load curve, one method is to take the mean of historical data for each of the 48 timestamps of day. Another statistical method is to take the peak value of the Kernel density estimation for each of the 48 timestamps of the day. Hypothetically, this peak value is the most probable value of energy consumption for the given half hour timestamp. However, these methods are highly susceptible to season variation and therefore gives less accurate reference load curve.

One naïve method is to take an average of the consumption load curve of day "D-1" with the consumption load curve having maximum energy consumption among the curves of days "D-2" and "D-5" (only weekdays). This reference load curve is used for orange alerts. The vice versa of this method is also formulated for green alerts. Both these reference curve keeps the effect of the near past historical consumption, therefore maintaining the effect of temperature and seasonality. These naive methods are devised to introduce a bias, so that it nudges the households makes a better effort in future. The advanced methods are still in the process of study.

MAIN REFERENCES

Albadi, M. H., and E. F. El-Saadany. 2007. “Demand Response in Electricity Markets: An Overview.” In 2007 IEEE Power Engineering Society General Meeting, 1–5.

Bivas, Pierre. 2011. “La production d’effacement : comment offrir des économies d’électricité à des millions de foyers.” Le journal de l’ecole de Paris du management n°90 (4): 8–14.

Hatton, Leslie, and Philippe Charpentier. 2014. “Système électrique français : estimation de l’effacement des clients résidentiels,” 14.

Lesgards, Valérie, and Laure Frachet. 2012. “La Gestion de La Demande Résidentielle d’électricité: Retour Sur 30 Ans d’expérimentations Mondiales.” La Gestion de La Demande Résidentielle d’électricité: Retour Sur 30 Ans d’expérimentations Mondiales, no. 607: 162, 164, 192-210 [21 p.].

Neenan, B. 2009. “Residential Electricity Use Feedback: A Research Synthesis and Economic Framework,” 126.

Thaler, Richard H., and Cass R. Sunstein. 2008. Nudge: Improving Decisions about Health, Wealth, and Happiness. Nudge: Improving Decisions about Health, Wealth, and Happiness. New Haven, CT, US: Yale University Press.

Keywords: Reference load curve, Indirect energy flexibility, Residential sector, Nudge

Session W2.7 (Online Track): Ensuring high quality building simulations

Paper ID 30162: Identifying grey-box models from archetypes of apartment block buildings

Marius Eide Bagle* (1), Philip Maree (2), Harald Taxt Walnum (1), Igor Sartori (1)

1: SINTEF Community, Norway; 2: SINTEF Digital, Norway

AIM AND APPROACH

A potentially large amount of flexibility resides in the space heating of residential buildings. To realize this potential, it is necessary to model heat demand with models that are accurate enough and suitable for real time control.

Well-suited for this purpose are grey-box models, which combine a relatively simple physical descriptions of the building with data-driven inference of key parameters. However, identification of grey-box models poses a challenge: alongside energy use and weather data, the indoor temperature must also be known. Such data are scarcely available. Furthermore, it is not given that measurements from normal operation of buildings provide datasets that are 'rich' enough to successfully drive the identification process, or if special test periods should be carried out. Such a test would require the manipulation of indoor temperatures, possibly in periods of non-occupancy, for several days; a challenging task on real buildings.

This paper presents a method that aims at overcoming this bottleneck by combining features of both white-box and grey-box modelling. A set of white-box models (specifically, IDA-ICE models) representing the Norwegian stock of apartment blocks is available, based on ca. 20 archetypes previously developed in the Tabula/Episcope project [1].

SCIENTIFIC INNOVATION AND RELEVANCE

To enable a large-scale rollout of predictive control systems in the building stock, developing robust methods of model identification is of key importance. Provided that load profiles in the IDA-ICE models are validated (a parallel research activity), it is legitimate to assume that the indoor temperature profiles from the IDA-ICE archetypes are also representative for the real building stock, thus bypassing the need for measurement data.

Under this assumption, the grey-box models are identified from datasets generated by the IDA-ICE archetypes in two modalities: under normal operation with a seasonal (three-month) duration, and under special test periods with a duration of one to two weeks. During the test periods the IDA-ICE archetypes are excited with trains of heating events, Pseudo Random Binary Sequence (PRBS), aiming at exploring a wide and rapidly changing set of indoor temperatures around the comfort zone, ca. between 20 and 24 °C. It is also of interest to investigate the effect of temporal resolution in the identification process.

PRELIMINARY RESULTS AND CONCLUSIONS

Preliminary results have been obtained using the software package CTSM-R [2], which uses an extended Kalman filter (EKF) to formulate a maximum likelihood estimation problem. This is combined with a quasi-Newton method to find the optimal set of parameters. With this setup, there are issues regarding the initial parameterization of the states, which essentially have to be found via a qualified guess. One weakness using the EKF is that it does not incorporate state constraints, nor does it utilize potential underlying non-linear dynamics. To such an extent, Moving Horizon Estimation (MHE) is presented as an alternative optimization-based approach to solving the maximum a-posteriori estimate. Recent advances in computational power makes MHE a viable solution as an asymptotically stable observer which has shown in other cases to provide improved state estimation and greater robustness to poor guesses of the initial state [3].

Results will be evaluated considering both the physical meaningfulness of the parameters, e.g. by ensuring that the energy balance of the model is unbiased [4], and that the parameter values are accompanied by reasonable confidence intervals. To enable a fair comparison between the algorithms, a fixed structure for the grey-box model will be decided a priori.

MAIN REFERENCES

[1] "Tabula/Episcope - Norwegian archetypes." [Online]. Available: https://episcope.eu/building-typology/country/no.html. [Accessed: 30-Jun-2020].

[2] H. Madsen et. al, "Continuous Time Stochastic Modeling in R User's Guide and Reference Manual," 2018.

[3] Haseltine, Eric L., and James B. Rawlings. "Critical evaluation of extended Kalman filtering and moving-horizon estimation." Industrial & engineering chemistry research 44.8 (2005): 2451-2460.

[4] R. De Coninck, F. Magnusson, J. Åkesson, and L. Helsen, "Toolbox for development and validation of grey-box building models for forecasting and control," J. Build. Perform. Simul., vol. 9, no. 3, pp. 288–303, 2016.

Keywords: Grey-box modelling, Optimization, Building physics, Moving Horizon Estimation

Paper ID 30199: Integrating thermal, energy, lighting, and acoustics in building design approach: Lesson learned from students assignments

Rizki A. Mangkuto*, Anugrah Sabdono Sudarsono, R.S. Joko Sarwono

Institut Teknologi Bandung, Indonesia

AIM AND APPROACH

This paper aims to report results and lesson learned from students assignments in integrating thermal, energy, lighting, and acoustics performance in a building design simulation project. The objectives are to observe the students approach in defining and obtaining the baseline, target, and optimum performance in all mentioned aspects, and to analyse the differences between the submitted design proposals.

The following hypothetical case was assigned to the students: A classroom with two glazed windows is situated on the second floor of a third-story building; the north side is exposed to the exterior, whereas the other three sides are adjacent to an unconditioned corridor. The students were asked to simulate and predict the baseline and target values of thermal, energy, lighting (all using Sefaira) and acoustics (using CATT v8) aspects of the classroom, and optimise them only by modifying the interior material properties. The submitted project reports were analysed statistically to observe the trend.

SCIENTIFIC INNOVATION AND RELEVANCE

An integrated building performance simulation (BPS) approach considering thermal, energy, lighting, and acoustics performance altogether in building design is not so common. Integrating the first three mentioned aspects is indeed popular and is supported by current simulation tools, but the acoustics aspect is often not considered at the same time, i.e. it is not integrated in the design approach. This paper offers an insight on how graduate students (in Building Physics), some of which were professional architect and engineers, integrated and optimised all of those four aspects altogether, in an attempt to modify a classroom using BPS. There are also observations on how the proposed designs are compared to each other.

The results would be of relevance for building practitioners and educators, since the findings suggest that such integrated design can be quite challenging and is not trivial, particularly for those who are not familiar with the integrated approach, despite already having a background in the building industry.

PRELIMINARY RESULTS AND CONCLUSIONS

Based on the submitted reports, it is found that only 50% of the students modified the interior materials by considering the four aspects at the same time, so that the results were optimum for all aspects. The remaining 50% chose to optimise the materials considering the first three aspects, as they could be done in Sefaira, and then modified the material with respect to acoustics aspect in CATT v8; without realising that the last step might also influence or change the thermal, energy, and lighting performance. While this seems like a typical beginner's mistake in BPS, it can also happen for instance to specialists, i.e. those who are experienced BPS users but do not master the four aspects altogether.

Among the aspects, lighting performance (sDA300/50%) is the most consistently predicted, with the coefficient of variance (CV) of 0.16 (baseline) and 0.07 (optimum). In opposite, thermal performance (annual overheating hours) is the least consistent, with CV of 1.68 (baseline) and 1.82 (optimum).

In conclusion, the findings can be applied to improve the content of BPS education by employing integrated building performance design approach.

MAIN REFERENCES

I. Beausoleil-Morrison (2019) Learning the fundamentals of building performance simulation through an experiential teaching approach, Journal of Building Performance Simulation, 12:3, 308-325.

P.P. Charles, C.R. Thomas (2009) Four approaches to teaching with building performance simulation tools in undergraduate architecture and engineering education, Journal of Building Performance Simulation, 2:2, 95-114.

E. Mendes, N. Mendes (2019) An instructional design for building energy simulation e-learning: an interdisciplinary approach, Journal of Building Performance Simulation, 12:3, 326-342.

Keywords: integrated design approach, optimisation, building performance, student assignment

Paper ID 30538: Statistical methodologies for verification of building energy performance simulation

Amin Nouri*, Jérôme Frisch, Christoph van Treeck

Institute of Energy Efficiency and Sustainable Building (E3D), RWTH Aachen University, Germany

AIM AND APPROACH

Building performance simulation tools are being increasingly deployed by researchers and professionals to predict the thermal behavior of buildings. In general, modeling and simulation techniques are employed to determine the thermal characteristics of the building. Validation methods are used to ensure the accuracy of simulation results. Several standardized procedures exist to assess building performance simulation tools, such as ASHRAE Standard 140 as well as the German VDI 6007 and VDI 6020 guidelines. This paper is conducted within a research project funded by the German Federal Ministry for Economic Affairs and Energy (BMWi), which addresses the development of quality standards for building and systems energy performance simulation. The objective of this project is to develop a validation methodology, to define standards for simulation applications and to transfer them into planning practice.

SCIENTIFIC INNOVATION AND RELEVANCE

Buildings account for approximately 40% of the final energy consumption and 36% of the greenhouse gas emissions in Europe (European Commission, 2019). The Paris climate agreement aims to keep the global temperature rise to well below 2 °C above pre-industrial levels and to limit the temperature increase to a maximum of 1.5 °C. In the last few years, building performance simulation tools have been taken into consideration for the scientific community as well as industrial society. However, studies have revealed that there are substantial discrepancies between predictions of building performance simulation tools and the measured energy consumption in practice, which is referred to as the "performance gap". Validation and verification procedures are essential elements within the development of building simulation tools to assess reliability of the simulation results as well as to find and eliminate eventual bugs in the simulation software.

PRELIMINARY RESULTS AND CONCLUSIONS

The first part of this paper presents the comparison of the different standards and guidelines in the context of building and HVAC system performance simulation. The second part describes a systematic set of test cases for building and HVAC systems based on the ASHRAE Standard 140 and discusses the simulation approach in Modelica/Dymola. The third part presents the development and implementation of a validation methodology, which verifies the plausibility of simulation results and excludes simulation errors to the greatest extent. In this paper, first results as well as comparative analyses will be discussed. Another aspect of the research project is the development of a Platform. The main objective of the Platform is to provide a facility for defining individual test cases, to create individual simulation code, to perform a comparative validation, and to evaluate the accuracy of the simulation tools.

MAIN REFERENCES

Wetter, M., van Treeck, C. (2017). „IEA EBC Annex 60: New Generation Computing Tools for Building and Community Energy Systems".

Strachan, P., Svehla, K., Heusler, I., Kersken, M. (2016). „Whole model empirical validation on a full-scale building". Journal of Building Performance Simulation.

Judkoff, R., Wortman, D., O'Doherty, B., Burch, J. (2008). „A methodology for validating building energy analysis simulations". NREL, Golden.

Keywords: Verification, Validation, Building and HVAC systems, Comparative Platform

Paper ID 30208: Easy-to-implement simulation strategies for annual glare risk assessment based on the European Daylighting Standard

Bruno Bueno* (1), Abel Sepúlveda (2), Christoph Maurer (3), Simon Wacker (4), Taoning Wang (5), Tilmann E. Kuhn (6), Helen Rose Wilson (7)

1: Fraunhofer ISE, Germany; 2: Taltech, Estonia; 3: Fraunhofer ISE, Germany; 4: Fraunhofer ISE, Germany; 5: Lawrence Berkeley National Laboratory LBNL, Berkeley CA, USA; 6: Fraunhofer ISE, Germany; 7: Fraunhofer ISE, Germany

AIM AND APPROACH

One of the most important functions of fenestration systems, which triggers user manipulation of the facade, is the protection from daylight glare. The new European daylight standard EN 17037 contains assessment procedures, criteria and information on glare evaluation. Still, mainstream building simulation programs include limited capabilities for glare risk assessment [1], which sometimes results in suboptimal façade design. The problem resides in the complexity and high requirements of state-of-the-art glare evaluation techniques, including those proposed in EN 17037 based on the DGP index [2,3]. In this study, we benchmark available methods for DGP calculation [4,5] and propose simulation strategies that could be easily implemented in simulation programs. This will foster the integration of glare risk assessment in the design process of buildings.

SCIENTIFIC INNOVATION AND RELEVANCE

For a reliable glare risk assessment, it is crucial to account for the direct-direct optical transmittance and cut-off angle of fenestration systems. In this study, we analyse different options to represent the optical behaviour of fenestration components in simulation environments. Bi-directional scattering distribution functions (BSDF) with different resolutions [6] are compared with other strategies such as peak extraction. These representations are based on datasets that have to be experimentally obtained [7] and digitally transferred to simulation environments. The simulation approach, the timestep sampling and the optical representation of the shading technologies determine their applicability in simulation environments and the possibility to be integrated in the design process of buildings.

PRELIMINARY RESULTS AND CONCLUSIONS

In this study, state-of-the-art glare evaluation methods are compared. We propose new strategies based on optical data such as low-resolution BSDF, which is becoming available in international databases, combined with recent approaches such as peak extraction. Efficient sampling strategies reduce the computational cost of dynamic raytracing simulations without loss of accuracy. DGP calculations based on these strategies could be implemented in building simulation environments.

MAIN REFERENCES

[1] Bueno B, Sepúlveda A. A specific building simulation tool for the design and evaluation of innovative fenestration systems and their control. Building Simulation Conference, Rome, 2-4 September; 2019.

[2] J. Wienold, J. Christoffersen, Evaluation methods and development of a new glare prediction model for daylight environments with the use of CCD cameras, Energy and Buildings (2006).

[3] J. Wienold, T. Iwata, M. Sarey Khanie, E. Erell, E. Kaftan, R.G. Rodriguez, J.A. Yamin Garreton, T. Tzempelikos, I. Konstantzos, J. Christoffersen, T.E. Kuhn, C. Pierson, M. Andersen, Cross-validation and robustness of daylight glare metrics, Lighting Research and Technology (2019).

[4] E.S. Lee, D. Geisler-Moroder, G. Ward, Modeling the direct sun component in buildings using matrix algebraic approaches: Methods and validation, Solar Energy (2018).

[5] M. Abravesh, B. Bueno, S. Heidari, T.E. Kuhn, A method to evaluate glare risk from operable fenestration systems throughout a year, Building and Environment (2019).

[6] G. Ward, M. Kurt, N. Bonneel. Reducing anisotropic BSDF measurement to common practice. In: Proceedings of the Eurographics 2014 Workshop on Material Appearance Modeling: Issues and Acquisition. Lyon, France: Eurographics Association; 2014, p. 5–8.

[7] P. Apian-Bennewitz. New scanning gonio-photometer for extended BRTF measurements. In: Proceedings of SPIE; 2010.

Keywords: glare, DGP, daylight, benchmark, dynamic

Paper ID 30235: A machine-learning framework for daylight and visual comfort assessment in early design stages

Hanieh Nourkojouri* (1), Zahra Sadat Zomorodian (1), Mohammad Tahsildoost (1), Zohreh Shaghaghian (2)

1: Shahid Beheshti University, Iran, Islamic Republic of; 2: Texas A&M University, College Station, United States

AIM AND APPROACH

This research is mainly focused on daylight assessment in early stages of design through a new framework which is based on machine learning algorithms. A dataset was primarily developed from 2880 simulations derived from Honeybee for Grasshopper. The simulations were done for a shoebox space with a one side window. The alternatives emerged from different physical features including room dimensions, interior surfaces reflectance, window dimensions and orientations, number of windows and shading states. The metrics used for daylight evaluations included UDI, sDA, mDA, ASE and sVD. Quality Views were analyzed for the same shoebox spaces via a grasshopper-based algorithm which was developed from the LEED v4 evaluation framework for Quality Views. The dataset which had finally daylight, visual comfort and quality views outputs indices was further analyzed with Artificial Neural Network algorithm written in Python. The developed model could be used in early design stages analyzes without the need for time consuming simulations in previously used platforms and programs.

SCIENTIFIC INNOVATION AND RELEVANCE

This work presents a concept for development of a new method to evaluate daylight assessment and views at the same time without the need for time conserving simulations. The mentioned method includes 5 indices for daylight and visual comfort evaluations and 2 indices for evaluation of quality views. Having all these 7 metrics together could give the designers a comprehensive vision for the designed space's performance in daylight and visual comfort. Furthermore, with the application of machine learning algorithms like ANN which is used in this research, these results could be delivered to the designers and stakeholders quickly with determination of some simple physical features of the designed spaces and there will be no need for vast knowledge in building physics modeling and simulations for architects and designers using them.

PRELIMINARY RESULTS AND CONCLUSIONS

In this research a model was developed for prediction of daylight, visual comfort and quality view assessment at the same time in a space. The model takes some simple physical features of the space as inputs and the estimated outputs include 7 metrics indicating the daylight and visual comfort performance of the defined space. The model is based on an optimized ANN algorithm which achieved a prediction accuracy of about 97% and MSE index of 0.002.

MAIN REFERENCES

Ayoub, M., 2020. A review on machine learning algorithms to predict daylighting inside buildings, Solar Energy, Vol 202, p249-275.

Mardaljevic, J., 2015. Climate-based daylight modelling and its discontents. Presented at the Simple Buildings Better Buildings? Delivering performance through engineered solutions, CIBSE Technical Symposium,London, April 16-17th

Ngarambe, J., Irakoze, A., Young Yun, G., Kim, G., 2020, Comparative performance of machine learning algorithms in prediction of indoor daylight illuminances, Sustainability 2020,12(11), 4471

Tregenza, P., Mardaljevic, P., 2017, Daylighting Buildings: Standards and the needs of designer, Lighting Research and Technology, Vol 50, p63-79.

LEED® green building program, v4 for Building design and construction.

Keywords: Daylight assessment, Visual Comfort, Quality Views, Parametric daylight simulation, Machine Learning

Paper ID 30240: Reducing heat island effect: a mathematical model of green roof design

Jing Hong* (1), Dennis Michael Utzinger (2)

1: Rivion; 2: University of Wisconsin- Milwaukee

AIM AND APPROACH

Green roofs ease the heat island effect. Optimizing green roof design helps achieve this goal more efficiently. This paper proposes an energy model of green roofs to estimate surface temperature and validates them with experimental evidence that measured on the green roof at the University of Wisconsin - Milwaukee. Two energy balance equations separately indicate the heat flux through the bare soil and vegetation-covered surfaces. The energy flux density and surface temperature of bare soil and vegetation-covered surface were modeled by the proposed mathematical models. In the meantime, the effects of color, soil depth, and plant type on the surface temperature were analyzed to conclude the optimal green roof design.

SCIENTIFIC INNOVATION AND RELEVANCE

In the proposed energy model, the calculation of radiation, convection, and evaporation on surface temperature is simplified to use fewer unknown variables. For that reason, this model only requires basic weather data and an initial soil temperature reading to estimate the green roof surface temperature in any equation-solving program.

PRELIMINARY RESULTS AND CONCLUSIONS

The energy balance models explain how surface color, soil depth, and plant types affect the surface temperature of a green roof. In conclusion, the green roof surface temperature can be reduced by lighter surface color, shallower soil depth, and plants with lower internal leaf resistance and larger leaf size.

MAIN REFERENCES

[1] Susca, Tiziana, Stuart R. Gaffin, and G. R. Dell'Osso. "Positive effects of vegetation: Urban heat island and green roofs." Environmental pollution 159, no. 8-9 (2011): 2119-2126.

[2] Sailor, David J. "A green roof model for building energy simulation programs." Energy and buildings 40, no. 8 (2008): 1466-1478.

[3] Duffie, John A., and William A. Beckman. Solar engineering of thermal processes. John Wiley & Sons, 2013.

[4] Berdahl, Paul, and Marlo Martin. "Emissivity of clear skies." Solar Energy 32, no. 5 (1984): 663-664.

[5] Watmuff, J. H., W. W. S. Charters, and D. Proctor. "Solar and wind induced external coefficients-solar collectors." Cooperation Mediterraneenne pour l'Energie Solaire (1977): 56.

[6] Farouki, Omar T. Thermal properties of soils. No. CRREL-MONO-81-1. Cold Regions Research and Engineering Lab, Hanover NH, 1981.

[7] Oke, Timothy R. Boundary layer climates. Routledge, 2002.

[8] Gates, David M. Biophysical ecology. Courier Corporation, 2012.

[9] Monteith, John, and Mike Unsworth. Principles of environmental physics: plants, animals, and the atmosphere. Academic Press, 2013.

[10] American Society of Heating, Refrigerating and Air-Conditioning Engineers. 2013 Ashrae Handbook: Fundamentals. Inch-pound ed. Atlanta, Ga.: Ashrae, 2013

[11] Oyj, Vaisala. "Humidity Conversion Formulas—Calculation formulas for humidity." Vaisala: Helsinki, Finland (2013).

Keywords: Green roof, Mathematical model, Simulation, Heat Island Effect, Calibration

Paper ID 30364: Semantic web ontologies for buildings objects and their performance data

Eleanna Panagoulia*, Zachary Lancaster, Tarek Rakha

Georgia Institute of Technology, United States of America

AIM AND APPROACH

The normative ontologies of tools, federated and tailored by the Architecture, Engineering, Construction and Operations industry (AECO), constrain collaboration and interoperability due to the creation of specific and bounded representational spaces, outside of which, software cannot operate. Although Building Information Modeling (BIM) provides a foundation for collaboration and data exchange within a common platform, professional practice still relies heavily on document sharing to circumvent the limits of proprietary software ontologies, especially when integrating performance analytics.

Current approaches of using monolithic data schemas to communicate information (i.e. IFC) constrain the domain of possible content, hence reducing the data expressivity and failing to capture the diversity of data in the built environment. To overcome this constraint we must provide access to information across platforms without presupposing information loss. This requires a modular, consistent, specific enough vocabulary, to describe any element, without at the same time being ambiguous.

This paper proposes one such vocabulary in the form of a semantic web-based ontology for representing building envelopes and their linked data. We demonstrate the use of this ontology in a platform agnostic exchange between envelope performance analysis and industry standard design software; enabling decision-making for building envelopes in the context of high-performance retrofitting design.

SCIENTIFIC INNOVATION AND RELEVANCE

The ability to represent building elements as semantically complete objects would contribute to higher performance architectures. However, the identification and classification of building objects depends largely on user interpretation and appears to operate without an explicitly documented or comprehensive ontology. This impedes the consistent linking of analytical outcomes and auxiliary data to envelope systems and prevents the creation of a common information repository. This lack of consistent understanding of a project, across all disciplines, results in miscommunication and thus, deviation between the actual and anticipated performance.

Previous research indicated that a single interface would be insufficient to handle the multifarious datasets associated with the building objects and their performance. This paper's innovation is in the employment of a semantic web approach that proposes the use of consistent and practical ontologies describing both building elements and building performance data, while also describing the relationship between building elements and their analytical data. This ontological definition focuses on a bi-directional exchange between platforms, and makes performance and simulation data available to practitioners.

PRELIMINARY RESULTS AND CONCLUSIONS

This research specifies two novel ontological definitions and a software interface that makes use of these definitions. The first one is defined as the Ontology for Building Objects (OBO), describes an explicit graph-based, rule-driven description of building envelopes as discrete elements and the relationships between them. While OBE focuses primarily on a description of geometric objects, the second, the Ontology for Building Performance Data (OBPD), is a minimal definition for building performance analytics, its data requirements, outputs and the relationship between an analysis and a building element or set of elements. While OBO is strongly aligned with OBPD, it is also aligned with other known building ontologies.

The paper showcases an application that operates as middleware between different data stores, the user and an interface for data serialization towards a common data environment. We demonstrate the above using Revit's API and a dataset common to energy audits of existing buildings. The result is an automated method of identifying components enabled by the OBO's building objects definition and linking them to performance analytics based on OBPD's relationship specification. We present a lossless exchange and mapping of data that provides the ability to represent non-native, analytical data directly in the BIM environment.

MAIN REFERENCES

[1] R. Volk, J. Stengel, and F. Schultmann, "Building Information Modeling (BIM) for existing buildings - Literature review and future needs," Automation in Construction. 2014.

[2] E. Kamel and A. M. Memari, "Review of BIM's application in energy simulation: Tools, issues, and solutions," Autom. Constr., vol. 97, no. October 2018, pp. 164–180, 2019.

[3] P. Pauwels, M. Poveda-Villalón, Á. Sicilia, and J. Euzenat, "Semantic technologies and interoperability in the built environment," Semant. Web, 2018.

[4] U. Knaack, "Potential for innovative massive building envelope systems – Scenario development towards integrated active systems," J. Facade Des. Eng., 2015.

[5] S. B. Sadineni, S. Madala, and R. F. Boehm, "Passive building energy savings: A review of building envelope components," Renew. Sustain. Energy Rev., 2011.

[6] A. Mahdavi and D. Wolosiuk, "A Building Performance Indicator Ontology : Structure and Applications", 2019.

[7] M. H. Rasmussen, M. Lefrançois, P. Pauwels, C. A. Hviid, and J. Karlshøj, "Managing interrelated project information in AEC Knowledge Graphs," Autom. Constr., 2019.

[8] M. Bonduel, M. Vergauwen, R. Klein, M. H. Rasmussen, and P. Pauwels, "A novel workflow to combine bim and linked data for existing buildings," eWork Ebus. Archit. Eng. Constr. - Proc. 12th Eur. Conf. Prod. Process Model. ECPPM 2018.

Keywords: BIM, Ontology, Semantic web, Data Interoperability, Environmental Performance

Paper ID 30401: A machine-learning framework for acoustic design assessment in early design stages

Reyhane Abarghooie (1), ZahraSadat Zomorodian (1), Mohammad Tahsildoost (1), Zohreh Shaghaghian* (2)

1: Shahid Beheshti University, Tehran, Iran; 2: Texas A&M University, College Station, United States

AIM AND APPROACH

Predicting acoustic performance by utilizing simulation tools is a widely used approach being favored over time-costly scale model studies. In this field, building acoustic simulation tools are complicated by several challenges, including the high cost of acoustic tools, the need of acoustic expertise, and the time-consuming process of acoustic simulation. This research presents a methodology for measuring acoustic comfort using a soft-sensing approach in the early design stages of the building. This work presents a proof of concept for a novel machine learning method to estimate a set of typical room acoustics parameters by using only geometrical information as input features.

The proposed model is trained and evaluated by using a novel dataset composed of acoustical simulation of a single room with 2916 different configurations. In the stimulation process features that include room dimensions, window size, material absorption coefficient, furniture, and shading type has been analyzed by using Pachyderm acoustic software. The mentioned dataset is used as the input of a machine-learning model based on Deep Neural Network (DNN). The machine learning model is fully-connected DNN with 5 hidden layers.

SCIENTIFIC INNOVATION AND RELEVANCE

Acoustic comfort is one of the most important design parameters that affect people's satisfaction and productivity. Therefore predicting the sound conditions in the building is noticeable in both the research and the industrial environments. Traditional methods to study the sound propagation inside the rooms can be divided into three approaches: the geometrical models, the wave-based models, and the statistical models. The parameter values that render the acoustical comfort of a room should be estimated based on the geometry of the virtual room. While the complete physical models are typically computationally too demanding, an approximation can be made by using simplified mathematical methods such as Sabin or Eyring. Formulas that can only calculate RT, and do not often hold in typical environments such as offices, schools, accommodations, commercial, and healthcare environments. In this way, we aim to estimate some of the acoustical indexes which, include Reverberation Time (RT), Early Decay Time (EDT), Speech Transmission Index (STI), Clarity (C80), and Definition (D50), where the estimated values should be accurate enough for the plausible rendering of acoustic virtual reality. The goal of this project is to introduce a model with short calculation time to estimate the mentioned indexes in the nominated spaces.

PRELIMINARY RESULTS AND CONCLUSIONS

This work presents a machine-learning-based method to estimate the room acoustic indexes. The model takes geometric and physical features of a room as input and output the estimated RT, EDT, STI, C80, and D50 indexes values as a function of frequency. All models were trained and evaluated in a dataset that contains parametric acoustic simulation.

The estimation was performed by using a machine-learning model based on a DNN. The baseline model achieves appropriate results by using geometrical features extracted from the 3D model of rooms. The stimulation model achieves a prediction accuracy of approximately more than 96% during the mentioned indexes.

MAIN REFERENCES

Rossing, T.D. (2014). Springer Handbook of Acoustics. New York, Ny Springer.

Falcón Pérez (2018). Machine-learning-based estimation of room acoustic parameters. Master's thesis. Aalto University, Espoo

J. Bianco, M., Gerstoft, P., Traer, J., Ozanich, E., A. Roch, M., Gannot, S. and Alban Deledalle, C. (2019). Machine learning in acoustics: Theory and applications. The journal of the acoustical society of America.

H. and Zhai, Z. (John) (2016). Advances in building simulation and computational techniques: A review between 1987 and 2014. Energy and Buildings, 128, pp.319–335.

Huang, B., Pan, Z., Liu, Z., Hou, G. and Yang, H. (2017). Acoustic amenity analysis for high-rise building along urban expressway: Modeling traffic noise vertical propagation using neural networks. Transportation Research Part D: Transport and Environment, 53, pp.63–77.

Lovedee-Turner, M. and Murphy, D. (2018). Application of Machine Learning for the Spatial Analysis of Binaural Room Impulse Responses. Applied Sciences, 8(1), p.105.

Keywords: Room acoustic, Parametric acoustic simulation, Machine Learning, Deep Neural Network

Paper ID 30438: A Python library for Radiance matrix-based simulation control and EnergyPlus integration

Taoning Wang* (1), Greg Ward (2), Eleanor S. Lee (1)

1: Lawrence Berkeley National Lab, United States of America; 2: Anyhere Software

AIM AND APPROACH

In the realm of building simulation, there are many reasons why accurate, efficient ray-tracing-based solar radiation and daylighting simulations are needed: 1) accurate thermal and visual comfort predictions rely on detailed maps of solar radiation and luminance on an occupant's body or field of view. 2) broader adoption of energy efficiency, indoor environmental quality, and healthy building standards require accurate modeling of innovative fenestration solutions, and 3) increased trends toward integrated, advanced building design and control solutions require efficient models to work seamlessly within co-simulation environments, such as Spawn-of-EnergyPlus. Radiance matrix-based simulation methods provide efficient and accurate ways to simulate annual/dynamic daylighting and solar radiation, particularly optically-complex operable shades, and dynamic glazings. However, real-world adoption could be more pervasive if the model setup and simulation workflow were not so difficult to learn and prone to user errors.

The goal of this work is to enable users and software developers to adopt advanced Radiance matrix-based simulation methods without extensive knowledge and experience. A Python-based open-source library frads was developed, along with a series of command-line programs, to automate and speed up the use of these advanced simulation methods.

SCIENTIFIC INNOVATION AND RELEVANCE

Ray-tracing algorithms have typically been too time-consuming for practical use in annual simulations. With matrix-based methods, the time needed to conduct annual simulations using ray-tracing tools has been reduced by several orders of magnitude. With the advent of version 9.3, EnergyPlus has exposed its core simulation engine to any external software environment, which enables runtime Radiance and EnergyPlus integration. The entire workflow is implemented as a command-line program, but software developers can use frads to incorporate the workflow into any third-party software.

frads consist of many standard Radiance operations allowing for automated model translation from EnergyPlus, matrices generation, and runtime data-exchange with EnergyPlus. frads also comes with several command-line programs for routine operations, and these programs also serve as examples of using the frads library. The chief among these programs is mrad, an executive program that automates the Radiance two- to six-phase simulation workflow. Users need to provide the model description (e.g., geometry, material, and location) and parts of the model that are parametrized. mrad will then choose the suitable matrix phase method based on user-specified speed and accuracy requirements and carry out matrix generation and multiplication workflow, outputting irradiance, illuminance or luminance results per time step.

PRELIMINARY RESULTS AND CONCLUSIONS

The framework for frads has been developed with further refinements currently underway. Early tests indicate that Radiance and EnergyPlus run-time integration requires minimal to no user intervention, enabling rapid widespread adoption. One of the key advantages of run-time data-exchange between Radiance and EnergyPlus is that it enables the evaluation of advanced model predictive control of building dynamic facade in a multizone building. During runtime, at each timestep, EnergyPlus simulation pauses; given a dynamic facade control signal (e.g., via Spawn/ Modelica or manufacturer component model), Radiance computes resulting facade energy transfer data, which are then used in EnergyPlus to complete the relevant heat balance calculations.

With these accurate simulation methods as part of the toolset, engineers and designers are able to differentiate facade product performance through simulation, which in turn spurs innovations from inventors and manufacturers. Integration of Radiance and EnergyPlus in a co-simulation environment further boosts the confidence among engineers and designers in the whole-building energy simulation results.

MAIN REFERENCES

Gehbauer, C., Blum, D. H., Wang, T., & Lee, E. S. (2020). An assessment of the load modifying potential of model predictive controlled dynamic facades within the California context. Energy and Buildings, 210, 109762.

Wang, T., Ward, G., & Lee, E. S. (2018). Efficient modeling of optically-complex, non-coplanar exterior shading: Validation of matrix algebraic methods. Energy and Buildings, 174, 464-483.

Lee, E. S., Geisler-Moroder, D., & Ward, G. (2018). Validation of the Five-Phase Method for Simulating Complex Fenestration Systems with Radiance against Field Measurements.

McNeil, A., & Lee, E. S. (2013). A validation of the Radiance three-phase simulation method for modelling annual daylight performance of optically complex fenestration systems. Journal of Building Performance Simulation, 6(1), 24-37.

Keywords: Daylighting, Energy, Co-simulation, Matrix-based

Paper ID 30470: Functional mock-up unit based generic threat injection framework for energy efficient and resilient buildings

Yangyang Fu, Xing Lu, Zheng O'Neill*

Texas A$M University, United States of America

AIM AND APPROACH

Fault detection and diagnosis in building energy and control systems relies much on accessible faulty data, which can be obtained from experiments or dynamic simulation. While experiments that inject faults to actual building energy and control systems are risky to building operators, high-fidelity dynamic modeling and simulation can provide flexibility of injecting all types of faults to the system in order to investigate their impact.

EnergyPlus as a popular building energy simulator has integrated plenty of faulty models. However, the development exposed several limitations. First, it is difficult to model control-related faults, such as inappropriate PID parameters and communication delays etc. Second, the injection of pressure-involved faults is not straightforward. For example, to model duct fouling, users need to adjust the pressure head, minimum and maximum air flowrate in the fan model. Third, faults cannot be injected in the middle of simulation. Forth, fault evolution patterns are not considered.

This paper presents a flexible FMU-based fault injection framework that can address the above-mentioned issues. This framework will consider a comprehensive list of typical faults in the HVAC system. The injection method will be mathematically presented. Component-level and system-level faulty modeling and simulation will be demonstrated in the case study.

SCIENTIFIC INNOVATION AND RELEVANCE

This paper presents a flexible and generic functional mock-up unit (FMU) based threat injection frame- work in Python. FMU is a simulation model that is compliant to the functional mock-up interface (FMI), which defines a standardized interface for a model so that such a model can be simulated in a different en- vironment for model exchange or co-simulation. The utilization of FMU in this paper is to provide a standardized model interface for different building energy simulators so that a generic threat library can be connected to it and perform threat evaluation. The proposed framework can address most of the identified research gaps in terms of threat modeling and simulation. Currently, EnergyPlus, TRNSYS and Modelica models can all be compiled to FMU models. Therefore, this proposed framework can be reused for all three simulators with minimum modifications when the different software structures of these three simulators should be taken into considerations.

PRELIMINARY RESULTS AND CONCLUSIONS

This paper presents a generic and flexible threat in- jection framework based on Functional Mock-up Unit (FMU). Several Python libraries are developed to address automating the generation of FMU wrap- per models, define threats and threat injection, and perform step-wise simulation for both baseline and threat-injected systems. The standalone definition of threats and the use of FMU as modeling base enables this framework to be easily expanded to other FMU- supported building energy simulators. The framework is demonstrated in a Modelica/FMU environment by modeling and simulating single-/multiple- order of threats. The simulation results show that the framework can provide flexibility of injecting different types of threats at different starting time for different duration.

MAIN REFERENCES

Kim, Janghyun, Stephen Frank, James E. Braun, and David Goldwasser. "Representing small commercial building faults in EnergyPlus, Part I: Model development." Buildings 9, no. 11 (2019): 233.

Kim, Janghyun, Stephen Frank, Piljae Im, James E. Braun, David Goldwasser, and Matt Leach. "Representing Small Commercial Building Faults in EnergyPlus, Part II: Model Validation." Buildings 9, no. 12 (2019): 239.

Granderson, Jessica, Guanjing Lin, Ari Harding, Piljae Im, and Yan Chen. "Building fault detection data to aid diagnostic algorithm creation and performance testing." Scientific data 7, no. 1 (2020): 1-14.

Li, Yanfei, and Zheng O'Neill. "A critical review of fault modeling of HVAC systems in buildings." In Building Simulation, vol. 11, no. 5, pp. 953-975. Tsinghua University Press, 2018.

Zhang, Rongpeng, and Tianzhen Hong. "Modeling of HVAC operational faults in building performance simulation." Applied Energy 202 (2017): 178-188.

Keywords: fault detection and diagnosis, Modelica, fault injection

Paper ID 30572: Real-time light simulation methodology for expedited comparison and optimization studies through agent-based photon modeling

Vishal Vaidhyanathan*, Jichen Wang

Carnegie Mellon University, United States of America

AIM AND APPROACH

Daylight Simulations for illumination analyses for tasks involving comparative evaluation and optimization are time taking due to their dependency on external simulation engines. Conventional daylight analysis workflows for quantitative analysis in early phase design decisions for multiple design iterations can be cumbersome and require running thousands of test cases to optimize a workspace for point in time illuminance. This research proposes a methodology for utilizing light simulation as a metric for optimization that uses agent based modelling of photon particles, mimicking their properties using certain policies to generate instantaneous daylight illumination results for comparative and qualitative analysis of architectural spaces. This algorithm removes external dependencies of simulation engines and runs on the CPU. Since it works with direct geometry collision and particle based methods, it works with non-orthogonal and non-linear geometries as well at real-time. A large analysis test mesh can also be executed in real time using multi-threading of particle groups. This enables quick performance optimization for real time design decision making. It also consists of a front end interface for non-intuitive users to perform comparative studies with ease where fitness graphs are visualized in real time.

SCIENTIFIC INNOVATION AND RELEVANCE

Daylight simulations for tasks involving comparative evaluation, optimization, etc. - are time taking. They require external simulation-engine dependencies and take a while to return results. While the required tasks are comparative in nature, and multiple design options are to be rapidly evaluated and ranked, the performance evaluation process must be expedited (almost real-time), easy to setup (non-intuitive) and computationally non-intensive. In such scenarios, it is also sufficient if results are not absolute but rather comparative and "rankable". Agent Based Modelling (ABMs) is an approach to model systems composed of autonomous and interacting agents that have a set of policies or behavioral rules and are constrained in an environment. Parallels can be drawn between ABM and the properties of light particles (Photons). This research project explores if ABM can be used to emulate light behavior to an extent that it is possible to compare multiple design options and make daylight performance directed design decisions, in an expedited way and in real-time. The amount of illuminance and its uniformity are the two most critical criteria to validate good lighting conditions and are thus evaluated in this tool. The main inputs for this tool are the sun angle and geometry surfaces.

PRELIMINARY RESULTS AND CONCLUSIONS

This methodology with light simulation and optimization is flexible and extremely user friendly where input parameters can be changed during the design process that gives quantifiable standards for comparing design options. The tool gives the designer the power to come up with design choices by themselves, where the evaluation tool is just a helper for decision making. By alternating the input parameters and the definition of geometries, it can also be applied in other kinds of design processes. With this agent based algorithm, one can simulate the light behavior with a rough accuracy and fast calculation. Given that there are already plenty of well established light simulation tools in the field, a comparison was made between this tool and ladybug for grasshopper. The results though not as accurate as ladybug, resulted in the run time being very short which was almost comparable to a real time calculation. Moreover the input settings are much simpler than other tools given that it can take any surface type objects as the geometry input. These benefits make this simulator a very good tool in daylight evaluation and optimization by exploring a large amount of possibilities with multiple geometrical inputs to find the optimized solution.

MAIN REFERENCES

1. Illuminating Engineering Society. IES Approved Method: Spatial Daylight Autonomy (sDA) and Annual Sunlight Exposure (ASE). New York: IES; 2012. (IES LM-83-12).

2. Chamilothori, Kynthia & Chinazzo, Giorgia & Rodrigues, João & Dan-Glauser, Elise & Wienold, Jan

& Andersen, Marilyne. (2019). Subjective and physiological responses to façade and sunlight pattern

3. United States Green Building Council (USGBC), retrieved 2020.

Keywords: Agent based Modelling, Lighting Design, Optimization, Computational Design

Paper ID 30753: Radiant spectral energy for simulation in the built environment

Joseph Del Rocco, Joseph T. Kider Jr.*

University of Central Florida, School of Modeling, Simulation, and Training

AIM AND APPROACH

Fast and accurate daylighting and energy performance simulations are crucial for real-time control systems (RCS) in the built environment. RCS include responsive facades, adaptive e-glass, and smart HVAC control systems. State-of-the-art building monitoring systems driving the RCS require spectral energy inputs to take full advantage of both light and heat of solar and sky radiation. Additionally, building designers also require spectral energy for building and fenestration design and material decisions relating to circadian rhythm daylighting. Unfortunately, modern daylighting and energy simulations often neglect spectral energy in efforts to reduce computation time, and therefore do not provide the full global-illumination and energy solution needed by fine-grained RCS. We present an accurate, interactive, physically-based approach that utilizes a Transition Portal radiosity engine (Kider et al., 2019) to compute a full global-illumination solution with spectral energy for building simulations in real-time. This approach allows building simulations to leverage the visible spectrum for more accurate daylighting and the full spectrum for energy analysis. This novel method is intended for real-time control systems in the built environment and building adaptation for circadian daylighting.

SCIENTIFIC INNOVATION AND RELEVANCE

Modern building performance simulation often ignores real-time spectral energy input for various reasons. The calculations can be time-consuming and the data needs to be transformed and marshaled through a diverse array of legacy and modern daylighting and energy software systems (modeling packages, fenestration software, OpenStudio, EnergyPlus, Radiance, etc.). However, daylighting and energy performance software must eventually use full-spectrum solar and sky radiation to provide the most accurate results. Doing so may even help to simplify the building performance pipeline. We have anticipated this and progressively worked towards a solution for fine-grained real-time control systems manipulating both light and heat from the same spectral inputs. We have developed a novel system that interfaces with our radiosity engine which processes the spectral inputs and produces spectral outputs and can interface with modern tools. Spectral energy is generated from sky conditions through a machine-learned model (Del Rocco, 2020) from both real-world skydome photographs and synthetic sky renders to demonstrate our system under varied times of day and sky conditions. We also propose and simulate a novel adaptive e-glass solution that filters heat during warm climates but allows it during cool climates, to maximize energy performance with natural heating.

PRELIMINARY RESULTS AND CONCLUSIONS

We performed daylighting simulations and energy analysis on three separate building spaces with the following real-time control systems: automated blinds, a kinetic facade, and various e-glass configurations. Our preliminary results show that our spectral energy system is definitely fast enough to be used by building monitoring systems to drive real-time control systems, as well as by building designers using parametric design to plan shading devices and materials. We also demonstrate that our proposed adaptive e-glass solution can properly filter heat energy from the upper spectra during warm climates (allowing it during cool climates) based on configured limits, contributing to an overall more energy-efficient building performance solution. A cyber-physical prototype building monitoring system with e-glass control is proposed, which monitors the skydome with an all-sky camera as the root input of spectral energy, and then simulates the propagation of the spectral energy input throughout the building efficiently. These experiments demonstrate the feasibility of a system that can handle spectral energy calculations for building performance simulations and control systems in real-time.

MAIN REFERENCES

Kider, J. T., Walter, B., Fang, S., Sekkin, E., & Greenberg, D. P. (2019). Transition Portal for daylighting calculations in early phase design. Energy and Buildings.

Del Rocco, J., Bourke, P. D., Patterson, C. B., & Kider, J. T. (2020). Real-time spectral radiance estimation of hemispherical clear skies with machine learned regression models. Solar Energy.

S. F. Rockcastle, M. Danell, L. Petterson, and M. L. Ámundadóttir (2020). The Impact of Behavior on Healthy Circadian Light Exposure Under Daylight and Electric Lighting Scenarios. ACEEE Summer Study on Energy Efficiency in Buildings 2020.

Balakrishnan, P., & Jakubiec, J. A. (2019). Spectral Rendering with Daylight: A Comparison of Two Spectral Daylight Simulation Platforms. IBPSA Conference.

Rockcastle, S., Ámundadóttir, M. L., & Andersen, M. (2019). The Case for Occupant-Centric Daylight Analytics: a Comparison of Horizontal Illumination and Immersive View.International Conference of the International Building Performance Simulation Association

Adaptive Lighting for Alertness (ALFA) Software. https://solemma.com/Alfa.html

Inanici, M., and ZGF Architects LLP (2015). LARK Spectral Lighting plugin to Grasshopper. https://faculty.washington.edu/inanici/Lark/Lark_home_page.html

Ward, G. J. (1994, July). The RADIANCE lighting simulation and rendering system. In Proceedings of the 21st annual conference on Computer graphics and interactive techniques.

Keywords: spectral energy, daylighting, natural heating, smart glazing, real-time

Paper ID 31051: Short-term forecasting of building energy consumption with deep generative learning

Yichuan X. Ma*

The University of Hong Kong, Hong Kong S.A.R. (China)

AIM AND APPROACH

Short-term building energy consumption forecasting is highly valuable from both a technical and an economic point of view. In this paper, a deep generative learning model (GAN-Plus) taking account of short-term future meteorological data is proposed to accurately forecast building energy consumption in the near future. A conventional machine learning model (multilayer perceptron, MLP) and a non-meteorology-based version of the proposed model (GAN-Zero) were developed and comparatively tested as baseline models. Multi-year hourly meteorological data and actual energy consumption measurements from two office buildings in Shanghai were used for modelling and testing.

SCIENTIFIC INNOVATION AND RELEVANCE

(1) A novel generative short-term forecasting framework was delineated.

(2) Customised conditional GAN-based model with a 1D-UNet generator under the generative short-term forecasting framework were developed and quantitatively evaluated under multiple granularity settings.

(3) Historical and future meteorological information was taken into consideration to improve forecasting accuracy.

(4) Cross-case generalisability, which is an underappreciated issue in the field of building energy forecasting, was discussed.

PRELIMINARY RESULTS AND CONCLUSIONS

State-of-the-art accuracy and decent cross-case generalisability of the proposed GAN-based models were demonstrated. The proposed model outperformed the chance level and all the baseline models, achieving accuracies of 85.48% with hourly granularity and 94.77% with daily granularity. MLP failed in the cross-case forecasting task with an hourly CV-RMSE notably larger than 30%. Though similar in terms of performance in hourly forecasting, GAN-Plus showed superior accuracy (with a difference > 1%) than GAN-Zero in daily forecasting. In the noise robustness analysis, GAN-Plus demonstrated strong capability in dealing with the uncertainties in the future meteorological clues, with less than 1.5% deterioration of accuracy with a noise level of 20%. The advantage and potential risks of using the proposed GAN-Plus model were further discussed.

MAIN REFERENCES

[1] Deb, C., F. Zhang, J. Yang, S. E. Lee and K. W. Shah (2017). "A review on time series forecasting techniques for building energy consumption." Renewable and Sustainable Energy Reviews 74: 902-924.

[2] Fan, C., Y. Sun, Y. Zhao, M. Song and J. Wang (2019). "Deep learning-based feature engineering methods for improved building energy prediction." Applied energy 240: 35-45.

[3] Goodfellow, I. J., J. Pouget-Abadie, M. Mirza, B. Xu, D. Warde-Farley, S. Ozair, A. Courville and Y. Bengio (2014). "Generative adversarial networks." arXiv preprint arXiv:1406.2661.

[4] Ma, Y. X. and C. Yu (2020). "Impact of meteorological factors on high-rise office building energy consumption in Hong Kong: From a spatiotemporal perspective." Energy and Buildings 228: 110468.

[5] Mirza, M. and S. Osindero (2014). "Conditional generative adversarial nets." arXiv preprint arXiv:1411.1784.

[6] Monfet, D. and P. Radu Zmeureanu PhD (2009). "Calibration of a building energy model using measured data." ASHRAE Transactions 115: 348.

Ronneberger, O., P. Fischer and T. Brox (2015). U-net: Convolutional networks for biomedical image segmentation. International Conference on Medical image computing and computer-assisted intervention, Springer.

Keywords: Deep learning, short-term forecasting, building energy consumption, generative adversarial network

Session W2.8 (Online Track): Buildings paving the way for the energy transition

Paper ID 30189: Economic and emission reduction potential of renewable based energy systems in an apartment building in cold climatic conditions

Hassam ur Rehman*, Ala Hasan, Francesco Reda

VTT Technical Research Centre of Finland, Finland

AIM AND APPROACH

Climate change is one of the biggest challenges, largely caused by emissions, and to address such issue, renewable energy can be used for buildings and districts energy demands. The aim is to design and simulate a novel renewable based energy system integrated with the buildings in the district and to compared the proposed system against the reference energy system. The energy system is designed to provide space heating, domestic hot water and electricity to the buildings in the community. TRNSYS dynamic simulation software is used to model the energy system. Costs and CO2 emission are calculated and compared for the novel renewable based energy system and reference energy system.

SCIENTIFIC INNOVATION AND RELEVANCE

The cities and districts are setting ambitious targets to make buildings and districts carbon neutral in an economical manner. The novelty is to design renewable energy based district heating system with storage in order to integrate buildings in the community with clean energy production source. The focus and novelty lies in the analysis of the CO2 reduction potential, rather then technical performance. In addition to this the emission reduction cost and relative emission reduction is compared to the reference district heating system, to show the benefits. The concept is proposed in order to reduce the CO2 emissions to reach the European Commission and cities targets of 2050.

PRELIMINARY RESULTS AND CONCLUSIONS

The renewable energy based energy system can provide energy to the buildings in the community. These systems have the potential to reduce the CO2 emissions from the present state-of-the-art district heating girds by more then 50%. The onsite energy fraction can vary between 1% to 97%. The emission reduction cost can vary between 2 €/kg CO2/yr to 6 €/kg CO2/yr. Such renewable based systems could lower the carbon dioxide emissions from the existing energy systems to meet the Europe emissions reduction and cities targets.

MAIN REFERENCES

[1] ur Rehman, Hassam, et al. "Towards positive energy communities at high latitudes." Energy conversion and management 196 (2019): 175-195.

[2] Hirvonen, Janne, et al. "Towards the EU emissions targets of 2050: optimal energy renovation measures of Finnish apartment buildings." International Journal of Sustainable Energy 38.7 (2019): 649-672.

Keywords: CO2 emission reduction, renewable based energy system, seasonal storage, communities; solar energy; energy simulations

Paper ID 30951: Integration of a flexible tensile material in a second skin façade system: a passive method to enhance the energy performances of the Italian building scenario

Giovanni Ciampi*, Yorgos Spanodimitriou, Michelangelo Scorpio, Antonio Rosato, Sergio Sibilio

Department of Architecture and Industrial Design, University of Campania Luigi Vanvitelli, Italy

AIM AND APPROACH

In Italy the building panorama is mostly represented by historical buildings or buildings built in the mid-twentieth century, when the performances of the building envelope were scarcely considered [1]. Thus, in order to provide a comfortable indoor space for occupants, a considerable amount of energy is consumed, especially for the heating and cooling. For these reasons, retrofit actions are continuously fostered by the government, in order to encourage the improvement of the overall building energy efficiency. These improvements can be classified into passive and active ones. The energy conservation measures involve active actions, such as electric lighting, equipment, heating, ventilation and air conditioning (HVAC), etc., oriented towards the use of modern facilities with higher efficiency and low energy consumption, while passive actions are oriented towards the reduction of the energy demand over the life of the building. The passive actions can be incorporated within the construction elements, like envelope insulation and fenestration, or they can be integrated as external elements that can significantly reduce the cooling demand. However, it is not always possible to replace the glazing or the insulation layers, considering the complexities coming from, for example, the historical relevance of a building or the cost for the privates.

SCIENTIFIC INNOVATION AND RELEVANCE

In these cases, a lightweight, non-impacting and inexpensive solution, such as the installation of an external second skin layer on the façade, can be seen as a foreseeable passive solution to improve the thermal performances of the building envelope [2].

In this paper, the energy and environmental impacts of passive retrofit actions on an existing building are evaluated in terms of primary energy saving, carbon dioxide equivalent emissions and simple pay-back period [3]. The study focused on office building and is aimed at both proposing a general operational methodology and highlighting a best practice for the Italian territorial context. The primary energy consumption, the carbon dioxide equivalent emissions and cost-effectiveness, associated to the current scenario of the case study, have been evaluated through the dynamic simulation software TRNSYS, across a whole year. The simulation results associated to a reference office building have been compared with those achievable by adopting a passive retrofit action on the building envelope. In particular, the refurbishment consists of the installation, without invasive interventions, as a second skin, of a light innovative flexible PVC-coated polyester fabric [4] on the façade of the reference building.

PRELIMINARY RESULTS AND CONCLUSIONS

Different arrangements and control strategies have been considered for the second skin, specifically tailored on the characteristics of the polyester fabric. Both the energy and environmental comparison have been carried out according to the Italian scenario. The performed comparison among the current status and the possible future scenarios allowed to evaluate the potential benefits in terms of primary energy saving, carbon dioxide equivalent emissions reduction, and cost-effectiveness, assessing the advantages of each proposed retrofit action. In particular, the results highlighted that a seasonal control strategy, along with a proper integration of an insulation layer in the second skin system, allowed for a reduction of the cooling energy associated to the whole building up to 30%.

MAIN REFERENCES

[1] M. Scorpio, G. Ciampi, A. Rosato, L. Maffei, M. Masullo, M. Almeida, S. Sibilio. Electric-driven windows for historical buildings retrofit: Energy and visual sensitivity analysis for different control logics, Journal of Building Engineering, Volume 31, September 2020, 101398. https://doi.org/10.1016/j.jobe.2020.101398

[2] F. Ascione, N. Bianco, G. Maria Mauro, D.F. Napolitano. Building envelope design: Multi-objective optimization to minimize energy consumption, global cost and thermal discomfort. Application to different Italian climatic zones. Energy 2019;174:359–74. https://doi.org/10.1016/j.energy.2019.02.182.

[3] G. Angrisani, E. Entchev, C. Roselli, M. Sasso, F. Tariello, W. Yaïci. Dynamic simulation of a solar heating and cooling system for an officebuilding located in Southern Italy. Applied Thermal Engineering, Volume 103, 25 June 2016, Pages 377-390. https://doi.org/10.1016/j.applthermaleng.2016.04.094

[4] Schueco. PVC-coated polyester Fabric. https://www.schueco.com/de-en/fabricators/products/facades/textile-facades/facid65 (accessed July 12, 2020).

Keywords: Passive building envelope, Flexible materials, PVC-coated polyester fabric, Building energy efficiency, Energy savings

Paper ID 30405: Study of optimal control strategies for an energy production system using thermal mass of buildings

Charbel Salameh* (1,2), Bruno Peuportier (2), Patrick Schalbart (2)

1: Accenta; 2: Centre Efficacité énergétique des Systèmes Mines Paristech

AIM AND APPROACH

During the operation phase, HVAC systems (heating, cooling, and domestic hot water) represents up to 50% of buildings total energy consumption, and 75% of GHG (Greenhouse Gas) emissions in France. Photovoltaic and solar thermal production, geothermal heat pumps, BTES (Borehole Thermal Energy Storage), and short-term thermal storage (in hot or cold water tanks) can be associated to reduce cost and related emissions.

In this context, real-time energy management strategies adapted to this type of complex systems is investigated. Determining an energy management strategy for such systems, taking into account buildings thermal capacity, requires adequate building and system models that can evaluate energy loads with good precision in a reasonable calculation time for real-time application.

Dynamic thermal simulation is used in order to evaluate the consumption and production profiles of typical buildings and the associated energy system. An appropriate optimisation technique with an objective function is applied in order to determine optimal strategies.

SCIENTIFIC INNOVATION AND RELEVANCE

Considering the large impact of the buildings energy consumption and GHG emissions, methods and solutions are developed in order to decrease the energy consumption and the environmental impact of the sector. This study presents an innovative solution that is based on thermal energy storage using BTES.

In this first approach, we consider using a geothermal borehole model with a Ground Source Heat Pump (GSHP) coupled to a building model including an occupancy model and meteorological data. The goal is to optimise the control in real-time of the energy production and storage system based upon the calculation of the heating/cooling needs of the building. This optimisation takes into consideration the cost of the energy production and the COP of the GSHP influenced directly by the energy exchanged with the BTES.

The novelty of this model lies in the simultaneous multi-system simulation and optimisation taking into consideration the energy storage in the building thermal mass and in the BTES. The study aims at minimising the cost and impact of the thermal energy production. In addition, the system efficiency and building thermal needs profile can be optimised by allowing a temperature set point variation within the limits of occupants' comfort.

PRELIMINARY RESULTS AND CONCLUSIONS

The optimisation procedure was applied to a multi-family residential building of 4300 m² in the south of France.

Taking into consideration an hourly electricity price and a maximal and minimal limit for the temperature set point in the building, the results show a load shifting of the energy demand by shifting the demand from peak hours to low-cost hours. This is made possible by storing the energy in the thermal mass of the building, increasing the temperature during off-peak hours to an acceptable limit within the respect of the occupants' comfort.

The optimisation takes into account a short-term horizon considering the building dynamic, and a long-term horizon for the BTES which can range from a minimum of one year to a twenty year period to ensure the temperature balance in the ground storage, thus optimising the efficiency of the heat pump.

Early results show a decrease of energy consumption cost between 10% and 15%.

MAIN REFERENCES

Favre B., Peuportier B., Application of dynamic programming to study load shifting in buildings, Energy and Buildings 82 (2014) 57–64

Frapin, M., Chaplais, F., Schalbart, P. and Peuportier, B., Optimal control of heating in a two-zone building using price decomposition-coordination method, 31st International Conference on Efficiency, Cost, Optimization, Simulation and Environmental Impact of Energy Systems, June 2018, Guimarães - Portugal.

Malisani, P., Favre, B., Thiers, S., Peuportier, B., Chaplais, F. and Petit, N. Investigating the ability of various buildings in handling load shiftings, Power Engineering and Automation Conference, Wuhan, septembre 2011

PEUPORTIER, B. et SOMMEREUX, I. B., 1990. Simulation tool with its expert interface for the thermal design of multizone buildings. In : International Journal of Solar Energy. 1 janvier 1990. Vol. 8, n° 2, p. 109-120. DOI 10.1080/01425919008909714.

Robillart, M., Schalbart, P., Chaplais, F., Peuportier, B., Model reduction and model predictive control of energy-efficient buildings for electrical heating load shifting, J. Process Control (2018), 11:3, 294-308, https://doi.org/10.1080/19401493.2017.1349835

Robillart, M., Schalbart, P., Peuportier, B., Derivation of simplified control rules from an optimal strategy for electric heating in a residential building. Journal of Building Performance Simulation, July 2017. doi: 10.1080/19401493.2017.1349835.

Keywords: Building, optimisation, Model Predictive Control, Ground Source Heat Pump, Thermal Energy Storage

Paper ID 30309: Effect of introducing hybrid power system for housing

Saori Nakao* (1), Yasuyuki Shiraishi (1), Yusuke Shigematsu (2)

1: The University of Kitakyushu, Japan; 2: Saibugas Co., Ltd., Japan

AIM AND APPROACH

In this research, we propose a three-cells cooperation system that combines a fuel cell system (FC) and a storage battery system (BT) in a photovoltaic power generation (PV) to utilize renewable energy as the main power source in a house.[1,2] FC can be operated according to the power generation status of renewable energy because it can supply power stably and generate power at any time. In addition, FC can also use the waste heat generated by power generation for hot water supply. BT can charge and discharge excess power of PV that has not been consumed by itself. In this report, as the previous step of the demonstration experiment to be conducted in the future, we conduct a case study of annual energy simulation for a house with a three-cells cooperation system. The purpose of this study is to clarify the optimum operation method of a three-cells cooperation system by comparing simulation cases from the viewpoint of the energy saving and the economic efficiency. The energy saving evaluation items are annual energy consumption, energy self-sufficiency rate of the house, and CO2 emissions, and the economic evaluation items are the operating cost and energy payback time (EPT) when introducing each device.

SCIENTIFIC INNOVATION AND RELEVANCE

Although renewable energies such as PV and wind power generation are environmentally friendly and abundant resources that can be easily obtained at low cost, there is a problem that the amount of power generation changes depending on the weather. Therefore, a stable power supply becomes possible by combining renewable energy with a device that can stably supply electric power or a device that can store them.[3,4] Tsuneoka et al. [5] proposed an operation method using heat pump water heater(HPWH) and BT in combination with PV, and showed the economic superiority of autonomous operation by introducing BT. Akimoto et al. [6] investigated the operation method by introducing electric vehicle (EV) to consume excess power of PV, and proposed operation methods that improve energy self-sufficiency rate of the house according to the frequency of EV use. In this way, methods of combining various devices with renewable energy have been proposed, but there are not many case studies that compare multiple operation methods from the viewpoint of energy saving and economic efficiency assuming actual operation. In this report, we conduct a case study of the operation method for a house with a three-cells cooperation system assuming actual operation.

PRELIMINARY RESULTS AND CONCLUSIONS

By introducing a three-cells cooperation system in the house, it has become possible to supply power without using external power as much as possible. For the reason, energy self-sufficiency rate of the house was increased, and annual energy consumption and CO2 emissions were decreased. Since the power purchased from the grid has been significantly reduced, the operating costs have been reduced as well. Based on the above results, the effectiveness of the three-cells cooperation system was shown from the viewpoint of both the energy saving and the operating costs.[1,2] Regarding EPT, although the initial cost of PV and FC has decreased, the initial cost of BT is still high, so it took a considerable number of years to recover them. However, it was suggested that EPT could be significantly shortened by using the Feed-in Tariff, that electricity company purchases electricity generated from renewable energy at a fixed price, and selling excess power of PV.

MAIN REFERENCES

[1] Y.Shigematsu et al, Optimal control of a household fuel cell system in a three-cells cooperation system Part1 Project summary for building new energy utilization model, Heating, Air-Conditioning and Sanitary Engineers of Japan (in press, in Japanese).

[2] S.Nakao et al, Optimal control of a household fuel cell system in a three-cells cooperation system Part2 Examination of energy saving effect and economic efficiency by numerical simulation, Heating, Air-Conditioning and Sanitary Engineers of Japan (in press, in Japanese).

[3] Jijian Lian et al, A review on recent sizing methodologies of hybrid renewable energy systems, Energy Conversion and Management, Vol.199.(2019), pp.1-23.

[4] T.Wakui et al, Optimal operations management of residential energy supply networks with power and heat interchanges, Energy and Buildings, Vol.151.(2017), pp.167-186.

[5] Y.Tsuneoka et al, Operation Method for Self-Consumption of Surplus Power Considering Comfort in a Zero Energy House, Heating, Air-Conditioning and Sanitary Engineers of Japan, No.10.(2017), pp.293-296 (in Japanese).

[6] M.Akimoto et al, Operational strategies for self-consumption considering the use of an electric vehicle ina net zero energy house, Journal of Environmental Engineering (Transactions of AIJ), Vol.770.(2020), pp.277-287 (in Japanese).

Keywords: Photovoltaic power generation, Fuel cell system, Storage battery system, Energy simulation, Optimization

Paper ID 31141: Activation of the building thermal mass to store PV surplus energy

Arno Dentel*, Christina Betzold

Technische Hochschule Nürnberg, Germany

AIM AND APPROACH

Since 2018, a terraced house complex with shared energy system is monitored and evaluated regarding PV self-consumption and efficiency. Beside heat pumps and photovoltaics (PV), different kind of storage units are integrated and used to store PV surplus. Thermal storage units for heating and domestic hot water as well as an electrical storage are integrated into the higher-level control system and are charged selectively. However, since the capacity limits have been reached, additional storage options are interesting. The thermal mass of the building offers additional potential to store PV surplus energy. This paper evaluates the operation of thermal mass activation for the terraced house complex in order to increase the PV self-consumption and decrease the grid consumption.

The simulation study of thermal mass activation is realized in TRNSYS. Increasing the room set point temperature of 20 °C by 2 K during PV production, aims to a reduction of the grid consumption by 47 % and an increase of PV self-sufficiency from 23 % up to 64 %. At the same time, the mean deviation of the room set temperature increased from 1.7 K to 2.0 K. The results are compared to real measured data from the terraced house

SCIENTIFIC INNOVATION AND RELEVANCE

In this work, a control strategy is investigated in simulation and real operation, which activates the thermal building mass by increasing the setpoint in the entire building. Figure 1 shows the process of the control strategy, which depends on the available PV power. As soon as a PV power of 1 kW is exceeded, the setpoint is increased by 1 K to 6 K, depending on the variant. If the PV power is less than 1 kW, the default setpoint, e.g. 20 °C, is set. In real operation the interface between control strategy and building energy management system is realized by a SQL Server Database.

PRELIMINARY RESULTS AND CONCLUSIONS

The simulation results show that the activation of the thermal building mass with PV power offers good potential for reducing grid consumption while maintaining thermal comfort. Likewise, the operating costs can be reduced both with and without the EEG levy but with the EEG levy, the cost savings drop again after 3 K. A setpoint increase of 2 K provides the best result. The energy and financial results do not improve thereafter. The thermal comfort only deteriorates strongly with very high setpoint increases of 6 K.

The assessment of the storage potential shows that a room temperature increase from 20 C to 22 °C leads to a thermal power of 6 kW to 8 kW over a period of 6 hours. After approx. 6 hours, the thermal power is reduced because some rooms have already reached their setpoint. Due to the room-by-room building model, there is no complete switching off of the thermal power, as some rooms are already heated again depending on their temperature.In addition, the activation of the thermal building mass was implemented in the control of the terraced house complex in November 2020.

MAIN REFERENCES

EEG, 2020: German Renewable Energies Law

University of Madison-Wisconsin, 2017. TRNSYS-TRaNsient SYstems Simulation, Solar Energy Laboratory,. Version 18.1.

Keywords: ZEB, thermal storage, building mass, model predictive control

Paper ID 30391: Impact of implementing air-conditioning systems on the school building stock in Brazil considering climate change effects: a bottom-up benchmarking

Matheus Soares Geraldi, Mateus Vinicius Bavaresco*, Veronica Gnecco, Enedir Ghisi, Michele Fossati

Federal University of Santa Catarina, Brazil

AIM AND APPROACH

The study of the building stock is a key knowledge to improve energy performance and enhance indoor environmental quality. In Brazil, energy policies are still in their first steps [1]. This study is based on a previous work that assembled and analysed evidence-based data of public schools in Brazil [2], which revealed that around 90.1% of them are not equipped with adequate air-conditioning. Those systems are responsible for enhancing indoor thermal conditions, but also increase the energy use intensity [3]. Meanwhile, air-conditioning is being gradually implemented in schools according to the uninformed culture that only those systems provide adequate thermal conditions.

This paper aims to estimate how much the average energy use intensity (EUI) of the building stock will increase due to implementing air-conditioning in schools. For this purpose, EnergyPlus simulation was employed using reference buildings. Parametric simulations were adopted to represent the actual stock conditions for a sample of 378 buildings, and a calibration process fitted the simulated and actual whole-building energy consumption. Then, the calibrated models were used to simulate scenarios of air-conditioning implementations. Finally, predicting the increasing of EUI allows the evaluation of energy efficiency strategies instead of just implementing air-conditioning in the near future.

SCIENTIFIC INNOVATION AND RELEVANCE

The scientific innovation of this study relies on the method employed, which takes advantage of building stock modelling techniques to estimate future trends of the performance of the building stock. Benchmarking is a useful practice to evaluate if a building is efficient or not by comparing it against its pairs. Developing a benchmarking practice requires methods that take into consideration stock data. It usually adopts reference buildings and considers simulation models to establish the benchmark by regressive models – the reference energy performance of a building typology according to determined conditions. Advances in benchmarking methods are available in the literature; however, they are usually related to high-granularity data, specific focuses or their countries or regions [4]. In Brazil, a static benchmarking method was developed for bank branches [5] and high-rise buildings [6]. However, a comprehensive benchmarking method that considers future trends for school building stock is pioneering.

The relevance is expressed by the importance of measuring the carbonisation effect of the implementation of air-conditioning systems in the public-school stock in Brazil. Furthermore, by estimating this impact at stock-level, it is possible to propose passive and smart strategies instead of the unconscious implementation of air-conditioning systems.

PRELIMINARY RESULTS AND CONCLUSIONS

So far, the building stock was characterised. It was observed that only 12.95% out of all classrooms of the stock have air-conditioning. Similarly, only 32.7% of the administrative rooms and 31.9% of the labs have air-conditioning. Two specific schools that had air-conditioning installed in 2016 and 2017 were analysed. The first school had and

EUI of 11.9 kWh/m^2 before the implementation and 34.6 kWh/m^2 after. The second case presented 8.1 kWh/m^2 before and 15.8 kWh/m^2 after. Both cases presented an expressive increasing in EUI, as expected.

The use of a standard design in a national perspective does not support the use of passive strategies for energy efficiency, which are inherently dependent on the building context [7]. Then, the educational department opts to implement air-conditioning to boost thermal condition as a general rule. However, the utilisation of passive strategies such as increasing ceiling insulation, smart utilisation of the natural ventilation and improving shading devices could improve the thermal condition and reduce cooling loads. Finally, there is the possibility to take advantage of this standard design and propose standard solutions according to the building context.

MAIN REFERENCES

[1] A.P. Melo, M.J. Sorgato, R. Lamberts, Building energy performance assessment: Comparison between ASHRAE standard 90.1 and Brazilian regulation, Energy Build. 70 (2014) 372–383. doi:10.1016/j.enbuild.2013.11.080.

[2] M.S. Geraldi, E. Ghisi, Mapping the energy usage in Brazilian public schools, Energy Build. 224 (2020) 110209. doi:10.1016/j.enbuild.2020.110209.

[3] T.S. Saraiva, E.M. da Silva, M. Almeida, L. Bragança, Comparative study of comfort indicators for school constructions in sustainability methodologies: Schools in the amazon and the southeast region of Brazil, Sustain. 11 (2019). doi:10.3390/su11195216.

[4] M.S. Geraldi, E. Ghisi, Building-level and stock-level in contrast : A literature review of the energy performance of buildings during the operational stage, Energy Build. 211 (2020) 109810. doi:10.1016/j.enbuild.2020.109810.

[5] E.H. Borgstein, R. Lamberts, Developing energy consumption benchmarks for buildings: Bank branches in Brazil, Energy Build. 82 (2014) 82–91. doi:10.1016/j.enbuild.2014.07.028.

[6] T. Alves, L. Machado, R.G. de Souza, P. de Wilde, Assessing the energy saving potential of an existing high-rise office building stock, Energy Build. 173 (2018) 547–561. doi:10.1016/j.enbuild.2018.05.044.

[7] P. de Wilde, Building Performance Analysis, John Wiley & Sons Ltd, 2018.

Keywords: Building Energy Performance, Reference buildings, Schools, Building stock.

Paper ID 30418: Borehole latent energy storage system integrated with solar thermal collectors and heat pumps

Parham Eslami Nejad*, Arash Bastani, Alain Nguyen, Etienne Saloux

CanmetENERGY, NRCan, Canada

AIM AND APPROACH

In this study, low temperature borehole thermal energy storage integrated with thermal solar collectors and heat pumps is investigated for the Drake Landing Solar Community. Each borehole in the field is equipped with two independent U-pipe circuits and saturated sand backfill. Heat pumps and solar collectors are connected to these U-pipe circuits independently. Simultaneous heat injection and extraction to/from the boreholes can be achieved through this configuration without the need for an above-ground short-term storage system.

SCIENTIFIC INNOVATION AND RELEVANCE

Contrary to the existing system (Sibbitt et al. 2012), the 52-home community is conditioned using reversible heat pumps. In the heating mode, which is the dominant in the extreme cold conditions of Okotoks (Alberta, Canada), heat pumps extract heat from the boreholes and freeze saturated sand backfill. Solar thermal energy is injected to balance out the ground thermal load and to melt frozen backfill for storing latent thermal energy. Implementing some PV modules is also investigated to partially compensate power used by the heat pumps.

PRELIMINARY RESULTS AND CONCLUSIONS

A detailed numerical model of the system has been developed to evaluate the new concept. This study summarizes different effects of the borehole latent energy storage (BLES) system on the overall system performance and characteristics. Although solar fraction is reduced and the heat pump net electricity consumption from the grid is positive, simulation results confirmed significant reduction of solar storage temperature, total required borehole depth, solar collector areas, and borehole field footprint. This concept can be used to develop a new generation of low temperature solar community with integration of BLES system and heat pumps.

MAIN REFERENCES

Sibbitt, B., McClenahan, D., Djebbar, R., Thornton, J., Wong, B., Carriere, J., Kokko, J., 2012. The Performance of a High Solar Fraction Seasonal Storage District Heating System – Five Years of Operation. Energy Procedia. 30, 856–65.

Keywords: Borehole latent energy storage, Solar assisted ground source heat pump, Solar thermal collectors, Simulations

Paper ID 30351: Simulation-based analysis of the load-shifting potential of Heat Pumps in Multi-Family Houses using co-simulation

Daniel Schmidt*, Sabine Hoffmann

TU Kaiserslautern, Germany

AIM AND APPROACH

Due to the growing expansion of renewable energies (RE), electricity grids will face increasing challenges in the future. A rising share of renewables is required because of the finite nature of fossil fuels and in order to achieve the climate targets set. The feed-in of electricity from RE is highly fluctuating and hard to control. Demand-Side-Management is a way to adjust the demand of electricity to its generation.

In our research project we focus on heat pumps (HP) as a Power-to-Heat solution. Based on the good efficiency, HPs are increasingly used to heat buildings. This approach is particularly interesting because heat can be stored more easily compared to electricity as most residential buildings have a thermal storage. By using thermal storage systems, generation of heat can be decoupled from demand in terms of time. This contribution examines how heat generation in multi-family houses (MFH) can be shifted to grid-friendly times using model predictive control (MPC) and HPs.

SCIENTIFIC INNOVATION AND RELEVANCE

In a first step, various simulation models were created to investigate the flexibility of heat generation in different buildings. The research focuses on typical MFHs from Germany, France, Belgium and Luxembourg. For each country a reference building was modelled and represented in TRNSYS. The building models were modified to represent different years of construction with according heating demands. In addition, different storage volumes were modelled in TRNSYS to vary the existing heat capacity. The MPC system was programmed in MATLAB. Taking weather data and grid signals into account, it designs an operating plan of a heat pump for a horizon of 48 hours. The simulation results show that the developed MPC in combination with HPs can represent one possible solution to the challenges of energy transition.

Furthermore, we will explain the chosen solution for co-simulation of models in MATLAB and TRNSYS using the BCVTB (Building Controls Virtual Test Bed) platform. Based on the experiences gained, it can be shown how the advantages of individual simulation programs can be used and how the range and relevance of investigations can be increased by combining different software.

PRELIMINARY RESULTS AND CONCLUSIONS

The results of the project specify the extent to which the electricity requirement for heat generation can be controlled in MFHs of different years of construction. A special focus is set on the flexibility of the different buildings. The simulation results show how the operating time of the heat pump in combination with the controller can be shifted to support the electricity grid.

Additionally, the results point out that it may also have a financial advantage for end consumers to use a control system to exploit the cheapest possible electricity signals. Variable tariffs could provide incentives for electricity consumption in the future. The contribution also evaluates the potential for financial savings.

Altogether, the outcome of the project shall prove that Power-to-Heat in buildings plays an important role in the energy transition. The solution presented cannot solve the major problem of fluctuating generation from renewables and the necessary expansion of the electricity grids, but it can certainly make a contribution to stabilizing the grids in the future.

MAIN REFERENCES

This contribution was carried out during the Interreg research project PtH4GR²ID (Power-to-Heat for the Greater Region's Renewables Integration and Development), in which researchers from Germany, France, Belgium and Luxembourg are collaborating. All the models used in this study were developed within the project and details and further information can be seen in the corresponding reports (see homepage: http://www.pth4gr2id.com/).

More detailed information on the control system used can also be found in the following publication:

Röhrenbeck, S.; Benzarti, A.; Wellßow, W. H.; Maar, K.; Hauffe, P.; Maul, J.; Pahn, M.; Tersluisen, A.; Gündra, H.: Prädiktive Betriebsoptimierung drehzahlvariabler Wärmepumpen in Kombination mit preisvariablen Stromtarifen. VDE Kongress, Mannheim, 2016

The programs used to perform the simulations are TRNSYS (Transsolar, http://www.trnsys.com/) and MATLAB (MathWorks, https://de.mathworks.com/). The BCVTB platform (Lawrence Berkeley National Laboratory, https://simulationresearch.lbl.gov/bcvtb/FrontPage) was used to connect the two software.

Keywords: Co-Simulation, Model Predictive Control, Heat Pumps, Demand-Side-Management

Paper ID 30448: Operational control of earth-to-air heat exchanger using reinforcement learning

Kento Tomoda* (1), Yasuyuki Shiraishi (1), Dirk Saelens (2,3)

1: The University of Kitakyushu, Fukuoka, Japan; 2: KU Leuven, Department of Civil Engineering, Building Physics & Sustainable Design Section, Leuven, Belgium; 3: EnergyVille, Genk, Belgium

AIM AND APPROACH

This research focuses on the development of an optimal control system for earth to air heat exchangers (EAHE). During the operational phase, EAHE control system should determine optimal settings for airflow rate ensuring that the indoor air quality and the heat exchange are maximized. The EAHE control system should determine optimal settings for airflow rate ensuring that the indoor air quality and the heat exchange are maximized. It is, however difficult to optimize the system by conventional control methods since the response of the control is noticeable only a few days later. Our main purpose was to explore the applicability of an optimal control for EAHE by reinforcement learning (RL) controls which do not require the construction of control rules. We conducted the analysis with a CFD based simulator [1] as the RL environment and Proximal Policy Optimization [2] as the RL algorithm. The state which was obtained from the CFD was defined such as outside air temperature and surface temperature in the system. Actions were the binary value whether introduce outside air through the EAHE or not. Rewards were the obtained pre-cooling quantity through EAHE and the occurrence of condensation at representative point in the system.

SCIENTIFIC INNOVATION AND RELEVANCE

In this study, we conducted RL controls for the summer season using CFD-based high-precision long-term performance prediction simulator developed by us [1] as an RL environment. Wei et al [3] built learning environments of RL with an HVAC system which was EnergyPlus model and investigated the controllability for HVAC systems by Q learning and Deep Q-Network (DQN). The research in [4] implemented HVAC control by DQN algorithm with the indoor thermal comfort and energy consumption as rewards. RL could be applied to complex problems with a large number of state variables, such as Building Automation System (BAS), and is attracting attention as a next-generation control method. Training phase on RL, it requires a huge number of simulation trials. However, EAHE simulation requires long term analysis since its huge heat capacity and complex geometry. For this reason, it is difficult to simulate for EAHE many times. Considering the implementation for BAS, a highly accurate simulator is required because it takes time for relearning if there is a deviation between the prediction results and the real world. However, there was no research about application of RL control to EAHE due to difficulty in establishing an appropriate RL environment of EAHE.

PRELIMINARY RESULTS AND CONCLUSIONS

As the learning progressed, the obtained pre-cooling quantity was increased and the condensation time ratio in the system which correlation with airborne fungi of fresh air after passing through the system tended to decrease. In the current implementation, it was suggested that the control by RL could maximize the obtained pre-cooling quantity while suppressing condensation. Since only RL control was mentioned in this paper, comparison conventional control methods with that is a subject for future study.

MAIN REFERENCES

[1] Kento Tomoda et al. "Annual Prediction Method for the Pre-Cooling/Heating Performance of Earth-to-Air Heat Exchanger using Underground Air Tunnel". Proceedings of the 11th International Conference on Industrial Ventilation. Vol.1. (2015). pp. 339-346.

[2] John Schulman et al. Proximal Policy Optimization Algorithms. arXiv preprint. arXiv: 1707.06347. (2015).

[3] Tianshu Wei et al. "Deep Reinforcement Learning for Building HVAC Control". Design Automation Conference. (2017). pp. 1-6.

[4] William Valladares et al. "Energy optimization associated with thermal comfort and indoor air control via a deep reinforcement learning algorithm". Building and Environment. Vol.155. (2019), pp. 105-117.

Keywords: CFD, Reinforcement Learning Control, Earth-to-air heat exchanger system

Paper ID 30617: Distribution of potential savings from urban-scale energy modeling of a utility

Brett Bass (1), Joshua New* (2)

1: The University of Tennessee; 2: Oak Ridge National Laboratory

AIM AND APPROACH

While urban-scale building energy modeling is growing increasingly mature in data sources, algorithms, and empirical validation, there is still a need for best practices, guidelines, and standards for industry-accepted decision metrics relevant to specific use cases. Case studies are needed to inform such efforts. In addition, successful applications are need to motivate investment by, and in partnership with, utilities to scale grid-interactive efficient building technologies, realize aspirations of smart homes and cities, and dynamically dispatch load (rather than generation) in a way that stabilizes and reduces the cost of critical energy infrastructure. In partnership with the Electric Power Board of Chattanooga, TN, OpenStudio and EnergyPlus models were created of over 179,000 buildings and empirically validated against 15-minute whole-building electrical consumption of each building. Eight energy/demand-related measures relevant to nine utility-defined use cases are evaluated to showcase statistical distributions over the entire building stock of potential savings for energy, demand, emissions, and cost.

SCIENTIFIC INNOVATION AND RELEVANCE

Most urban-scale building energy modeling projects lack scalability due to use of geographically-limited data sources (e.g. tax assessor's data). Lessons learned from categorizing and analyzing 37 data sources [1] have been contrasted with tax assessor data to define 19 fields that may be useful in developing a comparison matrix with examples comparing/contrasting traditional vs. non-traditional urban-scale data sources [2]. New data sources and algorithms were applied iteratively to create slightly-different digital twins of all buildings in the 1400 km2 service area of a utility. By comparing the simulated energy use of each digital twin to actual 15-minute energy data from each building, we were able to prioritize the value of data source and algorithm combinations that tended to result in energy models that more closely match the true electrical consumption. These baselines were then modified to quantify energy, demand, emissions, and cost savings prior to potential program rollout, targeted marketing, and business model decisions for utility-prioritized use cases [3]. This study will review contributions to the data, algorithms, taxonomy, and value of data layers for specific use cases relevant to a utility.

PRELIMINARY RESULTS AND CONCLUSIONS

While empirical validation is increasingly common, high-resolution measured energy data is still relatively difficult to attain and is typically done for dozens to ~200 buildings in previous literature. This study performs empirical validation of 179,000 buildings, providing a larger sample size with greater statistical confidence in the final results.

Instead of reporting average savings for energy (kWh), this study will focus primarily on the contributions involving the distribution (e.g. box-and-whisker plots) of savings not only for energy but also for electrical peak demand (kW), emissions (kg), and cost ($ US dollars).

Multiple building technologies are evaluated including roof insulation, envelope sealing, more efficient lighting, smart thermostats, more efficient HVAC or water heater, and different HVAC or water heater types. As an example, smart-thermostat for utility-signaled pre-conditioning of buildings by 4.4°C two hours prior to peak demand saves on average 27% of a building's electrical demand but varies significantly from 0 to 93% across the 179,000 buildings. Aggregation of building-specific savings to utility-scale savings offers reduced risk for the financing and implementation of utility-scale savings. Upon approval by the utility, savings will be presented as average/building or percent to allow rough estimation of potential savings by other utilities.

MAIN REFERENCES

[1] Yuan, Jiangye, New, Joshua R., Sanyal, Jibonananda, and Omitaomu, Olufemi (2015). "Urban Search Data Sources." ORNL internal report ORNL/TM-2015/397, July 31, 2015, 70 pages.

[2] New, Joshua R., Adams, Mark, Garrison, Eric, Bass, Brett, and Guo, Tianjing (2020). "Scaling Beyond Tax Assessor Data." ASHRAE/IBPSA-USA 2020 Building Performance Analysis Conference & SimBuild (BPACS), Chicago, IL, Aug. 12-14, 2020.

[3] New, Joshua R., Adams, Mark, Im, Piljae, Yang, Hsiuhan, Hambrick, Joshua, Copeland, William, Bruce, Lilian, Ingraham, James A. (2018). "Automatic Building Energy Model Creation (AutoBEM) for Urban-Scale Energy Modeling and Assessment of Value Propositions for Electric Utilities." In Proceedings of the International Conference on Energy Engineering and Smart Grids (ESG), Fitzwilliam College, University of Cambridge, Cambridge city, United Kingdom, June 25-26, 2018.

Keywords: Urban-scale building energy modeling, empirical validation, energy, demand, emissions

Paper ID 30724: Dynamic simulation of a radiant ceiling panel incorporating PCMs for building cooling and heating applications

Andres Gallardo*, Umberto Berardi

Ryerson University, Canada

AIM AND APPROACH

Experimental and numerical studies have demonstrated the benefits of thermally activated building systems (TABS) in terms of thermal comfort and energy efficiency. However, one downside of TABS is that the system has to be incorporated in the building from the design stage, which limits its application to new buildings. To encourage the application of high thermal mass radiant systems in refurbishment, the authors developed a radiant ceiling panel system with incorporated PCM (PCM-RCP). The focus of this study is on the control strategy to operate the system, to meet both heating and cooling demands of a typical office building. Whole-building energy simulations have been used for evaluating the performance of the PCM-RCP system operating under the proposed control strategy.

SCIENTIFIC INNOVATION AND RELEVANCE

The authors aim to provide criteria to effectively design, size, and control radiant ceiling panels with integrated PCM for building heating and cooling applications. This paper explicitly presents the control of the proposed system, that has been developed as part of an integrated process that also includes the design and dimensioning of the system. The proposed control strategy take into consideration previous findings that indicated that radiant cooling systems with PCM have an optimum performance when the cycle of melting and solidification is completed as much as possible. Based on the radiant cooling ceiling panel surface temperature that represents the melting temperature of the incorporated PCM, a control algorithm is proposed for year-round automated operation, which meets the comfort requirements, is simple to implement and leads to an energy-efficient operation.

PRELIMINARY RESULTS AND CONCLUSIONS

Results indicate that the control strategy can be regarded as effective as it can respond well to changes in cooling and heating demand, meet the thermal comfort requirements, is simple to implement, and can lead to an energy-efficient operation due to the capability to operate only at night-time. Results also show that for cooling dominated climates, a melting temperature of 21°C is desirable to increase the passive cooling power of the system. However, in climates where both cooling and heating are required, a melting temperature of 24°C results in better thermal performance.

MAIN REFERENCES

- Jobli, M. I., Yao, R., Luo, Z., Shahrestani, M., Li, N., & Liu, H. (2019). Numerical and experimental studies of a Capillary-Tube embedded PCM component for improving indoor thermal environment. Applied Thermal Engineering, 148, 466–477.

- Lehmann, B., Dorer, V., & Koschenz, M. (2007). Application range of thermally activated building systems tabs. Energy and Buildings, 39(5), 593–598.

- Raftery, P., Duarte, C., Schiavon, S., & Bauman, F. (2017). A new control strategy for high thermal mass radiant systems. In Proceedings of the 15th IBPSA Conference (pp. 755–764).

- Romaní, J., De Gracia, A., & Cabeza, L. F. (2016). Simulation and control of thermally activated building systems (TABS). Energy and Buildings, 127, 22–42.

- Tödtli, J., Gwerder, M., Lehmann, B., Renggli, F., & Dorer, V. (2007). Integrated design of thermally activated building systems and their control. 9th REHVA World Congress for Building Technologies - CLIMA.

Keywords: radiant ceiling panels, PCM, TABS, HVAC, thermal energy storage

Paper ID 31154: The energy master planning process for districts

Matthias Haase*

ZHAW, Switzerland

AIM AND APPROACH

The energy master planning process for districts requires an analysis of different scenarios, which include new construction to different levels of energy efficiency, major renovation of all or some buildings comprising building stock under consideration with Deep Energy Retrofit of these buildings, minor renovations with energy-related scope of work, or demolition of some old buildings. Such analysis requires building energy modeling. In this research work we collected models of representative buildings from several countries and compared them.

SCIENTIFIC INNOVATION AND RELEVANCE

Complex districts consist of buildings and outdoor spaces with specific needs. The use that different buildings and areas are put to affects energy consumption, whereas the different functional patterns and stakeholder groups influence energy use. They are also associated with specific requirements that make it relevant to consider different types of performance indicators.

In the scope of this analysis both ventilation indicators and requirements with a direct or an indirect effect on energy consumption in districts are identified. When defining the relevance of performance indicators; legal requirements (i.e. for work environment), ownership or authority over parts of the district (single buildings or a complex of buildings), and cultural context also come into play.

PRELIMINARY RESULTS AND CONCLUSIONS

Energy can be considered to follow function because energy in the end is used to meet requirements defined by the activities that take place in a district. In each district, requirements are diversified by the type of activities/functions (residences, commercial (shops, retail), service (schools, restaurants, cafes, etc.), by the sizes of tenants rental spaces, or by the type of spaces (public areas, offices, parking etc.). The different activities can be characterized by functional patterns for various groups; – opening hours for commercial buildings will differ from operational hours for technical services and lighting. Facility operation has to meet the requirements of staff in commercial and cultural or service buildings before they open to the public. In districts, many tasks are performed outside of opening hours which require maintaining health and safety for the workers. Examples are maintenance and cleaning, sanitation and supply infrastructures, mobility and transport. In relation to this, the ratio of full operation of HVAC and lighting vs. opening hours or service hours is one index that could be used as a performance indicator.

MAIN REFERENCES

Baer D, Haase M (2020) Energymaster planning on neighbourhood level: learnings on stakeholders and constraints fromtheNorwegian case ofYdalir, proceedings world sustainable built environment conference, Gothenburg, Sweden

BREEAM (2019) bre environmental assessment method. https://www.breeam.com/.

CASBEE (2019) Comprehensive assessment system for built environment efficiency. https://www.ibec.or.jp/CASBEE/english/.

EED—Directive 2012/27/EU of the European parliament and of the council of 25 October 2012 on energy efficiency. Energy Efficiency Directive. https://ec.europa.eu/energy/en/topics/energyefficiency/targets-directive-and-rules/energy-efficiency-directive.

EnergyPlan. Energy systems simulation tool. https://www.energyplan.eu/.

EPBD (2018) Directive (EU) 2018/844 of the European parliament and of the council of 30 May 2018 amending directive 2010/31/EU on the energy performance of buildings.

Fox K (2016) Energy master planning perspectives and best practices, presentation to the federal utility partnership working group, May 2016. Cincinnati, OH

Haase M, Lohse R (2020) Process of energy master planning of resilient communities for comfort and energy solutions in districts. In: IOP conference series: earth and environmental science, vol 352, number 1, IOP Publishing Ltd.

Huang Z, Yu H, Peng Z, Zhao M (2015) Methods and tools for community energy planning: a review. Renew Sustain Energy Rev 42(C):1335–1348 (Elsevier)

Jank R (2017) Annex 51: case studies and guidelines for energy efficient communities. Energy Build 154:529–537

LEED. Leadership in energy and environmental design. https://leed.usgbc.org/leed.html.

Keywords: Energy master planning, Neighborhood level, district models

Paper ID 31155: Case study of an NZEB (renovation) with 7 years measurement data – what can a designer learn from it?

Matthias Haase*

ZHAW, Switzerland

AIM AND APPROACH

A typical residential building from 1937 located near Wuerzburg in Germany, was deep retrofitted in 2013. Roof, façade and ceiling in the basement were highly insulated and thermal bridges were minimized. Windows were replaced with three-layered glazed windows with wood-aluminium windows. A compact unit (balanced ventilation system with integrated air-to-water heat pump) was installed together with an 8 KW PV system with a south-west orientation and 50 degrees angle (roof-integrated). The ventilation ducts were integrated into the existing chimneys. Residential appliances/white goods (Refrigeration, Laundry, Dishwashing) were installed/replaced by A+++ equipment. Cooking equipment was replaced by induction device. Lighting fittings were replaced with LED in the whole building.

The simulations of energy balance and load profiles (in IDA ICE) match well with measurements. Self-consumption and export of electricity to the grid was monitored. The results show variations in self-consumption.

SCIENTIFIC INNOVATION AND RELEVANCE

The paper presents Monitoring data from 7 years performance of a NZEB. This represents a very comprehensive retrofitting towards plus energy level. Energy consumption and production as well as usage patterns and load profiles were monitored over a period of one year.

Besides key performance indicators for energy production and energy use it is interesting to monitor different KPI for self-consumption. Energy costs is another useful KPI.

The results have implications for energy load profiles which represent a new type of grid load that needs further recognition in planning of renovation work.

PRELIMINARY RESULTS AND CONCLUSIONS

This paper reports a case study of deep retrofitting of a residential building towards plus energy level. Energy consumption and production as well as usage patterns and load profiles were monitored over a period of one year. The simulations of energy balance and load profiles match well with measurements. Self-consumption and export of electricity to the grid was monitored. The results show variations in self-consumption. Annual self-consumption varied between 17% (2015) and 22% (2020).

MAIN REFERENCES

Energy strategy for Europe (2012) Energy market country report 2011 – Belgium. European Commission.

BPIE 2013, Data Hub for the energy performance of buildings. Source: http://www.buildingsdata.eu/

EPBD. 2010. Energy Performance in Buildings Directive 2008/0223, European Commission, Brussels.

Kreditanstalt für Wiederaufbau (KfW), Energy efficient refurbishment (in German: Energieeffizientes Sanieren),updated in 2015, financial support program

Energieeinsparverordnung 2014, (EneV), http://enev-2014.info/, access date Jan 2016

Salom, J., Marszal, A.J., Widén, J., Candanedo, J. and Byskov Lindberg, K., (2014) Analysis of Load Match and Grid Interaction Indicators in NZEB with High-Resolution Data. IEA Task 40/Annex 52 Towards Net Zero Energy Solar Buildings.

Keywords: Self-consumption, energy costs, historic building

Session W2.9 (Online Track): Improving indoor environmental quality

Paper ID 30299: Smart ventilation control to minimize the infection risk with COVID-19 in office buildings

Zhihong Pang (1), Pingfan Hu (2,3), Xing Lu (1), Qingsheng Wang (2,3), Zheng O'Neill* (1)

1: J. Mike Walker '66 Department of Mechanical Engineering, Texas A&M University, College Station, TX, USA; 2: Mary Kay O'Connor Process Safety Center, Texas A&M University, College Station, TX, USA; 3: Artie McFerrin Department of Chemical Engineering, Texas A&M University, College Station, TX, USA

AIM AND APPROACH

The spread of severe acute respiratory syndrome coronavirus 2 (SARS-CoV-2), also known as COVID-19, has already taken on pandemic proportions on a global scale. Many businesses have been delayed or even shut down in support of disease control amid the pandemic, leading to a rising number of social and economic issues in many countries.

In the context of strong advocates for reopening and economic recovery, this study aims to propose a novel ventilation control strategy to facilitate a smooth shift for the common office buildings from the normal mode, as which they were designed and operated by default, to the pandemic mode. To achieve this goal, firstly, a parametric computational fluid dynamics (CFD) analysis will be conducted to simulate the multiple operation scenarios of a VAV system under different occupancy and thermal load conditions. Secondly, based on the results of the parametric analysis, a scientific correlation between the CO2 concentration with the infection risk will be established to monitor the infection risk. Thirdly, a smart and novel ventilation control strategy will be developed by virtue of the quantitative relationship. This control strategy will then be evaluated by EnergyPlus simulations in terms of energy performance and effectiveness of risk mitigation.

SCIENTIFIC INNOVATION AND RELEVANCE

Existing literature shows that ventilation control (sufficient fresh air) can help dilute the virus concentration in buildings and reduce the infection risk due to airborne transmissions. However, how to increase the ventilation rate through smart control strategies for public buildings during emergencies (e.g., pandemic) still remains unexplored.

As far as the authors are concerned, although there have been a few studies trying to investigate the minimum ventilation rate for disease control purpose, there is no similar research, which aims at directly associating the infection risk with the CO2 concentration using a comprehensive parametric CFD analysis. Besides, the proposed study is the first of its kind which takes both the energy performance and infection risk mitigation into the evaluation of a smart ventilation control. Since the AHU operation for most VAV systems in commercial buildings is relied on the CO2 sensing, the results of this study is more feasible to be implemented in the real buildings to bring actual benefits.

PRELIMINARY RESULTS AND CONCLUSIONS

We have started to develop the CFD cases to prepare for the parametric analysis. The model setting and control scenario definition are still under discussion and yet to be finalized.

We have deployed a comprehensive environmental sensor network which measures the CO2 concentration, temperature, relative humidity, and light intensity for multiple pointes in a space for two complex rooms in a campus building. The key operation data of the HVAC system, e.g., the discharge airflow and temperature of the VAV box, the outdoor airflow rate, etc., has also been extracted from the Building Automation System (BAS) to facilitate the CFD model calibration.

MAIN REFERENCES

Li, Yiping, et al. "Role of ventilation in airborne transmission of infectious agents in the built environment-a multidisciplinary systematic review." Indoor air 17.1 (2007): 2-18.

Yu, Ignatius TS, et al. "Evidence of airborne transmission of the severe acute respiratory syndrome virus." New England Journal of Medicine 350.17 (2004): 1731-1739.

Sundell, Jan, et al. "Ventilation rates and health: multidisciplinary review of the scientific literature." Indoor air 21.3 (2011): 191-204.

Chen, Qingyan. "Ventilation performance prediction for buildings: A method overview and recent applications." Building and environment 44.4 (2009): 848-858.

Zhang, Z., and Q. Chen. "Experimental measurements and numerical simulations of particle transport and distribution in ventilated rooms." Atmospheric environment 40.18 (2006): 3396-3408.

Greenough, Anne. "Role of ventilation in RSV disease: CPAP, ventilation, HFO, ECMO." Paediatric respiratory reviews 10 (2009): 26-28.

O'Neill, Zheng D., et al. "Energy savings and ventilation performance from CO2-based demand controlled ventilation: Simulation results from ASHRAE RP-1747 (ASHRAE RP-1747)." Science and Technology for the Built Environment 26.2 (2020): 257-281.

Keywords: Demand-Controlled Ventilation, COVID-19, Occupant Well-being, Computational Fluid Dynamics, Energy Efficiency

Paper ID 30204: Numerical investigation of a novel diffuse wall ventilation: thermal comfort and cross-infection control effectiveness

Chen Zhang*, Gaute Larsen Tveit, Truls Ramstad, Peter V. Nielsen, Rasmus Lund Jensen

Department of the Built Environment, Aalborg University, Denmark

AIM AND APPROACH

A novel ventilation concept named as diffuse wall ventilation is proposed in this study and investigated by numerical simulations. The ventilation concept is similar to the diffuse ceiling ventilation[1][2], instead of using the suspected ceiling as an air inlet, the entire wall or large part of the wall serves as air inlet. The diffuse wall is made of textile material covered with vertically mounted lamellas, and the space behind the textile is utilized as a plenum to distribute air. Due to the large inlet area, the ventilated air is supplied into the occupied zone with very low velocity, and the airflow pattern in the room is dominated by the thermal buoyancy generated by the heat sources. A CFD model is developed to simulate the performance of diffuse ceiling ventilation in an Annex 20 room with two seated occupants [3][4]. The validated model is used to evaluate the thermal comfort, such as draft rate and temperature gradient, and cross-infection risk between two occupants under different flow rate and temperature difference scenarios. Finally, a parametrical study is performed to evaluate the impact of diffuse wall configurations. Three inlet configurations are investigated: full-wall inlet, half-wall inlet, and lamella inlet.

SCIENTIFIC INNOVATION AND RELEVANCE

The diffuse wall ventilation concept is proposed initially as a part of an office renovation project, where the low ceiling and visible inlets and ducts are not accepted due to the space limitation and design preference, making the traditional ventilation concepts unsuitable. Even though the concept of supplying air from wall inlets have been widely researched and applied, such as displacement ventilation and stratum ventilation, supplying air from the entire wall has not been well documented and lack of knowledge on the ventilation performance. In the wake of the COVID-19 pandemic, increased awareness of a ventilation system's ability to remove contaminants and inhibit the spreading of diseases is expected. Cross infection risk between two occupants in the office with diffuse wall ventilation is therefore evaluated, in addition to thermal comfort and system capacity.

PRELIMINARY RESULTS AND CONCLUSIONS

The room with full-wall inlet is used as a baseline study, operated at three scenarios with supply temperature of 20, 17, 13 oC, and ACH of 8.3, 4.2, 2.5 h-1, respectively. The supply air falls towards the floor when it leaves the inlet and develops into a plane discharge flow when it reached the floor. The depth of the discharge flow depends on the supply air temperature. The discharge flow travels alone the floor and rises until meeting the heat sources, where it is comparable with displacement ventilation. The air velocity in the occupied zone is very low with diffuse wall ventilation, and max velocity is lower than 0.05 m/s in all cases. The temperature difference between the ankle and head height is less than 2 oC in all cases. Local air quality index at the inhaled air εexp is used to assess the cross-infection risk between two occupants. The simulation results indicate the case 1 (Ts= 20 oC and ACH= 8 h-1), provide the highest εexp of 2.2. The study of different diffuse wall inlet configurations shows that the room with half-wall inlet performs the best among three configurations in terms of thermal comfort and air quality.

MAIN REFERENCES

[1] C. Zhang and P. Heiselberg, "Diffuse ceiling ventilation," REHVA J., no. Special Issue Nordic Technology, pp. 78–82, 2019.

[2] C. Zhang, T. Yu, P. Heiselberg, M. Pominaowski, and P. Nielsen, "Diffuse Ceiling Ventilation: Design Guide," DCE Techincal Rep., no. 27, 2016.

[3] A. D. Lemaire et al., "Room air and contaminant flow, evaluation of computational methods, subtask-1 Summary Report," IEA Annex 20 Air Flow Patterns within Build., p. 82, 1993.

[4] P. V. Nielsen, "Analysis and Design of Room Air Distribution Systems," HVAC&R Res., vol. 13, no. 6, pp. 987–997, 2007.

Keywords: Diffuse wall ventilation, CFD simulation, Cross-infection risk, Thermal comfort

Paper ID 30354: Developing thermal comfort maps for naturally ventilated spaces

Vardan Soi* (1), Swati Puchalapalli (2), Rashmin M. Damle (3)

1: Masters in Technology, Building Energy Performance (MTech BEP), CEPT University; 2: Terraviridis; 3: CEPT University

AIM AND APPROACH

This paper presents an alternative methodology to calculate and visualize operative temperature (Top) for naturally ventilated (NV) spaces. It helps calculate Top at multiple points in space and time. The methodology calculates three environmental variables namely air-temperature (Tair), solar adjusted mean radiant temperature (Tmrt) & air-velocity (Vair) to determine thermal comfort. The methodology of the calculation can be divided into two parts. The first part relies on thermal simulation to obtain Tair and surface temperatures to calculate Tmrt. The second part relies on a simplified CFD simulation to obtain Vair. The weather data is binned for wind direction to optimize the number of CFD runs. The proposed method limits itself to iso-thermal airflow hence, accounts only for wind assisted ventilation.

The data from CFD & thermal Simulation is integrated with other calculation variables such as view factor & effective radiant field (ERF) to derive Tair, Tmrt & Vair spatio-temporally. Top across the thermal zone is calculated for grid points to generate point-in-time thermal plots. Further, Top across all the data points in space is translated into a spatio-temporal representation for a given set of thermal comfort acceptability criteria which indicates the percentage of times a grid point meets the comfort condition.

SCIENTIFIC INNOVATION AND RELEVANCE

The literature review shows that Top has been calculated both spatially & temporally earlier (Mackie2015,Herkel1999,Marino2018,Webb2006), however with only two comfort variables (Tair, Tmrt) and for air-conditioned spaces. Further, several Studies plot Top for NV spaces, but are either point-in-time or space and do not capture impact of Vair on Top (jakubiec2019,Levitt2013). Vair distribution is one of the major factors behind thermal non-uniformity in NV spaces and Indoor thermal comfort is influenced greatly by airflow behaviour (Gan1994,Fulpagare2013). Thus, this research attempts to integrate three comfort parameters and captures thermal diversity in an indoor NV space both spatially & temporally.

NV is one of the most relevant passive strategies not just from the viewpoint of energy savings but also from the perspective of health, comfort & well-being (Jones1999). Spatial non-uniformity & lack of air movement has been the top reasons for thermal dissatisfaction in naturally ventilated scenarios. (Brager&Baker2009) Even if thermal comfort acceptability range for NV is wider, understanding the spatio-temporal characteristics of Top would help to minimize thermal non-uniformity and maximise comfort across the year. Thus, the proposed methodology could provide 'schematic design stage' support to designers helping them understand architectural dimensions of thermal comfort and thereby make informed design decisions.

PRELIMINARY RESULTS AND CONCLUSIONS

A lecture room of an institutional building in Hyderabad, India has been selected to develop the methodology. Preliminary results indicate that the proposed methodology captures thermal non-uniformity in the space due to spatio-temporal variation of Tair, Tmrt & Vair. For instance, on a typical summer day (2ndMay,11am), Tair=37oC, Tmrt shows the impact of the high surface temperature of the exposed south & east walls. The wind speed (3.2ms-1) & direction (236o) for this timestep corresponds to CFD results of bin 12. The directional impact of Vair distribution counters the high Tmrt in the south & core zones thereby, reducing Top in these grid points. Similarly, Top was obtained at spatial resolution across all points-in-time and conditional statements were scripted to assign credit '1' when Top<30oC (IMAC, NV) else '0'. These values for every grid point were averaged across the temporal range to obtain a spatio-temporal map which indicates percentage of time when the grid point is comfortable.

Thus, the proposed methodology can capture the impact of geometric implications along with the thermal properties of materials into a meaningful spatiotemporal interpretation of indoor thermal comfort, enabling designers in making informed design decisions ranging from façade to interiors and energy to comfort.

MAIN REFERENCES

Mackie, C.(2015). Pan Climatic.

Herkel, Sebastian, Frank Schöffel, J.D.(1999). Interactive Three-Dimensional Visualization of Thermal Comfort.

Webb, A. L.(2013). Mapping comfort: An analysis method for understanding diversity in the thermal environment. Proceedings of BS 2013: 13th Conference of the International Building Performance Simulation Association,1642–1648.

Marino, C., Nucara, A., & Pietrafesa, M. (2015). Mapping of the indoor comfort conditions considering the effect of solar radiation.Solar Energy,113,63–77.

Jakubiec, J. A., Doelling, M. C., & Heckmann, O.(2017). A Spatial and Temporal Framework for Analysing Daylight, Comfort, Energy and Beyond in Conceptual Building Design. 15th IBPSA Conference, (January)

Levitt, B., Ubbelohde, M. S., Loisos, G., & Brown, N. (2013). Thermal Autonomy as Metric and Design Process. Building Lasting Change2013.

Gan, G. (1994). Numerical Method for a Full Assessment of Indoor Thermal Comfort. Indoor Air

Fulpagare, Y., & Agrawal, N. (2013). Experimental Investigation on Room Ait Flow Pattern & Thermal Comfort Quantification. International Journal of Engineering Sciences & Emerging Technologies,IJESET,6(1),120–132.

Jones, A. P. (1999). Indoor air quality and health. Atmospheric Environment, 1(C), 57–115.

ISO 7726: Ergonomic of the thermal environment - Instruments for measuring physical quantities.

Brager, G., & Baker, L. (2009). Occupant satisfaction in mixed-mode buildings. Building Research and Information,37(4),369–380.

Keywords: Spatio-temporal maps, Thermal Comfort, Natural Ventilation, MRT

Paper ID 30453: Multi-objective optimisation of Indian Jaali fenestration system for visual, thermal and perceptual performance using computational methods

Afshan Rehman*, Amulya Surapaneni

Carnegie Mellon University, United States of America

AIM AND APPROACH

Jaali is a term used for perforated screens with floral or geometric patterns native to Indo-Islamic architecture. Historically, it has been used as an effective way in providing shade and privacy for building occupants in hot arid and hot humid climates. By changing the depth and the size of apertures of the Jaalis, one can reduce glare, solar insolation and achieve unobstructed views. Although individual studies on the impact of Jaalis on visual comfort and thermal comfort exist, no research currently examines how Jaali patterns affect both visual and thermal comfort at the same time. This paper introduces an experimental study to evaluate these concepts by replacing a glass facade of a modern library located in Delhi, India with a Jaali screen. A qualitative sample study was conducted for the library hall using six Jaali screens to ascertain the Jaali pattern that best improves the quality of light in the library hall. The most preferred Jaali pattern from the survey was then studied to quantitatively assess its visual and thermal properties. To determine the optimum aperture size and depth of the Jaalis, daylight metrics such as ASE and sDA, and thermal metrics such as indoor solar radiation were used.

SCIENTIFIC INNOVATION AND RELEVANCE

People spend 90% of their time indoors and improving the indoor environmental quality increases the well-being and productivity of the occupants. For this, it is important to improve indoor environments quantitatively and qualitatively.

Jaali, an innovative latticed screen of ancient Indian architecture, affects both visual and thermal quality of a space. This paper introduces an experimental study that analyses the effect of shadows created by these Jaali patterns perceptually. Qualitative studies were used to analyse the perceptual performance of these patterns using surveys. Quantitative study was used to examine the effect of this Jaali pattern on both visual and thermal comfort by using visual programming simulation environments. Innovations in building simulations have paved a way for optimizing these intricate Jaali patterns for daylight and thermal comfort. The solid to void ratio and depth of Jaali screens were optimized parametrically to produce well-lit, thermally comfortable and glare-free indoor spaces.

The paper highlights an important research methodology to generate an optimized innovative Jaali pattern that can be used by architects and building performance engineers to develop building screens that improve the visual and thermal quality of indoor spaces.

PRELIMINARY RESULTS AND CONCLUSIONS

The first section of the results presents a survey, in which the respondents selected one preferred Jaali pattern. The subsequent sections show how the solid-void ratio and thickness of the this Jaali pattern was changed parametrically to arrive at a final optimised pattern that gives the best possible results for daylight and thermal analysis.

The simulation results showed that a deeper Jaali screen provided better indoor thermal performance by reducing the indoor solar radiation. However, the overarching aim of this research was to create a Jaali screen that satisfies both visual and thermal comfort. A 100mm deep screen was selected, as Jaali screens that are more than 100 mm deep were drastically reducing the spatial daylight. This paper generated a final Jaali screen of solid-void ratio of 70%-30% and a depth of 100mm using both qualitative and quantitative strategies. The result of the paper demonstrates that the solid-void ratio and depth of Jaali screens can be optimized to produce well-lit, thermally comfortable and glare-free indoor spaces. The paper highlights an important research methodology to generate an optimized Jaali pattern that can be used to develop building screens that improve the visual and thermal quality of indoors.

MAIN REFERENCES

1. Batool, A. (2014). "Quantifying Environmental Performance of Jali Screen Facades for contemporary buildings in lahore, pakistan." Available from ProQuest Dissertations & Thesis Global.

2. Sherif, A., Sabry, H., & Rakha, T. (2012). External perforated Solar Screens for daylighting in residential desert buildings: Identification of minimum perforation percentages. Solar Energy, 86(6), 1929–1940. https://doi.org/10.1016/j.solener.2012.02.029

3. Gandhi, D (2014). "Stone Jaali - Daylight Performance Analysis." Available from CARBSE article series.

4. Abdullahi, Y., & Embi, M. R. B. (2013). Evolution of Islamic geometric patterns. Frontiers of Architectural Research, 2(2), 243–251. https://doi.org/10.1016/j.foar.2013.03.002

(2018, April 7). Retrieved from https://sites.psu.edu/perforatedscreendesigner/

(2019, October 25). Retrieved from https://fluidhandlingpro.com/bernoulli-equation-and-the-venturi-effect.

5. Hamdani, M. & Bekkouche, Sidi Mohammed El Amine & Benouaz, Tayeb & Belarbi, Rafik & Mohamed Kamal, Cherier. (2014). Minimization of indoor temperatures and total solar insolation by optimizing the building orientation in hot climate. Engineering Structures and Technologies. 6. 131-149. 10.3846/2029882X.2014.988756.

6. Kurma, Naga Viswatej. (2017). Analyzing Sun Angles of Solar Time Dial for Design of Building Envelope Components for the Region of Andhra Pradesh, India. International journal of Emerging Trends in Science and Technology. 03. 5018-5028. 10.18535/ijetst/v4i3.06.

Keywords: Jaali, Daylight analysis, Thermal analysis

Paper ID 30474: Fine-grained measurement of the indoor built environment with robotic vacuum cleaners

Joseph McLaughlin* (1), Michal Young (1), Siobhan Rockcastle (1,2)

1: University of Oregon, United States of America; 2: Baker Lighting Lab

AIM AND APPROACH

This paper presents a method of measuring ambient conditions in the built environment with a fine spatial and temporal granularity using autonomous robot vacuum cleaners. Our method records the illuminance and temperature of a space with 10cm accuracy over sub-second time-steps. The result is a dense dataset of measurements, capturing features of the environment over both space and time.

The built environment exposes occupants to a range of variable lighting and thermal conditions that are inherently dynamic over space and time. Past studies on the effects of light and thermal variability reveal the impact that dynamic exposure to these attributes may have on occupant health, comfort, and satisfaction indoors (Kim et al., 2018; Danell et al., 2020). However, we still lack methods of collecting distributed data about environmental attributes and rely heavily on statically mounted sensors. To address the limitations of static sensors, recent trends have explored the use of wearables devices, which are useful in modeling the experiences of occupants within a space (Ulaganathan et al, 2019). Our approach is distinguished from these methods, employing a single sensor package to map the built environment with the fidelity necessary to model the interdynamics of the space itself.

SCIENTIFIC INNOVATION AND RELEVANCE

We attached a small computer with illuminance and temperature sensors to the exterior of an autonomous vacuum. We then constructed an indoor positioning system using a network of ultra-wideband radio transceivers to locate the vacuum cleaner as it moves through the space (Zafari et al., 2019). Measurements are collated on a remote host, providing network access to a dataset that is uniquely high-fidelity over both space and time.

This method offers several distinct advantages over typical methods of static measurement. Namely, our method provides the fidelity to make narrowly-localized assessments throughout a space. Further, our method can reveal "blind spots" missed by static measurements that rely on a fixed mounting location. This is done inexpensively and portably, using a single sensor package and a commercially available robot vacuum.

Our method may prove its utility in both research and industry applications, where the temporal and spatial data of indoor environmental attributes could drive building control systems, monitor real-time comfort conditions, or assess the baseline performance of a building without the cost of a hard-wired sensor network. Lighting and HVAC systems that utilize real-time data to adjust building performance would benefit from the immediate and continuous feedback that our method provides.

PRELIMINARY RESULTS AND CONCLUSIONS

We performed an initial study of our method, recording 30,000 illuminance measurements over the course of 8 hours in an approximately 44m2 residential living space. On hourly intervals, the autonomous vacuum cleaner navigated the space for 30 minutes while measurements were taken from an on-board computer. Initial evaluation confirms the highly variable characteristics of illuminance measurements over space and time. The results further illustrate several intricate interactions within the space such as the effects of tall objects obstructing the path of daylight throughout the day. Our full-length paper will present the hardware and software components of our method as well as an expanded set of pilot results to demonstrate how the ambient dynamics of the built environment affect occupants on a uniquely local scale.

MAIN REFERENCES

Ulaganathan, S., Read, S. A., Collins, M. J., & Vincent, S. J. "Influence of Seasons upon Personal Light Exposure and Longitudinal Axial Length Changes in Young Adults." Acta Ophthalmologica, vol. 97, no. 2, 2019, pp. e256–65. Wiley Online Library, doi: 10.1111/aos.13904.

Kim, J., Schiavon, S., Brager, G. "Personal comfort models – A new paradigm in thermal comfort for occupant-centric environmental control." Building and Environment, vol. 132, 2018, pp. 114-124,

Dannel, M., Amundadottir, M., Rockcastle, S. "Evaluating Temporal and Spatial Light Exposure Profiles for Typical Building Occupants. " SimAUD 2020, Vienna Virtual Conference, May 25 – 27, 2020

F. Zafari, A. Gkelias and K. K. Leung, "A Survey of Indoor Localization Systems and Technologies," in IEEE Communications Surveys & Tutorials, vol. 21, no. 3, pp. 2568-2599, 2019

Keywords: Modeling, Ambient information, Daylight performance

Paper ID 30494: Investigating the relationship between interpretability and performance for optimal rule-based control

Eikichi Ono* (1,2), Kuniaki Mihara (2), Takamasa Hasama (2), Yuichi Takemasa (2), Bertrand Lasternas (1), Adrian Chong (1)

1: National University of Singapore, Singapore; 2: Kajima Technical Research Institute Singapore, Singapore

AIM AND APPROACH

Recently, optimal controls such as model predictive control, or MPC, have become a hot and popular research topic. However, there are some hurdles for their application such as cost, reliability, tuning, maintenance, warranty and liability. On the other hand, some researches try to simplify optimal control algorithm based on rigorous optimization results. Robillart et al. (2018) have developed a regression model based on the optimization result of dynamic programming to reduce calculation cost. Drgoňa et al. (2018) have tried to approximate an optimization result of MPC by two learning method, regression tree (RT) and time delay neural network (TDNN). They point out that a RT model can be easily transformed to a rule-based control, but has some disadvantages such as accuracy and model complexity, compared to more complex algorithms like TDNN. However, there is no research that intends to optimize rules of rule-based controls in terms of control design support. Therefore, this research aims to develop an optimal rule generation method using machine learning techniques based on optimization results according to the structure of rule-based control logics.

SCIENTIFIC INNOVATION AND RELEVANCE

Regression can be a solution for simplification of optimal controls as Robillart et al. (2018) studied, but complex regression models are not suitable for rule-based controls. The idea of Drgoňa et al. (2018), using regression tree, may be good because it has a similar structure to rule-based control logics. However, they consider only binary decisions in the RT model. Similarly, May-Ostendorp et al. (2011) and Domahidi et al. (2014) have tried to simplify control logics based on optimization results, but they deal with only binary decisions. Because including only binary decisions leads to high complexity, redundancy, and less accuracy of logics, it should be better to study the structure of the regression tree in reference to actual rule-based controls. Thus, we try to improve the structure of the tree by mixing binary decisions and linear functions at its nodes and leaves, leading to more practical rules in terms of their accuracy and interpretability. Since the optimal rule-based control consists of simple logics, it has some advantages compared to complex control algorithms like MPC; e.g. it is easily understandable and adjustable for engineers and can be widely installed in not only new buildings but also existing buildings.

PRELIMINARY RESULTS AND CONCLUSIONS

We conduct a simulation case study which aims to maximize both thermal comfort and energy performance for an office space air-conditioned by a dedicated outdoor air system with ceiling fans (Mihara et al., 2019) considering thermal preference distribution of occupants. Since the system can control two variables of zone air temperature and air velocity, and occupants have different thermal sensation even in the same thermal condition, this optimization problem can be complicated and it is not easy to optimize rule-based control design. Firstly, the setpoints of zone air temperature, supply air temperature and ceiling fan speed are optimized by a genetic algorithm. The result shows that the optimal operation using ceiling fans and considering individual thermal preference can significantly increase thermal comfort of occupants and also reduce energy consumption, but cannot be easily interpretable for control design. Secondly, as a result of optimal rule generation by the improved regression tree model based on the optimization result, it is found that the model can generate more accurate and interpretable rules than the original regression tree model.

MAIN REFERENCES

[1] Domahidi, A., Ullmann, F., Morari, M. and Jones, C. N. (2014). Learning decision rules for energy efficient building control. Journal of Process Control, 24, 763–772.

[2] Drgoňa, J., Picard, D., Kvasnica, M. and Helsen, L. (2018). Approximate model predictive building control via machine learning. Applied Energy, 218, 199–216.

[3] May-Ostendorp, P., Henze, G. P., Corbin, C. D., Rajagopalan, B. and Felsmann, C. (2011). Model-predictive control of mixed-mode buildings with rule extraction. Building and Environment, 46, 428-437.

[4] Mihara, K., Sekhar, C., Takemasa, Y., Lasternas, B. and Tham, K. W. (2019). Thermal comfort and energy performance of a dedicated outdoor air system with ceiling fans in hot and humid climate. Energy & Buildings, 203, 109448.

[5] Robillart, M., Schalbart, P. and Peuportier, B. (2018). Derivation of simplified control rules from an optimal strategy for electric heating in a residential building. Journal of Building Performance Simulation, Vol. 11, No. 3, 294–308.

Keywords: Rule-based control, optimal control, interpretability

Paper ID 30800: Performance of aerogels as thermal insulation materials for buildings in tropical climates

Aminu Wali Bashir, Eliane Hayashi Suzuki*, Joao Roberto Diego Petreche, Brenda Chaves Coelho Leite

Universidade de São Paulo, São Paulo, Brazil

AIM AND APPROACH

This study aims to examine the influence and suitability of inserting aerogels on wall surfaces of buildings in the tropical regions with an emphasis on energy consumption reduction and thermal comfort improvement.

Building modeling and computational energy simulations were conducted on tropical climates of Lagos, Nigeria, and Belem-Para, Brazil. The selected locations have similar annual weather conditions, and the architecture of buildings is also similar. A single design was selected to represent the two locations: a residential building with thermal zones distributed on two floors.

As the local materials used for construction vary significantly, the simulation models were based on the materials commonly used in each location and with aerogel material inserted on the external wall surfaces of the buildings. The results of the simulations were compared to effectively conclude on the significance of the insulation material in each location. The average temperatures are almost constant throughout the year in both locations, therefore, simulations were run for a typical design day in summer. Zones' mean air & operative temperatures and air relative humidity values were measured and compared to the standard recommended range. The analysis of the results was carried out based on each floor.

SCIENTIFIC INNOVATION AND RELEVANCE

The two selected locations are both in coastal regions, with moderately high temperatures, high relative humidity, and high rainfall virtually throughout the year. The use of HVAC systems for cooling of buildings in these locations covers a large percentage of the total energy consumption. However, energy efficiency measures through the adoption of insulation materials are recognized to be tools that could affect the heat gain and/or loss within buildings and reduce energy demand. Aerogel has then been identified to be the most promising solution when insulation is required due to its very low thermal conductivity feature [1], [2], hence its selection for this study.

There are various publications on the application of aerogels in buildings [3], [4], [5], [6], but none considered analyzing its significance through modeling and building energy simulations and its consequences on tropical climates. Thus, this study intends to lay out a foundation for this type of methodology which, if proven to be significant, can serve as a base reference for insulating buildings in tropical climates (which covers almost 40% of the global landmass). Also, these strategies could improve thermal comfort in the buildings, especially in locations where energy supply is not constant.

PRELIMINARY RESULTS AND CONCLUSIONS

• Simulation with conventional materials

The measured mean air and operative temperatures for a typical summer day in the two selected locations were similar and could be comparable. Also, the values were above the recommended range in both locations especially for thermal zones on the first floor, basically due to the effect of solar radiation. The corresponding air relative humidity values recorded in Lagos were within the recommended range while the values for Belém were above the range virtually throughout the day.

• Simulations with aerogel material

The insertion of the insulation material appeared to have more significance in Belém than in Lagos. A decrease of 2.1% on first floor zones and up to 6.1% on ground floor zones was recorded in the measured temperatures in Belém, while for Lagos, the decrease was between 0.6% for first floor zones to 1.3% for ground floor zones. However, the decrease in the measured temperatures slightly increased the corresponding air relative humidity.

The insertion of Aerogel on wall surfaces of the models showed the potential of reducing the indoor temperatures in both locations. This means there will be less demand for mechanical cooling and at the same time improves thermal comfort.

MAIN REFERENCES

[1]. Allouhi, A., El Fouih, Y., Kousksou, T., Jamil, A., Zeraouli, Y., & Mourad, Y. (2015). Energy consumption and efficiency in buildings: Current status and future trends. Journal of Cleaner Production. https://doi.org/10.1016/j.jclepro.2015.05.139

[2]. Jelle, B. P. (2011). Traditional, state-of-the-art, and future thermal building insulation materials and solutions - Properties, requirements, and possibilities. Energy and Buildings, 43(10), 2549–2563. https://doi.org/10.1016/j.enbuild.2011.05.015

[3]. Baetens, R., Jelle, B. P., & Gustavsen, A. (2011). Review: Aerogel insulation for building applications: A state-of-the-art review. Energy & Buildings. https://doi.org/10.1016/j.enbuild.2010.12.012

[4]. Cuce, E., Cuce, P. M., Wood, C. J., & Riffat, S. B. (2014). Toward aerogel based thermal superinsulation in buildings: A comprehensive review. Renewable and Sustainable Energy Reviews. https://doi.org/10.1016/j.rser.2014.03.017

[5]. Roya, H. N., Umberto Berardi, Nosrati, R. H., & Berardi, U. (2018). Hygrothermal characteristics of aerogel-enhanced insulating materials under different humidity and temperature conditions. Energy and Buildings, 158, 698–711. https://doi.org/10.1016/j.enbuild.2017.09.079

[6]. Riffat, S. B., & Qiu, G. (2013). A review of state-of-the-art aerogel applications in buildings. International Journal of Low-Carbon Technologies. https://doi.org/10.1093/ijlct/cts001

Keywords: Aerogels, Building energy consumption, Energy efficiency, Insulation materials, Building energy simulation

Paper ID 30823: A combined CFD-based genetic algorithm and human body thermoregulation model for designing the indoor thermal environment

Wei Liu (1), Yu Xue (2), Dayi Lai* (3)

1: Division of Sustainable Buildings, Department of Civil and Architectural Engineering, KTH Royal Institute of Technology, Brinellvägen 23, Stockholm, 100 44, Sweden; 2: School of Civil Engineering, Dalian University of Technology, Dalian 116024, China; 3: Department of Architecture, School of Design, Shanghai Jiao Tong University, Shanghai 200240, China

AIM AND APPROACH

A thermally comfortable indoor environment is critical for ensuring the health and productivity of the occupants (Schellen et al., 2010). This study aims to develop a combined CFD-based genetic algorithm and human body thermoregulation model for the inverse design of indoor thermal environment. This investigation developed the CFD solver in OpenFOAM (Jasak et al., 2007), the genetic algorithm by GenOpt (Xue et al., 2019), and the human body thermoregulation model in Matlab (Lai and Chen, 2016). A shell file was used to run those three parts in need automatically. The developed model was validated by the measured human skin temperature from literature. The model was further applied to design the thermal environments in an office with displacement ventilation and an aircraft cabin with mixed ventilation for demonstration.

SCIENTIFIC INNOVATION AND RELEVANCE

This study implemented a new way for evaluating the thermal comfort level in the inverse design of indoor environment. On the one hand, the developed model considered the heterogeneous thermal environments, such as vertical air temperature gradient, non-uniform radiation, and spatially different air distribution, which improves the accuracy in estimating the thermal comfort level. On the other hand, the coupling would not only overcome the inaccurate CFD prediction due to fixed surface temperature of the occupants, but also lead to reliable estimation of the occupants' thermal comfort level.

PRELIMINARY RESULTS AND CONCLUSIONS

After validating the performance of the human thermoregulation model coupled with CFD in predicting skin temperature, we used the model to inverse design the indoor thermal environment in an office and an aircraft cabin. In the office case with displacement ventilation, the thermal comfort was maximized when the air supply air temperature and velocity was 25 oC, and 0.08 m/s. While in the aircraft cabin, the objective function was the minimum when the slot diffuser supplied 26 oC of air by 0.32 m/s. Coupling thermoregulation model with CFD inverse design provides a more detailed way to optimize the indoor thermal environment, and at the same time, local thermal discomfort can be reduced.

MAIN REFERENCES

Jasak, H., Jemcov, A., Tukovic, Z., et al., 2007. Openfoam: A c++ library for complex physics simulations, in: International workshop on coupled methods in numerical dynamics, IUC Dubrovnik Croatia. pp. 1–20.

Lai, D., Chen, Q., 2016. A two-dimensional model for calculating heat transfer in the human body in a transient and non-uniform thermal environment. Energy and Buildings 118, 114–122.

Schellen, L., van Marken Lichtenbelt, W.D., Loomans, M.G., Toftum, J., De Wit, M.H., 2010. Differences between young adults and elderly in thermal comfort, productivity, and thermal physiology in response to a moderate temperature drift and a steady-state condition. Indoor air 20, 273–283.

Xue, Y., Liu, W., Wang, Q., Bu, F., 2019. Development of an integrated approach for the inverse design of built environment by a fast fluid dynamicsbased generic algorithm. Building and Environment 160, 106205.

Keywords: thermal comfort, thermoregulation model, inverse design

Paper ID 30858: Simulation of COVID-19 ultraviolet disinfection using coupled ray tracing and CFD

Nathaniel Jones* (1), Paul Lynch (2), Joseph Hewlings (1), Justin Boyd (2), Ryan Seffinger (1), Dan Lister (2), Renee Thomas (1)

1: Arup, United States of America; 2: Arup, United Kingdom

AIM AND APPROACH

Ultraviolet Germicidal Irradiance (UVGI) is the effective technique of employing ultraviolet radiant energy to inactivate (and render non-infectious) disease causing bacteria, mold spores, fungi and viruses. In the context of COVID-19, UVGI may be employed in the upper unoccupied zone of a room to reduce viral concentrations by four orders of magnitude. However, the effect on air in the occupied zone is dependent on mixing via air exchange and temperature gradients. We present a computational fluid dynamics (CFD) based approach to model disinfection of respiratory droplets and aerosolized pathogens in the occupied zone. Our CFD model informs the specification and placement of UVGI lamps to mitigate risk of infection. We use the CFD model to validate some simple calculation methods for UVGI effectiveness (including those by Rudnick & First and Beggs & Sleigh) which are then integrated with the Wells-Riley model of airborne infection risk to assess the relative benefit of UVGI with and against other measures. While the world awaits a cure, this method allows us to make facilities safer and thus prevent the spread of COVID-19 and all subsequent threats from respiratory virus outbreaks.

SCIENTIFIC INNOVATION AND RELEVANCE

The innovation in our method is the use of ray tracing and CFD to model the interaction of UVGI energy with pathogens in the unoccupied space and its effect on the occupied zone below. Experimentally, this technique was first shown in 1892 to be effective in destroying Bacillus anthracis – the bacteria that causes anthrax. UVGI technology has been used extensively and successfully in the hospital and healthcare setting. Using our simulation workflow, we can expand UVGI disinfection to the workplace broadly, in educational facilities, entertainment venues, transportation hubs – anywhere large groups of people transit or gather that have been affected by COVID-19. We can combine CFD with less resource-intensive tools to tailor the level of analysis as appropriate – from rapid screening of infection control measures to fine-tuning of UVGI system design.

PRELIMINARY RESULTS AND CONCLUSIONS

We demonstrate a tool that provides real-time calculation of UVGI dosing using the visual programming language Grasshopper within the Rhinoceros 3D environment, and an extension to this that exports UVGI fluence in an air volume to CFD. The CFD results are validated against the Wells-Riley, Rudnick & First and Beggs & Sleigh equations for simple geometries, but CFD also allows analysis of complex spaces, including those with partitions or obstructions that limit the mixing of air.

MAIN REFERENCES

Beggs, Clive B. & Sleigh, Andrew. “A quantitative method for evaluating the germicidal effect of upper room UV fields.” Journal of Aerosol Science 33 (2002) 1681-1699.

Buonanno, Giorgio, Stabile, Luca and Morawska, Lidia. “Estimation of airborne viral emission: quanta emission rate of SARS-CoV-2 for infection risk assessment.” Environment International 141 (2020) 105794.

Malayeri, Adel Haji, et al. "Fluence (UV dose) required to achieve incremental log inactivation of bacteria, protozoa, viruses and algae." IUVA News 18.3 (2016): 4-6.

Noakes, C. J., C. B. Beggs, and P. A. Sleigh. "Modelling the Performance of Upper Room Ultraviolet Germicidal Irradiation Devices in Ventilated Rooms: Comparison of Analytical and CFD Methods." Indoor and Built Environment 13.6 (2004): 477-488.

Riley, Richard L., and Solbert Permutt. "Room air disinfection by ultraviolet irradiation of upper air." Archives of Environmental Health: An International Journal 22.2 (1971): 208-219.

Riley, Richard L., Solbert Permutt, and James E. Kaufman. "Convection, air mixing, and ultraviolet air disinfection in rooms." Archives of Environmental Health: An International Journal 22.2 (1971): 200-207.

Rudnick, Stephen N., and Melvin W. First. "Fundamental factors affecting upper-room ultraviolet germicidal irradiation—Part II. Predicting effectiveness." Journal of Occupational and Environmental Hygiene 4.5 (2007): 352-362.

Keywords: germicidal ultraviolet irradiance, COVID-19, UV-C lamps, disinfection, computational fluid dynamics

Paper ID 30864: Radiation control optimization for dynamic glazing in building façades in subtropical climate

Eliane Hayashi Suzuki*, Racine Tadeu Araujo Prado, Brenda Chaves Coelho Leite

Universidade de Sao Paulo, Brazil

AIM AND APPROACH

This study aims to discuss the participation of the window as responsible for room air temperature increase along the day in subtropical humid climate. For this, a high insulated glazing system was proposed for offices windows in order to investigate compositions with low U-factor. Previous study regarding Brazilian climates [1] showed that insulated glass unit (IGU) might increase energy consumption in office buildings. In this study, dynamic insulated glass unit with electrochromic layer was considered. Although chromogenic glazing technology seems to be more adequate for cold climates, there is a discussion whether it is appropriated for hot climates either. Simulations were carried out for a small office room in Sao Paulo/Brazil considering automation controls currently available for electrochromic glazing, based on occupancy and outdoor air dry bulb temperature, for example. Also, other parameters were tested for construction index control, being more adequate for hot climates. Rhinoceros, Grasshopper with Ladybug and Honeybee plugins, EnergyPlus 9.3 and Parametric PreProcessor tools were considered in modeling and simulation process. Analysis were conducted in terms of adaptive comfort based on ISO 7730 [2] and ASHRAE 55 standards [3], and energy consumption criteria.

SCIENTIFIC INNOVATION AND RELEVANCE

Chromogenic glazing performance has been exhaustively studied in temperate climates in order to face rigorous winter and great temperature difference along the year. In this research, an evaluation of electrochromic insulated glass unit for subtropical humid climate is proposed. In electrochromic technology there is a reversible change in materials optical properties related to ionization process, according to environmental parameters [4]. [5] had found that electrochromic glazing allows passive heating during the winter and solar gain reduction in the summer, saving energy up to 48%. In the case of Sao Paulo/Brazil, demand for cooling is higher than for heating, because even when the outside air temperature is low, office activities' internal thermal load is enough to heat the environment. So, in most of the time in the winter, there is cooling demand rather than heating. For this reason, electrochromic would be useful for radiation control in the summer, when it is necessary to block diffuse and beam incident solar radiation. As electrochromic glazing is not commercialized in Brazil, it should be studied the types of existing controllers in North America [6] and Europe, and how they apply in the subtropics.

PRELIMINARY RESULTS AND CONCLUSIONS

In hot climates, where irradiance level is relatively high along the year, radiation control of windows plays an important role for thermal comfort and energy efficiency. Results showed that there is potential in using electrochromic glazing in subtropical climate, since there would be a combination of natural ventilation coupled with HVAC system. Window is usually the weaker device for thermal performance in the building envelope when compared to other construction elements. In this way, electrochromic glazing may reduce energy transmission through the window and decrease envelope contribution to thermal load in hot days. However, it might happen an undesired passive heating of the space as the windows have high insulation level, thus increasing air supply total cooling energy from HVAC system. Data analysis along the year showed better combinations of glazing status control for thermal performance, e.g. total incident radiation in windows surface and seasonal schedule considering initial index – darker or lighter state. This case can also be applied for residential rooms, which become home office in covid-19 pandemic period and have the same work demands for comfortable and healthy places.

MAIN REFERENCES

[1] PINTO, M. M.; WESTPHAL, F. S. Energy performance of office buildings in Brazil using insulated glass units. 16th IBPSA International Conference and Exhibition. In: Proceedings... Rome: IBPSA, 2019.

[2] INTERNATIONAL STANDARD ORGANIZATION. ISO 7730: Ergonomics of the thermal environment - Analytical determination and interpretation of thermal comfort using calculation of the PMV and PPD indices and local thermal comfort criteria. Geneva: International Organization for Standardization, 2005.

[3] AMERICAN SOCIETY OF HEATING, REFRIGERATING AND AIR-CONDITIONING ENGINEERS. ASHRAE Standard 55: Thermal Environmental Conditions for Human Occupancy. Atlanta: ASHRAE, 2013.

[4] BAETENS, R.; JELLE, B. P.; GUSTAVSEN, A. Properties, requirements and possibilities of smart windows for dynamic daylight and solar energy control in buildings: A state-of-the-art review. Solar Energy Materials and Solar Cells, v. 94, n. 2, p. 87–105, 2010.

[5] LEE, E. S.; CLAYBAUGH, E. S.; LAFRANCE, M. End user impacts of automated electrochromic windows in a pilot retrofit application. Energy and Buildings, v. 47, p. 267–284, 2012.

[6] DUSSAULT, J. M.; GOSSELIN, L. Office buildings with electrochromic windows: A sensitivity analysis of design parameters on energy performance, and thermal and visual comfort. Energy and Buildings, v. 153, p. 50–62, 2017.

Keywords: Thermal comfort, natural ventilation, air conditioning, radiation control, dynamic glazing

Paper ID 31085: Influence of interventions against spread of COVID-19 on noise levels in student restaurant

Lukáš Zelem* (1), Vojtech Chmelík (2), Monika Rychtáriková (3)

1: STU Bratislava, Faculty of Civil Engineering, Department of Architecture; 2: STU Bratislava, Faculty of Civil Engineering, Department of Architecture; 3: KU Leuven, Faculty of Architecture,

AIM AND APPROACH

One of the main interventions against the spread of COVID-19 is so called social distancing, which has rather large consequence on activities held in the building interior. The direct impact can be seen in the limitations of the number of people present in an enclosed space at the same time in order to fulfil the necessary distance between persons.

However, these rather unpleasant restrictions with dramatical impact on functioning of the so-called HORECA (hotel-restaurant-café) services, turned out to have also a positive impact on the indoor acoustic comfort. The main difference can be seen in the decreased noise levels in restaurants, cafes and bars.

SCIENTIFIC INNOVATION AND RELEVANCE

Smaller number of talking people present inside a room means also lower noise levels. The fact that human vocal output depends on a situation and that people adopt their voice power to the surrounding noise (so called Lombard effect) can (in moderately occupied restaurants) result in even lower sound levels than one could presume [1]. This article focuses on the prediction of noise levels produced by talking crowd of people in the student restaurant at Civil Engineering faculty STU Bratislava. The ground dimensions of cantina are 21 × 23 m, height of ceiling 3.8 m and volume 1775 m3. Standard capacity of student restaurant (before epidemy of COVID-19) is 280 sitting places. The two situations are compared: (1) typical situation before the epidemy of COVID-19 with maximum number of people present in this room and (2) situation with maximum allowed number of people present, so that the requirements of Slovak government against the spread of COVID-19 [1] would be fulfilled. A prediction of the sound pressure level is performed by means of Odeon software, using adaptation algorithm for Lombard effect.

PRELIMINARY RESULTS AND CONCLUSIONS

Decreasing the number of people present in a room means also decreasing of sound sources and therefore noise levels. But in case of restriction due to spread of COVID-19 it doesn't mean only fewer people in enclosed space, but also larger distance between groups of people with maximum of two people at one table as well. It could lead to higher number of speaking persons in one moment , but it might avoid that Lombard effect would take place.

MAIN REFERENCES

[1] J. H. Rindel, „Verbal communication and noise in eating establishments," Applied Acoustics, %1. vyd.71, pp. 1156-1161, 2010.

[2] UVZ, „Úrad verejného zdravotníctva Slovenskej republiky," Slovenská republika, 14. july 2020. [Online]. Available: http://www.uvzsr.sk/index.php?option=com_content&view=article&id=4132:covid-19-zavery-z-ustredneho-krizoveho-tabu-sr-povinnos-nosenia-ruok-na-verejnosti-zatvorenie-obchodov-v-nedeu-vyleneny-nakupny-as-pre-seniorov-a-pod&catid=250:koronavirus-2019-ncov&Ite. [Cit. 27 júl 2020].

[3] L. Zelem, V. Chmelík, M. Rychtáriková a C. Glorieux, „Analysis of the acoustic behavior of people in a restaurant," rev. EuroRegio, Porto, 2016.

Keywords: COVID-19, noise, levels, restaurant

Session W3.1: Practice and industry related case studies

Paper ID 30482: Buoyancy-driven ventilation in a real operational building: Uncertainty quantification of the discharge coefficient

Lup Wai Chew, Chen Chen, Catherine Gorlé*

Stanford University, United States of America

AIM AND APPROACH

The ventilation rate across an opening is characterized by the still-air discharge coefficient, Cd, which can be estimated experimentally. Once determined, Cd is often specified as a constant in empirical models used for the designs of natural ventilation systems. However, under realistic operating conditions, the value of Cd can be affected by deviations from the still-air conditions in the local flow (Etheridge, 2011). As a result, the assumption of a constant value can introduce errors in estimating the natural ventilation rate, especially when considering complex buildings.

We aim to study uncertainty in Cd in a real operational building with a buoyancy-driven ventilation system using large eddy simulations (LES) in OpenFOAM. We focus on modeling the Y2E2 building at Stanford University, as measurements of indoor and outdoor temperatures are available (Lamberti & Gorlé, 2018). The LES model is first validated with measurements collected in the building. The validated model is then used to simulate nine scenarios with different initial and boundary conditions for the indoor and outdoor temperature. The results are post-processed to determine the uncertainty in Cd for the different windows, and this uncertainty is propagated through an empirical model to quantify the resulting uncertainty in the ventilation rate.

SCIENTIFIC INNOVATION AND RELEVANCE

Buildings made up 32% of the total global energy consumption in 2010 (IPCC, 2014). In tropical countries such as India, air conditioning can account for more than half the energy consumption of buildings (Berardi, 2015). Natural ventilation does not consume energy and offers mitigation to the unsustainable energy consumption in the building sector. However, the design of natural ventilation systems is extremely challenging with many unpredictable parameters including the outdoor conditions and occupant behavior (Chen, 2009; Etheridge, 2011). The design process is often simplified by assuming a constant Cd for openings (Etheridge, 2011), although Cd is expected to vary especially in complex buildings. Our case study of a real operational building quantifies the errors introduced by the constant Cd assumption when estimating the natural ventilation rate, especially with unsteady outdoor conditions. The finding thus suggests that incorporating a non-constant Cd, informed by LES, can improve the design process of natural ventilation systems.

PRELIMINARY RESULTS AND CONCLUSIONS

The LES model is built and meshed in the ANSYS Workbench package. The simulation domain consists of the target building and its surrounding. The outermost boundaries (2-3H away from the building in each direction, where H = 13 m is the building height) have zero-gauge-pressure boundary condition for pressure and zero-gradient boundary condition for all other parameters. The windows are not fully opened (i.e., at an inclined angle) and are therefore modeled as openings with a porous pressure jump to include the resistance due to the inclined angle. The building walls have no-slip and adiabatic boundary conditions. The floors and ceiling have no-slip and constant-temperature boundary conditions. Wall functions are employed to reduce the total number of grid points. The mesh has 2.6 million grid points, which is verified to have achieved mesh-independent results by comparing to the results obtained with a finer mesh of 8.6 million grid points. The simulated temperatures agree well with measurements. The next step is to use the validated model to simulate the nine cases, which are then used to quantify the uncertainty in Cd.

MAIN REFERENCES

Berardi, U. (2015). Building energy consumption in US, EU, and BRIC countries. Procedia engineering, 118, 128-136.

Chen, Q. (2009). Ventilation performance prediction for buildings: A method overview and recent applications. Building and environment, 44(4), 848-858.

Etheridge, D. (2011). Natural ventilation of buildings: theory, measurement and design. John Wiley & Sons.

IPCC, Intergovernmental Panel on Climate Change (2014), Climate change 2014: Mitigation of climate change, Chapter 9: Buildings.

Lamberti, G., & Gorlé, C. (2018). Uncertainty quantification for modeling night-time ventilation in Stanford's Y2E2 building. Energy and Buildings, 168, 319-330.

Keywords: CFD, large eddy simulation, full-scale building, flow rate

Paper ID 31063: Impact of landscaping elements on pedestrian wind simulations using OpenFOAM

Rubina Ramponi* (1), Viet Le (2), Jake Haskell (3), Adam Brooks (2)

1: Arup, Dublin, Ireland; 2: Arup, New York, United States of America; 3: Arup, Berlin, Germany

AIM AND APPROACH

Modern cities have experienced for years a trend towards rapid urbanization and development of tall buildings to accommodate a growing population. The recent pandemic, however, has highlighted the impact of urbanization on the outdoor environment and the importance of well-designed outdoor spaces.

When looking at the user experience of the outdoors, wind has an important role in the perception of comfort. Numerical CFD simulations help designers to assess windiness around buildings and to identify areas that need mitigations. Testing the effectiveness of these different mitigation measures is a key component of pedestrian wind assessments, although modelling their aerodynamic properties in CFD is often a challenge.

This paper illustrates a case study where landscape is used to mitigate adverse wind conditions in a public space. CFD simulations are performed using the open-source software OpenFOAM and user-defined functions were implemented to describe the vegetation as a source of momentum and turbulence. A sensitivity study to the porosity of the trees is first carried out to identify suitable Leaf Area Density (LAD) values. Then, the sensitivity of the results to the use of turbulence sources, often ignored in commercial studies, is evaluated and discussed.

SCIENTIFIC INNOVATION AND RELEVANCE

A number of literature studies discuss tree modelling using wind tunnel (Bitog et al., 2011; Gromke and Ruck, 2008; Manickathan et al., 2018) and CFD (Buccolieri et al., 2018; Gromke and Blocken, 2015; Mochida et al., 2008), where trees are generally represented as porous media and characterized by a drag coefficient (Cd) and a porosity index, such as the leaf area density (LAD).

Manickathan et al. (2018) recently published a comprehensive wind tunnel study that provides a better understanding of the tree aerodynamic properties and behavior when exposed to the wind/ This study supports further investigation in the accurate modelling of vegetation using numerical methods.

Most of commercial studies ignore the turbulence effect of vegetation and model trees only as a momentum source/sink using a default porosity model. This study illustrates an easy implementation of user-defined sources in OpenFOAM to account for both momentum and turbulence and illustrate the advantages of a more accurate modelling of vegetation for pedestrian level wind studies.

PRELIMINARY RESULTS AND CONCLUSIONS

The study illustrates the benefits of landscaping to improve wind comfort in a public open space. The results show that foliage and planting effectively shelter key occupant areas, with a positive impact on the pedestrian wind comfort within the vicinity of the trees. An accurate representation of these mitigations is therefore important when carried out wind studies.

A sensitivity analysis of the LAD values was carried out and the impact of the LAD is less dramatic than expected. A LAD of 1.1 m^2/m^3 produces already a significant amount of wind shelter and the addition of further planting leads to diminishing returns. A LAD of 2.2 m^2/m^3 only reduces the local velocity at pedestrian level by a further 15% and also does not mitigate the issue of the strong shear flow around the corner of the building which also strongly impacts the pedestrian comfort at ground level. The results of the sensitivity analysis to the turbulence sources will discussed in the full paper.

MAIN REFERENCES

Bitog, J.P., Lee, I.B., Hwang, H.S., Shin, M.H., Hong, S.W., Seo, I.H., Mostafa, E., Pang, Z., 2011. A wind tunnel study on aerodynamic porosity and windbreak drag. Forest Science and technology 7(1), 8–16.

Buccolieri, R., Santiago, J.-L., Rivas, E., Sanchez, B., 2018. Review on urban tree modelling in CFD simulations: Aerodynamic, deposition and thermal effects. Urban Forestry & Urban Greening 31, 212–220.

Gromke, C., Blocken, B., 2015. Influence of avenue-trees on air quality at the urban neighborhood scale. Part I: Quality assurance studies and turbulent Schmidt number analysis for RANS CFD simulations. Environmental Pollution 196, 214–223.

Gromke, C., Ruck, B., 2008. Aerodynamic modelling of trees for small-scale wind tunnel studies. Forestry (Lond) 81, 243–258.

Manickathan, L., Defraeye, T., Allegrini, J., Derome, D., Carmeliet, J., 2018. Comparative study of flow field and drag coefficient of model and small natural trees in a wind tunnel. Urban Forestry & Urban Greening 35, 230–239.

Mochida, A., Tabata, Y., Iwata, T., Yoshino, H., 2008. Examining tree canopy models for CFD prediction of wind environment at pedestrian level. Journal of Wind Engineering and Industrial Aerodynamics, 4th International Symposium on Computational Wind Engineering (CWE2006) 96, 1667–1677.

Keywords: Pedestrian wind comfort, Landscaping, CFD, Wind

Paper ID 30679: Full-scale validation of natural ventilation models in Stanford's Y2E2 building

Chen Chen, Catherine Gorlé*

Stanford University, United States of America

AIM AND APPROACH

The objective of the present study is to accomplish full-scale validation of buoyancy-driven natural ventilation models for Stanford's Y2E2 building. We use computational fluid dynamics (CFD) and uncertainty quantification (UQ) to identify optimal locations for temperature sensors under uncertain boundary and initial conditions. This will ensure that the measurements provide values representative of the volume-averaged temperature, while also characterizing spatial variability in the temperature field for validation of CFD results. The presentation will first introduce the set-up of CFD model, the characterization of the uncertain input parameters and the propagation method. Subsequently, we will present the results for the quantities of interest defined to select the optimal sensor locations. The results of the full-scale experiments will be presented together with a comparison to predictions obtained by the CFD model and a low-fidelity building thermal model.

SCIENTIFIC INNOVATION AND RELEVANCE

Natural ventilation can significantly reduce building energy consumption, but the complexity of the governing flow and heat transfer phenomena and the variability in the boundary and operating conditions make robust design a challenging task. Efficient modeling strategies that can account for this complexity and uncertainty could provide essential information during the design process, but validation in full-scale buildings should be pursued to quantify their predictive capabilities. In a previous study, a multi-fidelity computational framework with UQ was proposed to predict the volume-averaged indoor air temperature during night-time ventilation in one of the atria of Stanford's Y2E2 building. The framework combines a computationally efficient building thermal model to quantify uncertainty in the boundary and operating conditions with a more expensive CFD model to more accurately represent the complexity of the flow (Lamberti & Gorlé, 2018). Comparison of the results with building sensor measurements indicated that the sensor locations might result in data that is not be representative of the volume-averaged temperature; hence, a more careful experimental design is needed to support successful validation. The proposed method could more generally be used to design full-scale experiments for validation, and to decide on the location of building sensors used during operation.

PRELIMINARY RESULTS AND CONCLUSIONS

For the CFD-based design of experiments, five dependent uncertain parameters are first characterized by fitting historical building measurements to beta distributions. Using a Nataf transformation, these dependent uncertain parameters are transformed into a set of independent random variables, which are propagated through the CFD model using a polynomial chaos approach. Three quantities of interest, which indicate whether a specific location is representative of the volume-averaged, minimum, or maximum temperature are defined. The mean and standard deviations of these quantities of interest are used to decide on the optimal sensor locations. On each floor, the volume-averaged temperature will be obtained by averaging data obtained from 5 sensors; measurements representative of the minimum and maximum temperatures will be obtained from additional sensors close to windows and in hallways between offices, respectively. The full-scale experiment is currently ongoing, and the results will be used to validate both the building thermal model and the CFD model predictions.

MAIN REFERENCES

Lamberti, G., & Gorlé, C. (2018). Uncertainty quantification for modeling night-time ventilation in Stanford's Y2E2 building. Energy and Buildings, 168, 319-330.

Keywords: CFD simulations, atrium building, natural ventilation, validation, full-scale experiment

Paper ID 30161: Whole-building energy simulation analysis and optimization of residential building equipped with air-duct system in three different regions

Arash Zarmehr*, Joseph T. Kider .Jr

Institute for Simulation and Training, University of Central Florida, United States of America

AIM AND APPROACH

Building-integrated photovoltaics (BIPV) systems, on the outer layer of a building structure, have the ability to reduce electricity and materials costs, as well as pollution, by taking advantage of renewable energy sources (sunlight and wind), and changes in thermal resistance. A BIPV ventilation air-gap system and its effects on heating and cooling loads are presented in this work. We use computational fluid dynamics to model, analyze, and compare BIPV air-gap and attic ventilation strategies and their impact on building energy performance. Results suggest an attic natural ventilation increase by 2.8 mph and a mean temperature decrease by 11.2%. One novel contribution of this work includes a BIPV attachment, that converts the air-gap into a miniature wind catcher to further improve building performance. This design improves upon traditional air-gap architectures by increasing natural air velocity while decreasing photovoltaic and attic temperatures.

SCIENTIFIC INNOVATION AND RELEVANCE

The results of this study can introduce the benefits of using BIPV special designs to reduce the investment cost of using renewable energy and increase ROI for residential buildings. It can also be extended to commercial buildings. These benefits include increase productivity and efficiency of PV (which means more power production) and using natural ventilation to reduce building cooling and heating loads (which means less end-use consumption).

PRELIMINARY RESULTS AND CONCLUSIONS

Primary results show the benefit of this air-duct BIPV system to increase airflow velocity under PV panels and then direct the high-velocity airflow to the attic. Results suggest an attic natural ventilation increase by 2.8 mph and a mean temperature decrease by 11.2%.

MAIN REFERENCES

Zarmehr, A., and J. T. Kider, Jr. (2018). Modeling and simulation of parametric wind-catcher designs for natural ventilation in sustainable building skin architecture. Advanced Building Skins.

Gan, G. (2009). Effect of air gap on the performance of building-integrated photovoltaics. Energy 34 (7), 913–921.

Goverde, H., D. Goossens, J. Govaerts, F. Catthoor, K. Baert, J. Poortmans, and J. Driesen (2017). Spatial and temporal analysis of wind effects on PV modules: Consequences for electrical power evaluation. Solar Energy 147, 292–299.

Keywords: Building Integrated Photovoltaic, BIPV, Air-Duct, Natural Ventilation

Paper ID 31135: Development of a digital twin of the temperature distribution in datacenters

Eric Terry*

Actiflow

AIM AND APPROACH

The aim of the project was to provide datacenter operators with a non-invasive technology to monitor the real time 3D temperature distribution in the entire datacenter room. The developed solution is called 4DCOOL and uses both

real time temperature measurement data and a 3D CFD (Computational Fluid Dynamics) model. The principle is to define how and where the CFD computer model is uncertain and then use a mathematical optimization technique to nudge the model towards the data within the limits of the uncertainty. With the right approach, the model stays physically correct but is much closer to the sensor data.

SCIENTIFIC INNOVATION AND RELEVANCE

The 3D temperature monitoring tool is based on the fusion of CFD and real time measurement data. There are well established techniques to do the envisaged

fusion which have been common in the field of weather prediction for decades, but far less applied to indoor climate applications.

The energy consumption of large datacenters is high, and the total energy consumption of the entire datacenter market is rising faster than in any other industry. Cooling accounts for 25-40% of datacenter energy consumption. Competition in the market and CO2 reduction requirements are drivers in the market to reduce energy consumption, but this introduces significant risks for datacenter operators. Mainly for colocation datacenters, it is crucial to have a maximum uptime of the datacenter, and to respect at all times the maximum air temperature provided to the customer's servers as specified in the service level agreement (SLA) with the customer. As a result, the industry is rather conservative to change or optimize cooling in the datacenter room, especially without good monitoring tool. Therefore, it is important to provide datacenter operators with a non-invasive technology to monitor the real time 3D temperature distribution in the entire datacenter room.

PRELIMINARY RESULTS AND CONCLUSIONS

The 4DCOOL monitoring tool was tested through 2 pilot projects in 2 different datacenters in Belgium and the Netherlands. During the pilot projects, one set of sensors was used as input for the software, and a second set of sensors was used to validate the output of the tool. In both pilot projects, the tool provided a reliable image of the temperature distribution in the datacenter.

In both pilot projects, the insights from the developed tool led to a re-arrangement of the lay-out of the perforated tiles and the cooling unit settings, leading to cooling energy cost savings of more than 20%.

The result of our development project is a validated software tool called 4DCOOL that can be coupled to existing DCIM (Data Center Infrastructure Management) software which provides the user with a quasi-real-time image of the temperature distribution in the datacenter room. Furthermore, the software has the functionality to set up "what-if" scenarios for the operator to assess the

consequences of (i) expanding or moving of the heat load, (ii) adjusting the settings of the cooling units or (iii) rearranging the perforated tile configuration in the floor or ceiling.

MAIN REFERENCES

- Geir Evensen (2003). The Ensemble Kalman Filter: theoretical formulation and practical implementation, Ocean Dynamics (53): 343-367

- Koomey J.G. (2008). Worldwide electricity used in data centers, stacks.iop.org/ERL/3/034008

Keywords: CFD, digital twin, datacenter, monitoring

Session W3.2: Ensuring high quality building simulations

Paper ID 31043: Characterised sun path patches as a way to design better shading

Andrew Corney* (1), Vladimir Bajic (2)

1: Trimble SketchUp, United Kingdom; 2: Trimble SketchUp, USA

AIM AND APPROACH

Interviews and surveys identified that a large proportion of architects use shadow analysis as the primary (and often only) way of designing shading systems.

The aim of this project was to find ways to enhance this natural workflow by providing low friction ways to get a better understanding about sun quality. The aim is to improve shading outcomes by providing better information in established workflows.

The concepts were developed and studied as part of a beta program with 190 participants, mostly archtiects. The beta program started with a range of surveys and interviews asking participants about if and how they design shading systems.

A beta application with the proposed workflow was then developed, firstly as a web-app, and then as a SketchUp extension.

SCIENTIFIC INNOVATION AND RELEVANCE

Understanding sun path diagrams and shadows are incredibly important and included in most architectural design courses.

However in most climates the time of day and time of year, the cloudiness of the sky, intensity of the sun and outside temperature as well as the nature of the building affect whether or not shading is really useful or not. Annual simulations with analysis tools are outside the capabilities of most architects and even where they are available, the use is sporadic because it does not fit neatly in the architect's design workflow.

A key reason identified in our research (over years) is that any friction in the architectural design process drastically reduces the effectiveness of building simulation. The work here sought to provide a defensible and useful improvement to the information used, while minimising the level of friction added to the design process and we feel this is very relevant.

PRELIMINARY RESULTS AND CONCLUSIONS

So far in the beta program research has shown that a very large proportion of participants (mostly architects, all SketchUp users) use the shadow functionality in SketchUp to design shading systems without using any other hourly simulation input.

Although many participants reported not remembering how to use sun path diagrams, most were able to pass straightforward tests on what the information presented is. This meant that the division of the sun path into 3 types of characterised "patches" (overheating, warming or passive) could also be interpreted and applied to inform design.

We used reverse shadow projections for shading strategies onto sun path diagrams to create visual explanations as to the effectiveness of different shading approaches. These could be generated quickly and easily while using SketchUp and provided a way to dynamically see how important and effective shading devices are, without the need to take the model into dedicated simulation.

We hope results from prototypes being made available for design use will also be presentable at the conference.

MAIN REFERENCES

Interviews and Discussions with architects and specialists participating in our beta program from the United States and Europe.

Sun, Wind and Light: Architectural Design Strategies, M, DeKay, GZ Brown

Keywords: architects, sun path, shading

Paper ID 30376: Modelling solar shadings with metallic slats for optimal daylighting. What parameters should we focus on?

Bertrand Deroisy* (1), Marshal Maskarenj (2), Sergio Altomonte (2)

1: Belgian Building Research Institute, Brussels, Belgium; 2: Université Catholique de Louvain, Louvain-la-Neuve, Belgium

AIM AND APPROACH

Designing high performance buildings requires a proper consideration of solar exposures and indoor climate conditions. In the current context of the climate change it must be possible to guarantee adequate thermal and visual comfort even with future climate conditions. Overheating is obviously an increasing risk factor for the occupants, especially in urban settings due to heat island effect. Meanwhile daylight, other than providing suitable conditions for vision, affects our physiological and psychological health. Solar shading systems with tilting metallic slats are commonly used to control daylight provision and energy transfer to the interior space. However, a precise characterization of their performance has to include many parameters, which are not always available. Very often, component level metrics are used to compare solutions, but these are neither consistently related to a specific context nor they are reliable for more complex building envelope assemblies. This study identifies the factors that most relevantly impact on the robustness of performance outcomes of daylighting simulations. It focuses specifically on the scattering properties of metallic surfaces and the shape of the slat profile on simulation results.

SCIENTIFIC INNOVATION AND RELEVANCE

The main solar shading systems used in Europe include roller blinds and venetian blinds, consisting of stacked and guided metallic slats. Standardized methods for characterizing solar-optical properties of regular transparent materials are well established, but robust methods do not exist yet for light scattering, shading and daylighting systems, such as venetian blinds, which have specific angle-dependent properties. Simulations of daylight provision and thermal radiative transfer through complex building envelopes needs to integrate the spatial distribution aspects from the materials and surfaces finishes used in the models. A comparison of standard methods, up-to date building level simulation techniques and advanced optical simulation tools was done in this study. Exact geometrical models of the slats and the possibility to integrate measured BSDF data to describe the scattering properties of the slat surface were used in advanced optical simulations techniques. A detailed analysis of the sensitivity of a set of input parameters (sky model, optical properties, slat design, glazing type) on the global performance of the solar shading systems was done. The results of this study allow to identify the simulation and modelling parameters that should be primarily considered when evaluating the daylighting performance of a building envelope with solar shading systems.

PRELIMINARY RESULTS AND CONCLUSIONS

Out of the four simulation parameters considered, the factor that has the largest effect on the consistency of simulation outcomes is represented by the tilt angle of the slats. An adequate and accurate setting of the tilt angle, based on solar altitudes and internal requirements, is essential to guarantee comfortable conditions for the building occupants at any time. The light scattering properties of the slat surfaces have a non-negligeable impact on daylight provision. Real slats are often relatively specular. Modelling them as diffuse surfaces generally underestimate transmittance ratios when the system is in a relatively open position. The shape of the slats can also have a significant influence when special profiles are used. However the difference between a flat slat and a typical curved slat is not detectable with advanced optical simulation techniques. The current building simulation applications do not allow to accurately estimate the impact of special slat profile shapes on daylighting performance. Large differences were observed between the two simulation based methods for medium sun angles. In general, more precise models are required whenever relatively specular surfaces, or special slat profiles, are used in shading systems.

MAIN REFERENCES

Capperucci, Loonen R., Hensen J.L.M, Rosemann A.L.P. (2018). Angle-dependent optical properties of advanced fenestration systems - Finding a right balance between model complexity and prediction error. Building simulation 12, 113–127.

Inanici M., Hashemloo A. (2017). An investigation of the daylighting simulation techniques and sky modeling practices for occupant centric evaluations. Building and Environment 113, 220-231.

Konis T., Lee E.S. (2015). Measured daylighting potential of a static optical louver system underreal sun and sky conditions. Building and Environment 92, 347-359.

Kuhn T., (2017). State of the art of advanced solar control devices for buildings. Solar Energy 154, 112-133.

Nilsson A., Jonsson J. (2010). Light-scattering properties of a Venetian blind slat used for daylighting applications. Solar Energy 84, 2103-2111.

Tzempelikos A., Chan Y-C. (2016) Estimating detailed optical properties of window shades from basic available data and modeling implications on daylighting and visual comfort, Energy and Buildings 126, 396-407.

Uribe D., Vera S., Bustamante W., McNeil A., Flamant G. (2019). Impact of different control strategies of perforated curved louvers on the visual comfort and energy consumption of office buildings in different climates, Solar Energy 2019, 495-510.

Keywords: Solar shading, Daylighting, BSDF, Venetian blind

Paper ID 30229: A spectral model for longwave radiant heat transfer: influence of new generation polymers in BES

Edouard Walther, Antoine Hubert*

AREP L'hypercube, France

AIM AND APPROACH

In Building Energy Simulation (BES), the modeling of radiation relies on a dual-band model: longwave, infra-red radiant heat transfer is linearised and computed separately from shortwave, solar radiation. This robust technique originates from the optical properties of glass, the latter being opaque to longwave radiation.

In the recent year, the use of polymer materials such as ETFE or LDPE has become popular in stations, greenhouses or leisure halls (Giuliano et al. 2010). In comparison with glass, they exhibit attractive features such as a reduced weight or higher visible transmittance.

The dual-band model is consistent for standard glass but appears to be unadapted to the aforementioned materials. Indeed, they are partially transparent to longwave radiation, with transmissivities ranging from 20 to 80% depending on the wavelength, which particularly affects the "greenhouse effect".

The present work aims at creating a spectral model for radiation transfer in multiple bandwidths and evaluating the influence of the new generation polymer materials on the greenhouse effect. Determining the ability of classical BES models for the simulation of radiant heat transfer through polymers depending on their cutoff wavelength in the infra-red domain is also an objective of this work.

SCIENTIFIC INNOVATION AND RELEVANCE

A few references mention applications using ETFE (Cremers & Marx 2016), (Hu et al. 2016), (Cremers & Marx 2017), however, to the best of the authors' knowledge, the effect of longwave transmissivity on indoor/outdoor radiation seems to be ignored. In the consulted literature, only (Poirazis et al. 2009) point out the lack of information about transmissivity in the longwave domain for polymers like ETFE and highlight the need for an extensive radiative model with experimental validation.

It hence appears interesting to explore the actual influence of such optical properties on heat transfer within buildings, which a spectral model reliably takes into account. Indeed, depending on the value of transmissivity in the infra-red range, this phenomenon may possibly be negligible as suggested in (Poirazis et al. 2009).

PRELIMINARY RESULTS AND CONCLUSIONS

A single-room, "shoebox" house serves as a test case. The window transmittance model follows (Curcija et al. 2018) and the wall model is built after a 4R3C scheme (Fraisse et al. 2002). In order to confirm the accuracy of the building model, a cross-validation is led using the EnergyPlus software with mere glass on transparent the southern walls. The results obtained show that the spectral model compares well with the dual-band model of EnergyPlus.

Preliminary results have demonstrated that ETFE does not filter the infrared radiations as efficiently as glass does, which is beneficial for longwave radiant cooling, however, given the higher transmittance in the solar spectrum, the temperature in buildings with ETFE may exceed the glazed building's temperature.

A comparison of the ETFE with LDPE, which transmissivity is even higher in the longwave range is currently explored. The differences obtained are in favour of a spectral model for BES of buildings with longwave transparent polymers.

MAIN REFERENCES

Vox, Giuliano & Teitel, M. & Pardossi, Alberto & Minuto, A. & Tinivella, F. & Schettini, Evelia. (2010). Sustainable greenhouse systems. Sustainable Agriculture: Technology, Planning and Management. 1-80.

Poirazis, H., Kragh, M., & Hogg, C. (2009, July). Energy modelling of ETFE membranes in building applications. In 11th International IBPSA Conference, Glasgow, Scotland (Vol. 144).

Curcija, C., Vidanovic, S., Hart, R., Jonsson, J., & Mitchell, R. (2018). WINDOW Technical Documentation. Lawrence Berkeley National Laboratory.

Cremers, J., & Marx, H. (2016). Comparative study of a new IR-absorbing film to improve solar shading and thermal comfort for ETFE structures. Procedia Engineering

Cremers J, Marx H. A new printed and spatially transformed ETFE foil provides shading and improves natural light and thermal comfort for membrane structures'. PLEA 2017.

Poirazis H, Kragh M, Hogg C. Energy modelling of ETFE membranes in building applications. In11th International IBPSA Conference, Glasgow, Scotland 2009

Hu J, Chen W, Qiu Z, Zhao B, Zhou J, Qu Y. Thermal performances of ETFE cushion roof integrated amorphous silicon photovoltaic. Energy Conversion and Management. 2015

Fraisse G, Viardot C, Lafabrie O, Achard G. Development of a simplified and accurate building model based on electrical analogy. Energy and buildings. 2002

Keywords: Longwave radiant heat transfer, spectral model, ETFE, Polymers

Paper ID 30335: Modelling naturally ventilated double skin facade in Modelica

Alessandro Dama* (1), Jaime Varas del Ser (1), Ettore Zanetti (1), Francesco Casella (1), Olena Kalyanova Larsen (2)

1: Politecnico di Milano, Italy; 2: Aalborg University, Denmark

AIM AND APPROACH

In recent decades, Double Skin Facades (DSF) and their thermal performance have been subject of numerous studies in literature. Despite this, the availability of rapid, robust and accurate tools for evaluating the performance of naturally ventilated double skin facades is still very limited, since only few published models have been accompanied by a complete experimental validation under variable boundary conditions, i.e. temperatures, solar irradiance and wind. Furthermore, the integration and/or coupling of such models within building energy simulation tools remains a complex task due to the multiple functionalities of the transparent and ventilated façade interacting with the building environment.

To this purpose this paper presents the implementation and validation of a model for naturally ventilated DSF in Modelica. The aim is to provide an open and robust tool easily integrable in the recent development of Modelica building libraries. Modelica, in fact, is an object oriented and open source programming language that has gained attention in the last decade, thanks to its ability to standardize and simplify modelling and thanks to its high potential when working with multi-domain systems.

SCIENTIFIC INNOVATION AND RELEVANCE

An ongoing international cooperation, under IBPSA Project 1, aim at creating a freely accessible, editable, documented and validated Modelica simulation library to support the design and operation of buildings and districts [1]. This work would contribute to the building library developments.

Validations of the selected model for naturally ventilated DSF was already presented in [2] and [3]. The further advantage of its implementation in Modelica is, thanks to its modularity, the possibility to perform again the model validation under different choices of the boundary conditions, isolating different model domains. Moreover, in this study the validation was extended to the simulation of a building module with the south facing DSF, giving a proof o the integration of the DSF model with the zone thermal model in Modelica. The experimental database used was provided by a field study on a full scale DSF "The Cube" carried out in Aalborg, Denmark [4].

Finally, a sensitivity analysis was performed on the convection in the ventilated channel, on the glazing solar absorptions and on the thermal capacities. It gave insight on the most relevant choices for the model parameters.

PRELIMINARY RESULTS AND CONCLUSIONS

Preliminary results of the DSF model implementation in Modelica had confirmed its capability to predict the variability of the mass flow rate, mainly due to the variable wind conditions, and improved its accuracy in predicting the outlet temperature and the inward heat flux. Such improvements are likely due to a better coupling in Modelica of the thermal and fluid-dynamic problems. The sensitivity analysis shows the importance of an accurate and detailed optical characterization of the window system and the role of the correlation adopted for the convection inside the ventilated channel. Otherwise, thermal capacity of glazing does not influence significantly the prediction even using a simulation timestep of fifteen minutes.

MAIN REFERENCES

[1] https://ibpsa.github.io/project1/

[2] A. Dama, D. Angeli, O. K. Larsen, Naturally ventilated double-skin facade in modeling and experiments, Energy and Buildings 144 (2017) 17–29

[3] A. Dama, M. Dopudi, O. K. Larsen, Experimental Validation of a Model for Naturally Ventilated Double-Skin Facades in proceedings of 7th International Building Physic Conference, IBPC 2018, Syracuse, NY, USA

[4] O. Kalyanova, Empirical Validation of Building Simulation Software: Modelling of Double Facades Final Report Technical Report IEA ECBCS Annex43/SHC Task 34 Validation of Building Energy Simulation Tools Subtask E

Keywords: adaptive facades, convection, wind pressure, solar radiation, experimental validation

Paper ID 31119: Open-source photovoltaic model for early building planning processes: Modeling, application and validation

Laura Maier* (1), Michael Kratz (1,2), Christian Vering (1), Philipp Mehrfeld (1), Dirk Müller (1)

1: RWTH Aachen University, E.ON Energy Research Center, Institute for Energy Efficient Buildings and Indoor Climate, Aachen, Germany; 2: currently studying at ETH, Zurich, Switzerland

AIM AND APPROACH

Within buildings, a great potential to reduce CO2 emissions exists. One common solution is to integrate renewable energy sources (RES) into BESs which are called interconnected systems. In this regard, PV systems are a promising technology as they enable sector coupling on the one hand and support local electricity generation on the other hand.

In order to exploit the full potential of PV systems, they have to be systematically integrated into the local control system. In this context, the proper sizing of PV modules plays an important role. This decision is made at an early planning stage. However, the optimum sizing of PV systems is challenging due to dynamic boundary conditions such as weather and its interdependencies with the whole BES, i.e. mounting's influence. In this context, simulation models facilitate the process of estimating future operation of PV modules.

We contribute to a more simplified planning process by applying the following steps:

1. We develop an open-source Modelica PV model for wafer-based cells, which is based on manufacturer data only and is suitable for early stage design.

2. We validate the model with measured data to prove mounting's influence.

SCIENTIFIC INNOVATION AND RELEVANCE

None of the researched Modelica PV models cover all of the following aspects:

• Open-source access

• Parameters based on manufacturer data only

• Integration of the mounting´s influence

• Validation based on measurement data

In addition, we quantify the influence of ohmic losses on the DC power output.

PRELIMINARY RESULTS AND CONCLUSIONS

In order to evaluate the model accuracy, we compare the simulated electrical energy of selected days with the measured one. The simulation mostly overestimates the electricity generation. This is caused by effects such as ageing, ohmic losses (OL) or staining, which are neglected within the model. Apart from that, the highest relative error is observed for the roof system at around 16 %.

To understand higher model errors for some days, we analyze the DC power output of the roof system in more detail. Here, we compare the simulated and the measured power on the day with the highest model error. The absolute difference between the data sets increases with raising power output. When also taking into account the

OL, the model error is decreased to 7 %. Here, we estimate the OL using the simulated cell temperature and measured current.

For a simulative OLs estimation, the detailed interconnection of the implemented PV modules has to be known. This information is not necessarily given at an early planning stage and complicates the parameterization tremendously. As this contradicts the model's aim to be simple and used at an early stage of planning, OLs are neglected.

MAIN REFERENCES

Batzelis, E., Papathanassiou, S.. A Method for the Analytical Extraction of the Single-Diode PV Model Parameters (2016). IEEE Transactions on Sustainable Energy 7, 504-512.

Boyd, M. (2015). High-Speed Monitoring of Multiple Grid-Connected Photovoltaic Array Configurations.

Boyd, M. (2017). Performance Data from the NIST Photovoltaic Arrays and Weather Station. Journal of Research of the NIST 122.

Duffie, J.A., Beckman, W.A. (edited by). (2013). Solar engineering of thermal processes. Fourth edition. Wiley. Hoboken, NJ (USA).

King, D.L., Boyson, W.E., Kratochvill, J.A. (edited by). (2005). SANDIA REPORT SAND 2004-3535 Unlimited Release Printed December 2004 Photovoltaic Array Performance Model." (2005).

Müller, D., Lauster, M., Constantin, A., Fuchs, M., Remmen, P. (2016) AIXLIB- An open-source Modelica library within the IEA-EBC Annex 60 Framework. Proceedings from BauSIM2016. Dresden (Germany), 14-16 September.

Keywords: Photovoltaic system model, Modelica, Interconnected Building Energy Systems, Validation

Session W3.3: Buildings paving the way for the energy transition

Paper ID 30411: The influence of uncertainties in grid electricity primary energy conversion factors on multi-criteria trade-off solutions in façade design optimisation

Samuel Bruno de Vries*, Roel C.G.M. Loonen, Jan L.M. Hensen

Eindhoven University of Technology, Netherlands, The

AIM AND APPROACH

Decision-making in façade design requires balancing of various competing performance aspects, including visual comfort, daylighting performance, thermal comfort, investment and operation costs, and CO2 emissions. These decisions are influenced by assumptions about cost parameters, HVAC specifications, and carbon intensity of grid electricity and other fuels.

The goal of this paper is to investigate how the optimal trade-off solutions and corresponding performance of façade design (glazing and advanced solar shading) change as a result of variations in the technical and economic context.

These topics are investigated through a case-study focussing on a novel sun tracking vertical blind which can switch between solar reflecting and solar absorbing states. The performance of this system is assessed in relation to other conventional solutions for controlling the admission of solar heat gains and daylight.

A multi-scale and multi-domain simulation approach is used to evaluate performance in terms of daylight glare probability simplified, spatial daylight autonomy, primary energy consumption and total costs. The influence of possible changes in the technical and economic context are considered by testing the influence of different assumptions regarding, cost parameters, the efficiency of heating and cooling systems and the primary energy ratio (PER) of electricity on energy performance and total costs.

SCIENTIFIC INNOVATION AND RELEVANCE

The increasing share of renewables in the electricity mix will decrease the overall carbon emissions of electricity in relation to other fuels and cause larger temporal fluctuation in the emissions of grid electricity. The emissions associated with building electricity demand will therefore become increasingly time-of-use dependant and building energy performance predictions are therefore clouded by uncertainties regarding the future electricity mix. Addressing these uncertainties in the facade design process is a topic that is still largely unexplored.

In modern high-performance office buildings, interactions between glazing, automated solar shading controls, daylight dimmable lighting and occupants are defining for building energy performance. The façade design problem therefore requires a multi-physics approach and is interwoven with control optimisation.

This paper utilizes a novel simulation tool chain, that:

- Employs a multi-physics co-simulation approach that allows for multi-objective decision support in the selection of solutions for controlling the admission daylight and solar energy.

- Uses a multi-scale simulation approach to evaluate the effects of a complex multi-state shading system on whole building performance.

- Enables co-optimisation of solar shading control and façade design aspects.

- Allows for evaluating design alternatives in combination with different HVAC systems and electricity mix scenarios, and robust solutions to be identified.

PRELIMINARY RESULTS AND CONCLUSIONS

- This study identifies façade design and solar shading control solutions that offer beneficial trade-offs in visual comfort, daylighting, energy performance and building related costs. In particular, it uncovers high-performing solutions that are robust to uncertainties regarding costs, the future electricity mix and the efficiency of HVAC systems.

- The advanced vertical blind control strategy offers superior daylighting, visual comfort and energy performance compared to conventional solutions. This conclusion is robust to different assumptions regarding PER scenarios, HVAC concepts and glazing systems.

- The presence of a daylight dimming system is an essential condition for the sun-tracking vertical blind system to offer improvements in energy performance over conventional automated control approaches.

- More efficient cooling systems and improvements in the PER of electricity will decrease the relative importance of energy performance in relation to other performance aspects in the selection of glazing and solar shading systems.

- If daylight dimming systems and efficient cooling systems become ubiquitous, and more renewable electricity from PV gives rise to a favourable PER in the summer months, reducing solar heat gains will become less important in the selection of glazing and the control of solar shading systems. Effective daylighting becomes the most defining aspect in improving energy performance.

MAIN REFERENCES

Aslihan, Tavil, and S. Lee Eleanor. 2006. "Effects of Overhangs on the Performance of Electrochromic Windows." Architectural Science Review.

Atzeri, Anna Maria, Andrea Gasparella, Francesca Cappelletti, and Athanasios Tzempelikos. 2018. "Comfort and energy performance analysis of different glazing systems coupled with three shading control strategies." Science and Technology for the Built Environment 24 (5):545-58.

Cubi, Eduard, Ganesh Doluweera, and Joule Bergerson. 2015. "Incorporation of electricity GHG emissions intensity variability into building environmental assessment." Applied energy 159:62-9

Méndez Echenagucia, Tomás, Alfonso Capozzoli, Ylenia Cascone, and Mario Sassone. 2015. "The early design stage of a building envelope: Multi-objective search through heating, cooling and lighting energy performance analysis." Applied energy 154:577-91.

Pilechiha, Peiman, Mohammadjavad Mahdavinejad, Farzad Pour Rahimian, Phillippa Carnemolla, and Saleh Seyedzadeh. 2020. "Multi-objective optimisation framework for designing office windows: quality of view, daylight and energy efficiency." Applied energy

Shen, Eric, Jia Hu, and Maulin Patel. 2014. "Energy and visual comfort analysis of lighting and daylight control strategies." Building and Environment 78:155-70.

St-Jacques, Max, Scott Bucking, and William O'Brien. 2020. "Spatially and temporally sensitive consumption-based emission factors from mixed-use electrical grids for building electrical use." Energy and Buildings 224:110249.

Keywords: Façade, Optimisation, Uncertainty, Electricity emission factors, Solar shading control

Paper ID 30334: Analysis of HVAC retrofit layouts including solar cooling system with adsorption heat pump - Modelling, dynamic simulation and multi-criteria evaluation

Paola Colombo*, Giulia Filippini, Rossano Scoccia, Marcello Aprile, Mario Motta

Department of energy, Politecnico di Milano, Italy

AIM AND APPROACH

Within the perspective of increasing attention to the environmental issues and energy policies, the construction sector has a crucial role. Buildings are responsible of the 40% of the total energy consumptions and generate the 36% of greenhouse gases in European Union. In addition, the long lifecycle of buildings leads to a very small rate of replacement of the existing stock. Thus it is necessary to pursue the objective of building retrofitting, starting from the older buildings stock, which has the largest energy saving potential. The European R&D project Heat4Cool addresses the specific construction segment of residential buildings, proposing innovative, efficient and cost-effective solutions for the space heating and cooling retrofitting. Among the Heat4Cool pilot sites, this work deals with the HVAC systems retrofitting for a residential building located in Valencia (Spain). From the energy analysis of the pre-retrofit status, this paper summarizes the steps of definition, modelling and evaluation of renovation layouts of the heating and cooling system, with the goal to identify the optimal configuration in terms of energy consumption reduction and costs.

SCIENTIFIC INNOVATION AND RELEVANCE

For the specific pilot site under analysis, renovation layouts of the heating and cooling system, including solar thermally driven adsorption heat pumps, are defined, modelled and evaluated. In TRNSYS 17 environment, the renovation layouts are modelled and tested, with the aim to evaluate the impact of the use of different technologies in terms of energy consumptions, costs and environmental impact. Parametric simulations are performed for the implementation of solar assisted thermal driven adsorption chiller, exploiting its potentiality for the solar cooling. The optimal case, among the ones simulated, analysed and compared, is determined through a multi-criteria decision analysis that takes into account energetic, environmental, economic and financial parameters.

PRELIMINARY RESULTS AND CONCLUSIONS

The solar thermal system resulted able to cover till the 52% of the heating (SH+DHW) energy needs, while the use of solar assisted adsorption chiller showed its potentialities in the reduction of the energy consumptions for the cooling needs, providing a contribution on the total cooling energy production that reaches the 27%. All the simulation resulted in line with the PE reduction target (with values up to 67%), while the discounted payback periods are around 10 years under certain reference conditions.

MAIN REFERENCES

I. Daßler, W. Mittelbach, Solar cooling with adsorption chillers, in: Energy Procedia, Elsevier Ltd, 2012: pp. 921–929. https://doi.org/10.1016/j.egypro.2012.11.104.

H.T. Chua, K.C. Ng, A. Malek, T. Kashiwagi, A. Akisawa, B.B. Saha, Modeling the performance of two-bed, silica gel-water adsorption chillers, Int. J. Refrig. 22 (1999) 194–204. https://doi.org/10.1016/S0140-7007(98)00063-2.

M. Schicktanz, T. Núñez, Modelling of an adsorption chiller for dynamic system simulation, Int. J. Refrig. 32 (2009) 588–595. https://doi.org/10.1016/j.ijrefrig.2009.02.011.

V. Palomba, S. Vasta, A. Freni, Q. Pan, R. Wang, X. Zhai, Increasing the share of renewables through adsorption solar cooling: A validated case study, Renew. Energy. 110 (2017) 126–140. https://doi.org/10.1016/j.renene.2016.12.016.

Solar Energy Laboratory, TRNSYS - TRaNsient SYstem Simulation program.

E. Zanetti, R. Scoccia, S. Garone, M. Aprile, M. Motta, L. Mazzarella, Energy Saving Potentials of a Centralized Hybrid Heating System via Adaptive Model Predictive Control in a Northern Italy Residential Building, Proc. Build. Simul. 2019 16th Conf. IBPSA. 16 (2020) 2925–2932. https://doi.org/10.26868/25222708.2019.210631.

R. Scoccia, T. Toppi, M. Aprile, M. Motta, Absorption and compression heat pump systems for space heating and DHW in European buildings: Energy, environmental and economic analysis, J. Build. Eng. 16 (2018) 94–105. https://doi.org/10.1016/J.JOBE.2017.12.006.

Keywords: HVAC retrofitting, solar cooling, solar assisted adsorption chiller, TRNSYS

Paper ID 30200: Performance comparison of static and adjustable photovoltaic panels

Dawid Wolosiuk, Matthias Schuss, Ardeshir Mahdavi*

Vienna University of Technology, Austria

AIM AND APPROACH

Building-integrated photo-voltaic (PV) panels can increase the fraction of renewable sources in the energy mix. Roof installations of PV panels are thus encouraged in many countries [1]. Thereby, one of the relevant questions concerns the cost-benefit analysis of static (fixed) installations versus dynamic installations capable of solar tracking. Moreover, in addition to entirely static option and the solar tracking options, it is conceivable to conduct regular directional adjustments "manually". Whereas this latter option would be more expensive than conventional static installations (due to the needed mechanical gear and manual labour), it would require significantly fewer resources that the solar tracking variants. The present contribution applies a high-resolution modelling approach and cost-estimation routines to compare the energetic output and estimated installation and maintenance cost of static, fully dynamic, and multiple instances of recurrent manually executed directional adjustment of the PV panels.

SCIENTIFIC INNOVATION AND RELEVANCE

The energetic efficiency of PV systems can be enhanced via the selection of the optimal orientation of the panels. In many instances, this orientation is determined based on the calculation of energy again over typically longer periods of time (e.g., over a typical year). Regular adjustment of the orientation (e.g., via automated solar tracking) can further increase the magnitude of the solar energy harnessed. However, this comes with notable increase of the cost of installing and maintaining the panels. In certain instances, regular manually-based adjustments may offer an alternative that is both energetically attractive and feasible form the point of view of installation and operation costs. However, related decisions require information about the energetic benefits and installation and maintenance costs of adjustable solutions. Likewise, in order to conduct a reliable cost-benefit analysis, the added value of higher adjustment frequency must be weighed against the expenses associated with higher frequencies.

PRELIMINARY RESULTS AND CONCLUSIONS

A computational platform has been developed involving a generator for high-resolution sky models, algorithms for the dynamic calculation of incident solar radiation [2]. Thereby, different panel slopes and directions can be parametrically assessed. The platform is coupled with an existing application that computes, for different types of PV panels, the magnitude of generated electricity as a function of the incident solar radiation [3, 4]. The comparison of static and adjustable systems is performed for three different climatic regions and different energy management scenarios, involving autarkic and grid-connected buildings with and without local storage capacity.

MAIN REFERENCES

[1] A. Jäger-Waldau, "PV Status Report 2019". EUR 29938 EN, Publications Office of the European Union, Luxembourg, 2019.

[2] D. Wolosiuk and A. Mahdavi, "Application of ontologically structured data for building performance analysis". Proceedings of the 11th annual Symposium on Simulation for Architecture & Urban Design (SimAUD), 2020, pp. 297-302.

[3] J. S. Stein, W. F. Holmgren, J. Forbess and C. W. Hansen, "PVLIB: Open source photovoltaic performance modeling functions for Matlab and Python". 2016 IEEE 43rd Photovoltaic Specialists Conference (PVSC), Portland, OR, 2016, pp. 3425-3430

[4] W. F. Holmgren, C. W. Hansen, and M. A. Mikofski. "pvlib python: a python package for modelling solar energy systems". Journal of Open Source Software, 3(29), 884, 2018.

Keywords: solar energy, PV panels, solar tracking, performance

Paper ID 31144: The path towards a zero fossil fuel Building

Frederik Maertens*, Wim Boydens

boydens engineering, Belgium

AIM AND APPROACH

A competition has been launched to build a +/-80.000m² mixed-use building in Belgium. The client requirements were high; class A comfort, BREEAM daylight compliance, low energy consumption, light building structure, maintenance friendly and, above all, a flexible and functional building.

As a starting point, the idea of the architect was to create the new building with minimal impact on the existing infrastructure and neighbouring surroundings.

Next to that main idea, the cladding material should be resistant to flash rust (coming from the rails) and easy to maintain.

A curtain wall façade was the obvious starting point of the design. Simulations were used to optimize the amount and location of the glazed elements, enhancing thermal and visual comfort while keeping the glazing as cladding material.

SCIENTIFIC INNOVATION AND RELEVANCE

- The presentation of this case gives insight in the use of simulations for specific multidisciplinary design choices in designs of large buildings.
- The case illustrates how dynamic simulations can assist and add value to multidisciplinary integrated designs.
- The presentation shows a practical step by step methodology to optimize the façade with various combined simulation tools.

PRELIMINARY RESULTS AND CONCLUSIONS

The developed façade made it possible to reach a building demand of 13kWh/m² for heating and 6kWh/m² for cooling, a passive building definition; including the renovated building part. As end result of this combination of simulations a visually fully glazed façade (seen from the outside) but insulated and provided with internal solar shading could still result in optimal indoor comfort in a passive way .

MAIN REFERENCES

no reference studies used

Keywords: facade, daylight, energy, comfort

Paper ID 30647: Retrofit of a 1970's apartment building into energy plus, results of a prototype apartment

Eelke P. Bontekoe* (1), Wilfried G.J.H.M. van Sark (1), Liza Looijen (2), Paul Das (3), Joris van den Heiligenberg (3), Arno F. Peekel (4)

1: Utrecht University, The Netherlands; 2: Hogeschool Utrecht, The Netherlands; 3: Bos Installatiewerken, The Netherlands; 4: Utrecht Sustainability Institute (USI), The Netherlands

AIM AND APPROACH

To reduce the CO2 emissions of the building stock in the Netherlands, 1.5 million existing houses have to be turned into NZEB (Near Zero Energy Buildings) before the end of 2030 [1]. This paper describes a case study in which we investigate if this challenge can be taken a step further by retrofitting a 10 story apartment building from the 1970's in such a way that it becomes a positive energy building.

SCIENTIFIC INNOVATION AND RELEVANCE

A new building design is made, where the thermal envelope is insulated, and facades are being upgraded by means of new modular building elements. These elements perform very well on insulation, contain systems for low temperature heating, have integrated window blinds and building-integrated photovoltaic (BIPV) modules. Central heat pumps on the roof of the building provide heat for space heating and domestic hot water (DHW). Each apartment has a booster heat pump to boost the DHW temperature above 60 degrees Celsius. More than 1100 PV modules are placed on the roof as well as on the facades of the building. To accommodate a part of these modules a special steel framing is designed on top of the building .

The energy requirements of the retrofitted building design were calculated with the help of the NZEB tool, which is based on PHPP v9 [2]. The energy production of the PV system is modeled by means of the PVSites software tool [3]. To understand if the proposed concept works in real life, a pilot apartment has been realized where the energy demand and the comfort levels have been monitored for about 1 year.

PRELIMINARY RESULTS AND CONCLUSIONS

The results of monitoring temperature, CO2 and relative humidity in the apartment, proved that in general the comfort is well within the limits of the latest European standards [4]. Only during an exceptional heatwave, overheating was measured. Over the course of the year, the heat pumps for space heating and DHW required more energy than expected. This was mainly due to a malfunctioning space heater. In case this heater would function properly, the PVSites tool showed that the energy production for the whole building is larger than the energy demand.

We can conclude that with our prototype apartment we have proven that it is possible to retrofit an apartment building such that it produces more energy by its PV system than its energy demand. Lessons learnt from the pilot project are already implemented in the next step of the project, where the whole building is being retrofitted.

MAIN REFERENCES

[1] Ministerie van Economische zaken en Klimaat, "Klimaatakkoord," Klimaatakkoord, p. 250, 2019.

[2] Passive House Institute, "Passive House Planning Package v9 (PHPP9)." 2019.

[3] PVSites, "Homepage of PVSites software package," 2019. [Online]. Available: https://www.pvsites.eu. [Accessed: 17-May-2019].

[4] Council of the European Union, "EN 16798-1:2019 Indoor environmental input parameters for design and assessment of energy performance of buildings- addressing indoor air quality, thermal environment, lighting and acoustics," 2019..

Keywords: BIPV, NZEB, Renovation, case study

Session W3.4: Historical and heritage buildings

Paper ID 30702: Estimation of the thermal properties using modal identification method and optimal experiment design applied to historical building walls

Benjamin Kadoch* (1), Julien Berger (2)

1: Aix Marseille Université, CNRS, IUSTI UMR 7343, 13453, Marseille, France; 2: Laboratoire des Sciences de l'Ingénieur pour l'Environnement (LaSIE), UMR 7356 CNRS, La Rochelle Université, CNRS, 17000, La Rochelle, France

AIM AND APPROACH

Thermal properties of the walls require to be known precisely since they play a crucial role on the assessment of the building energy efficiency [1, 2]. The unknown properties can be estimated by solving parameter estimation problem [3]. This procedure aims at minimizing a cost function between the model numerical predictions and the experimental observations. Nevertheless, retrieving the unknown parameters needs a large number of computations of the heat transfer problem. To decrease this computational effort, model reduction techniques can be employed. Several methods are reported in the literature to model the physical phenomena in building walls [4]. The Modal Identification Method (MIM) can be a pertinent choice since it has been demonstrated successful applications for inverse problem [5]. A complementary technique is the optimal experiment design methodology [6, 7] which determine the length of the measurement observations. The optimal design is searched according to the conditions of the experiment. Carrying the measurement for the optimal experiment design ensure to estimate the unknown parameter with the highest accuracy. The aim of this paper is to propose a fast and accurate parameter estimation method. This methodology is then used to estimate the thermal diffusivity of historical buildings.

SCIENTIFIC INNOVATION AND RELEVANCE

An innovative methodology is proposed based on two concepts. First, a reduced order model based on MIM method is used to significantly reduce the computational time of the direct problem without loosing accuracy. The reduced model is based on a state space representation where the matrices are built during a learning step. The latter is based on a minimization procedure between the predictions of the reduced and complete models. An interesting point is that the model is built to compute the field of interest and its sensitivity to the unknown parameter. The sensitivity is known straightforwardly using the matrices of the reduced model. The second concept is the optimal experiment design methodology. It is employed to determine a reduced sequence of observations. Three advantages are enhanced with this approach. First, it reduces the inherent computational cost of a posteriori model reduction methods since the learning step is carried for a reduced sequence. Then, the chosen sequence is optimal to estimate the parameter with accuracy. In addition, the inverse problem is solved only for a few days of observations. It reduced again the computational effort to retrieve the parameter.

PRELIMINARY RESULTS AND CONCLUSIONS

A first case study is proposed to validate the MIM model reduction method. This MIM model is built with a signal based on 4 values of thermal diffusivity. Then, the MIM model shows a very satisfying accuracy and efficiency to simulate the direct problem over a wide range of diffusivity. Regardless the inherent cost of the learning step, the model cut by 5 the computational cost of the direct problem. The model is also evaluated in the framework of inverse problem with simulated experimental observations. The unknown parameter is estimated with an error lower than the measurement uncertainty. After this validation case, the whole methodology is applied to more realistic ones. The issue is to estimate the thermal diffusivities of several old building walls. These are monitored during four months with three sensors drilled inside. Results for one wall give that the estimated diffusivity, using a model of order 10, is three times higher than the one provided by standards. Moreover, the parameter is estimated with an algorithm 5 times faster. These preliminary results show that the methodology is efficient to calibrate the model with a reduced computational effort and give an accurate thermal diffusivity.

MAIN REFERENCES

[1] A. Jumabekova, J. Berger, and A. Foucquier. Sensitivity analysis in the framework of parameter estimation problem for building energy performance: a continuous derivative based approach. submitted, 2019.

[2] T. Busser, M. Pailha, A. Piot, and M. Woloszyn. Simultaneous hygrothermal performance assessment of an air volume and surrounding highly hygroscopic walls. Building and Environment, 148:677 – 688, 2019.

[3] J. Berger, H.R.B. Orlande, N. Mendes, and S. Guernouti. Bayesian inference for estimating thermal properties of a historic building wall. Building and Environment, 106(SupplementC):327–339, 2016.

[4] J. Berger, N. Mendes, S. Guernouti, M. Woloszyn, and F. Chinesta. Review of Reduced Order Models for Heat and Moisture Transfer in Building Physics with Emphasis in PGD Approaches. Archives of Computational Methods in Engineering, pages 1–13, 2016.

[5] M. Girault, D. Petit, and E. Videcoq. The Use of Model Reduction and Function Decomposition for Identifying Boundary Conditions of A Linear Thermal System. Inverse Problems in Engineering, 11(5):425–455, 2003.

[6] D. Ucinski. Optimal Measurement Methods for Distributed Parameter System Identification. CRC Press, New York, 2004.

[7] J. V. Beck and K. J. Arnold. Parameter Estimation in Engineering and Science. John Wiley and Sons, New York, 1977.

Keywords: Parameter estimation problem, Model Calibration, Inverse Heat Conduction Problem, Modal Identification Method, Reduced Order Model

Paper ID 31123: Living lab 'De Schipjes': a zero-fossil-fuel energy concept in the historic city center of Bruges

Jelger Jansen* (1,2), Frederik Maertens (3), Wim Boydens (3,4), Lieve Helsen (1,2)

1: KU Leuven, Department of Mechanical Engineering, Celestijnenlaan 300 box 2421, 3001 Leuven, Belgium; 2: EnergyVille, Thor Park 8320, 3600 Genk, Belgium; 3: boydens engineering, Kooidam 6, 8200 Bruges, Belgium; 4: Ghent University, Department of Architecture and Urban planning, Jozef Plateaustraat 22, 9000 Ghent, Belgium

AIM AND APPROACH

This contribution aims to show the role and importance of dynamic simulations in the research project of 'De Schipjes', a social housing neighborhood located in the historic city center of Bruges. Firstly, an introduction is given in which some background information is provided on Almshouses 'De Schipjes' and in which the importance of dynamic simulations in the project is clarified, being to design and operate a 100% renewable energy sources-based district heating network. Secondly, the methodology and results (in the form of figures and tables) of both simulation studies are briefly discussed in two different sections. Finally, a conclusion is drawn.

SCIENTIFIC INNOVATION AND RELEVANCE

This contribution illustrates the importance of dynamic building simulations to design a 100% renewables-based district heating network in an historic city center and to optimize the control strategy during design and commissioning. Furthermore, this contribution shows the feasibility of such a zero-fossil-fuel energy concept in a historic city center. This is relevant for other historic centers in Europe (and beyond), where this concept might be reproduced or serve as an inspiration.

PRELIMINARY RESULTS AND CONCLUSIONS

The contribution focuses on two different stages in the project of 'De Schipjes'. In a first stage of the project, building simulations allowed an extensive comparison of four energy system scenarios (e.g. by using performance indicators) which led to the design of a low-temperature fully renewable energy sources-based district heating network supplied by a ground source heat pump and solar thermal collectors. In a later stage, detailed simulations were performed to assess different rule-based control (RBC) algorithms impacting the energy performance of the energy system and thermal comfort. In this contribution, four promising adapted RBCs are mentioned and the simulation results of these RBCs in comparison with a reference RBC are provided.

MAIN REFERENCES

Aertgeerts, A. (2016). IWT Proeftuin De Schipjes: Theoretische benadering van het verbeteringspotentieel. Internal Report, KU Leuven, Belgium, 2016.

Boydens, W., Feyaerts, S., Vandermeulen, A., Helsen, L., Jansen, J. (2020). Control strategy assessment of a small GSHP sourced DH system with end user DHW booster heat pumps. 12th IEA Heat Pump Conference. Jeju (Korea), 11-14 May 2020 (Postponed to April 26-29, due to outbreak of Coronavirus). Accepted for oral presentation.

Feyaerts, S. (2019). Impact van de regeling op de performantie van een klein thermisch net voor godshuizen, De Schipjes, te Brugge. Master's thesis, KU Leuven, Belgium, 2019.

Jorissen, F., Reynders, G., Baetens, R., Picard, D., Saelens, D., Helsen, L. (2018). Implementation and verification of the IDEAS building energy simulation

library. Journal of Building Performance Simulation 11(6), 669-688.

Van Kenhove, E., Aertgeerts, A., Laverge, J., Janssens, A. (2015). Energy Efficient Renovation of Heritage Residential Buildings Using Modelica Simulations. 14th Conference of International Building Performance Simulation Association. Hyderabad (India), 7-9 December 2015.

Wetter, M., Zuo, W., Nouidui, T.S., Pang, X. (2014). Modelica Buildings library. Journal of Building Performance Simulation 7(4), 253-270.

Keywords: historical city center, RES-based district heating design, rule-based control

Paper ID 30268: Simulation-based Optimization of thermal Comfort in a heritage Manor House

Torsten Schwan*, Rene Unger

EA Systems Dresden GmbH, Germany

AIM AND APPROACH

The preservation of historical building structures, especially in urban areas, is a particular challenge for cities, municipalities and private investors due to the changing requirements of space usage and energy supply. The primary goal is to find the best possible compromise between the future usability of the building as well as the preservation and conservation of the historical building fabric. A novel approach is the integration of a closed inner shell in timber construction within the existing structures. On the one hand, this enables the complete conversion of the building with regard to the latest space utilization and design requirements. On the other hand, this allows the compliance with the latest energy standards to be implemented with minimal intervention in the historical building envelope.

However, this noble goal places enormous challenges on architects and engineers and requires adequate evaluation tools for finding an optimal, individual solution. This paper describes an approach for the use of a hybrid digital twin model [1] based on Modelica building models and high-resolution energy system measurement data to evaluate the existing situation and various optimization measures for indoor air comfort and energy efficiency in a heritage-protected urban cottage.

SCIENTIFIC INNOVATION AND RELEVANCE

This paper actually addresses two types of innovation in the construction sector. On the one hand, the way in which a historic building is to be adapted to existing space and energy standards. On the other hand, the paper shows a new approach in the building sector of using a hybrid digital twin model to analyze, evaluate and optimize various measures in an early stage of the renovation.

The hybrid approach in the creation of the digital twin model combines the use of high-resolution, electronic measurement technology in the building with the development of dynamic simulation models of the building and its energy supply [2]. The combination of both methods provides a maximum of model accuracy by calibrating the model with temporally high-resolution measurement data and thus offers the engineers a platform for finding and digitally testing valid optimization measures in the real building. Particularly with regard to the individuality of historical buildings, a methodology is thus shown that can be used in a variety of practical applications.

PRELIMINARY RESULTS AND CONCLUSIONS

The work steps shown here began shortly after the owner put the presented historical building into operation. For this reason, high-resolution measurement data of energy consumption and thermal room behavior from several weeks and various supply situations were already available at the start of the project.

The modelling and calibration of the models enabled the mapping of the individual building behavior with an error tolerance of less than 3%.

The cause of the original shortfall in thermal energy could be determined with the help of the models and three different optimization measures were tested and evaluated. The developed solution enabled a reduction of the energy demand by up to 57% while at the same time significantly improving thermal comfort in more than 99% of the year.

MAIN REFERENCES

[1] Chinesta, F.; Cueto, E.; Abisset-Chavenne, E.; Duval, J.; Khaldi, Fouad (2018). Virtual, Dig-ital and Hybrid Twins: A New Paradigm in Data-Based Engineering and Engineered Data. Archives of Computational Methods in Engineering.

[2] Schwan, T., Schmitt S.; Castellani, A. (2019). Calibration of HVAC system models with monitoring data – Digital Twin meets measurement data. ESI Forum in Germany, Berlin, Germany.

Keywords: Hybrid Digital Twin, heritage-protected building, measurement data, calibration, system optimzation

Paper ID 30339: A hybrid measurement-simulation approach to determine the reflectance map of a historic tapestry

John Mardaljevic (1), Eleonora Brembilla* (2), Stephen Cannon-Brookes (3), Nigel Blades (4)

1: Loughborough University, United Kingdom; 2: TU Delft, Netherlands; 3: UCL, United Kingdom; 4: National Trust, United Kingdom

AIM AND APPROACH

In historic buildings, curators and conservators are increasingly choosing to display rooms under daylight illumination conditions comparable to how those spaces were originally used.[1] All predominantly daylit spaces will experience considerable spatio-temporal variation in natural illumination. However, monitoring of light levels to control exposure is only carried out at a limited number of locations, perhaps just one per space/room.[2] Light-sensitive objects such as large tapestries therefore present particular challenges since the daylight dose across the tapestry could vary significantly from that recorded at the periphery of the tapestry. An earlier study in a historic setting employed high dynamic range imaging (HDRI) to measure cumulative daylight exposure by using numerous wallpaper patches as proxy illuminance targets.[3] Knowing the reflectance of the wallpaper, the incident illuminance at each target patch could be derived from the patch luminance recorded in the HDR image (taken every 10 minutes). It was then possible to reconstruct the prevailing daylight illuminance incident on the paintings hung on the walls by interpolation across the wallpaper patches using a Kriging algorithm. The new approach uses the tapestry itself as the target. This required a novel hybrid HDR measurement – lighting simulation approach.

SCIENTIFIC INNOVATION AND RELEVANCE

The setting for the study is the Volury Room at Ham House, Richmond-upon-Thames (UK). The Volury contains a set of three large seventeenth century tapestries that occupy most of the three (non window) walls. The technique to derive incident illuminance from camera measured HDR luminance requires knowledge of the diffuse reflectance of the target. For scenarios such as the Volury, it was necessary to devise a method to determine the per-pixel reflectance (or albedo map) for the entire tapestry in situ without disturbing it. The basis for this was controlled illumination of the tapestry by a LED studio lamp. Characterisation of the lamp photometry allows determination of the illumination field across the tapestry. However, the room-reflected component added significantly to the light incident on the tapestry. A novel hybrid multi-step approach using HDR capture combined with Radiance lighting simulation was devised to account for the room-reflected component of illumination from the lamp.[4] With the HDR camera fixed in place for long-term monitoring, HDR captures were made of the tapestries illuminated by the LED panels. The new approach opens up the possibility of camera-based measurement of daylight dose in historic showrooms.

PRELIMINARY RESULTS AND CONCLUSIONS

The LED lamp photometry (measured under controlled conditions) was imported into a Radiance simulation of the Volury based on an accurate 3D model of the room. The simulated illumination field from the LED incident on one tapestry was predicted for increasing number of ambient bounces to account for the room reflected component until convergence was achieved. The Radiance simulated illumination field was then applied to the camera captured HDR image of the tapestry (illuminated by the LED lamp) via a warping procedure. The warping was required to precisely align the simulated illumination field with the camera HDR image. Next, using the precisely-aligned illumination field, the albedo map for the tapestry was derived from the HDR image. This was repeated for the other large tapestry visible from the camera position. At corresponding points across the simulated illumination fields there was remarkable agreement (<10%) with in situ point measurements of illuminance across the two tapestries taken at the time of HDR capture (under LED illumination). The derived albedo maps for the two tapestries are presented. Thereafter, the albedo maps will be used to derive the cumulative long-term (several months) daylight illumination across the two tapestries from the (automated) HDR capture images.

MAIN REFERENCES

[1] National Trust. The National Trust Manual of Housekeeping. Anova Books, London, 2011.

[2] N. Blades, K. Lithgow, S. Cannon-Brookes, and J. Mardaljevic. New tools for managing daylight exposure of works of art: case study of Hambletonian, Mount Stewart, Northern Ireland. Journal of the Institute of Conservation, 40(1):15–33, 01 2017.

[3] J. Mardaljevic, S. Cannon-Brookes, N. Blades, and K. Lithgow. Reconstruction of cumulative daylight illumination fields from HDR: Theory, deployment and in-situ validation. Lighting Research & Technology, (accepted), 2020.

[4] G. Ward Larson, R. Shakespeare, J. Mardaljevic, C. Ehrlich, E. Phillips, and P. Apian- Bennewitz. Rendering with Radiance: The Art and Science of Lighting Visualization. San Francisco: Morgan Kaufmann, 1998.

Keywords: Daylight, HDR, Simulation

Session W3.5: Improving indoor environmental quality

Paper ID 30624: The impact of light distribution and furniture layout on meeting light exposure objectives in an office - a simulation case study

Megan Danell* (1), Steffen Hartmeyer (1), Lisa Petterson (2), Robert Davis (3), Marilyne Andersen (1), Siobhan Rockcastle (4)

1: Ecole Polytechnique Federale de Lausanne (EPFL), Switzerland; 2: SRG Partnership, Portland, OR United States of America; 3: Pacific Northwest National Lab, Portland, OR, United States of America; 4: University of Oregon, Eugene, OR United States of America

AIM AND APPROACH

While the building industry is beginning to embrace the impact of light on non-visual responses that drive human health, there are limited design guidelines for how to implement effective light distribution and achieve recommended circadian light exposure. Human-centric factors that impact eye-level light exposure in a private office include the seating location, view direction, and eye level of an occupant, which interact with the light distribution, intensity, and spectrum of any given light source. This study presents a simulation-based study that compares various light distribution patterns, furniture configurations, and seating/standing positions to illustrate the impacts on human health potential from a non-visual health perspective.

The space used in our case study is an existing architectural office located in Portland, OR. Multiple luminaires and light distribution patterns are compared under different furniture layouts. The ALFA (Adaptive Lighting for Alertness) plug-in for Rhino is used to simulate Equivalent Melanopic Lux (EML) values for a series of hourly and daily time steps [4]. Electric light sources are simulated using industry-standard IES files and results are compared for both seated and standing view positions.

SCIENTIFIC INNOVATION AND RELEVANCE

A number of simulation techniques have emerged in recent years as a means to predict how varying factors within a space affect non-visual light exposure for building occupants. These techniques include the simulation of vertical illuminance values in incremental measurements across one or more view directions [3, 4, 6]. Recent work has also compared the impact of daylight [1, 3, 4, 5] and electric lighting sources [4, 5, 6, 7] on the health of indoor occupants. A recent study compared the impact of various overhead light sources on vertical eye-level exposure and horizontal task-plane illuminance [2]. This paper builds upon these studies by comparing a broader range of occupant-centric and spatial conditions that can impact the lighting design of an office space.

The novelty of our research lies in the comparison of light distribution patterns, furniture layout, and ergonomics within a private office space. Because eye-level light exposure accounts for light reflected off of vertical surfaces, luminaires that distribute light onto vertical surfaces and are located closer to the eye are potential assets for improving healthy lighting conditions. This work has the potential to bridge research with practical lighting design recommendations to improve the health and well-being of occupants located in private offices.

PRELIMINARY RESULTS AND CONCLUSIONS

Our results reveal a variety of intriguing outcomes. Intuitively, the combination of electric light and daylight sources systematically outperforms scenarios that rely exclusively on electric light only. While 36 out of 48 simulated scenarios achieved the minimum threshold for the WELL Building Standard of 150 melanopic lux (assuming continuous exposure between 9am and 1pm), less than half achieved the recommended 250 melanopic lux, despite achieving recommended task-plane illuminance values.

Comparing results from the various light distribution scenarios (direct vs. direct/indirect ceiling-mounted luminaires and wall-wash luminaires), the fixtures that provided a significant wall wash component achieved the highest EML values. This indicates the potential for vertical light distribution to act as a source of healthy light exposure that has not yet been thoroughly studied. A comparison of various ergonomic positions also reveals the variability in EML exposure between standing and seated positions. Further development of this proposal would contribute to the currently limited design guidelines for implementing effective lighting design to achieve circadian light exposure through the use of furniture configurations, eye level, and lighting distributions.

MAIN REFERENCES

[1] Acosta, I., Leslie, R., & Figueiro, M. (2017). Analysis of circadian stimulus allowed by daylighting in hospital rooms. Lighting Research & Technology, 49(1), 49–61.

[2] Jarboe, C., Snyder, J., & Figueiro, M. (2020). The effectiveness of light-emitting diode lighting for providing circadian stimulus in office spaces while minimizing energy use. Lighting Research & Technology, 52(2), 167–188.

[3] Amundadottir, M., Rockcastle, S., Sarey Khanie, M., & Andersen, M. (2017). A human-centric approach to assess daylight in buildings for non-visual health potential, visual interest and gaze behavior. Building and Environment, 113, 5–21.

[4] Saiedlue, S., Amirazar, A., Hu, J., & Place, W. (2019). Assessing Circadian Stimulus Potential of Lighting Systems in Office Buildings by Simulations. ARCC Conference Repository.

[5] Danell, M., Amundaddottir, M. L., & Rockcastle, S. (2020). Evaluating Temporal and Spatial Light Exposure Profiles for Typical Building Occupants. SimAUD Conference Proceedings.

[6] Dai, Q., Huang, Y., Hao, L., Lin, Y., & Chen, K. (2018). Spatial and spectral illumination design for energy-efficient circadian lighting. Building and Environment, 146, 216–225.

[7] Rockcastle, S., Danell M., Petterson, L., & Amundadottir, M. (2020). The Impact of Behavior on Healhty Circadian Light Exposure Under Daylight and Electric Lighting Simulations. ACEEE Conference Proceedings.

Keywords: lighting distribution, circadian, human-centric, non-visual

Paper ID 30463: Evaluating the use of photobiology-driven alertness and health measures for circadian lighting design

Athina Ji-Hae Alight*, J. Alstan Jakubiec

University of Toronto, John H. Daniels Faculty of Architecture, Landscape, and Design, Toronto, Canada

AIM AND APPROACH

The aim of this project is to evaluate a novel daylighting and electric lighting design workflow that assesses a space based upon light's impact on human photobiology-driven alertness, and health. The method, being published in a separate submission to the conference, works by translating timeseries spectrally-resolved light simulation data into photobiologically driven measures mediated by the response of intrinsically-photosensitive retinal ganglion cells (ipRGCs) in the human eye. These measures are used as input to a dynamic photobiological framework that accounts for light history, timing, spectrum, and homeostatic body rhythms. These measures are then visualised in a novel manner to communicate the impacts of lighting on space occupants. To demonstrate the value of this process, the predicted non-visual biological effects of six design variables (artificial lighting schedules, artificial light spectrum, occupant location, window spectral transmittance, surface reflectance, and two space designs) are simulated. The design variables are also applied to the frameworks suggested by Mardaljevic et al. (2013), Amundadottir et al. (2017), the WELL standard (2018), and Konis (2019) to test how these models respond to variations in spectrum and light exposure and how they differ in the resulting evaluation of architectural design.

SCIENTIFIC INNOVATION AND RELEVANCE

Several frameworks have been developed to evaluate lighting design for non-visual biological effects, which this paper compares and evaluates. Unlike light for visual tasks, the non-visual/circadian system, is sensitive to the timing of light exposure. The circadian system is also responsive to shorter wavelengths than the visual system. Illuminance that is appropriate to perform visual tasks therefore may not be enough to entrain circadian rhythms or maintain alertness and performance. Insufficient light after waking or excess light before sleep onset disrupts circadian rhythms with harmful health consequences. Furthermore, circadian disruption is cumulative, and depends on circadian entrainment in the recent past. As a result of these complexities, our model takes the history, timing, intensity, and spectrum of light exposure into account. To the best of our knowledge, the workflow presented in this paper is the only one that predicts explicit biological effects of light and spectrum over time rather than circadian light potential. Through implementing our model, we will demonstrate the results of the model and how architectural design can directly impact an occupant's circadian health. Our comparative analysis will showcase how predicting alertness and health measures differs from previous work and impacts the evaluation of architecture.

PRELIMINARY RESULTS AND CONCLUSIONS

For the six design parameters and using our new model, the following alertness and health photobiological measures are calculated: subjective alertness measured by the Karolina Sleepiness Scale, mean reaction time, attention lapses, and performance at rote tasks, the amount of circadian phase shift per day, the time of peak melatonin concentration, and the percent of total melatonin suppression per day due to acute light exposure.

The comparative analysis with previous models is based on two aspects: (1) comparisons of quantitative lighting units and (2) assessing differences in design evaluation outcomes as positive or negative. A compilation of the circadian calculation and evaluation methods used in existing photobiological lighting design evaluative frameworks (Mardaljevic et al. 2013, Amundadottir 2017, WELL Standard 2018, Konis 2019) are compared in this manner.. Each framework's responsiveness to lighting types (daylight, electric, or hybrid), evaluation timeframe (instantaneous, daily, or annual), and lighting design interventions is also compared and discussed.

MAIN REFERENCES

Amundadottir, M. L., Rockcastle, S., Khanie, M. S., & Andersen, M. (2017). A human-centric approach to assess daylight in buildings for non-visual health potential, visual interest and gaze behavior. Building and Environment, 113, 5-21.

Circadian lighting design. (2018). https://standard.wellcertified.com/light/circadian-lighting-design.

International WELL Building Institute, WELL Building Standard Circadian Lighting Design Feature. https://standard.wellcertified.com/light/circadian-lighting-design Last accessed 6/3/2020.

Konis, K. (2019). A circadian design assist tool to evaluate daylight access in buildings for human biological lighting needs. Solar Energy, 191, 449-458.

Mardaljevic, J., Andersen, M., Roy, N., & Christoffersen, J. (2013). A framework for predicting the non-visual effects of daylight–Part II: The simulation model. Lighting Research & Technology, 46(4), 388-406.

Solemma.com. ALFA. https://solemma.com/Alfa.html.

Tekieh, T., Lockey, S. W., Robinson, P. A., McCloskey, S., Zobaer, M. S., & Postnova, S. (2020). Modelling melanopsin-mediated effects of light on circadian phase, melatonin suppression and subjective sleepiness.

Keywords: daylighting, circadian lighting, non-visual effects, design process

Paper ID 30278: Integrated analysis of daylight and solar access building requirements and performance in urban environments in Estonia

Francesco De Luca*, Abel Sepúlveda

Tallinn University of Technology, Tallinn, Estonia

AIM AND APPROACH

Daylight and solar access are essential aspects of the indoor environmental quality of buildings. Adequate quantity of daylight helps to perform tasks with ease and its distribution increases architectural quality [1]. Appropriate direct solar access helps the entrainment of the circadian rhythm and the improvement of physiological and psychological well-being of occupants [2]. Thus, in most countries, regulations prescribe minimum quantities of daylight and direct solar access [3, 4].

The new EU standard Daylight in Buildings [5], to be acquired in Estonia, for daylight requires a minimum Daylight Factor (DF) of 0.7% on 95% of the simulation plane and of 2.2% on 50% of the plane closer the window. For solar access in dwellings it requires a minimum of 1.5 hours of exposure to sunlight calculated during one day between February 1st and March 21st.

The present study investigates optimal dwelling room parameters for 1) the fulfillment of both DF and sunlight requirements of the new EU standard in urban environments, where sunlight provision is most critical, and 2) adequate daylight availability. The aim is to help local authorities in the acquisition of the new EU standard and to provide designers with guidelines for the fulfillment of both requirements.

SCIENTIFIC INNOVATION AND RELEVANCE

The new EU standard requires different daylight but same sunlight quantity for the different countries. There are no studies about the relation between the two requirements in Estonia. Additionally, recent studies proved the scarce reliability of the Daylight Factor metric in predicting daylight availability in the country [6,7].

The innovation of the study lies in the integrated analysis of building performance for daylight and solar access at northern latitudes in urban environments. Additionally, it contributes to the assessment of the reliability of the DF requirements through climate based daylight simulations using the metrics of the LM-83-12 method Spatial Daylight Autonomy (sDA) and Annual Sunlight Exposure (ASE) [8]. The first predicts reliably daylight availability, the latter potential visual discomfort.

The integrated analysis of DF and solar access, DF and sDA, ASE and solar access permits to assess the efficacy of the requirements of the new EU standard for Estonia and to develop guidelines for the design in urban environments.

For the study a parametric model is realized that generate variations of a sidelit room orientation, window size, shading (upper floor balcony) and surrounding environment. The parametric model automates the calculation of solar access and the different daylight simulations at each variation.

PRELIMINARY RESULTS AND CONCLUSIONS

Simulations for 144 room variations, 8 orientations, 9 window sizes and 2 shading states are performed for March 21st without surrounding buildings and in three urban environments in Tallinn.

Without surrounding buildings, DF, that doesn't depend on orientation, is fulfilled by 66.7% of 18 variations, solar access, sDA and ASE by 62.5%, 49.3% and 56.9% of 144 variations respectively. DF and sunlight together, and sDA and ASE together, as required by the EU standard and by the LM-83-12 method, are fulfilled by 27.8% and 13.9% of all variations respectively.

In the three urban environments, DF is simulated also for the 8 orientations due to the different external obstructions. On average, solar access, DF, sDA and ASE are fulfilled by 51.8%, 20.3%, 17.8% and 68.5%, DF and sunlight, and sDA and ASE together are fulfilled by 14.4% and 3.7% of all variations respectively.

Preliminary results show the different fulfillment of DF and sunlight, the influence of the urban environment and the difficulty to achieve the two pair of performance together, and the daylight overestimation by the DF requirement in Estonia. The paper will present detailed simulation results and parameters necessary to fulfil the different metrics singularly and in pairs as required.

MAIN REFERENCES

1 - Reinhart, C.F. 2014. Daylighting Handbook I. Fundamentals. Designing with the Sun. Building Technology Press, Cambridge (USA).

2 – Lockley, S.W. 2009. Circadian rhythms: influence of light in humans. In: Squire LR (ed), Encyclopedia of Neuroscience (Vol. 2), Cambridge, MA, USA: Academic Press, 971–988.

3 – Dogan, T. and Park, Y.C. 2019. A critical review of daylighting metrics for residential architecture and a new metric for cold and temperate climates. Lighting Research & Technology, 51, 206–230.

4 – Darula, S., Christoffersen, J. and Malikova, M. 2015. Sunlight and insolation of building interiors. Energy Procedia, 78, 1245–1250.

5 - European Commission 2018. EN 17037:2018 Daylight in Buildings.

6 - De Luca, F., Kiil, M., Simson, R., Kurnitski, J. and Murula, R. 2019. Evaluating daylight factor standard through climate based daylight simulations and overheating regulations in Estonia. Proceedings of 16th IBPSA International Conference and Exhibition (BS2019), 3968-3975.

7 – Sepúlveda, A., De Luca, F., Thalfeldt, M. and Kurnitski, J. 2020. Analyzing the fulfillment of daylight and overheating requirements in residential and office buildings in Estonia. Building and Environment, 180, 107036.

8 - Illuminating Engineering Society 2013. IES LM-83-12 Approved Method: IES Spatial Daylight Autonomy (sDA) and Annual Sunlight Exposure (ASE).

Keywords: Daylight, solar access, regulations, parametric design, urban environment

Paper ID 30387: Machine learning techniques for the daylight and electric lighting performance predictions

Chantal Basurto*, Oliver Paul, Jérôme H. Kämpf

Idiap Research Institute, 1920 Martigny, Switzerland

AIM AND APPROACH

Despite the technological advance in the field of energy efficient buildings, the achievement of adequate lighting interior environments it is still a tight corner spot. The latter, due to the complex interplay occurring at different levels of the building performance, involving energy related and occupant's comfort issues. Such as, achieving a right balance between an increased daylight penetration for the reduction of heating and lighting loads in winter, while minimizing the risk of glare for the occupants. Providing an adequate solar protection while achieving a sufficient daylight provision at task area is a similar quest for summer time. Therefore, in order to undertake the intrinsically linked energy efficiency and occupant's comfort goals, recent research endeavors involve an integrated assessment of daylight, electric lighting, blinds and lighting controls. Nowadays, such evaluations are mostly performed with the use of computer simulations, which, due to the complexity of the issue, are still highly demanding in terms of computing time and performance capabilities; besides of the human-hours invested on the interaction with distinctive tools and interfaces. In order to improve the response time of daylight and electric lighting performance-predictions, machine learning techniques based on existing daylighting evaluation methods, are employed using surrogate models.

SCIENTIFIC INNOVATION AND RELEVANCE

In order to achieve an optimal control of blinds and electric lighting, a predictor model is employed to evaluate the impact of a blind's position choice on the work-plane illuminance and of glare in the occupant's eye. Including the predictor model in a Model Predictive Control (MPC) is the ultimate goal, aiming to obtain a quasi-real-time optimization of the building parameters, to provide visual comfort to the user with less electric lighting. Ubiquity is the main feature of this work, since the predictor model is derived from year-round simulations generated by the RADIANCE based matrix multiplication methods, where all possible blinds positions and weather conditions are considered. Simulation cost is another relevant feature of this work, since, due to the longer time that RADIANCE simulations require to complete, the predictor model is rather based on a statistical surrogate model realized with an Artificial Neural Network (ANN). A database is first produced and employed for the training of the surrogate model. Its input parameters are the weather data (direct and diffuse irradiance), sun and blinds position and electric lighting intensity, while the output data are key performance indicators (average work-plane illuminance and DGP glare index), for specific users.

PRELIMINARY RESULTS AND CONCLUSIONS

The method is applied to an office building located in Martigny, Switzerland, where two specific rooms are modeled and used as a case-study to demonstrate the model's predictor capabilities. The two models were created using Sketchup, while their material properties (reflectance, transmittance and other physical parameters) were measured using a Minolta Chromameter (CR-200b) and Gloss-meter (GM-060). The models were then calibrated according to on-site illuminance measurements obtained during the summer 2020, where the accuracy was reported as below 20%, providing the target precision for our surrogate model. A nearby weather station is used to gather direct and diffuse irradiance on hourly basis for the whole year under study. The vast amount of data obtained from the year-round RADIANCE simulations was determinant for defining the function between the inputs and outputs. Different ANNs (FFN, LSTM, GRU and CNN) are tested and compared to provide satisfactory precision for both illuminance and DGP, at a negligible simulation cost due to the data obtained beforehand from the RADIANCE simulations. The electric lighting contributions to the illuminance on the work-plane are computed separately, and the potential glare from the luminaires neglected. The developed surrogate model is finally validated against actual RADIANCE simulations and real monitoring.

MAIN REFERENCES

Ayoub, M., 2020. A review on machine learning algorithms to predict daylighting inside buildings. Solar Energy 202, 249–275. https://doi.org/10.1016/j.solener.2020.03.104

McNeil, A., 2013. The Five-Phase Method for Simulating Complex Fenestration with Radiance (Tutorial). Lawrence Berkeley National Laboratory, Berkeley, CA.

McNeil, A., 2012. A validation of the RADIANCE Three-Phase Simulation Method for Modeling Annual Daylight Performance of Optically Complex Fenestration Systems. Journal of Building Performance Simulation 1–14.

McNeil, A., 2010. The three-phase method for simulating Complex Fenestration with Radiance. Lawrence Berkeley National Laboratory, Berkeley, CA.

Nault, E., Moonen, P., Rey, E., Andersen, M., n.d. Predictive models for assessing the passive solar and daylight potential of neighborhood designs: A comparative proof-of-concept study. Building and Environment 116, 1–16. http://dx-doi.org/10.1016/j.buildenv.2017.01.018

Ward, G., 1985. RADIANCE lighting simulation software.

Wienold, J., Christoffersen, J., 2005. Towards a new daylight glare rating. Presented at the LUX Europa: Lighting for Humans, Berlin, Germany.

Zhaoyang, L., Cheng, S., Qi, D., 2020. A daylight-linked shading strategy for automated blinds based on model-based control and Radial Basis Function (RBF) optimization. Building and Environment 177. https://doi.org/10.1016/j.buildenv.2020.106854

Keywords: Daylighting and electric lighting, Machine Learning, blinds and lighting controls, illuminance and glare index, RADIANCE matrix multiplication methods.

Paper ID 31121: Luminance distributions in consultancy: Simulations or measurements?

Thijs Willem Kruisselbrink*

Peutz BV, Netherlands, The

AIM AND APPROACH

The luminance distribution is suitable tool to assess the lit environment in a holisitic manner. The luminance distribution is either simulated or measured. Both approaches have different strenghts and weaknesses. This work presents an analysis of the strenghts and weaknesses of the two approaches in order to find the most suitable approach for specific cases in consultancy.

SCIENTIFIC INNOVATION AND RELEVANCE

The lit environment is an intangible, but rather relevant component of the built environment, impacting performance, comfort, health and well-being. Research has shown that the lit environment has a multidimensional character. Consequently, the lit environment cannot be fully grasped by singular metrics that are often utilized inconsultancy such as the daylight factor (DF). Preferably, multiple metrics, associated to e.g. amount,distribution or directionality of light, are utilized todescribe the lit environment. However, despite significant effort, the research community has not found asatisfactory and holistic metric to capture the lit environment as a whole.

Alternatively, the luminance distribution can be a suitable means to describe the lit environment, as it contains information on the majority of relevant metrics. Nevertheless, practical implementation of the luminance distribution, simulation or measurement, in consultancy is rather limited. In addition to the advantagesof the luminance distribution, multiple limitations areassociated to its use, both for simulations and mea-surements.

PRELIMINARY RESULTS AND CONCLUSIONS

Simulations or measurements of the luminance distributions are suitable for different consultancy cases. Simulations are more suitable to find a optimal solution for the lit environment while measurements are able to assess a problem associated to the lit environment.

MAIN REFERENCES

Inanici, M. (2006, 6). Evaluation of high dynamicrange photography as a luminance data acquisitionsystem.Lighting Research and Technology 38(2),123–134.

Kruisselbrink, T. (2020, 10).Practical and continu-ous luminance distribution measurements for light-ing quality. Ph. D. thesis, Eindhoven University ofTechnology, Eindhoven.

Kruisselbrink, T., R. Dangol, and A. Rosemann(2018, 6).Photometric measurements of light-ing quality: An overview.Building and Environ-ment 138, 42–52.

Ochoa, C. E., M. B. Aries, and J. L. Hensen (2012, 7).State of the art in lighting simulation for buildingscience: a literature review.Journal of BuildingPerformance Simulation 5(4), 209–233.

Pierson, C., M. Bodart, J. Wienold, and A. Ja-cobs (2017). Luminance maps from High DynamicRange imaging : calibrations and adjustments forvisual comfort assessment. InLux Europa, Ljubl-jana, Slovenia, pp. 147–151.

Reinhard, E., G. Ward, S. Pattanaik, and P. Debevec(2006, 8).High Dynamic Range Imaging: Ac-quisition, Display, and Image-Based Lighting (TheMorgan Kaufmann Series in Computer Graphics).San Fransisco: Morgan Kaufmann Publishers Inc.

Reinhart, C. and O. Walkenhorst (2001). Valida-tion of dynamic RADIANCE-based daylight simu-lations for a test office with external blinds.Energyand Buildings 33(7), 683–697.

Van Den Wymelenberg, K. (2012).Evaluating Hu-man Visual Preference and Performance in an Of-fice Environment Using Luminance-based Metrics.Ph. D. thesis, University of Washington.

Keywords: luminance distribution, simulation, measurement, lit environment

Session W3.7 (Online Track): Ensuring high quality building simulations

Paper ID 30512: Design and test of reduced grey-box models adapted to office buildings

Thibault Péan*, Soledad Ibañez Iralde, Jordi Pascual, Jaume Salom

IREC Catalonia Institute for Energy Research, Spain

AIM AND APPROACH

Simplified models (oftentimes called black and grey box models), allows for high precise results of the energetic behaviour of buildings, with a lower resources demand than more extensive white-box models. Given their potential, a large amount of research has been produced in recent years to develop such simplified models, with the end purpose of performing quick energy demand calculation, or for integration in predictive control schemes. In this line, offices and tertiary buildings represent a large and interesting area of application for such models, given that they have a much higher energy intensity than residential buildings, they tend to be refurbished more often and monitored more closely. The present work aims at developing and adapting simplified models to the specificities of office buildings, to obtain reliable energy calculations and predictions. Such models, provided that they are thoroughly validated and tested, can be used for fast energy demand calculations, integration into overall building management system (that most office buildings already have), predictive controls, or fault detection algorithm for predictive maintenance.

SCIENTIFIC INNOVATION AND RELEVANCE

A large part of the existing literature focuses on simplified models for residential buildings. Office buildings possess some distinct characteristics, which must be taken into account when designing a simplified model. The internal gains from equipment, people and solar irradiation take larger values for this building typology, therefore the cooling demand is usually more important than in residential buildings. This is particularly relevant in the Mediterranean area, where the cooling demand might then represent the critical demand, while in the same region the heating demand is more critical for residential buildings. Furthermore, offices always have mechanical ventilation, which is not the case of most houses in Southern Europe. Finally, the occupancy schedule is more steady and less stochastic, dictated by opening office hours, with an exact opposite occupancy compared to a standard dwelling (i.e. occupants are usually at work when they are not at home), and larger non-occupancy periods (weekends).

To take into account these specificities, a grey-box model was developed, taking as starting point the resistance capacitance (RC) model described in ISO 13790, and modified with more detailed infiltration and ventilation gains. The resulting model has a R4C2 structure.

PRELIMINARY RESULTS AND CONCLUSIONS

To test the proposed grey-box model, a study case of an office building situated in the Mediterranean area of Spain was chosen. A white-box model was first developed with knowledge of the construction elements, and adjusted with monitoring data from the real building in the heating and cooling seasons. The purpose of this intermediate white-box model is to be able to generate different datasets for the identification of the grey-box model. The white-box model was calibrated by adjusting the infiltration rate, the thermal capacity, and the shading coefficient of the windows, leading to a root mean square error of 0.8°C (normalized: 3.5%) between model and data.

The white-box model was used to generate different datasets, exciting the building with a pseudo-random binary signal on the space heating or cooling power. Using these datasets, the identification process aimed at finding the appropriate values of the thermal capacities of the two nodes of the model, and the transmission coefficients of the opaque and transparent construction elements. Using two different optimization process, the best values for these parameters were obtained, leading to a fit of 75% between white-box and grey-box model data, which is satisfactory given the simplicity of the chosen RC model.

MAIN REFERENCES

ISO. ISO 13790: Energy Performance of buildings - Calculation of energy use for space heating and cooling. (2008).

De Coninck, R., & Helsen, L. (2016). Practical implementation and evaluation of model predictive control for an office building in Brussels. Energy and Buildings, 111, 290–298. https://doi.org/10.1016/j.enbuild.2015.11.014

Reynders, G., Diriken, J., & Saelens, D. (2014). Quality of grey-box models and identified parameters as function of the accuracy of input and observation signals. Energy and Buildings, 82, 263–274. https://doi.org/10.1016/j.enbuild.2014.07.025

De Coninck, R., Magnusson, F., Åkesson, J., & Helsen, L. (2015). Toolbox for development and validation of grey-box building models for forecasting and control. Journal of Building Performance Simulation, (July), 1–16. https://doi.org/10.1080/19401493.2015.1046933

Bacher, P., & Madsen, H. (2011). Identifying suitable models for the heat dynamics of buildings. Energy and Buildings, 43(7), 1511–1522. https://doi.org/10.1016/j.enbuild.2011.02.005

Keywords: RC model, thermal dynamics, grey-box modelling, tertiary buildings

Paper ID 30610: Evaluation of existing infiltration models used in building energy simulation

Yeonjin Bae (1), Jaewan Joe (2), Seungjae Lee (1), Piljae Im* (1), Lisa C. Ng (3)

1: Oak Ridge National Laboratory, United States; 2: Inha University, South Korea; 3: National Institute of Standards and Technology, United States

AIM AND APPROACH

This study aims to evaluate the existing infiltration models in EnergyPlus by comparing their simulation results with the infiltration rate estimated from heavily monitored field measurements. A series of tracer gas decay and fan pressurization tests were performed in a full-scale, multizone, and unoccupied commercial building. Three infiltration models within EnergyPlus, one with various sets of coefficients, are used to simulate infiltration rates. The fan pressurization test results were converted to the design infiltration rate for use in each infiltration model. The simulation results were compared with the infiltration rate estimated from the field measurements.

SCIENTIFIC INNOVATION AND RELEVANCE

Infiltration is one of the major sources of uncertainty in building energy simulation. Although many infiltration models exist, their structures and assumptions vary, and often, they are inaccurate for commercial buildings as they were developed based on empirical datasets from residential building models. In this study, tracer gas decay and whole-building pressurization tests were performed in the test building, and the results were used in three different infiltration models available in EnergyPlus. Also, a simulation study was conducted with an empirically validated EnergyPlus model to investigate how the selection of the infiltration model influences the predicted building heating energy consumption.

PRELIMINARY RESULTS AND CONCLUSIONS

In this study, the measured and predicted infiltration rates were compared. The predicted infiltration rates were calculated based on six infiltration models (i.e., three infiltration models from EnergyPlus with various sets of coefficients). Out of six models, "DOE-2", "EffectiveLeakageArea," and "FlowCoefficient" based models show significantly large discrepancies withe the measurements. For example, the median value of the predicted infiltration rates using the "DOE-2" model was only15.4 % of the median value of the measured rates. However, the absolute predictive error in the infiltration rate was small because the building is relatively airtight. If the airtightness of the target building is low (i.e., leaky), then the absolute predictive error would also increase.

Also, the simulation results (compared with the infiltration rate estimated from the field measurements) showed that the predicted infiltration rate and the estimated heating energy consumption can be significantly affected by the infiltration model selection.

MAIN REFERENCES

ASHRAE. 2017. ASHRAE Handbook Fundamentals, American Society of Heating, Refrigerating, and Air-Conditioning Engineers.

ASTM International. 2017. ASTM International Standard E741-11. Standard Test Method for Determining Air Change in a Single Zone by Means of a Tracer Gas Dilution. West Conshohocken, PA: ASTM International.

ASTM. International. 2019. ASTM Standard E779. Standard Test Method for Determining Air Leakage Rate by Fan Pressurization. West Conshohocken, PA.

Emmerich S.J., McDowell T.P. and FAIA W.A. 2007. Simulation of the Impact of Commercial Building Envelope Airtightness on Building Energy Utilization/DISCUSSION. ASHRAE Transactions, 113, p.379.

Gowri K., Winiarski, D., and Jarnagin R. 2009. Infiltration modeling guidelines for commercial building energy analysis (No. PNNL-18898). Pacific Northwest National Lab (PNNL), Richland, WA.

Han G., Srebric J., and Enache-Pommer E. 2015. Different modeling strategies of infiltration rates for an office building to improve accuracy of building energy simulations. Energy and Buildings, 86: 288–95.

Im P., Joe J., Bae Y., New J. R. 2020. Empirical validation of building energy modeling for multi-zones commercial buildings in cooling season. Applied Energy, 261, 114374.

Younes C., Shdid C. A., and Bitsuamlak G. 2012. Air infiltration through building envelopes: A review. Journal of Building Physics, 35(3): 267–302.

Keywords: Infiltration models, Building energy simulation (BES), Tracer gas test, fan pressurization test

Paper ID 30689: Exploring the possibility of calibrating a whole-building model from the short-term monitoring of selected reference rooms

Ilaria Pittana* (1), Riccardo Albertin (2), Alessandro Prada (3), Francesca Cappelletti (4), Andrea Gasparella (2)

1: Dep. of Industrial Engineering, University of Padua, Italy; 2: Faculty of Science and Technology, Free University of Bozen-Bolzano, Italy; 3: Dep. of Civil, Environmental and Mechanical Engineering, University of Trento, Italy; 4: Dep. of Architecture and Arts Iuav University of Venice, Italy

AIM AND APPROACH

Accurate simulation models of existing buildings provide a reliable picture of building and system behaviour, useful for diagnostic purposes, for designing retrofit interventions or improving the control strategies. Reliable building models can be obtained through monitoring and calibration, aiming at minimizing the discrepancy between predicted and actual performance by fine-tuning the values of the simulation parameters. Analytical calibration methods may become a complex and time expensive process, especially when a large numbers of parameters has to be estimated as in the case of the largest buildings. In addition, overfitting issues can undermine the reliability of the calibration process.

This work explores the possibility of carrying out a model calibration through low-cost and short-term measurements without falling into overfitting issues. In particular, the proposed approach is based on the selection of representative spaces in the buildings, and the identification of multiple monitoring periods during which only a subset of simulation parameters needs to be calibrated. The model is extended to the entire building in a multi-stage and multi-level approach.

The described approach is applied to a Primary School building located in the town of Schio (northern Italy).

SCIENTIFIC INNOVATION AND RELEVANCE

This method has two main advantages that represent its main innovation:

- It uses the measurements inside a small portion of a building (i.e. 9 rooms) to calibrate the whole building, lowering the monitoring costs.

- It uses four short monitoring periods to calibrate different sets of inputs separately, reducing the number of parameters to be optimized each time, thus improving the calibration process.

The described approach is applied to a Primary School building located in the town of Schio (northern Italy). The air temperature and relative humidity in two reference rooms and in the 7 boundary rooms have been monitored, considering four periods, with and without occupancy and during the heating season or the free-floating season. The calibration is developed with a multi-stage approach that considers the four selected periods to calibrate the inputs for which the sensitivity of the air temperature is highest and with a multi-level methodology that extends the parameters calibrated for the two rooms to the entire building and perform a second calibration to estimate the missing parameters.

PRELIMINARY RESULTS AND CONCLUSIONS

Two classrooms have been calibrated during four subsequently periods (Period 1: non-occupied building, passive mode; Period 2: non-occupied building, heating sys¬tem on; Period 3: occupied, passive mode; Period 4: occupied, heating on). The calibrated inputs of the partial-model of the school have been extended to the model of the whole school building in order to build the whole-building initial model in the same periods and to calibrate the residual unknown inputs.

The comparison of the temperature profiles obtained in the different stages and in the different levels of the calibration process with the measured data available in 9 rooms are presented by means of some statistical indices (RMSDavrg, CV(RMSD)avrg and R2avrg). Finally, the measured energy consumption for heating is compared to the calculated one.

MAIN REFERENCES

ASHRAE. Guideline 14–2002, Measurement of Energy and Demand Savings. American Society of Heating, Ventilating, and Air Conditioning Engineers, Atlanta, Georgia; 2002.

Moriasi D. N., Arnold D. N., Van Liew M. W., Bingner R. L., Harmel R. D., Veith T. L. (2007). Model evaluation guidelines for systematic quantification of accuracy in watershed simulations, Transactions of the ASABE, Vol. 50(3).

Penna P., Cappelletti F., Gasparella A., Tahmasebi F., Mahdavi A., (2015). "Multi-stage calibration of the simulation model of a school building through short-term monitoring." Special Issue ECPPM 2014 - 10th European Conference on Product and Process Modelling".

Penna P., Prada A., Cappelletti F., Gasparella A., (2015). "Multi-objective optimization of Energy Efficiency Measures in existing buildings." Journal of information technology in construction, 20, 132-145. Yang Z., Becerik-Gerber B. (2015). A model calibration framework for simultaneous multilevel building energy simulation. Applied Energy 149 (2015).

Tahmasebi F., Zach R., Schuß M., Mahdavi A. (2012). "Simulation Model Calibration: An Optimization-Based Approach". Fourth German Austrian IBPSA Conference Berlin University of the Arts.

Tahmasebi F., Mahdavi A. (2013). "A Two-Staged Simulation Model Calibration Approach To Virtual Sensors For Building Performance Data." Proceedings of BS2013: 13th Conference of International Building Performance Simulation Association, Chambéry, France, August 26-28.

Keywords: Monitoring, Building simulation, Calibration, Validation

Paper ID 30760: A novel multi-domain model for thermal comfort which includes building indoor CO2 concentrations

Sarah Crosby* (1), Adam Rysanek (2)

1: Department of Mechanical Engineering, University of British Columbia, Vancouver, Canada; 2: School of Architecture and Landscape Architecture, University of British Columbia, Vancouver, Canada

AIM AND APPROACH

In a prior work, we applied Bayesian logistic regression to correlate and characterize the relationship between perceived thermal comfort, thermal indoor conditions, and non-thermal metrics of indoor environmental quality (IEQ) such as CO2 concentrations and indoor noise levels. That work made use of the COPE dataset, a field study from the early 2000s conducted by the National Research Council of Canada. The study collected objective and subjective IEQ measurements from approximately 800 occupants of open-plan offices in large Canadian and US cities. This work updates the findings of the prior study by adding over 100 new samples of IEQ measurements collected from over 100 occupants of office spaces at the University of British Columbia in 2019. Bayesian logistic regression of the expanded dataset reinforces observations made in our first paper that thermal comfort is correlated to measured values of indoor CO2 concentrations and speech intelligibility, in addition to parameters of temperature and mean-radiant temperature. This paper formulates a new predictive model of thermal comfort, derived from the Bayesian logistic regression of the COPE and UBC datasets, which can be used by building modellers to predict thermal comfort in office settings based on thermal conditions, ventilation rates, and noise levels.

SCIENTIFIC INNOVATION AND RELEVANCE

Several recent studies have identified the multi-perceptual and multi-contextual relationship of thermal comfort - that thermal comfort may be related to other indices of IEQ and vice versa. Our work is one of few studies to evaluate these relationships quantitatively and in a manner that can support future thermal comfort prediction. The buildings sector is facing several conflated challenges, particularly now in a post-COVID-19 world. Energy use should be minimized to support climate change objectives, but indoor air quality and well-being cannot be sacrificed - if anything, it should be improved as well. Our research has suggested that an open plan office with greater acoustic privacy and/or high amounts of fresh air can provide the same level of thermal comfort at higher/lower temperatures than an office with minimal acoustic privacy and 'typical' fresh air ventilation rates. Establishing these relationships in a manner in which building designers can account for these effects in building simulation is important. If low-cost interventions such as acoustic dividers can improve the acceptability range of indoor temperatures in office settings, facilities managers can achieve energy savings through warmer summertime temperatures and cooler winter temperatures.

PRELIMINARY RESULTS AND CONCLUSIONS

The Bayesian regression results inferred from the expanded UBC + COPE dataset revealed stronger evidence to suggest that measured indoor CO2 concentrations and speech intelligibility are independently correlated with perceived thermal comfort. The statistical significance of these results is validated using several Bayesian model validation techniques which, in turn, validates the robustness and significance of our prior findings. We synthesize the regression results into polynomial expressions (similar to the PMV-PPD and Adaptive Comfort models) which can in turn be used to predict thermal comfort in office spaces while taking into account non-thermal metrics of Indoor environmental quality. This new model could be used by building performance simulation experts to predict occupants' thermal comfort in office spaces. We demonstrate how this can contribute to, and also provide obstacles to, identifying potential building energy savings with respect to heating, cooling, and mechanical ventilation, depending on the climate.

MAIN REFERENCES

S. Crosby, G. Newsham, J. Veitch, and A. Rysanek. Correlations between thermal satisfaction and non-thermal conditions of indoor environmental quality: Bayesian inference of a field study of offices. Energy and Buildings (in review), 2020.

S. Crosby, and A. Rysanek. Extending the Fanger PMV model to include the effect of non-thermal conditions on thermal comfort. Proceedings of eSim Conference, Vancouver, BC, Canada. 2021 (in review).

A. Jamrozik, C. Ramos, J. Zhao, J. Bernau, N. Clements, T. Vetting Wolf, and B. Bauer. A novel methodology to realistically monitor office occupant reactions and environmental conditions using a living lab. Building and Environment, 130 :190–199, 2018.

J. Langevin, J.and Wen and P. Gurian.Modeling thermal comfort holistically: Bayesian estimation of thermal sensation, acceptability, and preference distributions for office building occupants. Building and Environment, 69: 206–226, 2013.

M. Schweiker, E. Ampatzi, M. S. Andargie, R. K. Andersen, E. Azar, V. M. Barthelmes, C. Berger, ..., and L. P. Edappilly. Review of multi-domain approaches to indoor environmental perception and behaviour. Building and Environment, 2020

J. J. McArthur and C. Powell. Health and wellness in commercial buildings: Systematic review of sustainable building rating systems and alignment with contemporary research. Building and Environment 171 (2020): 106635.

Keywords: Thermal Comfort, Indoor Environmental Quality, Bayesian Inference, Open-plan offices, Occupant well-being

Paper ID 30232: A gap-filling method for room temperature data based on autoencoder neural networks

Antonio Liguori* (1), Romana Markovic (2), Jérôme Frisch (1), Andreas Wagner (2), Francesco Causone (3), Christoph van Treeck (1)

1: E3D - Institute of Energy Efficiency and Sustainable Building, RWTH Aachen University, Mathieustr. 30, 52074 Aachen, Germany; 2: Building Science Group, Karlsruhe Institute of Technology, Englerstr. 7, 76131 Karlsruhe, Germany; 3: Department of Energy, Politecnico di Milano, Via Lambruschini 4, 20156 Milano, Italy

AIM AND APPROACH

The aim of this paper is to investigate the feasibility of autoencoder neural networks for reconstructing the missing indoor air temperature sequences that are obtained from room automation. For that purpose, three deep learning architectures that include feed-forward, 1D-convolutional and long short-term memory (LSTM) denoising autoencoders were implemented to learn the daily indoor air temperature patterns. The sequences of the monitored room automation data were undersampled to 30 minutes' frequency and the reconstructed gaps in measured data ranged between few hours (10 % of the daily values) and around 22 hours (90 % of the daily values).

The developed models were evaluated using the monitoring data collected in multiple buildings with significant differences in thermal mass and design. Finally, the model's performance was compared to the baseline models such as linear interpolation and mean inserting.

SCIENTIFIC INNOVATION AND RELEVANCE

Thanks to the growing number of installed meters in buildings (Rätz, 2019), data-driven models have experienced an increasing use in the area of building energy and environmental performance optimization (Ibeigi, 2020), (Zhou, 2020). However, one of the main limitations for data-driven modeling is the presence of errors and missing values in these data sets (Chong, 2016). In the related literature, there is still little relevant research about this issue and, for this reason, the adopted reconstruction techniques often lead to limited models' performance (Chong, 2016).

In this regard, the proposed autoencoder neural networks for the missing data reconstruction represent a promising approach to fill the latter research gap. The particular novelty of this method is that it benefits from the spatial and temporal correlation of the input features, which can lead to significant performance improvement. This is achieved by relying on the LSTMs for handling the temporal relationships and convolutional units for handling the spatial correlations within the models' input and output features.

PRELIMINARY RESULTS AND CONCLUSIONS

The results showed, that the temporal (day-to-day) correlation have more predictive power, when compared to the spatial correlation (dimensions within the same day). The proposed autoencoder neural network could effectively reconstruct missing indoor air temperature data with accuracy over 90 % in terms of root mean squared error. Furthermore, the models' performance did not drop with the increased duration of the missing sequences. Accordingly, the obtained results confirmed the findings from Fan et at. (2018) - namely, autoencoder neural networks can capture patterns of building energy and control data, which represents a significant practical potential for the inclusion of these models in the real-time building control. Comprehensive results and further conclusions will be presented in the main paper.

MAIN REFERENCES

Chong et al., Imputation of missing values in building sensor data, ASHRAE and IBPSA-USA SimBuild 2016 Building Performance Modeling Conference, Salt Lake City, UT (2016).

Fan et al., Analytical investigation of autoencoder-based methods for unsupervised anomaly detection in building energy data, Appl Energy 211 (2018) 1123-1135.

Ibeigi et al., Prediction and optimization of energy consumption in an office building using artificial neural network and a genetic algorithm, Sustain Cities Soc 61 (2020) 102325.

Rätz et al., Automated data-driven modeling of building energy systems via machine learning algorithms, Energy Build 202 (2019) 109384.

Zhou et al., Using long short-term memory networks to predict energy consumption of air-conditioning systems, Sustain Cities Soc 55 (2020) 102000.

Keywords: neural networks, autoencoders, indoor environmental quality, missing data, occupant behavior

Paper ID 30207: A synthetic population model for representing occupant behaviors in buildings

Handi Chandra Putra*, Tianzhen Hong

Lawrence Berkeley National Laboratory, United States of America

AIM AND APPROACH

The current development of occupant behavior research has come to a more elaborate framework of building occupant interaction. Researchers collected behavioral data but often found it a challenge to meet the minimum number of required data points and the data interoperability requirements. Researchers address the first issue with the synthetic population and the latter with data ontologies. The two solutions are complementary to each other. One of the known ontologies, Drivers-Needs-Actions-Systems (DNAs) ontology, has been used by building modelers to describe energy-related occupant behavior[1]. The expansion of DNAs ontology is intended to pave a synthetic occupant population pathway that further its use in multiple applications, including the emerging agent-based modeling (ABM) [2]. The synthetic population approach is useful for detailed characterization of occupant-agent and a group-of-agent in the ABM environment. Previously collected data on occupants, including ASHRAE Thermal Comfort DB II [3], IEA Annex 66 data set [4], drive the synthetic behavior generation. Occupants' demographic characteristics are drawn from multiple sources, including U.S. decennial census and microdata sample[5], as well as The National Household Travel Survey[6]. Case studies of residential and commercial buildings are used to present the workflow of DNAs framework expansion, synthetic population generation, and agent-based modeling.

SCIENTIFIC INNOVATION AND RELEVANCE

The need for a more detailed occupant behavior modeling is becoming prevalent in the context of human-building interaction. Occupant modelers borrow behavioral theories and models from other established-disciplines in the area. The Theory of Planned Behavior (TPB), for example, describes how social norms, energy usage habits, and economic concern drive the behavior of occupants to adjust thermostat settings in their rooms [7]. Hence, the expansion of DNAs ontology introduces new elements that fall into five categories: socio-economic, geographical location, activities, subjective actions, and individual and collective adaptive actions. The locus of control of individual occupants within the collectivity of occupants is challenging in synthetic population research. Demographers have tried to synthesize individuals and match them with synthesized households. The synthesized individuals and individual-groups are, then, validated with the marginal distribution of real census data and microdata [8]. The synthetic occupant generation follows the DNAs ontology to describe the occupants and uses an integrated population synthesis method of Bayesian network (BN) and multi-level iterative proportional fitting (IPF). The resulting synthetic data represent the probability distribution of a sample of real data.

PRELIMINARY RESULTS AND CONCLUSIONS

The proposed enhancements to the DNAs framework represent occupant behavior in greater detail, which can inform occupant data collection and synthetic occupant data generation efforts. The core components of the framework that draws upon adaptive behavioral constructs, namely Drivers, Needs, Actions, and Systems remain unchanged. The expansion includes geographical location characteristics of an occupant, such as climate zone, urban and rural regions. Relevant energy-related policies in place, as well as infrastructural investments of a municipality, are also behavioral determinants that are attributed to location. Moreover, energy-use behaviors are also varied based on building types; hence, the framework includes the building types. Both statistical construct synthetic occupant populations and reproduce the sample occupant data with well-fitting distributions. The synthetic population model is, then, utilized for running ABM and generating better occupant data for building energy modeling.

MAIN REFERENCES

[1] T. Hong, S. D'Oca, W. J. N. Turner, and S. C. Taylor-Lange, "An ontology to represent energy-related occupant behavior in buildings. Part I: Introduction to the DNAs framework," Building and Environment, vol. 92, pp. 764–777, Oct. 2015.

[2] C. Berger and A. Mahdavi, "Review of current trends in agent-based modeling of building occupants for energy and indoor-environmental performance analysis," Building and Environment, vol. 173, p. 106726, Apr. 2020.

[3] V. F. Licina et al., "Development of the ASHRAE Global Thermal Comfort Database II," BUILDING AND ENVIRONMENT, vol. 142, pp. 502–512, Sep. 2018.

[4] International Energy Agency, "Energy in buildings and communities program. Annex 66: definition and simulation of occupant behavior in buildings." [Online]. Available: http://www.annex66.org/.

[5] U. S. Census. Bureau, 2015: http://factfinder2.census.gov.

[6] F. H. A. (FHWA), "National Household Travel Survey," 2017: https://nhts.ornl.gov/.

[7] E. L. Hewitt, C. J. Andrews, J. A. Senick, R. E. Wener, U. Krogmann, and M. Sorensen Allacci, "Distinguishing between green building occupants' reasoned and unplanned behaviours," Building Research & Information, vol. 44, no. 2, pp. 119–134, Feb. 2016.

[8]K. Müller, "A generalized approach to population synthesis," 2017.

Keywords: ontology, synthetic population, agent-based modeling

Paper ID 30288: How do buildings adapt to changing occupancy? A natural experiment

Brodie W. Hobson* (1), Tareq Abuimara (1), H. Burak Gunay (1), Guy R. Newsham (2)

1: Department of Civil and Environmental Engineering, Carleton University; 2: Construction Research Centre, National Research Council Canada

AIM AND APPROACH

Modern office buildings regulate their indoor climate through a series of conservative setpoints and equipment schedules that are chosen early in the design phase to ensure adequate indoor air quality at maximum occupancy. However, the occupancy of most modern office buildings rarely exceeds 50%, yet operations are seldom optimized for actual occupancy. This lack of adaptability to changing occupancy exacts a toll on buildings' energy use by providing excessive building services in an inefficient manner. Largely vacant offices buildings left in the wake of the recent COVID-19 pandemic present a natural experiment to evaluate the extent of this problem in an extreme case. Data from the building automation systems of several office buildings were collected before, during, and after the exodus of office workers to determine the impact this change in occupancy had on building energy use, and to establish a timeline for any operational changes to ventilation, temperature setpoints, and artificial lighting. Wi-Fi data were collected to estimate the occupancy of each building over this period for comparison. Three-point univariate changepoint models were employed to estimate the change in energy use pre- and post-pandemic. Design and operational challenges to buildings' adaptability to occupancy are identified.

SCIENTIFIC INNOVATION AND RELEVANCE

The novel case study presented capitalizes on the unique oppourtunity created by the COVID-19 pandemic to identify current obstacles that hinder the adaptability of commercial building HVAC and artificial lighting systems to variable occupancy. The issue of adaptability is sure to become more prevalent in the global commercial building stock with persistent habits such as teleworking evolving at an accelerated pace. This unprecedented natural experiment also allows for the collection of data in buildings with near-zero occupancy on a much larger timescale than ever before, which will generate new insights into the role of occupants in buildings' energy use. Operational strategies and changes made by operators after the onset of the pandemic – and the impact on building energy use – are discussed and quantified. The lag time between recommended changes and the implementation of these interventions in building automation systems is characterized and examined in the broader context of day-to-day building operations. Recommendations are put forth to help inform design decisions such as equipment sizing, as well as new simulation-based design approaches that take this variable occupancy into account to help improve buildings' adaptability to variable occupancy in extreme cases and during day-to-day operations alike.

PRELIMINARY RESULTS AND CONCLUSIONS

Preliminary data shows that the energy use was largely unchanged for several weeks after the COVID-19 lockdown began, even though many of the buildings' occupancies had dropped dramatically in a single day, and to near-zero in less than one week. While a small decrease in energy use was observed due to reduced plug-in equipment and lighting loads in the initial weeks, operating schedules and temperature setpoints were largely unchanged for months after the initial lockdown. The overall energy use of many buildings increased after the onset of the pandemic as new ventilation requirements with higher outdoor air fractions were introduced for health purposes. With the current state of operational practices, buildings are not adaptable to changes in occupancy. While a global pandemic is an extreme case of variable occupancy, changes to system-level equipment schedules and setpoints to optimize energy use for zero occupants should be comparatively simple to the changes required for optimizing more numerous zone-level systems for more subtle changes in occupancy during typical operations. If the additional savings made possible by more granular occupancy-centric controls are to be realized post-pandemic, significant changes to how buildings are designed, simulated, and operated are needed.

MAIN REFERENCES

F. Haldi and D. Robinson, "The impact of occupants' behaviour on building energy demand," J. Build. Perform. Simul., vol. 4, no. 4, pp. 323–338, 2011.

H. B. Gunay, "Improving energy efficiency in office buildings through adaptive control of the indoor climate," Carleton University, 2016.

H. B. Gunay, M. Ouf, G. Newsham, and W. O'Brien, "Sensitivity analysis and optimization of building operations," Energy Build., vol. 199, pp. 164–175, 2019.

H. B. Gunay, W. O'Brien, I. Beausoleil-Morrison, and B. Huchuk, "On adaptive occupant-learning window blind and lighting controls," Build. Res. Inf., vol. 42, no. 6, pp. 739–756, 2014.

I. E. Bennet and W. O'Brien, "Office building plug and light loads: Comparison of a multi-tenant office tower to conventional assumptions," Energy Build., vol. 153, pp. 461–475, 2017.

M. Ouf, W. O'Brien, and H. B. Gunay, "On quantifying building performance adaptability to variable occupancy," Build. Environ., vol. 155, no. February, pp. 257–267, 2019.

W. O'Brien et al., "Introducing IEA EBC Annex 79: Key challenges and opportunities in the field of occupant-centric building design and operation," Build. Environ., p. 106738, 2020.

Keywords: occupancy, building operations, occupancy-centric control, building automation systems, HVAC

Paper ID 30433: Investigating thermostat setpoint preferences in Canadian households

Karthik Panchabikesan* (1), Mohamed Ouf (1), Ursula Eicker (1), Guy Newsham (2), Heather Knudsen (2)

1: Concordia University, Montreal, QC, Canada; 2: Construction Research Centre, National Research Council of Canada, Ottawa, Canada

AIM AND APPROACH

Unlike commercial building occupants, residential dwellers have more control over their indoor environment as they often have direct and sole access to thermostats. However, despite the diversity in their indoor temperature preferences, most building energy simulations, codes and standards assume identical thermostat setpoints for all residential buildings. To this end, this study aims to demonstrate the variations in temperature setpoints across Canadian households by analyzing thermostat data collected from ~13,000 residential buildings. The specific objectives of this study are to (1) determine the average heating, and cooling thermostat setpoints in residential buildings, (2) rank the importance of different attributes that influence setpoint preferences, and (3) extract distinct heating and cooling setpoint profiles in residential buildings.

Multiple analysis approaches were used in this study. First, statistical methods were used to estimate the average thermostat setpoints in different provinces. Then, a random forest ensemble learning model was used to rank the relative importance of different attributes on average heating and cooling setpoint temperatures. Finally, the k-Shape clustering technique was used to extract distinct daily heating and cooling setpoint temperature profiles from ~13,000 Canadian homes.

SCIENTIFIC INNOVATION AND RELEVANCE

Studies indicate that residential buildings are responsible for 33% of total electricity use in Canada, and space conditioning (which is controlled by thermostats) accounts for 64% of their energy use. Therefore, occupants' thermostat setpoint preferences significantly influence buildings' energy use and play a vital role in the efficient operation of HVAC systems. As thermostat setpoint assumptions are critical inputs for building performance simulations, the diversity in occupants' preferences across different climate zones and household characteristics should be accounted for to improve the simulation accuracy. Previously, data on occupants' actual thermostat setpoint preferences were not widely available and obtained primarily through self-reporting via surveys. However, today's new generation of smart thermostats enables centralized data collection and sharing at large scales, which paves the way for investigating temperature setpoint preferences, and variations in thermostat settings (concerning several factors such as location, building type, age, etc.) in thousands of residential buildings at once. Such data analysis provides an unprecedented opportunity for re-visiting fixed/simplistic assumptions related to thermostat setpoint schedules in building energy simulations to better reflect operational parameters in actual buildings.

PRELIMINARY RESULTS AND CONCLUSIONS

The results obtained in this study were compared with the building energy codes and standards and differences up to ~3°C were found between the code assumptions and the present study. Moreover, in building energy codes and standards, only daytime and nighttime thermostat setpoint temperatures are provided, ignoring the common practice of setting a different daytime setpoint when occupants are away from home. In this context, assuming appropriate thermostat setpoints considering more typical current occupant schedules is recommended to increase the accuracy of the building simulations. The random forest model results suggest that building energy code provisions might consider different setpoint assumptions based on climatic zones and building characteristics. The analysis of daily setpoint profiles indicates significant variations within households, which were grouped into four main clusters for heating and cooling setpoint temperature, respectively. The majority of these clusters show a daytime temperature setback in residential buildings which is rarely accounted for in building energy codes and standards. Overall, the results presented in this study could be a significant input to updating temperature setpoint assumptions used in the building energy codes, and in energy simulations considering occupant preferences in residential buildings.

MAIN REFERENCES

1. Huchuk, B., O'Brien, W., and Sanner, S. 2018. A longitudinal study of thermostat behaviors based on climate, seasonal, and energy price considerations using connected thermostat data. Building and Environment, 139:199-210.

2. National Research Council Canada, 2017, National Energy Code of Canada for Buildings 2017. Ottawa, ON.

3. Paparrizos, J. and Gravano, L. 2015. k-shape: Efficient and accurate clustering of time series. In Proceedings of the 2015 ACM SIGMOD International Conference on Management of Data, 1855-1870.

4. Peffer, T., Pritoni, M, Meier, A., Aragon, C., and Perry, D. 2011. How occupants use thermostats in homes: A review. Building and Environment, 46(12):2529-2541.

5. Ren, X., Yan, D., and Hong, T. 2015. Data mining of space heating system performance in affordable housing. Building and Environment, 89:1-13.

6. Santin, O.G., Itard, L., and Visscher, H. 2009. The effect of occupancy and building characteristics on energy use for space and water heating in Dutch residential stock. Energy and buildings, 41(11):1223-1232.

Keywords: Thermostat setpoint schedules, Occupant preferences, Building simulation, Residential buildings

Paper ID 30703: Use of district energy modelling and stakeholder engagement in developing decarbonisation strategies

Susan Pierce* (1,2), Lorenzo De Donatis (2), Fabiano Pallonetto (3), Giovanni Tardioli (2)

1: University College Dublin; 2: Integrated Environmental Solutions Ltd.; 3: IVI Institute - Maynooth University

AIM AND APPROACH

Decarbonisation of urban areas will play a vital role in tackling climate change and meeting future carbon emission reduction targets. Densely populated districts, towns and cities are among the most challenging to develop sustainably. This study aims to model and calibrate the largest university campus in Ireland, namely the University College Dublin (UCD), including the adjacent urban community area and drive it through a full decarbonisation pathway by 2050, analysing several possible scenarios focused on the electricity and heating sectors. First, a virtual model of the whole campus, together with its electricity and heating networks, have been created and calibrated using time-series metered data and building management system (BMS) data for the whole 2019; then, future development scenarios have been explored to identify the optimum pathway for decarbonisation, addressing possible changes in the control strategies of existing plants as well as the addition of new assets and renewable energy systems.

SCIENTIFIC INNOVATION AND RELEVANCE

Merging the goals of different stakeholders, objectives and budgets is one of the main challenges to face in this process; simulation at urban scale can help overcoming this by driving decision-makers towards justified routes by leveraging the uncertainty of specific choices. The hybrid approach used for the analyses has given the work the possibility to merge real and simulated data through the interaction of various pieces of software communicating together in order to give strength and reliability to the future development scenarios, being based on a virtual baseline model which behaviour is accurately matching the one of the real campus.

PRELIMINARY RESULTS AND CONCLUSIONS

The modeling of both the thermal and electricity networks enabled the evaluation of the current energy and carbon footprint of the campus, together with the quantification of the impact of possible future developments on carbon dioxide emissions. The analysis concludes that the campus can achieve a 10% reduction in carbon dioxide emissions without the need to install new generation units. The introduction of new biomass combined heat and power unit would increase this reduction to 17% compared to the baseline. Coupling this with the installation of 3 MW of solar photovoltaic, the calculated campus potential, a total reduction of 26% can be achieved.

MAIN REFERENCES

• Capros, P., Paroussos, L., Fragkos, P., Tsani, S., Boitier, B.,Wagner, F., Busch, S., Resch, G., Blesl, M. and Bollen, J. (2014), `Description of models and scenarios used to assess European decarbonisation pathways', Energy Strategy Reviews 2(3-4), 220-230.

• Horan, W., Shawe, R., Moles, R. and O'Regan, B. (2019), `Development and evaluation of a method to estimate the potential of decarbonisation technologies deployment at higher education campuses', Sustainable Cities and Society 47(January), 101464.

• Kowalski, K., Stagl, S., Madlener, R. and Omann, I. (2009), `Sustainable energy futures: Methodological challenges in combining scenarios and participatory multi-criteria analysis', European Journal of Operational Research 197(3), 1063{1074.

• Oh, Y. K., Hwang, K. R., Kim, C., Kim, J. R. and Lee, J. S. (2018), `Recent developments and key barriers to advanced biofuels: A short review', Bioresource Technology 257(February), 320-333.

Keywords: Decarbonisation, Community, Sustainability, Renewables

Paper ID 31057: Investigating the correlation dynamism between WiFi connection counts and camera-based occupancy counts

Nastaran Alishahi, Mohamed M. Ouf*, Mazdak Nik-Bakht

Concordia University

AIM AND APPROACH

Despite the large variations in building occupancy patterns, operation of the heating, ventilation, and air conditioning (HVAC) systems typically assumes a full or near full occupancy. Systems and control sequences that provide demand-controlled ventilation (DCV) aim to address this issue by adjusting building operations based on the real-time occupancy estimates. However, these systems typically rely on measured data such as CO2 sensors as a proxy for occupancy, which presents several limitations, namely significant cost, maintenance requirements, and inherent inaccuracies. Therefore, obtaining accurate real-time building occupancy information using non-intrusive and low-cost approaches offers great potentials to optimize the operation of HVAC systems.

This study aims to investigate the evaluation of building occupancy through the analysis of WiFi counts as a proxy and validate this approach using camera-based image recognition counters. Specific objectives of this study include (i) validating the relationship between these two data streams to establish a correlation between WiFi counts and actual building occupancy; and (ii) developing predictive models to predict daily occupancy patterns including peak occupancy, arrival and departure times, from the WiFi traffic data. Several clustering and regression methods were applied to WiFi and camera-based counts from a library building in Montreal, to achieve these objectives.

SCIENTIFIC INNOVATION AND RELEVANCE

Although WiFi counts have been previously proposed and tested as a proxy for real-time building occupancy, validation with ground truth actual occupant counts has been fairly limited. Since obtaining ground truth occupancy count data is challenging, many studies typically relied on short-term manual counting that can be costly, intermittent, inaccurate, and time-consuming. To this end, this study relied on using camera-based occupancy counters to establish a continuous data stream for validating the correlation between WiFi counts and actual building occupancy. Two different types of counters were used to evaluate the actual occupancy, i.e. optical and thermal sensors. In this regard, after testing the accuracy of the counters through manual counting at random points over time, we considered the readings from these counters as the ground reference values. We employed various statistical analysis methods to identify correlations between the WiFi traffic and actual occupancy counts.

Once this relationship between WiFi and occupant counts was established, WiFi count data can be collected non-intrusively at low cost (i) to estimate building occupancy patterns including peak occupancy as well as arrival and departure times for scheduling HVAC system operations, and (ii) to utilize real-time occupancy information to operate building systems, such as DCV, accordingly.

PRELIMINARY RESULTS AND CONCLUSIONS

Preliminary results show a strong correlation between camera-based counts and the number of WiFi connections on different days. Various linear correlation methods, as well as statistical tests, were used to investigate such correlations. K-means clustering was then employed to identify clusters of typical daily occupancy patterns based on WiFi counts. Four clusters were identified with peak occupancy ranging between 1,000 and 1,800 people, representing different occupancy patterns on Fridays, weekends, and two clusters for the other four weekdays. Seasonality effects were identified on the patterns of occupancy over the weekdays/weekends, which were contributed to the fact that the library belongs to an educational organization. Multiple linear regression models developed for each cluster to predict peak occupancy, earliest arrival, and latest departure times, reached an R2 ranging between 80%-90% in different clusters.

MAIN REFERENCES

Ashouri, A., Newsham, G. R., Shi, Z., & Gunay, H. B. (2019). Day-ahead Prediction of Building Occupancy using WiFi Signals. 2019 IEEE 15th International Conference on Automation Science and Engineering (CASE), 1237–1242. https://doi.org/10.1109/COASE.2019.8843224

Hobson, B. W., Gunay, H. B., Ashouri, A., & Newsham, G. R. (2020). Clustering and motif identification for occupancy-centric control of an air handling unit. Energy and Buildings, 223, 110179. https://doi.org/10.1016/j.enbuild.2020.110179

Hobson, B. W., Lowcay, D., Gunay, H. B., Ashouri, A., & Newsham, G. R. (2019). Opportunistic occupancy-count estimation using sensor fusion: A case study. Building and Environment, 159, 106154. https://doi.org/10.1016/j.buildenv.2019.05.032

Ouf, M. M., Issa, M. H., Azzouz, A., & Sadick, A.-M. (2017). Effectiveness of using WiFi technologies to detect and predict building occupancy. Sustainable Buildings, 2, 7. https://doi.org/10.1051/sbuild/2017005

Wang, Y., & Shao, L. (2018). Understanding occupancy and user behaviour through Wi-Fi-based indoor positioning. Building Research & Information, 46(7), 725–737. https://doi.org/10.1080/09613218.2018.1378498

Wang, Z., Hong, T., Piette, M. A., & Pritoni, M. (2019). Inferring occupant counts from Wi-Fi data in buildings through machine learning. Building and Environment, 158, 281–294. https://doi.org/10.1016/j.buildenv.2019.05.015

Keywords: Occupancy estimation, Occupant count, Wi-Fi data, HVAC

Session W3.9 (Online Track): Buildings paving the way for the energy transition

Paper ID 30141: Modelica-json: Transforming energy models to digitize the control delivery process

Michael Wetter* (1), Jianjun Hu (1), Anand Prakash (1), Paul Ehrlich (2), Gabe Fierro (3), Milica Grahovac (1), Marco Pritoni (1), Lisa Rivalin (4), Dave Robin (5)

1: Berkeley Lab, Berkeley, CA; 2: Building Intelligence Group, Portland, OR; 3: University of California at Berkeley, Berkeley, CA; 4: Facebook, Menlo Park, CA; 5: BSC Softworks, Atlanta, GA

AIM AND APPROACH

The Modelica Buildings library is used for composing HVAC and control models in the Spawn of EnergyPlus and the OpenBuildingControl project. These models contain a substantial amount of information about HVAC systems and control sequences.

This paper introduces a software that parses Modelica-based HVAC and control models to generate documents that aid in building procurement, construction, commissioning and operation. Generated documents include:

- A list of control sensors, actuators and points for use by control providers to bid on a project.

- A JSON representation of the control sequence for use as input to a process that translates the control sequences to an implementation specific to the selected control product line.

- An English language description of the control sequence for the control provider and the building operator.

- A semantic model, using the Brick data model, that aids in deploying data analysis applications and configuring building diagnostics and FDD tools.

This work, through the use of CDL, will complement ASHRAE's Standards for Communication (Standard 135, BACnet) and Semantic Interoperability (Standard 223P) with a new Standard for expressing the control logic. It will also provide a path to reuse design data to create models using ASHRAE Standard 223P (http://www.bacnet.org/WG/SI/index.html).

SCIENTIFIC INNOVATION AND RELEVANCE

The scientific innovation lies in the combination of a declarative model of the HVAC system (Modelica) and the use of control sequences expressed in the Control Description Language CDL [Wetter et al., 2018] to aid the digitization of the building control delivery process, and to enable the evaluation and optimization of control sequences.

Current practice involves the HVAC designer writing a verbose description of the control sequences, which a project technician interprets to write code for deployment of the sequence in a proprietary control system. This is followed by a manual process to validate and confirm the operation without the quality control procedures commonly used in modern software development. As a result, the performance of these sequences is highly dependent on programmer skills, programmer understanding of the building systems, proper commissioning and timely update in case of operational changes [Pritoni et al, 2020]. The presented process can allow manufacturers to program and test sequences centrally and distribute them to installers [Paliaga et al. 2020], who are only required to take simple configuration steps.

This presentation will focus on the tools that extract control sequences and their semantic information for integration in such a digitized control delivery process.

PRELIMINARY RESULTS AND CONCLUSIONS

A CDL-to-JSON translator and a process which produces English language documentation (HTML or Word) from a JSON representation have been developed. A prototype translator from CDL to the commercial control product line EIKON of Automated Logic Control has also been developed. A translator from Modelica and its CDL subset to Brick [Balaji et al, 2016], a semantic model that is used as input to the development of ASHRAE Standard 223P,is in development, and preliminary results are expected at the time when the full paper is due.

Advantages of this approach include:

- making energy efficient sequences available to any building, independently from the skill of the contractor

- reducing human errors and commissioning time

- ability to quickly scale advanced sequences to a large number of buildings overcoming constraints related to workforce training

- reduced cost and for the building owner

MAIN REFERENCES

Michael Wetter, Milica Grahovac and Jianjun Hu. Control Description Language. 1st American Modelica Conference, Cambridge, MA, USA, August 2018.

Bharathan Balaji, Arka Bhattacharya, Gabe Fierro, Jingkun Gao, Joshua Gluck, Dezhi Hong, Aslak Johansen, Jason Koh, Yuvraj Agarwal, Mario Berges, David E. Culler, Rajesh Gupta, Mikkel Baun Kjaergaard, Joern Ploennigs, Kamin Whitehouse.

Brick v1.0 - Towards a Unified Metadata Schema for Buildings.

3rd ACM International Conference on Systems for Energy-Efficient Built Environments (BuildSys), November 2016.

Marco Pritoni, Anand Prakash, David Blum, Kun Zhang, Hwakong Cheng, Gwelen Paliaga, Rui Tang, Jessica Granderson. Advanced control sequences and FDD technology. Just shiny objects, or ready for scale? ACEEE summer study proceeding. 2020. (Accepted)

Paliaga G., Singla R., Snaith C., Lipp S., Mangalekar D., Cheng H. Pritoni M. Re-Envisioning RCx: Achieving Max Potential HVAC Controls Retrofits through Modernized BAS Hardware and Software. ACEEE summer study proceeding. 2020. (Accepted)

Keywords: Controls, Modelica, semantic models

Paper ID 30643: Effect of selected passive cooling measures on overheating in residential buildings in Canada

Michal Bartko*, Abdelaziz Laouadi, Michael Lacasse

National Research Council, Canada

AIM AND APPROACH

Due to the climate change effects, summer outdoor temperatures are rising with increasing frequencies. Such climatic changes may result in overheating of buildings' interior spaces, thus causing thermal discomfort and heat-related issues to building occupants such as heat exhaustion, dehydration, heat strokes and even death.

A methodology to evaluate overheating in buildings was developed and overheating was analysed using building simulation tool. EnergyPlus® simulation software with its group of Air Flow Network object was used for this purpose. Archetype building models of several types of Canadian residential buildings were developed, including:

- Midrise Multi Unit Residential Building

- Single family detached house

- Row (town) house

This paper reports on overheating in single detached houses. Simulations were carried out for Ottawa, Ontario, Canada. Weather files for the historical period and seven scenarios of future projections of climate were considered for the simulation to analyse overheating

SCIENTIFIC INNOVATION AND RELEVANCE

The overheating effects were investigated for residential buildings and thermal comfort was evaluated using the index of transient standard effective temperature. Several passive cooling measures were evaluated including:

Building envelope thermal mass (heat capacity)

- light- outside vinyl cladding on wood stud wall

- medium- outside brick veneer on wood stud wall

Construction sets with effective R-values

- old (1980s) R15

- retrofit (2015) R24

- current (2015) R18

- future- net zero R27

Shading

- Exterior rollers/ shutters

- Interior blinds

- Reflective interior blinds

Cooling

- Fully air conditioned

- Partially air conditioned with relaxed cooling set-point

Ventilation

- Mechanical, night ventilation

- Natural, open windows when indoor temperature is higher than the outside temperature and the set-point temperature of 26°C

- Mixed ventilation, natural and mechanical.

PRELIMINARY RESULTS AND CONCLUSIONS

The results showed that during outdoor extreme heat events, the room temperatures exceed the comfortable levels for extended time periods, thus creating heat stress for occupants. The effectiveness of several passive cooling strategies, such as shading, and ventilation, and trade-off between energy efficiency and building thermal resiliency were evaluated.

The effect of thermal insulation (R value) seems very significant with more insulated buildings more prone to overheating. Increased ventilation proved very efficient in decreasing the indoor air temperature and shortening duration of exposure to high temperatures considerably. This includes both, the night ventilation by opening windows when outdoor conditions allow, as well as mechanical ventilation. Shading by exterior blinds and electrochromic and thermochromic windows decreased the cooling loads by 75, 50 and 40% respectively compared to the commonly used interior blinds.

The modelling results using future climate with anticipated temperature increase of 3.5°C (intense, year 2081) by year 2100 show increase of the overall cooling loads by 30 to 40%.

MAIN REFERENCES

ASHRAE 55, (2017) Thermal Environmental Conditions for Human Occupancy

CMHC* (1984), Canadian wood-frame house construction

CMHC* (2013), Canadian wood-frame house construction

Laouadi, A. et al. (2018), Climate resilience of buildings, overheating in buildings- literature review

Laouadi, A. et. al. (2020), Development of reference summer weather years for analysis of overheating risk in buildings. Journal of Building Performance Simulation

NRCan, (2011) Survey of Households Energy Use

Parekh, A. (2005), Development of archetypes of building characteristics libraries for simplified energy use evaluation of houses

*CMHC- Canadian Mortgage and Housing Corporation

Keywords: Residential buildings, climate change, heat waves, overheating, natural ventilation

Paper ID 30693: District heating: a practical solution for reducing fossil fuel dependency in Quebec's remote communities

Annie Pike*, Michael Kummert

Polytechnique Montreal, Canada

AIM AND APPROACH

For over 100 years, district heating (DH) has been a reliable and economical alternative to individual heating technologies. More recently, the ability to harness alternative forms of thermal energy has made DH a prominent solution for the decarbonization of energy systems. These systems enhance synergies between electrical and thermal energy generation through waste heat recovery and conversion of excess renewable electricity. The aim of the present study is to characterize the potential of DH in the context of Quebec's remote communities, which currently rely largely on fossil fuels to power and heat their buildings.

This paper first details the procedure used to develop a set of building archetypes appropriate to the region to accurately simulate the hourly heating demand of Whapmagoostui-Kuujjuarapik, the community selected for this study. Next, a series of energy models are developed to represent varying levels of renewable energy integration, with and without DH. These models are used to assess the supplementary fossil fuel reductions enabled by coupling the thermal and electrical grids via a DH network. Simulation results are presented in terms of the financial investment required to achieve a GHG emissions reduction target to highlight the point at which DH becomes economically feasible.

SCIENTIFIC INNOVATION AND RELEVANCE

Reducing fossil fuel dependency in remote communities is an important challenge in Canada and has been the focus of many studies in recent years. However, most of these studies focus entirely on power generation without consideration of the fossil fuel use associated with heating. Where heating solutions are considered, they are typically limited to individual residential systems. The present study demonstrates the strength of community-scale thermal energy infrastructure as a low-carbon alternative. An additional contribution of this paper is the methodology used to quantify the community heating demand. Buildings in remote communities, specifically residential units, have energy-use characteristics that can deviate significantly from "typical" buildings. Energy simulation results from existing building archetypes for cold-climate regions [1] are compared to archetypes created based on building envelope and occupancy-related parameters specific to Whapmagoostui-Kuujjuarapik.

PRELIMINARY RESULTS AND CONCLUSIONS

A set of six archetypes covering 85% of the residential building stock in Whapmagoostui-Kuujjuarapik have been developed and reveal that existing archetypes significantly overestimate the residential heating demand. Initial simulations of the DH model have shown that generator waste heat recovery without any renewable energy source could eliminate over 15% of the combined building energy GHG emissions in Whapmagoostui-Kuujjuarapik. Further simulations that combine DH with wind energy generation indicate that DH is a more efficient solution than battery energy storage for increasing renewable energy penetration, and thereby reducing GHG emissions. The final paper will include a discussion of the economic parameters associated with these findings and present the levelized cost of energy in each scenario for increasing emissions reduction targets.

MAIN REFERENCES

[1] "Residential Prototype Building Models Building Energy Codes Program." [Online]. Available: https://www.energycodes.gov/development/residential/iecc_models. [Accessed: 13-Jul-2020].

[2] E. Wilson, C. E. Metzger, S. Horowitz, and R. Hendron, "2014 Building America House Simulation Protocols," 2014.

[3] L. Guo et al., "Optimal design of battery energy storage system for a wind-diesel off-grid power system in a remote Canadian community," IET Gener. Transm. Distrib., vol. 10, no. 3, pp. 608–616, 2016.

[4] Hydro-Québec Distribution, "Suivi sur les causes de la consommation en 2e tranche d'énergie au tarif DN," 2019.

[5] A. Dalla Rosa, R. Boulter, K. Church, and S. Svendsen, "District heating (DH) network design and operation toward a system-wide methodology for optimizing renewable energy solutions (SMORES) in Canada: A case study," Energy, vol. 45, no. 1, pp. 960–974, Sep. 2012.

[6] U. Persson and S. Werner, "Heat distribution and the future competitiveness of district heating," Appl. Energy, vol. 88, no. 3, pp. 568–576, Mar. 2011.

[7] H. Lund, B. Möller, B. V. Mathiesen, and A. Dyrelund, "The role of district heating in future renewable energy systems," Energy, vol. 35, no. 3, pp. 1381–1390, 2010.

Keywords: District Heating, Renewable Energy, Waste Heat Recovery, TRNSYS

Paper ID 30704: Quantifying the effect of multiple demand response actions on electricity demand and building services via surrogate modeling

Na Luo, Jared Langevin, Handi Chandra Putra*

Lawrence Berkeley National Laboratory, United States of America

AIM AND APPROACH

Demand response (DR) and dynamic grid integration of building loads is playing an increasingly important role in ensuring grid reliability and resilience in the face of day-to-day stresses and emergency events [1]. Identifying the potential changes in building demand and indoor services given certain DR strategies may facilitate the real-time decision-making process for building operators who wish to participate in DR programs [2].

In this study, we explore a surrogate modeling approach to predict changes in core building electricity demand and services in office and retail buildings under a given set of outdoor environmental conditions and dynamic DR actions. The surrogate modeling approach consists of the following steps: 1) define assumptions for different DR scenarios and develop DR measures in OpenStudio/EnergyPlus [3]; 2) simulate DR measures across all climate zones, building types and vintages of interest; 3) compile results into synthetic database covering various outcomes; 4) develop simpler regression models of building services and electricity demand using the synthetic database; 5) quantify the effect of multiple DR scenarios on building services and electricity demand. To maximize the flexibility of the developed regression models to future updating in real field settings, each is implemented in a Bayesian inference framework [4].

SCIENTIFIC INNOVATION AND RELEVANCE

A comprehensive set of DR scenarios are designed and simulated in this work across influential office and retail building loads to maximize the relevance of developed models to high-impact DR decision-making in buildings. Specifically, the generated synthetic database covers a wide variety of conditions as listed below, yielding a set of predictive models that can be broadly applied in guiding flexible building operations.

- Six types of DR strategies are simulated - 1) global temperature adjustment (GTA), 2) GTA+pre-cooling, 3) outdoor air (OA) reduction, 4) lighting dimming, 5) plug load reduction (limited to office buildings), and 6) packages of #1-5. For each DR measure, we defined three levels of adjustments - low, moderate and high; and two levels of DR event duration - shorter (e.g., 4 hours) and longer (e.g., 8 hours).

- Two prototypical commercial building types are selected - 1) medium-size office building and 2) standalone retail building - to investigate the effects of DR on building services and electricity demand under different operation modes.

- The simulations are also conducted across 4 prototypical vintages - pre-1980, 1980-2004, 2004 and 2010, as well as 13 climate zones, referring to the ASHRAE climate zone map to capture the potential impact of outdoor environments and local building design codes.

PRELIMINARY RESULTS AND CONCLUSIONS

The surrogate models are able to effectively predict changes in whole building demand as well as indoor temperature, relative humidity, illuminance and CO_2 concentration under the different DR scenarios suggested above, across building types and vintages. Models are assessed using several metrics that cover their accuracy and robustness, including R^2, Absolute Relative Error (ARE), Mean Absolute Deviation Percentage (MADP), and Variance inflation factor (VIF). Moreover, by implementing the surrogate models in a Bayesian framework, we are able to quantify the uncertainty in our models' predictions in a straightforward manner, which is critical for communicating the risk of unacceptable service level changes from DR actions to building operators.

Our models yield important insights on the likely impacts of various DR strategies on building demand, for example: 1) GTA with precooling exerts the most obvious effect on demand shedding among five individual DR strategies, while OA reduction has the least impact; 2) the costs of pre-cooling in older buildings is notably greater than in newer buildings - particularly for the office; and 3) the potential demand shedding from DR packages is less than the direct summation of the savings from each measure, suggesting meaningful interactions across the operation of these measures when packaged.

MAIN REFERENCES

[1] Motegi, N., Piette, M. A., Watson, D. S., Kiliccote, S., & Xu, P. (2007). Introduction to Commercial Building Control Strategies and Techniques for Demand Response.

[2] Yin, R., Kara, E. C., Li, Y., DeForest, N., Wang, K., Yong, T., & Stadler, M. (2016). Quantifying flexibility of commercial and residential loads for demand response using setpoint changes. Applied Energy, 177(September), 149–164. https://doi.org/10.1016/j.apenergy.2016.05.090.

[3] Crawley, D. B., Lawrie, L. K., Pedersen, C. O., & Winkelmann, F. C. (2000). Energy plus: energy simulation program. ASHRAE Journal, 42(4), 49–56.

[4] PyMC Development Team. (2018). PyMC3 Documentation. Available online: https://docs.pymc.io/.

Keywords: Building demand flexibility, Synthetic database, Surrogate modeling, Building demand and services, Decision making tool

Paper ID 30835: Optimizing price-informed operation of a battery storage system in an office building

Xuechen Lei*, Ellen Franconi, Yunyang Ye

Pacific Northwest National Laboratory, United States of America

AIM AND APPROACH

Many of tomorrow's building technologies go beyond efficiency considerations and target increased load flexibility. Such demand flexibility measures (DFMs) and behind-the-meter distributed energy resources (DERs) can postpone or reduce electric load based on a price or other grid signal, which supports achieving grid-interactive efficient buildings (GEB) and energy resilience. However, considering DFMs and DERs in energy policy creates a new set of analysis challenges that needs to be addressed. For example, new prescriptive efficiency measures amended to current US model energy codes historically have been evaluated using an average, blended electricity rate, which accounts for demand charges. Post 2019, new code proposals can be evaluated using a representative time-of-use (TOU) rate. While this is a step in the right direction, more sophisticated analysis methods are needed to consider a variety of TOU rates and make impact assessments that account for price-informed control. To address these needs, this study couples an office prototype building simulation model with varying-in-sophistication battery storage operating strategies for different electricity TOU rates. The analysis compares the energy savings impact of operating a battery storage system following simpler rule-of-thumb methods versus a semi-optimized priced-informed operation.

SCIENTIFIC INNOVATION AND RELEVANCE

The effective evaluation of cost effectiveness for capital-intensive demand flexibility measures, such as energy storage technologies, requires a customized approach that accounts for variations in TOU tariffs that occur across utilities and regions. While advanced optimization methods may be warranted when assessing multiple interactive systems and dynamic pricing, energy policy studies do not usually require the same level of detail. Instead, they need to make performance assessments across multiple building types and climate zones, while maintaining enough customization to address variations in the most influential parameters. This paper investigates such considerations for battery storage to develop a practical and efficient analytical approach to be applied in energy code development. The research explores using a heuristic method to develop a semi-optimized, customized strategy that is practical to integrate into whole-building simulation analysis to assess national savings potential. Its heuristic-based battery operation strategy also has the potential for implementation in a building energy management and control system. To verify the effectiveness of the proposed semi-optimized approach, we compare it to two simpler rule-of-thumb battery operation strategies. We conduct the analysis for two demand-based TOU rates applied to a large and medium office building in three locations representing hot-humid, mixed-humid, and cool-humid conditions.

PRELIMINARY RESULTS AND CONCLUSIONS

Implementing the simpler rule-of-thumb methods instead of a semi-optimized approach reduces annual electricity cost savings and, therefore, the cost effectiveness of the battery system investment. For the ASHRAE TOU rate, the results show that a more optimized approach provides additional savings to be ~ $5 - $8 / kWh battery capacity. For the ConEd TOU rate, the additional savings are ~ $5 - $10 / kWh. Perhaps of equal importance though, is the impact that a simplified approach has on battery cycling and its life. For example, Li-Ion battery system cost is estimated at $360/kWh capacity with the battery storage cost comprising $190/kWh of the total (Mongrid et al 2019). Levelizing capacity cost over a 10-year life indicates a $19/kWh year cost. Comparing this normalized first cost to the incremental savings of the semi-optimized solution, as well as the additional benefit realized from improving the battery life by 40% to 70%, indicates a substantial added benefit for utilizing a semi-optimized, customized TOU approach for evaluation battery storage impact.

MAIN REFERENCES

• Franconi, E, J Lerond, C Nambiar, M Rosenberg, R Hart, and D Kim. 2020. Realizing Demand Flexibility with Commercial Building Energy Codes. PNNL-29604. Pacific Northwest National Laboratory. Richland, WA. (publication pending)

• VanCutsem, O, M Kayal, D Blum, and M Pritoni. 2019. "Comparisons of MPC Formulations for Building Control under Commercial Time-of-Use Tariffs." 2019 IEEE PowerTech Milan.

• Panagiota, G, K Foteinski, A Heller, and C Rode. 2017. "Intelligent Scheduling of a Grid-Connected heat Pump in a Danish Detached House." Proceedings from the 15th IBPSA Conference, San Francisco California. August 7-9, 2017.

• Luo, N, T Hong, H Li, R Jia, and W Weng. 2017. Data Analytics and Optimization of an Ice-Based Energy Storage System for Commercial Buildings. Lawrence Berkeley National Laboratory. Berkeley, CA.

• Lauro, F, F Moretti, A Capozzoli, and S Panzieri. 2015. "Model Predictive Control for Building Active Demand Response Systems." The 7th International Conference on Sustainability in Energy and Buildings. Madeira, Portugal. May 16 – 19, 2017.

Keywords: Priced-based control, energy storage, grid-interactive buildings, codes and standards

Session T1.1: Practice and industry related case studies

Paper ID 30218: Assessing building envelope thermal performance using in-situ measurements

Jade Deltour* (1), Karel De Sloover (1), Sebastien Pecceu (1), Nicolas Heijmans (1), Geert Bauwens (2)

1: BBRI, Belgium; 2: KULeuven, Belgium

AIM AND APPROACH

There are two ways to evaluate a building's energy performance: by calculation or by benchmarking its energy consumption. The calculations have the advantage of being independent of the actual climate and user's preferences, but have the disadvantages to not take into account (not at all or not enough) the quality of the works and aspects not covered by the calculations procedures. It is just the opposite for the benchmarking of consumptions : they include the quality of the works but are not climate and user independent, so cannot be qualified of "building" performance.

Nowadays, in-between solutions are being developed, to take the advantages of both approaches without having their inconvenient. Some focus on the building envelope with a test called "co-heating", that takes as long as 15 days, which is too long. Other use very long term monitoring of consumptions and indoor conditions, and to extract a climate and user independent energy performance among this big amount of data. In both cases, the development of those new techniques requires the intensive use of building simulations.

Building simulation have been used to develop a short (≈ 5 days) dynamic co-heating test to measure the heat loss coefficient (HLC) of a building.

SCIENTIFIC INNOVATION AND RELEVANCE

To select the most reliable protocol, we analysed existing and new protocols based on virtual datasets generated with a simple building simulation software (CAPSOL).

A protocol is a combination of a data collection process and a data analysis process. In the "regular co-heating protocol" the data is collected over a span of 2 weeks, while the building is thermostatically heated to approximately 25 °C (at least 10 °C higher compared with the outdoor temperature). Then, data is analysed through linear regression.

In order to reduce the measurement duration, there are two possible options. First, test conditions can be varied by periodically toggling the heating system on and off. Secondly, more advanced data analysis methods can be applied, using autoregressive with exogenous input (ARX) models or state space models (RC network) solved using differential equations.

The flexibility offered by building simulations allowed the evaluation of different measurement conditions (duration, heating excitation and building typology) combined with various analysis methods. More than 100 protocols and over 2.000 virtual data sets were generated. These measurement protocols (virtually generated) were then evaluated based on accuracy and reproducibility of results.

PRELIMINARY RESULTS AND CONCLUSIONS

This work would not have been possible without the use of virtual data. Simulations allowed us to test a lot of measurements protocols on different typologies, while also giving us a first idea of the reproducibility of these measurements.

The combination of a heating excitation according to a PRBS signal using a state space model for the data analysis seems to be the most stable tested protocol and allows the test duration to be reduced to 5 days (rather than 15 days for a regular co-heating test). Simulations highlighted the interest to limit the impact of the sun during measurements. Moreover, the building typology clearly impacts the results: it is more difficult to obtain an accurate and reproducible result as the insulation level of the building increases. Depending on the typology, the reproducibility varies between +/- 10% to +/- 20%. Finally, the selected protocol was applied on 6 new houses, the results of the experimental campaign were consistent with the values predicted by calculation. However, for these 6 cases, the calculated HLC under evaluated the measured HLC by approximately 14%.

MAIN REFERENCES

[1] CEN-TC89, N2044, Draft Standard from WG13 TG 5,Thermal performance of buildings – In-situ testing of completed buildings - Part 1: Data collection for whole building aggregate heat loss test

CEN-TC89, N2046, Draft Standard from WG13 TG 5, Thermal performance of buildings – In-situ testing of completed buildings - Part 2: Steady-state data analysis for whole building aggregate heat loss test

[2] J. Wingfield, D. Johnston, D. Miles-Shenton, M. Bell, Leeds Metropolitan University, Whole House Heat Loss Test Method (Coheating), (2010)

[3] G. Bauwens, S. Roels, PhD Theses, In Situ Testing of a Building's Overall Heat Loss Coefficient - Embedding Quasi-stationary and Dynamic Tests in a Building Physical and Statistical Framework, (2015)

[4] International Energy Agency, EBC Annex 58, Report of Subtask 3, part 2 :Thermal performance characterization using time series data – statistical guidelines,

[5] S. Thébault, R. Bouchié, En. & Build., Refinement of the ISABELE method regarding uncertainty quantification and thermal dynamics modelling, 178, 182–205, (2018)

[6] E.Mangematin, G. Pandraud, D. Roux, C. R. Physique, Quick measurements of energy efficiency of buildings , Volume 13, Issue 4, 383-390 (2012)

[7] K. R. Godfrey. Correlation methods. Paper, International Federation of Automatic Control, London, 1980

Keywords: virtual dataset, co-heating, state space model, measurement, identification

Paper ID 31056: Simulating gentle failure as an approach to building resiliency

Ibone Santiago Trojaola* (1), Susan Ubbelohde (1,2), George Loisos (1), Nathan Brown (1), Eduardo Pintos (1), Santosh Philip (1)

1: Loisos + Ubbelohde, Alameda CA USA; 2: University of California, Berkeley CA USA

AIM AND APPROACH

Increased frequency of events such as mudslides and wildfires, as well as the potential for earthquakes in California, have pushed projects to consider the building's resilience during an extended power outage and challenging thermal conditions with climate change. In addition to efforts to move annual performance towards zero energy, teams now look at strategies that maximize passive survivability for an indefinite period of time.

This paper describes the case study of an art-center/residence in California where the project recognized these environmental challenges and extended goals to address potential disasters and climate change scenarios, with time becoming an element in energy performance in response to these events.

Simulation of energy use and thermal comfort with a focus on passive survivability helped defining the project targets, developing an implementation plan and sizing the systems. Modeling results enabled stakeholders to make informed decisions and define expectations for a range of scenarios.

This case study includes an energy generation system consisting of several photovoltaic arrays and an energy storage battery system. In addition to installing these systems, the building was designed to enable the owners to automatically implement progressive resiliency stages managing power consumption and effectively reducing energy use as needed during outages.

SCIENTIFIC INNOVATION AND RELEVANCE

This paper describes a "gentle failure" approach to resilience in the context of multiple potential immediate disasters and long-term weather change. It reframes the idea of future proofing buildings by focusing on adaptability and diversity of strategies rather than oversizing systems or hardening the envelope to respond to more extreme conditions.

Every year the environment shows that resilience is more and more critical for every stage of design. Policies are taking the role of addressing this issue but there are still not many examples of how this research applies to and is integrated into practice. Literature on resiliency typically addresses urban-level planning but not how simulation supports the design process or the methodology of actually implementing potential strategies in practice. Similarly, publications about how to obtain data representing climate change scenarios to be used in building simulations are available but do not describe how to think through this range of possibilities and deal with results to design both the building and the systems in a non-normative manner.

This case study builds on previous research for essential services buildings by detailing the implementation of the resiliency plan with construction for the building scheduled to finish in July 2021.

PRELIMINARY RESULTS AND CONCLUSIONS

Resilience was achieved by including appropriately sized energy production and storage systems but also developing an implementation plan to allow the building to "sail itself" through power outages. Space-by space calculations with specific load and usage profiles based on occupant's preferences were used to develop energy budgets and design the electrical and mechanical systems for three resiliency stages (normal, reduced and minimal operation).

The envelope design was informed by building performance analysis results both in terms of thermal autonomy and energy use reduction. Using multiple iterations of an energy plus model, a series of design alternatives were tested to understand how the building performs with parametric variations of insulation in the roof and wall assemblies, thermal mass and glazing specifications.

Detailed, multivariable performance data over multi-day periods representing extreme weather events to describe potential future scenarios occurring more frequently due to climate change. The building's flexibility and adaptation surpasses the needs for zero energy performance and allows the project to be operated after being disconnected from the grid, relying on battery storage and PV panels as sources of energy. A performance monitoring system provides measured data of the actual use profile to revisit assumptions if needed.

MAIN REFERENCES

1-Roostaie S. et al (2019),' Sustainability and resilience: A review of definitions, relationships, and their integration into a combined building assessment framework' , Building and Environment, Volume 154, May 2019, Pages 132-144

2-Rajkovich N. et al (2019), 'Climate Change Resilience Strategies for the Building Sector: Examining Existing Domains of Resilience Utilized by Design Professionals', Article in Sustainability 2019, 11, 2888; doi:10.3390/su11102888

3-Dickinson, R. and Brannon, R. (2016) 'Generating Future Weather Files for Resilience', PLEA 2016: 36th International Conference on Passive and Low Energy Architecture.

4-Trogal, K. et al, Architecture and Resilience: Interdisciplinary Dialogues (2018) Taylor & Francis Ltd.

5-Sillmann, J. et al. (2017) 'Understanding, modeling and predicting weather and climate extremes: Challenges and opportunities', Weather and Climate Extremes.

Elsevier Ltd, 18(November), pp. 65–74. doi: 10.1016/j.wace.2017.10.003.

6-Santiago Trojaola, I. et al. (2017) ' Simulation in Support of Zero Net Energy and Resilient Design', Proceedings of the 15th IBPSA Conference San Francisco, pp 2333-2341.

7-Brian, R. (2014) "The Principles of Future-Proofing: A Broader Understanding of Resiliency in the Historic Built Environment." Journal of Preservation Education and Research, vol. 7 (2014): 31–49.

Keywords: Battery storage, PV system sizing, Zero Net Energy, Resiliency

Paper ID 30260: Energetic self-sufficiency of a greenhouse residence: a dynamic techno-financial feasibility study

Charlotte Verhaeghe* (1), Mateusz Bobier (1), Rik Berens (1), Amaryllis Audenaert (1), Stijn Verbeke (1,2)

1: University of Antwerp, Belgium; 2: VITO, unit Smart Energy and Built Environment, Belgium

AIM AND APPROACH

The dependence of high energy performance buildings on external energy supply options can further be reduced by increasing the building's energetic self-sufficiency (ESS), up to the level of complete energy autonomy. A techno-financial analysis is carried out for different stages of ESS, up to the 100% ESS-level, for a greenhouse residence in Belgium. The energy demand, retrieved by means of dynamic whole building simulation, is to a large extent covered by on-site renewable energy sources (RES), combined with a certain capacity of electrical energy storage as a buffer, both sized using external scripts. The financial feasibility of the ESS-variants is evaluated for a variety of both static and dynamic pricing mechanisms under three economic cornerstone scenarios. The results indicate that increased (annual) ESS requires highly oversized on-site RES and energy storage options during most part of the year, thereby making the investment financially unattractive. The hypothesis that the techno-financial feasibility of various ESS-levels is highly dependent on the considered boundary conditions (e.g. end-use, time, spatial -and economic boundaries) is furthermore confirmed.

SCIENTIFIC INNOVATION AND RELEVANCE

While there is a large amount of literature on technical and financial feasibility of (nearly or net) Zero Energy Buildings (ZEBs), there is less research on the comparison of dynamic technical and financial feasibility of the different stages of self-sufficiency towards the off-grid level. In this study dynamic whole building simulation tool EnergyPlus is coupled to external scripts for sizing energy storage technologies in relation to different levels of self-sufficiency. The dynamic- and in the future more realistic and relevant- pricing mechanisms are a direct result of the growing stress on the energy grid, which is increasingly characterized by intermittent renewable energy production. Therefore, it is relevant to analyse the effect of various energy pricing mechanism on the financial feasibility of the building project.

PRELIMINARY RESULTS AND CONCLUSIONS

The hypothesis that the techno-financial feasibility of various ESS-levels is highly dependent on the considered boundary conditions (e.g. end-use, time, spatial -and economic boundaries) is confirmed.

Furthermore, the dynamics of energy demand and generation profiles and energy pricing scenarios have an important effect on both the technical and financial analysis. Simple rules of thumb are not likely to be suitable in early design decisions, which indicates the need for (co-)simulations instead.

The results indicate that it is essential to allow for energy storage to both serve the building and the energy grid. Programmable energy management is furthermore a promising solution to increase financial attractiveness of increased ESS.

In addition, it was found that a zonal model to determine the effects of the greenhouse on the energy demand of the building is more accurate than a single zone model in EnergyPlus.

MAIN REFERENCES

[1] Marszal AJ, Heiselberg P, Bourrelle JS, Musall E, Voss K, Sartori I, et al. Zero Energy Building - A review of definitions and calculation methodologies. Energy Build 2011;43:971–9. https://doi.org/10.1016/j.enbuild.2010.12.022.

[2] Vale B, Vale R. The new autonomous house: Design and planning for sustainability. Thames & H. London: 2000.

[3] Torcellini P, Pless S, Deru M, Crawley D. Zero Energy Buildings: A Critical Look at the Definition. ACEEE Summer Study Pacific Grove, 2006, p. 15. https://doi.org/10.1016/S1471-0846(02)80045-2.

[4] Kolokotsa D, Rovas D, Kosmatopoulos E, Kalaitzakis K. A roadmap towards intelligent net zero- and positive-energy buildings. Sol Energy 2011;85:3067–84. https://doi.org/10.1016/j.solener.2010.09.001.

[5] Sun Y, Ma R, Chen J, Xu T. Heuristic optimization for grid-interactive net-zero energy building design through the glowworm swarm algorithm. Energy Build 2020;208:109644. https://doi.org/10.1016/j.enbuild.2019.109644.

[6] Georgakarakos AD, Mayfield M, Hathway EA. Battery storage systems in smart grid optimised buildings. Energy Procedia, vol. 151, Elsevier Ltd; 2018, p. 23–30. https://doi.org/10.1016/j.egypro.2018.09.022.

[7] Treado S, Delgoshaei P, Windham A. An Evaluation of Simulation-based approaches for building control system design and integration. Trans Control Mech Syst 2014;3:2544–51.

Keywords: Autonomous buildings, grid-interactive buildings, energy storage, dynamic technical and financial feasibility evaluation

Paper ID 30806: Energy prediction under changed demand conditions: robust machine learning models and input feature combinations

Thomas Schranz* (1), Johannes Exenberger (1), Christian Møldrup Legaard (2), Ján Drgoňa (3), Gerald Schweiger (1)

1: Graz University of Technology, Austria; 2: Aarhus University, Denmark; 3: Pacific Northwest National Laboratory, USA

AIM AND APPROACH

Deciding on a suitable algorithm for energy demand prediction in a building is non-trivial and depends on the availability of data. In this paper we compare four machine learning models, commonly found in the literature, in terms of their generalization performance and in terms of how using different sets of input features affects accuracy. This is tested on a data set where consumption patterns differ significantly between training and evaluation because of the Covid-19 pandemic. We provide a hands-on guide and supply a Python framework for building operators to adapt and use in their applications.

In this paper we investigate approaches to predict hourly electric energy usage of a five-floor, mixed-use academic building at the Graz University of Technology (TUG) accommodating offices, seminar rooms, laboratories and a lecture hall. We compare the performance of models using combinations of previously observed hourly energy consumption, occupancy data (approximated through registrations in the University's resource management system), weather data, features engineered from date and time (time of day, weekday, holiday) and hourly water consumption as input features.

SCIENTIFIC INNOVATION AND RELEVANCE

With this paper, we contribute to the state of the art in building energy forecasting by assessing the performance and the robustness of four machine learning algorithms (linear regression, random forest, fully-connected neural network, and recurrent neural network) with various sets of input features.

We analyze models in terms of their ability to predict short-term energy demand in a building where the consumption patterns differ significantly between training and test data because of the Covid-19 pandemic.

We provide guidelines for practitioners by

* examining how different lookback and prediction horizons influence the accuracy and robustness of the machine learning models for single-step energy demand prediction

* benchmarking models using additional input features, such as weather data, against models predicting future energy demand from past consumption values only.

* examining the potential of integrating water consumption data for data-driven energy prediction.

As a side-contribution, the Python machine learning framework, as well as the data used in the experiments described here is published on Github. This allows researchers and practitioners to reproduce the results presented in this paper and to adapt the framework for their purposes and applications.

PRELIMINARY RESULTS AND CONCLUSIONS

Results show that using features engineered from date and time affects prediction performance most significantly, regardless of the choice of model and lookback horizon. Additionally, we found that simple models, such as linear regression and random forests perform very well both in terms of generalization ability and robustness with respect to the choice of input features and lookback horizon. Especially the random forests showed exceptional generalization performance for all choices of input features.

Conversely, the neural networks performed well when predicting from previous consumption values alone, but were sensible to the choice of inputs. It stands to reason that this issue could be addressed with regularization techniques, such as dropout layers, L1 or L2 regularization. Besides, we found that the choice of input features and the choice of lookback horizons for the neural networks interacted with each other. Consequently, we did not find neural networks models to adequately fulfill the requirement of working well without extensive testing and tweaking.

MAIN REFERENCES

Bontempi, G., S. Ben Taieb, and Y.-A. Le Borgne(2013). Machine Learning Strategies for Time Se-ries Forecasting.

European Commission and Climate Action DG(2019).Going climate-neutral by 2050: a strategiclong-term vision for a prosperous, modern, com-petitive and climate-neutral EU economy.

Sun, Y., F. Haghighat, and B. C. M. Fung (2020).A review of the-state-of-the-art in data-driven ap-proaches for building energy prediction.

Rätz, M., A. P. Javadi, M. Baranski, K. Finkbeiner,and D. M ̈uller (2019).Automated data-drivenmodeling of building energy systems via machinelearning algorithms.

Keywords: Machine Learning, Energy Prediction, Model Robustness, Deep Learning

Paper ID 31147: Quantification of the cost of oversizing cooling installations in a case study under construction in Vietnam

Pedro Marques*, Mathias Vandecasteele, Kien Le Trung, Wim Boydens

Studiebureau Boydens, Belgium

AIM AND APPROACH

In countries where energy regulation is rather lax and young, the HVAC design team might lack the incentive to do a careful sizing of the heating and cooling loads in a building. Instead, predefined values in watts per square meter, generous “safety factors” or a combination of both are used.

Oversizing of HVAC installations occurs more often than one would expect - upon site visits in Vietnam, installations with several chillers are seen running one chiller only. In such cases, oversizing did not necessarily translate into malfunctioning on site, for the building occupants felt no issues – no difference in comfort between an oversized installation and a rightly sized one.

However, such design practice will unquestionably end up in a higher capital investment, lower plant efficiency, higher energy and maintenance costs when compared to a more carefully designed installation.

This document compares side-to-side the final figures of a as-built HVAC design with a dynamic cooling load simulation and quantifies the hidden cost of the oversizing for the target building.

SCIENTIFIC INNOVATION AND RELEVANCE

The prevention of oversizing is a fundamental step for more sustainable buildings, as the impact on energy is two-fold: reducing the installation and running costs of cooling plants allows for the difference to be shifted toward more sustainable solutions.

Besides the fundamental role of regulation to enforce compliance with industry-acknowledged norms, dynamic simulations can play a fundamental role as the time and funds invested in knowledge, licenses and experience can largely be offset by the savings in the capital investment on each project.

PRELIMINARY RESULTS AND CONCLUSIONS

In this case study, the estimation of 1.6 million USD, or 40 usd/m^2 is undervalued, as other cost affecting factors such as net floor area available for rent, cost of oversized technical spaces and ceiling clearance, are not quantified.

The oversizing is a problem that is difficult to identify, and even harder to quantify, but very real. The difficulty is that there might be no signs of discontent:

• The HVAC designer delivered a system that provides air conditioning as requested, and "better installing too much than not enough".

• The building manager has no major problems running the system. If one chiller goes out of service, plenty of backup is available.

• The design team technical knowledge is insufficient to assess the issue, either because the team lacks experience, or because past experience comes from analogous designs.

• No regulation is enforcing the reporting and benchmarking of the efficiencies of the cooling plant, or the general building consumption.

MAIN REFERENCES

nothing to declare

Keywords: oversizing, dynamic simulations, cooling installations

Session T1.2: Ensuring high quality building simulations

Paper ID 30230: Spatial distribution of thermal comfort: a case study in Paris' station

Edouard Walther* (1), Mateusz Bogdan (1), Fanny Peyre (2), Christian Inard (2)

1: AREP L'hypercube, France; 2: LASIE Université de La Rochelle, France

AIM AND APPROACH

In Building Energy Simulation (BES), the environmental variables for thermal comfort such as air velocity, radiant temperature, dry bulb temperature and relative humidity are nodal, meaning a single scalar value is computed for each thermal zone, regardless of their size. This long-used approach is valid for most of the built environment, however it may prove to be inappropriate in specific cases, such as widely open and/or glazed buildings. Radiation and convection being major contributors to the heat balance of the human body (Pickup and De Dear, 2000), the existence of indoor solar fluxes and of significant variations in the air velocity field cannot be overlooked for a proper evaluation of thermal comfort.

Commercial tools lack the spatial distribution of radiant temperature and air velocity fields for the computation of comfort levels in semi-open places. The aim of the present work is hence to provide a detailed distribution of thermal comfort based on an "enriched" nodal approach, with hourly solar distribution and air velocity fields. A proof of concept is proposed under the form of a case study, using the state-of-the-art, open source softwares: energyplus, coupled with Radiance and openFOAM.

SCIENTIFIC INNOVATION AND RELEVANCE

The method presented in (Walther et al. 2019) is chosen for the computation of natural ventilation, using CFD to obtain the detailed pressure boundary conditions. As a by-product, this approach yields a simulation set-up allowing to compute indoor velocity fields within open buildings with a moderate additional effort.

The indoor isothermal velocity field is obtained by transposing the approach of (Delpech et al. 2005) to the indoor environment: the distribution of air velocities is solved with an isothermal k-epsilon model for 12 incident wind directions, with their median meteorological velocity as inlet. Hourly velocities are reconstructed by field interpolation, selecting the closest precomputed fields. The amplitude is determined by means of scaling by the hourly weather data (Delpech et al. 2005).

The long wave contribution to radiant temperature is obtained directly from the nodal value of the BES. The shortwave contribution is calculated the daylight coefficients method, also known as "the Two-Phase" method (Subramaniam 2017).

The PET index (Walther & Goestschel, 2018) is chosen as comfort metric. In order to preserve both temporal and spatial variation of comfort, the Outdoor Thermal Comfort Autonomy (Nazarian et al. 2019) was used for the qualification of comfortable spaces.

PRELIMINARY RESULTS AND CONCLUSIONS

The presented approach allows for the computation of a spatial distribution of comfort indexes in indoor spaces. It provides a better understanding of the causes of local thermal (dis-)comfort, as each of the underlying phenomena (e.g. solar radiation, wind distribution) has been calculated.

The study of the comfort patterns obtained is interesting for architects as it becomes possible to differentiate "sub-zones" of occupation within a given volume and provides information at a higher level of detail than previously available to space designers.

Interestingly, the computation of the OTCA using different thermal comfort indexes yields rather different results.

As a perspective, one could argue that zonal models are a cheaper alternative to computational fluid dynamics. The coupling of such models with energyplus is however numerically challenging. The actual longwave heat transfer may significantly impact thermal comfort depending on the indoor environment. The detailed radiation computation with view factors on idealised individuals is currently explored.

MAIN REFERENCES

Delpech P., et al. 2005. Pedestrian wind comfort assessment criteria: A comparative Study. EACWE4. Prague: J. Naprestek & C. Fischer.

Nazarian, N., Acero, J. A., & Norford, L. (2019). Outdoor thermal comfort autonomy: Performance metrics for climate-conscious urban design. Building and Environment, 155, 145-160.

Pickup, Janelle, and Richard de Dear. "An outdoor thermal comfort index (OUT_SET*)-part I-the model and its assumptions." Biometeorology and urban climatology at the turn of the millenium. Selected papers from the Conference ICB-ICUC. Vol. 99. 2000.

Subramaniam S. (2017) Daylighting simulations with Radiance using matrix-besed methods, Berkeley Lab, (https://www.radiance-online.org/learning/tutorials/matrix-based-methods)

Edouard Walther, Antoine Hubert, Alexis Sauvageon, Mateusz Bogdan, Improving the Prediction of Natural Ventilation by Coupling CFD and BES : A Case Study in Strasbourg Train Station, IBPSA 2019.

Walther E, Goestchel Q. The PET comfort index: Questioning the model. Building and Environment. 2018 Jun 1;137:1-0.

Keywords: Indoor thermal comfort, Spatial distribution, OTCA

Paper ID 30535: Spectral and biological simulation methods for the design of healthy circadian lighting

J. Alstan Jakubiec*, Athina Alight

University of Toronto, Canada

AIM AND APPROACH

Medical and architectural research has identified that light has significant impacts on alertness, sleep quality, and productivity [1-4]. This has broadly been referred to as circadian lighting, and its effects are dependent on the time of exposure and are sensitive to blue, short wavelength light. Existing calculation methods and standards assess the impact of light on health outcomes [5-7]; however, they have a variety of weaknesses: lacking spectral information, estimating scaling factors to account for spectral power distributions, or relying on instantaneous calculations although light timing and exposure history are crucial in assessing the effects of light on health. No existing methodology calculates spectrally- and temporally-varying light exposure and maps its impact on biological and health-related outcomes. This paper will communicate such a simulation method using time-varying 81-channel spectrally specific lighting simulations [8] and apply the results to existing dynamic photobiology models from medical literature [9]. We predict reaction time, sleepiness, task performance, sleep time and melatonin levels in the bloodstream entirely as a factor of timed light exposure history. This paper shares how these calculation models are implemented, demonstrates them under some week-long scenarios, and proposes new metrics and communication methods to assess the impact of light on health.

SCIENTIFIC INNOVATION AND RELEVANCE

We will present a new scheme for calculating spectral daylight simulations on an annual basis combined with electric lighting which can be dimmed, controlled, and color-shifted on a schedule. Innovations over existing spectral calculation methods include: a physics-based simulation process for the estimation of spatially resolved sky color and spectrum, the inclusion of spectral window and opaque surface materials at a high resolution of 81-channels (5 nm steps) [8], and the calculation of modern non-image forming light quantities following CIE standard S 026 [10]. Furthermore, we present an implementation of a dynamic model of human arousal dynamics [9] which can predict the impact of light on sleep time, subjective sleepiness, task performance, reaction time, and blood plasma melatonin levels. The model is responsive to the time, color, and quantity of light exposure in addition to light history. Two cases are presented using seven days of light history for a modelled occupant. These new calculation methods and metrics will illustrate the actual benefits of lighting design for circadian health, which has not been accomplished using simulation before.

PRELIMINARY RESULTS AND CONCLUSIONS

A new simulation method for time-varying lighting spectra from daylight and controlled electric lighting is presented as well as its application to modelling human circadian biology. Detailed information will be presented on how light interacts with homeostatic and circadian bodily rhythms to impact health and alertness. We calculate the impact of light on sleep time, subjective sleepiness, task performance, reaction time, and blood plasma melatonin levels and present novel methods of evaluating these measures in architectural lighting design. By applying these new models, we present the means to answer longstanding questions regarding the specific health impacts of light exposure in the built environment. Two sample light exposure profiles are calculated over a period of 1 week to illustrate these effects.

In addition to this outcome, a second paper is being prepared to apply our method to a variety of parametric design variants and compare the outcomes to existing methods of evaluating circadian lighting.

MAIN REFERENCES

1. Rea, M. and M. Figueiro, Light as a circadian stimulus for architectural lighting. Lighting Research & Technology, 2018. 50(4): p. 497-510.

2. Figueiro, M.G., K. Gonzales, and D. Pedler, Designing with circadian stimulus. LD+A: The Magazine of the Illuminating Engineering Society, 2016. 8: p. 30-34.

3. Dijk, D.-J. and S.N. Archer, Light, sleep, and circadian rhythms: together again. PLoS Biol, 2009. 7(6): p. e1000145.

4. Lucas, R.J., et al., Measuring and using light in the melanopsin age. Trends in Neurosciences, 2014. 37(1): p. 1-9.

5. Konis, K., A novel circadian daylight metric for building design and evaluation. Building and Environment, 2017. 113: p. 22-38.

6. Inanici, M., M. Brennan, and E. Clark, Spectral daylighting simulations: Computing circadian light, in Building Simulation. 2015: Hyderabad, India.

7. IWBI, The WELL Building Standard. Washington, DC, Delos Living LLC, 2016.

8. Solemma LLC, ALFA – Adaptive Lighting for Alertness. 2020.

9. Postnova, S., S.W. Lockley, and P.A. Robinson, Prediction of cognitive performance and subjective sleepiness using a model of arousal dynamics. Journal of biological rhythms, 2018. 33(2): p. 203-218.

10. CIE, CIE S 026/E:2018 CIE System for Metrology of Optical Radiation for ipRGC-Influenced Responses to Light. 2018, Commission internationale de l'eclairage.

Keywords: lighting, circadian lighting, light and health

Paper ID 30698: An optimization of the switching temperature for an energy system with a double-source heat pump

Sara Bordignon*, Giuseppe Emmi, Marco Rossi, Michele De Carli, Angelo Zarrella

Department of Industrial Engineering - Applied Physics Section, University of Padova, Italy

AIM AND APPROACH

Enhancing the efficiency of the energy systems for the air-conditioning and domestic hot water production in buildings and reducing the related primary energy consumption through the exploitation of renewable energy sources have become of great interest in the late years.

This paper presents a dynamic analysis of a double-source energy system, aiming at optimally combine the air and the soil as heat source/sink for a heat pump.

In the field of multi-source energy systems, the dual-source heat pump represents a valuable solution to increase the system's efficiency, compared to a more common air-source heat pump and, at the same time, to reduce the size of the borehole heat exchangers field that should be coupled to a ground source heat pump [1]. These solutions can, therefore, decrease the initial investment and, depending on the plant configurations and on the boundary conditions (i.e. climatic area) [2], an optimisation of the energy system should be carried out to avoid unnecessary oversizing of the components and to further increase the efficiency of the heat pump [3].

SCIENTIFIC INNOVATION AND RELEVANCE

The dynamic energy model is developed in TRNSYS [4] environment, where a novel Type for the modelling of the heat pump is adopted. The purpose of the analysis is to investigate the performances of the system varying the switching temperature, that is the parameter which establishes the heat source to be used by the heat pump. The heat pump Type is a flexible model and can be used as a double-source heat pump. The outputs are based on the polynomials of the compressors, which can be modified by the user, while the fluid properties in the different points of the thermodynamic cycles can be obtained thanks to an internal link to REFPROP [5]. The heat pump is connected to two thermal energy storages from which heat is withdrawn for the air-conditioning and domestic hot water production respectively. Through this approach, it is possible to investigate and predict in detail the thermal and electrical behaviour of double source heat pump systems, which can be used in the context of both retrofitted and new buildings, under different weather conditions.

PRELIMINARY RESULTS AND CONCLUSIONS

The study consisted in the dynamic energy modelling of a double source system where a heat pump exploits both the air and the soil as heat source/sink, for heating, cooling and domestic hot water production of a building. It allows testing a novel TRNSYS Type for the heat pump, which can be used for different simulation applications. The analysis identifies an optimal switching temperature for the investigated case study, which allows increasing the efficiency of the system, reducing the size of the borehole heat exchangers field and decreasing the thermal drift of the ground temperature, due to unbalanced operating condition. The simulations are carried out for different climatic conditions, to make it possible a comparison of the optimal switching temperature for the different locations.

MAIN REFERENCES

[1] I. Grossi, M. Dongellini, A. Piazzi, and G. L. Morini, "Dynamic modelling and energy performance analysis of an innovative dual-source heat pump system," Appl. Therm. Eng., vol. 142, pp. 745–759, Sep. 2018.

[2] T. Catalina, J. Virgone, and E. Blanco, "Multi-source energy systems analysis using a multi-criteria decision aid methodology," Renew. Energy, vol. 36, no. 8, pp. 2245–2252, 2011.

[3] G. Hou, H. Taherian, L. Li, J. Fuse, and L. Moradi, "System performance analysis of a hybrid ground source heat pump with optimal control strategies based on numerical simulations," Geothermics, vol. 86, p. 101849, Jul. 2020.

[4] University of Wisconsin--Madison. Solar Energy Laboratory. "TRNSYS, a Transient Simulation Program." Madison, Wis. :The Laboratory, 1975.

[5] M. Lemmon, E.W., Bell, I.H., Huber, M.L., "M.O. NIST Standard Reference Database 23: Reference Fluid Thermodynamic and Transport Properties-REFPROP, Version 10.0." National Institute of Standards and Technology, Standard Reference Data Program, Gaithersburg, 2018.

Keywords: GSHP, TRNSYS, double-source heat pump, optimization, renewable energy

Paper ID 30165: Wind-driven air flow modelling in Modelica: verification and implementation in the IDEAS library

Klaas De Jonge* (1,3), Filip Jorissen (2), Lieve Helsen (2,4), Jelle Laverge (1)

1: Ghent University, Department of Architecture and Urban Planning , Belgium; 2: University of Leuven (KU Leuven), Department of Mechanical Engineering, Belgium; 3: FWO, Research Foundation Flanders; 4: EnergyVille, Thor Park, Waterschei, Belgium

AIM AND APPROACH

In building simulations, pressure driven airflow and heat flow modelling is often done in separate specialised programs. To include the mutual dependence of indoor temperatures and pressure driven airflow, simulations often require co-simulation using two distinct programs, where CONTAM is often used to compute the system flow rates and pressures. This requires the modeller to be skilled in multiple software packages, and co-simulation. Moreover, the resulting co-simulation framework is hard to extend and customize and an inefficient coupling can cause model inaccuracies or long computation times.

This paper presents a new implementation for pressure driven airflow in the open-source Modelica library IDEAS. It allows the modeller to model the heat transfer and interzonal, pressure driven airflow simultaneously, thereby using one software. This new implementation is validated dynamically by comparing the obtained results with a duplicate of the multi-zone model in the specialised software CONTAM.

SCIENTIFIC INNOVATION AND RELEVANCE

The new implementation will allow modellers to model the mutual dependence of indoor temperatures and pressure driven airflow simultaneously and thus faster. The default implementation does not require extensive additional input which will make it easier for modellers, not familiar with pressure driven airflow, to include this factor. Due to the extensible nature of Modelica, the models can be coupled to detailed HVAC simulations and other system models.

Towards an energy neutral building stock, understanding the different mechanisms at play is key. Combined heating/cooling and ventilation simulations are important to gain a better understanding of the interaction between these depending factors and their impact on energy use and indoor air quality, respectively.

PRELIMINARY RESULTS AND CONCLUSIONS

Results show almost perfect agreement between Modelica IDEAS and CONTAM when zone temperatures are comparable. This shows that the pressure driven airflow implementation is state-of-the-art and may be used as an alternative to co-simulations with CONTAM.

It is important to note that the default integration in IDEAS is limited to predefined airflow elements and relations. This can be a limiting factor for modelling specific airflow paths through a building (e.g. elevator shafts). However, the default implementation can be extended with custom airflow elements to model the additional airflow paths.

MAIN REFERENCES

[1] F. Jorissen, G. Reynders, R. Baetens, D. Picard, D. Saelens, and L. Helsen, "Implementation and verification of the IDEAS building energy simulation library," Journal of Building Performance Simulation, vol. 11, no. 6, pp. 669–688, Nov. 2018, doi: 10.1080/19401493.2018.1428361.

[2] W. S. Dols and B. J. Polidoro, "CONTAM User Guide and Program Documentation Version 3.2," National Institute of Standards and Technology, NIST TN 1887, Sep. 2015. doi: 10.6028/NIST.TN.1887.

[3] M. Wetter, "Multizone Airflow Model in Modelica," presented at the Modelica 2006, Vienna, Austria, Sep. 2006.

Keywords: Modelica IDEAS, CONTAM, Pressure driven airflow, Indoor Air Quality, Indoor temperatures

Paper ID 30840: Evaluation of energy and indoor environmental performance of mechanical ventilation systems under different climatic conditions

Dragos-Ioan Bogatu* (1), Ongun Berk Kazanci (1), Charalampos Angelopoulos (2), Daniel Coakley (2), Bjarne W. Olesen (1)

1: International Centre for Indoor Environment and Energy, Department of Civil Engineering, Technical University of Denmark, Nils Koppels Allé, Building 402, 2800 Kgs. Lyngby, Denmark; 2: Mitsubishi Electric R&D Centre Europe BV, 17 Firth Road (AC3), Houstoun Industrial Estate, Livingston, West Lothian EH54 5DJ

AIM AND APPROACH

One of the most significant global development challenges for the built environment is the mitigation of the impact of energy-intensive mechanical heating, ventilation and cooling systems (HVACS) in buildings. Growing population around the world and the fact that people tend to spend almost 90% of their time indoors [1], makes indoor environmental quality an important parameter to account for in new and existing buildings. Nowadays, more people rely on mechanical systems to maintain acceptable levels of Indoor Environmental Quality (IEQ). Mixed-mode buildings combine the use of mechanical and passive techniques to maintain thermally comfortable indoor environments. However, there are no clear guidelines on how to operate such mechanical ventilation systems in order to maintain acceptable levels of IEQ with the minimum possible energy consumption. The aim of this research was to perform dynamic thermal modelling simulations to evaluate the energy and indoor environmental performance of mechanical ventilation systems under different climatic conditions and building types. To examine this, the following objective was used: to identify the optimum operation mode of the mechanical system (bypass, heat recovery or enthalpy recovery) to achieve the energy savings whilst maintaining acceptable levels of IEQ.

SCIENTIFIC INNOVATION AND RELEVANCE

The evaluation of the performance of mechanical systems is often compared to the level of energy use in a building. For the energy use there is one KPI, kWh/m2 per year primary energy, which includes all buildings systems and energy carrier. However, this approach excludes the analysis of parameters for indoor environmental quality such as relative humidity, CO2 concentration, and operative temperature. This paper uses dynamic computer simulations for a residential and an office application to examine the performance of the mechanical ventilation systems under different outdoor conditions and under different indoor heating and cooling loads. The following cases will be analyzed for the climatic conditions of Edinburgh, Scotland; Copenhagen, Denmark; Zurich, Switzerland; and Palermo, Italy: i) Operate ventilation systems in heat recovery mode; ii) Operate ventilation systems in enthalpy recovery mode; and iii) Operate ventilation systems in bypass mode and increased ventilation for cooling (day- and nighttime). In all cases the following KPIs will be used: i) Yearly Room temperature and humidity distribution; ii) Yearly CO2 concentration distribution as an indicator of IEQ; and iii) Energy use of the mechanical systems in the different cases.

PRELIMINARY RESULTS AND CONCLUSIONS

It is expected that the enthalpy recovery mode will have the most effect in very humid and very dry climates. It is predicted that the combination of the passive modes that achieves higher energy savings will depend on the climate, period of the year and building type. It is expected that the enthalpy recovery mode will reach higher energy savings than the heat recovery mode since it allows extracting latent heat from humid air in addition to the recovery of sensible heat. Therefore, it is likely that the outdoor relative humidity and air temperature will influence the amount of energy savings obtained from the enthalpy recovery mode, compared to the heat recovery mode. The bypass mode will probably lead to higher energy savings in periods and locations where the outdoor temperature is lower than the indoor air temperature and there is a need for cooling (e.g. during nighttime in a hot summer period). Finally, it is expected that all the passive strategies analyzed in this study will reach acceptable CO2 concentrations during occupied hours.

MAIN REFERENCES

[1] N. E. Klepeis, "The National Human Activity Pattern Survey (NHAPS) - A Resource for Assessing Exposure to Environmental Pollutants," Lawrence Berkeley National Laboratory, 2001.

Keywords: Indoor Environmental Quality (IEQ), Mechanical Ventilation, Building Simulation

Session T1.3: Buildings paving the way for the energy transition

Paper ID 30266: PCM-enhanced building envelope for improved thermal comfort and energy efficiency in danish buildings

Morten Hagenau, Muhyiddine Jradi*

University of Southern Denmark, Denmark

AIM AND APPROACH

The building sector accounts for 40% of the Danish energy consumption profile with an equivalent share in the emission of greenhouse gasses [1]. With this large contribution, increasing the efficiency of this sector is key for reaching future energy goals. Considering the Danish building regulations [2], it can be noted that the requirements for new buildings have increased drastically, both on the level of the building envelope and the quality and specifications of energy systems. Alongside active energy improvements dealing with energy supply systems [3], passive techniques targeting the building envelope are demonstrated as viable options to improve the building performance [4]. In this study, a holistic investigation of integrating phase change materials within the building envelope is presented aiming to improve indoor thermal comfort and reduce energy consumption. A dynamic energy modeling and performance evaluation of building envelope enhanced with phase change materials under Danish conditions is carried out. Three case study real buildings are considered, where corresponding holistic dynamic energy models are developed and calibrated using actual collected data. Using the calibrated energy models, the impact of a PCM-enhanced building envelope in the three cases is simulated and evaluated on both thermal comfort and energy consumption.

SCIENTIFIC INNOVATION AND RELEVANCE

A large block of studies has investigated PCM applications in buildings from different perspectives, divided as active and passive applications [5]. Active PCM applications are PCM modules integrated within the building HVAC system as an active energy storage component. On the other hand, passive applications take the form of PCM-enhanced building envelope, where the PCM is embedded within the envelope constructions as an interdependent layer or mixed with another material. Nevertheless, the PCM potential in storing latent heat has proven to be one of the drivers of net zero energy building design [6]. The recent investigations have clearly shown the large potential of PCM-enhanced building envelope applications in reducing the overall cooling and heating load and hence the corresponding greenhouse gas emissions. As the danish building regulation is being progressively tightened concerning the building envelope and the overall energy performance of buildings, PCM integration in the building component provide a potential alternative to enhance the overall thermal envelope and reduce heating and cooling loads. In this context, the impact of using PCMs within building envelopes on the heating and cooling energy consumption of Danish buildings has not been extensively studied, in particular through using dynamic energy modeling and simulations.

PRELIMINARY RESULTS AND CONCLUSIONS

The evaluation of the impacts of implementing PCMs within the building envelope is carried out by employing dynamic energy models. As the PCM charging and discharging is a dynamic process affected by multiple factors, the use of such models is key towards a more accurate prediction of the PCM behavior. Three case study buildings are considered, a university office building, a daycare center and a residential single-family house. Using EnergyPlus, dynamic energy models are developed and calibrated using actual data. Then, the calibrated models were employed to simulate and evaluate the impact of a PCM-enhanced building envelope on both thermal comfort and energy consumption. The results demonstrate a reduction on energy consumption ranging from 5 to 33%, with a reduction in the peak summer temperature of up to 5°C. The results of this work will serve as a basis for passive PCM applications in Danish buildings and will aid the use of PCM-enhanced envelope in the design of future energy efficient and Danish net zero energy buildings. This is supported by the reported considerable energy consumption reduction, cooling, and heating loads reduction as well as the decrease in the average indoor temperature in the summer in the three buildings investigated.

MAIN REFERENCES

[1] Jradi, M., Veje, C., Jørgensen, B.N. (2017): Deep energy renovation of the Mærsk office building in denmark using a holistic design approach. Energy and Buildings 151, 306-319.

[2] Bygningsreglementet, The Danish Building Regulations (2020). https://bygningsreglementet.dk/

[3] Derradji, L., Errebai, F.B., M. Amara, M. (2017): Effect of PCM in improving the thermal comfort in buildings. Energy Procedia 107, 157-161.

[4] Rathore, P.K.S., Shukla, S.K. (2019): Potential of macroencapsulated PCM for thermal energy storage in buildings: a comprehensive review. Construction and Building Materials 225, 723-744.

[5] Iten, M., Liu, S., Shukla, A. (2016): A review on the air-PCM-TES application for free cooling and heating in the buildings. Renewable and Sustainable Energy Reviews 61, 175-186.

[6] Kenzhekhanov, S., Memon, S.A., Adilkhanova I. (2020): Quantitative evaluation of thermal performance and energy saving potential of the building integrated with PCM in a subarctic climate. Energy 192, 116607.

Keywords: Phase change material, PCM, dynamic modeling and simulation, EnergyPlus, thermal comfort

Paper ID 30740: Comparison of data-driven model-based and model-free control approaches for unlocking building energy flexibility

Anjukan Kathirgamanathan* (1,3), Eleni Mangina (2,3), Donal P. Finn (1,3)

1: School of Mechanical and Materials Engineering, University College Dublin, Ireland; 2: School of Computer Science, University College Dublin, Ireland; 3: Energy Institute, University College Dublin, Ireland

AIM AND APPROACH

Utilisation of energy flexibility in buildings, within a demand side management context, is increasingly being considered as a vital contributor to the future smart grid [1]. However, due to the heterogenous nature of the building stock, developing computationally tractable control-oriented models, which adequately represent the complex and nonlinear thermal-dynamics of individual buildings, is proving to be a major hurdle [2]. The current paper compares two predictive control techniques to harness energy flexibility in buildings; using a data-driven model-based approach and a model-free approach. The 'separation of variables' technique together with the random forest algorithm is tested as the model-based approach [3]. A 'Soft Actor Critic' reinforcement learning agent is used as the model-free approach.

A white-box EnergyPlus model of an archetype commercial building is used as the virtual testbed building for designing and evaluating the two control approaches. Energy flexibility is provided through the use of building passive thermal mass, active thermal energy storage, on-site photovoltaic generation, and electrical battery storage. Both control approaches are evaluated for ease of design, computational burden and efficacy of the control performance to utilise the energy flexibility on the basis of dynamic signals from the grid.

SCIENTIFIC INNOVATION AND RELEVANCE

Data-driven approaches show promising potential in unlocking energy flexibility in commercial buildings where the cost of deriving a first-principles physics based model is high or in some cases impracticable [4]. With commercial buildings being complex and exhibiting non-linear behaviour, building control based on physics based models has often been limited to being rudimentary and non-predictive and hence unable to fully harness the energy flexibility such buildings possess. A literature review of prior work, on the application of data-driven methods for building demand side management, has shown that two distinct data-driven control approaches exist: model-based and model-free [2]. To the best knowledge of the authors, there have been limited comparisons made of the two approaches on a benchmark case. Without such benchmarks, it is very difficult to compare the efficacy and scalability of the different techniques due to the varying buildings and dynamic boundary conditions used in individual studies [2]. The current research, which is being carried out as part of the IEA Annex 81 Data-Driven Smart Buildings [5], will aid in understanding the relative benefits and challenges of both control approaches as applied to an archetype building.

PRELIMINARY RESULTS AND CONCLUSIONS

A 12 storey, 46,000 m2 floor area, commercial office building was used as a case study. The model-based controller was developed using Python, with the Eppy package and an OPC bridge used for co-simulation together with EnergyPlus. In the model-free case, the Gym-Eplus package was used as a wrapper for the OpenAI gym environment and to interface with EnergyPlus. Preliminary work has applied the model-based approach (separation of variables using random forests) in the optimal control of the building under dynamic pricing signals and found that the controller is able to shift consumption thereby reducing the energy cost to the building. Separately, the model-free approach, using the 'Soft Actor Critic' reinforcement learning algorithm, has been applied to different buildings as part of the CityLearn competition [6]. The reinforcement learning technique was shown to be capable of smoothening the district demand profile. Ongoing work aims to apply both techniques to the same virtual testbed building and benchmark the performance of the controllers. This work is essential in aiding the widespread adoption of data-driven control approaches to the building demand side management problem.

MAIN REFERENCES

[1] K. O. Aduda, T. Labeodan, W. Zeiler, G. Boxem, and Y. Zhao, "Demand side flexibility: Potentials and building performance implications," Sustain. Cities Soc., vol. 22, pp. 146–163, 2016.

[2] A. Kathirgamanathan, M. De Rosa, E. Mangina, and D. P. Finn, "Data-driven Predictive Control for Unlocking Building Energy Flexibility: A Review," Renew. Sustain. Energy Rev., vol. In Press., 2020.

[3] A. Jain, F. Smarra, and R. Mangharam, "Data Predictive Control using Regression Trees and Ensemble Learning," 2017 56th IEEE Conf. Decis. Control, vol. 135, pp. 4446–4451, 2017.

[4] F. Smarra, A. Jain, T. De Rubeis, D. Ambrosini, A. D. Innocenzo, and R. Mangharam, "Data-driven model predictive control using random forests for building energy optimization and climate control," Appl. Energy, vol. 226, pp. 1252–1272, 2018.

[5] IEA EBC Annex 81, "Data-Driven Smart Buildings EBC ANNEX 81," 2019.

[6] J. R. Vázquez-Canteli, J. Kämpf, G. Henze, and Z. Nagy, "CityLearn v1.0: An OpenAI gym environment for demand response with deep reinforcement learning," BuildSys 2019 - Proc. 6th ACM Int. Conf. Syst. Energy-Efficient Build. Cities, Transp., pp. 356–357, 2019.

Keywords: building energy flexibility, data-driven, reinforcement learning

Paper ID 30143: Time-based economic hierarchical model predictive control of all-electric energy systems in non-residential buildings

Laura Maier*, Larissa Kühn, Philipp Mehrfeld, Dirk Müller

RWTH Aachen University, E.ON Energy Research Center, Institute for Energy Efficient Buildings and Indoor Climate

AIM AND APPROACH

Model predictive control (MPC) is a known approach to optimize building energy system (BES) operation. In this context, studies prove that the choice of control and prediction horizons are dependent on the regarded BES and its boundary conditions. In some cases, a mix of large and short prediction horizons is useful to exploit the BES´s full potential. This is especially true for seasonal storages or for economic boundary conditions, based on a long-term performance. This challenge is met by the concept of hierarchical MPC (HMPC). HMPC separates the control problem in different layers. In literature, these layers can be time-based and/or space-based. In this work, we focus on a time-based approach in order to incorporate seasonal pricing effects.

This work aims at assessing the potential of time-based HMPC based on mixed integer linear and quadratic programs (MILP and MIQP) from an eco-economic point of view. The use case is a non-residential BES basing on electricity only. The BES consists of two types of storages (battery, electric vehicles) and local electricity generation by photovoltaics (PV). We couple a MILP- and MIQP-based HMPC with a dynamic simulation using validated Modelica models to evaluate the BES´s operation under predefined eco-economic scenarios.

SCIENTIFIC INNOVATION AND RELEVANCE

In scientific literature, different MPC approaches for the optimization of BES operation have already been investigated. However, for the combination of PV, battery and electric vehicles, the main focus lies on residential buildings. Here, we focus on non-residential buildings by applying the deduced methodology to a business and innovation center in Berlin called FUBIC. FUBIC is a highly complex, interconnected BES. The simultaneous occurrence of heating and cooling demands as well as the integration of components whose efficiency depends on the boundary condition represent an additional challenge for the HMPC.

We contribute to research in this field by optimizing FUBIC´s operation with regard to defined eco-economic scenarios. We evaluate the suitability of a MILP and MIQP approach for time-based HMPC. Furthermore, we compare the HMPC to a conventional rule-based control concept and to a 1-layer MPC approach. Thus, we can assess the potential of a time-based HMPC concept. Based on this, we deduce factors, which support the use of time-based HMPC structures.

PRELIMINARY RESULTS AND CONCLUSIONS

Preliminary results indicate that the HMPC leads to lower operating costs and CO2 emissions in comparison to a 1-layer MPC approach and the conventional rule-based control.

However, we detect a gap in the large and short prediction horizon optimization. The large prediction horizon optimization differs greatly from the short one in some periods. This is caused by model simplifications in the upper layer on the one hand and the constraints, which are communicated between the layers, on the other hand. Consequently, for the proper implementation of a time-based HMPC, it is important to assess how and which constraints are communicated between the layers.

Apart from that, the economic boundary conditions and the storage size influence the value added by a time-based HMPC structure the most. Moreover, it becomes obvious that the 1-layer approach also results in lower operating costs and CO2 emissions compared to the rule-based control concept but the whole potential of MPC control is not exploited.

MAIN REFERENCES

S. Kim, J. Kim, K. Cho and G. Byeon, "Optimal Operation Control for Multiple BESSs of a Large-Scale Customer Under Time-Based Pricing," in IEEE Transactions on Power Systems, vol. 33, no. 1, pp. 803-816, Jan. 2018, doi: 10.1109/TPWRS.2017.2696571.

R. Kumar, M. J. Wenzel, M. J. Ellis, M. N. ElBsat, K. H. Drees and V. M. Zavala, "Handling Long Horizons in MPC: A Stochastic Programming Approach," 2018 Annual American Control Conference (ACC), Milwaukee, WI, 2018, pp. 715-720, doi: 10.23919/ACC.2018.8430780.

M. E. Raoufat, B. Asghari and R. Sharma, "Model predictive BESS control for demand charge management and PV-utilization improvement," 2018 IEEE Power & Energy Society Innovative Smart Grid Technologies Conference (ISGT), Washington, DC, 2018, pp. 1-5, doi: 10.1109/ISGT.2018.8403403.

Y. Shi, B. Xu, D. Wang and B. Zhang, "Using Battery Storage for Peak Shaving and Frequency Regulation: Joint Optimization for Superlinear Gains," 2018 IEEE Power & Energy Society General Meeting (PESGM), Portland, OR, 2018, pp. 1-1, doi: 10.1109/PESGM.2018.8586227.

K. Vatanparvar and R. Sharma, "Battery Optimal Approach to Demand Charge Reduction in Behind-The-Meter Energy Management Systems," 2018 IEEE Power & Energy Society General Meeting (PESGM), Portland, OR, 2018, pp. 1-5, doi: 10.1109/PESGM.2018.8586597.

Keywords: Modelica, MILP, Hierarchical MPC, Battery Energy System Storages

Paper ID 30774: Large-scale multicriteria sorting for the integration of photovoltaic systems in the urban environment

Martin Thebault (1), Gilles Desthieux (2), Christophe Ménézo* (1)

1: University Savoie Mont Blanc, LOCIE/FRESBE, F- 74944 Annecy-le-Vieux, France; 2: Haute école du paysage d'ingénierie et d'architecture de Genève (hepia), Institute for Landscaping Architecture Construction and Territory (inPACT), University of Applied Sciences Western Switzerland, Geneva, Switzerland

AIM AND APPROACH

This work proposes a multicriteria approach in order to evaluate the suitability of a building to be equipped with Photovoltaic (PV) systems. In the present case, technical (roofs complexity), economic (payback period), environmental (CO2 reduction), energetic (self-consumption) as well as social (heritage constraints) are considered. These criteria are evaluated for each building of the Great Geneva agglomeration, then a multicriteria methods, ELECTRE TRI, allows to sort these roof in three categories that corresponds to “Very High”, “High”, “Moderate” suitability.

SCIENTIFIC INNOVATION AND RELEVANCE

Sorting processes have been widely used in cities. They take the form of GIS tools called solar cadastres (IRENA, 2019). The solar cadastre of a city is a geographical interface, which makes it possible to know how much sunlight a given roof, building, home or a part of them receives. In most of the current cadastres, roofs are sorted based on the solar irradiation they receive i.e. the energy received yearly per square meter.

However, cities are complex environments and information about the solar potential is not sufficient when assessing the suitability of a roof to host a photovoltaic systems. Other key aspects must be considered such as, for instance, the presence of superstructure elements on the roof (Desthieux 2018), structural robustness, economic feasibility (Sommerfeldt 2017) or the heritage and aesthetic character of the given building (Florio 2018). Therefore, the sortings that have been proposed are limited as they do not address the multicriteria aspect of the suitability of a roof to host PV systems.

PRELIMINARY RESULTS AND CONCLUSIONS

This paper presents a sorting methodology of the buildings according to their suitability to be equipped with PV systems. This sorting is based on a multicriteria approach performed using the ELECTRE TRI method. The different criteria considered are, the self-sufficiency, the payback periods, the roof complexity, the reduction in CO2 emissions and the heritage constraints. Three classes of suitability are defined which are 'Very high', 'High' and 'Moderate' suitability.

The method is then applied to the case of the Geneva Agglomeration (Grand Genève). This French-Swiss region is composed of nearly 260000 buildings. The different parameters of the ELECTRE TRI models are defined based on discussion with experts (thresholds, weights, limits of the classes, ...).

The methodology successfully allows to evaluate the suitability of the roofs and provide a relevant tool to help decision makers into identifying the right building for PV system installation.

MAIN REFERENCES

IRENA, 2019. Solar simulators: Application to developing cities [WWW Document]. Publ.-SimulatorsAppl.--Dev.-Cities. URL /publications/2019/Jan/Solar-simulators-Application-to-developing-cities (accessed 10.17.19).

Desthieux, G., Carneiro, C., Camponovo, R., Ineichen, P., Morello, E., Boulmier, A., Abdennadher, N., Dervey, S., Ellert, C., 2018. Solar Energy Potential Assessment on Rooftops and Facades in Large Built Environments Based on LiDAR Data, Image Processing, and Cloud Computing. Methodological Background, Application, and Validation in Geneva (Solar Cadaster). Front. Built Environ. 4. https://doi.org/10.3389/fbuil.2018.00014

Thebault, Martin, Vincent Clivillé, Lamia Berrah, et Gilles Desthieux. 2020. « Multicriteria roof sorting for the integration of photovoltaic systems in urban environments ». Sustainable Cities and Society, 102259.

Figueira, J., Greco, S., Ehrgott, M., 2005. Multiple criteria decision analysis: state of the art surveys. Springer Science & Business Media.

Keywords: multicriteria, solar, urban, pluridisciplinar

Session T1.4: Improving indoor environmental quality

Paper ID 30827: Development of a 3D and high resolution dynamic thermal model of a room with sun patch evolution for thermal comfort applications

Teddy Gresse* (1), Lucie Merlier (2), Jean-Jacques Roux (1), Frédéric Kuznik (1)

1: Univ Lyon, INSA Lyon, CNRS, CETHIL, UMR5008, 69621 Villeurbanne, France; 2: Univ Lyon, UCBL, INSA Lyon, CNRS, CETHIL, UMR5008, 69622 Villeurbanne, France

AIM AND APPROACH

Rapidly varying environmental phenomena, such as solar radiation or airflows can significantly affect thermal comfort. Hence, this contribution presents the development and validation of an efficient dynamic thermal model designed to simulate a room for thermal comfort applications, using a 3D and high-resolution description of heat conduction in the envelope and surface balances. In particular, the model can handle short-time steps and calculates the sun patch, which corresponds to the projection of solar radiation through a window onto interior walls.

The model is validated using detailed experimental data of a low energy building called BestLab [1]. Surface temperature and indoor air temperature are especially compared, and the discrepancies are quantified.

SCIENTIFIC INNOVATION AND RELEVANCE

Building thermal models generally aim to predict the energy consumption of buildings. Thus, they consider thermal loads over a long period such as a year, and neglect or simplify some features of thermal transfers. Conduction in the building envelope is typically considered 1D, short-wave radiation is projected on the floor and typical hourly weather data are used.

Yet, buildings and indoor environmental conditions are exposed to rapid environmental variations. As a response, surface temperature distribution, indoor airflows and room air temperature may substantially but locally vary, which can affect the thermal comfort of inhabitants [2] [3]. Thus, relevantly addressing thermal comfort requires higher resolution simulations able to manage locally rapid environmental variations. In particular, the use of a 3D model that includes the calculation of the sun patch and relevant convective effects can significantly improve thermal comfort prediction thanks to a better prediction of the different indoor environmental quantities [4] [5].

PRELIMINARY RESULTS AND CONCLUSIONS

The developed model is already able to simulate the three-dimensional heat conduction in a typical cubic room with a refined mesh. The sun patch detection is also successfully implemented.

Regarding the reference validation case, the complex geometry and mesh of the test case are well reproduced. To complete this validation study, current developments focus on the implementation of varying boundary conditions using weather data that vary over short time-steps.

Next step is to couple this thermal model with Computational Fluid Dynamics (CFD) to improve the description of indoor convective transfers at the walls and obtain an accurate and detailed operative temperature map for thermal comfort prediction.

MAIN REFERENCES

[1] A. Rodler, "Modélisation dynamique tridimensionnelle avec tache solaire pour la simulation du comportement thermique d'un batiment basse consommation," 2014.

[2] E. Arens, H. Zhang and C. Huizenga, "Partial- and whole-body thermal sensation and comfort—Part II: Non-uniform environmental conditions," Journal of Thermal Biology, 2006.

[3] Y. Zhang and R. Zhao, "Relationship between thermal sensation and comfort in non-uniform and dynamic environments," Building and Environment, 2009.

[4] A. Rolder, J. Virgone and J.-J. Roux, "Sun patch impact for the evaluation of operative temperatures distributions," Engineering and Architecture (SCEA), 2014.

[5] A. Rodler, J.-J. Roux, J. Virgone, K. Eui-Jong and J.-L. Hubert, "Are 3D heat tranfer formulations with short time-step and sun patch evolution necessary for building simulation?," Conference paper, 2013.

Keywords: Building Energy Model, three-dimensional heat transfers, sun patch, thermal comfort

Paper ID 30515: Impact of building thermal inertia in different climates using energy dynamic simulation through a simplified description model

Roberto Rugani* (1), Marco Bigazzi (1), Fabio Fantozzi (1), Marco Marengo (2), Marco Picco (2), Giacomo Salvadori (1)

1: University of Pisa, Pisa, Italy; 2: University of Brighton, Brighton, United Kingdom

AIM AND APPROACH

This study is part of a wider research aimed at creating building type databases for simplified dynamic thermal models of buildings. Among all the factors that influence energy consumption and comfort, the control of internal temperature is undoubtedly the crucial issue [1][2]. The environment temperature variations caused by all the heat exchanges mainly depend on several parameters related to the way the building is designed [3]: the thermal-physical properties of the building materials are surely the main parameters. For many years the quality of building envelopes has been assessed only on the reduction of thermal transmittance [4]; however, recent researches have shown that, especially in order to reduce the need for cooling energy, not only thermal transmittance but also heat capacity should be carefully considered [5]. Thermal inertia causes temporal variations in the heat transfer of external conditions inside the building, and vice versa; although standards such as EN-ISO-13786 allow its impact to be evaluated using semi-stationary methods, the only way to correctly understand its effect is to use dynamic tools. The aim of this project is to evaluate the thermal inertia of the building envelope under different weather conditions, using a simplified dynamic building energy simulation screening tool [6].

SCIENTIFIC INNOVATION AND RELEVANCE

Several building envelopes have been conceived and compared in order to analyse the impact of thermal inertia. Therefore, a database regarding envelope's constructions and transparent surfaces has been developed to be exhaustive, flexible and modular. This database has then been implemented in a simplified dynamic simulation tool that has been used to evaluate the impact of thermal inertia of different building solutions in the internal comfort of an edifice and in its energy consumption. The tool is based on the well-known building performance simulation program EnergyPlus; and the associated simplified model used has been defined in previous studies by assessing the impact of simplifications on simulation results [7].

The goal of this study is to evaluate the behaviour of different envelopes in two significantly different climate conditions representative of the European climate (London and Brindisi) and for two end uses: residential and office. In order to do so, four comparisons have been identified: envelope with similar transmittance values, but different areal heat capacity; different insulation position in heavyweight envelopes; envelope with similar periodic transmittance values, but different internal heat capacity; and the impact of using PCM boards in lightweight envelopes.

PRELIMINARY RESULTS AND CONCLUSIONS

In order to compare the results of the dynamic energy simulations, a standard building have been defined at geometric level, conceived as a parallelepiped.

Comparisons are based on both thermal comfort and consumption. In particular, in order to evaluate the occupants' internal comfort, operative temperature has been chosen as a suitable parameter (EN-ISO-15251)[8].

From the results can be concluded that thermal inertia has a great relevance, especially in warm climates where energy savings achieved up to 8%. The important role of the internal heat capacity is highlighted by the results of the second and third comparisons, in which, assuming the same periodic transmittance, the envelope with the highest heat capacity has 12% less discomfort levels and 5% less cooling need than the others. In addition, a passive strategy such as a good night-time ventilation during summer has a much more positive impact on high-inertia envelopes, with energy savings up to 17%. However, lightweight envelopes have several advantages, ensuring an adequate level of transmittance with lower thicknesses. Therefore, the analysis has been extended to PCM boards as a solution to improve the dynamic behaviour of lightweight structures, with a tested result of only 1% higher cooling need than the heavyweight envelope.

MAIN REFERENCES

1. European Parliament Directive 2010/31/EU on the energy performance of buildings. Off. J. Eur. Union 2010.

2. European Parliament Directive (EU) 2018/844 amending Directive 2010/31/EU on the energy performance of buildings and Directive 2012/27/EU on energy efficiency. Off. J. Eur. Union 2018.

3. Pacheco, R.; Ordóñez, J.; Martínez, G. Energy efficient design of building: A review. Renew. Sustain. Energy Rev. 2012, 16, 3559–3573, doi:10.1016/j.rser.2012.03.045.

4. Stazi, F. Thermal Inertia in Energy Efficient Building Envelopes; Stazi, F.B.T.-T.I. in E.E.B.E., Ed.; Butterworth-Heinemann, 2017; ISBN 978-0-12-813970-7.

5. Leccese, F.; Salvadori, G.; Asdrubali, F.; Gori, P. Passive thermal behaviour of buildings: Performance of external multi-layered walls and influence of internal walls. Appl. Energy 2018, 225, 1078–1089, doi:10.1016/j.apenergy.2018.05.090.

6. Picco, M.; Marengo, M. A Fast Response Performance Simulation Screening Tool in Support Of Early Stage Building Design. Proc. 16th IBPSA Conf. 2019, 1296–1303, doi:10.26868/25222708.2019.210252.

7. Picco, M.; Marengo, M. Energy simulation in early stage building design: Simplified models and impact on results. Build. Simul. Appl. 2015, 2015-February, 119–126.

8. EN 15251:2007 Indoor environmental input parameters for design and assessment of energy performance of buildings addressing indoor air quality, thermal environment, lighting and acoustics; 2007;

Keywords: Thermal inertia, Phase Change Material, Building envelopes, Internal comfort, Simplified simulation tool

Paper ID 30885: Bedroom environmental discomforts, occupant behaviors, and sleep quality based on an online survey

Chenxi Liao* (1,2), Jelle Laverge (1), Mizuho Akimoto (3), Chandra Sekhar (4), Pawel Wargocki (2)

1: Research group Building Physics, Construction and Climate Control, Department of Architecture and Urban Planning, Ghent University, Belgium; 2: International Centre for Indoor Environment and Energy, Department of Civil Engineering, Technical University of Denmark; 3: Department of Architecture, Waseda University; 4: School of Design and Environment, National University of Singapore

AIM AND APPROACH

The purpose of this study was to examine the association between discomfort in bedrooms caused by noise, stuffy air, "too warm", "too cool" conditions, and sleep quality, and the association between occupant behaviors and environmental discomfort in bedrooms, as well as the association between occupant behaviors and sleep quality. An online questionnaire survey was conducted in the summer to investigate the sleep quality of people living in Belgium (temperate climate). It investigated the level of discomfort, if any, and sleep quality. Last but not least, the situation of bedroom environmental discomforts can be used for defining indoor environmental quality in building simulation.

SCIENTIFIC INNOVATION AND RELEVANCE

Bedroom comfort affects sleep quality, which is vital for humans health and next-day performance (Hirshkowitz et al., 2015; Opp, 2009). An increasing number of studies showed the importance of thermal comfort (Imagawa and Rijal, 2015; Lee and Shaman, 2017; Lei et al., 2017), air quality (Laverge and Janssens, 2012; Mishra et al., 2018; Strom-Tejsen et al., 2016), and acoustic comfort (Caddick et al., 2018) for good sleep quality. However, previous studies mainly focused on the effect of only one factor of bedroom environment on sleep quality, although thermal comfort, air quality, and acoustic comfort may influence sleep quality interactively, where at least for thermal comfort and indoor air quality (Xiong et al., 2020). In this study, the frequencies of bedroom environmental discomforts of "too warm", too cool", noise and stuffy air, sleep quality (the Pittsburgh Sleep Quality Index (PSQI)), as well as occupant behaviors, were investigated, via an online questionnaire survey in Belgium (temperate climate), from July to August 2020.

PRELIMINARY RESULTS AND CONCLUSIONS

A total of 83 responses was received. The respondents were 43 males and 40 females aged 27 – 32 years. Almost half of the respondents (47.7 %) had a PSQI score greater than 5, which was indicated as poor sleepers. A total of 87.8%, 68.9%, 32.4% and 18.9% of respondents were disturbed regularly or occasionally during sleep by "too warm" conditions, noise, stuffy air and "too cool" conditions, respectively. Responses of people who were disturbed by "too warm" conditions were significantly associated with poor sleep quality; 67.4% of PSQI scores were higher compared to those who were not disturbed ($p < 0.05$). The PSQI scores increased with the increasing number of bedroom environmental discomforts, the effect being close to significance (p-trend < 0.066). It was concluded that "too warm" conditions were the major bedroom environmental discomfort in summer in Belgium. People experienced poor sleep quality if they were disturbed by more than one of the bedroom environmental discomforts. This study investigated to what extent were people disturbed by thermal discomfort, noise and stuffy air during sleep in summer. More similar studies are required to be conducted in the other three seasons or regions.

MAIN REFERENCES

Caddick, Z. A., et al., 2018. A review of the environmental parameters necessary for an optimal sleep environment. Building and Environment. 132, 11-20.

Hirshkowitz, M., et al., 2015. National Sleep Foundation's sleep time duration recommendations: methodology and results summary. Sleep Health. 1, 40-43.

Imagawa, H., Rijal, H. B., 2015. Field survey of the thermal comfort, quality of sleep and typical occupant behaviour in the bedrooms of Japanese houses during the hot and humid season. Architectural Science Review. 58, 11-23.

Laverge J., Janssens A., 2012. Analysis of the influence of ventilation rate on sleep pattern. Indoor Air Conferences. Austin, TX: ISIAQ.

Lee, W. V., Shaman, J., 2017. Heat-coping strategies and bedroom thermal satisfaction in New York City. Science of the Total Environment. 574, 1217-1231.

Lei, Z. P., et al., 2017. Effect of natural ventilation on indoor air quality and thermal comfort in dormitory during winter. Building and Environment. 125, 240-247.

Mishra, A. K., et al., 2018. Window/door opening-mediated bedroom ventilation and its impact on sleep quality of healthy, young adults. Indoor Air. 28, 339-351.

Opp, M. R., 2009. Sleeping to fuel the immune system: mammalian sleep and resistance to parasites. Bmc Evolutionary Biology. 9.

...

Keywords: Bedroom environmental discomfort, Occupant behavior, Sleep quality, Survey

Paper ID 30524: Phase-change materials selection: numerical study based on design of experiments.

Gilles Baudoin*, Geoffrey van Moeseke

Université catholique de Louvain, UCLouvain, Belgium

AIM AND APPROACH

Researchers have shown an increased interest in phase-change materials (PCM) to modify thermal mass in buildings. PCM offer an opportunity to increase the thermal mass effect on a given temperature range. A key aspect of PCM selection is the melting temperature but the optimised melting temperature may significantly vary in the literature.

The melting temperature selection could be affected by the properties of the PCM-enhanced component, by the climate and by the building under investigation. To the best of our knowledge, the influence of the building under investigation has never been systematically studied. Therefore, this study aims to quantify the influence of the building parameters on the PCM selection, and more specifically on the optimal melting temperature.

Dynamic simulations of a test-cell in a temperate climate are currently being conducted with EnergyPlus. Various combinations of the building parameters were selected based on design of experiments. For each combination, a case with and without PCM were compared to determine the potential gains of using PCM on energy needs, both for cooling ΔEcool and heating ΔEheat. A metamodel was then constructed to link the energy savings ΔEcool and ΔEheat., with the building parameters in the form of a second order polynomial function.

SCIENTIFIC INNOVATION AND RELEVANCE

To our knowledge, it is the first time that optimal melting temperature is systematically studied for different building parameters. While previous studies were focusing on specific cases, this investigation will allow to generalise the results to a larger set of building configurations.

Knowing the effect of the building parameters on the PCM selection would also allow (i) to facilitate PCM selection in early design phase and (ii) to evaluate the impact of operational conditions on PCM selection. Moreover, this study will provide some support for the original idea of using two different melting temperatures to optimise thermal mass in the building sector. It would allow to identify the best PCM for heating and the one for cooling. In further investigations, combinations of these two PCMs could be compared with the addition of one PCM only.

PRELIMINARY RESULTS AND CONCLUSIONS

Initial results were obtained with eight different building parameters, e.g. wall insulation, and three different PCM-panels. They indicated that: (i) the achievable savings could be higher for cooling than for heating and (ii) an inadequate melting temperature selection could even lead to negative effect for heating. However, these results were not sufficient to identify the different influences of each building parameter on the PCM selection.

The next step is to add the melting temperature in the metamodel. The solution being investigated is to calculate the optimised melting temperature for each combination of parameters, and to build a metamodel for the optimised melting temperature. This metamodel would directly link the optimised melting temperature with the building parameters.

MAIN REFERENCES

S. E. Kalnæs, B. P. Jelle, Phase change materials and products for building applications: A state-of-the-art review and future research opportunities, Energy and Buildings 94 (2015) 150–176.

P. C. Tabares-Velasco, C. Christensen, M. Bianchi, Verification and validation of EnergyPlus phase change material model for opaque wall assemblies, Building and Environment 54 (2012) 186–196.

M. Saffari, A. de Gracia, S. Ushak, L. F. Cabeza, Economic impact of integrating PCM as passive system in buildings using Fanger comfort model, Energy and Buildings 112 (2016) 159–172.

G. Evola, L. Marletta, F. Sicurella, A methodology for investigating the effectiveness of PCM wallboards for summer thermal comfort in buildings, Building and Environment 59 (2013) 517-527.

M. Mäkelä, Experimental design and response surface methodology in energy applications: A tutorial review, Energy Conversion and Management 151 (2017) 630–640.

S.-G. Yong, J. Kim, J. Cho, J. Koo, Meta-models for building energy loads at an arbitrary location, Journal of Building Engineering 25 (2019) 100823.

P. Westermann, R. Evins, Surrogate modelling for sustainable building design – A review, Energy and Buildings 198 (2019) 170–186.

Keywords: phase-change materials, response surface method, thermal mass

Paper ID 31084: Study of the influence of temperature on the moisture buffering capacity of bio-based concretes

Igue Fathia Dahir* (1), Anh Dung Tran Le (1), Alexandra Bourdot (2), Promis Geoffrey (1), Sy Tuan Nguyen (3), Omar Douzane (1), Laurent Lahoche (1), Thierry Langlet (1)

1: University of Picardie Jules Verne, France; 2: Ecole normale supérieure Paris-Saclay, France; 3: University of Science and Technology – the University of Danang, Viet Nam

AIM AND APPROACH

The aim of this article is investigate the hygric performances of bio-based materials. The MBV value characterizes the ability of a material or multi-layer component to moderate the variation of indoor relative humidity (RH). In the literature, the moisture buffer value was determined at a constant temperature, normally at 23°C. However, in reality, the indoor temperature of the buildings is variable. Therefore, this study will examine the influence of temperature on the moisture buffer value (MBV). First, the physical models are presented. Second, the numerical models have been implemented in the Simulation Problem Analysis and Research Kernel (SPARK) suite to the complex problems. Then, the simulation tools are validated with the experimental results found in the literature. The study will be carried out on a building envelope made of palm and sunflower concretes (bio-based concretes). The boundary conditions of the studied wall are chosen according to the protocol proposed in the NordTest to calculate MBV value as function of temperature. The results showed that the increase in temperature induces an increase in the MBV value. Using this numerical model presented in this paper can predict and optimize the hygric performance of bio-based materials designed for building application.

SCIENTIFIC INNOVATION AND RELEVANCE

Bio-based concretes (such as hemp concrete, flax concrete…) are dedicated to natural construction which is a mean of achieving sustainable construction over time.

In the literature, the moisture buffer value (MBV) was determined at constant temperature, 23°C. However, in reality, the indoor temperature of the buildings is variable. The experimental results showed that the MBV value is impacted by the temperature. For example, MBV values of palm concrete measured at 23°C and 10°C are 2.96 and 2.03 (g.m-2. % RH-1) respectively (with a percentage deviation of 31.42 %). Therefore, it is necessary to carry out an in-depth study on the impact of temperature on the MBV value of bio-based materials.

The use of the presented numerical model which has been validated experimentally can predict and optimize the hygric performance of bio-based materials designed for building application.

PRELIMINARY RESULTS AND CONCLUSIONS

The numerical model to study the influence of temperature on the MBV value has been validated by comparing with the results found in the literature.

The preliminary results showed that the increase in temperature induces an increase in the MBV value and using the numerical model presented in this paper can predict and optimize the hygric performance of bio-based materials.

MAIN REFERENCES

Chennouf, N., Agoudil, B. 2018. « Hygrothermal characterization of a new bio-based construction material: Concrete reinforced with date palm fibers ». Construction and Building Materials (192): 382-394. doi.org/10.1016/j.conbuildmat.2018.10.089.

Janssen, H., Roels, S. 2009. « Qualitative and quantitative assessment of interior moisture buffering by enclosures ». Energy and Building 41 (4): 382-394. doi: 10.1016/j.enbuild.2008.11.007.

Philip, J.R., De Vries, D.A. 1957. « Moisture movement in porous materials under temperature gradients ». Transaction of American Geophysical Union 38 (2): 222–232. doi.org/10.1029/TR038i002p00222.

Rode, C., Peuhkuri, R., Lone, H., Time, B., Gustavsen, A., Ojanen, T., Ahonen, J., Svennberg, K. « 2005. Moisture buffering of building materials ». Nordic Innovation Centre Report: BYG-DTU R-126: 1601–2917.

Tran Le, A.D. 2010. « Etude des transferts hygrothermiques dans le béton de chanvre et leur application au bâtiment (sous-titre: simulation numérique et approche expérimentale) ». Thèse de doctorat, Reims : Université de Reims Champagne-Ardenne.

Keywords: Bio-based materials, moisture buffer capacity, temperature, humidity, simulation and modelling

Session T1.5: Climate change and bioclimatic design

Paper ID 30360: An integrated approach to quantify the potential local climate mitigation of a district energy network compared to individual air conditioning systems.

G-E Kyriakodis* (1,2), Emmanuel Bozonnet (1), Peter Riederer (2)

1: University of La Rochelle; 2: CSTB Sophia-Antipolis

AIM AND APPROACH

The transition from building to district scale is supported numerically by the nascent field of Urban Building Energy Modelling. However, the concurrent assessment of building energy demand and local climate conditions is sparsely studied explicitly, while there is a lack of integrated tools embedding the modelling of district energy systems. This article presents a developed coupled model to account for building energy needs, urban heat island and site-specific effects, along with the district energy system operation. To illustrate the approach, we examine the mitigation potential of a district network under projected climate conditions for a city with an oceanic climate.

SCIENTIFIC INNOVATION AND RELEVANCE

This paper demonstrates a coupled multi-model to account for urban building energy demand, UHI and site-specific effects as well as the integration of district energy systems to urban energy studies.

The developed methodology is based on a bottom-up approach where the explicit building footprints are enclosed by outdoor air cells of the urban canopy in which heat and mass transfer modelling techniques are employed under a zonal approach. The derived coupled urban canopy model is linked with a simplified urban boundary module and a vertical diffusion model. The spatial scale of the model corresponds to urban neighbourhoods while the temporal one to annual simulations given the limited execution time (23h/simulation). The model can be used either as a standalone in order to assess the urban thermal performance at various levels, or it can be readily decoupled to support a detailed meso-model.

PRELIMINARY RESULTS AND CONCLUSIONS

The neighbourhood case study under future climate conditions for a city with an oceanic climate, examined the mitigation potential of a district energy network as a free on-site heat emitter, compared to individual energy systems, such as AC units, in terms of outdoor thermal comfort and energy efficiency. The simulation outcomes assessed a mitigation capacity of the cooling network close to 2%, considering the CDH as an outdoor thermal comfort indicator related to local climate modifications. The latter is also associated with a mean air temperature reduction of 0.4oC during the cooling period. However, the mitigation potential of the district energy system is correlated with a cooling penalty due to the ground losses of the tube circuit of approximately 18 MWh which cannot be balanced by the reduction of the cooling demand at the zone level due to the elimination of the rejected heat. Nevertheless, the increased cooling demand when the UHI effect is considered limits the cooling penalty.

MAIN REFERENCES

1. Santamouris, M. (2020). Recent progress on urban overheating and heat island research. Integrated assessment of the energy, environmental, vulnerability and health impact. Synergies with the global climate change. Energy and Buildings, 207.

2. Gros, A., Bozonnet, E. and C. Inard (2014). Cool Materials Impact at District Scale—Coupling Building Energy and Microclimate Models. Sustainable Cities and Society 13, 254–266.

3. Kyriakodis, G-E., 'Development of a coupled simulation tool for urban building energy demand, district energy systems and microclimate modeling', PhD Thesis, 2020

4. Kyriakodis, G-E., et. al., "Quantifying the Impact of Urban Microclimate in Detailed Urban Building Energy Simulations.", IBPSA proceedings, 2019.

5. Riederer, P., Partenay, V., Perez, N., Nocito, C., Trigance, R. and T. Guiot (2015). Development of a Simulation Platform for the Evaluation of District Energy System Performances. Proceedings from BS2015: Building Simulation Conference. Hyderabad (India), 7-9 December 2015

6. Reinhart, C. F., et. al.,'Urban building energy modeling–a review of a nascent field'. Building and Environment, vol. 97, 2016

7. Perera, A., et. al., "Quantifying the impact of urban climate by extending the boundaries of urban energy system modeling," Applied Energy, vol. 222, pp. 847–860, 2018.

Keywords: UHI, UBEM, microclimate modeling, District energy networks, Synchronous coupling

Paper ID 30196: Building Performance Simulation to support tree planting for cooling need reduction: a machine learning approach

Massimo Palme* (1), Claudio Carrasco (2), Riccardo Privitera (3), Daniele La Rosa (3)

1: Universidad Católica del Norte, Chile; 2: Universidad de Valparaíso, Chile; 3: University of Catania, Italy

AIM AND APPROACH

Greening the city is recognised as a main strategy to improve cities liveability, outdoor environment and buildings' energy efficiency in summer (Ng et al. 2012). A lot of research has been conducted on the topic of green urban environment simulation, especially in terms of shadows (Balogun et al. 2014, Laband and Sophocleus 2009), however most of these works used simplified models not considering the boundary conditions introduced by the nature of the urban planning problem. Only few attempts to cope with concrete configurations have been done until the date. This work proposes a machine learning approach to predict, based on certain number of previously run simulations, the contribution of trees' shadows to cooling needs reduction in Mediterranean climates. This procedure can allow urban planners to evaluate a specific situation in terms of some easily observed parameters (building shape, type of trees, distance from the main façade, orientation, number of façades shadowed) and to obtain an estimation of cooling reduction or a classification in ranges of effectiveness (e.g. very poor, poor, medium, good, very good) of the configuration examined. Urban planners compare the resulting savings with other benefits and the cost of implementation of the green infrastructure.

SCIENTIFIC INNOVATION AND RELEVANCE

Machine learning refers to the capability of an algorithm to learn from some available data and to develop a strategy to predict continuous or categorical values of a variable from other variables used as predictors. There are many algorithms that can be used to learn from data, an in some cases best results are obtained by using a combination of different algorithms (ensemble). Machine learning has been already used in urban planning problems (Oh et al 2006, Liu et al. 2017, Ye et al. 2019), suggesting that the approach can be very helpful for stakeholders. In this work, four different algorithms (knn, random forest, glm and loess) are tested as well as the ensemble of all of them, both considering a continuous and a categorical evaluation. Software R is used to test the performance of the algorithms and to visualize the results. The sensitivity, specificity and global accuracy of the different algorithms used can be compared. The approach is capable to be extended by the inclusion of new observations or new parameters to be used as predictors, resulting in a very flexible too to estimate the saving that can be obtained in each configuration tested.

PRELIMINARY RESULTS AND CONCLUSIONS

First attempts in using this methodology show that if a prediction in terms of a continuous variable (cooling needs or energy savings of a specific configuration) is difficult (root mean square error of about 30%), a categorical evaluation can be obtained with an accuracy of about 95% with a single threshold value and about 70% with a five-categories classification. Such a result is useful for planners, who often take the decision in function of the question: is effective (in this configuration) to plant trees (or not)? Machine learning approach can answer to this question with a 95% of accuracy once established the threshold value (i.e. the saving that permits to equilibrate the cost-benefit analysis result). Among algorithms, random forest seems very interesting because of the high sensitivity (it assures that the minimum performance is achieved) and because the indication of the variable importance, that is a very useful result to be taken in consideration for future planning decisions.

MAIN REFERENCES

Balogun, A.A., Morakinyo, T.E., Adegun, O.B. (2014). Effect of tree-shading on energy demand of two similar buildings. Energy and Buildings 81, 305-315

Laband, D., Sophocleus, V. (2009). An experimental analysis of the impact of tree shade on electricity consumption, Arboriculture and Urban Forestry, 35, 197-202.

Liu, L., Silva, E., Wu, C., Wang, H. (2017). A machine learning-based method for the large-scale evaluation of the qualities of the urban environment. Computers, Environment and Urban Systems. 65, 113-125.

Ng., E., Chen, L., Wang, Y., Yuan, C. (2012). A study on the cooling effects of greening in a high-density city: An experience from Hong Kong. Building and Environment 47, 256-271

Oh, J., Hwang, J.E., Smith, S., Koile, K. (2006). Learning from main streets. A Machine learnign approach identifying neighborhood comercial districts. Innovations in Design and Decision Support Systems in Architecture and Urban Planning.

Ye, Y., Richards, D., Lu, Y., Song, X., Zhuang, Y., Zeng, W., Zong, T. (2019). Measuring daily accessed street greenery: A human-scale approach for informing better urban planning practices. Landscape and Urban Planning 191

Keywords: Machine learning, Green infrastructure, Cooling needs

Paper ID 30692: Development and analysis of a metric to manage overheating risks in residential energy codes

Rasoul Asaee*, Alex Ferguson

Natural Resources Canada, Canada

AIM AND APPROACH

The objective of this study was to propose a performance compliance requirement that will reduce the risks of overheating in homes by limiting design choices that are well understood to contribute to overheating.

We developed an archetype home to examine how solar gains for different glazing areas, different orientations, and various specifications for window solar gains. We also examined how predictions about cooling loads could vary between different building simulation platforms. We used 240 archetypes to evaluate the impact of the overheating test on home design. The 240 archetypes represent contemporary housing in eight major housing markets across Canada.

SCIENTIFIC INNOVATION AND RELEVANCE

In cold climates, health and economic concerns continue to focus attention on occupants' comfort and energy consumption in winter rather than summer. For years, builders tend to increase solar gain through windows as a low-cost means to reduce the heating energy consumption in buildings. As the buildings become more energy tight, the risk of overheating during the summertime increases substantially. If designers do not consider the implications of adverse solar gains in summer, these homes may require significant cooling loads, and in extreme conditions, the heat stress caused can lead to premature death. Code officials asked authors to develop a metric to prevent the overheating risks in new Canadian residential buildings. We used building performance simulation to develop a workable solution that is easy to understand for the industry and assessed the impact of the new metric on building design.

PRELIMINARY RESULTS AND CONCLUSIONS

Our results show that archetypes with small fenestration and door to wall ratio (FDWR < 10%), and low- to mid-gain windows generally comply with the requirement. As expected, archetypes that have higher windows area are less likely to comply with the overheating requirements. Likewise, the likelihood of compliance decreases in archetypes suing high-gain windows. For example, none of the archetypes with FDWRs above 30% complied with the requirement as designed; nor did any of the archetypes with very high gain windows (SHGC > 0.6).

This work led to the addition of an overheating metric for the first time in a Canadian building code. This requirement affects design choices in homes with high window areas. In such houses, the requirement can be met by choosing low solar-gain windows. Reducing window area, relocating east and west-facing windows to north and south facades, and installing overhangs also help homes comply. Authors recommended against permitting builders to install air-conditioning as an alternate means of compliance.

MAIN REFERENCES

Environment and Climate Change Canada (2016). Pan-Canadian Framework on Clean Growth and Climate Change: Canadas plan to address climate change and grow the economy. Environment and Climate Change Canada.

Parekh, A., R. Charron, S. Poirier, and L. Roux (2018). Testing of HOT2000 version 11 in accordance with ASHRAE standard 140-2014. In Proceedings of the eSim 2018 Conference. Montreal, QC (Canada), 9-10 May 2018.

Crawley, D. B., Lawrie, L. K., Pedersen, C. O., & Winkelmann, F. C. (2000). Energy plus: energy simulation program. ASHRAE Journal, 42(4), 49-56.

NRCan (2018b). EnerGuide in Canada, Natural Resources Canada: Office of Energy Efficiency. Available online at www.nrcan.gc.ca/energy/products/energuide/12523.

Keywords: energy code, overheating risk, residential building, energy simulation, solar gain

Session T1.6: Ensuring high quality building simulations

Paper ID 31086: Faster computation of g-functions used for modeling of ground heat exchangers

Jeffrey Spitler*, Jack Cook

Oklahoma State University, United States of America

AIM AND APPROACH

Temperature response functions, known as g-functions, are a computationally efficient method for simulating ground heat exchangers (GHEs), used with ground-source heat pump (GSHP) systems as part of a whole-building energy simulation. In fact, at present, there are no other methods that have sufficient accuracy and are fast enough to simulate a ground-source heat pump system in a whole-building energy simulation.

The concept, mathematical derivation and an implementation of a g-function calculation program were developed by Claesson and Eskilson (1985). More recently Cimmino (2018a, 2018b, 2019) developed an open-source g-function calculation tool known as pygfunction. This tool offers great flexibility for the user to compute g-functions for specific configurations of boreholes. However, for large borehole configurations (with ~1000 boreholes), the required time to compute a single g-function can take several hours, and the required RAM can be on the order of 100 GB, greatly exceeding most desktop PCs. In order to develop libraries of g-functions and training sets for machine learning approaches, we are computing hundreds of thousands of g-functions. This paper describes further development of Cimmino's methodology to speed the computation and reduce the memory requirements.

SCIENTIFIC INNOVATION AND RELEVANCE

Dusseault and Pasquier(2019) demonstrated use of artificial neural networks to compute g-functions. 500,000 g-functions computed for fields of up to 10 boreholes were used to train the neural network. Pasquier (2019) has estimated that a million g-functions might be needed train a neural network. As described by Spitler, et al. (2020) the computing time and memory scale approximately with the square of the number of segments used. Each borehole is divided into multiple segments treated as finite line sources and the accuracy depends on the number of segments used. To compute g-functions for ~1000 borehole configurations with deep (~400m) boreholes even the resources of our university's high-performance computing cluster are insufficient, as we have very few nodes with available RAM exceeding 256 GB.

Therefore, any effort to develop neural network training sets, or even libraries of g-functions that include very large borehole configurations will require development of an improved tool that is faster and uses less memory than pygfunction. This is also necessary for individual users wishing to compute custom g-functions on their desktop PC. The work described in this paper represents a significant improvement in both speed and memory requirements.

PRELIMINARY RESULTS AND CONCLUSIONS

At the time of this writing, we have obtained about a 2.5-fold increase in computational speed for borehole fields with 3000-5000 segments. (This work was done on a desktop PC with 16 GB RAM, limiting the size of the borefield that could be analyzed.) For the case with 3000 segments, the memory requirement is reduced by about 75%, but a more careful analysis will be done once we move back to the university's high-performance computing cluster. To date, the primary improvements have been obtained by switching from Python to C++, streamlining the code, using hash tables to store segment-to-segment interactions, and taking advantage of multi-threading. Additional, more substantive changes are planned, which should provide further improvement. Even with the existing improvements, g-functions can be computed much faster with much less memory, allowing larger borefields to be analyzed on desktop PCs and more widely available, lower-memory cluster nodes.

MAIN REFERENCES

Cimmino, M. 2018a. Fast calculation of the g-functions of geothermal borehole fields using similarities in the evaluation of the finite line source solution. Journal of Building Performance Simulation 11(6): 655-668.

Cimmino, M. 2018b. pygfunction: an open-source toolbox for the evaluation of thermal. eSim 2018, Montreál, IBPSA Canada. 492-501.

Cimmino, M. 2019. Semi-Analytical Method for g-Function Calculation of bore fields with series- and parallel-connected boreholes. Science and Technology for the Built Environment 25(8): 1007-1022.

Claesson, J. and P. Eskilson. 1985. Thermal analysis of heat extraction boreholes. Proceedings of 3rd International Conference on Energy Storage for Building Heating and Cooling ENERSTOCK 85, Toronto, Canada, Public Works Canada. 222–227.

Dusseault, B. and P. Pasquier. 2019. Efficient g-function approximation with artificial neural networks for a varying number of boreholes on a regular or irregular layout. Science and Technology for the Built Environment 25(8): 1023-1035.

Pasquier, P. 2019. E-mail. J. D. Spitler.

Spitler, J.D. J. Cook, X. Liu. 2020. Recent Experiences Calculating g-functions for Use in Simulation of Ground Heat Exchangers. Submitted to GRC Transactions.

Keywords: ground-source heat pumps, ground heat exchangers, borehole heat exchangers

Paper ID 30210: Evaluating the impact of neighbouring structures on a high-rise residential block's performance

Weijie Xu*, Carlos Jimenez-Bescos, John Kaiser Calautit

University of Nottingham, United Kingdom

AIM AND APPROACH

The work aims to analyse the impact of building locations and neighbouring buildings on its energy performance by undertaking parametric modelling. In this research, a 12-storey residential building was investigated using a validated Building Energy Simulation (BES) model. IESVE will be used to assess the energy performance of the residential building, which is a dynamic thermal simulation based on the modelling of the heat transfer processes between a building and its microclimate. The cities selected for this research are Copenhagen – Denmark, Beijing – China and Singapore, which represents three different climatic conditions: oceanic, humid continental and tropical climate, respectively. The effect of parameters such as surroundings' number and positions were discussed. In total, over 180 scenarios were modelled to understand the influence of the variation of different parameters.

SCIENTIFIC INNOVATION AND RELEVANCE

Urbanisation has increased rapidly in recent years, and more high-rise buildings are built to accommodate the increasing urban population. The energy used by buildings in the urban area accounts for over one-third of the total energy demand. The energy usage of high-rise buildings is much higher than the low-rise building. High-rise buildings affect the solar conditions and wind environment of the surrounding areas which can have positive and negative effects depending upon the site-specific circumstances. The use of multi-parameter optimisation approach for the building design can achieve considerable energy savings and cost reduction while improving thermal comfort. Factors or parameters that influence building energy consumptions includes type of energy used, altitude, location, fabric, building design, shape orientation, operation and technology usage and have been explored in the literature. A limited number of studies investigated the energy performance of high-rise residential buildings in different climatic conditions. Although previous works focused on the urban morphology effect, not many works assessed the influence on high-rise building energy performance. In addition, most previous studies focused on how the UHI phenomenon affect the heating and cooling performance of buildings. There is little study on the association between neighbouring high-rise structures and energy consumption.

PRELIMINARY RESULTS AND CONCLUSIONS

Surrounding buildings could positively and negatively influence the central building's performance, depending on the location and shading conditions. For a single building, the annual solar gain difference between Denmark and China was 46.2MWh, and the total energy load difference was 47MWh. Influence on the thermal comfort of the inhabitants was also evaluated. Under the same set conditions, there were great differences between PMV and PPD values among the three locations, so the indoor set conditions should be determined depending on the local environment. Multiple high-rise buildings with narrow spacing were observed to have the largest impact on the central building's energy performance. Overall, the results indicate that the design and layout are important issues to consider for high-rise buildings as they can influence the energy use up to 14.35% when the central building was surrounded by eight double-height buildings with 15m spacing. The present work has shown the importance of considering the influence of neighbouring structures when performing energy simulations of high rise buildings.

MAIN REFERENCES

Vallati, A., et al., On the Impact ofUrban Micro Climate on the Energy Consumption of Buildings.Energy Procedia, 2015. 82: p. 506-511.

Cheung, C.K., R.J. Fuller, and M.B. Luther, Energy-efficient envelope design for high-rise apartments.Energy and Buildings, 2005. 37(1): p. 37-48.

Raji, B., M.J. Tenpierik, and A. van den Dobbelsteen, An assessment of energy-saving solutions for the envelope design of high-rise buildings in temperate climates: A case study in the Netherlands.Energy and Buildings, 2016. 124: p. 210-221.

Mardookhy, M., et al., A study of energy efficiency in residential buildings in Knoxville, Tennessee.Journal of Cleaner Production, 2014. 85: p. 241-249.

Mardookhy, M., et al., A study of energy efficiency in residential buildings in Knoxville, Tennessee.Journal of Cleaner Production, 2014. 85: p. 241-249.

Zhai, Z.J. and J.M. Helman, Implications of climate changesto building energy and design.Sustainable Cities and Society, 2019. 44: p. 511-519.

Samuelson, H., et al., Parametric energy simulation in early design: High-rise residential buildings in urban contexts.Building and Environment, 2016. 101: p. 19-31.

Keywords: building performance, high-rise buildings, neighbouring buildings, building simulation, parametric analysis

Paper ID 30865: Where HVAC models fail - a conceptual framework for extending effective HVAC modelling into early concept design of net zero buildings

David Rulff*, Theodor Victor Christiaanse, Ralph Evins

University of Victoria, Canada

AIM AND APPROACH

The performance gap between measured building performance and simulated results in practice is a well documented issue, with error derived from a variety of sources (including misapplied tools and methods, uncertainty in design parameters, stochastic variability in operation and occupancy controls, etc.) and related to different stages of a building project life cycle (conceptualization, detailed design, construction and occupancy).[1] Traditional bottom-up, deterministic modeling frameworks for building HVAC systems are misaligned with a performance-based design paradigm, especially in early stages where broad design spaces and significant uncertainty are present.[2] For building energy modellers, the extent to which available conceptual frameworks can adapt to these changing conditions needed to be evaluated, with a consistent methodology that captures top-down functional requirements and uncertainty in a parametric analysis. The goal of this study is to engage in dialogue with the broader energy modelling community by outlining a variety of conceptual frameworks for representing HVAC systems from literature and practice, with particular emphasis on their influence on the reliability in applied building performance simulation.

SCIENTIFIC INNOVATION AND RELEVANCE

This study presents a summary and comparison of different conceptual frameworks for HVAC modelling with a focus on net-zero buildings. This has significant implications in the context of increasingly stringent performance-based Building Codes and Standards that aim to regulate the majority of building stock in many countries, in efforts to reduce the impact of building energy consumption and associated greenhouse gas emissions on climate change. Each framework is deconstructed into characteristic components that allow consistent comparison, including how they capture the over-arching design and process goals, functional requirements, performance requirements (and corresponding performance indicators); as well as their capacity to integrate verification methods, functional systems and sub-systems, and stochastic elements or agents.[2] Alignment with best practice building modeling methodologies, technical schema and practical considerations is explored through implementation of a series of test cases in EnergyPlus, drawn from literature on sources of HVAC modeling errors and industry testing protocols.[3][4] Recommendations for future research and refinement of a common conceptual framework for HVAC are compiled from the results.

PRELIMINARY RESULTS AND CONCLUSIONS

Promising HVAC modeling frameworks are found as part of larger building simulation research works, especially from three diverse areas of study: characterizing problem spaces and conducting feature engineering for black box machine learning methods,[5] describing system objects and functional schema for detailed calibration of high fidelity models,[6] and decomposition of building systems into functional units for best practice building simulation.[2] The different frameworks were used to define and implement test cases with increasingly complex HVAC system configurations in EnergyPlus, within the context of a performance-based design paradigm. Preliminary results indicate that uncertainty can be mitigated using statistical methods from calibration research, while limitations due to over-specification or excessive detail can be mitigated by consistent hierarchical decomposition that allows sub-elements and sub-systems to be collapsed into less granular functional units. The connection to the top-down performance requirements can be maintained through the same functional units, also helping preserve interpretability for machine learning methods that may be applied.

MAIN REFERENCES

[1] de Wilde, P. (2014). The gap between predicted and measured energy performance of buildings: A framework for investigation. Automation in Construction, 41, 40–49. https://doi.org/10.1016/j.autcon.2014.02.009

[2] Augenbroe, G. (2019). 10 The role of simulation in performance-based building. Building Performance Simulation for Design and Operation.

[3] Eisenhower, B., O'Neill, Z., Fonoberov, V. A., & Mezić, I. (2012). Uncertainty and sensitivity decomposition of building energy models. Journal of Building Performance Simulation, 5(3), 171–184. https://doi.org/10.1080/19401493.2010.549964

[4] Tian, W., Heo, Y., de Wilde, P., Li, Z., Yan, D., Park, C. S., Feng, X., & Augenbroe, G. (2018). A review of uncertainty analysis in building energy assessment. Renewable and Sustainable Energy Reviews, 93, 285–301. https://doi.org/10.1016/j.rser.2018.05.029

[5] Zhang, Y., O'Neill, Z., Dong, B., & Augenbroe, G. (2015). Comparisons of inverse modeling approaches for predicting building energy performance. Building and Environment, 86, 177–190. https://doi.org/10.1016/j.buildenv.2014.12.023

[6] Coakley, D., Raftery, P., & Keane, M. (2014). A review of methods to match building energy simulation models to measured data. Renewable and Sustainable Energy Reviews, 37, 123–141. https://doi.org/10.1016/j.rser.2014.05.007

Keywords: high fidelity systems modeling, conceptual framework, HVAC schema

Session T2.1: Practice and industry related case studies

Paper ID 30540: Developing a simplified methodology for simulating energy performance of buildings with thermally activated building systems

Mohsen Sharifi*, Josué Borrajo Bastero, Rana Mahmoud, Eline Himpe, Jelle Laverge

Ghent university, Belgium

AIM AND APPROACH

Detailed building models allow the HVAC designers to find optimal solutions. However, more complexity in modelling will also result in more efforts for the designers in the component sizing procedure specially when the optimal design of the system is highly depended on the control system. In Thermally Activated Building Systems (TABS), which are embedded pipes in the concrete of the building structure, the role of control is pivotal and thus the control has to be simulated during the design. Therefore, the building model and optimal control has to be coupled to an optimisation algorithm for optimal component sizing. This makes a computationally heavy pre-design procedure. This paper addresses the need for a for a simplified methodology for modelling building with TABS. We calculated the dynamic hourly heating and cooling loads of three case studies with white-box method and used them in simplified resistor-capacitor (RC) model developed by the grey-box approach and the parameters of RC model were estimated by inverse modelling approach. The methodology was applied on three different case studies and the results imply that the developed model by proposed methodology can be confidently coupled to an optimal control to be used for sizing HVAC components.

SCIENTIFIC INNOVATION AND RELEVANCE

1-An automated algorithm for modelling building with TABS

2-Taking the advantage of white-box modelling to have high accuracy

3- Dynamic thermal behavior of the building can be incorporated to the design procedure with a simple model

PRELIMINARY RESULTS AND CONCLUSIONS

Using built-in Modelica libraries and prepared code, the building demand is calculated accurately and easily. Then, a simplified model is derived by the proposed methodology and optimal sizing of HVAC components can be achieved in the early stage of design without expensive mathematical calculations. The first simulation results show that the simplified model can predict the dynamic thermal behaviour of the building accurately and the designer can be confident about the sizing of components when suing the proposed methodology.

MAIN REFERENCES

1- W. Boydens, D. Costola, A. Dentel, T. Dippel, L. Ferkl, A. Görtgens, L. Helsen, J. Hoogmartens, B.W. Olesen, W. Parijs, M. Sourbron, C. Verhelst, J. Verheyen, C. Wagner, REHVA Guidebook No. 20: Improved system design and control of GEOTABS buildings: Design and operation of GEOTABS systems, REHVA, Brussels, 2013. www.rehva.eu.

2- Sharifi, M., Mahmoud, R., Himpe, E., & Laverge, J. (n.d.). Interaction of GEOTABS and secondary heating and cooling systems in hybridGEOTABS buildings : towards a sizing methodology.

3- Sourbron, M. (2012). Dynamic thermal behaviour of buildings with concrete core activation. PhD thesis

4-Jorissen, F. (2018). Toolchain for Optimal Control and Design of Energy Systems in Buildings. (April). Retrieved from https://limo.libis.be/primo-explore/fulldisplay?docid=LIRIAS1652305&context=L&vid=Lirias&search_scope=Lirias&tab=default_tab&lang=en_US

Keywords: building energy simulation, TABS, simplified RC model, design

Paper ID 30377: Analysis of the thermal performance of a ground water storage cell with helical shaped pipe for ground source heat pumps

Marco Marigo* (1), Enrico Prataviera (1), Sara Bordignon (1), Michele Bottarelli (2), Angelo Zarrella (1)

1: Department of Industrial Engineering, University of Padova, Italy; 2: Architecture Department, University of Ferrara, Italy

AIM AND APPROACH

Ground Source Heat Pump (GSHP) systems are promising technologies that can reduce fossil fuel consumption and CO2 emission due to the air-conditioning in buildings. The advantages of this technology compared to air-source heat pumps are well-known. However, the installation of vertical ground heat exchanger is expensive, consequently alternative solutions are necessary.

In this work, a particular vertical ground heat exchanger for residential buildings is studied. It consists of a helical shaped pipe installed in a water tank (cell) buried into the ground. The water of the tank can be groundwater or greywater from the dwelling. The bottom of the cell is installed at about 3 m under the ground surface. Different cells with helical shaped pipe can be coupled in parallel to the heat pump that provides heating and cooling for the building.

The new system was installed in a residential building in Treviso (Italy) and its temperatures were monitored on the long-term. Numerical simulations in transient conditions via a commercial finite element software were carried out to analyse the thermal behaviour of the cell. The long-term performance of the system was also simulated through a numerical model named CaRM (Capacity Resistance Model), which was modified to simulate this system.

SCIENTIFIC INNOVATION AND RELEVANCE

Ground heat exchangers for GSHP can be vertically or horizontally oriented. For vertical ground heat exchangers, deep vertical bores need to be drilled increasing the initial cost of the installation. In addition, deep bores (usually 100 m long) are not feasible in some locations due to geological and legal constraints. Horizontal ground heat exchangers are installed at shallow depths (usually less than 2 m) requiring significantly more surface. Moreover, decreasing the depth of installation, they are more affected by weather variations.

In the presented heat exchanger, the higher thermal capacitance of the water in the tank is exploited in order to store energy, limit the peak loads and establish suitable control strategies. The heat transfer between ground and tank is continuous, and it is not related to the operation of the system. The lower depth allows the installation without the necessity of legal permissions and the maintenance is easier. The initial cost for the installation is lower than a drilling for a common vertical borehole heat exchanger.

PRELIMINARY RESULTS AND CONCLUSIONS

A model in a finite element software was built to analyse the thermal behaviour of the system in the short term. In particular, the transient response to a heat load profile of the whole system was investigated analysing the heat exchanged between cell and ground, in both heating and cooling mode. A sensitivity analysis on the main parameters (e.g. geometry and material of the tank, pitch of the helical shaped pipe, installation depth, ground thermal properties) was carried out studying the effect on the thermal performance of the system. Then, a capacitance-resistance model was also developed for long-term simulations.

Simulation results are in good agreement with field measurements. They show the relationship between the temperature change and the volume of water in the cell. In addition, the new system is also compared with a common double U-tube configuration in terms of thermal performance on the long-term and from the economic point of view. The economic analysis points out the lower cost of the new solution.

MAIN REFERENCES

Warner J, Liu X, Shi L, Qu M, Zhang M. A novel shallow bore ground heat exchanger for ground source heat pump applications - Model development and validation. Applied Thermal Engineering 2020; 164:114460.

Zhang M, Liu X, Biswas K, Warner J. A 3D numerical investigation of a novel shallow bore ground heat exchanger integrated with phase change material. Applied Thermal Engineering 2019; 162:114297.

Bonamente E, Aquino A, Cotana F. A PCM thermal storage for ground-source heat pumps: simulating the system performance via CFD approach. Energy Procedia 2016; 101:1079-1086.

Lund JW. Geothermal heat pumps - an overview, Geo-Heat Centre Quarterly Bullettin 2001; 22:1–8.

De Carli M, Tonon M, Zarrella A, Zecchin R. A computational capacity resistance model (CaRM) for vertical ground coupled heat exchangers. Renewable Energy 2010; 35:1537-1550

Zarrella A, De Carli M. Heat transfer analysis of short helical borehole heat exchangers. Applied Energy 2013; 102:1477-1491

Keywords: Ground Source Heat Pumps, Borehole heat exchanger, Heat pump, Heat transfer

Paper ID 30717: Borehole thermal energy storage integration: a case study

Jonas Cleiren* (1), Freek Van Riet (2), Kristof Smits (1), Wolf Nys (1), Roel Vandenbulcke (1), Ivan Verhaert (3)

1: Hysopt NV, Antwerp, Belgium; 2: Noven NV, Ghent, Belgium; 3: EMIB research group, University of Antwerp, Antwerp, Belgium

AIM AND APPROACH

In the past decades, a large number of geothermal energy storage systems like Borehole Thermal Energy Storage (BTES) and Aquifer Thermal Energy Storage (ATES) have been widely applied. However, these systems are rather complex to design and prone to design errors, often resulting in suboptimal systems mainly on a hydraulic and control related point of view. In order to obtain a well-functioning system despite the complexity, a case study is conducted to address various optimisation possibilities.

To optimise a BTES system, a case study is conducted using the simulation software Hysopt because it takes both the thermal, as well as the hydraulic and control behaviour into account. In the case study, the initial design of a BTES system is optimised by simulating variants with different thermal, hydraulic and control design choices and comparing the resulting performances with each other. The analysis takes into account different KPIs like the energy consumption/cost, CO2 emission and investment cost.

SCIENTIFIC INNOVATION AND RELEVANCE

Geothermal energy storage systems are called complex because of the combination of multiple types of production units, the possible desired thermal balance of the ground, the different operating conditions and the control strategy. Because of the complexity, extensive professional knowledge is required to design and correctly simulate geothermal energy storage systems. Even when doing so, there will always remain opportunities for improvement.

The conducted case study starts with the optimisation of the distribution network with the end-units to decrease the return and/or supply temperature. Afterwards the hydraulic configuration and control strategy of the energy centre is optimised to increase the contribution of the heat pump, which results in reduced energy consumption and CO2 emissions. The remaining excess heat and/or cold is stored in the BTES. Depending on the difference between extracted and injected heat, the ground can be thermally imbalanced. To address this imbalance, a few alternatives are proposed to improve the thermal balance. In addition, optimisations to keep the investment cost as low as possible are also proposed.

PRELIMINARY RESULTS AND CONCLUSIONS

The conducted case study, including simulation of thermal, hydraulic and control behaviour, resulted in different design options. Depending on the priority of the KPIs, the most suitable design can be chosen.

The different design choices are mostly focused on hydraulic configurations and control strategies, mainly because these aspects are often neglected. In the case study, the energy consumption can be reduced by 40%, the energy cost by 30% and the CO2 emission also by 30%. Furthermore, some design choices lowered the investment cost and improved the thermal balance of the soil.

From the analysis of the simulation results, it can also be concluded that the performance of the system with geothermal energy storage is very sensitive to small adjustments in the hydraulics and controls. Therefore, it is recommended to choose a more robust design, for instance one with relatively simplistic controls, to reduce the error sensitivity.

MAIN REFERENCES

R. Vandenbulcke, "Hydronic Simulation and Optimisation", University of Antwerp, 2013.

F. Van Riet, "Hydronic design of hybrid thermal production systems in buildings", University of Antwerp, 2019.

F. Van Riet, R. Vandenbulcke, J. Cleiren, and I. Verhaert, "Hydronic Optimisation Of Hybrid Heating Systems: A Methodology Based On Base Circuits", IBPSA Building Simulation 2019.

F. Van Riet, H. El Khaoui, F. Hulsbosch, G. Steenackers, and I. Verhaert, "Exploring the novel software Hysopt: a comparison of hydronic heat distribution systems of an apartment building", 12th REHVA World Congress CLIMA 2016.

J. Cleiren, "Hydronic design of heat- and cold production systems with geothermal seasonal energy storage", University of Antwerp, 2018.

X. Q. Zhai, M. Qu, X. Yu, Y. Yang, and R. Z. Wang, "A review for the applications of integrated approaches of ground-coupled heat pump systems", Renew. Sustain. Energy, vol. 15, no. 6, pp. 3133-3140, 2011.

L. Gao, J. Zhao, Q. An, J. Wang, and X. Liu, "A review on system performance studies of aquifer thermal energy storage", Energy Procedia, vol. 142, pp. 3537-3545, 2017.

M. Bloemendal, T. Olsthoorn, and F. Boons, "How to achieve optimal and sustainable use of the subsurface for Aquifer Thermal Energy Storage", Energy Policy, vol. 66, pp. 104-114, 2014.

Keywords: Geothermal energy storage, BTES, Heat pump, Hydronics, Control strategies

Paper ID 31150: Towards a new simulation based hybridGEOTABS design methodology!

Alexander Jean Lucienne Berquin*

Boydens Engineering, Belgium

AIM AND APPROACH

Providing an easy-to-use method to determine the optimal system powers of hybridGEOTABS in a predesign stage. The new method is based on a design moment analysis and correlation analysis of a simulated database.

SCIENTIFIC INNOVATION AND RELEVANCE

It could be possible to size the different components of hybridGEOTABS in a predesign stage based on correlations between peak demands and optimal system powers.

PRELIMINARY RESULTS AND CONCLUSIONS

It was shown that it is difficult to define specific design conditions to size the hybridGEOTABS components. However, main circumstances could be distinguished for which the system powers are peaking.

Furthermore, It is clear that high correlations exist between peak demands and the optimal system powers in heating and cooling mode. This was also investigated for different climates.

MAIN REFERENCES

Berquin, A. (2020). Towards a new hybridGEOTABS design methodology! Universiteit Gent.

Franziska Bockelmann, Stefan Plesser, & Hanna Soldaty. (2017). REHVA Guidebook No. 20—Advanced System Design and Operation of GEOTABS Buildings. REHVA.

Mahmoud, R., Himpe, E., Delghust, M., & Laverge, J. (2019). D2.2 A set of parametric geometries for the (sub)typologies studied. UGent.

Sharifi, M., Himpe, E., & Laverge, J. (2019). D2.1 An automated baseload search algorithm for GEOTABS sizing. Universiteit Gent.

Keywords: hybridGEOTABS, design moment analysis, correlation analysis, simulated database, predesign

Session T2.2: Ensuring high quality building simulations

Paper ID 30191: Investigation of the temperature-dependent conductivity towards improving energy performance of building

Anna Wieprzkowicz*, Dariusz Heim

Lodz University of Technology, Poland

AIM AND APPROACH

The aim of the study is to investigate the effect of the temperature-dependent thermal conductivity on energy demand. The analysis was performed for the basic test building composed of a single rectangular zone, based on the Case 900 from ANSI/ASHRAE Standard 140-2011 (ANSI/ASHRAE 2011). The analysis was conducted for three locations characterized by different climate conditions: Norway, Bergen - cold climate without a dry season and with cold summer; Romania, Bucharest - temperate climate without a dry season and with warm summer; United Arab Emirates, Abu Dhabi – hot desert climate.

Ten functions of temperature-dependent thermal conductivity were developed and tested using non-linear properties subroutine implemented in ESP-r. The results were compared with the constant values of thermal conductivity: the maximum and minimum values included in predefined functions.

Results were analysed in terms of heating and cooling energy demand, the temperature of the insulation layer and dynamically changing thermal conductivity values as well as parameters characterizing the dynamics of thermal properties switching.

SCIENTIFIC INNOVATION AND RELEVANCE

Dynamic properties of building components are highly expected regarding overall energy performance (Jelle, Gustavsen, and Baetens 2010) and future construction technology (Berardi et al. 2018). In case of heat transfer through building construction two thermal properties are technically possible to be adjusted: capacity and conductivity, or combined both as a so-called heat retention index

The technical solution of dynamic insulation materials (DIM), also named controllable insulation materials (CIM) is mainly based on pore (gas filles) structures. It should be noted that only radical changes in the conductivity can lead to the reduction in energy consumption.

One of the parameters which determine the overall performance of DIM is control strategies. The initial work e.g. (Park, Srubar, and Krarti 2015) have investigated a binary R-value control in which the thermal resistance of the wall is either "on" or "off". Further research, (Rupp and Krarti 2019) investigated the improved strategy by using wall surface temperatures, the heating/cooling set-point temperature, and the temperature in the middle of the wall. It is found that adding a period during which the R-value of the wall can vary continuously within a defined range has the potential to modestly decrease the heating and cooling energy consumption.

PRELIMINARY RESULTS AND CONCLUSIONS

The results of building energy performance equipped with dynamic insulation materials were presented in the paper. The numerical model was developed in ESP-r considering the different function of thermal conductivity versus temperature. The sudden and smooth changes in conductivity values were considered in case of heating/cooling energy demand. Analyses were performed for three geographical locations: Bergen, Bucharest and Abu Dhabi. For each location, the energy savings were determined and the best function was finally defined. The biggest difference in cooling demand was 50% for Bucharest and 25% for Abu Dhabi. In both cases, the rapid changes in lambda values gave better energy performance of the dynamic insulation in case of analysed building. It was also concluded that periodic increase of thermal conductivity has no beneficial effect of heating energy demand - in a heating season high thermal resistance is crucial.

MAIN REFERENCES

Jelle, Bjørn Petter, Arild Gustavsen, and Ruben Baetens. 2010. "The Path to the High Performance Thermal Building Insulation Materials and Solutions of Tomorrow." Journal of Building Physics 34 (2): 99–123. doi:10.1177/1744259110372782.

Berardi, Umberto, Lamberto Tronchin, Massimiliano Manfren, and Benedetto Nastasi. 2018. "On the Effects of Variation of Thermal Conductivity in Buildings in the Italian Construction Sector." Energies 11 (4). MDPI AG. doi:10.3390/en11040872.

Park, Benjamin, Wil V. Srubar, and Moncef Krarti. 2015. "Energy Performance Analysis of Variable Thermal Resistance Envelopes in Residential Buildings." Energy and Buildings 103 (July). Elsevier Ltd: 317–325. doi:10.1016/j.enbuild.2015.06.061.

Rupp, Shawn, and Moncef Krarti. 2019. "Analysis of Multi-Step Control Strategies for Dynamic Insulation Systems." Energy and Buildings 204 (December). Elsevier Ltd: 109459. doi:10.1016/j.enbuild.2019.109459.

Keywords: thermal conductivity, energy demand, building materials

Paper ID 30828: 3D drone-based time-lapse thermography: a case study of roof vulnerability characterization using photogrammetry and performance simulation implications

Tarek Rakha* (1), Yasser El Masri (1), Kaiwen Chen (1), Pieter De Wilde (2)

1: Georgia Institute of Technology, United States of America; 2: University of Plymouth, United Kingdom

AIM AND APPROACH

This paper investigates airborne time-lapse thermography using drones as a method of translating 3D envelope CAD models generated via RGB and IR photogrammetry into whole Building Energy Modeling (BEM) software. The goal is to develop a novel 4D approach that incorporates defects detected at varying situations and instances of time into comprehensive thermal profiles for envelopes. We propose the development of novel 3D thermography models constructed from time lapse IR image data collected using UAS to inform BPS envelope modeling inputs. A case study is presented for a courtyard building on a North American campus, where the research team employed a drone to fly in two path types, perimeter polygon and in a strip. The flights repeated in 2-hour intervals starting at 9 AM until 5 PM (5 flights). In this work, we are focusing on the roof as a building element that is typically assumed to perform uniformly and is neglected in as-built inspections due to the impossibility of perceiving the entire component without being significantly higher than it.

SCIENTIFIC INNOVATION AND RELEVANCE

Multiple built environment applications have made use of thermography, mainly focusing on defect identification using perspectives from the IR spectrum. However, standard IR readings are typically undergone in singular points in time, when in several cases, such as varying pressure differences or latent heat gain, anomalies can only be revealed at specific times of the day, possibly in different seasons of the year. The pursuit of a time-series based envelope diagnosis is not new; it started in the early 80's under the terminology "Transient Thermography," which evolved into "Time Sequential Thermography" in the late 90's, and finally to "Time Lapse Thermography" in the mid-2010's onward. Therefore, this paper's scientific contribution is evolving this concept through two proposed innovations: 1) developing a novel Time Lapse Thermography inspection methodology that employs drones to inspect building envelopes at different times during the day that characterize an envelope's performance comprehensively without being constricted spatially; 2) employing drone-collected and geolocated RGB and IR images to build 3D envelope CAD models using photogrammetry, and introducing the concept of 3D thermography to identify envelope defects. The work's relevance is the translation of such models into simulation software for more accurate and faster existing building envelop modeling.

PRELIMINARY RESULTS AND CONCLUSIONS

We presented in this paper a novel workflow for 3D envelope modeling using aerial time-lapse IR data collection using UAS. A comprehensive roof thermal profile was developed for a case study building employing photogrammetry software Agisoft Photoscan, which generated temporal IR inspections of a building's roof using multiple 3D thermography CAD models. The goal was to develop a building inspection framework that utilizes drones equipped with IR cameras to collect data time series, which in turn informs envelope performance modeling to accurately depict thermal resistance as well as anomalies for more accurate BPS. The paper concludes that on-site building envelope inspections are significantly enhanced by a time-based passive thermography audit approach that has spatial liberty due to the use of UAS and discussed the potential for integrating this technology effectively in BEM to inform both facilities management and retrofitting design. Future work should explore full envelope 4D CAD modeling in concert with thermography. Advances in this field are expected to leap the process of building envelope energy audits forward to become ubiquitous and informative to decision makers aiming to retrofit and manage our existing built environments to become significantly more highly performing.

MAIN REFERENCES

Edis, Ecem, Inês Flores-Colen, and Jorge De Brito. 2015. "Time-Dependent Passive Building Thermography for Detecting Delamination of Adhered Ceramic Cladding." Journal of Nondestructive Evaluation 34 (3): 1–16. https://doi.org/10.1007/s10921-015-0297-5.

Fox, Matthew, David Coley, Steve Goodhew, and Pieter De Wilde. 2015. "Time-Lapse Thermography for Building Defect Detection." Energy and Buildings 92: 95–106. https://doi.org/10.1016/j.enbuild.2015.01.021.

Gharawi, Mohanned Al, Yaw Adu-Gyamfi, and Glenn Washer. 2019. "A Framework for Automated Time-Lapse Thermography Data Processing." Construction and Building Materials 227: 116507. https://doi.org/10.1016/j.conbuildmat.2019.07.233.

Grinzato, E., V. Vavilov, and T. Kauppinen. 1998. "Quantitative Infrared Thermography in Buildings." Energy and Buildings 29 (1): 1–9.

Rakha, T., & Gorodetsky, A. (2018). Review of Unmanned Aerial System (UAS) applications in the built environment: Towards automated building inspection procedures using drones. Automation in Construction, 93, 252-264.

Hobbs, Chris. 1992. "Transient Thermography." Sensor Review 12 (1): 8–13.

Hoyano, Akira, Kohichi Asano, and Takehisa Kanamaru. 1999. "Analysis of the Sensible Heat Flux from the Exterior Surface of Buildings Using Time Sequential Thermography." Atmospheric Environment 33 (24–25): 3941–51.

Ibarra-Castanedo, Clemente, Stefano Sfarra, Matthieu Klein, and Xavier Maldague. 2017. "Solar Loading Thermography: Time-Lapsed Thermographic Survey and Advanced Thermographic Signal Processing for the Inspection of Civil Engineering and Cultural Heritage Structures." Infrared Physics and Technology 82: 56–74.

Keywords: Envelope, Unmanned Aerial Systems (UAS), Drones, Thermal Imaging, 3D Modeling

Paper ID 30585: Dynamic modelling and comparison between transient step response of capacitive hygrometers and chilled mirrors

Ettore Zanetti*, Rossano Scoccia, Marcello Aprile, Mario Motta

Politecnico di Milano, Italy

AIM AND APPROACH

This manuscript presents the results of an experimental study carried out to compare the transient responses of three capacitive hygrometers with respect to three chilled mirrors. Several experiments were carried out changing the values of air flow rate, temperature, and relative humidity in the test chambers. The results of these experiments were used to derive different models of the sensors, from black box data driven models, to grey box models. These can be used to check and eventually reconstruct transient data operation of desiccant evaporative cooling heat exchangers which are inherently transient. The sensors were tested in the ReLab research group facility [1], two climatic chambers were used to simulate the different conditions, the sensors were installed in a long cylindrical duct with a 16cm diameter. The step analysis was carried out by having a fan actively driving air inside the duct and by rapidly moving one end of the duct from one chamber to the other at different conditions until a steady state was reached.

SCIENTIFIC INNOVATION AND RELEVANCE

Desiccant Evaporative Cooling (DEC) systems have seen an increased interest in academia [2] and commercial applications [3] for Heating ,Ventilation and Air Conditioning (HVAC) applications. The core components of these systems are the direct or indirect evaporative cooler coupled with a desiccant component. Traditionally the process air goes through the desiccant component, which can be a rotary enthalpic wheel filled with silica gel or other desiccant materials, and after being dehumidified and heated up is cooled via the direct or indirect evaporative cooler [4]. However, in the last years more compact solutions that incorporate the evaporative cooling and the desiccant action at the same time were developed as shown in [5],[3]. These new heat exchanger designs do not allow a continuous operation as the enthalpy wheel, they are transient systems where the silica gel dehumidify the air moisture until being full of water, then they are regenerated and the cycles can repeat. This transient behavior on a macroscale raises the problem of having a reliable measurement of the moisture content at the outlet of the heat exchanger. This manuscript shows the transient response of capacitive and chilled mirror hygrometers, trying to develop a dynamic model to simulate the transient response.

PRELIMINARY RESULTS AND CONCLUSIONS

The experiments were carried out for three air flow rates (100-360-500 kg/h), three values of temperatures (20-30-35) and three (20-35-60) values of relative humidity. The preliminary result analysis shows that the chilled mirrors are faster for coupled Temperature and relative humidity step changes, while for just a relative humidity change the two instruments perform in a similar fashion. This is due to the type of capacitive hygrometer tested that has a heavy metal head, drastically increasing its thermal inertia. Starting from the experimental data the dynamic models for the instruments will be developed and validated against experimental data.

MAIN REFERENCES

[1] P. di M. Department of Energy, "ReLab." [Online]. Available: http://www.relab.polimi.it/laboratorio/laboratorio/.

[2] Y. Yang, G. Cui, and C. Q. Lan, "Developments in evaporative cooling and enhanced evaporative cooling - A review," Renew. Sustain. Energy Rev., vol. 113, no. May, p. 109230, 2019.

[3] M. Beccali, P. Finocchiaro, M. Motta, and B. Di Pietra, "Monitoring and Energy Performance Assessment of the Compact DEC HVAC System 'Freescoo Facade' in Lampedusa (Italy)," Eurosun Conf. Proc., pp. 1–8, 2018.

[4] X. N. Wu, T. S. Ge, Y. J. Dai, and R. Z. Wang, "Review on substrate of solid desiccant dehumidification system," Renew. Sustain. Energy Rev., vol. 82, pp. 3236–3249, Feb. 2018.

[5] T. S. Ge, Y. Li, R. Z. Ã. Wang, and Y. J. Dai, "A review of the mathematical models for predicting rotary desiccant wheel," vol. 12, pp. 1485–1528, 2008.

Keywords: dynamic response, capacitive hygrometer, chilled mirror, dessicant system

Paper ID 30203: Improving the Reliability of theoretical approaches for hygrothermal characterization and modeling of building envelopes

Imane Oubrahim* (1,2,3), Thierry Duforestel (1,3), Rafik Belarbi (2,3)

1: EDF R&D, TREE EDF Lab Les Renardières, 77818 Moret-sur-Loing, France; 2: LaSIE UMR 7356, CNRS, La Rochelle Université, Avenue Michel Crépeau, 17042 La Ro- chelle Cedex 1, France; 3: 4ev Lab CNRS, Université de La Rochelle, Electricité de France EDF, Avenue Michel Cré- peau, 17042 La Rochelle Cedex 1, France

AIM AND APPROACH

The prevailing thermal methods turned out to be insufficient to deal with recent hygrothermal development in the building industry. Furthermore, the recent advances in dynamic modeling and advanced hygrothermal measurements also show their limitations when applied to these problems (Duforestel 2015). To summarize the situation, the simulated results have been in disagreement with the ones measured, chiefly in dynamic configurations. The analysis of these results led us to two essential weaknesses in the current methods: an underestimation of the vapor permeability of materials (linked to an experimental bias), and ignoring the hysteretic nature of moisture sorption in hygrothermal models. Therefore, a national project (SmartRéno) aims to complement our theoretical and experimental background to produce reliable tools for hygrothermal simulation, by investigating and redefining four main characteristics of building materials: The water vapor diffusion coefficient, the gas and liquid relative permeabilities and the sorption hysteresis.

Accordingly, this article will, first, present the precedent works that led us to this project. Then, the problematics to which we expect to respond and the methodology to be used to achieve the objectives will be detailed. Finally, we will put forward the first results of the work in relation to the objectives of this project.

SCIENTIFIC INNOVATION AND RELEVANCE

All of the work in this paper must lead to a complete corpus of hygrothermal characterization methods, accurate enough to meet the needs of hygrothermal studies, and to an updated heat and mass transfer simulation tool (SYRTHES, developed at EDF R&D) integrating these changes.

We will investigate the impact of changing each coefficient at a time in order to analyze and explain the impact of all individual changes on the overall behavior of the simulation tool. Then, we will integrate all the coefficients revisited in the tool and finally validate the latter by comparing it to existing static and dynamic experimental results.

PRELIMINARY RESULTS AND CONCLUSIONS

A sorption hysteresis model has been progressively identified based on the literrature. It has been integrated into the simulation tool SYRTHES for coupled heat and mass transfers. A comparison between experimental and simulated results has been carried out with various versions of the SYRTHES model:

- With an average curve of the mais sorption curves;

- Pure Mualem model;

- Pure linear model;

- Mixed model.

This comparison made it possible to highlight the impact of taking this phenomenon into account on the different calculated potentials.

A new experimental method for measuring water vapor diffusion coefficient is being tested on a varied range of materials used for the construction and renovation of building envelopes.

It is based on the principle of the cup test method with an adaptation to measure the difference in gas pressure between the two faces of the tested sample.

These tests are also being simulated in order to estimate the impact of the transfer characteristics on the expected results. It is expected that this new testing process will allow to determine the water vapor diffusion coefficient with an higher accuracy and will also allow the measurement of the gas relative permeability.

MAIN REFERENCES

-Mualem, Y. (1973). Modified approach to capillary hysteresis based on similarity hypothesis.

Water Resources Research, Vol. 9, No. 5.

- Mualem, Y. (1974). A conceptual model of hysteresis. Water Resources Research, Vol. 10, No. 3.

- Duforestel, T. (2015). Des transferts couplés de masse et de chaleur à la conception bioclima-

tique : recherches sur l'efficacité énergétique des bâtiments. Habilitaion à diriger des recherches, Faculté des Sciences et Technologies, Département de Mécanique.

- EDF, I. Rupp, C. Peniguel, Syrthes, version 4.3.6: https://www.edf.fr/groupe-edf/qui-sommes-nous/activites/notre-communaute-scientifique/syrthes.

- Projet SmartRéno (mars 2019 – juin 2021): https://smart-reno.recherche.univ-lr.fr/.

Keywords: Sorption Hysteresis, modeling, renovation, water vapor diffusion coefficient, gas relative permeability

Session T2.3: Buildings paving the way for the energy transition

Paper ID 30614: Thermal inertia analysis of family house heating system: towards higher energy flexibility

Vojtěch Zavřel*, Elizaveta Andreeva, Jakub Šimek, Martin Barták

CTU in Prague, Faculty of Mechanical Engineering, Czech Republic

AIM AND APPROACH

The aim of this research is to quantify energy flexibility potential of a typical family house with a heat pump driven heating system. The current research is in framework of a larger research project. The ultimate goal is to introduce and implement a supervisory control tool that maximize the energy flexibility in residential and office buildings and satisfy requirements of the future smart grid without compromising thermal comfort of occupants. The presented research can be considered as a preparatory phase to understand the dynamics of the controlled system.

The thermal inertia of a family house is studied using building energy simulation (BES) tool TRNSYS. The BES tool enables to create theoretical conditions to study step response of the heating system with a heat pump. The virtual experiments focus on the cooling down period of the indoor environment with respect to various setting of the wall structure and storage tank volume.

The research can greatly benefit from the physics-based nature of the BES tool, that support the deep understanding of the influential parameters of the heating system. The results of this study are used as background materials by control engineers for further development of the supervisory control.

SCIENTIFIC INNOVATION AND RELEVANCE

The energy sector in central Europe has been undergoing transformation towards carbon neutrality. For instance, Germany intends to provide at least 35% of gross electricity consumption through renewable energy sources by 2020 and 80% by 2050. The application of intermittent renewable energy sources (RES) is increasing and additional strategies must be considered to accommodate these sources. Building energy flexibility is relatively new phenomena allowing the controlled demand response, when the users can take active role at the energy market and control their demand upon the dynamic energy pricing. The premise of current research is that if the demand response strategies are applied at large scale, it would ultimately help to accommodate more intermittent renewable energy sources into the electrical grid and reduce the carbon footprint. To recall, the current research supports development of new strategies aiming to maximize the building sector flexibility and thermal storage capacity.

PRELIMINARY RESULTS AND CONCLUSIONS

The preliminary results revealed promising inter-daily storage potential of the building thermal inertia, that may support flexible operation of the heat pump without compromising indoor thermal comfort. The model represents a typical 2 stories family house. The theoretical step-response experiments were defined for constant boundary conditions of 0 °C outdoor temperature and no radiation. The heating system was set to maintain the representative indoor temperature at 25 °C and tank temperature at 55 °C. At the beginning of the experiment the heating system was switched off. The course of the representative temperatures during the cooling period indicated the time-constant of the system. Various building structures and tank configuration were investigated under the described boundary conditions. The discharge time of the tested storage tanks (range of 170 to 1000 l) was observed in range of 3 to 12 hours. The representative indoor temperature drops about 5 °C within 50 to 90 hours depending on the storage tank and construction settings. It can be concluded that the thermal inertia of the building and heating system can provide sufficient inter-daily storage capacity for heat pump flexible operation related to the dynamic grid conditions (e.g. dynamic price signal, grid CO2 intensity).

MAIN REFERENCES

S. Ø. Jensen et al., “IEA EBC Annex 67 Energy Flexible Buildings,” Energy Build., vol. 155, pp. 25–34, Nov. 2017.

Le Dréau, J. and Heiselberg, P. (2016) ‘Energy flexibility of residential buildings using short term heat storage in the thermal mass’, Energy, 111, pp. 991–1002. doi: 10.1016/j.energy.2016.05.076

S. Verbeke and A. Audenaert, “Thermal inertia in buildings: A review of impacts across climate and building use,” Renewable and Sustainable Energy Reviews, vol. 82. Elsevier Ltd, pp. 2300–2318, 01-Feb-2018.

K. Foteinaki, R. Li, A. Heller, and C. Rode, “Heating system energy flexibility of low-energy residential buildings,” Energy Build., vol. 180, pp. 95–108, 2018.

R. Li and S. You, “Exploring potential of energy flexibility in buildings for energy system services,” CSEE J. Power Energy Syst., vol. 4, no. 4, pp. 434–443, 2019.

Keywords: energy flexibility, thermal inertia, demand response, family house

Paper ID 30346: Comparing Machine Learning based Methods to standard Regression Methods for MPC on a virtual Testbed

Felix Bünning* (1,2), Corentin Pfister (1,2), Ahmed Aboudonia (2), Philipp Heer (1), John Lygeros (2)

1: Urban Energy Systems Laboratory, Empa, Switzerland; 2: Automatic Control Laboratory, ETH Zürich, Switzerland

AIM AND APPROACH

Model Predictive Control (MPC) can significantly reduce the energy consumption of buildings [1]. To reduce the modelling effort and related costs, data-driven methods for building MPC have been applied and tested recently in the area of residential buildings [2,3]. The approaches are based on Machine Learning (ML) methods such as Artificial Neural Networks (ANN) and Random forests, which have shown game-changing results in various fields, like games, speech and image recognition, in the recent past. However, our experiments with MPC on real residential buildings suggest that simple methods based on linear regression can lead to comparable or better results than advanced ML-based methods. To verify this observation, we apply the different methods to the residential building cases of the virtual testbed BOPTEST [4], which is designed to benchmark MPC algorithms for buildings, and analyze their performance in terms of prediction accuracy, control performance and sample efficiency.

SCIENTIFIC INNOVATION AND RELEVANCE

Data-driven generation of building models that are suitable for MPC is highly relevant for both science and industry, as the manual generation of physics-based models can be cost-prohibitive for the widespread application of MPC in buildings. Methods such as ANN could be a compelling choice for model generation as they are universal function approximators and have shown promising prediction performance in various fields. But similarly to the case of reinforcement learning in locomotion tasks [5], it is not clear whether advanced ML methods always show advantageous performance compared to simpler methods such as linear regression. This issue should be analyzed as many ML methods also come with noticeable disadvantages such as loss of linearity/convexity and physical interpretability.

PRELIMINARY RESULTS AND CONCLUSIONS

Our preliminary results in experiments suggest that two simple methods, one based on identifying a single state linear model with least-squares linear regression and one ARX model, both match or outperform models based on input-convex ANN and random forests in terms of prediction accuracy. We apply the same models to the BOPTEST simulation framework and analyze the different methods in terms of prediction accuracy, control performance, sample efficiency and other model properties (convexity etc.). The numerical results are not available yet, but will give a better understanding of advantages and disadvantages of the various approaches.

MAIN REFERENCES

[1] Sturzenegger, David, Dimitrios Gyalistras, Manfred Morari, and Roy S. Smith. "Model predictive climate control of a swiss office building: Implementation, results, and cost–benefit analysis." IEEE Transactions on Control Systems Technology 24, no. 1 (2015): 1-12.

[2] Smarra, Francesco, Achin Jain, Tullio de Rubeis, Dario Ambrosini, Alessandro D'Innocenzo, and Rahul Mangharam. "Data-driven model predictive control using random forests for building energy optimization and climate control." Applied energy 226 (2018): 1252-1272.

[3] Bünning, Felix, Benjamin Huber, Philipp Heer, Ahmed Aboudonia, and John Lygeros. "Experimental demonstration of data predictive control for energy optimization and thermal comfort in buildings." Energy and Buildings 211 (2020): 109792.

[4] Blum, David, Filip Jorissen, Sen Huang, Javier Arroyo, Kyle Benne, Yanfei Li, Valentin Gavan et al. "Prototyping the BOPTEST framework for simulation-based testing of advanced control strategies in buildings." (2019).

[5] Mania, Horia, Aurelia Guy, and Benjamin Recht. "Simple random search provides a competitive approach to reinforcement learning." arXiv preprint arXiv:1803.07055 (2018).

Keywords: Model Predictive Control, Machine Learning, Identification, Data Predictive Control

Paper ID 30432: Advanced Control of Dynamic Facades and HVAC with Reinforcement Learning based on standardized co-Simulation

Christoph Gehbauer (1), Andreas Rippl* (1,2), Eleanor Lee (1)

1: Lawrence Berkeley National Laboratory; 2: University of Applied Sciences Technikum Wien

AIM AND APPROACH

The aim of this paper is to demonstrate the use of standardized co-simulation models to develop an advanced controller for dynamic facade systems. The objective of the controller is to reduce the cost of electricity for the building owner and improve occupant comfort, while reducing stress on the electric power grid. Buildings equipped with dynamic facades and advanced control systems are able to shed and shift load, to enable further increase of Distributed Energy Resources (DER).

The controller is developed using reinforcement learning (RL), a discipline of machine learning which gained large interest after the success of DeepMind [2] and availability of low-cost computing hardware. RL is a technique where a controller "learns" to map situations to actions, based on a numerical reward given from an environment. This process utilizes high-fidelity simulation models to learn general policies, e.g., how to best operate buildings with dynamic facades. The models are developed in Modelica and Radiance, and are coupled through the standardized Functional Mockup Interface (FMI) [5] for co-simulation. FMI allows the rapid replacement of training models, and since neither FMI nor RL are domain specific, in theory, the framework could be used for developing any control strategy based on RL.

SCIENTIFIC INNOVATION AND RELEVANCE

Future building control strategies will have to consider the interdependency of technologies within buildings, such as dynamic facades, HVAC, and artificial lighting, as well as within the community, such as time-varying generation from DER. The FMI provides the flexibility for this variety of technologies, with its defined co-simulation interface. It is agnostic to simulation domains and complexities, which makes it particularly useful in the coupling of buildings with DERs.

The Lawrence Berkeley National Laboratory (LBNL) is currently conducting multiple projects which focus on auto-adaptive controls [3], with model predictive control (MPC) being one promising technology. However, due to high computational costs for calculating the actions by the MPC controller at each timestep, e.g., every 5 minutes, the hardware costs are typically higher than with simple heuristic control. Other than MPC control, RL requires less computing power at the edge device, e.g., at the controllable facade, since the computational work is done prior to deployment, during the training of the controller with the simulation environment. Once deployed, occupancy feedback or sensing data can be used to continue training the controller for the specific building.

PRELIMINARY RESULTS AND CONCLUSIONS

Using RL for control has the advantage of lower commissioning costs, since it ships pre-trained, and the control is able to adapt itself during operation based on feedback from sensors in the building. This suggests a higher resilience to changes in occupancy, building use and systems over the lifetime of the building. Furthermore, RL control is able to adapt to possible model mismatches between the simulation and the real building over time, an area where MPC-based systems have difficulties.

The co-simulation setup with FMI enables a flexible simulation environment where a rapid replacement of simulation models is possible. During the development of the controller the complexity of the system can be constantly increased. Thus, the progress of the development is not stopped by lengthy adjustments of the simulation environment.

MAIN REFERENCES

[1] Barth-Maron, G., Hoffman, M.W., Budden, D., Dabney, W., Horgan, D., TB, D., Muldal, A., Heess, N., Lillicrap, T., 2018. Distributed Distributional Deterministic Policy Gradients. ArXiv180408617 Cs Stat.

[2] Silver, D., Schrittwieser, J., Simonyan, K., Antonoglou, I., Huang, A., Guez, A., Hubert, T., Baker, L., Lai, M., Bolton, A., Chen, Y., Lillicrap, T., Hui, F., Sifre, L., van den Driessche, G., Graepel, T., Hassabis, D., 2017. Mastering the game of Go without human knowledge. Nature 550, 354–359. https://doi.org/10.1038/nature24270

[3] Gehbauer, C., Blum, D.H., Wang, T., Lee, E.S., 2020. An assessment of the load modifying potential of model predictive controlled dynamic facades within the California context. Energy Build. 210, 109762.

[4] Lillicrap, T.P., Hunt, J.J., Pritzel, A., Heess, N., Erez, T., Tassa, Y., Silver, D., Wierstra, D., 2015. Continuous control with deep reinforcement learning. ArXiv150902971 Cs Stat.

[5] Modelica Association, 2014. Functional Mock-up Interface. URL https://fmi-standard.org/ (accessed 7.11.20).

Keywords: reinforcement learning, machine learning, co-simulation, functional mock-up interface

Paper ID 30566: Analysis of the impact of predictive models on the quality of the Model Predictive Control for an experimental building

Arash Erfani* (1), Xingji Yu (2), Tuule Mall Kull (3), Peder Bacher (4), Tohid Jafarinejad (1), Staf Roels (1), Dirk Saelens (1)

1: Katholieke Universiteit Leuven, Belgium; 2: Norwegian University of Science and Technology, Trondheim, Norway; 3: Tallinn University of Technology, Tallinn, Estonia; 4: Technical University of Denmark, Lyngby, Denmark

AIM AND APPROACH

In this paper, application of different data-driven models are assessed in a Model Predictive Control (MPC) framework. The studied building is an experimental building located in Holzkirchen, Germany. The dwelling is equipped with a floor heating system in which hot water is supplied by an air-to-water heat pump. A simulation model of this building is developed in Modelica, which serves as the emulator. Afterwards, different data-driven models were developed to characterize the thermal behavior of the building. These models range from single-state grey-box models to more advanced artificial intelligence models such as Artificial Neural Networks (ANN) and Support vector Machine (SVM). In the next step, MPC has been formulated to assess the performance of the predictive controller based on each of the developed models. Performance of the controller is quantified based on Key performance indicators (KPIs). The KPIs used in this study are thermal discomfort level and electricity cost. T he models are compared in terms of specific KPIs as well. These KPIs include the coefficient of determination (R2) and Normalized Root Mean Square Error (NRMSE) fit.

SCIENTIFIC INNOVATION AND RELEVANCE

In line with decarbonization of the energy system, share of Renewable Energy Sources (RES) in the energy system is bound to increase. The uncertain nature of renewable energy sources affects all the energy consumers. Reliable utilization of RES demands good forecast of the availability of these sources. Integrating forecasts in common controllers in buildings is not readily possible if not impossible. Hence, applying more advanced control strategies is inevitable. Amongst various control strategies suggested for optimization of building's thermal performance, MPC is very promising. In spite of the abundance of these studies, MPC is still not commonly applied in buildings . One of the issues hampering easy and cost-efficient implementation of MPC in buildings is developing a predictive model for the building. This study aims at comparing different modeling techniques, which could be used to identify a building's thermal characteristic . Contrary to the common studies in this field, we take into account the multiple-step-ahead prediction error associated with each model as a KPI. In addition, two different indicators for model's accuracy are considered to quantify the models quality, and then it is determined which of these indicators is more suitable to score predictive models used in an MPC framework.

PRELIMINARY RESULTS AND CONCLUSIONS

In this study, various data-driven approaches have been developed to characterize thermal behavior of a dwelling. This study aims at assessing the impact of different modeling approaches for a building in an MPC framework. The models developed for this study are AutoRegressive with eXogenous inputs (ARX), grey-box with orders ranging from 1 to 3, black-box state space models with different orders, Support Vector Machine (SVM) and Artificial Neural network (ANN). Comparing model's KPI with the ones of the MPC shows that which of the model's KPIs reflect model's performance in the application better. The results of applying MPC with different models for 2 weeks in the heating season are presented. These results show that a model with an acceptable coefficient of determination could lead to an MPC with KPIs, which are reasonably close to the KPIs obtained by a high order model with better fit. The other conclusion of this work is the importance of taking multiple step-ahead prediction error and not just one-step ahead error into account while developing the model.

MAIN REFERENCES

1- A. Afram, F. Janabi-Sharifi, Theory and applications of HVAC control systems - a review of model predictive control (MPC),Build Environ, 72 (2014), pp. 343-355,

2- Bourdeau M., Zhai X.Q., Nefzaoui E., et al. Modelling and forecasting building energy consumption: a review of data-driven techniques

Sustainable Cities Soc., 48 (2019), Article 101533

3- Maarten Sourbron, Clara Verhelst & Lieve Helsen (2013) Building models for model predictive control of office buildings with concrete core activation, Journal of Building Performance Simulation, 6:3, 175-198,

4- A. Erfani, A. Rajabi-Ghahnaviyeh, M. Boroushaki, Design and construction of a non-linear model predictive controller for building's cooling system

Building and Environment, 133 (2018), pp. 237-245,

5- T.Y. ChenApplication of adaptive predictive control to a floor heating system with a large thermal lag

Energy Build., 34 (1) (2002), pp. 45-51

6- Z. Vana, J. Cigler, J. Siroky, E. Zacekova, L. FerklModel-based energy efficient control applied to an office building

J. Process Control, 24 (6) (2014), pp. 790-797,

Keywords: Model predictive Control, Building Modeling, Data-driven Characterization, Thermal Behavior Identification, Artificial Intelligence

Session T2.4: Improving indoor environmental quality

Paper ID 30198: Missing data imputation for diagnostic purposes in building system : application to an office setting

Houda Najeh*, Stéphane Ploix

G-SCOP laboratory, University of Grenoble Alpes,France

AIM AND APPROACH

In a building system, the diagnostic result is calculated from a set of tests ; each defined by a bunch of data, a behavioral constraint and a set of possible explanations in case of anomaly. To make a test, data are required from different sensors. However, missing data is a common issue in buildings. Because a detection tests relies on several variables and that all the data should be available along a given time horizon, performong tests could become an issue.

In the literature, many techniques have been proposed to impute missing data. However, detection tests could detect anomalies with usual types of imputation : it is indeed necessary that the imputed data don't generate false alarms. It is all the more important for diagnosis from first principle approach where detection of anomalies has a strong impact on the diagnosis.

This paper discusses the issue of imputation for diagnostic purposes i.e generate data so that satisfy test constraints. The proposed imputation algorithm is based on satisfying genetic algorithm which search iteratively for imputed data satisfying behavioral while preserving as much as possible statistical properties.

The efficiency of the method is evaluated on measurements obtained from a real building: an office setting.

SCIENTIFIC INNOVATION AND RELEVANCE

Imputing missing data for diagnosis in building is discussed in this paper. In a building system, the diagnostic result is calculated from a set of tests, each defined by a bunch of data, a behavioral constraint and a set of explanations in case of anomaly. The data gaps are the most important fault type in building. Throwing away data could lead to wrong diagnostic results due to the reduced sample size. So, rather than deleting the incomplete data sets, another approach is to impute the missing values.

At the begining, the limits of the imputation methods in the literature are highlithted. The innovation of this paper results in the fact that the detection tests could detect anomalies on the imputed data. The imputed data should not generate false alarms. The contribution is the imputation for diagnosis, ie under positive test constraints.

The proposed algorithm for imputation of missing data is a genetic algorithm based method that impute solution over the iterations.

It is based on four steps: random creation of the initial population, selection of individuals, mutations and crossover operations, computation of the selection criteria. The data are imputed with respect to behavioral constraint satisfaction and preservation of statistical properties.

PRELIMINARY RESULTS AND CONCLUSIONS

- This paper tackles the issue of data gaps imputation for diagnoses purposes and proposes a new methodology based on genetic algorithm techniques for imputing missing data.

- Genetic Algorithm is used to improve the accuracy of the imputed solution over the iterations.

- Missing data are imputed under constraints of satisfaction of behavioral constraints of a test of detection and conservation of statistical properties for the time series of data after imputation process

- A real case study is used for validation

MAIN REFERENCES

[1] Priya, R. D., & Kuppuswami, S. (2012). A genetic algorithm based approach for imputing missing discrete attribute values in databases. WSEAS Transactions on Information Science and Applications, 9(6), 169-178.

[2] Al-Douri, Y., & Hamodi, H. (2017). Data imputing using genetic algorithms (GA): A case study of cost data for tunnel fans. In International Symposium on Mine Planning and Equipment Selection, Luleå, August 29-31 2017 (pp. 205-208). Luleå tekniska universitet.

[3] Patil, D. V., & Bichkar, R. S. (2010). Multiple imputation of missing data with genetic algorithm based techniques. IJCA Special Issue on" Evolutionary Computation for Optimization Techniques, 74-78.

[4] Shahzad, W., Rehman, Q., & Ahmed, E. (2017). Missing data imputation using genetic algorithm for supervised learning. International Journal of Advanced Computer Science and Applications, 8(3), 438-445.

[5] Najeh, H., Singh, M. P., Ploix, S., Chabir, K., & Abdelkrim, M. N. (2020). Automatic thresholding for sensor data gap detection using statistical approach. In Sustainability in Energy and Buildings (pp. 455-467). Springer, Singapore.

Keywords: Imputation, missing values, genetic algorithm, building system, diagnosis

Paper ID 30727: Approaches toward multi-aspect indoor-environmental performance indicators

Ardeshir Mahdavi*, Christiane Berger

Department of Building Physics and Building Ecology, TU Wien

AIM AND APPROACH

In earlier contributions [1, 2], a conceptual differentiation was suggested between three kinds of building performance simulation results. Some performance indicators, such as heating and cooling loads, are not directly relevant to buildings' indoor-environmental performance. A second class of indicators, such as task illuminance levels or room air CO2 concentrations are relevant to occupancy requirements (e.g., visual comfort, indoor air quality). But they do not have direct phenomenal correlates: People do not see illuminance or sense CO2 concentration. But these performance indicators may be linked to a third class of indicators that is not only relevant to human occupancy, but also correlates directly with phenomenal experience. Examples of such indicators are luminance of light sources, indoor air temperature, and sound pressure level. Certain characteristics of these last two categories are relevant to the present contribution. Specifically, indicators in these categories are frequently obtained for isolated aspects of indoor environment, i.e., thermal, visual, olfactory, or auditory. Aggregation of the values obtained for such isolated aspects into the value of a unified indoor-environmental quality indicator remains a challenge. The present treatment explores the potential of three strategic paths toward formulation and operationalization of multi-aspect indoor-environmental performance indicators.

SCIENTIFIC INNOVATION AND RELEVANCE

Recent state of the art reviews [3, 4, 5] in the framework of IEA Annex 79 [6] have documented the paucity of in-depth knowledge concerning the cross-modal and combined effects of multiple indoor-environmental factors on building inhabitants' perception and behavior. A further insight emerges from an ongoing review [7] of national and international indoor-environmentally relevant multi-aspect codes, standards, and guidelines: The procedures employed in these resources to aggregate quality assessments in separate domains into a unified evaluation scale appear to be rather opaque, if not arbitrary. The results of these investigations underline the need for the formulation and pursuit of strategies to address the challenge of defining aggregate (multi-domain) indoor-environmental performance indicators. Advances is this area are directly relevant and beneficial toward more effective scenarios of performance simulation deployment for building design and operation support.

PRELIMINARY RESULTS AND CONCLUSIONS

The present contribution outlines and exemplifies three strategic paths to address the aforementioned challenge, namely the development of multi-domain performance indicators. Illustrated briefly, these three paths are as follows: i) The relative weight of single-domain performance indicators toward the derivation of a unified indoor-environmental performance indicator is pursued following a Bayesian strategy involving iterative refinement of judgements elicited from expert panels. The weights matrix is structured in relation to a variety of building types and contexts; ii) Existing and future large-scale survey results from building users provide the information basis of the second path. Thereby, subjective single-domain evaluations of various aspects of sample environments are mapped into a unified formalism with calibrated coefficient for contributing single-domain variables; iii) The third path, which is at once the most rigorous and the most demanding, requires an empirically grounded understanding of the emergence of the unified field of perception from the vantage point of physiology, neuroscience, and cognitive psychology. The aforementioned reviews [3, 4, 5] reveal the inclusive state of the art in this area. The contribution is thus expected to provide a systematic map of necessary short-term, mid-term, and long-term efforts toward establishing computable cross-domain indoor-environmental performance indicators.

MAIN REFERENCES

1. Mahdavi, A. 2004. Reflections on computational building models. Building and Environment. Volume 39, Issue 8, August 2004. ISSN 0360-1323. pp. 913 – 925.

2. Mahdavi, A. 2011. The Human Dimension of Building Performance Simulation. Driving better design through simulation (keynote). Proceedings of the 12th Conference of The International Building Performance Simulation Association, Sydney, Australia. ISBN: 978-0-646-56510-1; pp. K16 - K33.

3. Schweiker, M. et al. 2020. Review of multi-domain approaches to indoor environmental perception and behaviour. Building and Environment. 176. 106804. doi: 10.1016/j.buildenv.2020.106804.

4. Mahdavi, A. et al. 2020a. Understanding multi-aspect indoor-environmental exposure situations: Past insights and future needs. The 16th Conference of the International Society of Indoor Air Quality & Climate (Indoor Air 2020) COEX, Seoul, Korea (to be published).

5. Mahdavi, A. 2020b. Explanatory stories of human perception and behavior in buildings. Building and Environment, 168; 106498.

6. IEA EBC Annex 79. https://annex79.iea-ebc.org/.

7. Mahdavi, A. et al. 2020b. Necessary conditions for multi-domain indoor environmental quality standards (to be published).

Keywords: Building simulation, indoor-environment, multi-domain performance indicators

Paper ID 30147: Comparison of overheating risk in nearly zero energy dwelling based on three different overheating calculation methods

Shady Attia* (1), Ramin Rahif (1), Abdulrahman Fani (1), Mohamed Amer (2)

1: Univeristé de Liège, Belgium; 2: Renowatt, Liege, Belgium

AIM AND APPROACH

This study aims to investigate the vulnerability of dwellings to overheating risk in Belgium. The objectives consist of establishing an energy simulation model of a verified case study within the Walloon Region and assessing its overheating risk based on the three overheating indicators. A five years monitored nearly zero energy house is used as a reference building. A validated building performance simulation model is created and validated in EnegryPlus. By exploring a large body of the literature and standards, we decided to select three overheating assessment methods. The three overheating risk indicators comprise 1)the Belgian EPBD overheating indicator, 2) the Passive House overheating indicator, and 3) the ambient warmness degree and indoor overheating degree indicators developed by Hamdy et al. (2017). Finally, the potential of ventilative cooling is assessed in the mitigation of overheating risk. The abovementioned method is applied to a lightweight single-family nearly zero energy with a total surface area of 174 m2 located in Eupen, Belgium.

SCIENTIFIC INNOVATION AND RELEVANCE

This study applies state-of-the-art method for overheating assessment of new timber construction in Belgium. The innovation lies in the comparative approach that provides an accurate evaluation for a highly insulated and airtight building. The study combines in-situ measurements of building thermo-physical characteristics and energy use monitoring with building performance simulation.

Our research contributes to overheating evaluation methods in residential buildings. The research methodology used in this research is part of the IEA Annex 80 activities on resilient cooling for buildings. The method of Hamdy et al. (2017) is based on EN 16798 Part 1 and 2. It runs an aggregated overheating value every time step based on the operative temperature, taking into account the solar radiation and thermal mass effects. The outcomes can be used to assess the cooling systems' performance and effectiveness of passive adaptation measures in a changing climate. Overall, this paper provides an essential basis to improve indoor thermal conditions and climate change resilient design of buildings, which are within the scope of BS 2021 conference.

PRELIMINARY RESULTS AND CONCLUSIONS

Results confirm a 216% difference in the overheated hours between the EPBD method and the used method of Hamdy et al. 2017. As an additional contribution of the study relates to ventilative cooling, which is assessed within the simulation model, proving an 82% potential as a mitigation strategy.

The results of the proposed study emphasize the need to improve the current legislative framework of the Walloon Region by adopting the set of long-term thermal discomfort indicators that allow the classification of the building regarding future projections, and the necessity of adapting dynamic calculations instead of static based estimation.

To conclude, this research gives an insight into whether nearly zero energy timber buildings are able to suppress the increase in average outdoor temperature in future climates. It also reveals the potential of ventilative cooling as a mitigation strategy and predicts to what extend its potential will decrease as global warming increases.

MAIN REFERENCES

Butcher, K., and Craig, B. "Chartered Institution of Building Services Engineers," Environ. Des. CIBSE Guide Eighth Ed. Ed CIBSE Guide Chart. Inst. Build. Serv. Eng. Lond., 2015.

Carlucci, S., & Pagliano, L. (2012). A review of indices for the long-term evaluation of the general thermal comfort conditions in buildings. Energy and Buildings, 53, 194-205.

Carlucci, S., Bai, L., de Dear, R., & Yang, L. (2018). Review of adaptive thermal comfort models in built environmental regulatory documents. Building and Environment, 137, 73-89.

CEN, EN 16798-1: Energy performance of buildings–ventilation for buildings–Part 1: Brussels, 2019.

Goethals, K., Coppieters, L., Delghust, M., & Janssens, A. (2013). Evaluation of the quasi-steady-state overheating indicator for passive houses. In 11th REHVA World Congress. Society of Environmental Engineering.

Hamdy, M., Carlucci, S., Hoes, P. J., & Hensen, J. L. (2017). The impact of climate change on the overheating risk in dwellings—A Dutch case study. Building and Environment, 122, 307-323.

Khovalyg, D., Kazanci, O. B., Halvorsen, H., Gundlach, I., Bahnfleth, W. P., Toftum, J., & Olesen, B. W. (2020). Critical review of standards for indoor thermal environment and air quality. Energy and Buildings, 109819.

Keywords: thermal discomfort, ambient warmness degree, Indoor overheating degree, EPBD, near-zero energy

Paper ID 31040: Influence of zoned use and occupancy on Daylight Autonomy in high-rise mass housing in Colombia

Alejandra Jiménez Gómez*, Alejandro Naranjo Ortiz, Elizabeth Parra Correa, Lucas Arango Díaz, Jorge Hernán Salazar Trujillo

Universidad Nacional de Colombia, Colombia

AIM AND APPROACH

The objective of this research focuses on evaluating the autonomy of natural light (DA) in a low-budget housing space in Colombia, based on the considerations of its use and occupation. Physic-spatial characteristics of various dwellings were defined to carry out computational simulations of natural light autonomy (DA) through the Grasshopper plug-in for Rhino. Consider the next parameters: a) window/wall ratio, b) type of glass, c) position of the window and d) depth of balconies. Besides, it included a proposal for evaluating the autonomy of natural light (DA) in domestic spaces in the Colombian tropics according to the combined light output of different controlled areas in the same environment, as a result of the various visual tasks carried out in it.

SCIENTIFIC INNOVATION AND RELEVANCE

In the Colombian tropics, in low-budget housing, it is common for the social zone to be used as a workspace. This particularity calls for studies of the light output of this type of space. The novelty and relevance of this work focus on evaluating the light performance of low-budget homes in the Colombian tropics, as well as on the proposal of a method to lightly evaluate spaces based on the division into zones. However, based on the singularity of use and occupation of these homes, it is decided to propose a different use of the dynamic metrics given the context and assuming that it is a benefit for energy consumption, contrary to a single assessment for the entire space in which due to the nature of the space under study would cast a negative in terms of its lighting since it is based on the hypothesis that it is more relevant to make evaluations by zones in each space than to assess the entire space. In this way, it would be possible to identify, for this type of space, which areas are more suitable according to the visual task.

PRELIMINARY RESULTS AND CONCLUSIONS

The preliminary results show that the valuation of space from the areas that compose it allows us to identify how many opportunities for grouping or occupation there are and what tasks can be carried out to achieve a higher light output. It is believed that in tropical conditions with high daytime variability of solar radiation conditions it is desirable to use the standard metrics differently and hope to perhaps define additional metrics. The proposed method demonstrates potential, as it demonstrates the importance of zoning spaces according to their use and occupation, in addition to the good use and exploitation of the elements that make up the facade to obtain an adequate entrance and distribution of natural light in the space.

MAIN REFERENCES

Bin Dalmouk, M., & Al-Sallal, K. A. (2020). Modeling Daylight in Adjacent Spaces of the Courtyard under Clear Sky Conditions. Proceedings of Building Simulation 2019: 16th Conference of IBPSA, 16, 990–995. https://doi.org/10.26868/25222708.2019.210307

Elzeyadi, I. M. K., & Batool, A. (2017). Veiled Facades: Impacts of Patterned-Mass Shades on Building Energy Savings, Daylighting Autonomy, and Glare Management in Three Different Climate Zones Solar screens environmental performance. October. https://doi.org/10.1016/0360-1315(82)90012-4

Fela, R. F., Utami, S. S., Mangkuto, R. A., Suroso, D. J., & Mada, G. (2019). The Effects of Orientation , Window Size , and Lighting Control to Climate-Based Daylight Performance and Lighting Energy Demand on Buildings in Tropical Area Department of Nuclear Engineering and Engineering Physics , Faculty of Engineering , Universitas. 1075–1082.

Potočnik, J., & Košir, M. (2020). Influence of commercial glazing and wall colours on the resulting non-visual daylight conditions of an office. Building and Environment, 171(December 2019). https://doi.org/10.1016/j.buildenv.2019.106627

Keywords: Daylight autonomy, living place, light performance, parametric computational simulation, architecture

Paper ID 30677: Assessing the current and future risk of overheating in London's care homes: The effect of passive ventilation

Ioanna Tsoulou (1), Nishesh Jain (1), Eleni Oikonomou (1), Giorgos Petrou* (1), Alastair Howard (2), Rajat Gupta (2), Anna Mavrogianni (1), Ai Milojevic (3), Paul Wilkinson (3), Michael Davies (1)

1: University College London, United Kingdom; 2: Oxford Brookes University, United Kingdom; 3: London School of Hygiene and Tropical Medicine, United Kingdom

AIM AND APPROACH

Climate change increases the likelihood of warmer summers and heat waves with adverse effects on thermal comfort and human health [1]. Older people and those who are physically frail are more vulnerable to heat, [3], including care home residents. The World Health Organization projects that heat-related morbidity and mortality will increase appreciably between 2030 and 2050 [6]. However, the heat risks to older people and how those risks may be reduced by building adaptations have not been quantified [2].

In this paper, we present findings from a study of thermal environments in an old care home and a modern care home in London, UK, using empirical observations and building temperature simulations. Indoor temperature and relative humidity data were obtained through continuous monitoring during the summer of 2019 and used to calibrate thermal simulations (EnergyPlus V8.9 via the DesignBuilder Graphical user Interface) of the indoor environment with and without passive cooling measures such as shading and natural ventilation.

SCIENTIFIC INNOVATION AND RELEVANCE

Our research represents a novel method to evaluate heat-related risks to health in care settings under a representative set of climate change scenarios. We combine temperature and humidity measurements into a heat index as an indicator of thermal (dis)comfort and heat stress [4, 5] to supplement dry bulb temperature data. The results of our thermal models add to evidence on the effect of passive ventilation in care homes with contrasting building characteristics. Coupled with perspectives from public health officials, care home practitioners and stakeholders involved in built-environment preparedness for vulnerable groups, the research will help inform public health guidance on protection against heat risks in care homes. It will also be relevant to building standards and guidelines with respect to the effect of low-energy and easy-to-implement building adaptations in reducing the risk of overheating.

PRELIMINARY RESULTS AND CONCLUSIONS

Our work on thermal modeling utilized CIBSE's test reference year (TRY) and design summer year (DSY1) weather files based on UK Climate Projections 2009 (UKCP09) and under the high emissions, 50% percentile scenario. Preliminary results showed that the mean daily indoor heat index in the two case study care homes is likely to increase by 2 °C by the 2050s and by 4 °C by the 2080s. Furthermore, they suggest that night-time window opening has a high impact in reducing heat index in both care home settings. Yet, the overall effectiveness of shading and passive ventilation in reducing overheating was highly varied, depending on the age of care home (and related insulation/heat loss), floor number, orientation, duration and time of day of window opening. The knowledge gained from the study will help care providers and relevant stakeholders from the built environment and policy development in adapting old and modern care homes for a warming climate.

MAIN REFERENCES

1. IPCC. (2014). Climate change 2014: Synthesis report. Contribution of Working Groups I, II and III to the Fifth Assessment Report of the Intergovernmental Panel on Climate Change. . In R. K. P. a. L. A. Meyer (Ed.). Geneva, Switzerland. from https://www.ipcc.ch/report/ar5/syr/

2. Kingsborough, A., Jenkins, K., & Hall, J. W. (2017). Development and appraisal of long-term adaptation pathways for managing heat-risk in London. Climate Risk Management, 16, 73-92.

3. Kovats, R. S., & Hajat, S. (2008). Heat stress and public health: a critical review. Annu. Rev. Public Health, 29, 41-55.

4. Quinn, A., Tamerius, J. D., Perzanowski, M., Jacobson, J. S., Goldstein, I., Acosta, L., & Shaman, J. (2014). Predicting indoor heat exposure risk during extreme heat events. Science of the total environment, 490, 686-693.

5. Tsoulou, I., Andrews, C. J., He, R., Mainelis, G., & Senick, J. (2020). Summertime thermal conditions and senior resident behaviors in public housing: A case study in Elizabeth, NJ, USA. Building and Environment, 168, 106411.

6. World Health Organization. (2014). Quantitative risk assessment of the effects of climate change on selected causes of death, 2030s and 2050s.

Keywords: Overheating, Care Homes, Passive Ventilation, Thermal Modeling, Thermal Comfort

Session T2.5: Climate change and bioclimatic design

Paper ID 30762: A modeling framework for building-integrated biotic Carbon Sequestration (CS) techniques: Towards mitigating climate change

Jayati Chhabra*, Tarek Rakha

Georgia Institute of Technology, United States of America

AIM AND APPROACH

While several methods have been previously developed to assess the climatic impacts of a building's landscaping elements, built environment practice still lacks a framework to determine the carbon sequestration (CS) potential of such architectural features. This paper proposes a modeling framework to estimate the amount of carbon that can be sequestered by a structure that utilizes landscaping elements to enhance environmental performance. The proposed approach focuses specifically on three techniques that have already been demonstrated and evaluated for carbon absorption in practice: green roofs; vertical greenery systems (VGS); and algae facades/curtains. The proposed modeling framework accounts for site and climate-based variations in solar radiation across the globe, as well as different plant types, species, the type of photobioreactors in the case of micro-algae, and their energy conversion efficiency ratios. The model aims to enable architects, engineers, and high-performance building specialists to plan, design and inform their decisions on the area and orientation of building elements that need to be covered by a single or a combination of techniques at an early design stage to sequester carbon, in an effort to not just decrease the environmental impact of buildings, but also contribute to mitigating the effects of climate change.

SCIENTIFIC INNOVATION AND RELEVANCE

Anthropogenic climate change is a reality of our time. Apart from the target of reducing greenhouse gas emissions, it is important that the built environment starts contributing to carbon sequestration (the process of capturing carbon from the atmosphere and storing it securely) at a massive scale. With limited tools, guidelines, and resources regarding the implementation of CS techniques in buildings, it can be a challenging task for architects and designers to incorporate such techniques in their projects. This research, thus, proposes an innovative simulation-based work-flow to quickly measure and analyze the CS potential of built environment projects through the application of various techniques appropriate for their building based on region, cost, and area. Weather files of any location are used to calculate the solar radiation and Photosynthetically Active Radiation (PAR). Consequently, based on the energy conversion ratio (ECR) and roof, facade or window area where the technique is to be applied, total solar energy transformed into usable biomass is computed. Finally, the estimated dry weight of the biomass from the energy is used to determine the sequestered carbon content, which is specifically relevant to the advancement of climate change conscious architectural and built environment praxis.

PRELIMINARY RESULTS AND CONCLUSIONS

The output from this paper showcases the CS potential of the three techniques in the context of five major Koppen classified climate zones – tropical, dry, moderate, continental, and polar. The work-flow for algae facades is validated against measured data from collected information in practice from the BIQ house in Germany and Photo.synth.etica by EcologicStudio. A work-flow is formulated specifically for green roofs and VGS from previously published literature by (Getter et al. 2009), (Whittinghill et al. 2014), (Pulselli et al. 2014), and (Kuronuma et al. 2018). The primary focus is on the varying effect of solar radiation, which ranges from 788.036 kWh/m2 (Antarctica) to 2166.897 kWh/m2 (India) in the tested sites. Results show that micro-algae with a theoretically maximum ECR of 8% can sequester up to 20.8 kgCO2/m2 annually (equivalent to the CO2 amount emitted by two cars in a day). Green roofs and VGS, can sequester carbon using an ECR of 0.4%-0.7%, even though the theoretical maximum for plants ranges from 4.6% - 6%. The paper concludes by discussing a process for strategic planning of CS techniques during the design stage, which may lead to buildings contributing to the efforts of climate change mitigation.

MAIN REFERENCES

Amir, Atikah F, Foong S Yeok, and Abdul M A Rahman. 2014. "Estimation of Annual Carbon Sequestration in Psophocarpus Tetragonobulus Used as Biofacade in Tropical Environment," 31–37.

https://doi.org/10.17758/ur.u1214310.

ecoLogicStudio. 2019. "SYSTEM PhotoSynthetica." 2019. https://www.photosynthetica.co.uk/system.

Getter, Kristin L., D. Bradley Rowe, G. Philip Robertson, Bert M. Cregg, and Jeffrey A. Andresen. 2009. "Carbon Sequestration Potential of Extensive Green Roofs." Environmental Science & Technology 43 (19): 7564–70. https://doi.org/10.1021/es901539x.

Kuronuma, Takanori, Hitoshi Watanabe, Tatsuaki Ishihara, Daitoku Kou, Kazunari Toushima, Masaya Ando, and Satoshi Shindo. 2018. "CO2 Payoffof Extensive Green Roofs with Different Vegetation Species." Sustainability (Switzerland) 10 (7): 1–12. https://doi.org/10.3390/su10072256.

Pulselli, R. M., F. Saladini, E. Neri, and S. Bastianoni. 2014. "A Comprehensive Lifecycle Evaluation of Vertical Greenery Systems Based on Systemic Indicators." WIT Transactions on Ecology and the Environment 191:1017–24. https://doi.org/10.2495/SC140862.

Whittinghill, Leigh J., D. Bradley Rowe, Robert Schutzki, and Bert M. Cregg. 2014. "Quantifying Carbon Sequestration of Various Green Roof and Ornamental Landscape Systems." Landscape and Urban Planning 123: 41–48. https://doi.org/10.1016/j.landurbplan.2013.11.015.

Wurm, Jan, and Martin Pauli. 2016. "SolarLeaf: The World's First Bioreactive Façade." Architectural Research Quarterly 20 (1):73–79.

https://doi.org/10.1017/S1359135516000245.

Keywords: Carbon Sequestration, Climate Change, Landscaping, Mitigation, Simulation

Paper ID 30315: Generating future-urban weather files for building performance simulations: case studies in London

Agnese Salvati*, Maria Kolokotroni

Brunel University London, United Kingdom

AIM AND APPROACH

The indoor environmental quality of buildings may be significantly compromised in the next decades due to the overlapping effects of climate change and urban warming. Climate change will determine an increase of air temperature which will be amplified in cities, due to the heat island (UHI) effect and the reduced wind speed in urban fabric. Considering that urban population and population ageing are increasing, the ability to predict the thermal behaviour of buildings under future-urban weather conditions is crucial to prevent serious health risks to the most vulnerable population groups due to building overheating.

This study presents a methodology to generate future-urban weather files to perform building performance simulations. The methodology considers the global climate change predictions based on the IPCC scenarios for 2050 for London and the local climate modifications determined by different urban areas across the city. A naturally ventilated residential buildings is modelled in EnergyPlus to assess the impact of changing outdoor boundary conditions on the building heating and cooling demand. The results are compared to highlight the relative impact of climate change and urban effects on buildings across the city of London.

SCIENTIFIC INNOVATION AND RELEVANCE

While many methods have been developed to generate future weather files, much less literature is available on methodologies to include urban climate modifications in future weather files. This study contributes to fill this gap, by presenting a methodology to generate future-urban weather files for London, that can be applied also to other cities.

First, future weather files are generated by dynamic downscaling of global climate models predictions (using the CORDEX datasets) and performing a bias-correction of the modelled data using historical observations. Second, the local climate modifications determined by the varying characteristic of the urban fabric across the city are included by using the Urban Weather Generator (UWG) and the URBVENT wind models. To this aim, the test building is assumed to be located in three different areas across the city: 1) the mixed-use and dense city centre, 2) a typical urban residential area and 3) a sub-urban low-density area. Historical observations of air temperature in the three areas are used to calibrate the UWG models. The results allow to understand the relative and overlapping impact of climate change and urban effects on the heating and cooling demand of residential buildings in the future (2050).

PRELIMINARY RESULTS AND CONCLUSIONS

The climate projections for London show a higher increase of air temperature in summer months than winter months. Similarly, the UHI intensity is stronger in summer than in winter. Therefore, the combined effect of climate change and UHI intensity will strongly modify the annual share of the building cooling and heating needs.

In the present weather conditions, for the building located in the suburban location where the UHI is negligible, the annual energy demand consists of 62% for heating and the 38% for cooling. In the future climate scenario, the share will become 53 % for heating and 47% for cooling in the suburban location and 47% for heating and 53% for cooling in the typical residential area of London. A higher impact is expected for the densest urban area.

In conclusion, the preliminary results indicate that urban climate is responsible for a significant increase of the cooling demand and needs to be included in the assessment of the future energy performance of urban building in London. The modelling procedure can be applied also to other cities and could be useful for a more accurate prediction of the impact of climate change on building in urban areas.

MAIN REFERENCES

Machard, A., Inard, C., Alessandrini, J.-M., Pelé, C., & Ribéron, J. (2020). A Methodology for Assembling Future Weather Files including Heatwaves for Building Thermal Simulations from the European Coordinated Regional Downscaling Experiment (EURO-CORDEX) Climate Data. Energies, 13, 3424.

Salvati, A., Palme, M., Chiesa, G., & Kolokotroni, M. (2020). Built form, urban climate and building energy modelling: case-studies in Rome and Antofagasta. Journal of Building Performance Simulation, 13(2), 209–225.

Salvati, A., Monti, P., Coch Roura, H., & Cecere, C. (2019). Climatic performance of urban textures: Analysis tools for a Mediterranean urban context. Energy and Buildings, 185, 162–179.

Salvati, A., & Kolokotroni, M. (2019). Microclimate Data For Building Energy Modelling: Study On ENVI-Met Forcing Data. In V. Corrado & A. Gasparella (Eds.), Proceedings of the 16th IBPSA Conference (pp. 3361–3368). Rome, Sept. 2-4, 2019.

Kolokotroni, M., Ren, X., Davies, M., & Mavrogianni, A. (2012). London's urban heat island: Impact on current and future energy consumption in office buildings. Energy and Buildings, 47, 302–311.

Kolokotroni, M., & Giridharan, R. (2008). Urban heat island intensity in London: An investigation of the impact of physical characteristics on changes in outdoor air temperature during summer. Solar Energy, 82(11), 986–998.

Keywords: Future-urban weather files, Urban climate, Climate change, Urban Heat Island, Urban Weather Generator

Paper ID 30862: A bootstrap method to investigate the variability of overheating risk against the future climate uncertainty in dwellings

Cheng Cui*, Rokia Raslan, Ivan Korolija

Institute for Environmental Design and Engineering, University College London, London, UK

AIM AND APPROACH

Future overheating risk in dwellings can be potentially mitigated by minimising the variability of overheating hours against uncertain future climate via robust optimisation. However, the estimation of this variability value through the utilisation of percentile-based probabilistic weather data has yet to be sufficiently investigated. In recognition of the knowledge gap in deriving robustness information via the percentile-based probabilistic weather data to facilitate robust optimisation for building design, this paper aims to apply Bayesian bootstrapping to investigate the variability of domestic overheating risk, against the future climate uncertainty.

SCIENTIFIC INNOVATION AND RELEVANCE

1. Quantification of the estimation accuracy of the overheating variability against climate uncertainties under different weather scenarios;

2. Proposed use of the skew normal distribution as an alternative for an adequately accurate estimation of overheating variability in the case of limited computational resources.

PRELIMINARY RESULTS AND CONCLUSIONS

The results indicate significant overheating risk in regulation-compliant houses. An increased degree of difficulty is also suggested in obtaining accurate estimations when considering time periods further in the future and when using higher emissions assumptions. In addition, the skew normal distribution can be used for a simpler and faster estimation, but the underlying uncertainties must be strengthened throughout its implementation.

MAIN REFERENCES

van Hooff, T., B. Blocken, J. L. M. Hensen, andH. J. P. Timmermans (2014, December). On thepredicted effectiveness of climate adaptation mea-sures for residential buildings.BUILDING ANDENVIRONMENT 82, 300–316.

Wright, J., E. Nikolaidou, and C. J. Hopfe (2016,September). Exhaustive search: Does it have arole in explorative design? InProceedings of the3rd IBPSA-England Conference BSO 2016, GreatNorth Museum, Newcastle, UK, pp. 8. IBPSA-England.

Chernick, M. R. (2008).Bootstrap Methods: A Guidefor Practitioners and Researchers. Hoboken, N.J.:Wiley-Interscience.

Keywords: uncertainty, climate change, retrofit, resilience, overheating

Session T2.6: Ensuring high quality building simulations

Paper ID 30778: Dimensionality reduction for calibration of dynamic building simulation models

Aurelien Bres*

AIT Austrian Institute of Technology, Austria

AIM AND APPROACH

The calibration of dynamic building simulation with subhourly data can involve comparing a high number of simulated and measured values, whereby the predictability and informativeness of values at a single time step may be limited. It can be assumed that dimensionality reduction, i.e. transforming the simulation output data to a space of lower dimension while retaining significant features, has the potential to improve the computational efficiency of calibration.

This contribution aims at investigating the use of linear and nonlinear dimensionality reduction techniques for the calibration of dynamic building simulation models involving uncertainties pertaining to both building and heating systems. Two calibration cases are considered: in a first case ("open loop"), heat generation is assumed to be known and fixed and the response of building and storage components is calibrated. In a second case ("closed loop"), heat generation in closed-loop control is also subject to calibration.

The techniques used for dimensionality reduction are principal component analysis (linear) one the one hand, and on the other hand nonlinear feature extraction based on artificial neural networks. A Bayesian calibration method is then applied to the corresponding outputs with reduced dimension, and its performance is assessed with synthetic data.

SCIENTIFIC INNOVATION AND RELEVANCE

Calibration of building performance simulation is a key approach for ensuring high-quality models and an active area of research. In the last decades, calibration practice has generally moved from monthly to daily, hourly and subhourly values. This has made the challenge of dealing with (high-dimensional) multivariate output data increasingly relevant.

Chong et al. (2017) addressed Bayesian calibration of building energy models with high temporal resolution, and reduced the size of the output by selecting a representative subset based on methods of information theory. Higdon et al. (2008) used linear dimensionality reduction in combination with Bayesian calibration of models with high-dimensional output. Remmen et al. (2019) reported a massive reduction in computational costs when performing Bayesian calibration of building simulation on statistical indices rather than using whole time series, but noted that some (time-varying) parameters could not be calibrated with this method.

The proposed contribution is the first to compare linear and nonlinear dimensionality reduction techniques in the context of Bayesian calibration for dynamic building simulation.

The corresponding improvement in calibration efficiency is of high relevance for calibration endeavours based on high resolution building monitoring data.

PRELIMINARY RESULTS AND CONCLUSIONS

The applicability of linear versus nonlinear dimensionality reduction is shown to be dependent on the calibration case: in the open loop case, linear dimensionality reduction makes it possible to explain a high share of the variance in system response with a few linear components, based on which efficient calibration is possible. In the closed loop case, system behavior at individual points in time is less predictable, and significant components are not identified by linear dimensionality reduction. In this case, nonlinear dimensionality reduction based on artificial neural networks still makes it possible to find features of the output with a significant dependency on the inputs. In both cases, dimensionality reduction is shown to yield more informative values for calibration than daily aggregated values, as demonstrated by a reduction in parameter uncertainty with synthetic data experiments.

MAIN REFERENCES

Chong, A., Lam, K. P., Pozzi, M., & Yang, J. (2017). Bayesian calibration of building energy models with large datasets. Energy and Buildings, 154, 343-355.

Higdon, D., Gattiker, J., Williams, B., & Rightley, M. (2008). Computer model calibration using high-dimensional output. Journal of the American Statistical Association, 103(482), 570-583.

Kent, M. G., Schiavon, S., & Jakubiec, J. A. (2020). A dimensionality reduction method to select the most representative daylight illuminance distributions. Journal of Building Performance Simulation, 13(1), 122-135.

Reddy, T. A., Maor, I., & Panjapornpon, C. (2007). Calibrating detailed building energy simulation programs with measured data—Part I: General methodology (RP-1051). Hvac&R Research, 13(2), 221-241.

Remmen, P., Schäfer, J., & Müller, D. (2019). Refinement of Dynamic Non-Residential Building Archetypes Using Measurement Data and Bayesian Calibration.

Keywords: dimensionality reduction, Bayesian calibration, multivariate outputs, calibration

Paper ID 30852: Reducing the energy performance gap on building stock level using actual energy data

Paula van den Brom*, Laure Itard

Delft University of Technology, The Netherlands

AIM AND APPROACH

The aim of this paper is to reduce the energy performance gap on building stock level. One of the most important explanations for the energy performance gap are invalid assumptions for building components and building installation settings.When existing buildings are simulated it is often not possible to determine the insulation rate of the building without destructive research, or to determine the average indoor temperature without measuring in many houses. Because of these difficulties, assumptions are made for the input of building energy simulation models. In this paper annual actual energy consumption data of multiple dwellings is used to calibrate building energy simulation models on building stock level. For the calibration, which can be expressed as an optimisation, machine learning algorithms are used. The method is based on the concept that the building energy simulation model can determine a more correct input by "learning" from measured annual energy data.

SCIENTIFIC INNOVATION AND RELEVANCE

Building energy simulation models are an important tool, not only for building design but also for policy making. Previous research has shown that there is a significant gap between actual energy consumption, and the energy consumption calculated by building energy simulation models. Many researchers, practitioners, and policymakers mainly impute this energy performance gap to occupant behaviour. One would expect this gap to be less at building stock level because occupant behaviour would be averaged. However, the performance gap is known to be high at a building stock level too, indicating a more structural problem in building energy simulation models. Being able to correctly assess and predict energy use in the building stock, it is essential to realise national and international energy saving targets. Until now it has been possible to reduce the energy performance gap on individual level using real energy data of an individual building to calibrate a building energy simulation model. However, this is not effective on a building stock level. In this paper a calibration method on building stock level is introduced to make building energy simulation outcomes a more reliable tool for policymakers.

PRELIMINARY RESULTS AND CONCLUSIONS

The preliminary results are promising and show that using actual energy data can indeed reduce the energy performance cap significantly. The paper will also give an indication which optimisation algorithms are the most effective and how much data is required to execute the calibration effectively.

MAIN REFERENCES

Hamilton, I.G., et al., Energy efficiency in the British housing stock: Energy demand and the Homes Energy Efficiency Database. Energy Policy, 2013. 60: p. 462-480.

de Wilde, P., The gap between predicted and measured energy performance of buildings: A framework for investigation. Automation in Construction, 2014. 41: p. 40-49.

Majcen, D., L. Itard and H. Visscher, Actual and theoretical gas consumption in Dutch dwellings: What causes the differences? Energy Policy, 2013. 61: p. 460-471.

Guerra Santin, O., Occupant behaviour in energy efficient dwellings: evidence of a rebound effect. Journal of Housing and the Built Environment, 2013. 28(2): p. 311-327.

Menezes, A.C.A.A.C., Predicted vs. actual energy performance of non-domestic buildings: Using post-occupancy evaluation data to reduce the performance gap. Applied Energy, 2012. 97: p. 355-364. Majcen, D., L.C.M. Itard and H. Visscher, Theoretical vs. actual energy consumption of labelled dwellings in the Netherlands: Discrepancies and policy implications. Energy Policy, 2013. 54: p. 125-136.

EU, amending Directive 2010/31/EU on the energy performance of buildings and Directive 2012/27/EU on energy efficiency, in L 156/75, E. Parlement, Editor. 2018, Offical Journal of the European Union: Strassbourgh. p. 17.

Keywords: energy performance gap, calibration, machine learning, building stock

Paper ID 30111: Using a large floorplan dataset to study the impact of simulation zoning simplifications

Aurelien Bres*

AIT Austrian Institute of Technology, Austria

AIM AND APPROACH

Zoning is a central concept in most building energy modelling tools. A simulation zone represents a spatial unit in which heat balance is calculated, based on the assumption of homogeneous air temperature. The way in which spaces are grouped into zones, which we refer to as simulation zoning, is a key aspect of model complexity. In comparison to a detailed zoning with one zone per room, zoning schemes grouping rooms e.g. according to their orientation or use may result in fewer zones, thus in faster run times and a more understandable model, but at the price of some deviations in results. While such deviations have been quantified in individual cases, they can be expected to depend strongly on individual building characteristics and especially on their respective floor plans. This contribution aims at investigating the impact of simulation zoning on results in a more systematic way, making use of a large dataset of floor plans. Simulation models are created and run for more than hundred distinct floor plans and with different zoning schemes. Differences between results (heat demand and temperatures) with detailed zoning and with simplified zoning are calculated for all these models and statistically analysed.

SCIENTIFIC INNOVATION AND RELEVANCE

This contribution will provide quantitative answers to questions of high relevance for dynamic thermal simulation practitioners and researchers, many of which frequently need to balance accuracy and computational requirements: How much accuracy can be expected to be lost by grouping rooms into simulation zones according to a given zoning scheme (e.g. one zone per floor), and what is the probability that the corresponding error exceeds a certain acceptable level (e.g. 2 K in the maximum temperature or 5% in the yearly heat demand)?

To our knowledge, most of the previous investigations of the impact of simulation zoning on results (e.g. [1,3,5]) have used single building models, and none of them has used more than 25 different floorplans [2]. Thus, the proposed contribution will be the first that may claim to provide a statistically relevant assessment of the impact of simulation zoning for general floorplans.

PRELIMINARY RESULTS AND CONCLUSIONS

Automated preparation of EnergyPlus models from annotated floorplans in the form of vector graphics (from the dataset of Kalervo et al. [4]) has been implemented. Based on this, 300 simulations have already been carried out for 100 floorplans and 3 zoning schemes (one zone per room, zones per orientation, one zone per floor), under the assumption of uniform internal loads. Some of the preliminary results for the coarsest zoning (one zone per floor) are: mean bias error for yearly heating demand between -7.2 % and 2.0 % (median of the absolute value 1.7 %), mean error for maximum zone air temperature between -2.7 K and -0.5 K (median -1.2 K).

The results are similar for many floorplans, but also show some significant variability and a few outliers. Also, the significance of the error magnitudes may be viewed as generally higher with certain indicators (such as maximum temperature) in comparison to others (such as yearly heating demand). The distributions of possible errors due to zoning simplifications estimated in this contribution have the potential to enable risk-aware zoning decisions in thermal simulation practice.

MAIN REFERENCES

[1] De Souza, C. B. and S. Alsaadani (2012). Thermal zoning in speculative office buildings: discussing the connections between space layout and inside temperature control. In First Building Simulation and Optimization Conference, pp. 417-424.

[2] Dogan, T., E. Saratsis, and C. Reinhart (2015). The optimization potential of floor-plan typologies in early design energy modeling. In Proceedings of the 14th IBPSA Building Simulation Conference, pp. 1853-1860.

[3] Heo, Y., G. Ren, and M. Sunikka-Blank (2016). Investigating an adequate level of modelling for energy analysis of domestic buildings. In 3rd Asia Conference of International Building Performance Simulation Association.

[4] Kalervo, A., J. Ylioinas, M. Häikiö, A. Karhu, and J. Kannala (2019). Cubicasa5k: A dataset and

an improved multi-task model for oorplan image analysis. In Scandinavian Conference on Image Analysis, pp. 28{40. Springer.

[5] Rivalin, L., D. Marchio, P. Stabat, M. Caciolo, and B. Cogné (2014). Influence of building zoning on annual energy demand. In International High Performance Buildings Conference.

[6] Shin, M. and J. S. Haberl (2019). Thermal zoning for building HVAC design and energy simulation: A literature review. Energy and Buildings, volume 203.

Keywords: zoning, floorplans, simplifications, model complexity

Session T2.7 (Online Track): Ensuring high quality building simulations

Paper ID 30145: Natural ventilation simulation for application to environmental planning in early design stage

Nozomu Ota*, Sei Ito

Shimizu Corpration, Japan

AIM AND APPROACH

The performance of natural ventilation varies depending on the local characteristics, surrounding buildings, building shape, ventilation openings arrangement and method of opening and closing, etc., so time and effort are required to determine the energy efficiency effect. Normally it is necessary to quantitatively determine the performance in the early design stage when the building form is determined, but there have been almost no tools that can be used within the limited time and resources of the early design stage. For example, OpenStudio, TRNFlow, and IES are capable of advanced simulation, including natural ventilation, but these are for precise analysis in the latter stages of design and are difficult to use in the early stage.

Therefore coupled indoor-outdoor natural ventilation simulation that enables environmental designers to easily and rapidly investigate energy efficiency schemes has been developed for application to early design stage natural ventilation schemes. The tool has been developed into 3 phases that can be utilized in stages as the design progresses. Phase 1 is site environment analysis, Phase 2 is an examination of multiple proposals, and Phase 3 calculates the annual energy efficiency.

SCIENTIFIC INNOVATION AND RELEVANCE

A Rhinoceros 3D model and wind pressure coefficients obtained from CFD analysis are coupled to the data in each phase, enabling seamless investigation reflecting the site properties from the early design stage.

- In Phase 1, the outdoor airflow analysis is performed considering the surrounding buildings by coupling with CFD software. Natural Ventilation Potential map indicating the openings positions to promote natural ventilation are produced by weighting the wind pressure coefficients with wind direction frequency, and This is the output that gives notice to the environmental designer.

- In Phase 2, multiple or optimization analysis using evaluation indexes by air conditioning reduction time is performed with a tool produced using Grasshopper.

- In Phase 3, the energy efficiency effect is calculated using Visual NETS, a unique Thermal and airflow network simulation program. A ventilation automatic opening and closing function is provided as standard to enable designers to easily and accurately calculate the energy efficiency effect. Also, By minimizing the amount of calculation, adopting the skyline method for solving simultaneous equations, and using the binary output of the result, Visual NETS has achieved 20 times speed than EnergyPlus in the model of 400 rooms. The tool has been widely applied to large-scale buildings.

PRELIMINARY RESULTS AND CONCLUSIONS

From a comparison between actual measurements and simulation for a 9,000 m2 total floor area office building in Tokyo, it was confirmed that temperatures and ventilation rates were generally well reproduced.

Also, a case study was performed for a medium-scale office building, which demonstrated the effectiveness of the simulation. As a result of preparing the potential map in Phase 1 and analyzing multiple proposals in Phase 2, a strong correlation was found between the average wind pressure coefficient of the natural ventilation potential map and the air conditioning reduction time, which confirmed the effectiveness of the natural ventilation potential map. Also in Phase 3, the optimized ventilation opening arrangement increased the cooling load reduction effect by more than 10% compared with an orthodox arrangement of ventilation openings. This suggests that investigating multiple proposals in the early design stage when the ventilation opening scheme can be changed comparatively easily contributes to increasing the energy efficiency effect, and it also confirmed the effectiveness of this simulation.

MAIN REFERENCES

- Marco Picco, Marco Marengo1(2019) A Fast Response Performance Simulation Screening Tool in Support Of Early Stage Building Design, Building Simulation 2019, Roma Italy.

- Nari Yoon, Ali Malkawi (2017) Predicting the Effectiveness of Wind-Driven Natural Ventilation Strategy For Interactive Building Design, BuildingSimulation 2017, San Francisco.

- Ali Mohammadzadeh, Miroslava Kavgic, Ali Al-janabi (2019) Energy Management System (EMS): The Impact of Natural Ventilation and Shading Controlon Thermal Performance of University Building in Winnipeg, Canada, Building Simulation 2019, Roma Italy.

- H. Okuyama,(1999) Thermal and airflow network simulation program NETS, Building Simulation'99, Kyoto, Japan, pp. 1237–44

- Sei Ito, Yasunori Akashi, Jongyeon Lim (2017)Study on Thermal Load Calculation for Ceiling Radiant Cooling Panel System, Building Simulation 2017, San Francisco.

- Y.Miyakawa, A.Matsuda, T.Kato(2001) Parallel Processing of Space Cholesky Factorization by Generalized Skyline Method, Information Processing Society of Japan, vol. 42, No. 4, pp. 762–770

Keywords: Natural Ventilation, Early Design Stage, Rhinoceros, CFD, Air Flow Network

Paper ID 30429: Control logic and parameters in a VAV system considering unevenly distributed internal loads and damper characteristics

Akari Nomura* (1), Shin Yamamoto (2), Shohei Miyata (1), Yasunori Akashi (1), Masashi Momota (3), Takao Sawachi (4)

1: Department of Architecture, School of Engineering, The University of Tokyo, Japan; 2: Taisei Corporation, Tokyo, Japan; 3: Tokyo Denki University, Tokyo, Japan; 4: Building Research Institute, Ibaraki, Japan

AIM AND APPROACH

Energy conservation in heating, ventilation, and air conditioning (HVAC) systems has been a major issue due to their high energy consumption percentage of the total energy used in buildings [1]. Therefore, various energy-saving controls related to automated controls such as variable air volume (VAV) control, variable water volume (VWV) control, and chiller unit control have been adopted. However, it is not easy to investigate appropriate control logics and parameters in the design phase and quantify their energy-saving effect [2]. In the operation phase, rooms are often utilized differently from conditions assumed in the design phase; however, reliable energy conservation is strongly demanded under such circumstances. Therefore, an advanced simulation tool that can properly evaluate energy-saving controls is needed even in the design phase [3,4].

In this study, we built a detailed and comprehensive simulation program for an HVAC system including automated controls, and investigated appropriate control logics and parameters for a VAV system depending on unevenly distributed internal loads and damper characteristics through case studies. Because the quantitative energy-saving effect obtained by this simulation program has been difficult to calculate by conventional ones, it is expected to provide useful information for the energy-saving building certification system that provides incentives.

SCIENTIFIC INNOVATION AND RELEVANCE

This study contributes to the new body of knowledge by presenting some of the probable operational issues of VAV control systems that have been neglected in the design phase and at the start of the operation, such as unevenly distributed internal loads among zones, using a simulation. Another significant achievement is clarifying the effects of control logics and parameters on the performance of energy-saving controls related to automatic controls based on simulation results.

To calculate these impacts at the building level, we developed a dynamic simulation in Python with the following three features.

i) Calculate the power of pumps and fans theoretically based on physical models considering pressure distribution

ii) Incorporate the control models, for example, proportional-integral (PI) control and VAV control such as static pressure set-point reset, and easy to change these control logics and parameters

iii) Calculate whole building state from building envelope to VAV dampers

Despite the current high energy performance of VAV control systems, their energy-saving potential can still be improved [2]. This study has great potential for VAV control systems with higher energy efficiency when combined with relevant studies such as automated fault detection and diagnosis.

PRELIMINARY RESULTS AND CONCLUSIONS

For an existing experimental RC building with six rooms, we built a simulation in which almost the same control logic as the real one was incorporated, with a 5-second interval. As a base case, the calculation with the input of a summer representative day resulted in accord with the control logic for each part, such as chiller, airflow, air handling unit, VAV system, and indoor environment. From these results, the validity of the simulation was demonstrated.

From the case studies conducted using the simulation, we obtained the following results.

Case 1) Unevenly distributed internal loads; when loads in some rooms were larger than in the other rooms and the flow rate set-point stuck to the maximum, it adversely affected the controllability of the other rooms with lower loads. A suitable set of control logics and parameters was further investigated.

Case 2) Control logic adjusted to suit equipment characteristics; when the range of the suitable static pressure was adjusted corresponding to the damper characteristics, the control of supply air fans was more stabilized.

These findings suggest an appropriate set of control logics and parameters for expected operating conditions can improve both energy efficiency and indoor air quality in VAV control systems.

MAIN REFERENCES

[1] Pérez-Lombard, L., Ortiz, J., & Pout, C. (2008). A review on buildings energy consumption information. Energy and buildings, 40(3), 394-398.

[2] Okochi, G. S., & Yao, Y. (2016). A review of recent developments and technological advancements of variable-air-volume (VAV) air-conditioning systems. Renewable and Sustainable Energy Reviews, 59, 784-817.

[3] Motomura, A., Akashi, Y., Lim, J., Zhang, W., Miyata, S., Sawachi, T., & Akamine, Y. (2019). Evaluation method for energy saving effects by VAV/VWV control in buildings Part 3-4 [in Japanese]. Proceeding from The Society of Heating, Air-Conditioning and Sanitary Engineers of Japan conference 2019. Sapporo (Japan), 18-20 September 2019.

[4] Yamamoto, S., Akashi, Y., Miyata, S., Zhang, W., Akamine, Y., & Sawachi, T. (2020). Evaluation method for energy saving effects by VAV/VWV control in buildings Part 6 [in Japanese]. Proceeding from The Society of Heating, Air-Conditioning and Sanitary Engineers of Japan conference 2020. Online, 16-18 September 2020.

Keywords: HVAC system, Variable air volume(VAV), Parameter, Control logic, Integrated simulation

Paper ID 30510: Development of test procedure for the evaluation of building energy simulation tools - phase III addition of subsystem test toward systematic diagnostics for HVAC system simulation-

Sei Ito* (1), Eikichi Ono (2), Harunori Yoshida (3), Hiromasa Yamaguchi (4), Eisuke Togashi (5), Kazuki Yajima (6), Hiroshi Ninomiya (7), Koichi Shinagawa (8)

1: Shimizu Corporation, Japan; 2: National University of Singapore, Singapore; 3: Professor Emeritus of Kyoto University, Japan; 4: Kansai Electric Power Co., Inc., Japan; 5: Kogakuin University, Japan; 6: SHINRYO CORPORATION, Japan; 7: Nikken Sekkei Ltd., Japan; 8: Nihon Sekkei, Inc., Japan

AIM AND APPROACH

Energy performance simulation of HVAC systems is a key element for system performance evaluation in a design phase, energy conservation code compliance, a green building certification program such as LEED, commissioning of the systems in a construction and operation phase, etc. Under such circumstances, the need for reliability evaluation of simulation tools is increasing. ASHRAE Standard 140 is a well-known test procedure for said purpose, but building fabrics and HVAC systems commonly used in Japan is out of scope. Therefore, aiming for the development of a test procedure focusing on those issues, a technical committee of the Society of Heating, Air-Conditioning and Sanitary Engineers of Japan (SHASE) was established and published a guideline of the procedure in 2016, the outline of which was presented at Building Simulation 2017. At Building Simulation 2019 we also presented the simulation trial results of the expanded test cases by several users, where we showed that test results indicated some discrepancies in energy consumption between tools and test participants. In this report, we report the further investigation about the developed method to identify the causes of the discrepancies with the results of additional test trials.

SCIENTIFIC INNOVATION AND RELEVANCE

The test results of HVAC system simulation reflect the tool's model characteristics and input errors or lack of modeling knowledge of users, which may cause the discrepancy in the results between tools. Because of such complexity, it is difficult to identify the cause of the discrepancy by using the annual simulation results of the whole HVAC system.

The air-side HVAC equipment analytical verification tests on ASHRAE Standard 140 uses an approach to test simulation results of several HVAC subsystem types mainly based on evaluation of thermal load by changing the input conditions step-by-step under steady state conditions. Although our approach is basically the same as we expand the approach by adding the evaluation of the state and energy performance of the system to focus on characteristics of equipment, control strategy and the system. We divide the whole system into three subsystems which are cooling water, heat source, and air handling unit subsystem. A set of test cases with different input conditions for each subsystem test can allow us to detect and diagnose faults in the tested tool and mistakes in input by users by evaluating whether the expected qualitative difference between the cases can be obtained on a step-by-step basis.

PRELIMINARY RESULTS AND CONCLUSIONS

A simulation test trial was conducted for the three subsystems. For example, regarding the cooling water subsystem test, there are six test cases that differ in the input conditions from the base case, which are outside wet bulb air temperature, load factor, control set points of outlet temperature and temperature difference of cooling water. By comparing the results of the cases, we could confirm whether the cooling tower performance and the control of the variable flow rate of the cooling water pump, etc. were properly modeled and simulated. In addition, we found that this subsystem test can suggest the probable causes of the difference in results such as modeling faults in tools and user mistakes and misunderstanding in input parameters of equipment and control logic characteristics by comparing the step-by-step results of a tool qualitatively. In conclusion, the subsystem test has been shown to have the following three advantages; 1) to be able to compare the characteristics of the tools and clarify their characteristics, 2) to be able to provide the tool developer with a test method that is useful for finding program faults, and 3) to help tool users discover input errors and misunderstandings with regard to modeling.

MAIN REFERENCES

ASHRAE (2017). ASHRAE Standard 140-2017, Standard Method of Test for the Evaluation of Building Energy Analysis Computer Programs. Atlanta, GA: ASHRAE.

R. Judkoff and J. Neymark (1995). International Energy Agency Building Energy Simulatin Test (BESTEST) and Diagnostic Method, National Renewable Energy Laboratory.

SHASE (2016). SHASE-G 1008-2016, Guideline of Test Procedure for the Evaluation of Building Energy Simulation Tool. Society of Heating, Air-Conditioning and Sanitary Engineers of Japan.

J. Neymark, M. Kennedy, R. Judkoff (2017). Airside HVAC Air-Distribution System Model Test Cases for ASHRAE Standard 140. Proceedings of the 15th International Building Performance Simulation Association Conference, Building Simulation 2017, San Francisco: pp.644-653.

E. Ono, S. Ito, H. Yoshida (2017). Development of Test Procedure for the Evaluation of Building Energy Simulation Tools. Proceedings of the 15th International Building Performance Simulation Association Conference, Building Simulation 2017, San Francisco: pp.380-388.

E. Ono, S. Ito, H. Yoshida (2019). Development of Test Procedure for the Evaluation of Building Energy Simulation Tools –Phase II Expansion of Evaluation Targets and Results of Simulation Trials-. Proceedings of the 16th International Building Performance Simulation Association Conference, Building Simulation 2019, Rome: pp.4578-4585.

Keywords: Energy performance simulation, Evaluation procedure, Diagnostics, Subsystem test

Paper ID 30686: Transfer learning based inverse modeling to identify unknown building properties

Yun-Dam Ko, Cheol-Soo Park*

Seoul National University, Korea, Republic of (South Korea)

AIM AND APPROACH

Due to lack of detailed building energy data (e.g. thermal properties of building envelopes, infiltration, occupant behavior, hourly sub-metered energy data etc.), most building energy benchmarking systems rely on EUI (Energy Use Intensity, kwh/m2.yr) for energy benchmarking. Precisely, it is closer to 'energy use benchmarking' than 'performance benchmarking'. In addition, the current benchmarking system is conducted for buildings of a same use type, e.g. office, education, hospital, etc.

In this regard, this study aims to present a data-driven methodology to identify building characteristics (e.g. thermal properties of opaque and transparent envelopes, efficiencies of mechanical systems, operational information [setpoint temperature, operation hours]) from energy use data (e.g. monthly gas and electricity use), which will lead to performance benchmarking.

For this purpose, Domain Adaptive Transfer Learning (DATL) is selected. A DATL model will be trained with virtual data and be used for benchmarking for 100,000 existing buildings located in Seoul, South Korea. This proposed approach will provide identification of the aforementioned unmeasured/uncertain building characteristics and will be used to categorize buildings into peer groups by 'similar' performance characteristics.

SCIENTIFIC INNOVATION AND RELEVANCE

To overcome the lack of the detailed building energy data mentioned in "Aim and Approach", the authors propose the DATL based approach. It is expected that this approach enables to identify unmeasured important features based on measured energy use, and it could be a good alternative when detailed building data are not available.

As a result, the proposed approach can provide more rational building energy 'performance' benchmarking. While the conventional approach (i.e. energy use benchmarking) cannot identify the reason for high energy consumption, the proposed approach, performance benchmarking enhanced by the transfer learning algorithm, is expected to inform decision makers of the degree of energy performance of a given building, e.g. bad envelopes, low efficiencies of mechanical systems, etc.

PRELIMINARY RESULTS AND CONCLUSIONS

Firstly, EnergyPlus reference buildings will be sampled using Latin Hypercube Sampling for 'source domain' data. A DATL model will be trained from the generated source domain data. This model will be used to identify thermal properties of building envelopes and efficiencies of HVAC systems, given monthly gas and electricity use data. The developed model will be tested with available data gathered from 100,000 existing buildings located in Seoul, South Korea.

Then, the authors will categorize buildings into peer groups in terms of similar 'energy performance' using a clustering algorithm, e.g. K-means. Case studies will be conducted to compare benchmarking results of our proposed approach with two conventional approaches (by use type and by daily load profile).

MAIN REFERENCES

Luo, X., Hong, T., Chen, Y. and Piette, M.A. (2017), Electric load shape benchmarking for small- and medium-sized commercial buildings, Applied Energy, 204, 715-725.

Gao, X. and Malkawi, A. (2014), A new methodology for building energy performance benchmarking: An approach based on intelligent clustering algorithm, Energy and Buildings, 84, 607-616.

Park, J.Y., Yang, X., Miller, C., Arjunan, P. and Nagy, Z. (2019), Apples or oranges? Identification of fundamental load shape profiles for benchmarking buildings using a large and diverse dataset, Applied Energy, 236, 1280-1295.

Westermann, P., Deb, C., Schlueter, A. and Evins, R. (2020), Unsupervised learning of energy signatures to identify the heating system and building type using smart meter data, Applied Energy, 264, 114715

Zhan, S., Liu, Z., Chong, A. and Yan, D. (2020), Building categorization revisited: A clustering-based approach to using smart meter data for building energy benchmarking, Applied Energy, 269, 114920

Keywords: Transfer learning, Building energy benchmarking, Clustering, Machine learning

Paper ID 31060: Meta-modelling of operation schedules of commercial buildings based on measured electricity demand data

Yohei Yamaguchi*, Fumiya Enokihara, Yoshiyuki Shimoda

Osaka University, Japan

AIM AND APPROACH

In simulations of the energy demand of commercial buildings, the assumptions regarding the building operations characterise the magnitude and temporal variation of the energy demand. However, the operating conditions of building stock are rarely available; thus, building stock and urban building energy models are often unable to simulate realistic time variations in energy demand. To address this issue, this paper presents a method for extracting an operation schedule from the time-series data of an electricity demand. A building's electricity demand is considered as the sum of the electricity consumption of all ofall the appliances used in the building. The appliances are divided into four groups based on two factors: 1) the existence of operations by building occupants, and 2) the existence of seasonal changes in electricity consumption. In the proposed method, a load disaggregation method is applied to extract the load component driven by the operations of building occupants without seasonal changes as a week-length profile. The profile is further summarised using a few characterisation parameters. Finally, joint probability distributions of the characterisation parameters are developed to assign operation conditions to the simulated buildings. This method is applied to a few thousand commercial buildings in Japan.

SCIENTIFIC INNOVATION AND RELEVANCE

The methods for considering realistic operation conditions in building stock energy models can be divided into the following three groups: (1) those using stochastic activity-based models to stochastically simulate occupancy conditions based on survey-based data, (2) those using building occupancy conditions extracted based on personal location data, and (3) those using operation schedules derived based on time-series electricity demand data measured in buildings (Bianchi et al., 2020). This study establishes a method for the third group. The time-series electricity demand data is first disaggregated by using a load profile disaggregation method to extract the average profile of a one-week period, i.e. 24 hours × 7 days = a 168-hour profile; in this profile, the seasonal variation is removed. The average week-length profile represents the magnitude of the building operation at each time of the day of the week. Then, the week-length profile is used to extract parameters representing the relationships between the times of day, days of the week, and magnitudes of the profile. Finally, parameters derived from sample buildings are used to develop statistical distributions of the parameters as a representation of the operation conditions of buildings within a building group.

PRELIMINARY RESULTS AND CONCLUSIONS

This paper first describes the detailed processes of the demand disaggregation, parameterisation of the operation conditions, and development of joint probability distributions of the parameters, i.e. the start time and duration of the operation. The results show that the extracted operation schedules exhibit general characteristics for each business sector. Most office buildings have a sharp rise in the morning, a slow decrease during evening hours, and low values during night hours and weekend days. The distributions of the characteristics of the school buildings are similar to those of offices. A hotel's start times are concentrated in the early morning, whereas the operation duration is much longer than those of the office and school. A hospital also has an early start time, but the operation duration decreases with a delay in the start time. A similar trend is observed for retail. A restaurant shows a diverse pattern for the start time and duration of operation, characterising the differences in the opening hours. The results demonstrate that the method is useful for capturing the variations in the operation conditions of buildings. The statistical distributions can be used to assign building operation conditions to the reference building models used in building stock models.

MAIN REFERENCES

Bianchi, C., Zhang, L, Goldwasser, D, Parker, A, Horsey, H. (2020). Modeling occupancy-driven building loads for large and diversified building stocks through the use of parametric schedules. Applied Energy; 276:115470. doi:10.1016/j.apenergy.2020.115470.

Keywords: Building operation, commercial building, meta-modelling, time-series energy demand

Session T2.8 (Online Track): Buildings paving the way for the energy transition

Paper ID 30159: Further development and validation of the "PROFet" energy demand load profiles estimator

Kamilla Heimar Andersen*, Synne Krekling Lien, Karen Byskov Lindberg, Harald Taxt Walnum, Igor Sartori

SINTEF AS, Norway

AIM AND APPROACH

Recent developments in policy highlight the importance of utilizing end-user flexibility to support the decarbonization of the energy system [1], [2]. Energy flexibility of buildings is also a research topic within the Norwegian research center for Zero Emission Neighbourhoods (FME-ZEN), where a ZEN definition is being developed considering several key performance indicators (KPI) [3]. The energy flexibility KPIs evaluate how well a building/a neighborhood can respond to signals from the grid and manage its demand, storage, and local generation to optimize its response to such signals while satisfying user comfort [4], [5]. To establish such KPIs, standard load profiles (i.e., unaware signal profiles) should be used as a reference to evaluate against on the flexibility potential.

Previous work [6] has developed a model, here named PROFet, for forecasting aggregated weather dependent load profiles of buildings, based on energy measurements from buildings connected to district heating, collected in a database here named trEASURE. The database and the model have hourly resolution and treat energy use for heating and electric specific purposes separately.

The aim of this study is to present further developments of the trEASURE database and to validate the typical load profiles from the PROFet model with out-of-sample datasets.

SCIENTIFIC INNOVATION AND RELEVANCE

First, the measurements database trEASURE has been extended to include more than 300 entries representing ca. 2.4 million m2 of floor area, subdivided into 11 building categories, both residential and (mostly) non-residential buildings. Second, the heating measurements have been pre-treated with a decomposition of the domestic hot water (DHW) part, using a hybrid seasonal/energy signature method [7]. Third, the energy efficiency of the buildings has been inferred by comparing the temperature dependency of the space heating measurement data with reference values from building standards. This resulted into three levels of energy efficiency: efficient - representing low-energy and passive house buildings; intermediate - representing newer buildings built according to the 2010 code, TEK10, which is also regarded as an ambitious yet realistic target for energy-efficient renovations [8]; and regular - for all other buildings.

Besides, the typical load profiles from the PROFet model had not yet been validated with out-of-sample datasets; this is presented in this paper.

PRELIMINARY RESULTS AND CONCLUSIONS

New sets of coefficients have been generated by the PROFet model with respect to the initial work [6], thanks to the extension and the improvements in the underlying trEASURE database. The resulting load profiles for given locations and years are compared with out-of-sample datasets, i.e., energy measurements that are not included in the trEASURE database. The out-of-sample datasets are limited to few building categories, such as apartment block, office, and school, and to the regular energy efficiency standard. This is due to difficulties in creating large-good quality datasets, with respect to data availability (incl. of metadata) and the need for time-consuming data cleansing routines. It should be noted that since the PROFet model is meant to generate load profiles valid at an aggregated level, also the out-of-sample datasets need to be conspicuous to have a meaningful comparison. The validation is based on both graphical comparisons for typical days and on statistical indicators, following the ASHRAE Guideline 14 [9].

The resulting load profiles for a standardized Oslo climate will also be presented graphically and as tabulated values, and it will be discussed how these profiles can be used to develop the energy flexibility KPIs in the Norwegian ZEN definition.

MAIN REFERENCES

[1] European Commission, "Clean Energy for All Europeans - The Winter Package.," 2018.

[2] IEA, "Status of Power System Transformation 2019, Technology report — May 2019," 2019.

[3] M. K. Wiik, S. M. Fufa, I. Andresen, H. Brattebo, and A. Gustavsen, "A Norwegian zero-emission neighborhood (ZEN) definition and a ZEN key performance indicator (KPI) tool," IOP Conf. Ser. Earth Environ. Sci., vol. 352, no. 1, 2019.

[4] R. G. Junker et al. (2018) Characterizing the energy flexibility of buildings and districts, Applied Energy, Vol. 225, pp. 175-182.

[5] IEA EBC (2019) Characterization of Energy Flexibility in Buildings, deliverable of Annex 67 Energy Flexible Buildings.

[6] K. B. Lindberg, S. J. Bakker, and I. Sartori, "Modelling electric and heat load profiles of non-residential buildings for use in long-term aggregate load forecasts," Util. Policy, vol. 58, no. March, pp. 63–88, 2019.

[7] S. K. Lien, D. Ivanko, and I. Sartori, "Domestic hot water decomposition from measured total heat in Norwegian buildings (Paper submitted for review)," in Build-Sim Nordic 2020.

[8] Enova (2012) Potensial- og barrierestudie – Energieffektivisering I norske bygg, Enova report 2012:01 (in Norwegian).

[9] ASHRAE (2014) Measurement of Energy, Demand, and Water Savings, ASHRAE Guideline 14-2014.

Keywords: ZEN, energy, flexibility, building, load profiles

Paper ID 30910: The efforts towards development of an energy efficiency upgrade platform

Mahnameh Taheri*, Loic Jacob, Colin Parry, Sahar Mirzaie, Agnieszka Hermanowicz, Ioanna Vrachimi, Alan Wegienka

arbnco Ltd., United Kingdom

AIM AND APPROACH

The building sector can have a significant contribution in emission reduction by energy retrofit actions. Thus, policymakers are making building energy performance a main concern in their sustainability decision makings. However, the building industry usually relies on simplified assessment tools and methods, which does not address optimized energy and cost saving strategies at a large scale. This is due to, among other reasons, the required expertise, time, and effort as well as complexity of the analysis [1]. This contribution presents the efforts towards development of an energy assessment platform to assist policy makers, building engineers, energy managers, etc., to evaluate cost effective retrofit actions and provide the best return from investments for commercial buildings. Portfolio assessments commonly fall into three phases of benchmarking, detailed investigation, and scaling of findings across portfolio [2]. The methodology involves, clustering the buildings based on their energy performance, selecting the most influential group for further analysis, and performing detailed investigations for the selected group. This study covers development of i) a GUI for convenient easy collection of building data, ii) a recommendation engine connected to a database of retrofit and renewable options with associated energy and cost savings, and iii) an investment decision engine.

SCIENTIFIC INNOVATION AND RELEVANCE

Rapid and reliable energy performance assessments have been one of the main concerns of the building science community. Improving the building energy performance predictions and addressing the shortcomings of the conventional approaches encouraged the development of new techniques. As mentioned above, the presented platform enables the users to a) benchmark their buildings' energy consumption, b) audit their building energy features, and c) explore different energy retrofit, renewable, and investment options. At the screening level, the analysis is based on limited user input, including, minimum twelve months of metered energy consumption data together with building location, area, and year of build. The core of the analysis behind the platform is sophisticated load shape analysis and Machine Learning (ML) methods. Load shape analysis is used here for the purpose of benchmarking based on a set of predefined efficiency factors [3]. Moreover, a novel application of a disaggregation algorithm for commercial whole building energy data is used to disaggregate the consumption data into categories of weather dependent, scheduled, and base loads. Load shape analysis, benchmarking building energy efficiency, and energy disaggregation are the backbone of the development of the enhanced investment recommendation engine.

PRELIMINARY RESULTS AND CONCLUSIONS

The project presented here provides a portfolio-scale energy assessment decision-making service. The result is a set of energy efficient alternatives for building elements and systems which improves the energy performance, considering the user-specified constraints. The process starts with an initial portfolio screening, to identify worst performing buildings with the greatest energy saving potential for further detailed analysis. For this step, a GUI is developed for a convenient and easy collection of more detailed user-dependent data. An inexperienced user can provide the very basic building data using the interface. Some unspecified building input parameters are then estimated based on typical characteristics of a building, such as type, age, location. The platform suggests actions, including energy efficiency retrofits, related to building fabric, lighting, and HVAC efficiency, renewables, i.e., PV and wind, and storage options. Then the financial feasibility of different investment alternatives is scoped. The financial metrics for these assessments include net present value, internal rate of return, simple payback, discounted payback period, etc. The data-based approach presented in this paper is used to evaluate different components of a buildings' energy profile and opens up the possibility of identifying the biggest retrofit opportunities with minimal specialist time over large building stocks.

MAIN REFERENCES

1. Lee, S.H., Hong, T., Sawaya, G., Chen, Y., Piette, M.A. 2015. DEEP: A Database of Energy Efficiency Performance to Accelerate Energy Retrofitting of Commercial Buildings. ASHRAE Winter Conference, January 2015, Chicago.

2. Franconi E. M. and Bendewald M. J. 2014. Analyzing Energy-Efficiency Opportunities across Building Portfolios. American Council for an Energy-Efficient Economy, ACEEE, Summer Study on Energy Efficiency in Buildings.

3. Taheri, M., Rastogi, P., Parry. C., Wegienka, A. 2019. Benchmarking Building Energy Consumption Using Efficiency Factors. Proceedings of the 16th International IBPSA Conference, Building Simulation 2019. September 2019, Rome, Italy. DOI: 10.26868/25222708.2019.210575.

Keywords: Energy efficiency, Retrofit analysis, Software development

Paper ID 30173: Alternative modelling approaches to energy performance certificates

David Jenkins*, Peter McCallum, Sandhya Patidar, Sally Semple

Heriot-Watt University, United Kingdom

AIM AND APPROACH

The outputs required from building simulation are impacted by the requirements of building energy policy, and the direction of that policy towards low-carbon energy systems. To model total energy consumption of a building is a different task to characterising the demand of that building at a transient level; to do so at scale is an additional level of complexity. With the success and ubiquity of Energy Performance Certificates (EPC) across Europe, there is tendency to use this form of building assessment as the main vehicle for characterising building energy demand. However, there is growing evidence of EPCs being applied to areas for which they were not designed to serve. By comparing a range of alternative building energy modelling techniques with the current methodologies underlying EPCs, this study proposes future directions for standardised energy assessment of residential buildings and the important role of modelling choices within this. Alternatives to traditional steady-state models for standardised assessment are proposed, with a framework for critiquing new methods proposed. The work will be placed within the wider framework of the Energy Performance in Buildings Directive (EPBD), discussing the extent that alternative methods may fit within this (noting existing responses to the EPBD across Europe).

SCIENTIFIC INNOVATION AND RELEVANCE

There is an information gap between the modelling used in research and that applied for standardised energy assessments of buildings. This paper aims to overcome the assumption that state-of-the-art building simulation and/or statistical modelling is inherently too complex and inefficient for use with EPC-type assessments. The work particularly focusses on two areas of modelling: i) Theoretical dynamic simulation that is applied at scale within an efficient interface and ii) statistical modelling of empirical data, by way of a clustering analysis to allow for classification of different homes/households by their measured energy demand characteristics. The dynamic simulation is based on currently available simulation software but applied through bespoke techniques that efficiently gather data on buildings from GIS/OS, open-source energy data, and existing EPC assessment data. The statistical model approach applies models developed by the authors that apply k-means clustering to proxy parameters of energy use via Principal Component Analysis, alongside disaggregation techniques to better categorise aspects of transient demand. Placing such techniques within a framework that is commensurate with EPC-type assessments raises the prospect of bridging the gap between large-scale generation of energy ratings of buildings (used in policy to encourage market transformation) and more detailed building simulation.

PRELIMINARY RESULTS AND CONCLUSIONS

The work proposes a dynamic simulation approach and statistical empirical method that could be used as a starting point for a new form of standardised energy assessment. However, in doing so, the paper also proposes standard criteria that any new method of energy assessment (for generating some form of energy rating) could be critiqued against. These criteria reflect the need for future energy assessments to align with some definition of reality, better understand the importance of (transient) demand flexibility, accommodate new technologies, be suitable for punitive action on homeowners (noting the direction of building energy policy across Europe), are suitable for extrapolation and standardisation, and are able to set a standard for quality of input information. By discussing the proposed modelling methods against these metrics of success, the paper will demonstrate what is specifically possible with the novel models generated by the authors, but also set a framework under which future energy assessments could be judged against. Initial results indicate that the traditional assumption that simple steady-state models are the only option for large-scale energy assessment of residential buildings is no longer valid, and the modelling community should work to update this approach to EPCs.

MAIN REFERENCES

Beckel, C., Sadamori, L., Staake, T., & Santini, S. (2014). Revealing household characteristics from smart meter data. Energy, 78, 397–410. https://doi.org/10.1016/j.energy.2014.10.025

Hardy, A., & Glew, D. (2019). An analysis of errors in the Energy Performance certificate database. Energy Policy, 129, 1168–1178. https://doi.org/10.1016/j.enpol.2019.03.022

Lomas, K. J., Beizaee, A., Allinson, D., Haines, V. J., Beckhelling, J., Loveday, D. L., Porritt, S. M., Mallaband, B., & Morton, A. (2019). A domestic operational rating for UK homes: Concept, formulation and application. Energy and Buildings, 201, 90–117. https://doi.org/10.1016/j.enbuild.2019.07.021

McCallum, P., Jenkins, D. P., Peacock, A. D., Patidar, S., Andoni, M., Flynn, D., & Robu, V. (2019). A multi-sectoral approach to modelling community energy demand of the built environment. Energy Policy, 132, 865–875. https://doi.org/10.1016/j.enpol.2019.06.041

Semple, S., & Jenkins, D. (2020). Variation of energy performance certificate assessments in the European Union. Energy Policy, 137, 111127. https://doi.org/10.1016/j.enpol.2019.111127

Keywords: EPC, Assessment, Dynamic energy modelling, standardisation

Paper ID 30733: Predicting the effects of building energy conservation policies: Modeling decision-making of building owners and tenants to maximize their own profits

Atsushi Funabiki* (1), Yuki Oto (2), Shohei Miyata (1), Yasunori Akashi (1)

1: Department of Architecture, Graduate School of Engineering, The University of Tokyo, Japan; 2: Obayashi Corporation

AIM AND APPROACH

At the 21st Conference of the Parties (COP21) to the United Nations Framework Convention on Climate Change (UNFCCC) in 2015, the Paris Agreement was adopted as an international agreement on environmental issue [1]. In response to this agreement, Japan has also set a goal to reduce carbon dioxide emissions. To achieve this goal, the spread of environmentally friendly buildings is required urgently, and support from a political perspective is essential to make buildings become more energy efficient.

To make policy which maximizes energy conservation efficiently, it is necessary to predict the effects of the policies. However, the interactions of building owners and tenants on building performance have not been considered in this context. Therefore, in this study, we built a simulation model for an office building area which consists of building owners and tenants as agents and predicts agents' decision makings based on their performances. The purpose of this study is to contribute to the planning of the optimal policies by predicting the effects of energy conservation policies including their combination. In this paper, Multi-Agent Simulation (MAS) which is one of the bottom-up simulation was employed.

SCIENTIFIC INNOVATION AND RELEVANCE

In addition to, Europe and the United States, Japan is also promoting the introduction of energy-saving policies, with subsidies, tax incentives, loans and other forms of assistance. In Japan, there are subsidy programs for buildings announced in Building Energy Efficiency Act of the Ministry of Land, Infrastructure and Transport and several other ministries [2-4]. In the real world, it is difficult to evaluate energy conservation policies in buildings by means of demonstrations, because of the huge economic and time costs involved, and it is generally verified by using simulations. Previous studies have modeled decision-making with residents and building owners as agents, aiming to predict greenhouse gas emissions from new construction and renovation, and facility upgrade, and to manage facility space efficiently [5-8]. By contrast, there are few simulations that incorporate multiple stakeholder decision-making, including subsidies for tenants, and few that examine the synergistic effects and counteracting effects of combinations of policies, therefore, it is difficult to consider in advance whether current policies are effective. It is novel and meaningful to incorporate multiple agents and policies into the simulation model constructed in this study to enable the prediction of energy performance and evaluation of energy-saving policies for buildings.

PRELIMINARY RESULTS AND CONCLUSIONS

In this paper, the objective policies are The Building Energy Efficiency Act, new construction subsidies and renovation subsidies. The behavioral logic of the agents in the simulation is as follows. The building owners change rents, rebuild or renovate to maximize theire profits. Tenants, on the other hand, move to new offices when they meet their preferences. At this time, the tenant chose a building that maximizes its own utility based on the rent and building performance of the building. The main four results of previous simulations are as follows. 1: Large subsidies are needed to make a significant impact on new construction subsidies. 2: Renovation subsidies can have an early effect, but the effects may converge in the long run. 3: the appropriate percentage of subsidies may vary depending on the combination of policies. 4: To progress the energy conservation, it is necessary to raise tenants' awareness of energy conservation, and although it will take time, it will have a significant effect. Further studies are needed to investigate policies to help tenants give more weight to energy efficiency in their preferences for building choices.

MAIN REFERENCES

[1] UNCFCC, The Paris Agreement.

https://unfccc.int/files/essential_background/convention/application/pdf/english_paris_agreement.pdf.

[2] Ministry of Land, Infrastructure and Transport, Building Energy Efficiency Act.

https://www.mlit.go.jp/common/001134876.pdf.

[3] Ministry of Economy, Trade and Industry, Agency for Natural Resources and Energy,

https://www.meti.go.jp/main/yosan/yosan_fy2019/pr/en/shoshin_taka_12.pdf.

[4] Ministry of the Environment, http://www.env.go.jp/earth/earth/ondanka/mat31y_01-03.pdf.

[5] Nägeli, C., Jakob, M., Catenazzi, G., Ostermeyer, Y., 2020. Towards agent-based building stock modeling: Bottom- up modeling of long-term stock dynamics affecting the energy and climate impact of building stocks, Energy and Buildings, 211, 109763.

[6] Liang, X., Yu, T., Hong, J., Shen, G.Q., 2019. Making incentive policies more effective: An agent-based model for energy efficiency retrofit in China. Energy Policy, 126, 177-189.

[7] Jia, M., Srinivasan, R.S., Ries, R., Weyer, N., Bharathy, G., 2019. A systematic development and validation approach to a novel agent-based modeling of occupant behaviors in commercial buildings, Energy and Buildings, 199, 352-367.

[8] Kavgic, M., Mavrogianni, A., Mumovic, D., Summerfield, A., Stevanovic, Z., Djurovic-Petrovic M., 2010. A review of bottom-up building stock models for energy consumption in the residential sector, Building and Environment, 45, 1683-1697.

Keywords: Building energy conservation policies, Agent-based modeling, Multi-agent system, Owner and tenant behaviors

Paper ID 31125: Urban climate simulation: coupling of mesoscale meteorological model with building-resolved neighborhood CFD simulation

Jan Carmeliet* (1), Aytaç Kubilay (1), Dominik Strebel (1), Dominique Derome (2)

1: ETH Zurich, Switzerland; 2: Université de Sherbrooke, Canada

AIM AND APPROACH

Cities show a heat island effect, where night temperatures are markedly higher in urban areas than rural ones. Cities are thus particularly affected when struck by heatwaves, which are, amongst others, climate extremes occurring at increasing frequency under the course of climate change. It is imperative to understand by how much the urban climate during heat waves could be mitigated using different solutions. Such solutions can be permanently in place, like evaporative cooling pavement materials or trees, or seasonal, like awnings and sun shading devices, to in-sync with heat waves, such as watering and irrigation interventions.

However, the simulation of the local urban climate at the scale of a few buildings to a full neighbourhood must consider the following physical processes, local wind dynamics, urban ventilation, solar radiation entrapment and local shading, evapo-transpiration of vegetation, heat and mass transport in porous media lining the streets, wind-driven rain deposition and run-off, and local anthropogenic heat sources.

To do so, we model local urban climate by coupling mesoscale meteorological simulation results to building-resolved computational fluid dynamics, coupled to radiation, heat-air-moisture transport model and building energy simulation. This paper shows the potential of this approach from mesoscale to

building- and material-scale.

SCIENTIFIC INNOVATION AND RELEVANCE

xMeteorological models provide boundary conditions that are realistic, as they use local topography and integrate land-use models. However, their local precision is limited by many factors: coarse resolution, parametrization of different phenomena and computational limitations. Coupling of meteorological models, to provide the boundary conditions of CFD models, which resolve a full neighborhood spatially, is an on-going research field, where multiple approaches exist with different advantages and computational cost. At neighborhood scale, our CFD model is coupled to a radiation exchange model and a heat and mass transport model, allowing to take into account all the required physics. The development of methodologies incorporating all relevant urban elements under realistic climate loading should be accompanied by interface with a comfort and physiological models of occupants, as well as methods for health impact assessment. These assessments have to be done under different scenarios of future climate. The validation of such coupling is complex and could use, for example, large datasets of careful urban measurement campaigns. In this paper, we present a validation exercise for the components of the modeling approach and present avenues for full coupling validation.

The approach presented provides capacities to assess the potential of urban mitigation solutions for heat waves.

PRELIMINARY RESULTS AND CONCLUSIONS

We present the sub-models and the couplings strategies of the modeling framework and provide one example of its application for an urban square in Zurich in terms of the impacts of using alternate solutions of pavements and adding vegetation.

Sustainable mitigation solutions to local urban heat islands, especially in case of heat waves, are expected to require a combination of measures such as the use of ventilation lanes, shadowing, evaporative cooling, vegetation, reflective surfaces. Realistic climate boundary conditions are necessary to deal with such extreme cases and analyses at local scale are required in order to assess the local heat islands and assess correctly the impact of these mitigation measures. The proposed coupled simulation approach is a carefully developed framework, providing a physically sound and validated representation of the complex local urban climate.

MAIN REFERENCES

Ferrari A, Kubilay A, Derome D, Carmeliet J. 2020. The Use of Permeable and Reflective Pavements as a Potential Strategy for Urban Heat Island Mitigation. Urban Climate 31: 100534.

Kubilay A, Derome D, Carmeliet J. 2018. Coupling of Physical Phenomena in Urban Microclimate: A Model Integrating Air Flow, Wind-Driven Rain, Radiation and Transport in Building Materials. Urban Climate 24: 398–418.

Kubilay A, Derome D, Carmeliet J. 2019a. Impact of Evaporative Cooling Due to Wetting of Urban Materials on Local Thermal Comfort in a Street Canyon. Sustainable Cities and Society 49: 101574.

Keywords: Urban scale, building-resolved, meteorological models, CFD

Session T2.9 (Online Track): Practice and industry related case studies

Paper ID 30138: Case study of evaporative cooling technologies for Australian office buildings

Paul Bannister* (1), Hongsen Zhang (2), Stephen White (3)

1: DeltaQ Consulting Service Pty Ltd, Australia; 2: Enerefficiency Pty Ltd, Australia; 3: Commonwealth Scientific and Industrial Research Organisation, Australia

AIM AND APPROACH

This paper summarises work, undertaken by the authors as part of the PRIME Net Zero Energy HVAC Technology project, investigating the potential for a range of evaporative cooling technologies to reduced HVAC energy consumption for office buildings in Australia.

A medium-sized office building model was simulated IES a range of configurations using weather files of typical locations across the 8 Building Code of Australia (Volume 1) climate zones in Australia. The HVAC types used in the simulation include:

(1) Conventional AHU-VAV system, which was used as the baseline.

(2) Model (1) plus direct/indirect evaporative cooling.

(3) Model (1) plus dewpoint cooler and direct evaporative cooling.

(4) Model (1) plus dewpoint cooler only

(5) Model (1) plus direct/indirect evaporative cooling and desiccant wheel.

SCIENTIFIC INNOVATION AND RELEVANCE

The use of evaporative cooling technologies provides significant opportunities for reduced energy use particularly in cooler and less humid climates, but they have achieved little penetration in the Australian office sector. The purpose of the study is to better understand the potential energy savings associated with a range of conventional evaporative cooling technologies (direct and indirect evaporative cooling) and less common approaches (dewpoint coolers, indirect evaporative cooling in combination with desiccant wheel). For this study they have been assessed in the context of otherwise conventional Australian office design and in particular are based on the use of a variable air-volume air-conditioning system.

PRELIMINARY RESULTS AND CONCLUSIONS

It was found that

(1) Configurations (3) and (4) offer the best energy/greenhouse savings (13-44%, dependent on climate zone) most for climate zones with the exception of Climate Zones 1 (Darwin) & 8 (Thredbo).

(2) The reactivation energy required for configuration (5) overshadowed the electricity savings achieved. If "free" reactivation heat were to be available, then this configuration provided the best outcomes in Climate Zone 1. Furthermore in Climate Zone 8 it was found that the chiller could be wholly replaced by desiccant wheel and evaporative cooling system potentially reducing the system capital cost.

Some additional simulations were run which demonstrated that further optimisation of control sequences could further extend the benefits of model (4) to 55% in the most favourable climate zone (Climate Zone 3 - Alice Springs).

This preliminary study has indicated significant potential for evaporative technologies in less humid and cooler climates in particular. Further work is required to understand the full potential of evaporative cooling technologies in more detail, particularly with respect to their application in other HVAC systems (especially those using higher supply air temperatures, such as underfloor systems) and other building types.

MAIN REFERENCES

To be provided.

Keywords: Evaporative cooling, office buildings, dewpoint cooler

Paper ID 30497: Optimization of electric lighting in a hospital emergency room - A case study

Zhafirah Ajrina*, Desliana Putri Pratiwi, Rizki Armanto Mangkuto, Revantino Revantino

Institut Teknologi Bandung, Indonesia

AIM AND APPROACH

Lighting design has become an important major factor in building construction planning. Optimum lighting is required to achieve an adequate visual performance while reducing visual discomfort, particularly, when executing tasks that require high accuracy. One of the most critical examples of lighting design is lighting in the healthcare facilities or hospitals, especially in emergency room. It is one of the parts in a hospital that provides early treatment for patients who suffer illness and injury, which can endanger their life-being. Therefore, it should be designed to obtain specific room requirement and lighting standard, since it can effectively build the room environment to influence physical and mental health, thus provides comfort of not only the patients, but also the medical staff. This paper aims to propose an optimized electric lighting design for Karya Dharma Kembang Tanjung Hospital, in this case at the emergency room, by means of modelling and simulation. Computational model of the room was built in DIALux Evo, and the simulation was executed with and without daylighting scene of four room types (lobby, examination room, treatment room, and corridor). Four different variables were observed and simulated for obtaining the optimum result.

SCIENTIFIC INNOVATION AND RELEVANCE

Lighting optimization in healthcare area was mostly done using several similar input variables, namely the polar diagram luminaires and reflectance walls. This research considers the variety of room type as well as the lighting standard in the emergency room, which is the most critical space in a hospital. This study also focuses on the treatment room to observe the local lighting and the influence of the visual discomfort from the doctor's perspective. This would be relevant to solve one of the problems in current pandemic situation, where many patients are found in the emergency room. In this case, simulation may help providing several lighting design solutions, which can enhance the accuracy of examination by the medical staff.

PRELIMINARY RESULTS AND CONCLUSIONS

The average horizontal illuminance and uniformity in the overall room type, in the case without daylight, has achieved the requirement. However, the task lighting condition in the examination and treatment room need to be further adjusted in order to enhance the visual performance. In this case, local lighting is applied to the area of patient's bed (mattress). The optimum lighting design for examination room is achieved using Phillips TBS165 G 2xTL5-28W HFS C6 luminaires with wall reflectance of 0.8. The selected option is based on the highest average illuminance value of those rooms compare to other types of luminaires, thus resulted good uniformity of 0.76. Then, the enhancement of lighting in treatment room is conducted by rotating the luminaire by 40° of normal from the right side of mattress orient to the doctor, it yields the smallest vertical illuminance and UGR values. However, the optimization of luminaire angle is only occurred when the observation area is along the patient's mouth up to the abdomen. Further investigation is needed when the observation area is more extensive to ensure good lighting performance and visual comfort in the room.

MAIN REFERENCES

1. A. H. Primanti, R. A. Mangkuto, D. P. Sari, and B. A. Kautsar, "Optimization of artificial lighting in physiotherapy room of Hasan Sadikin Hospital," AIP Conf. Proc., vol. 2088, no. 1, p. 040001, Mar. 2019.

2. G. F. M. Lima, J. Tavares, I. S. Peretta, K. Yamanaka, A. Cardoso, and E. Lamounier, "Optimization of lighting design usign Genetic algorithms," in 2010 9th IEEE/IAS International Conference on Industry Applications - INDUSCON 2010, Sao Paulo, pp. 1–6, Nov.2010.

3. "Hospital Lighting Guide and Standards". BFW Inc.

4. Kementerian Kesehatan. 2012. Pedoman Teknis Bangunan Rumah Sakit Ruang Gawat Darurat, Direktorat Bina Pelayanan Penunjang Medik dan Sarana Kesehatan, Kementerian Kesehatan RI.

Keywords: luminaire angle, average illuminance, uniformity, hospital

Paper ID 30746: Reallocation of heating consumption and costs based on decoupling of household heat metering data

Yuying Li (1), Fan Yang (1), Peng Xue* (1), Rui Zhao (2), Jingchao Xie (1), Jiaping Liu (1)

1: Beijing University of Technology, China, People's Republic of; 2: Tianjin Thermal Power Co., Ltd., Tianjin, China

AIM AND APPROACH

Implementing household heat metering and charging is a significant way to realize building energy conservation in northern China, but the current heat metering has two contradictions for promotion. The first fact is the corner users have a higher heating consumption than the center users with its bigger area of external envelope, and the second is the heating costs of high-temperature users will increase sharply due to the phenomenon of adjacent room heat transfer (ARHT). This study selects a residential building in Tianjin, China as a research object, and field measurement and energy simulation are adopted to reveal every process of heat transfer including ARHT. The heating load and ARHT are further reallocated for each unit, and a new heat charging scheme is then proposed. The results provide a case study for the fair allocation of heat charging, which is vital to the implementation of household heat metering in China.

SCIENTIFIC INNOVATION AND RELEVANCE

This study establishes a physical building model by EnergyPlus based on real design documents, and Python programming is employed to analyze the heat transfer process and create the indoor temperature decoupling model. It helps to clearly identify the user's heating behavior and precisely calculate the additional heating consumption and heating costs of ARHT. Comprehensively considering the influence of the corner user's large heating consumption and ARHT problem on the theoretical heating consumption, this research also calculates the heating consumption at different locations and ARHT with various temperature differences and occupancy rates. Finally, the reallocation of heating consumption and heat charging is proposed.

PRELIMINARY RESULTS AND CONCLUSIONS

After correcting the heating consumption of the whole building, the heating consumption of corner unit on the top floor decreases 33% of it before reallocation, and the heating cost also reduces by about 20%. The heating consumption of center users increase 20%, and heating costs increase above 10% than before. The ARHT increases with the increase of the temperature differences between adjacent rooms. When the heat transfer only occurs to one adjacent room, the ARHT of a unit could take up around 25% of the total heating consumption with the temperature difference of 10 degrees. With a rather low occupancy rate of the whole building, the heat loss will be twice of the basic heating consumption of an ordinary room with a higher vacant adjacent room numbers. The ratio of ARHT to the real total heating consumption could stay around 10% and even reach up to 50%, and the heating costs of high-temperature users could increase 70% during one heating season. This quantitative research provides a case study for heat charging, which pushes the heat fee policy towards further fair allocation.

MAIN REFERENCES

[1] Canale L, Battaglia V, Ficco G, Puglisi G and Dell'Isola M. Dynamic Evaluation of Heat Thefts Due to Different Thermal Performances and Operations between Adjacent Dwellings. Applied Sciences, 2020, 10(7), 2436.

[2] Cholewa T, Siggelsten S, Balen I, Ficco G. Heat cost allocation in buildings: Possibilities, problems and solutions. Journal of building engineering, 2020, 101349.

[3] Ling J, Li Q, Xing J. The influence of apartment location on household space heating consumption in multi-apartment buildings. Energy and Buildings, 2015, 103: 185-197.

[4] Lukic N, Nikolic N, Timotijevic S, Tasic, S. Influence of an unheated apartment on the heating consumption of residential building considering current regulations—Case of Serbia. Energy and Buildings, 2017, 155, 16-24.

[5] Michnikowski P. Allocation of heating costs with consideration to energy transfer from adjacent apartments. Energy and Buildings, 2017, 139: 224-231.

[6] Siggelsten S. Reallocation of heating costs due to heat transfer between adjacent apartments. Energy and Buildings, 2014, 75, 256-263.

[7] Xue P, Yang F, Zhang Y, Zhao MJ, Xie JC, Liu JP. Quantitative study on adjacent room heat transfer: Heating load and influencing factors. Sustainable Cities and Society, 2019, 51, 101720.

Keywords: household heat metering, temperature decoupling model, heating consumption simulation, adjacent room heat transfer, heat charging

Paper ID 30933: Feasibility study for combined mist evaporation cooling with carbon dioxide enrichment for greenhouse cultivation

Craig Edward Farnham* (1), Mami Oishi (2), Jihui Yuan (3)

1: Osaka City University, Japan; 2: Research Institute of Environment, Agriculture and Fisheries, Osaka Prefecture; 3: Toyohashi Univeristy of Technology

AIM AND APPROACH

Greenhouse farming in many regions of the world faces the issue of overheating. When natural ventilation nor mechanical ventilation is sufficient to prevent overheating, a low-cost solution is evaporative cooling, often with wetted media and strong mechanical ventilation. Fine mist sprays from high pressure nozzles can evaporate completely without this strong ventilation, and may be sufficient with natural ventilation to prevent overheating, as well as prevent air saturation which would block further mist evaporation, thus limiting the cooling effect and causing undesirable wetting of the plants and fruit.

Increased carbon dioxide (CO_2) concentrations may help increase crop yields by up to 30%. In "plant factories" (controlled environments for optimized plant growth) CO_2 enrichment is one growth-promoting technique. However, plant factories are costly to create and operate. In a typical low-cost greenhouse in a warm climate, is it practical to use CO_2 enrichment if combined with evaporation cooling and low ventilation rates?

In this feasibility study, a model of a greenhouse used in eggplant cultivation is developed in Ansys FLUENT 14.0. The boundary conditions are based on on-site measurements including air exchanges, mist spray rates, and continuous temperature and humidity logging.

SCIENTIFIC INNOVATION AND RELEVANCE

The Fluent Discrete Phase Model(DPM) is used to model both the water mist spray and the CO2 injection. The precise modeling of the CO2 flow near the injection pipes is not important to the overall model. Thus, modeling and meshing these many small inlets is avoided by using the DPM model, taking particles of solid CO2 which are set to near-instantly "combust" into CO2 gas, while assuring this does not have a significant effect on the overall thermal model. The model mesh is set finest at the mist spray injectors, where accurately representing the rapid evaporation is of key importance.

The steady state model is used to explore the parameters of mist spray flow rate (full capacity or half), air exchanges (high and low), outdoor air conditions (hot and dry or warm and humid) to find the optimal spray for the given condition. The CO2 distribution is also found for these conditions.

If the low ventilation rates for mist spray evaporation cooling can also permit CO2 enrichment in greenhouses, this may present a low-cost alternative form of "plant factory" which could be of use in warmer climates including in developing countries to increase cultivation yields.

PRELIMINARY RESULTS AND CONCLUSIONS

The test site is using a set mist spray schedule. The objective is to determine the effect of the single mist cooling control regime on the product quality. That is, only one test condition can be done per growing season in each greenhouse. Thus, it is desired to use simulation to test various mist spray parameters to optimize cooling for the climate at the site. Those results can then be tested in the following growing season. Further, the CO2 concentration and distribution can be modelled along with the mist spray flow and cooling effect, and visualized to indicate potential dead zones or accumulations.

At the time of writing this abstract, the mist spray model results (particle paths, air temperatures) are well-matched with on-site measurements on hotter dry days, while less so on humid days. The key uncertainties are the actual natural ventilation rate and the insolation. We are attempting to deduce air exchanges from the change in absolute humidity within the greenhouse given a constant spray rate. Basic air exchange calculations show CO2 enrichment to a significant level should be possible, but the levels and spatial distribution is still being processed.

MAIN REFERENCES

S.B.Idso, B.A.Kimball, M.G.Anderson, J.R.Mauney. Effects of atmospheric CO2 enrichment on plant growth: the interactive role of air temperature. Agriculture, Ecosystems & Environment, 20(1), pp. 1-10. (1987)

S. Murakamia, Y. Fujimoto, Y. Hayashi, H. Fuchikami, and T. Hattori. Energy cost minimization in plant factories considering weather factors using additive Bayesian networks. Journal of International Council on Electrical Engineering 8(1), pp. 128–135, (2018)

A. Perdigones, J. Garcia, A. Romero, et al. Cooling strategies for greenhouses in summer: Control of fogging by pulse width modulation. Biosystems Engineering 99(4), pp.573-586, (2008)

A. Ganguly and S. Ghosh, A Review of Ventilation and Cooling Technologies in Agricultural Greenhouse Application, Iranica Journal of Energy & Environment 2(1): pp.32-46, (2011)

Keywords: mist, cooling, greenhouse, evaporation

Paper ID 30523: Seasonal performance and potential of heat pumps in Catalonia

Thibault Péan*, Joaquim Romaní, Elena Fuentes, Jaume Salom

IREC Catalonia Institute for Energy Research, Spain

AIM AND APPROACH

Heat pump systems are considered a key element to the decarbonisation of our energy systems, given their ability to supply the thermal loads with (potentially renewable) electricity at a high efficiency. To understand better the impact of a large rollout of heat pumps to provide space heating and domestic hot water services in the residential sector, it is important to analyse in details their performance not only in steady-state conditions, but also on an annual basis, based on specific climates and emission systems, which greatly affect their efficiency. The present study therefore aims at evaluating the seasonal coefficient of performance (SCOP) through four different methods, including simplified calculations, normative methods included in European standards, and detailed annual simulations. These methods are compared for the different climate zones present in Catalonia (Spain), to understand the potential regional differences. Further than the SCOP, the impact of different emission systems is also studied, especially considering the replacement of gas boilers with heat pumps, and evaluating whether the previously installed emitters can be reused with the heat pump, which work at lower supply temperature. Answering these questions will form a basis to support the regional governmental decisions regarding energy plans and sustainable development.

SCIENTIFIC INNOVATION AND RELEVANCE

A thorough comparison of the seasonal performance factor of residential heat pumps is especially relevant, because the different calculation methods may vary from a simple equation correcting the nominal COP, to a more detailed approach based on the number of hours where the heat pump operates annually in a given climate. This study will thus enable to understand the deviations that can be expected in seasonal efficiency, because of the evaluation method, the geographical and climatic disparities, the heat pump control implemented and the final heat emitters. To perform the annual dynamic simulations, a new and detailed dynamic model of air-to-water heat pump has been used.

The question of the refurbishment is also particularly important, if policies are to incite to a large rollout of heat pumps. The most common heating system in Spanish homes consist of a gas boiler supplying water radiators. A boiler can easily be replaced by a heat pump, but changing the radiators for a circuit of fancoils or a radiant floor supposes additional construction works and investments. If the current emission system can be kept and still function correctly at lower temperature, this will reduce the installation costs and encourage a larger heat pump deployment.

PRELIMINARY RESULTS AND CONCLUSIONS

The first results indicate that the regulation of the supply temperature produced by the heat pump, linked to the emission system, has a larger impact on the seasonal efficiency than the climatic differences within Catalonia. With low temperature systems like radiant floor, the SCOP reaches 3.0 to 4.1, while it stays around 2.2 to 3.0 for high temperature systems like radiators. The use of heating curves in supply temperature control, recommended by manufacturers, produce a significant improvement in seasonal efficiency, highlighted by the yearly simulations performed in seven climate zones. The methods from the standard tend to overestimate the seasonal efficiency, as they are based on steady-state performance maps provided by the manufacturers in ideal conditions, while the reality of the building dynamics and the actual control of the heat pump have a slightly negative impact on the performance.

Regarding the replacement of gas boilers with heat pumps without changing the emission system, the first results indicate that such configuration would work in moderate climates. In colder climates, it would cause an important number of hours outside the comfort zone, because of the inherent limits of an emission system designed to work at high temperatures, but operating at lower temperature.

MAIN REFERENCES

UNE-EN 14825:2018. Acondicionadores de aire, enfriadoras de líquido y bombas de calor con compresor accionado eléctricamente para la calefacción y la refrigeración de locales. Ensayos y clasificación en condiciones de carga parcial y cálculo del rendimiento. Madrid, España, 2018.

Instituto para la Diversificación y Ahorro de la Energía (IDAE), "Prestaciones Medias Estacionales De Las Bombas De Calor Para Producción De Calor En Edificios," 2007.

Riederer, P., Partenay, V., & Raguideau, O. (2009). Dynamic test method for the determination of the global seasonal performance factor of heat pumps used for heating, cooling and domestic hot water preparation. IBPSA 2009 - International Building Performance Simulation Association 2009, (April), 752–759.

T. Péan, "Heat pump controls to exploit the energy flexibility of building thermal loads," Universitat Politècnica de Catalunya, 2020.

Péan, T., Costa-Castelló, R., & Salom, J. (2019). Price and carbon-based energy flexibility of residential heating and cooling loads using model predictive control. Sustainable Cities and Society, 50. https://doi.org/10.1016/j.scs.2019.101579

Pospíšil, J., Špiláček, M., & Kudela, L. (2018). Potential of predictive control for improvement of seasonal coefficient of performance of air source heat pump in Central European climate zone. Energy, 154, 415–423. https://doi.org/10.1016/j.energy.2018.04.131

Keywords: heat pumps, refurbishment, seasonal performance

Session T3.1: Practice and industry related case studies

Paper ID 31055: Using simulation, visualization and monitoring to address barriers to high-performance building strategies

Nathan Brown* (1), Ibone Santiago (1), M Susan Ubbelohde (1,2), George Loisos (1), Santosh Philip (1)

1: Loisos + Ubbelohde, United States of America; 2: University of California, Berkeley

AIM AND APPROACH

Design strategies for high performance often require a tuned system where envelope and HVAC strategies are complementary. However, normative approaches tend to encourage the design of oversized HVAC equipment as well as its use independent of envelope systems, thus potentially undermining high-performance strategies. One contributing factor is the use of compliance simulations containing envelope assumptions and behavior that are often not representative of the specificity of high-performance projects, which reduces their usefulness as a design tool. Other factors include risk-aversion of the design team and stakeholder preconceptions of thermal comfort factors as related to HVAC systems and not to the envelope.

This paper presents such methods, developed through practice, to use simulation, visualization, and monitoring as tools to quantify, communicate, and document performance. We present 3 case studies, one in depth, to illustrate the use of these methods to dramatically change the size and concept of mechanical systems. In addition, we present the use of developing a monitoring plan in order to further assuage the fear that HVAC designers will be held accountable for thermal variation outside the control of mechanical systems.

SCIENTIFIC INNOVATION AND RELEVANCE

The transition to a zero-energy, zero-carbon future has to include significant shifts in building practices deployed today. We have much of the knowledge required: much research has been published on innovative approaches to integrating building envelope and mechanical system design. We also have powerful resources around simulation and optimization techniques.

A significant barrier is the entrenchment of normative practice in everyday practice among design teams.

This paper adds to the body of knowledge of how to deploy modeling, visualization and monitoring as pieces of the bridge from normative to high-performance practice.

PRELIMINARY RESULTS AND CONCLUSIONS

We illustrate this approach first through projects to condition classroom spaces using tempered outside air rather than a packaged HVAC system. We then present a detailed case study of an art center/residence designed with high mass and earth-coupled thermal strategies. Initial simulations demonstrated the significant energy reduction benefits of these strategies.

In each of these cases, project particularities emerged as challenges to high-performance, low-energy approaches, including wide variations in occupancy profiles, stakeholder thermostat setting preferences informed by non-representative previous experience, and resiliency scenarios. These challenges required a plan for "who owners will call when something goes wrong."

Modeling and visualization provided a way to test scenarios being considered and help the team and stakeholders understand performance differences between each scenario. This allowed team members, especially mechanical system designers, to managing expectations and delineating areas of responsibility. Development of a monitoring plan provided a second layer of risk management since it provided a plan to disaggregate potential future issues with performance. In this way, performance issues can be more easily disaggregated as issues related to envelope, mechanical system, or building operation. Monitoring thus formed an additional insurance policy so that thermal performance can be compared to expectations (simulated results).

MAIN REFERENCES

Levitt, Brendon, et al. "Thermal autonomy as metric and design process." CaGBC National Conference and Expo: Pushing the Boundary–Net Positive Buildings. 2013.

Nytsch-Geusen, C., Kaul, W,. Rädler, J., Westermann, L., Shenoy, V., Balekai, P. (2019). The Digital Twin as a Base for the Design of Building Control Strategies

http://www.ibpsa.org/proceedings/BS2019/BS2019_210389.pdf

Jones, N, L., Chaires, I., Goehring, A. (2019). Detailed Thermal Comfort Analysis from Preliminary to Final Design

http://www.ibpsa.org/proceedings/BS2019/BS2019_210875.pdf

Walikewitz, N., Jänicke, B., Langner, M., Meier, F., Endlicher, W. (2015). The difference between the mean radiant temperature and the air temperature within indoor environments: A case study during summer conditions. Build Environ 84, 151-161.

Raftery, P., C. Duarte, S. Schiavon, and F. Bauman. 2017. A new control strategy for high thermal mass radiant systems. www.escholarship.org/uc/item/5tz4n92b

Keywords: high-performance building, simulation, visualization, monitoring

Paper ID 30242: In-situ validation of a new sizing methodology for combined production and distribution for domestic hot water and space heating

Senne Van Minnebruggen*, Ivan Verhaert

University of Antwerp,Faculty of Applied Engineering, Antwerp, Belgium

AIM AND APPROACH

Recent studies show that collective heating and cooling have the potential to decrease CO2-emissions and offer flexibility to increase the amount of renewable energy sources in the overall energy supply [1][2]. In order to achieve these advantages a proper design is crucial. The first step in the design of an optimal collective heating system is sizing. Within Belgian context, currently no standards or design rules exist for the sizing of collective heating systems that provide hot water for space heating purposes as well as for domestic hot water[3]. However, in practice a diversity of 'customized design rules' exists. Designers use a mix of foreign design rules and own rules of thumb or fabricant specific design guides. The lack of transparency disables a proper comparison and prevents new insights into either space heating of DHW demand to be incorporated. Eventually, this leads to poorly designed systems and discussions between different stakeholders after installation. This raises the question: 'How to properly and transparently size collective heating systems?' By comparing different design methods and analysing heat meter data the objective of this paper is to validate a recently introduced design rule [3], which tackles the issue of transparency and compatibility with other standards.

SCIENTIFIC INNOVATION AND RELEVANCE

In order to validate this new design rule a validation methodology has been developed to define the peak heat consumption of the installation on the basis of data from heat meters. Subsequently, the methodology determines a characteristic that shows the possible combinations of production capacity and heat storage to cover the peak heat consumption. However, some challenges arise in order to analyse the peak heat consumption. Firstly, the poor quality of the data from heat meters needs to be addressed. To cope with this a datacleaning is performed. Secondly, to obtain the peak heat consumption one has to take the influence of the outdoor conditions and the data resolution (measurement interval size) into account. In order to deal with the data resolution the methodology uses insights from a recent design methodology for DHW systems based on tap patterns[4]. Furthermore, to cope with the influence of the outdoor conditions, the relation between the peak heat consumption and the outdoor temperature is investigated. As a result the maximum or worst case peak heat consumption profile can be determined and, consequently, the corresponding production capacity – heat storage characteristic. The outcome of the design rule can then be tested based on this characteristic.

PRELIMINARY RESULTS AND CONCLUSIONS

In order to test the methodology and consequently validate the design rule, the validation methodology was applied to two case studies. From the result of the case studies follows that the method is able to obtain the desired characteristic. Nevertheless, the design rule could not be validated. Based on the analysis of the heat consumption of both case studies over the considered measurement period, some deviations in heat consumption were found. As a result of these deviations, the relationship between peak heat consumption and outdoor conditions could not be determined unequivocally. It is therefore possible that the result obtained underestimates the required capacity. In order to validate the design rule in a representative manner and to refine the validation methodology, it is advisable to apply the validation methodology to several case studies. Whereas these case studies should meet some requirements/advices as set out in this paper. Also, due to the diversity in design rules, an overview and evaluation of these different design rules proves to be necessary.

MAIN REFERENCES

[1] D. Connolly et al., "Heat Roadmap Europe: Combining district heating with heat savings to decarbonise the EU energy system," Energy Policy, vol. 65, pp. 475–489, Feb. 2014, doi: 10.1016/J.ENPOL.2013.10.035.

[2] H. Lund et al., "4th Generation District Heating (4GDH): Integrating smart thermal grids into future sustainable energy systems," Energy, vol. 68, pp. 1–11, Apr. 2014, doi: 10.1016/J.ENERGY.2014.02.089.

[3] I. Verhaert, "Design methodology for combined production and distribution for domestic hot water and space heating," E3S Web Conf., vol. 111, p. 01089, 2019, doi: 10.1051/e3sconf/201911101089.

[4] I. Verhaert, B. Bleys, S. Binnemans, and E. Janssen, "A Methodology to Design Domestic Hot Water Production Systems Based on Tap Patterns," CLIMA2016 - Proc. 12th REHVA World Congr., 2016.

Keywords: Collective heating system, Heat consumption analysis, Heat meter readings

Paper ID 31156: Comparison of Heating Power calculated through standard and building simulation models

Paul Van den Bossche* (1), Jeroen Van der Veken (1), Sébastien Pecceu (1), Stijn Verbeke (2)

1: BBRI, Belgium; 2: EMIB, UAntwerpen, Belgium

AIM AND APPROACH

With the shifting focus from increasing renewable energy production towards energy distribution issues and finding a continuous match between supply and demand, the need to define a correct required heating power in buildings also increases. Whereas calculations in the past mainly focused on comfort and 'enough power at any moment', a better understanding of the profiles of the required heating power is necessary to bring it in line with renewable energy availability. In addition, the installation costs of highly efficient systems such as heat pumps are more dependent on the installed power compared with classic boiler systems. However, experience from practice tend to show that the EN12831-1 overestimates the necessary heating power.

SCIENTIFIC INNOVATION AND RELEVANCE

Key Innovations

- Comparison of calculated, measured and simulated heat power for about 1000 cases

- Simulations based on realistic winter conditions

Research Implications

The original climatic data (period 2010-2020) showed clear correlations between dry bulb temperature, wind speed and solar radiation, all impacting the heat balance. When studying dynamics at extreme (cold) periods, these correlations need to be maintained, while commercially available climatic data the air temperatures are often not correlated with the wind data.

PRELIMINARY RESULTS AND CONCLUSIONS

Both measurements and simulations show an overdimensioning factor that ranges from rather small and acceptable (10 %, a minimum safety margin that could be justified for sizing purposes) to very significant (> 100 %), which is probably not acceptable from economic and efficiency related perspective.

What we can conclude from the simulation results is that between -3°C and 15°C the required power follows the trend of the linear heating curve that connects the calculated standard power at design temperature and 0 at 20°C. However, below these temperatures the simulated retrieved power appears to flatten out; thereby increasing the discrepancy with the calculations according to the procedures in the standard.

Several explanations may be at the origin of this effect. Firstly, in the Belgian climate no days are observed with, at the same time, the lowest temperatures, maximum wind speed and without solar gains. The standard implicitly assumes simultaneous occurrence of these factors, but this is not observed in the actual weather data.

Looking at our measurement data, this 'plateau effect' is even increased, probably by changing occupant behaviour and automatic control interventions.

MAIN REFERENCES

CEN - European Committee for standardization (2017) EN 12831-1 : Energy performance of buildings - Method for calculation of the design heat load - Part 1: Space heating load (EN 12831-1:2017)

Dols, W. S., & Polidoro, B. J. (2015). Contam user guide and program documentation version 3.2 (No. Technical Note (NIST TN)-1887).

EnergyPlus version 9.4.0; Lawrence Berkeley National Laboratory, U.S. Department of Energy and others.

NBN - Bureau voor normalisatie (2020). NBN EN 12831-1 ANB:2020. Belgian annexe to EN 12831-1:2017, in Dutch and French.

Peccue, S., Van den Bossche, P., Impact of Building Airtightness on Heat Generator and Heat Emission Equipment Sizing. AIVC Conference Athens September (2020 postponed) 2021.

Verbeke, S., Thermal inertia in dwellings, PhD dissertation, Universiteit Antwerpen (2017)

Keywords: Heating power, EN12831-1, climatic data

Paper ID 30430: Definition and validation of a thermal zone model in Modelica language with Holzkirchen Twin Houses experiment

Rossella Alesci* (1), Ettore Zanetti (1), Rossano Scoccia (1), Alberto Leva (2), Mario Motta (1)

1: Department of Energy, Politecnico di Milano, 20156 Milano, Italy; 2: Dipartimento di Elettronica, Informazione e Bioingegneria , Politecnico di Milano, 20133 Milano, Italy

AIM AND APPROACH

This manuscript presents the validation of the Modelica library "BuildingsAndPlants" thermal zone model. The Twin Houses experimental data and a TRNSYS model [1] were used as comparison. The "BuildingsAndPlants" library allows to easily integrate new models representing new advanced technologies (such as ventilated façades, double skin façades etc.) or new HVAC systems models in thermal zone model. In fact, until now, this process is difficult and time-consuming using the already available software.

In the "BuildingsAndPlants" library the use of components from the Modelica Standard library is reduced and all the heat transfer and fluid dynamics are modelled inside the library. In addition, it tends to minimize the use of inheritances and other properties that allow to write the code in a slightly more redundant but simpler way making it easier for the users to retrace the equations at the base of the models.

SCIENTIFIC INNOVATION AND RELEVANCE

This work focuses attention on providing a modular thermal zone model where each element can be changed. This allows to test different models (model classes, level of physic complexity) and to choose each time the most appropriate for the case under exam. Hence, the scientific innovation of this work is related to the different focus of the "BuildingsandPlants" library.

The thermal zone model defined in the "BuildingsandPlants" library can be adopted as baseline for the analysis of the behavior of new components such as new construction technologies or new HVAC systems. In fact, thanks to the modularity and the understandability of this library it will be very easy for an expert in building physics, to incorporate new components in the existing thermal zone model. In this way, each user can just add the new sub model which he wants to study, without the need to model each time the entire thermal zone from the beginning.

In addition, the understandability of this library will reduce the time required to learn how to use and modify it.

PRELIMINARY RESULTS AND CONCLUSIONS

The preliminary results of the thermal zone model described in this paper are obtained applying the case study of the Twin Houses in Holzkirchen [2].

In particular, in free floating conditions the air temperature trend is close to the one registered by the sensor, with a maximum temperature difference of 1°C and a time shift between the two curves that is almost zero along the entire simulation period.

On the other hand, at constant temperature, the required thermal power is underestimated of the 46% with respect to the value provided by the sensor.

Thus, one may conclude that, in free floating conditions, the model simulates in the correct way the behavior of the reference building, also at a dynamic level, while, when the electric heater is switched on, there is an underestimation of the required thermal power that is related to the hypothesis of full mix on which the model is based. This conclusion is also confirmed by another article [3] which analyzed the same case study.

In addition, the same model was created in TRNSYS and adopted to compare the results. In this case, the difference between the two integrals of the thermal power is of the 10%.

MAIN REFERENCES

[1] Solar energy laboratory, "TRNSYS 18 Getting Started," Trnsys 18, vol. 1, pp. 1–9, 2014.

[2] P. Strachan, I. Heusler, M. Kersken, and M. J. Jiménez, "Empirical Whole Model Validation Modelling Specification: Test Case Twin_House_Experiment_2 IEA ECB Annex 58 Validation of Building Energy Simulation Tools Subtask 4."

[3] G. Masy et al., "Lessons learned from heat balance analysis for holzkirchen twin houses experiment," Energy Procedia, vol. 78, pp. 3270–3275, 2015.

[4] W. Li et al., "Modeling urban building energy use: A review of modeling approaches and procedures," Energy, vol. 141, pp. 2445–2457, Dec. 2017.

[5] ASHRAE, "FUNDAMENTALS," in ASHRAE HANDBOOK, Atlanta,GA, 2013.

[6] A. Sodja and B. Zupančič, "Modelling thermal processes in buildings using an object-oriented approach and Modelica," Simul. Model. Pract. Theory, vol. 17, no. 6, pp. 1143–1159, 2009.

[7] P. Fritzson, Principles of Object Oriented Modeling and Simulation with Modelica 3.3. IEE Press, Wiley, 2015.

[8] A. Dama and F. Lastaria, "Explicit versus implicit method for radiative heat transfer in gray and diffuse enclosures," Int. J. Heat Mass Transf., vol. 55, no. 13–14, pp. 3829–3833, 2012.

[9] M. Wetter, "Modeling of Heat Transfer in Rooms in the Modelica ' Buildings ' Library," 2013.

Keywords: Validation, Modelica, BuildingsAndPlants, dynamic simulation

Paper ID 30626: Calibration of a multi-residential building energy model – Part II: Calibration using surrogate-based optimization

Van Long Lê* (1), Charlotte Marguerite (1), Charlotte Beauthier (1), Olivier Fontaine de Ghélin (1), Cécile Goffaux (1), Loïc de Moffarts (2)

1: Cenaero, Belgium; 2: Thomas & Piron Group, Belgium

AIM AND APPROACH

The present study, co funded by European Regional Development Fund and Thomas & Piron Group, aims at developing an energy toolbox for building energy load prediction and building fault detection and diagnosis. The idea is to use a calibrated energy model as an assistant to guarantee the energy performance of the building.

The research approach adopted in the present study is a stepwise energy model calibration. The building energy model is initialized based on the data from design and construction phases of the building and using the OpenStudio/EnergyPlus software. This energy model is then subjected to a sensitivity analysis (details of methodology will be described in a separated paper, i.e. Calibration of a multi-residential building energy model PartI: Cluster-Based sensitivity analysis) with the aim of identifying the most important parameters for the energy consumption. These parameters will then be considered as variables of an optimization-based calibration. A metamodel-based evolutionary algorithm is used to minimize the discrepancies between simulation results and real consumption data collected from smart meters. The Normalized Mean Bias Error (NMBE) and the Coefficient of Variation of the Root Mean Square Error (CvRMSE) will be calculated for each apartment to assess the accuracy of calibrated energy model.

SCIENTIFIC INNOVATION AND RELEVANCE

The present study deals with the calibration of a white-box energy model of a multi-family, multi-story building. The latter has 40 apartments having 1 to 3 bedrooms and a single-family house with 3 bedrooms. The energy consumption data are collected by smart meters installed for each building apartment. The real weather data are provided by the Royal Meteorological Institute of Belgium. As the calibrated energy model will be used to predict energy load as well as for fault detection and diagnosis at building and apartment levels, the energy model considers a large number of parameters. Even after a reduction step (selection of the most influent parameters through a sensitivity analysis), the total number of parameters for the optimization-based calibration is still large. Moreover, the white-box energy model simulation is quite expensive in terms of computational time. Therefore, the metamodel technique (e.g. Radial Basis Function, Kriging) is used for reducing computational costs of the white-box energy model simulation and the evolutionary algorithm, implemented within the Cenaero inhouse optimization platform, is used for finding the global optimum of the constrained optimization problem.

PRELIMINARY RESULTS AND CONCLUSIONS

The preliminary results show that the accuracy of energy model is largely improved compared with the baseline energy model of the building. The NMBE values calculated are within the recommended value range, which indicates that the discrepancies between the simulation results and the measurements are reasonable. However the CvRMSE is still quite high for several apartments and additional work is needed to obtain a value that fits better to the recommendations for model calibration (e.g. within the range validated by a performance measurement and verification protocol such as the ASHRAE guideline 14, International Performance Measurement and Verification Protocol).

MAIN REFERENCES

Crawley, Drury B., Linda K. Lawrie, Frederick C. Winkelmann, W. F. Buhl, Y. Joe Huang, Curtis O. Pedersen, Richard K. Strand, et al. 2001. "EnergyPlus: Creating a New-Generation Building Energy Simulation Program." Energy and Buildings. https://doi.org/10.1016/S0378-7788(00)00114-6.

Beauthier, Charlotte, Paul Beaucaire, Caroline Sainvitu, 2017. "A Surrogate-Based Evolutionary Algorithm for Highly Constrained Design Problems." In GECCO 2017 - Proceedings of the Genetic and Evolutionary Computation Conference Companion, 1614–21. New York, NY, USA: Association for Computing Machinery, Inc. https://doi.org/10.1145/3067695.3082538.

Li, Wancheng, Zhe Tian, Yakai Lu, and Fawei Fu. 2018. "Stepwise Calibration for Residential Building Thermal Performance Model Using Hourly Heat Consumption Data." Energy and Buildings. https://doi.org/10.1016/j.enbuild.2018.10.001.

Fabrizio, Enrico, and Valentina Monetti. 2015. "Methodologies and Advancements in the Calibration of Building Energy Models." Energies. https://doi.org/10.3390/en8042548.

Sun, Kaiyu, Tianzhen Hong, Sarah C. Taylor-Lange, and Mary Ann Piette. 2016. "A Pattern-Based Automated Approach to Building Energy Model Calibration." Applied Energy 165 (March): 214–24. https://doi.org/10.1016/j.apenergy.2015.12.026.

Goldwasser, David, Brian Ball, Amanda Farthing, Stephen Frank, and Piljae Im. 2018. Advances in Calibration of Building Energy Models to Time Series Data: Preprint. Golden, CO: National Renewable Energy Laboratory. NREL/CP-5500-70865.

Keywords: energy model calibration, surrogate-based optimization, sensitivity analysis

Session T3.2: Ensuring high quality building simulations

Paper ID 30771: FMI Co-Simulation between 2D/3D component models and HVAC/control models

Andreas Nicolai*, Andreas Söhnchen

TU Dresden, Germany

AIM AND APPROACH

Detailed construction and building component models, including hygrothermal porous material transport models, can be used to model a large variatiy of modern energy transfer and storage systems. These include heated concrete slabs, shallow soil heat collectors, heated wall layers, combined photovoltaic and construction panels etc. The interaction with connected HVAC systems/energy distribution systems and/or control models is, however, often limited in such tools. Using three distinct application cases, the article describes the extension of the hygrothermal transport model DELPHIN with an FMI co-simulation interface and the setup of a coupled simulation with external models. The article covers the some details of the implemenation, but also derived best-practice approaches on FMI co-simulation algorithms and parameters, suitable for the tested application scenarios.

SCIENTIFIC INNOVATION AND RELEVANCE

Description of tested, practice oriented co-simulation cases, using a detailed sub-model (building component model/hygrothermal transport model) and external control/energy distribution model. Description and demonstration of consequenced from Co-Simulation algorithm and parameter selection. Derivation of best-practice recommendations for setting up such coupled simulations, and defined interfaces and coupling parameters.

PRELIMINARY RESULTS AND CONCLUSIONS

Co-Simulation support and use successfully demontrated in quite different use cases. Illustration, that supporting even the most basic variant of the FMI co-simulation standart enhances functionality of isolated models significantly. Discussion of co-simulation parameters show, that for application cases in building energy simulation/energy supply system simulation, a simple Gauss-Seidel co-simulation type with moderate time step selection is usually sufficient.

MAIN REFERENCES

FMI co-simulation standard

Nicolai, A. and Paepcke, A.Entwicklung der Kopplungstechnologie von Komplexmodellen für Bauteil-, Raum- und Gebäudesimulation mit Modelica-basierten Anlagen-, Regelungs- und Nutzermodellen, 2018, Technischer Report

Nicolai, A. and Paepcke, A.; Co-Simulation between detailed building energy performance simulation and Modelica HVAC component models, 2017, 12th International Modelica Conference, Prague

Keywords: FMI, co-simulation, component models, HVAC, control models

Paper ID 30236: Disaggregation of digital meter data for synthetic load profile generation

Toon Bogaerts* (2,3), Stef Jacobs (1), Sara Ghane (2,3), Freek Van Riet (1), Wim Casteels (2,3), Siegfried Mercelis (2,3), Ivan Verhaert (1), Peter Hellinckx (2,3)

1: Energy and Materials in Infrastructure and Buildings, University of Antwerp, Belgium; 2: IDLab, University of Antwerp, Belgium; 3: Imec, Belgium

AIM AND APPROACH

Building simulations require accurate Synthetic Load Profiles (SLP) of electricity consumption to research interaction with the grid or coping strategies for appliance-induced overheating. This is only possible by means of empirically validated user behavior profiles, i.e. based on in situ measurements. Logging for each single appliance separately is, however, expensive and labor-intensive. This means that centrally measured data should be disaggregated into the data of individual appliances.

The aim of this research is therefore to evaluate two event-based Non-Intrusive Load Monitoring (NILM) techniques for data disaggregation for appliance recognition: classification trees and timeseries analysis using deep learning . Moreover, the compatibility of these both techniques with low temporal resolution of the measurements is verified.

Finally, based on the results, the applicability of the NILM-techniques on the digital energy meters in Belgium is discussed.

A public labelled dataset is considered as case study. The dataset contains one-week measurements with state transitions of the individual appliances as labels. The dataset is separated in order to train, validate and test the NILM-techniques.

SCIENTIFIC INNOVATION AND RELEVANCE

According to the European Commission, digitalization of the energy system is a necessity for the transition towards a sustainable future. In this context, the roll out of digital energy meters in Europe has started. The next step is to translate the digital meters into "smart" meters. While great efforts have been made in previous research -including to the level of commercially available tools-, existing sources lack objective evaluation of disaggregation techniques. Therefore, no reliable tools exist for central measurements for generating Synthetic Load Profiles.

Therefore, this paper discusses the development and evaluation of different NILM-techniques for user behavior profile applications. Moreover, it compares different temporal resolutions to take into account different read-out frequencies. Indeed, e.g. only the more recent digital energy meters in Belgium are equipped with the high-frequency read-out S1 gate, while previous versions are limited to a frequency of 0.1 Hz.

To establish disaggregation of appliances, we look to use beyond the state of the art time series classification techniques such as appliance fingerprinting, timeseries classification and feature extraction combined with classification trees. More specifically, the use of Long Short Term Memory cells will be used to analyze time series.

PRELIMINARY RESULTS AND CONCLUSIONS

Preliminary research shows that an optimized decision tree classifier is able to identify appliances in a similar fashion as statistical methods. Feature engineering improved the performance of the tree. We look to further improve these result by using recurrent neural networks such as LSTM's to extract complex features from the timeseries data. These features can be passed to a fully connected multilayer perceptron classifier to distinguish the different appliances. Furthermore, the influence of the sample rate did not affect the decision tree. This relation will be further analyzed to estimate the importance of the s1 gate. Finally, we will look into the possibility of continues labeling of timeseries with a relation between the hidden state of the recurrent neural network and the states of appliances. With these techniques we look to achieve 5%-10% better performance in means of accuracy beyond statistical methods.

Before the discussed algorithms can be applied for the generation of Synthetic Load Profiles, future work should focus on their applicability on buildings with other user behaviour and appliances than represented by the used dataset.

MAIN REFERENCES

• Anderson, K. et al., BLUED: A Fully Labeled Public Dataset for Event-Based Non-Intrusive Load Monitoring Research, in proceedings of ACM SustKDD'12, 2012

• NGUYEN, M., et al. A novel feature extraction and classification algorithm based on power components using single-point monitoring for NILM. In: 2015 IEEE 28th Canadian Conference on Electrical and Computer Engineering (CCECE). IEEE, 2015. p. 37-40.

• KELLY, Jack; KNOTTENBELT, William. Neural nilm: Deep neural networks applied to energy disaggregation. In: Proceedings of the 2nd ACM International Conference on Embedded Systems for Energy-Efficient Built Environments. 2015. p. 55-64.

Keywords: Non Intrusive Load Monitoring, Smart meters, Data disaggregation

Paper ID 30932: Measurement of the building envelope thermal performance in collective housings

Lorena de Carvalho Araujo* (1,2), Simon Thébault (1), Laurent Mora (2), Thomas Recht (2)

1: CSTB, France; 2: Univérsité de Bordeaux, France

AIM AND APPROACH

Building energy efficiency is a key factor in reducing CO2 emissions and assuring the comfort level for inhabitants. Governments have been valorizing the energy performance standards through thermal regulations and economic incentives. These are often based on results from building simulation softwares, achieved during building design stage. However, the real thermal performance can significantly deviate from the predicted one [1]. It is important to have reliable performance indicators to assure new building quality and to estimate the gains accruing after renovation works. The application of an in-situ method after construction or retrofitting phases enables the measurement of such indicators, as the whole heat loss coefficient (HLC) [2] and the heat loss coefficient by transmission (Htr) [3]. Collective housing counts for an important part of building stock, for this reason, mature technologies to measure its thermal performance are necessary. The current paper studies the applicability of a short duration test for identifying the HLC and Htr in collective housings and how to optimize the test protocol.

SCIENTIFIC INNOVATION AND RELEVANCE

There are different available methods in the literature for measuring the building envelope thermal performance (for instance: average method, energy signature, PSTAR, EBBE, co-heating, ISABELE, QUB, and others) [2],[3],[4],[5],[6]. They present variations concerning the mathematical approach, the duration, the protocol modalities and the applicability [7],[8]. Among those, methods like energy signature and EBBE can be applied to collective housing. However, they use static models, presenting long measurement periods that usually lasts for more than one season extending upto a period of few years. Presently, there are not short duration tests that have been validated regarding the HLC and Htr estimation quality for this building typology. The propose of the current paper is to study a dynamic approach using grey box models, that allows the reduction of test protocol for identifying the thermal performance of collective housing's envelope. In addition, variations of the test protocol, allows the study of optimal test conditions applied virtually to a medium collective housing. Furthermore, the relevancy of test protocol is verified by its application in-situ in a real building. This research suggests an alternative to evaluate the whole building heat loss coefficient of collective housings, with measurements duration shorter than one week.

PRELIMINARY RESULTS AND CONCLUSIONS

We modelled in a thermal dynamic simulation software (Pléiades + COMFIE) a collective housing from the Residence Figuières Vignettes in Feyzin, France. It presents 1300 m² divided in four floors and sixteen apartments, 21 thermal zones and the thermal properties of components are in a level of a retrofitted building. 336 variations from a protocol inspired by ISABELE method were applied to this model in order to study the impact of several key parameters of the protocol (duration, heating power, set point temperature, preheating) in the quality of Htr estimation.

The test duration and the temperature difference from the beginning and the end of the test were the most influent parameters in the quality of the Htr indicator. For durations equal or superior to four days of measurement, the tests presented a Htr bias inferior to 15% for moderate internal temperature variation.

Besides the virtual experiments, we applied a test into a real collective housing composed of three apartments located in Sallanches, France. The overall heat loss coefficients level of this building was measured using the SEREINE method for a period of one week. After two days of test, the results are stable and present an uncertainty inferior to 15%.

MAIN REFERENCES

[1] Wang, Liping & Mathew, Paul & Pang, Xiufeng. Uncertainties in energy consumption introduced by building operations and weather for a medium-size office building. Energy and Buildings. (2012).

[2] Bauwens G. In Situ Testing of a Building's Overall Heat Loss Coefficient – Embedding Quasi-stationary and Dynamic Tests in a Building Physical and Statistical Framework. Doctoral Thesis. (2015) .

[3] Thébault, Simon. Contribution à l'évaluation in situ des performances d'isolation thermique de l'enveloppe des bâtiments. Thèse de doctorat. (2017).

[4] Cohen M. et al., EPILOG Livrable n° 2 Rapport de synthèse sur la méthodologie employée avec tests sur cas d'étude théoriques par simulation. PACTE. (2017).

[5] Nordström, Gustav, Helena Johnsson, and Sofia Lidelöw. "Using the energy signature method to estimate the effective U-value of buildings." Sustainability in Energy and Buildings. Springer, Berlin, (2013).

[6] Wingfield, Jez, et al. "Whole house heat loss test method (Coheating)." Leeds Metropolitan University (2010).

[7] IEA Annex 58. Reliable Building Energy Performance Characterisation Based on Full Scale Dynamic Measurements. (2015)

[8] Roels, Staf, et al. "On site characterisation of the overall heat loss coefficient: Comparison of different assessment methods by a blind validation exercise on a round robin test box." Energy and Buildings 153 (2017).

Keywords: heat loss coefficient, collective housings, in-situ test

Paper ID 30327: Cooling demand reduction approaches for typical buildings in a future city district in mid-Sweden

Sana Sayadi*, Abolfazl Hayati, Jan Akander, Mathias Cehlin

Universuty of Gävle, Sweden

AIM AND APPROACH

The increase in population and living standards, as well as global warming and heatwaves due to climate change, have created a challenge to meet the cooling demand in buildings. Using currently available sources of energy endangers future energy security[1]. Therefore, implementing new approaches to reduce energy requirements in buildings to pave the path for energy transition is an area of interest. This study aims to analyze and minimize the cooling requirement for a multifamily building through simulations in a new city district in mid-Sweden. Buildings must meet the Near Zero Energy Building (NZEB) requirements based on the new Swedish National building regulations [2]. This study first explores the cooling demand of the building by means of simulations with IDA Indoor Climate and Energy (IDA-ICE) software, then investigates the effect of different mechanical ventilation strategies, window properties and orientations. The characteristics are aligned with Key Performance Indices (KPIs) which are based on the proposed list from IEA Annex 80: Resilient cooling of buildings. Climate files of normal and extreme conditions are considered for the simulations [3]. After implementing the changes in the building, results and their effect on cooling demand is investigated.

SCIENTIFIC INNOVATION AND RELEVANCE

Fulfilling the latest building regulations and implementing the most energy-efficient characteristics in the building, aligned with Annex 80's proposed KPIs help meeting the NZEB requirements. Performance of the multifamily building with focus on robustness and resilience for a future city district has to be considered. Implementing optimum building and window specification and using different climate files help fulfilling the future resilient NZEB buildings. Today's residential buildings, mainly in Sweden, have been designed to fulfill heating requirements but are seldom designed and equipped with systems for space cooling. This study investigates the future cooling requirement in terms of building regulation requirement and future climate conditions containing heat-waves.

PRELIMINARY RESULTS AND CONCLUSIONS

The cooling demand is expected to rise as the climate changes, therefore, buildings should be resilient to the future heat-waves. The chosen model meets NZEB requirements by implementing different optimized characteristics of a building aligned with Annex 80's KPIs and Swedish National Building regulations. The results envision the minimum cooling demand through optimum combination of the building's specifications in the new city district in mid-Sweden to provide the required comfort for the residents

MAIN REFERENCES

[1] Z. X. Jing, X. S. Jiang, Q. H. Wu, W. H. Tang, and B. Hua, "Modelling and optimal operation of a small-scale integrated energy based district heating and cooling system," Energy, 2014, doi: 10.1016/j.energy.2014.06.030.

[2] Boverket, "Konsekvensutredning BFS 2020:4 Boverkets föreskrifter om ändring i verkets byggregler (2011:6) – föreskrifter och allmänna råd, BBR, avsnitt 5 och 9", [In Swedish] Report on proposed NZEB building regulations version BBR 29, 2020.

[3] A. Machard, C. Inard, J.-M. Alessandrini, C. Pelé, and J. Ribéron, "A Methodology for Assembling Future Weather Files Including Heatwaves for Building Thermal Simulations from the European Coordinated Regional Downscaling Experiment (EURO-CORDEX) climate data," Energies, 2020, 13 (13), p. 3424, doi: 10.3390/en13133424.

Keywords: Building simulation, multi-family buildings, energy efficiency measure, future weather

Paper ID 30848: A critical review and cost-benefit analysis of green roofs in the United States: an optimization-based energy performance simulation study

Deva Shree Saini* (1,2), Tarek Rakha (1)

1: School of Architecture, Georgia Institute of Technology, Atlanta, GA, USA; 2: Hastings Architecture, Nashville, TN, USA

AIM AND APPROACH

Green roofs are often identified as energy-efficient techniques which, through their various mechanisms, contribute to a comfortable indoor environment. A significant number of published literature has investigated the thermal performance of a green roof under various climatic conditions and building parameters. Some of these studies monitored reduction in the thermal flux and roof surface temperature, thereby presenting it as an effective solution for energy savings, but this reduction might not translate into an equivalent drop in energy consumption. Other reviews that focused on the direct energy impacts of a green roof may be perceived as misleading in terms of its energy improvement potential due to their lack of representation of actual thermal parameters. The objective of this paper is to assess the energy performance claims made by previous publications to cast a light on possible exaggerations in green roof energy savings, by using a simulation-based model as the support that considers realistic settings. This was done by modeling a green roof assembly on reference office buildings developed by the US Department of Energy (DOE) for four climate zones. Simulations were carried out in EnergyPlus, and they responded to the variations in climate and the green roof parameters.

SCIENTIFIC INNOVATION AND RELEVANCE

This study's innovation is the focus on demonstrating the energy savings potential of green roofs as compared to the traditional insulation of a building. The paper poses that green roofs are not effective measures for reducing energy consumption if the base roof's R-value is equivalent to that of the entire green roof assembly. Therefore, their deployment should not be based solely on their energy savings features but should include other benefits, such as improved air quality, carbon sequestration potential, storm-water management and, reduction of urban heat islands. Contemporary codes and standards for building insulation mandate compliance to recommended values that are typically high enough to have thermal efficiency approximately equal to or greater than provided by a green roof. Considering their high installation, operation, and maintenance costs, the average payback period with such a small amount of savings is considerably high. This paper will also demonstrate a Life Cycle Cost Analysis (LCCA) of a green roof, and convert benefits other than energy savings into a monetary unit to better reflect their economic potential. The paper will discuss an outlook beyond the cost comparison as well, as green roofs can outperform conventional roofs from an ecological and social point of view.

PRELIMINARY RESULTS AND CONCLUSIONS

Multiple scenarios of green roof assemblies were developed based on their identified parameters using a brute-force approach. The annual energy savings for the best combination of parameters for four climate zones were as follows: Arid (Phoenix)-2.8%, Mediterranean (Los Angeles)-2.25%, Tropical (Atlanta)-1.36%, Temperate(Chicago)-1.5%. The simulation study shows limited energy-saving results. The less pronounced results can be attributed to high R-Value of the insulation which prevents the upward heat flow by reducing the thermal gradient created by a vegetated roof. Using the set of design parameters obtained from the preliminary study, optimization will be carried out using GenOpt, a generic optimization program coupled with EnergyPlus, to obtain a set of parameters which will minimize the energy consumption. The outcomes are majorly expected to indicate low insulation values for high energy savings. This study, therefore, aims to bring to notice the significance of a key parameter, insulation thickness, which critically impacts a green roof's performance by identifying literature that may have inadvertently inflated energy-saving outputs by failing to mention the thermal properties of base roof insulation and the other related factors. This research will present competitively beneficial aspects of green roofs as key factors in establishing it as a sustainable measure.

MAIN REFERENCES

Castleton, H. F., Stovin, V., Beck, S.B., & Davison, J. B. (2010). Green roofs; building energy savings and the potential for retrofit. Energy & Buildings, 42(10), 1582-1591.

Gagliano, A., Detommaso, M., Nocera, F., & Evola, G. (2015). A multi-criteria methodology for comparing the energy and environmental behavior of cool, green and traditional roofs. Building and Environment, 90, 71-81.

Jaffal, I., Ouldboukhitine, S.E., & Belarbi, R. (2012). A comprehensive study of the impact of green roofs on building energy performance. Renewable Energy, 43, 157-164.

Kotsiris, G., Androutsopoulos, A., Polychroni, E., & Nektarios, P. A. (2012). Dynamic U-value estimation and energy simulation for green roofs. Energy and buildings, 45, 240-249.

Niachou, A., Papakonstantinou, K., Santamouris, M., Tsangrassoulis, A., & Mihalakakou, G. (2001). Analysis of the green roof thermal properties and investigation of its energy performance. Energy and buildings, 33(7), 719-729.

Olivieri, F., Di Perna, C., D'Orazio, M., Olivieri, L., & Neila, J. (2013). Experimental measurements and numerical model for the summer performance assessment of extensive green roofs in a Mediterranean coastal climate. Energy and Buildings, 63, 1-14.

Keywords: green roof, energy-savings, insulation, optimization

Paper ID 30481: LSTM hourly solar irradiance prediction using local measurements and weather forecasts

ByungKi Jeon*, EuiJong Kim

Inha Univ., Korea, Republic of (South Korea)

AIM AND APPROACH

Predicting solar irradiance is important for various predictive control strategies in buildings. we proposed a model to predict the local irradiance using limited input data. The model is data-driven one using Long short-term memory(LSTM). To supply with abundant datasets, the weather data from other unrelated climates or regions are used.

SCIENTIFIC INNOVATION AND RELEVANCE

The proposed model has the following three advantages over the existing solar prediction model. First, the proposed LSTM model trains the model using only the data provided by the weather forecast system, which can be easily obtained with the API. Second, long-term measured data in the target area are not required. As shown in previous studies, most studies require long-term measured weather information in an area to predict local irradiance. Third, the proposed deep learning model does not require additional learning to update the model once it is built. Since the existing deep learning irradiance prediction model uses data measured in a specific region, it is common to update the model by learning the measurement data periodically to improve the performance of the model.

PRELIMINARY RESULTS AND CONCLUSIONS

The proposed model was verified by experiments carried out by local measurement of solar irradiances. The proposed model showed an error of learning performance RMSE 17.4W/m^2, prediction performance 18W/m^2 compared to existing weather data testsets. The measurement comparison also showed similar results with the simulation cases.

MAIN REFERENCES

According to the study of Qing, it was noted that these solar irradiance prediction models are difficult to use for the purpose of controlling residential buildings or small and medium-sized buildings. The reason is that the solar irradiance models can be effectively used only by large-scale operators because past solar irradiance and meteorological data, input data of the learning model, require expensive equipment to measure for a long time in one region and are continuously collected. This means that the existing solar irradiance prediction models need to be improved to apply the aforementioned prediction control algorithm such as MPC. Therefore, Qing's research developed a solar irradiance model that predicts solar irradiance by improving the simple solar irradiance prediction model using only the weather forecast system that is easily available through the Long short term memory (LSTM) deep learning algorithm, which is advantageous for predicting time-series data. This work is similar to this reference, but less measurement and weather information are required for both learning and predicting with slightly higher accuracy.

Qing, X., & Niu, Y. (2018). Hourly day-ahead solar irradiance prediction using weather forecasts by LSTM. Energy, 148, 461-468.

Keywords: solar irradiance, long term short term memory, weather prediction

Paper ID 31030: A critical review of the effectiveness of the Sustainability Tracking, Assessment & Rating System (STARS) framework on campus sustainability

Lauren E. Doocy, Arash Zarmehr*, Joseph T. Kider Jr

University of Central Florida, United States of America

AIM AND APPROACH

Ratings, certifications and assessments provide sustainability goals for design, construction, and operation for the built environment. Standards such as LEED, Energy Star, BREEM, and WELL offer benchmarks for buildings. Recently there has been a push to expand these certifications to small communities and university campuses representing buildings as a part of a larger interconnected system. Modern standards such as the Leadership in Energy and Environmental Design (LEED) for Cities and Communities, the Sustainability Tracking, Assessment, and Rating System (STARS), and WELL community allow individual parts of the built environment to function as a whole. These programs are voluntary, self-reporting frameworks to help small communities and universities track and measure their sustainability progress across measures of energy, water, transportation, and comfort. However, if a built space does not efficiently serve its occupants, the sustainability discussion becomes irrelevant asserting that these standards need to be both prescriptive and performative to increase the health and wellness of occupants in addition to sustainability goals. This study evaluates LEED for communities, STARS, and WELL community's ability to rate sustainability with the occupants in mind and analyzes several university campuses' attempts to meet these certifications. We then propose more occupant-centric sustainability measures to enhance human impact.

SCIENTIFIC INNOVATION AND RELEVANCE

This paper compares three rating systems (LEED community, STARS, and WELL community) across several universities. These campuses are a small network of buildings, all fulfilling different functions, and intended to allow students to live, work, and study efficiently. University campuses present an interesting case study because they resemble small communities and act as a living laboratory for small scale simulation of large scale urban spaces. Their resemblance, size, and impact on society, campuses have the opportunity to create change locally and provide insights for larger urban spaces. Each standard hcas different goals. For example, STARS evaluates universities on their overall sustainability including academics, engagement, operation, and planning, the efficiency of the built environment of the campus will have the greatest impact on overall campus sustainability. Similar to LEED certification criticism, STARS has its own shortcomings. While building sustainability initiatives can have extremely positive environmental impacts, impractical buildings are useless and expensive efforts if left unused by the occupants. This paper takes a critical look at the standard and deep look at the energy and water usage on a case-study campus determining guidance to policymakers for code implementation and enforcement.

PRELIMINARY RESULTS AND CONCLUSIONS

Our preliminary results first compare and contrast the three sustainability certifications for small communities and discuss their impact on university campuses and occupant health and wellbeing. For example, STARS puts emphasis on the academic and environmental engagement of students. However, much of the campus environmental impact is overshadowed by the achievement of less pressing initiatives, such as the presence of environment focused courses and on-campus environmental engagement. Campus operational capacity should suggest campus buildings working at maximum efficiency. Further, depopulation of the campus should suggest campus buildings will see a decreased workload. Since the standard university has periodic depopulation of campus, campus buildings should respond accordingly. Without considering building and operation efficiency, STARS cannot accurately evaluate university campus sustainability. We currently have analyzed 100 universities with the stars rating system and are expanding to look at the LEED and WELL standards. Finally, we have taken a deep dive into a case-study university's individual and campus building, energy, and water data to provide insights and suggestions to help provide a better implementation of these standards across campus.

MAIN REFERENCES

Habib M Alshuwaikhat and Ismaila Abubakar. 2008. An integrated approach to achieving campus sustainability: assessment of the current campus environmental management practices. Journal of Cleaner Production 16, 16 (2008), 1777-1785.

Peggy F. Bartlett and Geoffrey W Chase. 2004. Sustainability on campus: Stories and strategies for change. MIT Press.

Rebeka Lukman, Abhishek Tiwary, and Adisa Azapagic. 2009. Towards greening a university campus: The case of University of Maribor, Slovenia. Resources Conservation and Recycling 53, 11 (2009), 639-644.

Guy R Newsham, Sandra Mancini, and Benjamin J Birt. 2009. Do LEED-certified buildings save energy? Yes, but... Energy and Buildings 41, 8 (2009), 987-905.

Monika Urbanski and Walter Leal Filho. 2015. Measuring sustainability at universities by means of the Sustainability Tracking, Assessment and Ratings System (STARS): early findings from STARS data. Environment, Development and Sustainability 17, 2 (2015), 209-220

Keywords: Urban Sustainability, Campus Sustainability, Higher Education, Sustainability Networks, Community Assessment

Paper ID 30615: Building services engineering concept to create plus energy industrial hall buildings

Tobias Blanke* (1), Bernd Prof. Dr. Döring (2), Joachim Göttsche (1), Markus Hagenkamp (1), Vitali Reger (3), Markus Prof. Dr. Kuhnhenne (3)

1: FH Aachen, Solar Institut Jülich, Germany; 2: FH Aachen, Germany; 3: RWTH Aachen, Chair for Sustainable Metal Building Envelopes, Germany

AIM AND APPROACH

New buildings should be nearly zero energy buildings according to the EPBD since 2020. In future new buildings, it will be necessary to optimize the system technology and the use of renewable energies in the design process. In addition, an integrated approach, evaluation and optimization regarding energy demand and energy supply will be necessary. So-called plus-energy buildings even generate more energy than they need for their own requirements on an annual average. This poses special challenges to hall buildings with their high room volumes.

The paper presents a building services engineering concept (BSEC) for industrial buildings to bring them up to plus-energy standards. The BSEC is explained using a sample building, its function is demonstrated by numerical simulations. The concept is based on two key elements: a high-quality building envelope to ensure low transmission heat losses and a heat pump [1], which is connected to the ground via deep-foundation steel piles [2]. The heat pump is supported by solar-thermally activated steel curtain walls, which both supply heat to the ground and serve for cooling at night [2]. The heat is supplied to or removed from the hall via surface heating and cooling elements in the trapezoidal profile of the ceiling.

SCIENTIFIC INNOVATION AND RELEVANCE

The combination of innovative steel products and different buildings services will be shown. The BESC will be explained and the calculation results presented. Besides dimensioning aids for the BESC will be given.

PRELIMINARY RESULTS AND CONCLUSIONS

Plus energy standard for industrial hall buildings are possible. The explained system shows a good combination of the different building service components.

MAIN REFERENCES

[1] Plusenergiegebäude 2.0 in Stahlleichtbauweise; Vitali Reger Dipl.-Ing., Univ.-Prof. Dr.-Ing. Markus Kuhnhenne, Prof. Dr.-Ing. Helmut Hachul, Prof. Dr.-Ing. Bernd Döring, Dr.-Ing. Thiemo Ebbert, Tobias Blanke, Dr. rer. nat. Joachim Göttsche; 2019; https://doi.org/10.1002/stab.201900034

[2] Nutzung erneuerbarer Energien durch thermische Aktivierung von Komponenten aus Stahl Vitali Reger Dipl.-Ing., Univ.-Prof. Dr.-Ing. Markus Kuhnhenne, Prof. Dr.-Ing. Helmut Hachul, Prof. Dr.-Ing. Bernd Döring, Dr.-Ing. Thiemo Ebbert, Tobias Blanke; 2020; https://doi.org/10.1002/stab.202000031

Keywords: industrial hall buildings, plus energy concept, innovative steel products

Session T3.4: Improving indoor environmental quality

Paper ID 30542: Implementation of an innovative hybrid low-energy cooling strategy controlled with a fuzzy logic algorithm in a dwelling: numerical and experimental approaches

Aurélie Foucquier*, Simon Bal-Fontaine, Ophélie Ouvrier-Bonnaz, Etienne Wurtz

CEA LITEN, France

AIM AND APPROACH

In cold and temperate climates, the current technologies allow now to reach a good efficiency considering the energy losses. Buildings are highly insulated and their airtightness increases a lot. However, whereas those saving solutions are particularly efficient for heating considerations, they can penalize a lot the summer comfort. Thus, lots occupants resort to active cooling system leading finally to a significant increase of the global energy consumption. Thus, the objective of our work is to propose a low-energy cooling solution combining intelligently passive and soft systems in an individual dwelling to reach both the thermal comfort and the energy consumption. More specifically, a natural ventilation device and a shallow geothermal solution have been combined to constitute an overall cooling system. The feasibility of this cooling solution has been established during an experiment operated near Bordeaux in France along two successive summers. However, doubt remains regarding the resilience of such a solution function of the diversity of the French climates. To be able to study the robustness of this cooling solution, a numerical approach has been chosen. Especially, a co-simulation combining a FMU EnergyPlus of the dwelling, a geothermal simplified model and a monitoring of the cooling system has been implemented.

SCIENTIFIC INNOVATION AND RELEVANCE

A numerical approach has been employed in order to reach the robustness of an innovative low-energy cooling system regardless of the climate in an individual dwelling. Especially, it deals with a co-simulation combining a FMU EnergyPlus and Python programming. Indeed, the EnergyPlus model reproduces the dwelling behaviour allowing natural ventilation to operate. Python is used to model the shallow geothermal cooling model and to implement the monitoring of the overall cooling system. This overall cooling system combines the natural ventilation and the shallow geothermal solution, monitored by fuzzy logic to prioritize one among the other when specific indoor and outdoor conditions are reached. The shallow geothermal model is modelled by a RC model that has been calibrated from measurements collected during experimental summer campaigns near Bordeaux in France. Particularly, this geothermal solution consists in a water tank of about 1500L located in the crawlspace of a dwelling and connected to a cooling floor. The tank is then cooled by conduction through the ground. A first experimental campaign has shown that the ground alone is not sufficient to cool enough the water tank. Thus, a fan supplying air from outdoor has been added into the crawlspace to bring another cooling source.

PRELIMINARY RESULTS AND CONCLUSIONS

From the simulation approach, the overall cooling system has been tested in eight different geographical locations in France during summer. Several kinds of climate have been selected from the colder to the warmer. In order to evaluate the performance of the overall cooling system to reach thermal comfort by maintaining a good energy efficiency, several indicators have been proposed: hot and cold thermal discomfort, energy performance and intermittencies. This last aspect must be taken into account to avoid both the acoustic discomfort and the system erosion. Moreover, for some indicators, different performance targets required to be defined according to the climate. Indeed, the level of hot and cold discomfort cannot be considered in a same manner for each geographical location given that the degree of the occupants' acceptability of a thermal discomfort varies with the climate. As an example, the habitants from the South of France, where temperatures in summer is hot, will tolerate higher temperatures and will probably not support colder temperatures compared with the habitants from the North of France, where temperatures are usually really lower. Considering this aspect, results have shown good performance indicators of this innovative low-energy cooling solution regardless of the climate.

MAIN REFERENCES

H. C. Spindler, L. K. (2009). Naturally ventilated and mixed-mode buildings-Part II : Optimal Control. Building and Environment, 44, 750-761

M.J. Sorgato, A. M. (2016). The effect of window opening ventilation control on residential building energy consumption. Energy and Buildings, 133, 1-13.

Keywords: summer comfort, passive and soft cooling, climate, resilience, dwelling

Paper ID 30926: Development of adiabatic and isotherm humidifier models to compare health, energy and water use

Lien De Backer*, Elisa Van Kenhove, Lieve Vandorpe, Jelle Laverge

Ghent University, Belgium

AIM AND APPROACH

Nowadays when humidification is needed, steam (isotherm) humidification is mandatory in Belgium since it is safe in preventing legionellosis. The research question is whether adiabatic humidification, which should be more energy efficient than isotherm humidification [1], can be used without Legionella hazard risks. Previously, a biological Legionella growth model that estimates Legionella concentrations in domestic hot water applications is developed in Modelica [2]. To answer the research question, the aim of this paper is to broaden this bacterial growth model to other water containing systems, in this case humidifiers. Firstly, Modelica simulation libraries are screened for existing humidifier models. Unfortunately, none of these models are suitable for the purpose of this research. Since these models do not contain a water circuit, they cannot model Legionella growth. Therefore, a new model for adiabatic and isothermal humidification has been developed [3]. The model consists of two parts. The first part contains components to model moist air flowing through the humidifier, such as a preheater, humidifier and post-heater[4]. The second part contains the water system. In this part, the mass flow rate of water in the humidifier is modelled using components such as a water tank, a circulation pump and pipes.

SCIENTIFIC INNOVATION AND RELEVANCE

Existing humidifier models only consider mass flow of moist air. Since these models do not contain mass flow equations for water, it is not possible to estimate Legionella growth as this requires knowledge as about the water temperature and renewal of water in the tank. These developed models allow to study the potential of total energy savings of an adiabatic humidifier installed in an air handling unit taking into account the Legionella risk. The newly developed simulation model can be used stand-alone or can be connected to a building model. The model can also be used in other situations, besides the healthiness of adiabatic humidification, the healthiness of adiabatic cooling in summer, misting machines in grocery stores and cooling mist sprayers can be examined. Moreover, the model can be transformed into a cooling tower model by replacing the post-heater with a heat exchanger between the cooled air and the water circuit which cools the building, and by modifying the control.

PRELIMINARY RESULTS AND CONCLUSIONS

The adiabatic and isotherm (steam) humidifier models are developed and tested in two simulated cases, namely a fictive museum and a fictive office building. Energy use, water use and Legionella concentration are compared for both humidifiers for a period of two weeks during winter and mid-season. For each season, two variations were studied. In the first variation, the building is open during two weeks. In the second variation, the building is reopened after one week during which no visitors or employees were present. The simulated cases and scenarios show that it is possible to safely operate an adiabatic humidifier provided that the correct measures, i.e. sufficiently emptying the tank are taken or flushing the system with hot water. These measures have to be taken will strongly depend on case by case and the corresponding demand for water, stagnation of water in the water tank. This indicates that the legislation, prohibiting adiabatic humidification, may be too strict and opens doors to further in-depth research. It should be emphasized that these findings still need to be confirmed by validation of the Legionella part of the simulation model in a laboratory experiment, once the COVID19 measures allow.

MAIN REFERENCES

[1] M.-S. Jo, J.-H. Shin, W.-J. Kim, and J.-W. Jeong, "Energy-Saving Benefits of Adiabatic Humidification in the Air Conditioning Systems of Semiconductor Cleanrooms," Energies, vol. 10, no. 11, p. 1774, Nov. 2017.

[2] E. Van Kenhove, L. De Backer, A. Janssens, and J. Laverge, "Simulation of Legionella concentration in domestic hot water: comparison of pipe and boiler models," J. Build. Perform. Simul., vol. 12, no. 5, 2019.

[3] L. Vandorpe, "Adiabatic vs. Isotherm Humidification –Analysing Health, Energy and Water Use," Ghent University, 2020.

[4] R. Lazzarin and L. Nalini, Air Humidification - Technical, Health and Energy Aspects tle. Carel, 2006.

Keywords: Legionella pneumophila, humidification, health, energy and water use, simulation

Paper ID 31107: Gas phase air cleaning effects on ventilation energy use and the implications of CO2 concentration as an IAQ indicator for ventilation control

Dragos-Ioan Bogatu*, Ongun Berk Kazanci, Bjarne Wilkens Olesen

Intl. Center for Indoor Environment and Energy, Technical University of Denmark, Denmark

AIM AND APPROACH

Gas phase air cleaning can be used to improve the Indoor Air Quality (IAQ) or lower the required ventilation rate for the same IAQ and thus reduce the energy use for preheating/cooling and from transporting the outside air. This study aims to quantify the energy savings resulting from the use of air cleaning technology and to clarify the use of CO2 concentration as an IAQ indicator for control and evaluation according to present standards.

Dynamic building simulations were used to capture the relative differences in energy savings and indoor air quality (IAQ) between multiple clean air delivery rates. An office space was modelled where air flow rate was supplied through a generic heating, ventilation and air conditioning (HVAC) system with and without heat recovery. The required air flow rate was determined according to Method-I presented in EN 16798-1:2019 and adjusted according to the air cleaner's clean air delivery rate (CADR). The yearly simulations were run for different locations, i.e., Copenhagen, Denmark; Zürich, Switzerland; Palermo, Italy; and Tokyo, Japan and under two different building classes, low and very-low polluting buildings.

SCIENTIFIC INNOVATION AND RELEVANCE

Due to its energy savings potential from both preheating/cooling of outside air and fan energy, air cleaning technology represents a current topic of research and interest. This study quantified the energy savings relative to multiple CADRs along with the impact on the IAQ, thus offering an overview of potential benefits of air cleaning technology. A possible alternative to energy savings without the need for change in building materials was assessed by investigating different building polluting levels, i.e. low and very-low polluting buildings. Furthermore, the energy savings resulting from a reduction of air flow rate associated only to the removal of building related pollutants was presented, indicating new possible research areas with respect to air cleaning technology.

Additionally, this study complements standard EN16798-1:2019 and the related technical report EN16798-2:2019. By following the methods described, it highlights critical aspects when designing ventilation systems with air cleaning technology. Furthermore, the impact of the CO2 concentration on the evaluation of the resulting IAQ and its use as a demand control ventilation (DCV) indicator when employing air cleaning were analysed.

PRELIMINARY RESULTS AND CONCLUSIONS

The fan energy use reduced with an increase in the CADR. The heat exchanger's presence reduced the total energy use by saving energy from preheating and cooling the outside air no matter the CADR. The reduction in energy use was however significantly dependent on the geographical location.

The same IAQ level was achieved for a reduced ventilation flow rate since the air cleaner removed the pollutants associated with the occupants and the building. However, an increase in the CADR led to higher CO2 concentrations, especially for very-low polluting buildings, due to the inverse proportional relationship between CO2 and air flow rate. Therefore, in case of DCV where CO2 concentration is the indicator the set point must be adjusted, i.e. the CO2 concentration must be increased for a decrease in the required air flow rate, in order to maintain the same IAQ. The same CO2 concentration must then also be used for the space's IAQ evaluation stage to avoid categorizing the resulting IAQ to a lower level than the real one. Finally, the results showed that gas-phase air cleaners could be used in low-polluting buildings to reach an energy use equal to very-low polluting buildings while maintaining the same IAQ level.

MAIN REFERENCES

CEN. (2019). DS/EN 16798-1:2019. Energy performance of buildings - Ventilation for buildings - Part 1: Indoor environmental input parameters for design and assessment of energy performance of buildings addressing indoor air quality, thermal environment, lighting and acoustics.

CEN. (2019). DS/EN 16798-2:2019. Energy performance of buildings - Ventilation for buildings - Part 2: Interpretation of the requirements in EN 16798-1.

Keywords: air cleaning, energy performance, indoor air quality, ventilation, CO2 control

Session T3.5: Ensuring high quality building simulations

Paper ID 30214: Influence of internal control simplifications in heat pump system modelling for energy flexibility evaluations

Maarten Evens* (1,2), Alessia Arteconi (1,2,3)

1: KU Leuven, Belgium; 2: EnergyVille, Belgium; 3: Università Politecnica delle Marche, Italy

AIM AND APPROACH

Modulating electric driven heat pumps can support the energy transition. This technology provides the possibility to couple the heat and electricity sector. In combination with thermal energy storages, heat pumps offer the ability to adapt their load profile in order to provide flexibility to different stakeholders [1]. Testing control strategies of flexible heat pumps can be done in simulations, field trials or hardware-in-the-loop (HIL) [2]. HIL can also be used for model validations and to improve modelling quality. The modelling approaches for the heat pump internal control strategies differ, while the impact remains uncertain. Firstly, this paper addresses the influence of these internal strategies via yearly simulations with one minute time steps on a Flemish residential building in Modelica [3]. Control blocks were gradually added between simulations and the influence of each control block was investigated. Secondly, a predictive rule-based control with spot prices was used for flexible heat pump operation. A generally used heat pump model for flexibility testing was compared with a more accurate model able to represent the internal control strategies of the heat pump. Therefore, the modelling influence of internal control strategies on the ability of the heat pump for flexibility provision was also investigated.

SCIENTIFIC INNOVATION AND RELEVANCE

The main objective of this paper is to develop a model of a heat pump for residential applications able to represent in an accurate way the heat pump behaviour, without neglecting transient effects and constraints due to the limitations related to the internal heat pump control. Existing literature shows different modelling approaches to represent the real heat pump behaviour via performance map based models, while the impact of simplifications in the heat pump internal control strategies remains uncertain. In addition, currently market-available heat pumps do not offer direct modulation control to the end-user and internal time limits need to be respected. To develop flexible heat pump control strategies, those effects have to be considered, because they influence the exact behaviour of the heat pump. Furthermore they are crucial to keep end-user comfort, to obtain the requested consumption profile and to maintain the high efficiency of the total equipment.

PRELIMINARY RESULTS AND CONCLUSIONS

The goal of this paper was twofold. Firstly, the influence of simplifications in the heat pump internal control strategies was investigated. It was shown that non-negligible effects and profile modification were introduced. Different indicators such as energy use, extracted power, temperature levels, time limits, number of cycles and the use of the internal back-up heater were compared. Comparison between simplified heat pump control approaches and the most accurate heat pump model of this paper, showed that between 42 – 86 % of the year, deviations in the electrical heat pump power were within +/- 10 %. Most deviations were caused by domestic hot water cycles, but were introduced by different internal control strategies. Secondly, the heat pump behaviour under a flexible control strategy when using spot prices was analysed. Hence, it could be concluded that the internal control strategies influence the real heat pump behaviour and that these control strategies should be included to represent the real capabilities for flexibility provision. In future work, the presented heat pump model will be tested with advanced control strategies with heat pumps in a hardware-in-the-loop approach.

MAIN REFERENCES

[1] A. Arteconi and F. Polonara, "Assessing the demand side management potential and the energy flexibility of heat pumps in buildings," Energies, vol. 11, no. 7, pp. 1–19, 2018, doi: 10.3390/en11071846.

[2] W. El-Baz, L. Mayerhofer, P. Tzscheutschler, and U. Wagner, "Hardware in the loop real-time simulation for heating systems: model validation and dynamics analysis," Energies, vol. 11, no. 11, 2018, doi: 10.3390/en11113159.

[3] F. Jorissen, G. Reynders, R. Baetens, D. Picard, D. Saelens, and L. Helsen, "Implementation and verification of the IDEAS building energy simulation library," J. Build. Perform. Simul., vol. 11, no. 6, pp. 669–688, Nov. 2018, doi: 10.1080/19401493.2018.1428361.

Keywords: Heat pump, Flexibility, Control, Hardware-in-the-loop, Modelica

Paper ID 31012: The effect of the external driving forces modelling on the calculated building energy need through the use of dynamic simulation

Giovanna De Luca*, Franz Giorgio Maria Bianco Mauthe Degerfeld, Ilaria Ballarini, Vincenzo Corrado

Politecnico di Torino, Turin, Italy

AIM AND APPROACH

In literature, different calculation models and assumptions on the physical phenomena related to the interaction between the outdoor environment (air, sky, solar radiation, etc.) and the outdoor building envelope are provided. The above assumptions primarily concern the simplifications applied to the numerical models and the temporal variation of the parameters describing the interaction between the components. The present work analyzes the influence of the assumptions specified in the standardized hourly method of EN ISO 52016-1 [1] on the thermal energy needs of a building. The analysis is carried out by means of a full detailed dynamic calculation model (EnergyPlus). The topics evaluated include: i) the extra thermal radiation to the sky; ii) the external surface heat transfer coefficient; iii) the determination of the solar irradiance on the building envelope components and iv) the shading of the building envelope components. To assess the above-mentioned phenomena different sky models are considered, as well as both constant and time-varying values of the input parameters, if applicable. To guarantee a general validity of the outcomes, different building categories and weather conditions are considered. In more detail, a residential and an office building are analyzed in different Italian climate zones.

SCIENTIFIC INNOVATION AND RELEVANCE

The EN ISO 52016-1 [1] hourly method is based on assumptions and simplifications chosen so as to allow a sufficient accuracy in the outcomes with a low amount of input data required [2]. This is an attempt to overcome the low accessibility to detailed input data, often required by the detailed calculation models. However, the effects and errors related to the assumptions of the EN ISO 52016-1 standard [1] have not been sufficiently investigated yet. Zakula et al. [3] deepened the differences in the building energy performance by adopting fixed or variable windows physical properties. Ballarini et al. [2] analyzed the discrepancy between the results of the EN ISO 52016-1 method and of the EN ISO 13790 model compared to the detailed dynamic calculation of Energy Plus. However, not all the assumptions introduced by EN ISO 52016-1 [1] have been evaluated in the past studies. Within this framework, a sensitivity analysis investigating the effects of the simplifications of EN ISO 52016-1 [1] is carried out in the present work. The analysis performed allows to understand to what extent the calculation assumptions and the modelling options of the considered parameters can influence the simulation results and the accuracy of the models.

PRELIMINARY RESULTS AND CONCLUSIONS

The results of this study highlight the variation in the accuracy of a dynamic model results when different modelling simplifications and assumptions of the external driving forces are considered. The results show that a detail decrease in the definition of input parameters can influence the results in a not negligible way. For this reason, it would be preferable to choose a more detailed approach. In the same time, switching from constant to hourly input parameters would significantly increase the complexity of the model. The analysis also showed that the results are strictly linked to the building use and to the envelope properties, such as the thermal insulation or the window-to-wall ratio. In fact, in certain circumstances the outcomes are not sensitive to the simplifications applied to the models. Thus, the assumptions and simplifications applied in the definition of the input parameters should be further explored as to guarantee both a minor loss of accuracy of the model and the simplicity of the assessment, considering the different characteristics of the buildings and their related energy behaviors.

MAIN REFERENCES

[1] EN ISO 52016-1. Energy performance of buildings - Energy needs for heating and cooling, internal temperatures and sensible and latent heat loads - Part 1: Calculation procedures. (2017)

[2] D. van Dijk, "EN ISO 52016-1: The New International Standard To Calculate Building Energy Needs for 1 Heating And Cooling, Internal Temperatures And Heating And Cooling Load", Proceedings of 16th IBPSA Building Simulation Conference BS2019, 2-4 September 2019, Roma (Italia).

[3] T. Zakula, M. Bagaric, N. Ferdelji, B. Milovanovic, S. Mudrinic, K. Ritosa, "Comparison of dynamic simulations and the ISO 52016 standard for the assessment of building energy performance", Applied Energy, volume 254 (2019).

[4] I. Ballarini, A. Costantino, E. Fabrizio, V. Corrado, "The Dynamic Model of EN ISO 52016-1 for the Energy Assessment of Buildings Compared to Simplified and Detailed Simulation Methods", Proceedings of 16th IBPSA Building Simulation Conference BS2019, 2-4 September 2019, Roma (Italia).

Keywords: Building energy performance, Outdoor driving forces, Building envelope, Simplified dynamic models

Paper ID 30583: Validation of spectral simulation tools for the prediction of indoor daylight exposure

Clotilde Pierson* (1,2), Mariëlle Aarts (2), Marilyne Andersen (1)

1: EPFL, Switzerland; 2: TU/e, Netherlands

AIM AND APPROACH

The discovery of ipRGCs in the human eye [1-3] has led to a new area of research studying the effects of ocular light exposure on the so-called non-visual response to (or non-image forming effects of) light. An increasing number of studies raise awareness on the importance of a proper exposure to light, as they demonstrate how it impacts alertness, sleep, health and other neurobehavioral processes.

Considering that we spend the vast majority of our time indoors, it becomes urgent to get a better understanding of our 'light hygiene' within the built environment and to anticipate the implications of design decisions affecting access to daylight. To this end, new simulation workflows relying on spectral simulation tools and models able to link given properties of light with neurophysiological responses are needed, in addition to laboratory and field studies.

This research aims at evaluating spectral simulation workflows to study the non-visual response to daylight. The goal is to assess the validity of spectral simulation tools in predicting building occupants' exposure to light, considering both the spectral distribution and intensity over time. The methodology consists of a comparison of data measured in an office-like laboratory against spectrally simulated data.

SCIENTIFIC INNOVATION AND RELEVANCE

Two spectral simulation tools, ALFA [4] and LARK [5], are available today to simulate light exposure inside the built environment. What remains unclear is how well they represent reality, i.e. actual spectral irradiances experienced indoors. With the fast-growing interest for research on the non-visual response to light, the validation of these spectral simulation tools is highly needed to ensure the accuracy of their results and build reliable knowledge in this area of research.

Since the principal use of these spectral simulation tools is the study of non-image forming effects of light, the comparison of measured and simulated data in this case is done mainly in terms of non-visual response.

The prediction of the non-visual response to light through simulations requires two main steps:

- simulating light exposure at the eye (spectral power distribution and intensity) over a certain period of time,

- anticipating the non-visual response that the light exposure may induce, based on existing models.

The non-visual response model that will be principally considered in this study is one that links a time-series of light exposure to two kinds of non-visual responses, namely the direct response (nvRD) and the circadian response (nvRC) [6], though alpha-opic irradiances [7] will also be analyzed.

PRELIMINARY RESULTS AND CONCLUSIONS

Preliminary comparisons between spectral irradiances measured and simulated inside an office-like laboratory using ALFA software are promising. With either a clear or an overcast sky, the spectral power distribution appears to be similar throughout the course of the day for a selection of three different view directions. It is not yet clear, however, how the simulation tools perform for intermediate sky conditions, for which the sky model cannot be accurately representative of reality. Regarding absolute spectral irradiance values, the simulated values generally differ from the measured ones in the range of 0 to 50%.

To determine the extent to which differences between measured and simulated spectral irradiances may actually matter in the context of non-image forming effects of light, this difference is converted into differences in non-visual responses with the help of the nvRD and nvRC models [6]. While one can expect these models' outcomes to be similar under uniform skies, how much they differ under intermediate skies is to be further evaluated. At last, a main take-away from such a study is to open the discussion about the significance of a difference when it comes to non-visual responses, so as to define a threshold for valid spectral simulation tools.

MAIN REFERENCES

[1] Brainard GC, Hanifin JP, Greeson JM, et al. Action Spectrum for Melatonin Regulation in Humans: Evidence for a Novel Circadian Photoreceptor. The Journal of Neuroscience 2001; 21: 6405–6412.

[2] Berson DM, Dunn FA, Takao M. Phototransduction by Retinal Ganglion Cells That Set the Circadian Clock. Science 2002; 295: 1070–1073.

[3] Hattar S, Liao HW, Takao M, et al. Melanopsin-containing retinal ganglion cells: architecture, projections, and intrinsic photosensitivity. Science 2002; 295: 1065–1070.

[4] ALFA - Adaptive Lighting for Alertness, 2020, Solemma LLC, USA.

[5] Lark Spectral Lighting, 2015, Mehlika Inanici (University of Washington) and ZGF Architects LLP, USA.

[6] Ámundadóttir M. Light-driven model for identifying indicators of non-visual health potential in the built environment. 2016. [PhD thesis, EPFL]

[7] Lucas RJ, Peirson SN, Berson DM, et al. Measuring and using light in the melanopsin age. Trends in Neurosciences 2014; 37(1):1–9.

Keywords: spectral simulation, daylight, non-visual response, validation study

Paper ID 30252: Developing a test framework for assessing building thermal resilience

Shabnam Homaei*, Mohamed Hamdy

Norwegian University of Science and Technology, Norway

AIM AND APPROACH

Building performance can be affected by foreseen and/or unforeseen events such as changing environment (e.g. extreme weather conditions due to climate change) or changing requirements (e.g. new regulations/policies). Thus, buildings should be able to react to these events in order to last their performance and functionality. In this regard, the concept of resilience is recently gaining ground in the context of building design. However, there should be a clear definition and a commonly agreed assessment framework for resilience before it can be fully implemented in the building design. The most important concepts, which should be used to provide a clear and comprehensive definition of resilience and its assessment framework can be summarized in four questions: "resilience of what?", "resilience to what?", "resilience at what stage?", and "resilience for what?" This paper reviews resilience definition and different approaches toward resilient building design in the existing literature. The paper also introduces a conceptual framework for assessing building energy resilience by answering the four abovementioned questions for a case study building in cold climate countries.

SCIENTIFIC INNOVATION AND RELEVANCE

Most of the literature in the context of resilience are focusing on defining resilience and its assessment framework in larger scales such as communities, cities, and urban scales against disruptive events such as hurricanes, flooding, and earthquakes. This study defines the concept of resilience in the building scale and its energy systems, which can help designers and decision-makers to design resilient buildings in the design or renovation phases. Furthermore, simulation results can be used as input data for the proposed framework in order to assess the energy resilience of buildings with a focus on different key performance indicators. This will ensure designers about the simulation results and their functionality not only for current but also for the future conditions.

PRELIMINARY RESULTS AND CONCLUSIONS

The proposed framework states that to be energy resilient, the building and its energy systems should be able to "prepare and plan for", "absorb", "recover from", and "more successfully adapt" to adverse events in the operation phase. This framework can be used by designers and decision-makers in the design or renovation phase of the buildings in order to protect them against the foreseen and/or unforeseen events. The results show possible evaluations for the energy resilience of the cases study building in cold climate countries.

MAIN REFERENCES

[1] Nik, V. M., Perera, A. T. D., & Chen, D. (2020). Towards climate resilient urban energy systems: A review. National Science Review.

[2] Sharifi, A. (2019). Resilient urban forms: A review of literature on streets and street networks. Building and Environment, 147, 171-187.

[3] Sharifi, A., & Yamagata, Y. (2016). Principles and criteria for assessing urban energy resilience: A literature review. Renewable and Sustainable Energy Reviews, 60, 1654-1677.

[4] Meerow, S., Newell, J. P., & Stults, M. (2016). Defining urban resilience: A review. Landscape and urban planning, 147, 38-49.

[5] Jonathan Mak, W. H., & John Clarkson, P. (2017). Towards the design of resilient large-scale engineering systems.

[6] Moazami, A., Carlucci, S., & Geving, S. (2019, September). Robust and resilient buildings: A framework for defining the protection against climate uncertainty. In IOP Conference Series: Materials Science and Engineering (Vol. 609, No. 7, p. 072068). IOP Publishing.

Keywords: Resilience, Building energy, Preparation, Absorption, Recovery and Adaptation.

Session T3.6: The role of occupants

Paper ID 30813: Indoor climate profiles in Danish apartments based on high temporal resolution measurements

Ricardo Forgiarini Rupp* (1), Jørn Toftum (1), Gianluca Trotta (2), Rune Korsholm Andersen (1)

1: International Centre for Indoor Environment and Energy, Department of Civil Engineering, Technical University of Denmark; 2: Department of the Built Environment, The Faculty of Engineering and Science, Aalborg University Copenhagen

AIM AND APPROACH

The aim of this work is to define profiles of the indoor climate in Danish multifamily residential buildings. Measurements of indoor temperature, relative humidity and CO2 concentration were conducted every five minutes for more than a year in 45 apartments using internet of things devices. Cluster analysis was employed to derive the indoor climate profiles. The relationship between indoor and outdoor climate was analysed through piecewise linear regression.

SCIENTIFIC INNOVATION AND RELEVANCE

Building simulation relies on good-quality input data in order to provide good predictions of indoor climate and energy. Understanding occupancy patterns is crucial to perform accurate building simulation, which contributes to design better buildings with good indoor environment and low energy use. In this work, a large field campaign in residential buildings in Denmark using IoT devices was performed in order to acquire information on indoor environmental quality. The adoption of IoT devices allowed the occupants to visualize in real-time the indoor environmental conditions in their homes and also provided a rich database for scientific purposes. To our knowledge, the database is the most detailed Danish sample of residential buildings on indoor environmental parameters. Results may improve building simulation models, helping to reduce the gap between predicted vs. actual energy use in residential buildings.

PRELIMINARY RESULTS AND CONCLUSIONS

Almost 7 million measurements of environmental parameters were collected. Considerable discrepancies in indoor environmental parameters were observed across the apartments. Overheating and high CO2 concentration (above 1,000ppm) were observed in a number of apartments. Daily profiles of indoor climate were made and the relation between indoor and outdoor temperatures was investigated. The CO2 concentrations provided the possibility to make an overall assessment of occupancy. Occupants typically left the home in the morning and arrived back in the late afternoon, but different profiles were identified in the database. A comprehensive database on indoor environmental quality was built and analysed in this work, which provided a better understanding of indoor environmental quality in residential buildings at a national level.

MAIN REFERENCES

Fabi, Valentina, Rune Vinther Andersen, and Stefano Paolo Corgnati. (2013). Influence of Occupant's Heating Set-Point Preferences on Indoor Environmental Quality and Heating Demand in Residential Buildings. HVAC&R Research 19(5), 635–45.

Nguyen, J. L., Schwartz, J., & Dockery, D. W. (2014). The relationship between indoor and outdoor temperature, apparent temperature, relative humidity, and absolute humidity. Indoor Air 24(1), 103–112.

O'Brien, W., et al. (2020). Introducing IEA EBC annex 79: Key challenges and opportunities in the field of occupant-centric building design and operation. Building and Environment 178, 106738.

Yan, D., Hong, T., Dong, B., Mahdavi, A., D'Oca, S., Gaetani, I., Feng, X (2017). IEA EBC Annex 66: Definition and simulation of occupant behavior in buildings. Energy and Buildings 156, 258–270.

Yoshino, H., Hong, T., Nord, N. (2017). IEA EBC annex 53: Total energy use in buildings—Analysis and evaluation methods. Energy and Buildings 152, 124–136.

Keywords: Heating, Behaviour, Temperate climate, Residential

Paper ID 30817: Development of an interactive virtual reality simulation environment with a thermal feedback for the user

Christoph Nytsch-Geusen* (1), Werner Kaul (1), Kushagra Mathur (1), Lucas Westermann (1), Martin Kriegel (2)

1: UdK Berlin, Germany; 2: TU Berlin, Germany

AIM AND APPROACH

This contribution describes the first experimental results of the development of a virtual reality (VR) simulation environment, in which the user is immersively embedded in a 3D visualization of the building model and experiences the simulated room climate of the associated thermal simulation model physically as thermal feedback. The user of the simulation environment can also interact with the room model in VR and thus influence the room energy balance. To implement this approach, a Modelica-based thermal building model was coupled with the 3D visualization environment Unity in real time. The user can change their room position in the VR model and operate the interactive room elements, for example by opening a window or changing a setpoint of the room thermostat. These interactions cause a change in the physical state in the thermal building model, whereupon e.g. the calculated indoor air temperature is increased or decreased. This calculated current room air temperature is generated in the VR simulation environment for the user by an air conditioning device and made physically experienceable for him.

SCIENTIFIC INNOVATION AND RELEVANCE

Previous simulation tools used in the design and planning process for indoor climate analysis and building energy efficiency are not immersive. The user considers the room and the building as objects to be analyzed or optimized with a certain distance in the energetic simulation tool. However, the perception of the indoor climate is inherently an immersive sensual process. A VR simulation environment in which the user perceives the 3D space visually on the one hand, but also receives thermal feedback on the other and can significantly influence the room energy balance through interaction, opens up the potential to convey realistic experiences about the dynamics of the indoor climate of rooms and buildings.

PRELIMINARY RESULTS AND CONCLUSIONS

The development of the interactive VR simulation environment takes place within the research project "EnOB: GEnEff - A new type of evaluation of building-energy efficiency and an innovative demonstration by means of simulation methods and virtual reality". The article describes the software-technical coupling of the interactive virtual reality environment with the thermal room model as well as a first experimental setup for the physical perception of the simulated room air temperature. For this purpose, a simple application scenario is presented, in which the interaction alternatives for the user are explained and described in the experiment.

MAIN REFERENCES

Nytsch-Geusen, C., Möckel, J., Rädler, J., Thorade, M. 2017. BuildingSystems_VR - A new approach for immersive and interactive building energy simulation. Proceedings Building Simulation Conference 2017, San Francisco, USA.

Nytsch-Geusen, C.; Mathur, K. 2020. Entwicklung einer Virtual Reality-Umgebung zur interaktiven themischen Raumsimulation. Sucessfully submitted for BAUSIM 2020, Graz, Austria.

Keywords: VR simulation environment, immersive indoor climate experience, user interaction

Paper ID 30238: Impact and comparison of standard-based, stochastic, and tracked user behaviour in open space offices: A Swiss case study

Verena Marie Barthelmes*, Florian Guillaumat, Dolaana Khovalyg

Thermal Engineering for the Built Environment Laboratory (TEBEL), École polytechnique fédérale de Lausanne (EPFL), Switzerland

AIM AND APPROACH

A large corpus of literature has shown that the role of occupants and their interactions with the building envelope and systems represent a key factor of uncertainty when predicting building energy use with simulation tools [1-3]. During different stages of design, it is common practice to implement fixed standard user profiles [4] for taking into account human-related factors, such as occupants' presence or natural ventilation rates. The aim of this paper is to contribute to gaining further knowledge on (i) how reliably current national standards reflect real user behaviour and (ii) how much the deviation between standard and real user profiles might impact predictions of heating/cooling needs and indoor environmental comfort. The study is based on two-week measurement campaigns during the heating and cooling season in an open space office located in Lausanne, Switzerland. Occupant-related building simulation input starting from the reference model is based on (a) Swiss national standards [5] and (b) behavioural patterns (e.g., occupancy, window control behaviour) directly tracked and modeled from human-building interactions in the case study building and monitored potential influencing variables related to the global environment (thermal, olfactory, visual, acoustic) [6].

SCIENTIFIC INNOVATION AND RELEVANCE

The relevance of this paper relies on a critical analysis of the mismatch between standard profiles and real or modeled user behaviour patterns in order to further quantify uncertainties related to occupant input in building simulation programs. Rather than looking into the total impact of only assumed behavioural patterns on building energy use (e.g., different work styles [7]), the aim of this paper is to gain a deeper understanding to what extent more sophisticated behavioural models could reduce the uncertainty related to human-related factors in the simulation outcomes with respect to standardized and deterministic input. While previous studies mostly focused on determining probabilistic relationships between single-domain variables (e.g., thermal environment) and window control behaviour, the present contribution is aimed at including a more comprehensive set of multi-domain variables in the analysis.

PRELIMINARY RESULTS AND CONCLUSIONS

Preliminary results show significant discrepancies between predicted heating and cooling demands when running the model with standard-based and tracked/modeled building occupant-related simulation input. The results highlight the potentials and limitations of current national standards for space usage data for energy and building installations, and to what extent they could be enhanced with more sophisticated (or probabilistic) occupant behaviour modeling techniques. The modeling approach further highlights the most suitable predictors for window control behaviour based on the measurements taken in the case study building. The analysis of cause-effect relationships between behaviours and multi-domain environmental variables (includingacoustical parameters) allows for a brief discussion on gaps in current building energy simulation tools (BEST) with regard to the implementation of a larger set of environmental variables that might influence the behaviour of building occupants.

MAIN REFERENCES

[1] Hopfe, C.J., and Hensen, J.L.M. 2011. Uncertainty analysis in building performance simulation for design support. Energy and Buildings, vol. 43, no. 10, pp. 2798–2805.

[2] International Energy Agency (IEA) Energy in Buildings and Communities Programme. Annex 79 - Occupant-Centric Building Design and Operation. Available online: http://annex79.iea-ebc.org/

[3] Masoso, O.T., and Grobler, L.J. 2010. The dark side of occupants' behaviour on building energy use. Energy and Buildings, vol. 42, no. 2, pp. 173–177.

[4] O'Brien, W. et al. 2020 An international review of occupant-related aspects of building energy codes and standards, Building and Environment, Volume 179, 106906, ISSN 0360-1323.

[5] Schweizerische Ingenieur- und Architektenverein SIA, SIA 2024:2015 Space usage data for energy and building installations, 2016. Zurich.

[6] Barthelmes, V.M., Karmann, C., Licina, D., Andersen, M., and Khovalyg. D. 2020. A multi-domain data collection strategy for capturing relationships between occupant behaviour, comfort, indoor environment, and energy use in offices. In: Proceedings of the 11th Windsor Conference 2020 "Resilient Comfort", April, 2020.

[7] Hong, T., and Lin. H.W. 2012. Occupant behavior: impact on energy use of private offices. Proceedings of the 1st Asia conference of International Building Performance Simulation Association, Shanghai, China, November 25–27.

Keywords: Occupant behaviour, open space office, standards for space usage, probabilistic models, multi-domain

Paper ID 30847: Comparative window use study: habits versus deliberative actions in residential buildings

Silke Verbruggen*, Marc Delghust, Jelle Laverge, Arnold Janssens

Research Group Building Physics, Ghent University, Belgium

AIM AND APPROACH

The aim of the study is to evaluate the impact of window use habits on residential energy use and indoor climate. In this study the results of a survey querying the habits of 500 occupants in Belgium is used to simulate the habitual window use and to assess the impact of this habitual behaviour on the heating energy use, indoor temperature and indoor air quality. First the habitual window use detected in the survey is translated in window use profiles, based on stochastic occupancy and activity profiles. These window use profiles are applied as input for energy simulations in Modelica. Next, these results are compared to the results of a second set of simulations that apply typical weather-based models to simulate the window use.

SCIENTIFIC INNOVATION AND RELEVANCE

Many occupants behave in a habitual way with their building controls, such as windows, solar shades and thermostat [1]. However, this is neglected in most user behaviour models. Most window use models are primarily based on weather variables consequently they will predict similar window opening behaviour in dwellings that are situated in the same climate. In this study the importance of considering these window use habits is assessed.

The development of new types of window use models based on these habits is recommended since energy simulations using weather based models will underestimate the variance in window use, therefore neglecting those households that open the windows very often or very little.

PRELIMINARY RESULTS AND CONCLUSIONS

The translation of the habits indicated in the survey to window use profiles revealed that the average window opening duration of the surveyed households is similar as the ones predicted with a simple weather based model [2]. Only the opening time in the bedroom is a lot higher with the habit-based window use profiles (summer 52%, winter 13% compared to 26% and 6%). This is the consequence of the fact that many occupants leave windows open at night which will not be predicted by the weather-based models.

The average opening duration might be in correspondence with weather-based models, however, the variance in opening times is severely neglected with the weather based models. In all rooms the habit based models have a range of 0 to 100% opening time, while this range is a lot smaller for the weather based models with a maximum of only 40%.

The lack of extreme window use will have a serious effect on the energy use and indoor climate predicted with energy simulations.

MAIN REFERENCES

[1] S. Verbruggen, M. Delghust, J. Laverge, A. Janssens, Inclusion of window opening habits in a window model based on activity and occupancy patterns, in: Clima 2019, 2019: p. In Review.

[2] J. Maeyens, A. Janssens, A stochastic ventilation model regarding leakage and user behaviour, in: 2nd Int. Conf. Build. Phys., Leuven, 2003: pp. 737–748.

Keywords: Window use, Occupant behavior, residential energy use, habits

Paper ID 30962: Designing and experimenting nudge signals to act on the energy signature of households for implementing indirect energy flexibility

Muhammad Salman Shahid* (1), Benoît Delinchant (1), Béatrice Roussillon (2), Frédéric Wurtz (1), Daniel Llerena (2), Adélaïde Fadhuile (2), Nils Artiges (1)

1: G2Elab, 21 Rue des Martyrs, CS 90624, 38031 Grenoble CEDEX 1, France; 2: GAEL, 1241 Rue des Résidences, 38400 Saint-Martin-d'Hères, France

AIM AND APPROACH

To mitigate energy network constraints, energy flexibility can be implemented either through direct flexibility (optimal dispatch or demand response) or through indirect flexibility (urge the consumers to modify their energy consumption within a degree of their comfort). The aim of this study is to analyze the potential of indirect energy flexibility for residential sector. For this purpose, an experiment is set up with 175 French households. Generally, the peak energy consumption in a household occurs in the evening. Whereas the maximum renewable energy production occurs during the afternoon of hot days. Therefore, customized SMS alerts are sent to the households to either implement peak shaving or load shifting on alert day "D". These SMS are in accordance with the predefined list of appliances that the households use (or not use) during alert period. The orange alerts were sent on day "D-1" to implement peak shaving in the evening of alert day "D" during winter, whereas the green alerts are sent on day "D-1" for load shifting from evening to afternoon on alert day "D" during summer. As a feedback, each household visualizes its effort in the form of an image of consumption load curve superposed by reference load curve.

SCIENTIFIC INNOVATION AND RELEVANCE

An experiment is setup with 175 French households to test a hypothesis regarding indirect energy flexibility. The proposed hypothesis is that the knowledge of energy intermittency would influence energy consumption behavior of households and use of "nudge signals" would bring a positive reinforcement in their energy consumption behavior. For this purpose, an automated system of sending SMS for potential alert day "D" is implemented. The orange alerts are based on the forecasted network congestion in national grid during winter whereas; the green alerts are based on the forecasted national load curve of residential sector for next 15 days. The forecast is made on the historical data of national load curve of residential sector (to be discussed in detail in another paper). At day "D+2", the consumption load curve is compared against the reference load curve (to be discussed in detail in another paper) to visualize the impact of nudge signal. The comparison image is also viewed by concerning household as a feedback. The study is relevant in terms of testing indirect energy flexibility at a small scale to mitigate the energy imbalance caused by the intermittence of renewable energy production. This study might then be prospectively extrapolated at bigger scale.

PRELIMINARY RESULTS AND CONCLUSIONS

As part of preliminary results, an automated experimental setup (programmed in "python" language) is put in place with the procedures of calculating the reference load curve and prediction on national load curve of residential sector (both to be discussed in detail in separate papers). 9 orange alerts were sent towards the households during the winter of 2019-2020. The average daily load curve of treated group is compared with control group (both groups having similar energy consumption characteristics). Econometrically, it is observed that the cumulative energy saving is around 10% in response to the sent orange alerts (also known as nudge signals). Another indicator compares the consumption load curve against the reference load curve for both alert days and non alert days. It depicts that for most of the households, the energy consumption is reduced during the alert days as compared to the energy consumption during the non-alert days. The result of green alert is still awaited. The preliminary results conclude that the knowledge of energy intermittency influences the energy consumption behavior of certain households. However it is still to be determined whether it is caused by a certain bias (e.g. mindset of reducing GHG emissions at individual level etc.)

MAIN REFERENCES

Albadi, M. H., and E. F. El-Saadany. 2007. "Demand Response in Electricity Markets: An Overview." In 2007 IEEE Power Engineering Society General Meeting, 1–5.

Bivas, Pierre. 2011. "La production d'effacement : comment offrir des économies d'électricité à des millions de foyers." Le journal de l'ecole de Paris du management n°90 (4): 8–14.

Hatton, Leslie, and Philippe Charpentier. 2014. "Système électrique français : estimation de l'effacement des clients résidentiels," 14.

Lesgards, Valérie, and Laure Frachet. 2012. "La Gestion de La Demande Résidentielle d'électricité: Retour Sur 30 Ans d'expérimentations Mondiales." La Gestion de La Demande Résidentielle d'électricité: Retour Sur 30 Ans d'expérimentations Mondiales, no. 607: 162, 164, 192-210 [21 p.].

Neenan, B. 2009. "Residential Electricity Use Feedback: A Research Synthesis and Economic Framework," 126.

Thaler, Richard H., and Cass R. Sunstein. 2008. Nudge: Improving Decisions about Health, Wealth, and Happiness. Nudge: Improving Decisions about Health, Wealth, and Happiness. New Haven, CT, US: Yale University Press.

Tsemekidi Tzeiranaki, Sofia, Paolo Bertoldi, Francesca Diluiso, Luca Castellazzi, Marina Economidou, Nicola Labanca, Tiago Ribeiro Serrenho, and Paolo Zangheri. 2019. "Analysis of the EU Residential Energy Consumption: Trends and Determinants." Energies 12 (6): 1065.

Keywords: Indirect Energy Flexibility, Nudge, Residential Sector, Feedback, Positive Reinforcement

Session T3.7 (Online Track): Ensuring high quality building simulations

Paper ID 30312: Development of energy use benchmark data with respect to locations, housing types, gross area and construction year from household energy standing survey in Korea

Na Hyeon Lee*, Dong Hyun Seo

Chungbuk National University(CBNU), Korea, Republic of (South Korea)

AIM AND APPROACH

The aim is to present residential energy use features that can be used in various fields by analyzing monthly consumption of electricity and fuels on the basis of principal variables such as location, housing type, gross area and construction year.

[Approach]

- Architectural properties and residential energy monthly consumption data of sampled houses are collected from the Year 2016 microdata of the Household Energy Standing Sample Survey (HESS). Quality control process is carried out to remove the outlier consumption data.

- Energy consumption is analyzed and energy use features are presented through systematic analysis criteria based on the locations, housing types, gross area, and construction year, which are considered to be principal factors of energy consumption.

- A literature review on residential energy use features is analyzed. And then, the authors identify the current research practices and determines the reliability of HESS energy consumption data which is generated in this research.

SCIENTIFIC INNOVATION AND RELEVANCE

- According to statistics from the EIAS of Ministry of Land, Infrastructure, and Transport in 2017, residential buildings account for about 44% of total energy consumption in Korea. Therefore the government has been enforced with a strong building efficiency policy over the past decade. Despite the efforts of the government, there is a lack of data to identify the energy consumption landscape of residential buildings in Korea.

- This study contributes to presenting detailed residential energy use features through a new analysis system that has not been used in Korea by selecting location, housing type, gross area, and construction year as the main variables. Detailed residential energy use features data to be presented as a result of this study are expected to be more useful to researchers and policymakers as well as to the general public.

PRELIMINARY RESULTS AND CONCLUSIONS

Monthly fuel and electricity consumption are analyzed based on the four major variables such as location, housing type, gross area, and construction year. By doing so detailed energy use benchmark data are derived for residential buildings in Korea.

MAIN REFERENCES

- Lee, N., Kim, H., Seo, D. (2019). Analysis of Residential energy Use Features with Respect to Location, Housing Type, Gross Area and Construction Year from Household Energy Standing Survey, Journal of KIAEBS, 13(6), 536-549.

- Raffio, G., Isambert, O., Mertz G., Schreier C., Kissock K. (2007). Targeting residential energy assistance, Proceedings of Energy Sustainability Conference, 489-95.

- Filippini, M., Hunt, L. (2012). US Residential Energy Demand and Energy Efficiency: A Stochastic Demand frontier Approach, Energy Economics 34, 1484-1491.

Keywords: Residential energy, HESS, Micro data

Paper ID 30379: Integration of probabilistic methods and parametric tools for performance-based building design decision making

Fatemeh Shahsavari, Wei Yan*

Texas A&M University, United States of America

AIM AND APPROACH

Building performance simulation (BPS) tools are useful in the field of building design optimization but mostly fail to deal with uncertainties. The commonly-used BPS tools, such as EnergyPlus and TRNSYS, act solely upon deterministic sets of input data, disregarding the associated uncertainties (Suna et al., 2014). Solving optimization problems in the field of performance-based building design with deterministic approaches may oversimplify reality and lead to overestimation or underestimation of building performance (Yao, 2014).

Tian et al. (2018) identified the key sources of uncertainties in building energy simulation related to climate change, occupant behavior, and degradation of the HVAC system and building envelope materials over time. Uncertainty analysis methods such as Monte Carlo support robust optimization to find the optimum design alternative subjected to existing uncertainties.

The aim of this research is to deploy the capabilities of design parametric tools to integrate uncertainty analysis methods such as Monte Carlo in performance-based design decision making. The probabilistic methods versus deterministic methods allow considering different attitudes towards risk in design projects. Also, the probabilistic methods allow the calculation of expected value, expected utility, and confidence intervals to help designers make informed design decisions based on the project needs and existing uncertainties.

SCIENTIFIC INNOVATION AND RELEVANCE

The probabilistic methods provide designers with additional information including the mean value and variance of the simulation result for different design alternatives. Also, other factors including expected value and expected utility inform designers about the benefit-cost of each design alternative under different possible conditions. This extra information allows designers to make informed decisions with regard to the requirements of the design project. For instance, occupant thermal comfort may be a higher priority in design compared to construction cost in a health care facility design project, while the order may be different in a storage facility design project.

PRELIMINARY RESULTS AND CONCLUSIONS

Considering uncertainties in design input parameters with associated probability density functions and using Monte Carlo sampling methods allow performing building simulations for a range of inputs and getting the probability distribution of simulation output. We have tested a probabilistic method with a test case of three design alternatives for a modular classroom design and concluded that the ranking order of design alternatives obtained from the probabilistic methods may be different from what deterministic method suggests. Our previous work shows major discrepancies between deterministic and probabilistic results in building energy simulations.

This research will expand the previous experiment with more complicated design scenarios and studying the effect of different probability distribution types of design input parameters on the results. The probabilistic results may provide designers with better insight into design consequences under different possible conditions.

MAIN REFERENCES

Suna, K., Yana, D., Hongb, D., & Guoa, S. (2014). Stochastic modeling of overtime occupancy and its application in building energy simulation and calibration. Building and Environment, 79(2010), 1–12.

Tian, W., Heo, Y., de Wilde, P., Li, Z., Yan, D., Park, C. S., Feng, X., & Augenbroe, G. (2018). A review of uncertainty analysis in building energy assessment. Renewable and Sustainable Energy Reviews, 93(January 2017), 285–301. https://doi.org/10.1016/j.rser.2018.05.029

Yao, J. (2014). Determining the energy performance of manually controlled solar shades: A stochastic model based co-simulation analysis. Applied Energy, 127, 64–80. https://doi.org/10.1016/j.apenergy.2014.04.046

Keywords: Building performance simulation, uncertainty analysis, design decision making, expected value, expected utility.

Paper ID 30426: Energy prediction impact of the Space Level Occupancy schedule for a Primary School

Yingli Lou (1), Yunyang Ye (2), Wangda Zuo* (1,3), Jian Zhang (2)

1: University of Colorado, Boudler, United States of America; 2: Pacific Northwest National Laboratory, United States of America; 3: National Renewable Energy Laboratory, United States of America

AIM AND APPROACH

This research is aimed to explore whether it's necessary to develop heterogeneous occupancy schedules for building energy models. Both homogeneous and heterogeneous occupancy schedules can describe occupant status in the building. On one hand, the homogeneous occupancy schedule focuses on the total number of occupants in the building, and occupancy schedule is the same for all rooms [1,2]. On the other hand, the heterogeneous occupancy schedule describes occupant status in each room of the building so that each room has its unique occupancy schedule, which may improve the prediction accuracy [2,3]. Collecting data for the homogeneous schedule is easier than the heterogeneous schedule because it only requires an occupancy sensor in building entrance, while the heterogeneous schedule needs multiple sensors and it is labor intensive to identifying those data. Therefore, it is important to quantify the potential accuracy difference between these two schedules so that practitioners can select the schedule according to their needs.

This research uses primary school building as case study and develops homogeneous and heterogeneous occupancy schedules for it. Then, energy saving of primary school model [4] from ASHRAE Standard 90.1-2004 to 90.1-2016 using both schedules are calculated. Energy saving difference between these two schedules is compared.

SCIENTIFIC INNOVATION AND RELEVANCE

In the conventional view, occupancy has minimal impact on building energy prediction for buildings without occupancy-based control (OBC). Therefore, most researches employ a homogeneous schedule to represent building occupancy [5]. However, occupancy does impact the required heating and cooling loads of a building [6,7], which will then impact the building energy performance. Thus, this research evaluates the energy impact of occupancy schedules for buildings without OBC. This research develops homogeneous and heterogeneous occupancy schedules and compared their energy predictions.

Building energy models are critical for various energy related applications, such as building energy performance evaluation, energy efficiency measures selection, and large-scale building energy prediction. Determining the impact of heterogeneous occupancy schedule on building energy modeling will provide a guideline for practitioners in building energy modeling to seek a balance between modeling accuracy and labor time.

PRELIMINARY RESULTS AND CONCLUSIONS

In this case study, Energy savings of primary school from ASHRAE Standard 90.1-2004 to ASHRAE Standard 90.1-2016 using homogeneous and heterogeneous occupancy schedules are shown in Table 1.

Table 1. Energy saving of primary school from ASHRAE Standard 90.1-2004 to ASHRAE Standard 90.1-2016

Climate zones	Homogeneous occupancy schedule (%)	Heterogeneous occupancy schedule (%)	Difference (%)
1A	44.29	45.55	-1.26
2A	43.67	44.53	-0.86

2B 36.57 37.66 -1.09

3A 41.34 41.95 -0.61

3B 36.85 37.60 -0.75

3C 42.81 44.33 -1.51

4A 32.97 32.69 0.27

4B 39.76 39.86 -0.09

4C 43.50 42.36 1.14

5A 47.06 47.42 -0.35

5B 40.75 39.96 0.79

5C 43.38 43.82 -0.44

6A 47.36 47.35 0.01

6B 45.33 45.46 -0.13

7A 47.68 47.56 0.12

8A 44.16 43.31 0.86

Occupancy has impact on building energy prediction for buildings without OBC. And the impact of occupancy on energy saving evaluation is more significant in hot areas (climate zones 1, 2, and 3).

MAIN REFERENCES

[1]. Oldewurtel, F., Sturzenegger, D., & Morari, M. (2013). Importance of occupancy information for building climate control. Applied energy, 101, 521-532.

[2]. Chen, Y., Hong, T., & Luo, X. (2018, February). An agent-based stochastic Occupancy Simulator. In Building Simulation (Vol. 11, No. 1, pp. 37-49). Tsinghua University Press.

[3]. Agarwal, Y., Balaji, B., Gupta, R., Lyles, J., Wei, M., & Weng, T. (2010, November). Occupancy-driven energy management for smart building automation. In Proceedings of the 2nd ACM workshop on embedded sensing systems for energy-efficiency in building (pp. 1-6).

[4]. DOE. (2020). Commercial prototype building models. https://www.energycodes.gov/development/commercial/prototype_models

[5]. Ye, Y., Lou, Y., Zuo, W., Franconi, E., & Wang, G. (2020). How Do Electricity Pricing Programs Impact the Selection of Energy Efficiency Measures?-A Case Study with US Medium Office Buildings. Energy and Buildings, 110267.

[6]. Kwok, S. S., & Lee, E. W. (2011). A study of the importance of occupancy to building cooling load in prediction by intelligent approach. Energy Conversion and Management, 52(7), 2555-2564.

[7]. Kwok, S. S., Yuen, R. K., & Lee, E. W. (2011). An intelligent approach to assessing the effect of building occupancy on building cooling load prediction. Building and Environment, 46(8), 1681-1690.

Keywords: Building Energy Model, Building Occupancy, Building Energy Saving Evaluation

Paper ID 30440: Impact of ELA calibration methods on building energy model fidelity

Mohanned M Althobaiti* (1,2), Godfried L Augenbroe (1)

1: Georgia Institute of Technology, Atlanta, USA; 2: King Saud University, Riyadh, KSA

AIM AND APPROACH

As building performance is increasingly improved and building energy consumption decreases, a greater percentage of the total energy loss of a building occurs through envelope leakage. This leakage is characterized by the effective leakage area or ELA, which is a proxy parameter to what is essentially a complex flow phenomenon through cracks driven by pressure differences. Current building performance simulation (BPS) uses software modules that approximately calculate envelope infiltration, but the literature shows that their calibration and validation is still unsatisfactory. In fact, calibration and validation of BPS models is still an important subject of study in our quest to improve the fidelity of simulation-based predictions in various applications. The high level of interaction and subsumption between parameters can result in a model that approximates the measurements well (and thus meets the ASHRAE auditing threshold) but whose "best estimates" of parameters are unreliable. This can be a problem in performance contracting when limits have been agreed on certain parameters such as ELA and U-value. It can also be problematic in the use of the model for certain performance assessments. This paper introduces underlying issues discusses results of direct and indirect calibration with different model fidelities.

SCIENTIFIC INNOVATION AND RELEVANCE

The study focuses on the calibration of building energy models of existing buildings. It does so by conducting calibration for different experiments, i.e., with different data resolutions, and for different model fidelities. The calibration is anchored around ELA and its impact on "best estimates" of other parameters is verified. The study develops a new framework to address calibration and validation for different combinations of data and model fidelity, where each combination leads to probability distributions of the calibration parameter set. For each combination the ultimate aim is to determine the fitness of the resulting building energy model for given application studies such as building energy benchmarking, fault detection, unmet hour verification, etc.

The results are meaningful for better understanding façade infiltration and better understanding of the limits of calibrated models. The paper focuses exclusively on existing buildings, but its findings may lead to large scale data sets of calibrated ELA values in existing buildings, that may find their way into better ELA quantification in energy models of new designs.

PRELIMINARY RESULTS AND CONCLUSIONS

A representative case study is conducted to demonstrate the approach. To explore the role of model fidelity, two tools are chosen, i.e., a reduced-order energy calculator and EnergyPlus, which represents the high-fidelity tool. The results demonstrate the effectiveness of the proposed calibration process in approaching the true values of the calibration parameters with focus on ELA.

Calibration techniques have encountered great improvement in recent years and are well supported by ASHRAE and numerous theoretical and field studies. The criteria laid down in the ASHRAE guideline are useful, but more work is needed to understand the special nature of calibration of building energy models. Such criteria stipulate ranges of admissible error in the total estimated energy consumption of a building but do not address underlying aspects such as uncertainty embedded in the models or initial guesses about the value of building a model. The paper discusses the criteria that need to be met in order to guarantee that a calibration not only meets the minimum required CVRMSE of the predicted outcomes, but also provides a close enough approximation of the best estimate of ELA obtainable with an experimental (tracer gas) set-up.

MAIN REFERENCES

Augenbroe, G., Heo, Y., & Choudhary, R. (2011). Risk Analysis of Energy-Efficiency Projects Based on Bayesian Calibration of Building Energy Models. 12th Conference of International Building Performance Simulation Association, (pp. 2579-2586). Sydney.

Campolongo, F., Cariboni, J., & Saltelli, A. (2007). An Effective Screening Design For Sensitivity Analysis of Large Models. Environmental Modelling & Software, 22(10), 1509-1518.

Chong, A. (2018). Datasets, Bayesian Calibration of Building Energy Models for Large Datasets. Carnegie Mellon University. Thesis.

Costola, D., Blocken, B., Ohba, M., & Hensen, J. (2010). Uncertainty in airflow rate calculations due to the use of surface-averaged pressure coefficients. Energy and Buildings 42 881–888.

de Wilde, P., van der Vordan, M., Augenbroe, G., & Kaan, H. (2001). The Need for Computational Support In Energy-Efficient Design Projects In The Netherlands. Seventh International IBPSA Conference, (pp. 513-519). Rio de Janeiro.

de Wit, M. S. (2001). Uncertainty in Predection of Thermal Comfort in Buildings. Delft: Delft University of Technology.

Judkoff, R., & Neymark, J. (2006). Model validation and testing: the methodological foundation of ASHRAE Standard 140.

Reddy, T. A. (2006). Literature review on calibration of building energy simulation programs: Uses, problems, procedures, uncertainty, and tools. ASHRAE Transactions, 112(1), 226–240.

Keywords: Uncertainty analysis, Energy modeling, Effective leakage area, Model fidelity

Paper ID 30466: Development of Gray-box modeling technique for HVAC&R systems with automatic calibration process

Hyemi Kim* (1), Sumiyoshi Daisuke (2), Youngjin Choi (3)

1: Ph.D. Student, Faculty of Human-Environment Studies, Kyushu University, Japan; 2: Associate Prof., Faculty of Human-Environment Studies, Kyushu University, Japan; 3: Assistant Prof., Department of Architectural Engineering, Kyonggi University, Repubpic of Korea

AIM AND APPROACH

Key challenges faced with the simulation tools in the building performance simulation(BPS) sectors include: (1) to create a detailed physical-based energy model using measured data, (2) to create a compatible model that data is collected in real time (3) to lack the technician or energy manager with the building simulation knowledge [1]. This study develops an automatic simulation tool of Heating, Ventilation, Air Conditioning, and Refrigeration (HVAC&R) systems based on physics characteristics using the measured data. However, the existing simulation modeling process is costly and time-consuming to collect data in buildings. There are some problems about the limitations of the items that correspond to the physical formula [2]. Moreover, the techniques of systematic calibration are required to reconcile the discrepancies between the predicted and measured data [3].The purpose of this paper is to develop the gray-box modeling method in HVAC&R systems entailed with the automated calibration process, and to validate the reproducing performance of the simplicity systems. At first, we create each equipment model, and implement the modeling of the whole primary systems. Each device model selects a physical-based parameter and those are automatically tuned by machine learning. The reproducibility of the simplified HVAC&R system model using TRNSYS is verified.

SCIENTIFIC INNOVATION AND RELEVANCE

This study systematically founds an automatic modeling procedure of HVAC&R systems in central air conditioning systems. The developed modeling method has two typical characteristics that introduce Grey-box and automatic calibration methods. The existing research [1-4] of two methods have modeled a whole building energy simulation, not about HVAC systems. In addition, the most of research have developed an optimized algorithm or implemented FDD (Fault Detection and Diagnostics) about the specific system or equipment. Those are insignificant to grasp the status of operating system using a systematic modeling method that can cope with numerous systems. This study has the innovations: (1) to make a machine learning model of the physics-based on each equipment, (2) to systematize and categorize the modeling procedure so that the results of (1) can be extended to the system scope to respond various systems and equipment, (3) to improve the accuracy of the physical parameters by automatizing the stage of calibrating it one by one and making the whole steps accurate. The developed modeling method can identify each device's operation in HVAC&R systems by the physics principle. Afterward, this method will be able to be adapt during the developing optimized algorithm and FDD procedure of each equipment.

PRELIMINARY RESULTS AND CONCLUSIONS

This paper develops the modeling method of air-cooled systems in the initial stages of the whole study. The equipment adopted by the system are heat-source devices (e.g. turbo chiller, heat pump), heat storage tank, inverter pump, heat exchanger, etc. The development process of the modeling method of each device verifies the accuracy by using the various measured data. The modeling method in this study calculates the parameters based on the physics considering the minimized input data of each equipment, and it categorizes characteristics that are different from each manufacturer. Next, the selected parameter is tuned using the machine learning and calibrates by each hour. Each modeling output value is substituted into the input value of the other equipment, and modeling of the simplified air-cooled system is performed with TRNSYS. As a result, while the modeling of the whole system is performed, detailed operation situation will be grasped using the basic input variables. The HVAC&R system modeling method developed in this study uses fewer input variables and considers measured data items compared to the existing simulation method. Although the new data is unstable out of the range of the training data, and the output of the high accuracy is expected.

MAIN REFERENCES

The main references in this study are as follows. Hong et al. and Coakley et al. are the researches about the calibration for reducing discrepancies during the steps of building modeling and simulating. Harish et al. and Afroz et al. are papers about the classified modeling method with White-box, Black-box, and Grey-box. More than 100 existing studies are organized and analyzed in each research. Based on these researches, this study specified the modeling target, method and process, and calibration method through comparison with the relevant studies.

[Reference]

1) Tianzhen Hong et al., Building Simulation: Ten Challenges, Building Simulation, March 2018

2) V.S.K.V. Harish et al., A review on modeling and simulation of building energy systems, Renewable and Sustainable Energy Reviews, 56 (2016), pp.1272–1292

3) Daniel Coakley et al., A review of methods to match building energy simulation models to measured data, Renewable and Sustainable Energy Reviews,37 (2014), pp.123–141

4) Zakia Afroz et al., Modeling techniques used in building HVAC control systems: A review, Renewable and Sustainable Energy Reviews, 83 (2018), pp.64–84

Keywords: Automatized, Calibrate, Building Performacne Simulation, ANN, Grey-box modeling

Paper ID 30648: Development of a metamodel to assess building thermal performance for naturally ventilated residential buildings

Rodolfo Kirch Veiga, Letícia Gabriela Eli, Amanda Fraga Krelling, Marcelo Salles Olinger, Ana Paula Melo, Roberto Lamberts*

Federal University of Santa Catarina, Brazil

AIM AND APPROACH

Building energy simulation programs allow the assessment of the impact of different air conditioning systems into the total building final energy consumption, however, it requires high computational cost and effort. On the other hand, metamodels are quick and easy-to-use tools that support experts in decision-making to optimize the thermal performance of buildings. The main objective of this study is to develop a metamodel to predict the thermal performance of multifamily residential buildings. The thermal performance will be evaluated according to a percentage of hours where the operating temperature of the room is within a range of interest (PHFT). To develop the metamodel it will be necessary to define a parameterizable model. The development of the parameterizable model will consist of a sequence of simplifications made on the original base model. The simplifications that will be carried out have as main objective to reduce the amount of descriptions necessary to simulate the room of interest. Simplifications will be assessed through statistical analysis and heat balance in some critical cases. The computer simulation program EnergyPlus was adopted to obtain the PHFT for each case. The study emphasized the residential buildings located in São Paulo, Brazil.

SCIENTIFIC INNOVATION AND RELEVANCE

The development of accurate models for the prediction of the thermal performance of buildings has been intensified regarding the user thermal conditions needs and, consequently, the high energy demand of buildings, directly related to the expressive energy consumption of artificial air conditioning system. Metamodels are quick and easy-to-use tools, capable of identifying low-cost solutions that lead to energy efficiency. These tools can assist not only in the early design stages of new building projects and retrofit projects, but also facilitate the development of thermal performance approaches. Therefore, the development of a metamodel that accurately describes the thermal performance of residential building is a relevant challenge, since it can be applied to an extensive group of buildings. In this way, a tool that can be widely applied becomes viable, indirectly generating an increase in the energy efficiency of the residential building sector.

PRELIMINARY RESULTS AND CONCLUSIONS

The preliminary results showed that the proposed simplifications can compromise the heat balance, and, consequently, the percentage of hours where the operating temperature of the room is within a range of interest (PHFT). It was observed that it is a consequence related to the consideration of adiabatic surfaces. Therefore, in order to continue the process of simplifications on the original base model and to achieve a parameterizable model, it was necessary to adopt new measures that consider the heat exchanges on the room surfaces of interest. The thermal performance of each simplification will be compared with the base model based on the difference in PHFT indicated by root-mean-square error (RMSE) and mean absolute error (MAE) of the descriptive statistics of the reductions in comparison with the base model and illustrative graphics. The simplification process will be interrupted when a parameterizable model is found whose thermal performance is similar to the original base model. In this study, the Keras library - PHYTON language, was adopted as a function for artificial neural networks. The global sensitivity analysis of Sobol will be applied in order to identify the influence of the inputs data on the output data.

MAIN REFERENCES

AÏT-SAHALIA, Yacine; XIU, Dacheng. Principal component analysis of high-frequency data. Journal of the American Statistical Association, v. 114, n. 525, p. 287-303, 2019.

AMASYALI, Kadir; EL-GOHARY, Nora M. A review of data-driven building energy consumption prediction studies. Renewable and Sustainable Energy Reviews, v. 81, p. 1192-1205, 2018.

DEB, Chirag et al. A review on time series forecasting techniques for building energy consumption. Renewable and Sustainable Energy Reviews, v. 74, p. 902-924, 2017.

HAASE, M.; AMATO, A. An investigation of the potential for natural ventilation and building orientation to achieve thermal comfort in warm and humid climates. Solar energy, v. 83, n. 3, p. 389-399, 2009.

MOHD NAWI, Nazri; ATOMIA, Walid Hasen; REHMAN, Mohammad Zubair. The effect of data pre-processing on optimized training of artificial neural networks. 2013.

VEHTARI, Aki; GELMAN, Andrew; GABRY, Jonah. Practical Bayesian model evaluation using leave-one-out cross-validation and WAIC. Statistics and Computing, v. 27, n. 5, p. 1413-1432, 2017.

WANG, Zeyu; SRINIVASAN, Ravi S. A review of artificial intelligence based building energy use prediction: Contrasting the capabilities of single and ensemble prediction models. Renewable and Sustainable Energy Reviews, v. 75, p. 796-808, 2017.

Keywords: Building Simulation, Accuracy, Simplified model

Paper ID 30725: A knowledge-based framework for building model performance verification

Yan Chen, Jeremy Lerond*, Xuechen Lei, Michael Rosenberg, Draguna Vrabie

Pacific Northwest National Laboratory, United States of America

AIM AND APPROACH

Building simulation is widely used by researchers, regulators, and engineers to quantify building energy performance. One application is building energy codes benchmarking, that is, creating code compliant building energy models and simulating their performance. To obtain confident benchmarking results, it is critical to verify that code requirements are properly implemented and simulated. Currently, this is done in a manual and ad-hoc manner. Such a verification process is tedious, error-prone and time-consuming for several reasons: 1) the number of models to be checked is very large, 2) modelers check models differently, which might cause difficulties for quality control, and 3) manual checks can be highly repetitive if not automated. To solve these challenges, we propose a data-driven performance verification framework, which conducts automated output-based verification of building energy code requirements (especially control requirements which have been verified via time-series output). This framework has an extensible knowledge base that contains programmatic interpretation and implementation of code requirements. While this framework was developed for building energy code benchmarking it can be extended and used for other situations such as code compliance performance modeling evaluation, and other control performance verification.

SCIENTIFIC INNOVATION AND RELEVANCE

In the past, whole building energy modeling quality assurance (QA) and quality control (QC) have mainly been done using (or a combination of) input, output summary, or comparative output-based verifications. Most of these approaches tend to minimize the model review time while still being able to identify potential modeling errors within building energy models. The main downside to these approaches is that 1) they lack standardization, and 2) they do not typically evaluate the performance of the model at a granular level (e.g. simulation timestep) and hence might not be able to identify potential system control issues which in turns could have an impact on the simulation results. The proposed framework provides a transparent, reproduceable and automated way of checking the implementation of building system control strategies in a model based on simulation timestep data. The knowledge base developed as part of the framework uses a wide range of methods to verify simulation outputs: from simple rule-based verifications to more sophisticated approaches that leverage applied machine learning techniques.

PRELIMINARY RESULTS AND CONCLUSIONS

We present two example uses of this framework where the implementation of ASHRAE Standard 90.1 requirements are checked: daylighting dimming control and integrated economizer control. Detailed discussion includes how the framework automatically instruments an EnergyPlus model, collects necessary data from various input and output files, and uses different types of algorithms to check the implementation of the code requirements. The examples also illustrate how the framework can provide verification results with quantitative and visual outputs to help users better understand how the requirements are implemented in the model.

MAIN REFERENCES

BEMLibrary. n.d. Review Checklists and References. https://www.bemlibrary.com/index.php/practicioners/resources/review-checklists-references/.

Butzbaugh J.B., C.A. Antonopoulo. n.d. Introducing the Automated Fault Detection & Diagnostics Technology Challenge. PNNL-SA-153582.

Butzbaugh J.B., C.A. Antonopoulos, A. Tidwell. n.d. Automatic Fault Detection & Diagnostics Market Analysis. PNNL-30077.

Chris Balbach, David Bosworth. 2016. "Automated methods for improving energy model qualkity assurance and quality control." Building Performance Modeling Conference. Salt

Lake City, UT: ASHRAE and IBPSA-USA. 385-392.

COMNET. 2012. COMNET, Standardizing energy modeling for buildings. https://comnet.org/.

Kara Vega, Shelley Beaulieu, Maria Karpman. 2014. "Lessons Learned: Performing QC on Energy Models." ACEEE Summer Study on Energy Efficiency in Buildings. 339-350.

NREL. 2016. BuildingSync XML Schema. https://buildingsync.net/index.html .

Keywords: Simulation, Knowledgebase, Control performance verification, Quality control, Quality assurance

Paper ID 30732: Solar radiation nowcasting through advanced CNN model integrated with ResNet structure

Lei Chen* (1), Yangluxi Li (1), Hu Du (1), Yukun Lai (2)

1: Welsh school of Architecture, Cardiff University, United Kingdom; 2: School of Computer Science and Informatics, Cardiff University, United Kingdom

AIM AND APPROACH

Although a range of solar radiation forecasting methods have been developed for predicting photovoltaic generation, only a few of them pay attention to the solar radiation forecasting for building energy demand. From the perspective of building performance modelling, solar radiation forecasting needs to meet several critical requirements including high spatial resolution (1m-2km) and high temporal resolution (5-60mins), the accurate value of Direct Normal Irradiance (DNI) and Diffuse Horizontal Irradiance (DHI), which differs the requirement for predicting photovoltaic generation. As the geometric sum of DNI and DHI, accurate prediction of Global Horizontal Irradiance (GHI) with high spatial-temporal resolution tends to be the prerequisite for accurate prediction of DNI and DHI.

This research aims to construct a hybrid nowcasting model to predict GHI in high spatial-temporal resolution.

In this article, the authors adopt an advanced Convolutional Neural Network (CNN) model with Residual Neural Network (ResNet) structure to identify the cloud image information and predict the GHI at 10 minutes intervals merely using cloud images captured by a ground-based sky camera. On this basis, several ResNet structures are compared to achieve the optimal nowcasting model for GHI.

SCIENTIFIC INNOVATION AND RELEVANCE

Solar irradiance has a key role in predicting renewable thermal and electrical generation, informing demand response and ensuring high-quality real-time building simulations. In the past several decades, a series of solar irradiance forecasting methods have been developed and they could be classified as Numerical Weather Prediction Methods (NWP), Statistical Methods, Top-Down Methods, Bottom-Up Methods and Hybrid Methods [1]. Each method has its unique feature in term of spatial and temporal resolutions [2], for example, the spatial horizon of Bottom-Up Methods ranging from 1 meter to 2 kilometres and its temporal horizon ranging from 5 to 30 minutes. In recent years, researchers have explored various approaches [3], models [4] and data acquisition tools [5] for achieving high spatial-temporal resolution solar irradiance forecasting.

The key innovations of this study include:

1.Classifying solar radiation forecasting methods and reviewing recent representative articles.

2.Proposing a hybrid method using CNN-ResNet model to nowcast solar irradiance.

3.Only using ground-based cloud images as input to predict solar irradiation.

4.Comparing several ResNet structures to achieve the optimal nowcasting model for GHI.

PRELIMINARY RESULTS AND CONCLUSIONS

This study developed a hybrid nowcasting method for Global Horizontal Irradiance (GHI) through using an advanced CNN model with ResNet structure to identify the ground cloud images information collected by an open database – NREL dataset. Several ResNet structures are compared to achieve optimal nowcasting model for GHI. The results show that the ResNet structure could efficiently capture the significant information of cloud images and the ResNet-152 has the optimal performance on solar irradiance nowcasting. Final result also indicates that a low prediction error and high relative coefficient values are got. The accomplishment of this model apparently decreases the difficulty of solar irradiance forecast because only need to use sky image. Nevertheless, taking consideration of this model generalization capability, future relative investigations had better to be performed to improve precision under other sky condition.

MAIN REFERENCES

[1] Chen, L., Du, H. and Li, Y., 2019, September. Scoping Low-Cost Measures to Nowcast Sub-Hourly Solar Radiations for Buildings. In IOP Conference Series: Earth and Environmental Science (Vol. 329, No. 1, p. 012041).

[2] Diagne, M., David, M., Lauret, P., Boland, J. and Schmutz, N., 2013. Review of solar irradiance forecasting methods and a proposition for small-scale insular grids. Renewable and Sustainable Energy Reviews, 27, pp.65-76.

[3] Alonso, J., Batlles, F.J. and Portillo, C., 2015. Solar irradiance forecasting at one-minute intervals for different sky conditions using sky camera images. Energy Conversion and Management, 105, pp.1166-1177.

[4] Chu, Y., Li, M., Pedro, H.T. and Coimbra, C.F., 2015. Real-time prediction intervals for intra-hour DNI forecasts. Renewable Energy, 83, pp.234-244.

[5] Du, H., Jones, P., Segarra, E. L. & Bandera, C. F., 2018. Development Of A Rest Api For Obtaining Site-Specific Historical And Near-Future Weather Data In Epw Forma. Building Simulation and Optimization 2018, 2018 Emmanuel College, University Of Cambridge.

[6] Perez, R., Ineichen, P., Maxwell, E., Seals, R., & Zelenka, A., 1991. Dynamic models for hourly global-to-direct irradiance conversion. In Solar world congress: Proceedings of the biennial congress of the international solar energy society (Vol. 1, No. part II, pp. 951-956).

Keywords: Solar Irradiance Nowcasting, Global Horizontal Irradiance, High Spatial-Temporal Resolution

Paper ID 30874: Intelligent building envelopes energy evaluation: an integration of double-skin facades with earth-tubes

Payman Sadeghi*

Drury University, United States of America

AIM AND APPROACH

Ecological Architecture inclusively addresses all the matters that should and could be considered including but not limited to economic, social, or environmental dimensions. The integration of an ecological worldview, thus, should be recognized as a fundamental goal in reaching high-performance solutions that are formed as environmental systems rather than as standalone objects. Building envelope, if conceived as a responsive skin, can serve as a critical component of an ecological building. It is strictly linked to the environmental, social, and economical aspects of buildings and their context. Therefore, it should act in a creative, flexible, dynamic, and adaptive manner. A Double-Skin Facade (DSF) is product of this viewpoint: a multilayered formation to provide daylight, indoor air quality, natural ventilation, thermal and visual comfort, acoustics isolation, and aesthetic. In this domain, energy is surely a crucial concern too. This paper's focus is to examine ways to optimize the energy demand_ specifically Load Intensity (LI)_ of buildings with a DSF as an integrated, intelligent, and high performance building envelope system.

SCIENTIFIC INNOVATION AND RELEVANCE

From environmental and social standpoints, Double-Skin Facades (DSFs) are presumed as high-performance systems that integrate ecological solutions. Yet, DSFs' economic performance is debated mainly due to the overheating issue in hot seasons. That is, a DSF's goal to minimize buildings' energy use in hot climatic conditions is controversial, presenting it occasionally as not the most economically viable solution. In this study, it was hypothesized that the energy demand of a building with a naturally ventilated DSF, inspired by Persian wind-catchers, would be improved via linking it into the earth via earth-tubes, inspired by the concept of Roman hypocausts. The study speculates that this combination could resolve the DSF's overheating issue leading to a lower building load intensity, which would then increase the economic performance of a DSF system. To conduct the inquiry, a simulation method was applied that employed both TRNSYS and CANTAM softwares to model a nonresidential building with integrated DSF and earth-tube systems in three main climatic conditions of the United States. Combination of alternative design options, then, were assessed to compare their load intensity and identification of the optimized proposal.

PRELIMINARY RESULTS AND CONCLUSIONS

Consequently, the energy impact of alternative design options such as climate, DSF depth, shading, ventilation strategy, and earth-tubes integration was examined. The results showed that the assimilation of earth-tubes into a DSF system always increases the system's energy performance with no overheating issue provided that optimum DSF depth is envisioned. However, the integrated system demonstrated higher positive influence on buildings' energy performance in cold climates compared to moderate and hot climates where the impact is lower. Also, the energy demand in the integrated system is primarily linked to the building's ventilation strategy not the DSF's. In contrast to certain studies on DSFs suggesting that the integration of a DSF system will always increase cooling loads, a DSF system can considerably lower cooling loads on conditions that appropriate shading devices, construction types, DSF depths, and building ventilation strategies are implemented. After all, the study outcomes suggest additional emphasis on the role and the need for systematic communication with buildings science. To conclude, the research process has elaborately showcased how an early theorized design solution can still perform different from what had been known or read in the literature, highlighting the importance of evaluation.

MAIN REFERENCES

Bartok, J. (2012). Geothermal Heat for Greenhouses, Retrieved from: http://www.extension.org/pages/27790/geothermal-heat-for-greenhouses#.UwO-SvldVqU, Date Accessed: February 22, 2014.

Choi, W., Joe1, J., Kwak., & Huh, J. (2012). Operation and Control Strategies for Multi-Story Double Skin Facades during the Heating Season. In Energy and Buildings. 49. 454–465.

Gratia, E., & De Herde, A. (2004). Natural cooling strategies efficiency in an office building with a double-skin façade. Energy and buildings, 36(11), 1139-1152.

Kalyanova, O., Heiselberg, P., Felsmann, C., Poirazis, H., Strachan, P., & Wijsman, A. (2009, July). An empirical validation of building simulation software for modelling of double-skin facade (DSF). In Proceedings of the 11th IBPSA Conference, July (pp. 27-30).

Poizaris, H. (2004). Double Skin Facades for Office Buildings.

Wong, P. C., Prasad, D., & Behnia, M. (2008). A New Type of Double-Skin Façade Configuration for the Hot and Humid Climate. Energy and Buildings, 40(10), 1941-1945.

Xu, L., & Ojima, T. (2007). Field Experiments on Natural Energy Utilization in a Residential House with a Double Skin Façade System. Building and Environment, 42(5), 2014-2023.

Zhou, J., & Chen, Y. (2010). A Review on Applying Ventilated Double-Skin Facade to Buildings in Hot-Summer and Cold-Winter Zone in China. Renewable and Sustainable Energy Reviews, 14(4), 1321-1328.

Keywords: High-Performance Buildings, Double-Skin Façade (DSF), Earth-Tubes. Intelligent Building Skin, Simulation, and Energy Efficiency

Paper ID 30595: Whole building validation for simulation programs including synthetic users and heating systems: experiment and results

Matthias Kersken* (1), Paul Strachan (2)

1: Fraunhofer Institute for Building Physics IBP, Germany; 2: Energy Systems Research Unit, Department of Mechanical and Aerospace Engineering, University of Strathclyde, Glasgow, UK

AIM AND APPROACH

Dynamic Building Energy Simulation (BES) tools are increasingly used in the design of buildings, building services and controls. Especially when considering controls of complex buildings, the tools in use need to reliably respond the occupants' various inputs, represent the buildings' service systems and most importantly, capture the interactions between both correctly. Within IEA EBC Annex 58 a side-by-side full scale validation experiment on Fraunhofer IBP's Twin Houses was undertaken and made public, investigating BES tools predictive capabilities, focusing particularly on climate and the building envelope [1]. Within the current IEA EBC Annex 71 a new experiment was conducted focusing on occupants' influence and heating systems with controls. To guarantee a known but realistic occupancy, synthetic users with probabilistic elements were included [2]. During the experimental design, a sensitivity analysis was conducted to identify the most critical influences on the validation [3]. After these items were identified, either the experiment was changed or suitable instrumentation was installed. Using the detailed experimental specification and provided boundary conditions, 13 modelling teams using 8 different programs submitted "blind validation" results for the main part of the experiment. Updated submissions to correct modelling errors were made after the full experimental dataset was released.

SCIENTIFIC INNOVATION AND RELEVANCE

Because it is relatively cheap and fast the method of inter model comparison is widespread for the validation of building energy performance simulation software. Well known examples for inter model comparison are BESTEST and CEN 13791. Also, several in situ measurement datasets are available, but most of them are created using very simple test boxes [6]. In addition, often the datasets underlying published empirical validation studies are not available, so the validation can not be repeated with updated versions of the tools validated previously or other tools. This contribution presents a whole building empirical validation, including a publicly available dataset, based on detailed measurements and extensive documentation, including occupancy and different heating systems in a side-by-side configuration, allowing for direct comparison of different heating systems.

PRELIMINARY RESULTS AND CONCLUSIONS

The experiences from Annex 58 clearly show that the publicly available dataset itself is a very valuable outcome of such an experiment. Together the Annex 58 datasets [7] and [8] were downloaded about 300 times since their publication. The current Annex 71 dataset (including documentation) is also published [9] already. Both datasets are suitable for validation, training and educational purposes. This paper briefly summarises the detailed experimental specification and datasets recorded over a 4-month winter period for both electric and underfloor heating systems in two two-storey houses with realistic time–varying occupancy-related heat gains. The experiment consisted of several phases of increasing complexity. The paper will then show selected comparisons of the modelling predictions with the measured data, emphasising areas of good agreement, but also areas where agreement was poorer, indicating deficiencies in the modelling.

MAIN REFERENCES

[1] P. Strachan, K. Svehla, I. Heusler und M. Kersken, „Whole model empirical validation on a full-scale building," Journal of Building Performance Simulation, Bd. 9, Nr. 4, pp. 331-350, 09 2015.

[2] G. Flett und N. Kelly, „An occupant-differentiated, higher-order Markov Chain method for prediction of domestic occupancy," Energy and Buildings, Nr. 230, pp. 219-230, 2016.

[3] E. Mantesi, K. Mourkos, C. Hopfe, R. McLeod, P. Vatougiou, M. Kersken und P. Strachan, „Deploying Building Simulation to Enhance the Experimental Design of a Full-scale Empirical Validation Project," in Proceedings BuildingSimulation2019, Rome, Italy, 2019.

[4] P. Strachan, „Model validation using the PASSYS Test cells," Building and Environment, Bd. 28, Nr. 2, 1993.

[5] „Annex 58 - BES-Model Validation DATA: doi: 10.15129/8a86bbbb-7be8-4a87-be76-0372985ea228," [Online]. Available: https://pure.strath.ac.uk/portal/en/datasets/twin-houses-empirical-dataset-experiment-1%288a86bbbb-7be8-4a87-be76-0372985ea228%29.html.

[6] „Annex 58 - BES-Model Validation DATAset 2: doi: 10.15129/94559779-e781-4318-8842-80a2b1201668," [Online]. Available: https://pure.strath.ac.uk/portal/en/datasets/twin-houses-empirical-validation-dataset-experiment-2(94559779-e781-4318-8842-80a2b1201668).html.

[7] M. Kersken und P. Strachan, Twin House Experiment IEA EBC Annex 71 Validation of Building Energy Simulation Tools - Specifications and dataset - (DOI: 10.24406/fordatis/76) http://fordatis.fraunhofer.de/handle/fordatis/161, 2020.

Keywords: Empirical Validation, Validation Dataset, Heating Systems, Stochastic Occupancy

Paper ID 30356: A methodology for generating a synthetic local urban climate weather profile for building energy simulations in hot arid areas

Mohamed elnabawi Mahgoub* (1), Neveen Hamza (2)

1: Applied Science University, Bahrain; 2: Newcastle University, UK

AIM AND APPROACH

The paper examines the thermal performance of different shading designs for an urban alley in the hot arid context. Computational fluid dynamics (CFD) programs can provide detailed distributions and calculation of the air velocity and temperature distribution if all of the heat factors are set as dynamic boundary conditions and heat generators. However, this method is computationally very expensive and time demanding to provide predictions for a season or year. Alternatively, an integration of building energy modeling (BEM) and Computational fluid dynamics (CFD) can eliminate many of these assumptions, as the two models exchange the appropriate boundary conditions. Therefore, Designbuilder as BEM handles the external surface temperature for the main building surrounding the semi- enclosed areas, while Fluent as CFD simulates the street air flow and air temperature. Therefore, the paper main aim is to evaluate the operation of an alternative approach through the coupling between the Computational fluid dynamics (CFD) with the building energy modeling (BEM) in order to propose guidelines to mitigate the UHI effect and improve the urban microclimate. The numerical results were validated and compared to the experimental ones.

SCIENTIFIC INNOVATION AND RELEVANCE

• The coupling of building energy simulation and CFD simulations has been used to improve energy modeling performance (Zhai and Chen 2005; Tian and Zuo 2013).). As most of the available studies investigate the effect of outdoor environment on the building performance and indoor environment. However, the current paper examines the building performance on the outdoor microclimate and urban thermal comfort by calculating an outdoor comfort index

• Most studies assumed the reliability and accuracy of the building energy modelling (BEM) and Computational fluid dynamics (CFD) programs and did not conduct or offer a well validation result to support their findings (Zhai et al. 2002 and Tian et al, 2018). Few of the studies provided the validation of either program against experimental data (Du et al. 2015; Tian, Sevilla, Zuo, et al. 2017; Zuo, Wetter, et al. 2016) and small-scale experiments (Novoselac 2004; Wang 2007). Nevertheless, it needs future research efforts to develop small-scale experiments with high-quality measurements to validate the co-simulation (Tian et al, 2018). The current study provides a detailed validation based on the mean air temperature distribution, surface temperature and mean radiant temperature

PRELIMINARY RESULTS AND CONCLUSIONS

• The study has validated a coupled simulation assessment framework against data obtained from the field measurement. The validations verify that the process of can provide reasonable and reliable predictions for mean air temperature distribution for the urban alley and surface temperature for the surrounding buildings. the simulation outcomes showed a consistent trend with the observed ones and that the temperature differences are within 0.8°C to 1.2°C.

• Measurements and modelling of the outdoor thermal conditions has to be seen as a tool for urban planning and mitigation of conditions to increase utilization of outdoor urban spaces, especially in a context of globally increasing air temperatures and UHI in summer.

• The coupled simulation produces more accurate and detailed results than the separate simulations because; first, CFD receives more precise and real-time thermal boundary conditions and can predict the dynamic outdoor environment conditions that are important for the assessment of outdoor air quality and thermal comfort, while BES obtains more accurate convection heat from enclosures and can provide more accurate estimation of dynamic thermal behaviours of building.

MAIN REFERENCES

• Abdulrahman, A and Sharples, S (2014) 'Analysing and optimizing the performance of hybrid housing design in the Middle East'. In: 2nd IBPSA-England Conference, UCL.

• Allegrini, J., Carmeliet, J. (2018) Simulations of local heat islands in Zürich with coupled CFD and building energy models. Urban climate, 2018; 24: 340-359.

• Barbason, M., and Reiter, S., (2014) 'Coupling building energy simulation and computational fluid dynamics: Application to a two-storey house in a temperate climate',

• Blocken, B., Stathopoulos, T., Carmeliet, J. and Hensen, J.L.M. (2011) 'Application of computational fluid dynamics in building performance simulation for the outdoor environment: an overview', Journal of Building Performance Simulation, 4(2), pp. 157-184.

• Chao, Y., and Ng, E., (2012) 'Building porosity for better urban ventilation in high-density cities A computational parametric study', Building and Environment 50;176e189

• W. Tian. X. Han, W. Zuo, M. Sohn 2018. "Building Energy Simulation Coupled with CFD for Indoor Environment: A Critical Review and Recent Applications." Energy and Buildings, 165, pp.184-199.

• Yi, Y.K., Feng, N. Dynamic integration between building energy simulation (BES) and computational fluid dynamics (CFD) simulation for building exterior surface. Build. Simul. 6, 297–308 (2013).

Keywords: Computational fluid dynamics (CFD), building energy modeling (BEM), outdoor thermal comfort

Session T3.8 (Online Track): Climate change and bioclimatic design

Paper ID 30234: Impacts of climate change on mechanical ventilative cooling in high-rise buildings

Haohan Sha, Dahai Qi*

Université de Sherbrooke, Canada

AIM AND APPROACH

This study aims to investigate the impacts of climate change on the energy performance of mechanical ventilative cooling (VC) in high-rise buildings in Canada. Based on an optimal control strategy of VC, the investigation approach includes a hybrid model combining a machine learning model and a coupled ventilation and chiller cooling energy model. A high-rise institutional building located in Montreal is selected as a case study.

Using the building cooling load measured from the Building Automation System (BAS) in the case study, different machine learning models were trained to predict building cooling and compared. The most accurate model was selected. Meanwhile, the energy model of ventilative and chiller cooling that was validated by the BAS data was used to predict the energy consumption. With the application of the machine learning model and optimal control strategy, the energy savings of VC can be predicted by the energy model. The hourly future weather data in 2050s and 2080s were generated under two scenarios (mild and extreme climate change), and then used in the machine learning model to estimate the building cooling load. With future building cooling load, the future energy performance of the mechanical VC can be evaluated.

SCIENTIFIC INNOVATION AND RELEVANCE

There has been growing concern about climate change over the last decades. However, little research investigates the impacts of climate change on the energy performance of mechanical VC in high-rise buildings. The previous studies focus on the impacts of climate change on building energy consumption in different regions, e.g. Canada [1], China [2], U.S.[3]. etc. Considering the energy performance of VC can be affected by climate [4], climate change may influence the energy performance of mechanical VC. Therefore, the investigation of the energy performance of VC in future climates is required.

This study proposes a hybrid model that combines the machine learning technique and the energy model to predict the energy savings of VC in future climates. Machine learning techniques have been proved to accurately predict the building cooling load or energy consumption [5,6]. However, as a data-driven method, it is difficult for machine learning models to make predictions by changing their internal parameters [7], e.g. changing nominal fresh airflow rates of the mechanical ventilation system to reach VC. Through combining the coupled ventilation and chiller energy model, the hybrid model can predict the energy savings of VC under different settings, e.g. different nominal fresh airflow rates.

PRELIMINARY RESULTS AND CONCLUSIONS

The energy savings of VC under two climate change scenarios throughout the year were investigated. From the preliminary results, this study obtains four major conclusions.

1. Five different machine learning models were used to predict the building cooling load: linear regression, ridge regression, elastic net, support vector regression, and multilayer perceptron regression. Among these techniques, the multilayer perceptron regression model shows the highest accuracy to predict the building cooling load.

2. Based on the future hourly weather data of 2050s and 2080s, the available hours of VC (outdoor air conditions lower than the temperature and humidity setpoints) between June to September reduce, but the available hours of VC in shoulder season (March to May, October, and November) increase.

3. The energy savings of VC changes with climates. Generally, the changes in energy savings match the changes of the available hours of VC. The energy savings increase in the shoulder seasons (March to May, October and November). However, the energy savings decrease in the summertime (June to September).

4. There is an optimal nominal ventilation flow rates for energy savings of mechanical VC, and this optimal nominal ventilation flow rate does not change with climates.

MAIN REFERENCES

[1] U. Berardi, P. Jafarpur, Assessing the impact of climate change on building heating and cooling energy demand in Canada, Renew. Sustain. Energy Rev. 121 (2020) 109681.

[2] K.K.W. Wan, D.H.W. Li, W. Pan, J.C. Lam, Impact of climate change on building energy use in different climate zones and mitigation and adaptation implications, Appl. Energy. 97 (2012) 274–282.

[3] H. Wang, Q. Chen, Impact of climate change heating and cooling energy use in buildings in the United States, Energy Build. 82 (2014) 428–436.

[4] Y. Chen, Z. Tong, A. Malkawi, Investigating natural ventilation potentials across the globe: Regional and climatic variations, Build. Environ. 122 (2017) 386–396. https://doi.org/10.1016/j.buildenv.2017.06.026.

[5] C. Fan, F. Xiao, Y. Zhao, A short-term building cooling load prediction method using deep learning algorithms, Appl. Energy. 195 (2017) 222–233.

[6] C. Fan, Y. Sun, Y. Zhao, M. Song, J. Wang, Deep learning-based feature engineering methods for improved building energy prediction, Appl. Energy. 240 (2019) 35–45.

[7] K. Amasyali, N.M. El-Gohary, A review of data-driven building energy consumption prediction studies, Renew. Sustain. Energy Rev. 81 (2018) 1192–1205.

Keywords: climate change, ventilative cooling, machine learning

Paper ID 30490: Simulations beyond the building, identifying climate adaptation scale jumping potentials to district and city level. Research by design for the city of Monterotondo (Italy)

Maria Beatrice Andreucci*, Luciano Cupelloni, Fabrizio Tucci

Department of Planning, Design, Technology of Architecture, Sapienza University of Rome, Italy

AIM AND APPROACH

Centres for public and private services, as well as hubs for local and regional knowledge production, innovation and infrastructure, small and medium-sized cities (SMC) constitute the building blocks of metropolitan areas, and lend character and distinctiveness to their environmental, social and economic landscapes.

Several SMC are currently doing substantial efforts in adapting their 'urban envelope' to increase community resilience towards the impact of climate change.

On this basis, the objective of the proposed contribution is to illustrate an applied research case study regarding the effectiveness of simulation tools aimed at supporting climate adaptation design strategies at different scales for the 'satellite city' of Monterotondo, belonging to the metropolitan area of Rome. Critical issues and opportunities for smart energy development, along with the potential of renewable energy resources, sustainable transportation and modal shift will be presented and discussed, together with the detailed illustration of simulation results supporting integrated pilot actions in significant sectors, i.e. for the identification of the best operational methods in new and existing districts and neighbourhoods; the construction of new urban infrastructures, and the retrofit of old ones.

SCIENTIFIC INNOVATION AND RELEVANCE

The recently affirmed 'Smart Energy City', as a concept, ambitions, deployed technologies and programmes, rapidly gained traction in Monterotondo from 2012 onwards.

In 2016, the new Civic Tower was built, following the principle of sustainable design for construction, management and maintenance of buildings. Multiple advantages of the LEED 'Platinum' certified (2019) building originated more specifically from the implementation of bioclimatic design, sustainable water management solutions, the use of recycled or renewable materials, and a regenerative and self-sufficient production of energy.

Current priorities of Monterotondo's urban agenda regard the identification of scale jumping potentials to district and city level sustainability, and encompass testing and implementing state-of-the-art technologies and solutions for improving energy efficiency that supports both solutions and application development for citizen services. Among the various interventions which have been proposed and tested through this research by design: (i) a new eco-neighbourhood of 4 ha, planned to accommodate the growing population of the Roman hinterland; (ii) the revamping of the industrial area, expected to become a positive energy district, as well as an effective tool of propulsive local growth; (iii) the architectural and urban regeneration of the historical city centre; (iv) the adaptive re-use of the former bricks factories.

PRELIMINARY RESULTS AND CONCLUSIONS

The contribution will first introduce different simulation tools and the applications developed at multiple scales to support climate adaptive transformations for the city of Monterotondo. Subsequently, the expected results of the selected interventions will be presented in detail, and the multiple co-benefits for people and the environment will be discussed.

With the acknowledgement of value of local stakeholders, latest evidence of current approaches to scale jumping from green building to district and city level will be showcased.

Climate change adaptation and mitigation measures will be identified and presented, highlighting envisioned strategies and innovative solutions benefitting the whole community, within the framework of the broader UN Sustainable Development Goals.

Specific evidence will be given to evidence-based bioclimatic design as a trigger for smart communities, seeking to assess different urban systems vis-à-vis their multiple contribution to improving the quality of life of citizens, and of the overall urban built environment, in a climate change scenario.

The research, exploring different environmental technological solutions, and urban green blue infrastructure as an integrated design approach to urban regeneration at different scales, aims at contributing to the dissemination of the relevance and replicability of socio-technological innovations for a low-carbon future in small and medium-sized cities.

MAIN REFERENCES

Butturi, M.A., Lolli, F., Sellitto, M.A., Balugani, E., Gamberini, R., Rimini, B. (2019). Renewable energy in eco-industrial parks and urban-industrial symbiosis. A literature review and a conceptual synthesis. Applied Energy, 255, 113825.

Coccolo, S., Kampf, J., Scartezzini, J.L., Pearlmutter, D. (2016). Outdoor human comfort and thermal stress. A comprehensive review on models and standards. Urban Climate, 18, 33-57.

Monti, A. (2016). Energy Positive Neighborhood and Smart Energy Districts Methods, Tools, and Experiences from the Field. Cambridge: Academic Press.

Olgyay, V. (1962). Design with climate. Princeton, NJ, USA: Princeton University Press.

Park, S., Tuller, S.E., Jo, M. (2014). Application of Universal Thermal Climate Index (UTCI) for microclimatic analysis in urban thermal environments. Landscape and Urban Planning, 125, 146-155.

Roudsari, M., Pak, M., Ladybug (2013). A parametric environmental plugin for Grasshopper to help designers create an environmentally conscious design. Proceedings of BS2013: 13th Conference of the IBPSA, Chambery, France, August 26-28, 3128-3135.

Santamouris, M. (2014). On the energy impact of urban heat island and global warming on buildings. Energy and Buildings, 82, 100-113.

Wu, J. Zhang, Y., Meng, Q. (2013). Calculation method of sky view factor based on rhino-grasshopper platform. Proceedings of BS 2013: 13th Conference of the IBPSA, Chambéry, France, 258-266.

Keywords: Climate change adaptation, evidence-based design, adaptive reuse, environmental technological design, nature-based solutions

Paper ID 30622: Life cycle efficiency of solar shading systems: a proof-of-concept

Nazanin Rezaei Oghazi (1,2), Thomas Jusselme* (1), Elise Barri (1), Sergi Aguacil Moreno (3), Marilyne Andersen (2)

1: Energy Institute, University of Applied Science of Western Switzerland (HEIA-FR, HES-SO), Fribourg, Switzerland;; 2: Laboratory of Integrated Performance in Design (LIPID), EPFL Lausanne, Switzerland; 3: Building2050 group, Ecole polytechnique fédérale de Lausanne (EPFL), Fribourg, Switzerland

AIM AND APPROACH

Solar control techniques are essential for any bioclimatic approaches. Aside from their comfort benefits for the glare control, they are commonly implemented into buildings to decrease cooling needs. However, these techniques require embodied energy, and in some cases, they also consume operational energy during their use phase for their automation. Hence, the carbon neutrality objective at the 2050 horizon makes choosing solar control techniques based on the building operation phase only less and less relevant. A life cycle assessment approach would need to be used to quantify the overall energy and carbon performance. To that end, recent research has defined a new performance indicator: the Life Cycle Efficiency Ratio (LCER) [1], whose first application to solar control systems is the point of this paper. After a literature review that allowed to highlight previous research about the benefits of such systems in climatic contexts similar to Switzerland, their operational and embodied carbon impacts are being investigated by assessing different solar shading strategies and calculating their respective LCER. Finally, a Swiss low-carbon case-study, the smart living lab building in Fribourg, is used to evaluate the LCER more accurately and specifically within the Swiss context.

SCIENTIFIC INNOVATION AND RELEVANCE

Daylighting and carbon performances are two -still- independent and very active research topics. The first is crucial for user comfort, while the second is one of the major issues that humankind will have to face during the 21st century. However, assessing the carbon emissions of daylighting strategies has never been performed so far. This paper proposes a proof-of-concept methodology together with a case study, based on the solar control systems, to rank their performance according to daylight performance as well as carbon metrics. This preliminary study would be the first to propose a methodological framework that can hopefully be used for other building components affecting these two performance indicators. This work also gives valuable design guidance when it comes to choosing efficient solar control systems under carbon budget constraints. This contribution thus couples the climate change issue with solar shading, a fundamental bioclimatic strategy, by relying on building performance simulations with Energy Plus.

PRELIMINARY RESULTS AND CONCLUSIONS

The literature review pinpoints research on the impact of shading techniques on the operational energy consumption, while revealing that embodied impacts have generally been overlooked except for two papers. The first [2] concludes that an overhang system increases the CO2 emissions, due to the strong building materials used to reinforce the outdoor structures and the low latitude of Hong Kong, reducing the shading effect of external overhangs. The second [3] concludes that within the subtropical Brazilian climate, three out of four shading strategies had a positive impact on the overall building lifecycle, with the PVC roller shutter having the highest positive energy balance. These examples highlight the importance of a lifecycle approach with solar control systems that could have either positive or negative impacts. In the present study, based on a very different climate and environmental context, a preliminary analysis has been completed with the calculation of LCER for different shading strategies based on the operational benefits of literature-based case studies [4-6], that were all adapted to the Swiss context. The results show high discrepancies with both positive and negative impacts. Ongoing research using the case study will give a comprehensive understanding of solar control LCER within the Swiss context.

MAIN REFERENCES

[1] Brambilla, A., Bonvin, J., Flourentzou, F., Jusselme, T., 2018. Life cycle efficiency ratio: A new performance indicator for a life cycle driven approach to evaluate the potential of ventilative cooling and thermal inertia. Energy and Buildings 163, 22–33. https://doi.org/10.1016/j.enbuild.2017.12.010

[2] Huang, Y., Niu, J., Chung, T., 2012. Energy and carbon emission payback analysis for energy-efficient retrofitting in buildings—Overhang shading option. Energy and Buildings 44, 94–103. https://doi.org/10.1016/j.enbuild.2011.10.027

[3] Invidiata, A., Ghisi, E., 2016. Life-cycle energy and cost analyses of window shading used to improve the thermal performance of houses. Journal of Cleaner Production 133, 1371–1383. https://doi.org/10.1016/j.jclepro.2016.06.072

[4] Nielsen, M.V., Svendsen, S., Jensen, L.B., 2011. Quantifying the potential of automated dynamic solar shading in office buildings through integrated simulations of energy and daylight. Solar Energy 85, 757–768. https://doi.org/10.1016/j.solener.2011.01.010

[5] Tzempelikos, A., Athienitis, A.K., 2007. The impact of shading design and control on building cooling and lighting demand. Solar Energy 81, 369–382. https://doi.org/10.1016/j.solener.2006.06.015

[6] Brotas, L., Rusovan, D., 2013. Parametric Daylight Envelope. Presented at the PLEA 2013 : Proceedings of the 29th Sustainable Architecture for a Renewable Future Conference, Munich, Germany, p. 6.

Keywords: solar control, life cycle efficiency, carbon emissions

Paper ID 30627: Building thermal simulation-based climate classification of India

Manoj Kumar Sundaram*, Nusrat Ali, Rajasekar E

IIT Roorkee, India

AIM AND APPROACH

Climate classification of a region is used for various applications based on parameters like temperature, precipitation, vapour pressure, solar radiation, and seasonality. For building energy efficiency, it is one of the main steps that forms the basis. There must be a climate classification intended specifically for building energy efficiency since older classifications like Köppen Geiger was developed for agriculture and other related applications. A dedicated climate zone map will help in building codes and energy guidelines development. To depict the relation between climate and building energy performance combination of climate clustering with building performance simulation provide robust tools (Walsh et al., 2017). This paper aims to develop a climate classification map for BEE purpose based on building energy simulation results derived across 50km x 50km grid location across 1157 points covering the Indian peninsular. The daily discomfort hours will be used to cluster climate zones with GIS tools. The derived climate cluster can be used to create bioclimatic design guidelines for buildings.

SCIENTIFIC INNOVATION AND RELEVANCE

This study will develop a climate zone map of India which will be used exclusively for bioclimatic building design. The results will help in formulating climate zone specific guidelines for new and existing building design. Existing Indian climate classifications are performed through supervised techniques. One of the classifications currently being in use for the National Building Code of India (2005) subdivides the country into five zones based on monthly maximum temperature and relative humidity only. Though the classification is meant for building design the number of zones is less as it fails to address climatic variations throughout the country. This is especially demonstrated in the “composite” climate zone that encompasses a huge expanse of lowland areas, which is otherwise proven in several studies to have climate differences (Naveen Kishore and Rekha, 2018), (Khambadkone and Jain, 2017), (Rijildas and Rajasekar, 2016)

For fewer cluster groups in a smaller area country, percentage mean misclassified areas were used in a study to identify overlap of discomfort hours (Walsh et al., 2018). Another study used building heating and cooling load to validate a hierarchical clustering (Xiong et al., 2019). The paper will attempt to create classification based on the building simulation results.

PRELIMINARY RESULTS AND CONCLUSIONS

A building energy simulation are done for a case government approved naturally ventilated residential building model in energy plus for 1157 locations (50x50 Km grid) using TMY data of past 30 years till 2010 extracted from Meteonorm. The case building is a two-bedroom apartment flat. The buildings' windows are operated based on temperature control for naturally ventilation. The roof, west wall and floor are considered adiabatic. The degree discomfort hours (DDH) is extracted by the difference of zone air temperature from base neutral temperature calculated using running mean zone temperature for the whole year in all locations. Python based data management code is created to extract the expected output of cooling discomfort hours and heating discomfort hours. Further the results will be geospatially mapped across the locations and analysed to create cluster zones using GIS techniques to create a climate zone cluster map for India. This zone map will represent regions where specific design strategies can be used to achieve thermal comfort for occupants with low energy demand.

MAIN REFERENCES

Walsh A, Cóstola D and Labaki LC (2017) Review of methods for climatic zoning for building energy efficiency programs. Building and Environment 112: 337–350. DOI: 10.1016/j.buildenv.2016.11.046.

Walsh A, Cóstola D and Labaki LC (2018) Performance-based validation of climatic zoning for building energy efficiency applications. Applied Energy 212(December 2017). Elsevier: 416–427. DOI: 10.1016/j.apenergy.2017.12.044

Rajasekar E, Thakur G and Zeleke B (2020) Thermal Adaptation and Sustainable Housing in Cold Climates. In: Encyclopedia of Renewable and Sustainable Materials. Elsevier Ltd., pp. 244–258. DOI: 10.1016/b978-0-12-803581-8.11412-2.

Naveen Kishore K and Rekha J (2018) A bioclimatic approach to develop spatial zoning maps for

comfort, passive heating and cooling strategies within a composite zone of India. Building and

Environment 128(November 2017): 190–215. DOI: 10.1016/j.buildenv.2017.11.029.

Xiong J, Yao R, Grimmond S, et al. (2019) A hierarchical climatic zoning method for energy efficient building design applied in the region with diverse climate characteristics. Energy and Buildings 186: 355–367. DOI: 10.1016/j.enbuild.2019.01.005.

Keywords: Climate Zone, Thermal simulation, DDH, Guidelines

Paper ID 30691: Evaluation of daylighting strategies based on their embodied carbon emissions: a first methodological framework and case study

Nazanin Rezaei Oghazi (1,2), Thomas Jusselme* (2), Marilyne Andersen (1)

1: Laboratory of Integrated Performance in Design (LIPID), Ecole Polytechnique Fédérale de Lausanne (EPFL), Switzerland; 2: Energy Institute, University of Applied Science of Western Switzerland (HEIA-FR, HES-SO), Fribourg, Switzerland

AIM AND APPROACH

The impact of daylighting strategies on a building's carbon emissions have been assessed mostly based on the building's use phase and their operational benefits so far, thereby overlooking embodied carbon emissions linked to material production, construction, maintenance and end of life. The emergence of Life Cycle Assessments (LCA) is likely to shift the focus from operational energy impacts to the life-cycle carbon impacts. Applying Design Space Exploration (DSE) to LCA has already demonstrated a great potential to mitigate carbon emissions of future buildings early in the design process. However, there is no methodological framework yet for early-stage DSE applicable to both daylight and life-cycle carbon performance. Such a framework would allow designers to better understand the relationships between these performance indicators, which would promote low-carbon daylighting solutions. From a careful literature review, we were able to identify some of the key challenges ahead if we want to apply DSE for LCA and for daylight together, as well as some promising techniques that may be able to address them. As an initial step, we need to assess the feasibility of their combination through a proof-of-concept investigation using a case study: a residential unit for the Solar Decathlon Europe (SDe) 2021 competition.

SCIENTIFIC INNOVATION AND RELEVANCE

This paper aims to address the issue of reaching carbon neutrality objectives together with human comfort by coupling two inherently different performance criteria i.e. CO2 emissions and daylight. The is dependent on material quantity and type while the former is a spatial and dynamic phenomenon.

However, a review of the literature shows that the embodied carbon emissions of daylighting strategies, indirectly but strongly influenced by the material choices and by adopted construction techniques, have not really been assessed so far. Accordingly, this paper presents a first, proof-of-concept method which aims to rank daylighting strategies based on their life-cycle carbon emissions. The approach consists of providing a multi-criteria platform to co-assess adequate illuminance as well as carbon emissions of design options. To keep the number of degrees of freedom reasonable, it relies on the static the Daylight Factor (DF) metric and on the Global Warming Potential (GWP). The proposed method shall provide useful insights about the interactions between these two performance indicators such as, for instance, how much the GWP of given design options may change when the DF objective is increased from 1 to 2%.

PRELIMINARY RESULTS AND CONCLUSIONS

The critical review of previous studies suggest that the application of DSE for daylight performance assessment is a time-consuming and computationally expensive process due to the following reasons: (a) the spatial nature of daylighting simulations and thus number of combinations to assess, further increased in the case of dynamic simulations i.e. for climate-based metrics [1], [2], (b) the high number of design parameters in DSE when addressing both daylight and carbon performance [3], [4] and (c) the wide range values that design parameters may ultimately have when considered at early design phases [5]. To cope with the aforementioned challenges, we proposed to adapt DSE with two complementary techniques i.e. sensitivity analysis and Artificial Neural Networks (ANN). To conduct a DSE, we took the advantage of interoperability of BIM and Building Performance Simulation (BPS) tools i.e. EnergyPlus and DIVA with a visual scripting program i.e. Grasshopper and data visualization techniques, namely, parallel coordinate plots. Based on the ongoing investigations on the façade variations for SDe 2021 design brief, we can identify which daylighting strategy is likely to have the lowest carbon emissions, which design parameter may have the highest influence and how an increase in daylight performance may affect the GWP.

MAIN REFERENCES

[1] C.-L. Lorenz, A. B. Spaeth, C. Bleil de Souza, and M. S. Packianather, "Artificial Neural Networks for parametric daylight design," Archit. Sci. Rev., pp. 1–12, 2019.

[2] H. Samuelson, S. Claussnitzer, A. Goyal, Y. Chen, and A. Romo-Castillo, "Parametric energy simulation in early design: High-rise residential buildings in urban contexts," Build. Environ., vol. 101, pp. 19–31, 2016.

[3] T. Østergård, R. Jensen, and S. E. Maagaard, "Early Building Design: Informed decision-making by exploring multidimensional design space using sensitivity analysis," Energy Build., vol. 142, Mar. 2017, doi: 10.1016/j.enbuild.2017.02.059.

[4] T. Jusselme, E. Rey, and M. Andersen, "Data-driven method for low-carbon building design at early stages," École Polytechnique Fédérale de Lausanne (EPFL), 2020.

[5] R. Rezaee, J. Brown, J. Haymaker, and G. Augenbroe, "A novel inverse data driven modelling approach to performance-based building design during early stages," Adv. Eng. Inform., vol. 41, p. 100925, 2019.

Keywords: daylight, carbon emissions, Design Space Exploration, early stages of design

Paper ID 30765: Assessment of a weather-based climate classification with building energy simulation

Giovanni Pernigotto* (1), Andrea Gasparella (1), Jan LM Hensen (2)

1: Free University of Bozen-Bolzano, Italy; 2: TU Eindhoven - Eindhoven, The Netherlands

AIM AND APPROACH

According to the Energy Performance of Buildings Directive (EPBD) 2018/844/EU, a new high-performance and smart building is supposed to "maintain energy performance and operation [..] through the adaptation of energy consumption", "adapt its operation mode in response to the needs of the occupant" and show "flexibility and load shifting capacities". The design of such a building, however, is not able to leave the external weather conditions out of consideration. Furthermore, the same holds also at a larger scale, for example when it comes to define expected targets at national or European level achievable from new building constructions and building energy renovations according to the proposed European paradigm. In this context, climate classification can play a significant role to minimize errors and improve representativeness. However, different approaches can be implemented (Bai et al., 2020; Walsh et al., 2017; Yang et al., 2020).

SCIENTIFIC INNOVATION AND RELEVANCE

In this research, we analysed a weather-based method for climate classification proposed by the authors in previous studies (Pernigotto and Gasparella, 2018; Pernigotto et al., 2019), with the goal of assessing strong points and weaknesses in the framework of building energy performance analyses at European scale. The approach is based on the evaluation through statistical indicators of some variables descriptive of the weather quantities most relevant for building energy performance (i.e., dry bulb temperature, water vapour partial pressure, global horizontal solar irradiance), processing these inputs through hierarchical clustering, and selecting the representative locations for each cluster by means of the Kolmogorov-Smirnov test.

As a next step, in the present research, we aimed to see if robustness and representativeness can be achieved not only as far as the weather quantities are concerned but also in terms of simulated building energy performance, similarly to other approaches in the literature (Walsh et al., 2018).

PRELIMINARY RESULTS AND CONCLUSIONS

The proposed methodology gave results substantially different from the conventional classification systems adopted worldwide, such as the Köppen-Geiger classification (Peel et al., 2007) or the heating degree-days systems (ASHRAE, 2020). In particular, homogeneity of classes was found improved and overlapping issues removed. This allowed also to improve the quality of the statistical properties of the sample of climates and, thus, the robustness of the selection procedure for representative locations.

As a whole, the proposed weather-based clustering can be considered a “general purpose” climate classification, which can be applied to multiple contexts related to building simulation. While it proved to be suitable when heating demand is concerned, the presence of redundant results makes the computational effort suboptimal in case of the evaluation of cooling demand. Consequently, if the aim of a researcher or a policy maker is to assess only the energy performance for cooling at European scale, simplifications can be applied. On the contrary, when the focus is put on heating or global energy demand, the adoption of representative climates obtained through the proposed methodology can be considered satisfactory.

MAIN REFERENCES

ASHRAE. (2020). ANSI / ASHRAE Standard 169-2020, Climatic Data for Building Design.

Bai, L., Yang, L., Song, B., & Liu N. (2020). A new approach to develop a climate classification for building energy efficiency addressing Chinese climate characteristics. Energy 195, 116982.

Pernigotto, G., & Gasparella, A. (2018). Classification of European Climates for Building Energy Simulation Analyses. Proceedings of the V International High Performance Buildings Conference at Purdue. West Lafayette (Indiana, U.S.), 9-12 July 2018.

Pernigotto, G., Walsh, A., Gasparella, A., & Hensen, J.L.M. (2019). Clustering of European Climates and Representative Climate Identification for Building Energy Simulation Analyses. Proceedings of Building Simulation 2019. Rome (Italy), 2-4 September 2019.

Walsh, A., Cóstola, D., & Labaki, L.C. (2017). Review of methods for climatic zoning for building energy efficiency programs. Building and Environment 112, 337-350.

Walsh, A., Cóstola, D., & Labaki, L.C. (2018). Performance-based validation of climatic zoning for building energy efficiency applications. Applied Energy 212, 416-427.

Yang, L., Lyu, K., Li, H., & Liu, Y. (2020). Building climate zoning in China

using supervised classification-based machine learning. Building and Environment 171, 106663.

Keywords: climate classification, weather, climate clustering, representative climates

Paper ID 30782: Development of the thermal performance index for office building envelope in Asian hot-humid climate zones considering global warming

Sheng-Fen Chang (1), Weian Chen* (1), Ruey-Lung Hwang (1), Wen-Mei Shih (1), Kuo-Tsang Huang (2)

1: National Kaohsiung Normal University, Kaohsiung, Taiwan; 2: National Taiwan University, Taipei, Taiwan

AIM AND APPROACH

The Taiwan-EEWH green building certification system is dedicated for the hot-humid climate region and had launched the overseas version (EEWH-OS) in 2016[1]. However, when promoting to other countries, especially ASEAN countries, there are two problems inherent in the evaluation of building envelope performance should be solved. Firstly, take office buildings as an example, the ENVLOAD in Taiwan and the OTTV in some ASEAN countries are developed on the basis of the local climate that the formulas cannot be directly applied to other countries[2-7]. Secondly, these formulas were all established based on the weather data before 2000; however, due to the trend of climate change becomes dramatic recently, it is essential to adjust the formulas through the predicted future weather data. A total of 360 office building models were generated by Latin Hypercube Sampling. Building energy simulation program, EnergyPlus, was used to predict the cooling loads for these office building models that located in 8 cities within five timelines: current(2010s), 2030s, 2050s, 2070s and 2090s under the scenarios of RCP2.6, RCP4.5, RCP8.5. Specifically speaking, Taipei, Hong Kong, Hanoi, Bangkok, Kuala Lumpur, Singapore, Bandar Seri Begawan and Manila are selected as the target cities.

SCIENTIFIC INNOVATION AND RELEVANCE

In order to eliminate the problem that resulted from the climate parameters inherent in the ENVLOAD or OTTV formulas, this paper standardize the impact of climate conditions on the transmitted cooling load through building envelops by the performance-rating-based approach., In addition, a new index of envelope thermal performance, named PR is developed for commercial building in hot-humid climate region in Asia as well. Instead of ENVLOAD, using PR as the performance index of building envelope in the Taiwan-EEWH can not only contributing to the promotion of EEWH-OS, but assist architects in adjusting the design of the building envelope to better deal with climate change.

PRELIMINARY RESULTS AND CONCLUSIONS

By using the multiple regression analysis, the unified PR formula for Asian hot-humid climate region under current timeline is $PR=7.9\cdot(1-WWR)\cdot Uw+3.3\cdot WWR\cdot Uf+114.0\cdot WWR\cdot SC\cdot CF-18.2$. The performance rating (PR) is the percentage of the proposed building's annual sensible cooling load in the perimeter area that exceeds a specific baseline building. The PR benchmark value which corresponding to the in force ENVLOAD benchmark value for passing the EEWH-certification is 30%. According to the building locations and RCP scenarios, the simulation results show that sensible cooling loads of the benchmark building for EEWH-certification increase by 2%-7% for 2030s, 5%-15% for 2050s, 5%-24% for 2070s, and 5%-32% for 2090s. To maintain the EEWH-certificated buildings for the same sensible cooling load as in the 2010s, the strict benchmark for PR values have been formulated for the four timelines mentioned above, which are respectively 22%-25%, 17%-21%, 12%-18% and 9%-15%. The results of sensitivity analysis show that WWR (window to wall ratio), SC1 (shading coefficient of glass) and SC2 (shading coefficient of exterior shading device) make the greatest influence on PR. Besides, the significance of SC1 and SC2 decrease over time, and Uw (U-value of external walls) and Uf (U-value of glass) increase.

MAIN REFERENCES

[1] Architecture and Building Research Institute. Ministry of the Interior. Evaluation manual for green building in taiwan.2015 OS enition,2015.(In Chinese).

[2] Architecture and Building Research Institute. Ministry of the Interior. Evaluation manual for green building in taiwan.2015 BC enition,2015.(In Chinese).

[3] Authority, B. (1995). Code of Practice for Overall Thermal Transfer Value in Buildings 1995. Building Authority, Hong Kong.

[4] Chou, S. and Y. Lee (1988). "A simplified overall thermal transfer value equation for building envelopes." Energy 13(8): 657-670.Dürrenmatt DJ, Wanner O.

[5] Chow,W. K. and Chan, K. T., 1995. Parameterization study of the overall thermal transfer value equation for buildings, Applied Energy, 50 (1995): 247-268.

[6] Guidelines for Energy Efficiency in Buildings. (1989, December). Ministry of Energy, Telecommunications and Posts, Malaysia.

[7] Handbook on Energy Conservation in Buildings and Building Services. (1979). Building and Construction Authority, Public Works Department, Singapore.

Keywords: envelope thermal performance index, climate change, hot-humid climate, green building

Paper ID 30930: Solutions for efficient and environmental friendly cooling of data centers

Andrea Rocchetti* (1), Luca Socci (1), Giovanni Pierazzuoli (2)

1: University of Florence, Department of Industrial Engineering, Italy; 2: Seeley International Europe

AIM AND APPROACH

Starting from the 2000s, demand for data processing has grown dramatically. This request has resulted in an increasing demand of electricity for data centers, characterizing them as "hotspots" having an high environmental impact. Electricity in data centers is used not only for Information and Technology equipment, but is also dedicated to the cooling system. Cooling system has to dissipate thermal load generated by servers (temperature control), and also regulate humidity. Cooling system are divided into "liquid systems" and "air systems". In air system the cooling air entering the servers has to respect some thermo-hygrometric parameters: dry bulb temperature, dew point temperature, relative humidity (1) (2). In our work we would like to demonstrate that the simultaneous use of direct air side free cooling, air distribution systems (ADS) dedicated to servers and technologies based on indirect evaporative cooling leads to electricity savings, so to a decrease in environmental impacts. A case study has been investigated to test our numerical simulation.

SCIENTIFIC INNOVATION AND RELEVANCE

The realization of a controlled air distribution system is fundamental. In a traditional ADS there is a mixing between cold air from the cooling system and the hot air exiting the server. In a dedicated ADS, the cooling air is sent directly to server level, and the warm air is taken near to exit of servers (3).

The use of free cooling is very important to achieve reductions of electricity consumption. It's definitely convenient to use outdoor air as "cooling power carrier" if it has correct values of temperature and humidity (4). Outdoor air could be mixed with recirculated air to adjust thermo-hygrometric parameters.

The implementation of indirect evaporative cooling alongside traditional cooling technologies (direct expansion or chilled water coil) could be very interesting. A part of the design air flow can be treated by an indirect evaporative cooler (IEC): the air enters in the "dry channel" (primary air) where it's cooled at constant specific humidity; at the exit, part of the air is taken back (secondary air) and enters in the "wet channel", where is cooled and humidified by the water. So, the secondary air can cool the primary ones thanks to a heat exchange surface.

PRELIMINARY RESULTS AND CONCLUSIONS

With a dedicated ADS, we can work with higher temperature for the cooling air (20-25 °C, instead 13-18 °C) with a fall in the design air flow (about 40 %), and a consequent reduction in the consumption of energy for cooling.

Rising the maximum temperature (until a certain limit) at which we accept the utilization of outdoor air leads to a decrease in electric consumption. As a result of our analysis, optimization in free cooling temperature levels can halve the consumption of cooling chiller.

Another saving is achievable with an IEC. This kind of cooler requires a minimum quantity of electricity (is based on water consumption) and the saving depends by the percentage of design air flow treated by the cooler. In our study, we have found a saving variable between 10 and 20%, depending by the simulation scenario.

In addition, an IEC system has good environmental performances if we evaluate it from the Water-Energy Nexus perspective (electric energy saving leads also to water resource saving).

This system could be improved using a desiccant system upstream of the IEC: if the air is drier, IEC is more effective and can treat a higher percentage of the air flow.

MAIN REFERENCES

1. Capozzoli A, Primiceri G, Cooling systems in data centers: state of art and emerging technologies, Energy Procedia,Vol 83, 2015, pp 484-493, https://doi.org/10.1016/j.egypro.2015.12.168

2. Ebrahimi K, Jones G F, Fleischer A S, A review of data center cooling technology, operating conditions and the corresponding low-grade waste heat recovery opportunities, Renewable and Sustainable Energy Reviews, Vol 31, 2014, pp 622-638, https://doi.org/10.1016/j.rser.2013.12.007

3. Neil R, The different types of air distribution for IT environments, Schneider Electric – Data Center Science Center, White Paper 55 Rev 4, 2017,

4. Zhang H, Shao S, Xu H, Zou H, Tian C, Free cooling of data centers: A review, Renewable and Sustainable Energy Reviews, Vol 35, 2014, pp 171-182, https://doi.org/10.1016/j.rser.2014.04.017

Keywords: energy nexus approach, low consumption cooling, data center.

Paper ID 31046: Building performance simulation as a guide to design decision making: an analysis of architecture students' design process.

Larissa Pereira de Souza (1,2,3), Roberto Lamberts* (1,2), Carlos Eduardo Verzola Vaz (1,3)

1: Federal University of Santa Catarina; 2: Laboratory of Energy Efficiency in Buildings; 3: Advanced Modeling Group

AIM AND APPROACH

This research aims to analyze the design process of four groups of architecture students using building performance simulation to assist in design decision making. To do so, they learned how to use two Ladybug tools, in which they evaluated the site location potentials with Ladybug and then simulated the building performance with Honeybee. In the exercise, they had a simple one-room residential volume with building surroundings. They could change building parameters such as building's dimensions, materials, colors, orientations, and location on-site, as well as windows' dimensions, materials, and location on building to improve building performance. Each group received a different city with a very different climate, but the baseline simulation file and the end report were the same for each group. First, they analyzed the site location potentials regarding bioclimatic strategies (such as sun and wind availability). Then they performed a reference simulation in which they made no changes in the building to start analyzing and making decisions in design. Every later analysis, proposal, and changes in design were described before and after each simulation.

SCIENTIFIC INNOVATION AND RELEVANCE

Many architects agree that building performance simulation should be embedded in the design process. Unfortunately, thermal physics and design teaching are, in most cases, two separate courses with different teaching methods in architectural education. It results in segregated knowledge, and that leads to non-concurrent project development. Without the knowledge, building designers rarely apply bioclimatic strategies and test if it leads to the desired effect in the building. It is why most building performance analyses are carried out in the final stages of design when better changes to improve building performance are not available. Architectural education could overcome this problem by testing new ways and software that could be more effective to be part of the design process. Over the past few years, some researchers developed teaching experiences to combine performance simulations and design. In the end, they reinforce that there is no effective method for teaching simulation to architects. That said, the need for more experiments in the area is essential for its development.

PRELIMINARY RESULTS AND CONCLUSIONS

The study had a total of 17 groups, in which four of them were chosen to be analyzed in the present study. The groups had cities located in different climate zones by the Koppen-Geiger climate classification. The cities chosen were Cairo, located in a hot desert climate; Petrolina, which is considered a hot semi-arid climate; Warsaw, located in an oceanic climate; and Chicago, located in a humid continental climate. All groups used the relationship between adaptative comfort improvement and HVAC consumption decrease as a guide to evaluating the changes made in the design. Except for the group studying the Warsaw city, the other three groups reached near 80% of adaptative comfort at the end of the project by reducing more than 70% of cooling (first group) or heating (second group) consumption. Also, all groups needed from 4 to 6 design changes to reach such a design configuration. The main differences between groups were de changes in the design process. While the first three groups started by changing orientation and site location dimensions to block or use the sun with the surroundings, the Chicago group improved the materials first, which they described as the most relevant change in a cold climate.

MAIN REFERENCES

AIA, The American Institute of Architects. 2019. Architect's Guide to Building Performance - Integrating Performance Simulation in the Design Process. http://content.aia.org/sites/default/files/2019-06/Energy_Design_Modeling_Guide_v4.pdf.

Alsaadani, Sara, and Clarice Bleil De Souza. 2019. "Performer, Consumer or Expert? A Critical Review of Building Performance Simulation Training Paradigms for Building Design Decision-Making." Journal of Building Performance Simulation 12(3): 289–307. https://doi.org/10.1080/19401493.2018.1447602.

Beausoleil-Morrison, Ian. 2019. "Learning the Fundamentals of Building Performance Simulation through an Experiential Teaching Approach." Journal of Building Performance Simulation 12(3): 308–25. https://doi.org/10.1080/19401493.2018.1479773.

Bleil De Souza, Clarice. 2013. "Studies into the Use of Building Thermal Physics to Inform Design Decision Making." Automation in Construction 30: 81–93. http://dx.doi.org/10.1016/j.autcon.2012.11.026.

Bleil de Souza, Clarice, and Simon Tucker. 2015. "Thermal Simulation Software Outputs: A Framework to Produce Meaningful Information for Design Decision-Making." Journal of Building Performance Simulation 8(2): 57–78. https://doi.org/10.1080/19401493.2013.872191.

Shi, Xing et al. 2016. "A Review on Building Energy Efficient Design Optimization Rom the Perspective of Architects." Renewable and Sustainable Energy Reviews 65: 872–84. http://dx.doi.org/10.1016/j.rser.2016.07.050.

Tucker, Simon, and Clarice Bleil de Souza. 2016. "Thermal Simulation Outputs: Exploring the Concept of Patterns in Design Decision-Making." Journal of Building Performance Simulation 9(1): 30–49. https://doi.org/10.1080/19401493.2014.991755.

Keywords: Rhinoceros Grasshopper, Thermal simulation design, Teaching building simulation, Architectural education, BPS training

Paper ID 31052: Dynamic thermal performance of structurally optimized concrete floor slabs

Eduardo Gascon Alvarez*, Caitlin T. Mueller, Leslie K. Norford

MIT, United States of America

AIM AND APPROACH

As energy efficiency in buildings becomes increasingly urgent in the context of the global climate crisis, opportunities for considering interaction and integration of design strategies and systems are emerging as an important area of research. This paper specifically looks at the interaction of concrete floor systems in buildings as both structural systems with embodied energy and thermal systems with the potential to reduce operational energy consumption. Previous research has shown that shaping floor slabs through structural optimization can lead to substantial embodied energy reductions compared to typical flat slabs, resulting in ribbed structures that distribute carbon-intensive material in an efficient manner [1].

The research presented here offers new methodologies and results related to the dynamic thermal behavior of such shaped slabs, and specifically considers the ability of the thermal mass and geometric shape of the slab to store heat and offset energy demands related to cooling. In order to quantify these benefits, an "equivalent thermal slab" model is proposed. This model simplifies the complex three-dimensional heat transfer behaviour of a shaped slab into a one-dimensional flat slab that can be input into whole-building simulation engines such as EnergyPlus to quantify energy savings potential.

SCIENTIFIC INNOVATION AND RELEVANCE

One of the main innovations of our work is the application of the mentioned "equivalent thermal slab" model. Through thermal FEA simulations, the "exchanged energy" metric is computed for a given geometry and boundary conditions. This metric represents the thermal storage capacity of a specific slab for a 24h period and is used as a measure of equivalence between shaped and flat slabs. CFD simulations allow validating this assumption by accounting for the solid-fluid interactions. These simulations become necessary to fully understand how the slab geometry affects the heat exchange with the surrounding air and surfaces through convection and radiation. Recent literature review papers show how the surface heat transfer coefficient can substantially vary depending on the room conditions, especially for convection [2] .

Our work, similarly to recent research on the coupling of thermal mass and ventilation [3], understands that it is particularly relevant to study the thermal potential of structural elements as a path towards a more efficient building industry. Previous research on dynamic thermal performance of building components have focused on multilayer wall assemblies and the differences between lightweight and massive solutions [4] but no similar work has been found for slabs.

PRELIMINARY RESULTS AND CONCLUSIONS

Our assessments of the benefits of thermal mass on thermal comfort and operational energy have been informed by studies that span structural optimization, heat transfer and building physics.

At a component level, our first FEA simulations show how decreasing the distance between fins - i.e. increasing the exposed surface - translates into higher exchanged energy values. According to research on extended surfaces with fins [6], there is an optimal spacing below which performance drastically drops due to airflow restrictions. We expect to find this value and other geometrical correlations with our ongoing CFD simulations.

Also, early design explorations reveal the possibility of finding shaped slabs that have a 40% embodied energy reduction with respect to its thermally equivalent flat slab. This becomes particularly relevant considering that previous work on lifecycle analysis of thermal mass usually accounts for an initial surplus of embodied energy that is compensated with the operational savings throughout the building's lifespan [5].

Finally, our first whole-building simulations show that up to 11% cooling operational energy reductions can be achieved while still accounting for structurally optimized solutions. These savings are climate-dependent and are being obtained on a case-by-case basis.

MAIN REFERENCES

[1] Ismail, Mohamed and Mueller, C. "Minimizing Embodied Energy of Reinforced Concrete Floor Systems in LEDCs through Shape Optimization," Engineering Structures, 2020 (under review).

[2] Shinoda, Jun, et al. "A review of the surface heat transfer coefficients of radiant heating and cooling systems." Building and Environment 159 (2019): 106156.

[3] Craig, S. The optimal tuning, within carbon limits, of thermal mass in naturally ventilated buildings. Building and Environment 165 (2019) 106373

[4] Kosny, Jan, et al. "Dynamic thermal performance of concrete and masonry walls." Proceedings, Thermal Performance of the Exterior Envelopes of Buildings VII (1998): 629-643.

[5] Hacker, J. N., De Saulles, T. P., Minson, A. J., & Holmes, M. J. Embodied and operational carbon dioxide emissions from housing: A case study on the effects of thermal mass and climate change. Energy and Buildings 40(3) (2008) , 375–384.

[6] Jones, Charles D., and Lester F. Smith. "Optimum arrangement of rectangular fins on horizontal surfaces for free-convection heat transfer." J. Heat Transfer(1970): 6-10.

Keywords: Thermal mass, low carbon, dynamic thermal performance, structural optimization

Paper ID 31124: Simulation of the local urban climate and its mitigation during heat waves

Dominique Derome* (1), Aytaç Kubilay (2), Dominik Strebel (2), Andreas Rubin (2), Jan Carmeliet (2)

1: Université de Sherbrooke, Canada; 2: ETH Zurich, Switzerland

AIM AND APPROACH

Buildings in an urban setting are exposed to conditions that are markedly different from stand-alone buildings and building in rural areas. The presence of adjacent buildings affects wind patterns, exposure to wind-driven rain, solar radiation and shading, and more. Further, the urban heat island effect is compounded by climate change and has led to untenable living situations in cities undergoing a heat wave (e.g. 50 deaths in Montreal 2018 heat wave).

We expanded and validated a wind and wind-driven rain model obtaining a modeling suite that can capture all local urban physics, such as solar and thermal radiation, heat and mass transport in the materials lining the urban environment and in vegetation, including photosynthesis and opening of stomata. The model suite is operational with the couplings between the sub-models validated. This model suite allows to study the local climate in neighborhoods, under varying weather conditions, and determine the thermal comfort of pedestrians (Universal

Thermal Comfort Index). This modeling framework is used to assess which climatic loads to be used for long-term prediction of the building envelope hygrothermal performance.

Here we present the capacity to consider different mitigation solutions and consider their impacts on local urban environments undergoing a heat wave.

SCIENTIFIC INNOVATION AND RELEVANCE

Using this newly developed computational framework, studying the urban climate in its complex geometry and physics leads to new insights that can support the design of mitigation solutions. We present three avenues of investigation.

Vegetation plays a fundamental role in regulating urban environments. The incorporation of vegetation, such as trees and grass, is performed using a porous medium approach. As air flows through the foliage, a multiscale approach, the heat and moisture balance at all leaves is modelled. Evapotranspiration, shading and wind sheltering effects of vegetation on urban climate can be evaluated. We present examples of the effects of the trees and grass areas and an in-depth assessment of the impact of trees on local climate and comfort.

Porous materials are usually lining the urban environment. The incorporation of different evaporative cooling pavement solutions is performed with a parametric study varying the composition and properties of pavements. We look at different scenarios for the artificial wetting of porous pavement and the optimization of wetting events.

The geometry of the urban environments, different street canyons, different orientations of street canyons and squares, yields different wind patterns, solar radiation loads, rain deposition, etc. which combine in different urban climate profiles.

PRELIMINARY RESULTS AND CONCLUSIONS

The recently developed simulation framework has the capacity and potential to study the complex interactions of urban local environments. We present here different mitigation solutions for urban heat island effect and heat waves, assessed in terms of pedestrian comfort for realistic situations. Vegetation, namely trees, evaporative cooling pavements and neighborhood morphologies are shown here to illustrate this potential.

This framework can eventually be part of design methodologies for conceiving buildings and urban environments that will be able to mitigate the effects of urban heat islands, which should be compounded by climate change.

MAIN REFERENCES

Ferrari A, Kubilay A, Derome D, Carmeliet J. 2020. The Use of Permeable and Reflective Pavements as a Potential Strategy for Urban Heat Island Mitigation. Urban Climate 31: 100534.

Kubilay A, Derome D, Carmeliet J. 2018. Coupling of Physical Phenomena in Urban Microclimate: A Model Integrating Air Flow, Wind-Driven Rain, Radiation and Transport in Building Materials. Urban Climate 24: 398–418.

Kubilay A, Derome D, Carmeliet J. 2019a. Impact of Evaporative Cooling Due to Wetting of Urban Materials on Local Thermal Comfort in a Street Canyon. Sustainable Cities and Society 49: 101574.

Keywords: Urban local climate, all physics modeling, mitigation solutions

Session T3.9 (Online Track): Improving indoor environmental quality

Paper ID 30427: Assessing the impact of outdoor air pollution on natural ventilation potential in major Indian cities

Kopal Nihar*, Alex Nutkiewicz, Rishee K. Jain

Urban Informatics Lab, Stanford University, Stanford, USA

AIM AND APPROACH

Natural ventilation (NV) is an effective passive cooling strategy to meet growing energy demands by leveraging natural wind flow. However, NV inevitably leads to higher air exchange rates. In places with high concentrations of outside air pollutants, NV can increase these pollutants in the indoor environment leading to adverse occupant health impacts. This problem is particularly important for developing nations in hot-humid climates like India that are forecasting significant rise in cooling demand [1] as well as deterioration of air quality in major cities [2].

We utilize publicly available data from OpenAQ to assess the impact of outdoor air pollution on NV potential (NVP). We look at outdoor PM2.5 concentration for major Indian cities i.e. New Delhi, Mumbai and Bangalore for 2019-20. We model a typical residential building representative of informal settlements in India using EnergyPlus. This is combined with a differential equation model to evaluate indoor exposure of occupants to outside pollutants. We consider two scenarios for utilizing NV supplemented by mechanical systems - one in which air pollutants constrain NV availability and the other without such considerations - and the results compared to quantify the potential reduction in NV usability and corresponding energy penalty.

SCIENTIFIC INNOVATION AND RELEVANCE

Few works in the field have evaluated the impacts of outdoor air pollutants (particularly PM) on NVP but those that have undertaken this problem have focused primarily on commercial buildings. Authors of [3], [4] investigate the impact of ambient air pollution on energy savings through NV for commercial buildings in major cities of China and USA respectively.

Our approach aims to extend this to the residential sector with a particular focus on informal settlements; as occupants in such settlements are highly susceptible to the health risks of poor air quality. With over 881 million people living in informal settlements across the world [5], a minor improvement in space cooling and/or air pollution exposure has significant implications. Our initial analysis focuses primarily on large cities in India as many cities (particularly New Delhi and its neighbors) dominate the list of world's most polluted cities and have significant populations of informal settlement dwellers [6]. To accomplish this, we propose a modeling approach that integrates highly granular publicly available air pollution data with a coupled EnergyPlus and differential equation air-flow model to evaluate how NVP for space cooling changes when air pollution exposure is taken into account.

PRELIMINARY RESULTS AND CONCLUSIONS

We simulate the proposed model for the month of August'19 in New Delhi, Mumbai and Bangalore. We evaluate the impact on NV against a standard threshold of 25 μg/m3 average daily PM2.5 level as per World Health Organization (WHO) Air Quality guidelines. We observe that favourable hours for NV consistently decrease across all cities when PM2.5 levels increase. Maximum impact is observed for New Delhi where NVP is reduced by 75% whereas minimum impact is observed in Bangalore where NVP is reduced by only 20%. The corresponding reduction in cooling energy saving potential is observed to be 10% for New Delhi and 3% for Bangalore. Therefore, our results indicate that outdoor PM2.5 levels have a significant impact on NVP especially in highly polluted cities like New Delhi. As such, our work points to the synergistic opportunities that exist for reducing both air pollution and energy consumption in informal settlement residential dwellings across major Indian Cities. We conclude that paving a path towards a more sustainable energy future will require balancing tradeoffs between occupant thermal comfort, air pollution exposure and building energy usage in urban areas across the world.

MAIN REFERENCES

[1] Kumar, S., Sachar, S., Kachhawa, S., Goenka, A., Kasamsetty, S., George, G. (2018). Demand Analysis of Cooling by Sector in India in 2027. New Delhi: AEEE

[2] Guttikunda, S. K., Goel, R., & Pant, P. (2014). Nature of air pollution, emission sources, and management in the Indian cities. Atmospheric environment, 95, 501-510

[3] Tong, Z., Chen, Y., Malkawi, A., Liu, Z., & Freeman, R. B. (2016). Energy saving potential of natural ventilation in China: The impact of ambient air pollution. Applied energy, 179, 660-668

[4] Chen, J., Brager, G. S., Augenbroe, G., & Song, X. (2019). Impact of outdoor air quality on the natural ventilation usage of commercial buildings in the US. Applied Energy, 235, 673-684

[5] UN-HABITAT. SLUM ALMANAC 2015 2016 Tracking Improvement in the Lives of Slum Dweller; 2016:1–82.

[6] IQAir. 2019 World Air Quality Report: Region and City PM2.5 Ranking

Keywords: Natural ventilation, Air pollution, Informal Settlements, Building energy simulation, Indoor air quality

Paper ID 30547: Simulating the impact of ventilation on viral infection probability from aerosol transmission in enclosed spaces

Olivia Nile Sobek (1), Parag Rastogi* (1), John Allison (1), Graeme Jephson (1), Chris Pyke (2), Alan Wegienka (1)

1: arbnco ltd, United Kingdom; 2: Arc Skoru

AIM AND APPROACH

This study uses simulation to assess the impacts of occupant density and ventilation rates as control measures to reduce the risk of aerosol transmission of viruses (e.g., influenza, rhinovirus, SARS, and COVID-19) in large and small offices. The simulation outputs are selected to correspond with in-situ CO2 sensors and control points for implementation in offices around the world. The results of the simulation can be used to set targets for CO2 and other parameters that can be easily measured by low-cost sensors to manage risk of infection due to aerosol transmission.

We used CONTAM 3.2 to simulate small and large workspaces from very low to very high occupancy, representing socially-distanced office use and events such as conferences respectively. Each simulation was run with a standard AHU using ventilation rates from CIBSE Guide A [1] for the length of a typical workday. Occupants were assumed to generate CO2 at rates typical of office work. Using an augmentation of the Wells–Riley equation [2], we coupled these CO2 concentrations with known quantum generation rates of influenza, rhinovirus, and SARS, and preliminary rates of COVID-19 [3]. Finally, we assumed a number or fraction of infected people to model the probability of indoor airborne transmission.

SCIENTIFIC INNOVATION AND RELEVANCE

While studies have used the augmented Wells-Riley equation to model the risk of viruses at different ventilation levels e.g., [2], few have systematically looked at occupant density or ventilation as forms of control strategies or examined office spaces. By concentrating on offices and varying the number of occupants under four ventilation rate categories [1], we are able to model the impact of proposed social-distancing and ventilation control measures to reduce airborne infection of COVID-19, and similar viruses like SARS-CoV-1. This has implications beyond COVID-19, as similar strategies and behaviour applies for regular seasonal viruses like influenza and rhinovirus. This study does not suggest that aerosol transmission is the only transmission route, or that it is the most important. Rather, the idea is that all else held constant, infection probabilities are reduced with these actions.

This data can be used to suggest evidence-driven back-to-work protocols for buildings based on their ventilation systems and number of occupants. This could help tune ventilation rates or occupancy reductions to lower infection probability while maximising safe utilisation. Further, combining these parameters with continuous monitoring could allow infection probabilities to be calculated in real-time and immediate corrective actions from an HVAC system to lower infection probabilities.

PRELIMINARY RESULTS AND CONCLUSIONS

The results of this study clearly show a trend of decreased probabilities of viral infections with increased ventilation and decreased occupant density. There is a clear trend of increased viral infection probability with an increase in the number of infected people, and a decrease in the probability with an increase in ventilation. Increasing ventilation rates from 5 L/s/person to 20 L/s/person a decrease in the probability of infection from 2.3% to 43% can be seen depending on the virus and occupant density. Increasing the number of infected people within the same ventilation rate increased the probability anywhere from 5.5% to 50%. When the occupant density was halved with all other factors kept the same, the infection probability decreased between 0.1% and 6%. These results show the effectiveness of current social-distancing and ventilation-based virus control strategies, suggesting that by incorporating them into their back-to-work protocols an office space could substantially lessen the risk of infection to its occupants. By quantifying the probabilities of these strategies' effectiveness in typical office settings our study suggests the potential to model and simulate more complicated infectious circumstances and use real-time CO2 readings, to give building-specific advice on ventilation rates and occupancy.

MAIN REFERENCES

[1] CIBSE, Environmental design: CIBSE guide A, Eighth edition. London: Chartered Institution of Building Services Engineers, 2015.

[2] S. N. Rudnick and D. K. Milton, 'Risk of indoor airborne infection transmission estimated from carbon dioxide concentration', Indoor Air, vol. 13, no. 3, pp. 237–245, Sep. 2003, doi: 10.1034/j.1600-0668.2003.00189.x.

[3] H. Dai and B. Zhao, 'Association of infected probability of COVID-19 with ventilation rates in confined spaces: a Wells-Riley equation based investigation', Emergency Medicine, preprint, Apr. 2020. doi: 10.1101/2020.04.21.20072397.

Keywords: ventilation, aerosol transmission, CONTAM simulation, social-distancing, COVID-19

Paper ID 30554: Interactive Sankey diagram for visualizing real-time and/or simulation-derived energy flows

Claude Demers-Bélanger* (1), Adam Rysanek (2)

1: University of British Columbia, Canada; 2: School of Architecture and Landscape Architecture, University of British Columbia, Canada

AIM AND APPROACH

As UBC keeps adding buildings and students while putting forward a desire to be one of the greenest university campuses in Canada and in the world, energy management (UBC's Energy & Water Services, 2020) and decision-making toward energy become a major issue. The decision-makers need access to relevant information in a timely matter. This is an issue that not only UBC faces, but many institutions and district energy systems globally. This paper presents a new data analysis tool for visualizing district energy flows at varying scales, in real-time, via an interactive Sankey energy flow diagram. In the pilot application, the tool makes use of UBC's energy and water services (EWS) data collected with smart meters across the university's portfolio of over 150 buildings. The visualization tool allows users to inspect individual building energy use, by end-use, as well as download energy use data as visualized

SCIENTIFIC INNOVATION AND RELEVANCE

Sankey diagrams have been a common way to visualize flows of information. From economic flows (Bakenne, Nuttal, & Kazantzis, 2016) to energy sources (Cullen & Allwood, 2010), these charts have been able to help decision-makers better understand the relationship between dynamic systems driven by the flow of resources from source to end-use. Studies have shown that with better visualization, decision-makers can improve overall building performance diagnostics without having to resort or rely on building automation systems. Sankey diagrams can provide transparent carbon accounting of building energy use across end uses, building types, and time. Sankey diagrams also make access to measured building performance data more accessible, which can support practitioners seeking to calibrate building energy mdels. (Coakley, Raftery, & Keane, 2014). Though representing energy flows via a Sankey diagram is now new, we argue that an interactive Sankey diagram that is connected to a real-time building metering database, or a synthetic simulation-derived database, can improve one's understanding of measured and/or predicted building performance.

PRELIMINARY RESULTS AND CONCLUSIONS

The data is pulled from a time series specific data base (InfluxDB) and the web app will be delivered in JavaScript using D3.js module to create the Sankey diagrams. A full working example of the tool as well as a downloadable Github repository will be provided at the full submission of this paper.

MAIN REFERENCES

Bakenne, A., Nuttal, W., & Kazantzis, N. (2016). Sankey-Diagram-based insights into the hydrogen economy of today. International Journal of Hydrogen Energy, 7744-7753.

Coakley, D., Raftery, P., & Keane, M. (2014). A review of methods to match building energy simulation models to measured. Renewable and Sustainable Energy Reviews, 123-141.

Cullen, J. M., & Allwood, M. J. (2010). The efficient use of energy: Tracing the global flow of energy from fuel. Energy Policy, 75-81.

Lee, D., Cha, G., & Park, S. (2016). A Study on Data Visualization of Embedded Sensors for Building Energy Monitoring using BIM. INTERNATIONAL JOURNAL OF PRECISION ENGINEERING AND MANUFACTURING, 807-814.

Motegi, N., & Piette, M. (2002). Web-based energy information systems for large commercial buildings. National Conference on Building Commissioning.

Niu, S., Pan, W., & Zhao, Y. (2015). A BIM-GIS Integrated Web-based Visualization System for Low Energy Building Design. 9th International Symposium on Heating, Ventilation and Air Conditioning (ISHVAC) and the 3rd (pp. 2184-2192). Procedia Engineering 121.

UBC's Energy & Water Services. (2020, July 14). Energy Management. Retrieved from UBC'S Energy & Water Story: https://energy.ubc.ca/about-us/ubcs-story/energy-management/

Keywords: Visualization, Analytics, Energy flows, Real-time data.

Paper ID 30591: Indoor environmental quality (IEQ) assessment using laser-assisted data acquisition (LADA) of building geometry

Simon Wyke, Lasse Rohde*, Rasmus Lund Jensen

Department Of The Built Environment, Aalborg University, Denmark

AIM AND APPROACH

A new Indoor environmental quality (IEQ) assessment tool has been developed, to perform fast and inexpensive IEQ-evaluations, and performance dissemination that is more understandable for all stakeholders (Steen Larsen et al., 2019). This paper investigates how data for this new IEQ assessment tool can be collected efficiently through laser assisted data acquisition (LADA) of building geometry.

We tested LADA for the newly developed IEQ-evaluation tool on two flat units, and exported the acquired data from a handheld laser meter, used as a high-tech measuring tape, to the IEQ-tool, using data processing software, in which building geometries were modelled in 2D and exported into the Comma Separated Value format (CSV).

SCIENTIFIC INNOVATION AND RELEVANCE

IEQ is currently receiving increased attention in the building industry. However, IEQ is a multifaceted topic consisting of both thermal, visual, acoustic and air quality aspects, and their potential impact on occupant comfort, health and well-being (Rohde, Larsen, Jensen, & Larsen, 2019). This multifaceted nature makes IEQ difficult to comprehend for many building owners and occupants, and makes overall IEQ assessments a complicated, time-consuming and costly task.

Various building data are needed in order to evaluate the IEQ-potential of a building. It can be time-consuming and expensive to acquire, model and store such data in a repository e.g. a spreadsheet schema or building information model (BIM). Building models based on manual data input, additionally, often lack consistency and reliability (Gerrish, Ruikar, Cook, Johnson, & Phillip, 2016).

PRELIMINARY RESULTS AND CONCLUSIONS

Data acquisition for IEQ can be based on automated processes, including use of both photogrammetry and laser meters. Many studies focus on how to apply photogrammetry point-clouds to generate as-build models (Barazzetti, 2016). However, in this study we used a handheld laser meter to do laser assisted data acquisition (LADA), to limit the amount of data which needs to be acquired, to the exact needs of an IEQ-evaluation tool (Swanström Wyke, Svidt, Jønsson, Rohde, & Lund Jensen, 2019).

The test revealed that it was possible to acquire accurate geometry-data for IEQ-evaluation based on LADA, with low-effort and sufficient accuracy for the relevant IEQ-simulations.

MAIN REFERENCES

Barazzetti, L. (2016). Parametric as-built model generation of complex shapes from point clouds. Advanced Engineering Informatics, 30(3), 298–311. https://doi.org/10.1016/j.aei.2016.03.005

Gerrish, T., Ruikar, K., Cook, M., Johnson, M., & Phillip, M. (2016). Engineering , Construction and Architectural Management Article information : To cite this document : Trends of Skills and Productivity in the UK Construction Industry, 15(4), 372–382. https://doi.org/10.1108/eb021087

Rohde, L., Larsen, T. S., Jensen, R. L., & Larsen, O. K. (2019). Framing holistic indoor environment: Definitions of comfort, health and well-being. Indoor and Built Environment, 0(0), 1–19. https://doi.org/10.1177/1420326X19875795

Swanström Wyke, S. C., Svidt, K., Jønsson, K., Rohde, L., & Lund Jensen, R. (2019). Laser-Assisted Data Acquisition Of Building Geometry: Selection and use of laser meters and data processing software for IV20. Aalborg, Denmark.

Keywords: Data Acquisition, Laser Measuring, Indoor Environmental Quality (IEQ)

Paper ID 30632: Towards safer work environments during the COVID-19 crisis: A Study of different floor plan layouts and ventilation strategies coupling open foam and airborne pathogen data for actionable, simulation-based feedback in design

Zoe De Simone*, Patrick Kastner, Timur Dogan

Cornell University, United States of America

AIM AND APPROACH

As work environments struggle to reopen during the current COVID-19, it is crucial to establish tools for effective decision making. While a strong emphasis has been placed on determining general guidelines to reduce the risk of airborne viral spread (1),(2),(3),(4) there is a lack of open and easy to use simulation workflows to quantify the risk of airborne pathogens indoors on a case-to-case basis as floor-plans, furniture layouts, and ventilation inlet-outlet locations have a significant impact on indoor air movement.

We couple Computational Fluid Dynamics (CFD) passive scalar transport (5) with an energy model and viral decay function (6) to quantify the exposure to airborne pathogens by integrating this approach in the popular Rhino3d/Grasshopper CAD environment. We further demonstrate its application in a comparative study that investigates how plan and furniture layout, including small scale interventions such as plexiglass partitions and distancing measures, natural and hybrid ventilation strategies can impact the movement of airborne pathogens in indoor environments. The aim of the project is to enable stakeholders to assess existing workplaces, and to enable designers to create spaces that are more resilient in an epidemic.

SCIENTIFIC INNOVATION AND RELEVANCE

Spatial proximity, interpersonal contact, superficial contact and respiratory droplets are the four recognized means of indoor viral spread (7). While coughing and sneezing produce larger airborne particles that decay within 1s (8), small particles, produced when breathing, decay within 8 to 14 minutes (9). Once airborne, small droplets dehydrate, slowing their fall(10) and decay at rates ranging between 30 to 177 min depending on microclimate boundary conditions(11), including ventilation rates (8), thermal and radiation environment.

To model environmental factors such as air movement, thermal and uv exposure of small viral aerosol particles we introduce a new metric that uses the combination of passive scalar aerosol tracer, and an energy model coupled viral decay function (6).

Although ventilation reduces pathogen concentration, airflow patterns and direction can increase the risk of transmission between occupants. Reported transmission routes at a restaurant (12) demonstrate the importance of simulating seating arrangements and ventilation systems simultaneously. Our tool enables users to analyze the effectiveness of spatial layouts and ventilation strategies, providing architects with ways of validating muti-factor design strategies. Finally, we demonstrate the fidelity of the new metric by conducting case studies comparing venitation and floor plan layouts.

PRELIMINARY RESULTS AND CONCLUSIONS

The current health crisis suggests a clear need for easy-to-use health risk modeling tools for architects and engineers to evaluate the design of interior spaces. While existing guidelines offer general instructions regarding increased ventilation and distancing measures such as workstation configurations and plexiglass partitions (13), only few easy to use tools for ad-hoc evaluations of workplace configurations exist. In addition, the combination and efficacy of such guidelines has not been analyzed for a design space with multiple variables, such as floor plan layouts and ventilation direction. To understand the efficacy of existing spatial and ventilation guidelines on the mitigation of an airborne virus, the project investigates a series of typical office space layouts and ventilation systems(14).

We implement a prototypical design evaluation workflow that couples OpenFOAM, EnergyPlus, Eddy3D with the SARS-CoV-2 Airborne Decay Calculator (6). This allows users to simulate potential pathogen exposure levels in different scenarios considering layouts and ventilation strategies. A series of case studies evaluate floor plans with open space and partitioned spaces with mixed, displacement and natural ventilation. Preliminary results show significant differences between the test cases and suggesting promising fidelity of the proposed simulation methodology.

MAIN REFERENCES

1. ASHRAE, Indoor air quality guide. 2009.

2. HVAC in the context of COVID-19. 2020.

3. Stewart, Schoen, Mead, Olmsted, Sekhar, Vernon, ASHRAE Position Document on Infectious Aerosols. 2020.

4. WHO, Getting your workplace ready for COVID-19: How COVID-19 spreads, 2020.

5. Kastner , Dogan, A cylindrical meshing methodology for annual urban computational fluid dynamics simulations. 2020

6. Department of Homeland Security. Estimated Airborne Decay of SARS-CoV-2 2020.

7. Dietz , Horve, Coil , Fretz , Eisen , Van Den Wymelenberg, Novel Coronavirus (COVID-19) Pandemic: Built Environment Considerations ... 2020.

8. Somsen, van Rijn, Kooij, Bem, Bonn, Small droplet aerosols in poorly ventilated spaces and SARS-CoV-2 transmission. 2020.

9. Stadnytskyi , Bax, Bax , Anfinrud, The airborne lifetime of small speech droplets and their potential importance in SARS-CoV-2 transmission. 2020.

10. Wells W. ON AIR-BORNE INFECTION. 1934.

11. Smither, Eastaugh, Findlay, Lever. Experimental aerosol survival of SARS-CoV-2 in artificial .... Emerg Microbes Infect. 2020.

12. Lu, Gu , Li , Xu , Su , Lai , COVID-19 Outbreak Associated with Air Conditioning in Restaurant, Guangzhou, China, 2020

13. CDC. Communities, Schools, Workplaces, & Events. 2020.

14. Atkinson J, Chartier Y, Pessoa-Silva CL, Jensen P, Li Y, Seto W-H. Concepts and types of ventilation. 2009.

Keywords: COVID-19, Age of air, passive scalar, ventilation, CFD

Paper ID 30826: Computational analysis of aerothermal characteristics of skygardens and comparative assessment of different attenuators

Murtaza Mohammadi*, Paige Wenbin Tien, John Kaiser Calautit

University of Nottingham, United Kingdom

AIM AND APPROACH

The main aim of this study was to analyse the impact of vegetation on the aero-thermal conditions in a skygarden located on a high-rise building. The aerodynamic response of skygarden vegetation must be understood in detail to assist designers in the selection and arrangement of species and buffer elements for creating a conducive environment for occupants. The current study utilises Computational Fluid Dynamics (CFD) to analyse nine different vegetation configurations inside a high-rise skygarden to determine the aero-thermal performance. Simulations were carried out using the Steady Reynolds Averaged Navier–Stokes (SRANS) equation, where the vegetation was modelled as a porous zone with cooling capacity [1,2,3].

SCIENTIFIC INNOVATION AND RELEVANCE

Green spaces can greatly improve the quality of life, especially in urban areas, by influencing the health and well-being of citizens. It acts as a filter which reduces pollutants, dust and other harmful particles from the surrounding air, and are frequently planted as a barrier between emission source and adjacent areas to improve air quality. Additionally, vegetation can improve the local thermal condition by enabling shading and evaporative cooling [4,5]. Thus, building applied vegetative measures, such as green walls and roofs, are increasingly gaining popularity due to multifold benefits.

One such architectural feature, the skygarden, is an important design intervention for improving the social, economic and environmental values of a building. Many high-rise buildings feature a skygarden as a social space, allowing occupants to connect and experience outdoor freshness within a semi-enclosed environment [6]. Moreover, the current pandemic has forced designers to re-examine the spatial planning principles of built spaces, mainly due to health and safety concerns. Vegetation, including hedges, shrubs and trees can act as surfaces for deposition of aerosol particles [7], which are considered to be the primary mode of microbe transmission [8, 9]. Our study analyses the various attenuation effects of trees, primarily wind speeds and temperature.

PRELIMINARY RESULTS AND CONCLUSIONS

The results indicate that trees can attenuate high wind speeds in the skygarden for most of the configurations. Especially from the center to the leeward edge of the skygarden, wind speeds are within the comfort range at occupants' height. Near the windward edge, however, some amplification occurs due to constriction created in the air flow stream. The trees are able to extract heat from air and create cooling effect in their wake, up to 1°C. The effect is stronger with higher number of trees and reduced air speed.

Although the absolute performance of the skygarden improves with a higher number of trees, when taking into account the ground cover of trees, the relative performance of a single-row configuration is found better than a double row of trees. Thus, the design of the skygarden, requires an optimal balance between the benefits from the trees and the spatial cost of plantation.

MAIN REFERENCES

1. Kichah, A., Bournet, P.-E., Migeon, C., Boulard, T., Measurement and CFD simulation of microclimate characteristics and transpiration of an Impatiens pot plant crop in a greenhouse

2. Gromke, C., Blocken, B., Janssen, W., Merema, B., van Hooff, T., Timmermans, H., CFD analysis of transpirational cooling by vegetation: Case study for specific meteorological conditions during a heat wave in Arnhem, Netherlands

3. Manickathan, L., Defraeye, T., Allegrini, J., Derome, D., Carmeliet, J., Parametric study of the influence of environmental factors and tree properties on the transpirative cooling effect of trees

4. Imran, H. M., Kala, J., Ng, A.W. M., Muthukumaran, S., Effectiveness of vegetated patches as Green Infrastructure in mitigating Urban Heat Island effects during a heatwave event in the city of Melbourne

5. Besir, A. B., Cuce, E., Green roofs and facades: A comprehensive review, Renewable and Sustainable Energy Reviews

6. Pomeroy, J., The Skycourt and Skygarden, Greening the Urban Habitat

7. S. Janhäll, "Review on urban vegetation and particle air pollution - Deposition and dispersion,"

8. WHO, "DAMPNESS AND MOULD WHO GUIDELINES FOR INDOOR AIR QUALITY," 2009.World Health Organization

Keywords: Skygarden, wind comfort, high-rise building, computational fluid dynamics (CFD), vegetation

Paper ID 30855: The impact of PCM applications on thermal comfort in standardized preschool designs in 2 Brazilian climatic zones

Lorenzo Olivo Filippini*, Gabriela Sartori, Maurício Carvalho Ayres Torres

Universidade Federal do Rio Grande do Sul, Brazil

AIM AND APPROACH

The Federal Brazilian Government's Proinfância Program was developed in 2007 to ensure children's access to preschools. Preschool buildings' standardized plans were developed to speed up licitation processes. Since its creation, more than 5 billion Brazilian Reais had been invested, while more than 5,600 preschools of all types had already been built or were being built until 2017. According to the ABNT NBR 15.220-3:2005 criteria, Brazil is divided into eight bioclimatic zones. The standardized preschool building plan must observe the distinct climatic zones' characteristics to comply with thermal performance minimum levels. The thermal performance of buildings is directly associated with the occupant's thermal comfort. For thermal comfort maintenance, the use of Phase Change Materials (PCM) is a possible strategy that might be adopted to increase the thermal inertia of buildings envelope and reduce the thermal amplitude of indoor environments. This study aims to assess the impact of the PCM use on the thermal comfort in the preschool classrooms, according to the ASHRAE Standard 55-2013 Adaptive Model. Thermal simulations were performed using EnergyPlus, considering three different types of PCMs in four distinct applications for two bioclimatic zones in Brazil.

SCIENTIFIC INNOVATION AND RELEVANCE

Since there is no standard about thermal comfort in schools in Brazil, this research is a pioneer for assessing thermal comfort in naturally ventilated standardized preschool building plans in two specific bioclimatic zones in Brazil - 1 and 8 - using the ASHRAE Standard 55-2013 Adaptive Model. This study is also a pioneer for assessing the impact of using PCM to increase thermal comfort in the preschool standardized plan. The adoption of PCM might promote the improvement of thermal comfort levels in Brazilian preschool buildings, as well as, reduce the energy consumption in public buildings. Considering that more than 5,600 schools based on standardized designs had been built all over the country, the improvement of thermal comfort levels can impact a great number of children and save a significant amount of energy. Also, the results achieved in this study might encourage the development of a thermal comfort regulation for preschool buildings in Brazil, granting good quality learning spaces for children.

PRELIMINARY RESULTS AND CONCLUSIONS

The materials improved the thermal comfort by decreasing the operative temperature to different levels in each application in the prolonged occupancy spaces. The results indicate that, in the warmest Brazilian bioclimatic zone - zone 8 -, the impact of the PCM application was more relevant, reducing the thermal discomfort levels regarding heat. While in the Brazilian bioclimatic zone with more intense winter - zone 1 - the PCM application showed less relative reduction of the thermal discomfort levels. Results show that the average relative reduction in the total number of days in thermal discomfort conditions was approximately 7%, for zone one, and approximately 57% for zone eight. In the absolute number of days in thermal discomfort levels, those values represent an average reduction of 11,5 days, for zone 1, and 15 days, for zone 8. Considering both of these zones, it is perceived the capacity that the chosen PCMs have to passively improve the thermal comfort aspects towards the preschool occupants. Thus, PCMs might be an interesting alternative to air conditioning systems, especially in warmer climates.

MAIN REFERENCES

AL-JANABI, Ali; KAVGIC, Miroslava. Application and sensitivity analysis of the phase change material hysteresis method in EnergyPlus: A case study. Applied Thermal Engineering, v. 162, p.1-19, nov. 2019. Elsevier BV. http://dx.doi.org/10.1016/j.applthermaleng.2019.114222.

MAGENDRAN, Suhanyaa S. et al. Synthesis of organic phase change materials (PCM) for energy storage applications: A review. Nano-structures & Nano-objects, v. 20, p.1-18, out. 2019. Elsevier BV. http://dx.doi.org/10.1016/j.nanoso.2019.100399.

PAROUTOGLOU, Evdoxia et al. A PCM based cooling system for office buildings: a state of the art review. E3s Web Of Conferences, v. 111, p.1-8, 2019. EDP Sciences. http://dx.doi.org/10.1051/e3sconf/201911101026.

TABARES-VELASCO, Paulo Cesar; CHRISTENSEN, Craig; BIANCHI, Marcus. Verification and validation of EnergyPlus phase change material model for opaque wall assemblies. Building And Environment, v. 54, p. 186-196, ago. 2012. Elsevier BV. http://dx.doi.org/10.1016/j.buildenv.2012.02.019.

WEINLÄDER, Helmut; BECK, Andreas; FRICKE, Jochen. PCM-facade-panel for daylighting and room heating. Solar Energy, v. 78, n. 2, p. 177-186, fev. 2005. Elsevier BV. http://dx.doi.org/10.1016/j.solener.2004.04.013

Keywords: Thermal simulation, Adaptive model, thermal comfort, thermal performance

Paper ID 30974: Building meta-optimization: A study on the reuse of previous simulation data to reduce computational costs

Lucas Camilotti*, Nathan Mendes, Roberto Zanetti Freire

Pontifical Catholic University of Paraná, Brazil

AIM AND APPROACH

The proposed study aims to evaluate the benefits of data reuse in-between different simulation procedures during a whole-building energy optimization process. A simulation can be broken down into distinct tasks, and some of them may provide the same results when similar input parameters are adopted. Considering the scenario of a parametric optimization problem, where similar design vectors can be generated, the potential of data reuse between different simulations is present. As simulations are usually treated individually, without the exchange of information among them, this can cause a waste of time and resources by performing redundant tasks that could be avoided. To confirm our statements, we use the Domus software, a certified Brazilian tool for building energy analysis, as a case study considering a practical application of the proposed approach. One of the costliest stages during simulations using Domus is the view factor calculation [1], which is used to estimate the radiation which leaves a surface that strikes another, a calculation that depends exclusively on their geometry. By applying the proposed approach to building energy modeling optimization, the view factor obtained from previous simulations could be reused, providing significant improvement in time during the optimization process.

SCIENTIFIC INNOVATION AND RELEVANCE

Many attempts to reduce the execution time of costly optimization problems associated with whole-building simulation are constantly being performed. Methodologies such as parallelization and scheduling of simulations in a distributed environment [2,3], or the use of surrogate models for both the approximation of the objective function and the reduction on the number of simulations [4,5], have shown successful results in this field. The proposed methodology can provide yet another point of improvement, enabling yet another layer of reduction in overall optimization time. An evaluation of the proposed methodology in junction with the two aforementioned ones can provide useful insight in how well the three approaches work alongside one another. Moreover, most building simulation tools were not designed oriented to parametric optimization as a premise during the development stage, and as a result, minimal efforts were directed to provide integration points and functionalities for generic optimization tools. Based on that, the methodology has the additional benefit of encouraging simulation program developers to provide more fine-grained control over simulations, giving more freedom to the optimization tools utilized.

PRELIMINARY RESULTS AND CONCLUSIONS

The methodology was applied over a wide number of benchmark building design optimization problems, with selected design variables ranging from completely geometry-related, to completely unrelated, to a mix of both. The optimization processes were carried out multiple times for each problem, in order to obtain an average result and eliminate possible outliers. Preliminary results showed that the methodology has no weaknesses, aside from implementation complexity. Improvements showing a considerable reduction in overall optimization time were reported, with worst-case scenarios showing no changes, but also no penalties. Additionally, the view factor data reuse was carried out in cases of equal values for the whole-building, and an evaluation considering individual thermal zones separately has yet to be done, to verify if an even finer-grained control will prove to be beneficial for the whole-building optimization process.

MAIN REFERENCES

1. Augusto LD de C, Giacomet B, Mendes N. Numerical method for calculating view factor between two surfaces. In: IBPSA 2007 - International Building Performance Simulation Association 2007. 2007. p. 269–74.

2. Yang C, Li H, Rezgui Y, Petri I, Yuce B, Chen B, et al. High throughput computing based distributed genetic algorithm for building energy consumption optimization. Energy and Buildings. 2014;76:92–101.

3. Labib R, Baltazar J-C. Using Python to Automate the Preparation and Execution of Thousands of Daylighting and Glare Simulations on a Cloud Parallel Computing environment for Time-efficient Simulations. In: Proceedings of Building Simulation 2019: 16th Conference of IBPSA. 2020. p. 1545–51.

4. Carreras J, Pozo C, Boer D, Guillén-Gosálbez G, Caballero JA, Ruiz-Femenia R, et al. Systematic approach for the life cycle multi-objective optimization of buildings combining objective reduction and surrogate modeling. Energy and Buildings. 2016;130:506–18.

5. Barnes EC, McArthur JJ. Building Energy Use Surrogate Model Feature Selection – A Methodology Using Forward Stepwise Selection and LASSO Regression Methods. In: Proceedings of Building Simulation 2019: 16th Conference of IBPSA. 2020. p. 3078–85.

Keywords: computational cost, data reuse, meta-optimization, view factor

Paper ID 31098: Assessing performance of passive or low-energy resilient cooling technologies for pre-1980 medium office buildings in Phoenix, Arizona and Chicago, Illinois

Nari Yoon* (1), Sang Hoon Lee (1), Edward Arens (2), Hui Zhang (2), Ronnen Levinson (1)

1: Lawrence Berkeley National Laboratory, USA; 2: University of California, Berkeley, USA

AIM AND APPROACH

Our research evaluates a comprehensive set of key performance indicators (KPIs) for passive or low-energy building technologies (hereinafter, "strategies"). The strategies include cool envelope materials to reduce solar heat gain, natural ventilation to remove excessive indoor heat, and personal comfort systems to allow the cooling set point to be increased. They seek to provide resilient cooling for buildings, and thereby help building occupants adapt to increasingly frequent extreme heat events.

We used the whole-building energy simulation tool EnergyPlus to model prototype buildings before and after application of each strategy. Annual hourly HVAC energy use and indoor environmental conditions were used to calculate KPIs for each floor, for each multi-story perimeter/core zone, and for the whole building. We computed more than 50 KPIs and evaluated diagnostics (unit tests) to check their validity.

Major KPIs reported in this study include changes in annual heating, ventilation, and cooling energy uses; changes in seasonal and annual very-hot, warm, cool, and very-cold exceedance hours that gauge the number of discomfort occupied hours; and changes in seasonal and annual weighted exceedance hours, which are the sums of positive values of predicted mean vote (PMV) exceedance over occupied hours.

SCIENTIFIC INNOVATION AND RELEVANCE

We assessed how each strategy could make "building occupants" more resilient to hot weather, boosting comfort, health, and productivity; and make "building cooling systems" more resilient to hot weather, improving their ability to meet cooling load. These strategies could reduce illness and death in disadvantaged communities where residents lack air conditioning, and benefit any community subject to scheduled or unscheduled power outages. Our analysis of the selected strategies can thus inform to what extent a building and occupants may resist the consequences, reduce the impact, and recover from extreme weather conditions.

When expanded to a wide range of building types, vintages, sizes, and climates, our KPI approach can identify strategies that provide high cooling resilience under given conditions.

PRELIMINARY RESULTS AND CONCLUSIONS

We examined three-story medium office prototype buildings, each built prior to 1980, in Chicago, Illinois (cold winter / hot-humid summer) and Phoenix, Arizona (hot/dry both seasons).

- The annual fan-plus-cooling energy savings always exceeded the annual heating energy penalty.

- HVAC savings varied linearly with some parameters, but not with others. For example, cool-roof savings were proportional to increase in roof albedo, while ceiling-fan savings rose non-linearly with increase in average air speed.

- The annual comfort KPIs varied by region within the building. For example, in an uncooled three-story building, the thermal-comfort improvement (reduction in weighted warm exceedance hours) on the top floor was about 17-19 times that on the bottom floor when applying the cool-roof strategy.

- The annual HVAC savings yielded by ceiling fans and cool roofs in Phoenix were 3-4 times those in Chicago, and the annual thermal-comfort improvements made by these strategies in Phoenix were 2.5-3 times those in Chicago.

The comprehensive KPIs quantified cooling resilience of a building, providing detailed insights on how the selected strategies influence the energy and comfort performance of building zones and the whole building. We also demonstrated the ability to check building simulation with unit tests.

MAIN REFERENCES

ASHRAE. (2017). ANSI/ASHRAE Standard 55-2017: Thermal Environmental Conditions for Human Occupancy. https://www.ashrae.org/technical-resources/bookstore/standard-55-thermal-environmental-conditions-for-human-occupancy

Borgeson, S., & Brager, G. (2011). Comfort standards and variations in exceedance for mixed-mode buildings. Building Research & Information, 39(2), 118–133. https://doi.org/10.1080/09613218.2011.556345

Rosado, P. J., & Levinson, R. (2019). Potential benefits of cool walls on residential and commercial buildings across California and the United States: Conserving energy, saving money, and reducing emission of greenhouse gases and air pollutants. Energy and Buildings, 199, 588–607. https://doi.org/10.1016/j.enbuild.2019.02.028

Hoyt, T., Arens, E., & Zhang, H. (2015). Extending air temperature setpoints: Simulated energy savings and design considerations for new and retrofit buildings. Building and Environment, 88, 89–96. https://doi.org/10.1016/j.buildenv.2014.09.010

Yoon, N., Norford, L., Malkawi, A., Samuelson, H., & Piette, M. A. (2020). Dynamic metrics of natural ventilation cooling effectiveness for interactive modeling. Building and Environment, 180, 106994. https://doi.org/10.1016/j.buildenv.2020.106994

Keywords: Resilient cooling, cool building technologies, key performance indicator (KPI), energy simulation

Paper ID 31132: Optimum daylighting and thermal comfort simulation framework for school buildings in hot arid climate

Muhammad Adel Ahmed Mahmoud (1), Medhat Dorra (1), Khaled Nassar (2), Ayman Hassan El Hakea* (3), Mohamed Shaltot (4)

1: Department of Architecture, Cairo University, Egypt; 2: Construction Engineering Department, The American University in Cairo, Egypt; 3: Construction and Building Engineering Department, Arab Academy for Science, Technology and Maritime Transport, Cairo, Egypt; 4: EGEC Qatar for Engineering Consultations, Doha, Qatar

AIM AND APPROACH

While daylighting and thermal comfort design requirements in schools have significantly evolved, their associated metrics are still lagging behind. Most of the well-established relevant literature has long focused on illuminance, Daylight Factor (DF) and thermal charts. Nevertheless, modern Building Information Modeling (BIM)-enabled tools have provided optimal sustainable building design and performance. New metrics have emerged, namely Daylit Area % (DLA), Mean Daylight Factor (µDF), Daylight Autonomy (DA) and Useful Daylight Index (UDI), and unmet temperature and humidity hours. Moreover, there was a lack in previous research on schools in hot arid climate. Thus, this research proposes a BIM-enabled early design multi-objective simulation and optimization model for daylighting and thermal comfort in prototypical school buildings, whose aim is reaching the best combination of Window-to-Wall Ratio "WWR" and building orientation. The model features a Design Database Module (DDM), a Performance Criteria Database (PCD), and an Analytical Hierarchy Process Module (AHPM), used to estimate Criteria Relative Weights (CRW's). Next, scenarios are generated and checked against the site constraints and building codes, and only valid scenarios are considered using a Scenario Analysis Engine (SAE). For each valid scenario, a Combined Criteria Score (CCS) is obtained, and the optimum scenario is selected by the user.

SCIENTIFIC INNOVATION AND RELEVANCE

While traditionally, school building performance simulation was mostly performed for a few number of performance measures, this paper presents an AHP tool that enables the inclusion of as many measures as possible. This has been demonstrated by the authors through the adoption of five different daylighting metrics and two thermal comfort measures. On another hand, while the daylighting simulation was performed by integrating Rhinoceros® and DIVA® software tools, thermal simulation was carried out by integrating OpenStudio® and EnergyPlus® software packages. AHP criteria weights methods as well as three of the adopted software tools were scarcely used in previous related work.

PRELIMINARY RESULTS AND CONCLUSIONS

For the purpose of model implementation and testing, the authors considered three school building case studies, all situated in hot arid climates: Egypt, Saudi Arabia and Qatar. In general, the daylighting and thermal simulation results tended to provide better CCS results upon increasing WWR, with less dependence on building orientation. While the verification of the AHPM was done using consistency indices and ratios, the validation of the PCD and the model output was performed using surveys filled-in by experts in the field of architectural design. It was concluded that while the model proved as a useful early-design tool for school buildings in hot arid climates, it could well be extended to include other performance metrics such as energy and cost performance, and could also be used with other types of prototypical designs, such as residential, administrative, service buildings retail and congregational buildings.

MAIN REFERENCES

Anderson, D. R., Sweeny, D. J., Williams, T. A., Camm, J. D., & Martin, R. K. (2014). An Introduction to Management Science: Quantitative Approach to Decision Making (13th ed.). Mason, OH, US: South Western Cengage Learning.

ASHRAE. (2010). ASHRAE Standard 55-2010: Thermal Environment Conditions for Human Occupancy. Atlanta, GA, US: ANSI.

Mahmoud, M. A., Mashaly, I. A., Rashed, Y. M., & Nassar, K. (2016). A new dynamic climate-based daylight metric for sustainable building design in hot climates. Modern Environmental Science and Engineering, 1(2).

Reinhart, C. F. (2014). Daylighting Handbook 1: Fundamentals, Designing with the Sun. (R. Stein, Ed.) United States: Cristopher Reinhart.

Keywords: Building Performance Simulation, Sustainable Design, School Buildings, Hot Climate, Multi-objective Optimization

Paper ID 30653: End-use disaggregation in commercial buildings with the building automation system trend data

Narges Zaeri (1), H. Burak Gunay* (1), Araz Ashouri (1,2), Ian Wilton (2)

1: Carleton university, Ottawa, Canada; 2: National Research Council Canada, Ottawa, Canada

AIM AND APPROACH

End-use submetering is essential for energy management in large commercial and institutional buildings. However, most existing buildings lack adequate submetering even for major end-uses. End-use disaggregation techniques offer an untapped opportunity to supplement deficiencies in a metering network. This study presents an end-use disaggregation method for commercial buildings by using building automation system (BAS) data. The BAS trend data provide contextual information about the operational state of major energy-consuming systems and equipment such as fans, pumps, air handling unit (AHU) heating and cooling coils, and chillers. The method applies a series of multiple linear regression models disaggregating bulk metered heating, cooling, and electricity use data into different end-uses by using BAS data as predictors. The results demonstrate that the method can accurately disaggregate hourly building level electricity, heating, and cooling use into their end-use categories.

SCIENTIFIC INNOVATION AND RELEVANCE

In this study, the energy demand modeling procedure using two estimation methods was established by the following steps: (1) pre-processing of raw data, (2) feature selection, (3) model development, and (4) model evaluation and error calculation.

Two regression models were developed for each of the three energy meters: Sci-kit-learn linear-model (Python library) and the genetic algorithm(GA).

This analysis is performed using measured hourly energy consumption data of an academic office building in Ottawa, Canada. Hourly operational data for energy consumption was collected from the meter and automation system as well as WiFi access point data. Unlike most of the previous studies, this disaggregation problem utilized automation systems such as AHU fan state and occupancy estimation as well as meter data for energy reading to train the regression models.

PRELIMINARY RESULTS AND CONCLUSIONS

In this study, linear regression models that disaggregate bulk electricity, cooling, and heating meter data into end-uses of the higher spatial and categorical resolution were developed. The models use BAS trend data that indicate the operational state of the major energy-using equipment (e.g., AHU supply fan state). The models were trained with the Sci-kit-learn and the genetic algorithm. Their accuracy and ability to disaggregate end-uses were demonstrated upon data from an academic office building in Ottawa, Canada.

For electricity, the end-uses considered were fans and pumps, lighting and plug loads, and chillers. For heating and cooling, they were AHU heating and cooling coils, reheat coils, and other perimeters heating devices.

From the results of this study, it was observed that low-frequency meter data could be disaggregated at a reasonable accuracy with provided BAS data.

MAIN REFERENCES

1.Burak Gunay, H., Z. Shi, I. Wilton, and J. Bursill (2020). Disaggregation of commercial building end uses with automation system data. Energy and Buildings 223, 110222.

2. Sh. Walker, W. Khan, K. Katic, W. Maassen, Wim Zeiler." Accuracy of different machine learning algorithms and added-value of predicting aggregated-level energy performance of commercial buildings". Energy and Building 209, (2020)

3.W. O'Brian, A. Abdelalim, and H. B Gunay," Development of an office tenant electricity use model and its application for right-sizing HVAC equipment," Journal of Building Performance Simulation, vol.12, pp37-55,2019/01/02.

4. Eva Garcia-Martin, Crefeda Faviola Rodrigues, Graham Riley, HakanGrahn. "Estimation of energy consumption in machine learning".Journal of Parallel and Distributed Computing Volume 134, December 2019, Pages 75-88.

5.C.Fan, Y.Ding, "Cooling load prediction and optimal operation of HVAC systems using a multiple nonlinear regression model, Energy Build 19 7 (2019)7–17,doi: 10.1016 /J.ENBUILD. 2019.05.043

Keywords: Disaggregation, Use of bulk metered data, Leveraging building automation system, Regression-based approach

Paper ID 30211: The impact of using the application of a CNN-based approach for equipment usage detection on building energy

Shuangyu Wei, Paige Wenbin Tien*, Yupeng Wu, John Kaiser Calautit

University of Nottingham, United Kingdom

AIM AND APPROACH

The present work will develop a coupled approach based on computer vision and deep learning to detect and classify the usage of the equipment in a building space in real-time and concurrently. We propose a technique based on the Faster regions with CNN features (R-CNN) to process the obtained to detect and recognise the usage of equipment in an office space. The model is trained and deployed to a standard camera, and field tests were carried out in an office space. During the field test, the capabilities of the proposed method are evaluated by trying to detect how the occupants interact with personal computers. Different evaluation metrics suggested by previous works were employed. The field tests were carried out in an actual working space in a building at the University. The acquired real-time information is used to generate equipment usage profiles which can then be used to control HVAC operation and as input for building energy simulation (BES) models. In this study, the influence of the coupled approach on the energy demand is evaluated by simulating the case study building in BES with the typical profile and deep learning generated profile.

SCIENTIFIC INNOVATION AND RELEVANCE

As the use of equipment in office buildings increases, accurate equipment usage detection is valuable for energy consumption reduction in the built environment. Using the collected equipment usage information, building energy management system can automatically adjust the operation of heating, ventilation, and air-conditioning systems to meet the actual demands in different conditioned spaces in real-time. Previous studies highlighted that the use of conventional control strategies in office buildings such as "static" operation schedules could cause large energy waste in particular during unoccupied hours. Common equipment load detection techniques, such as power meters and survey, are unable to provide comprehensive and real-time equipment information necessary for demand-driven strategies. Building up from our previous research, a deep learning-based equipment load detection approach which employs artificial intelligence (AI) enabled cameras is proposed to obtain the accurate heat gain prediction and enable HVAC systems to meet the requirement of thermal comfort.

PRELIMINARY RESULTS AND CONCLUSIONS

The experimental results presented a detection accuracy of equipment detection of 83.33%. To investigate the impact of the proposed approach on building energy performance, the case study building was modelled and simulated. The simulation results presented that up to 35.95% reduction of internal heat gains could be potentially achieved with the use of deep learning influenced equipment detection profiles in comparison with the use of static schedules. It highlights the benefits of using real-time deep learning detection method to provide profiles to HVAC systems to achieve demand-driven controls which can minimize unnecessary building energy consumption while maintaining a comfortable indoor environment.

MAIN REFERENCES

1. Menezes AC, Cripps A, Buswell RA, Wright J, Bouchlaghem D. Estimating the energy consumption and power demand of small power equipment in office buildings. Energy and Buildings. 2014/06/01/ 2014;75:199-209. doi:https://doi.org/10.1016/j.enbuild.2014.02.011

2. Gunay HB, O'Brien W, Beausoleil-Morrison I, Gilani S. Modeling plug-in equipment load patterns in private office spaces. Energy and Buildings. 2016/06/01/ 2016;121:234-249. doi:https://doi.org/10.1016/j.enbuild.2016.03.001

3. Mahdavi A, Tahmasebi F, Kayalar M. Prediction of plug loads in office buildings: Simplified and probabilistic methods. Energy and Buildings. 2016/10/01/ 2016;129:322-329. doi:https://doi.org/10.1016/j.enbuild.2016.08.022

4. Ren S, He K, Girshick R, Sun J. Faster R-CNN: Towards Real-Time Object Detection with Region Proposal Networks. IEEE Transactions on Pattern Analysis and Machine Intelligence. 06/04 2015;39doi:10.1109/TPAMI.2016.2577031

Keywords: Deep learning, equipment detection, building energy savings, built environment, HVAC

Session T4.1: Ensuring high quality building simulations

Paper ID 30184: Collecting data for urban building energy modelling by remote sensing and machine learning

Philip Gorzalka* (1), Oana M. Garbasevschi (2,5), Jacob Estevam Schmiedt (3), Ariane Droin (2), Magdalena Linkiewicz (4), Michael Wurm (2), Bernhard Hoffschmidt (3)

1: Institute of Solar Research at German Aerospace Center (DLR), Jülich, Germany; 2: German Remote Sensing Data Center (DFD) at German Aerospace Center (DLR), Oberpfaffenhofen, Germany; 3: Institute for the Protection of Terrestrial Infrastructures at German Aerospace Center (DLR), Sankt Augustin, Germany; 4: Institute of Optical Sensor Systems at German Aerospace Center (DLR), Berlin-Adlershof, Germany; 5: ifo Institute for Economic Research, Munich, Germany

AIM AND APPROACH

We present a method that provides input data for urban building energy modelling (UBEM), especially for situations in which the existing information about the target area is insufficient. Minimum requirements for its applicability are

1) a set of footprints of the buildings in the area, containing information on their usage,

2) archetypes connecting information on building age, usage and size to its energetically relevant physical properties.

We show how different data gaps can be filled through remote sensing and machine learning (ML) approaches: The 3D building geometry (including semantics) is derived automatically from aerial images using photogrammetry. Windows are detected on the façade textures. For the case of missing age data, we present a ML approach that connects age information to 3D features and urban metrics. The ML model is trained on regions with accurate age information. We classify the state of retrofit using aerial infrared thermography in combination with dynamic building simulation. The simulation uses the 3D building polygons to model façade shading and view factors for radiation exchange.

We apply our method to a test quarter to demonstrate its usefulness and evaluate its accuracy.

SCIENTIFIC INNOVATION AND RELEVANCE

The importance of UBEM for planning of urban energy systems and for evaluating policy measures to reduce greenhouse gas emissions has been widely acknowledged in the literature. A wide range of tools has been developed to support these goals. However, for a good accuracy of results they require high-quality input data. As available data for UBEM is often sparse and regionally different, our approach helps to make UBEM tools applicable to more areas. At the same time, using remote sensing data ensures that input data is up to date. Finally, its high grade of automatization makes our approach highly scalable.

PRELIMINARY RESULTS AND CONCLUSIONS

The application of our method to a test quarter shows that the derivation of a 3D polygonal geometry from aerial imagery works reliably. Window detection on façade textures suffers from low image quality and visibility issues, but it outperforms default values for the window-wall ratio. Deriving building age classes from other cities through the ML approach is tolerably accurate. There is room for improvement, for example by taking additional features into account and using training examples with a more balanced distribution of age classes. The estimation of retrofit states using aerial infrared thermography adds valuable information for the energy demand calculation. We point out possible improvements of the data recording and evaluation that will further improve the reliability of the results.

By comparing energy demands that are calculated using different stages of data enrichment in our approach and using different simulation methods, the usefulness of introducing remote sensing data is evaluated.

All in all, the presented method is a promising way to fill gaps in UBEM input data and/or to ensure models are based on up-to-date information.

MAIN REFERENCES

Estevam Schmiedt, J., and Schiricke, B. (2020), 'A Projection And Clipping Method To Calculate Direct, Diffuse, And Reflected Irradiation', in V. Corrado, E. Fabrizio, A. Gasparella et al. (eds.), Proceedings of Building Simulation 2019: 16th Conference of IBPSA (Building Simulation Conference proceedings: IBPSA), 4777–84.

Frommholz, D., Linkiewicz, M., Meissner, H. et al. (2017), 'Reconstructing Buildings with Discontinuities and Roof Overhangs from Oblique Aerial Imagery', Int. Arch. Photogramm. Remote Sens. Spatial Inf. Sci., XLII-1/W1: 465–471.

Garbasevschi, O.(2020), 'Large-scale building age classification for urban energy demand estimation: Geo and satellite data for building age identification', Master's thesis, Delft University of Technology. TU Delft Repository. http://resolver.tudelft.nl/uuid:ba4511dd-4ba0-43eb-b6c8-ce3896cd5dc3

Remmen, P., Lauster, M., Mans, M. et al. (2018), 'TEASER. An open tool for urban energy modelling of building stocks', Journal of Building Performance Simulation, 11/1: 84–98.

Keywords: urban building energy modelling, aerial infrared thermography, photogrammetry, machine learning, energy retrofit

Paper ID 30345: Gh-urban: a Grasshopper plugin for employing CityGML data in dynamical building simulations and the utilization of energy Application Domain Extension

Farzan Banihashemi*, Hari Krishna Gadi, Hannes Harter, Werner Lang

Technical University of Munich, Germany

AIM AND APPROACH

Energy simulations in the urban context are useful tools for decision-makers and policy-making. To consider the buildings in dynamical building simulations different data requirements need to be fulfilled. The information stored in CityGML data and Application Domain Extension (ADE) can be useful data to build simulation models. Through the 3d city models, relevant semantics and geometries can be retrieved for urban simulations.

This paper presents Gh-urban, an open-source python-based plugin developed to utilize CityGML data for dynamical building simulations in Grasshopper. The tool can read and modify relevant attributes and geometries of 3d city models for building simulations in Grasshopper. The acquired data will be enriched in the Energy ADE for different use cases. For proof of concept, the connection of the Urban Modeling Interface (umi) and the Gh-urban plugin is investigated in an urban district in Munich as a case study.

SCIENTIFIC INNOVATION AND RELEVANCE

In the case study, the methodology and the tool for generating relevant CityGML Energy ADE data for the urban simulations are demonstrated. Based on the building age class and shape of the buildings a data bank for materials and components has been created and integrated into the plugin. Furthermore, different refurbishment and densification scenarios were generated via the developed tool for policy making and energy optimization.

The integration of the Gh-urban plugin and dynamical building simulations demonstrates an innovative approach to modify CityGML data and store the relevant energy simulation results and boundaries in the Energy ADE. The enriched CityGML data can be utilized and viewed by different decision-makers for better policy implications.

PRELIMINARY RESULTS AND CONCLUSIONS

The integration of CityGML data through the developed plugin with a dynamical urban simulation demonstrates the usefulness of 3d city models for optimizations on the urban scale. However, CityGML data are mainly available in the lower level of details (LOD 1 and 2). Therefore, data enrichment through the plugin is required. The developed tool and methodology can be applied to different simulation programs based on Grasshopper. Additionally, the enriched semantics of the buildings can be applied for other use cases, such as life cycle assessments on the urban scale.

MAIN REFERENCES

Jusuf, S., Mousseau, B., Godfroid, G.,Hui, V. (2017). Integrated modeling of CityGML and IFC for city/neighborhood development for urban microclimates analysis. Energy Procedia. 122. 145-150. 10.1016/j.egypro.2017.07.329.

Agugiaro, G., Benner, J., Cipriano, P. et al. (2018). The Energy Application Domain Extension for CityGML: enhancing interoperability for urban energy simulations. Open geospatial data, softw. stand. 3, 2. https://doi.org/10.1186/s40965-018-0042-y

Biljecki, F., Stoter, J., Ledoux, H., Zlatanova, S., Çöltekin, A. (2015). Applications of 3D City Models: State of the Art Review. ISPRS Int. J. Geo-Inf., 4, 2842-2889.

Biljecki, F., Kumar, K., Nagel, C. (2018). CityGML Application Domain Extension (ADE): overview of developments. Open geospatial data, softw. stand. 3, 13 . https://doi.org/10.1186/s40965-018-0055-6

Remmen, P., Lauster, M., Mans, M., Osterhage, T., Mueller, D. (2016). CityGML Import and Export for Dynamic Building Performance Simulation in Modelica, Building Simulation and Optimization conference (BSO2016), Newcastle upon Tyne, UK.

Reinhart, C., Dogan, T., Jakubiec, J., Rakha, T., Sang, A. (2013). UMI - An urban simulation environment for building energy use, daylighting and walkability. Proceedings of BS 2013: 13th Conference of the International Building Performance Simulation Association.

Keywords: CityGML, Energy ADE, Grasshopper

Paper ID 31113: Global sensitivity analysis for large scale building energy models: importance of building stock size and convergence

Matthias Yvan C. Van Hove*, Marc Delghust, Jelle Laverge

Ghent University, Department of Architecture and Urban Planning, Building Physics Research Group, Belgium

AIM AND APPROACH

Though uncertainty and sensitivity analysis methods have been widely applied in the context of building energy modelling (BEM), only a limited amount of studies have investigated their performance at scale. This paper aims at broadening the knowledge of global sensitivity analysis application at building stock level by applying the Sobol' SA at an internally developed building stock model. Further, the influence of the building stock size on the GSA model results will be addressed as well as the robustness of the GSA indices.

SCIENTIFIC INNOVATION AND RELEVANCE

Few research into uncertainty and sensitivity analyses at urban-scale building energy models has happened so far, let alone studies in which multiple techniques are compared with each other. There is an urgent need to broaden the research done in this field and come up with studies where these uncertainty and sensitivity tools are applied at urban scale to analyse their performance, applicability and feasibility. In order to address this, this research evaluates three different sensitivity analysis methods for a quasi-steady state building stock model with a set of 1076 calibrated detached houses. The study will compare the Sobol, DGSM and OAT sensitivity analysis techniques against each other by means of clear graphs and statistical metrics, give insights about the needed stock dimensions, handle convergence and speak about computation burden.

PRELIMINARY RESULTS AND CONCLUSIONS

The building stock simulations are still running. At the moment, the first 22000 buildings stocks (i.e., 1076 detached buildings) are calculated and evaluated using the Sobol sensitivity analysis method. These results clearly show an input parameter ranking with first, second and total sensitivity indices. In order to come to conclusions of the comparison between multiple sensitivity analysis techniques, we are waiting for the model evaluations of the other methods (i.e., DGSM and OAT). Computational cost comparison will be compared using the number of model evaluations needed and the point of convergence for each method. The non-parametric Kendall-Tau correlation test is used to evaluate the parameter ranking of the three sensitivity analysis methods and since there are three, a Spearman Test can be added as well. If all goes well and efficient, the tests can be done on two more different stocks or a combination of the three (i.e., detached, semi detached and terraced houses).

MAIN REFERENCES

Saltelli, A., P. Annoni, I. Azzini, F. Campolongo, M. Ratto, and S. Tarantola (2010). "Variance based sensitivity analysis of model output. Design and estimator for the total sensitivity index." Computer Physics Communications, 181(2):259-270

Sobol, I. M. and S. Kucherenko (2009). "Derivative based global sensitivity measures and their link with global sensitivity indices." Mathematics and Computers in Simulation, 79(10):3009-3017

Zhao, F., Lee, S. H., & Augenbroe, G. (2016). Reconstructing building stock to replicate energy consumption data. Energy and Buildings, 117, 301–312.

Keywords: Building stock modelling, Sobol' sensitivity analysis, Uncertainty Analysis, Quasi-steady-state model, IEA EBC Annex 70

Paper ID 30672: Uncertainty analysis of life cycle assessment input parameters on city quarter level

Hannes Harter* (1), Bruno Willenborg (2), Patricia Schneider-Marin (1), Farzan Banihashemi (1), Michael Vollmer (1), Kierdorf Daniel (1), Werner Lang (1), Thomas H. Kolbe (2)

1: Institute of Energy Efficient and Sustainable Design and Building, Technical University of Munich, Germany; 2: Chair of Geoinformatics, Technical University of Munich, Germany

AIM AND APPROACH

When talking about sustainable building or city development, it is mandatory to extend the analysis and evaluation of the ecological and energetic performance of buildings to a life-cycle based approach. Life Cycle Assessments (LCA) allows e.g. the identification of savings potentials and to support political decision-making regarding regulations. Especially regarding the political aspects, it is important to calculate LCAs for large building stocks. In previous studies a methodological approach for an automated, static LCA calculation of large residential building stocks has been developed and applied to a case study with over 115.000 buildings, using 3D city models based on CityGML-format. At urban district level, however, there are often data gaps in relation to relevant input data for calculations that are closed by making assumptions. These assumptions lead to uncertainties in calculation and interpretation of results. In order to know which parameters (around 100) have the greatest influence on LCA results, an uncertainty analysis is carried out using a case study with more than 300 buildings. The determined parameters also represent a key for sustainable development. By identifying the parameters, recommendations for action for municipalities can also be drawn up in relation to data acquisition for the implementation of such calculations.

SCIENTIFIC INNOVATION AND RELEVANCE

In addition to the high relevance of the thematic aspect of sustainable building, district and city development, the described approach offers a possibility to optimize and improve the district and city-related life cycle-based energy and emission-related simulation. It thereby contributes to the state of research. Through the results, which provide valuable information about the parameters, e.g. u-values (describing energetic standard of building), heat generator for heating and domestic hot water or different climatic conditions (reflecting climate change), that have the greatest influence on the calculation of the LCA results, input parameters and thus simulations can be optimized. Furthermore, one can significantly improve the LCA calculation results of large building stocks. Informing municipalities / city planners / geoinformatics / etc. about which parameters can have a high impact on LCA results, specific data can be acquired with high accuracy, which in turn can have a positive effect on not only the LCA calculation but also further simulations (e.g. thermal comfort in the interior). The development of strategies to minimize the life cycle-based influence of buildings in terms of resource consumption and their emissions can serve to demonstrate new scientific and sociopolitical goals and to develop general recommendations for action.

PRELIMINARY RESULTS AND CONCLUSIONS

Initial studies show that uncertainty and resulting variance in input parameters for LCA calculation at the district level can have a significant influence on the spread of the results. This can result in different interpretations and recommendations for action, which in turn can lead to different actions. The missing information on heat generators currently installed in existing buildings or missing information on renovation status of buildings on a city level can be cited as an example. Both missing information are replaced by the use of developed building typologies, e.g. depict country-specific, but no local character, which lead to large differences in the calculation results. Even if or precisely because these gaps in information cannot be closed in a timely manner, it is important to know which sensitivities or uncertainties are associated with simulations and results that are input-limited so that this information can be incorporated into the interpretation of the results. It can thus be stated that a sensitivity analysis of the input parameters for a city district-related LCA simulation represents a necessary and important step in the long way to a sustainable building stock.

MAIN REFERENCES

P. Schneider-Marin, H. Harter, K. Tkachuk, and W. Lang, "Uncertainty analysis of embedded energy and greenhouse gas emissions using BIM in early design stages," Sustain., vol. 12, no. 7, 2020.

A. Ciroth, G. Fleischer, and J. Steinbach, "Uncertainty calculation in life cycle assessments: A combined model of simulation and approximation," International Journal of Life Cycle Assessment, vol. 9, no. 4. Springer Verlag, pp. 216–226, 2004.

A. Mastrucci, A. Marvuglia, U. Leopold, and E. Benetto, "Life Cycle Assessment of building stocks from urban to transnational scales: A review," Renewable and Sustainable Energy Reviews, vol. 74. Elsevier Ltd, pp. 316–332, 01-Jul-2017.

E. Igos, E. Benetto, R. Meyer, P. Baustert, and B. Othoniel, "How to treat uncertainties in life cycle assessment studies?," Int. J. Life Cycle Assess., vol. 24, no. 4, pp. 794–807, Apr. 2019.

Keywords: Uncertainty Analysis, Life Cycle Assessment (LCA), Sustainable Building, CityGML, Urban Simulation

Paper ID 30573: Determination of household area and number of occupants for residential buildings based on census data and 3D CityGML building models for entire municipalities in Germany

Sally Köhler*, Matthias Betz, Keyu Bao, Verena Weiler, Bastian Schröter

Center for Sustainable Energy Technology, Stuttgart University of Applied Sciences, Stuttgart, Germany

AIM AND APPROACH

This paper presents a method for linking 3D building models based on CityGML standards with German census data to derive information on household area and the number of occupants per household for all types of residential buildings (e.g., apartments, terraced houses). Acquiring this information for entire neighborhoods is an important preparatory step for calculating occupancy-dependent demand profiles, e.g. for electricity or water within the modelling of urban energy systems [1]. First, the number of floors and the heated area per building is calculated based on the geometric information contained in the 3D building models. Secondly, the number of households per building is estimated using the average household area of the respective building type. Having information on the heated area and the number of households per building, household areas are assigned based on a frequency density distribution (FDD) and a greedy algorithm. The FDD is derived from data of the 2011 census [2] and gives the number of households depending on their size in 20 m² intervals. Lastly, occupants are allocated to the "sized" households using a second FDD, which links the household area to the number of occupants. These steps are applied for all residential buildings within the CityGML-file.

SCIENTIFIC INNOVATION AND RELEVANCE

Simulating time-resolved, occupancy-dependent demand profiles at building level for entire neighborhoods enables energy planers to consider aspects like sector coupling or networks/grid implications of "new" technologies such as electric vehicles or heat pumps more accurately [3]. To achieve this, more detailed information about buildings and their occupants is required than is typically available, for example for data protection reasons. Defining the level of granularity and detail needed, in order to create meaningful results, is a much-discussed topic in this area of expertise. The demonstrated method – although only a preparatory step for later applications – balances a high degree of accuracy and necessary simplifications for subsequent energy simulations on neighborhood scale. The data required for applying this method is usually available and the proposed approach requires little computational effort, yet the additional information available allows to calculate e.g. the electricity or water demands more accurately and in a differentiated manner for individual buildings. Furthermore, a similar method is under development for office buildings to calculate the number of workplaces, which is an important parameter for generating commercial building demand profiles, and which would thus extend the scope to non-residential buildings.

PRELIMINARY RESULTS AND CONCLUSIONS

The described approach has already been tested on three case studies with varying population densities. Comparing the calculated number of occupants of a municipality or a city quarter with statistical data, the results show an average deviation of -5%, ranging mostly between -15% and +15%.The method has been implemented inside the workflow-driven energy simulation platform SimStadt [4] [5]. Therein, the number of households and the number of occupants can be displayed; additionally, the respective household area in m^2 and their number of occupants are generated for each residential building type. This data can then be accessed by other calculation workflows in the SimStadt platform.

MAIN REFERENCES

[1] S. Köhler, M. Betz and U. Eicker, "Stochastic generation of household electricity load profiles in 15-minute resolution on building level for whole city quarters," in 16th IAEE European Conference, Ljubljana, 2019.

[2] Zensusdatenbank Zensus 2011, "https://ergebnisse.zensus2011.de/," 2011.

[3] P. Mancarella, G. Andersson, J. A. Pecas-Lopes and K. R. W. Bell, "Modelling of integrated multi-energy systems: drivers, requirements and opportunies," in Proc. Power Systems Computation Conference (PSCC), Genoa, 2016.

[4] R. Nouvel, K. Brassel, M. Bruse, E. Duminil, V. Coors, U. Eicker and D. Robinson, "SimStadt, a new workflow-driven urban energy simulation platform for CityGML city models," in CISBAT 2015 Proceedings, Lausanne, 2015.

[5] V. Weiler and U. Eicker, "Individual domestic hot water profiles for building simulation at urban scale," Proceedings IBPSA Building Simulation Conference, Rom, 2019.

Keywords: 3D building models, Interrelation of household area and number of occupants, Energy simulation

Session T4.2: Ensuring high quality building simulations

Paper ID 30687: Embedded single-board controller for Double Skin Facade : a co-simulation virtual test bed

Giovanni Gennaro* (1,2), Francesco Goia (3), Giuseppe De Michele (2), Marco Perino (1), Fabio Favoino (1)

1: Politecnico di Torino, Italy; 2: Eurac Research, Italy; 3: Norwegian University of Science and Technology, NTNU, Norway

AIM AND APPROACH

Dynamic transparent facades are multi-functional building systems able to change their thermophysical properties (e.g. g-value) in response to external stimuli or control logics, in order to meet different requirements such as energy efficiency and occupant comfort (thermal, visual, IAQ, acoustic). During building operation, in order to optimize the façade configurations according to the various and sometimes conflicting requirements, the design and implementation of the control method and system becomes important to ensure the achievement of the desired performance. The aim of this paper is to describe a framework for the implementation of real-time embedded controller for transparent dynamic facades, applied to a simple case study of a small office environment equipped with a Double Skin Façade (DSF). The core of the controller is the Raspberry Pi 4, a low-cost platform with powerful processors and thanks to its small size it can be easily integrated inside the façade together with the embedded sensors. Several execution and computational tasks can be conducted on this system efficiently, such as data analysis, building energy balances estimation and façade actuators control. Based on the experience and best practices found in literature both a calibrated white-box model developed in EnergyPlus and a rule-based controller have implemented.

SCIENTIFIC INNOVATION AND RELEVANCE

Different control strategies could be implemented for DSF: from simpler rule-based one, to more complex model-based control strategies. The latter could be based on either reduced or physical models, and for both models, depending on the computational time compared to the control timestamp, either an embedded controller (integrated with the façade) or an external processor (exchanging sensed data and actuated variables with a local controller) could be adopted. The present work investigates the potential of embedding model-based and rule-based control strategy in an integrated controller within the façade, relying on low-cost IoT sensors and processors. For this sake, a real case study (i.e. DSF mock-up mounted on an outdoor facility of the Polytechnic of Turin with an operable venetian blind) is used to compare the performance of the two control strategies adopting the following steps: (i) the white model of the DSF (EnergyPlus based) is created and calibrated to experimental data; (ii) the model simulation is synchronized with real-time climate data and embedded in the controller; (iii) for each control time step the white model is adopted to perform control decision making to optimize a certain performance objective; (iv) this process is compared to rule-based decision based on best practices.

PRELIMINARY RESULTS AND CONCLUSIONS

The development of a framework to test embedded controller for dynamic transparent facades presents different challenges and requirements: (i) the update of weather data acquired from a local climate station at each control time step, in order to perform model simulations with the real boundary conditions; (ii) the real-time integration of the measured data with the model; (iii) the computational time for the co-simulation of the models must be less than the time step of the control; (iv) the accuracy of simulated results. The main feature of this framework is the flexibility in terms of control strategies, limited only by the computational performance of the controller, which allows to implement control strategies based on physical, reduced models and simple rules. The computational performance of the Raspberry Pi allows to run several simulations (such as the number of the configuration that the dynamic facades can assume) within each control time step, to predict the optimal DSF configuration, providing very accurate results. Moreover, the integration in a single control system of sensors, control algorithms and actuators could be a robust solution for an embedded controller a dynamic façade, providing a more accurate prediction and decision-making support to the higher-level building control system.

MAIN REFERENCES

1. Loonen R.C.G.M., Favoino F., Hensen J.L.M., Overend M., Review of current status, requirements and opportunities for building performance simulation of adaptive facades. Journal of Building Performance Simulation 10 (2017).

2. Catto Lucchino E., Goia F., Lobaccaro G., Chaudhary G., Modelling of double skin facades in whole-building energy simulation tools: A review of current practices and possibilities for future developments. Building Simulation 12 (2019).

3. Aftab M., Chen C., Chau C.K., Rahwan T., Automatic HVAC control with real-time occupancy recognition and simulation-guided model predictive control in low-cost embedded system. Energy and Buildings 154 (2017).

Keywords: Dynamic transparent façades, Double Skin Façade, Façade control, Energyplus, Raspberry Pi

Paper ID 30390: Pyrano – A Python package for LiDAR-based solar irradiance simulations

Ádám Bognár*, Roel C.G.M. Loonen, Jan L.M. Hensen

Eindhoven University of Technology, The Netherlands

AIM AND APPROACH

Solar irradiance is a key input for modeling photovoltaic (PV) system performance and the influence of solar heat gains on a building's energy balance. In the BPS domain, efficient techniques for including the effects of obstructions and reflections have been developed by daylight modeling researchers, but these are often only used to calculate daylight metrics in interior spaces. However, with some adjustments, such methods could also be applied on external built surfaces to calculate the solar irradiance input for PV or solar heat gain simulations.

The goal of this paper is to introduce Pyrano, a new Python package for simulating solar irradiance on external built surfaces. Pyrano consists of five modules:

- A geometry pre-processor that handles irradiance sensor-point placement over EnergyPlus surfaces.

- A LiDAR point-cloud pre-processor for applying geometrical transformations on the LiDAR point cloud to align its coordinate system with the EnergyPlus geometry.

- A Python wrapper to execute certain Radiance sub-programs, such as epw2wea, gendaymtx and dctimestep.

- An input-output module for results visualization and connecting inputs and outputs between solar irradiance and PV simulation software.

- A module for calculating flux-transfer coefficients from pre-processed LiDAR point clouds for efficient matrix-based (sub)hourly annual solar irradiance simulations.

SCIENTIFIC INNOVATION AND RELEVANCE

If a building is situated in the built environment, its solar access is often influenced by reflections or shading by vegetation or other buildings. To be able to take this into account the geometry and reflectance properties of the surroundings need to be known, however, this input is often hard to acquire. The software presented in this work utilizes the 2.5 phase solar irradiance modeling method which was developed with a focus on the requirements of PV system and building solar heat gain simulations in urban context. It allows for taking into account shading and reflections based on the raw LiDAR point-cloud of the surroundings without the need for generating 3D surfaces nor conducting ray-tracing.

Simulation workflows for modeling PV together with a building situated in an urban environment are fragmented. Modeling of urban PV systems requires multidisciplinary knowledge from the modelers about simulating solar irradiance, building physics and electrical systems. Building energy modelers are usually architects, building engineers, or engineers specialized in BPS, rarely electrical engineers. This might have slowed the adoption of including PV for BPS investigations. In an attempt to address this issue, Pyrano bridges the gap between EnergyPlus (building energy), Radiance (irradiance) PVMismatch (PV power) simulations.

PRELIMINARY RESULTS AND CONCLUSIONS

Pyrano was developed in a way that the inputs it uses are compatible with the inputs used by the state of the art ray-tracing software Radiance. This makes it easy to validate the results of the software. The simulations with the proposed method were compared to Radiance simulations with various realistic case studies, showing less than 4% deviation in the simulated annual solar irradiance.

The full-paper will explain the underlying modeling methods and will demonstrate Pyrano with a case study of calculating solar heat gains and PV yield of a building situated in a dense urban environment.

Pyrano is free and open-source. The source code and tutorials are available at: https://gitlab.tue.nl/bp-tue/pyrano. The python package can be installed from: https://pypi.org/project/pyrano/.

MAIN REFERENCES

Subramaniam, S., 2017. Daylighting Simulations with Radiance using Matrix-based Methods.

Ward Larson, G., Shaskespeare, R., 2003. Rendering with Radiance. Morgan Kaufmann Publishers.

Reinhart, C.F., 2001. Daylight Availability and Manual Lighting Control in Office Buildings – Simulation Studies and Analysis of Measurements.

Mardaljevic, J., 2000. Daylight Simulation: Validation, Sky Models and Daylight Coefficient. De Montfort University Leicester.

Tregenza, P.R., Waters, I.M., 1983. Daylight coefficients. Light. Res. Technol. 15, 65–71.

Keywords: Matrix-based simulations, Solar irradiance, Photovoltaics, Solar heat gain, LiDAR point-clouds

Paper ID 30133: A practical approach for modelling PV off-grid systems in EnergyPlus using post-processing of data to identify black out days

Valentina Tomat*, Alfonso P. Ramallo-González, Antonio F. Skarmeta-Gómez

University of Murcia, Spain

AIM AND APPROACH

Photovoltaic (PV) installations are considered a key element in the fight against energy waste, allowing users to self-produce the energy needed according to their specific demand. Their market spread increased in the last years, because of both the technical advances and more affordable costs. [1]

In remote areas, where the connection to the grid is not possible and it would bring a high cost to bring the power supply, the off-grid PV system is an excellent solution. [2][3] Nevertheless, the sizing of a standalone system in the literature is mainly defined through intuitive methods, numerical methods and analytical methods [4] while, when it comes to model a dynamic simulation, most software does not provide specific tools. Common energy simulation software like EnergyPlus and TRNSYS allow to design grid-tied systems and hybrid systems (grid-tied systems with a battery back-up to avoid outages), but there is almost no literature in how to model off-grid systems. This project aims to propose an equivalent method to simulate the off-grid photovoltaic system with the most popular software: EnergyPlus.

SCIENTIFIC INNOVATION AND RELEVANCE

The main idea behind this paper is to use the simulator EnergyPlus to understand whether or not it is possible to model a standalone system starting from the design of a hybrid system, considering that the two systems are quite similar in design and components. The installation itself hardly provides enough energy to cover the demand throughout the year, because of weather unpredictability and because of unusual peak-demands. In a hybrid system, this scenario is solved by the eventual connection to the grid, that assures energy providing no matter the situation. In off-grid systems, when the installation does not cover the energy demand, black-out days occur.

Considered the similarity of the two situations, simulating a hybrid system, we can expect two main cases: In the first one, the building results to be autonomy all over the year, making the grid connection superfluous (when importation from facility does not occur at all, the building is considered able to operate also in off-grid conditions without outages [5]); in the second case, the model presents electricity coming from the utility: this energy can be converted in black-out days of an equivalent off-grid PV system.

PRELIMINARY RESULTS AND CONCLUSIONS

To the best of the authors' knowledge, the latter case has never been presented in a scientific paper. To test this method, the case study of an isolated house in the Region of Murcia, Spain, is presented. The building is modelled through SketchUp and Openstudio, while the dynamic energy simulation is obtained through the widely used software EnergyPlus. Results are given considering the number of days in which the battery charge at the end of the day is lower than 10%, i.e. the probability of a blackout day is high. Ten different scenarios are presented, differing in terms of electricity demand profile, PV peak power and battery storage capacity. Afterwards, the same scenarios are simulated with PVGIS, a tool implemented by the European Commission [6] to validate the results obtained with EnergyPlus. PVGIS is set to present the same electricity schedule and the same installation that has been used in EnergyPlus, to allow a more precise comparison. The comparison between the two methods shows good accuracy, inasmuch as the percentage of days varies between 0.5% and 11.9%, i.e., approximating by excess, between 2 and 44 blackout days of difference in the prevision.

MAIN REFERENCES

[1] Barbose, G. L., Darghouth, N. R., Millstein, D., Spears, M., Wiser, R. H., Buckley, M. & Grue, N. (2015). Tracking the sun VIII: the installed price of residential and non-residential photovoltaic systems in the United States.

[2] Zahedi, A., 2006. Solar photovoltaic (PV) energy; latest developments in the building integrated and hybrid PV systems, Renewable Energy, 31 (5) (2006) 711-718.

[3] Bekele, G., Tadesse, G., 2012. Feasibility study of small Hydro/PV/Wind hybrid system for off-grid rural electrification in Ethiopia, Applied Energy, 97 (2012) 5-15.

[4] Khatib, T., Mohamed, A., Sopian, K., 2013. A review of photovoltaic systems size optimization techniques, Renewable and Sustainable Energy Reviews 22 (2013) 454-465.

[5] Brumana, G., Franchini, G., Perdichizzi, A., 2017. Design and Performance Prediction of an Energy+ Building in Dubai, ScienceDirect Energy Procedia 126 (2017) 155-162.

[6] Huld, T., Müller, R., Gambardella, A., 2012. A new solar radiation database for estimating PV performance in Europe and Africa, Solar Energy 86 (6) (2012) 1803-1815.

Keywords: photovoltaic, off-grid, EnergyPlus, simulation, battery

Paper ID 30287: Initial validation of the one-diode photovoltaic model for the flexible panels

Dominika Knera*, Dariusz Heim, Michał Krempski-Smejda

Lodz University of Technology, Poland

AIM AND APPROACH

The flexible photovoltaic panels (FPV) became more and more popular in the building applications. In comparison with traditional, mainly crystalline silicon, the thin-film panels characterise by lightweight, low production cost or suitability for curved surfaces. The most popular thin-film PV are the Cadmium telluride (CdTe)/Cadmium sulphide (CdS) [1], as well as amorphous silicon and CIS/CIGS technologies [2]. In presented analysis flexible CIGS and semi-flexible crystalline silicon photovoltaic panels were tested experimentally in two configurations: free-standing (FSPV) and integrated with the wall (WIPV). Two aims of the study were formulated: to compare the performance of both PV panels and to verify the existing one-diode equivalent models implemented in ESP-r.

The experiments were conducted during selected days. Both PV configurations were tested parallel by measurements of power flow as well as the temperature of the modules. Additionally, the basic weather data were also registered, e.g. total and diffuse solar radiation, air temperature. The models of both installations (FSPV & WIPV) was developed in ESP-r using two one-diode photovoltaic models: Kelly's and Watsun-PV model. The simulation results were compared with experiments.

SCIENTIFIC INNOVATION AND RELEVANCE

Flexible PV technology has the potential to grow with a wide spectrum of application [3]. The advantage of them in comparison to stiff panels is visible in case of uneven surfaces. Additionally, building integrated flexible PV (BIPV) panels are usually much lighter than traditional. Those features make flexible PV more and more popular in building construction sector.

The existing PV models in ESP-r were validated for traditional crystalline PV modules equipped with glass sheets and aluminium frame [4]. The original part of this study is model validation for flexible thin-film PV under normal and extreme temperature. For FSPV the panel temperature slightly rises above the ambient air temperature because the effect of air cooling is high. When the PV panel is tightly joined with the building wall (WIPV) the risk of overheating rises rapidly and the effect of temperature on PV efficiency is much more visible. Additionally, the spectral characteristics of incident solar radiation will be analysed considering the visual and infrared part monitored on-site.

PRELIMINARY RESULTS AND CONCLUSIONS

The models of both constructions (FSPV & WIPV) and two photovoltaic technologies (c-Si and CIGS) were defined in ESP-r. The simulations were conducted using two one-diode photovoltaic models: Kelly's and Watsun-PV model. The initial results show the temperature and power flow for analyzed PV modules. The difference between power production and temperature is recognized. The experiments were done for clear and cloudy skies conditions. The final conclusion will be formulated based on the comparison between experimental and numerical results.

MAIN REFERENCES

[1] Visa I, Burduhos B, Neagoe M, Moldovan M, Duta A. Comparative analysis of the infield response of five types of photovoltaic modules. Renew Energy 2016;95:178–90. https://doi.org/https://doi.org/10.1016/j.renene.2016.04.003.

[2] Pandey AK, Tyagi V V., Selvaraj JA, Rahim NA, Tyagi SK. Recent advances in solar photovoltaic systems for emerging trends and advanced applications. Renew Sustain Energy Rev 2016;53:859–84. https://doi.org/10.1016/j.rser.2015.09.043.

[3] Ramanujam J, Bishop DM, Todorov TK, Gunawan O, Rath J, Nekovei R, et al. Flexible CIGS, CdTe and a-Si:H based thin film solar cells: A review. Prog Mater Sci 2019:100619. https://doi.org/10.1016/j.pmatsci.2019.100619.

[4] Mottillo M, Beausoleil-Morrison I, Couture L, Poissant Y. A comparison and validation of two photovoltaic models. Can. Sol. Build. Conf., 2006.

Keywords: solar energy, façade PV, power flow modelling

Session T4.3: Buildings paving the way for the energy transition

Paper ID 30180: Validated combined first and last year borefield sizing methodology.

Wouter Peere* (1), Damien Picard (1,2), Iago Cupeiro Figueroa (1), Wim Boydens (2,4), Lieve Helsen (1,3)

1: Department of Mechanical Engineering, University of Leuven (KU Leuven), Leuven, Belgium; 2: Boydens Engineering, Dilbeek, Belgium; 3: EnergyVille, Thor Park, Waterschei, Belgium; 4: Department of Architecture and Urban planning, University of Ghent, Ghent, Belgium

AIM AND APPROACH

This paper presents a new easy to use sizing method for borefields. Starting from the ASHRAE three-pulse sizing method (Ahmadfard, 2018a), there are mainly two variants found in literature: sizing based on the first year of operation (Carcel, 2016) and sizing based on the last year of operation (Ahmadfard, 2018b). In this paper we combine both methods in order to come up with an overall more correct and complete sizing method. This combined sizing method is then validated against an existing sizing method using earth energy designer (Heelström, 2000) and a dynamic simulation in Modelica (Laferrière, 2020) for both a heating and cooling dominated case from the point of view of the borefield.

SCIENTIFIC INNOVATION AND RELEVANCE

The combination of these two sizing methods, using both the first and last year of operation, is novel. The obtained method is fast and easy to implement and gives for specific borefield loads a more correct result than the existing individual methods. The correct sizing of a borefield is relevant due to the high cost associated to these investments. Keeping targets against climate change in mind, geothermal heating and cooling systems, using borefields, are more used nowadays. To make this technology attractive, the cost needs to be correct and not overstated. This method gives designers a new and quick tool to (more) correctly calculate the borefield size.

PRELIMINARY RESULTS AND CONCLUSIONS

This method is thoroughly validated using earth energy designer for both a cooling dominated and a heating dominated case from the point of view of the borefield. The validation results show that the current individual methods do not give the correct result in some cases (typically leading to undersized borefields) while this combined method does so. Moreover, this novel method is validated by a dynamic simulation using the IDEAS-library in Modelica to check whether the fluid temperature restrictions are kept. The presented combined sizing method sizes the field within these limits, whereas in some cases the current sizing methods on either the first or last year of operation don't.

MAIN REFERENCES

M. Ahmadfard. A Comprehensive Review of Vertical Ground Heat Exchangers

Sizing Models With Suggested Improvements. PhD thesis, École Polytechnique

de Montréal, 2018.

Mohammadamin Ahmadfard and Michel Bernier. Modifications to ASHRAE's

sizing method for vertical ground heat exchangers. Science and Technology for

the Built Environment, 24(7):803–817, 2018.

Patricia Monzó Cárcel, Michel Bernier, José Acuña, and Palne Mogensen. A

monthly based bore field sizing methodology with applications to optimum

borehole spacing. ASHRAE Transactions, 122, 01 2016.

G. Heelström, B. Sanner, Earth Energy Designer (EED), User's Manual, version 2 (2000).

Laferrière, A., Cimmino, M., Picard, D., Helsen, L. with Cimmino, M. (corresp. author) (2020). Development and validation of a full-time-scale semi-analytical model for the short- and long-term simulation of vertical geothermal bore fields. Geothermics, 86, 101788-101788. doi: 10.1016/j.geothermics.2019.101788 Open Access

Keywords: Borefield sizing method, ASHRAE, three-pulse, first and last year of operation, validation

Paper ID 30770: Efficiency of a heated air curtain under cross-jet temperature gradients

Claudio Alanis Ruiz* (1), Twan van Hooff (2,1), Bert Blocken (2), GertJan van Heijst (3)

1: Building Physics and Sustainable Design Section, KU Leuven, Belgium; 2: Building Physics and Services Unit, Eindhoven University of Technology, The Netherlands; 3: Fluids and Flows, Eindhoven University of Technology, The Netherlands

AIM AND APPROACH

Air curtains, which are essentially devices that generate an aerodynamic barrier, are used to reduce infiltration through entrance doors where transit is frequent. While isothermal jet air curtains (only driven by momentum) are commonplace, the use of air curtains with buoyant jets is fairly normal for certain applications. An example of the latter are heated air curtains at the entrance of buildings in cold and moderate climates, which are subjected to temperature differences between indoor and outdoor environments. In these air curtains an interaction occurs between natural convection, jet buoyancy and jet momentum. Despite the frequent occurrence of such scenarios, little has been done in investigating this interaction and its influence on air-curtain performance. Therefore, this study aims to provide an analysis of the added effect of jet temperature and ambient temperature on the overall flow behavior and the efficiency of a heated air curtain at the entrance of a building. The analysis is performed numerically using validated computational fluid dynamics (CFD) simulations. Two performance indicators are considered: (1) separation efficiency (related to mass transport), and (2) thermal efficiency (related to heat transport).

SCIENTIFIC INNOVATION AND RELEVANCE

Air infiltration is usually associated with increased energy demand in buildings. Moreover, problems concerning thermal comfort and indoor air quality can in many instances also be attributed to air infiltration. The infiltration of unconditioned air through access doors significantly contributes to these potential problems, especially in buildings where transit through these doors is frequent. For such reasons, air curtains are often implemented at these locations in order to facilitate the transit of people, vehicles and materials while minimizing energy losses, decreasing the transport of outdoor air pollutants to the inside and reducing thermal discomfort (e.g., due to wind draft and vertical air temperature differences). By virtue of the value that air curtains can provide in building applications, understanding their behavior and improving their performance is of great interest. The present contribution provides insights on the interaction of ambient conditions and air-curtain operating conditions, which has not been frequently studied in detail before. The analysis is conducted for a variety of representative combinations of ambient and air curtain operating conditions and overall performance is evaluated in terms of both heat and mass transfer.

PRELIMINARY RESULTS AND CONCLUSIONS

The results indicate that from energy conservation and pollution control perspectives, the use of a heated air curtain is not favorable over the use of an isothermal air curtain, since in every instance the heated air curtain yields additional heat losses and infiltration. Nevertheless, from a thermal comfort standpoint, it could be that the adoption of heated air curtains is beneficial to counteract potential factors of thermal discomfort at building entrances such as cold air drafts.

MAIN REFERENCES

Alanis Ruiz C., van Hooff T., Blocken B., and van Heijst G.J.F. 2018. CFD analysis of the effect of pressure gradients on the separation efficiency of a generic air curtain. Proceedings of Roomvent&Ventilation 2018, Helsinki, Finland: REHVA, pp. 241-246.

Brinks P., Kornadt O., and Oly R. 2015. Air infiltration assessment for industrial buildings. Energy and Buildings, pp. 663 - 76.

Khayrullina A., van Hooff T., Blocken B., and van Heijst G.J.F. 2017. PIV measurements of isothermal plane turbulent impinging jets at moderate Reynolds numbers. Experiments in Fluids, Vol. 58:31.

Lstiburek J., Pressnail K., and Timusk J. 2002. Air pressure and building envelopes. Journal of Building Physics, Vol. 26(1) pp. 53-91.

Younes C., Abi-Shdid C., and Bitsuamlak G. 2012. Air infiltration through building envelopes: A review. Journal of Building Physics, pp. 267-302.

Keywords: air curtain performance, separation efficiency, thermal efficiency, infiltration, computational fluid dynamics (CFD)

Paper ID 30530: Development of the hybridGEOTABS design methodology for feasibility study and pre-design

Eline Himpe* (1), Mohsen Sharifi (1), Rana Mahmoud (1), Filip Jorissen (3), Lieve Helsen (3), Wim Boydens (2), Jelle Laverge (1)

1: Ghent University, Belgium; 2: Boydens Engineering, Belgium; 3: KU Leuven, Belgium

AIM AND APPROACH

GEOTABS refers to the combination of a geothermal heat pump with thermally activated building systems, and is applied in low temperature heating and high temperature cooling of buildings. TABS is a radiant heating and cooling system and is beneficial in terms of thermal comfort and energy efficiency. When combined with a geothermal heat pump, it allows to make efficient use of low grade renewable energy sources. In order to maintain the thermal comfort when sudden and significant changes in heating or cooling loads appear and to maintain the thermal balance of the geothermal source, a secondary heating and cooling emission and generation system can be added, together with model predictive controls to optimise the operation of the hybrid system. This concept is called 'hybridGEOTABS'. In context of the hybridGEOTABS Horizon 2020-project, we developed a methodology to allow a near-optimal design for such hybridGEOTABS buildings, that is applicable in the early design stages. This paper sets out the main lines of the research study that has been undertaken to arrive to this methodology.

SCIENTIFIC INNOVATION AND RELEVANCE

The design hybridGEOTABS buildings brings along some challenges such as: how to decide on the sizing of the GEOTABS and secondary heating and cooling systems? How to incorporate the effects of the large thermal inertia of the TABS into the sizing of the systems? How to take into account the effect of the control on the sizing of the key components of the system? And most of all, how to do all of this while avoiding excessive engineering costs and detailed simulation studies from the very early design stages onwards?

In this project an integrated design solution is developed, that takes into account the aforementioned aspects to come to a near-optimal sizing result with the precision required in early design stages. Behind the resulting easy-to-use guidelines and materials, is found an intensive study and cooperation that incorporates the use of dynamic building energy simulations of the EU building stock, the development of load splitting algorithms and simulation studies of the effect of control on the design.

PRELIMINARY RESULTS AND CONCLUSIONS

This paper will walk the reader along the set-up and main outcomes of the aforementioned sub-studies and synthesise them into the development of a hybridGEOTABS design methodology. The main findings of this methodology are presented.

The resulting methodology is translated into a tool and guideline and allows the HVAC-designer to assess the feasibility of hybridGEOTABS in time similar to what is required in the design of more classical HVAC-solutions, and to come to a pre-sizing of the main system components. Before GEOTABS was mainly applied into a specific range of building designs and climates, by experienced designers. This newly developed methodology makes it possible to investigate the feasibility of hybridGEOTABS for a much wider range of buildings and building properties across EU, and makes the concept more accessible towards HVAC-designers and policy developers.

MAIN REFERENCES

[1] W. Boydens, D. Costola, A. Dentel, T. Dippel, L. Ferkl, A. Görtgens, L. Helsen, J. Hoogmartens, B.W. Olesen, W. Parijs, M. Sourbron, C. Verhelst, J. Verheyen, C. Wagner, REHVA Guidebook No. 20: Improved system design and control of GEOTABS buildings: Design and operation of GEOTABS systems, REHVA, Brussels, 2013. www.rehva.eu.

[2] E. Himpe, M. Vercautere, W. Boydens, L. Helsen, J. Laverge, GEOTABS concept and design : state-of-the-art, challenges and solutions, in: Proceedings of the REHVA Annual Meeting Conference Low Carbon Technologies in HVAC, 2018. http://hdl.handle.net/1854/LU-8565219 (accessed March 15, 2020).

[3] Model Predictive Control and Innovative System Integration of GEOTABS in hybrid Low Grade Thermal Energy Systems - hybridGEOTABS (EU H2020 project "MPC-GT" n°723649), Www.Hybridgeotabs.Eu. (2016). www.hybridgeotabs.eu.

[4] M. Sharifi, R. Mahmoud, E. Himpe, J. Laverge, Integrated Sizing Methodology for a hybridGEOTABS Building, ASHRAE Transactions. 125 (2019) 222–230.

Keywords: GEOTABS, HVAC design, MPC, H2020, hybridGEOTABS

Paper ID 30539: To be or not to be a hybridGEOTABS: energy performance of hybridGEOTABS buildings in the EU

Rana Mahmoud*, Mohsen Sharifi, Eline Himpe, Jelle Laverge

Ghent University, Belgium

AIM AND APPROACH

HybridGEOTABS technology plays an important role in providing clean energy for the building sector due to its high performance and use of renewable energy sources. However, for a building to be elected for the use of this technology it needs to follow a strict criterion in terms of energy performance. As stated by the GEOTABS guidebook [1]; it shall be a high insulated building with reduced window to wall ratio, using suitable orientation, with installed shading system and low internal heat gains. Therefore, limiting the implementation of this technology on buildings with different configurations.

In this paper we investigate the feasibility of hybridGEOTABS for a much wider range of configurations. We propose design guidelines for the possible building configurations that can allow higher shares of GEOTABS (primary system) and the residual can be covered by a secondary system. As a result, we can support the designer in the preliminary design stages to identify the eligibility of his building to be a hybridGEOTABS building or not. Our approach [2], [3] is based on the analysis of building stock dynamic simulations. We investigate the energy demand of a large spectrum of building geometrical and building physical properties combinations covering EU climatic zones.

SCIENTIFIC INNOVATION AND RELEVANCE

Our main focus is on non-residential buildings typologies namely offices, schools and elderly homes. One of the main difficulties faced when studying non-residential building stock is the scarcity of the available data. Our methodology allows the using of linear data found in the building energy certificates and transform these data into multi-zone energy models that can be dynamically simulated. The approach adopted in this research considers the different functions requirements and profiles based on extensive research regarding each typology. By implementing this method [3], we are able to obtain an hourly estimation of the energy demand of the buildings. The building energy models vary in terms of geometrical characteristics such as; surface area, volumes, window to wall ratios, as well as physical properties such as; internal heat gains profiles, possible orientation, thermal mass materials, varying the use of shading systems and the thermal performance of the envelope. The analysis is based on 40,000 cases of offices, 50,000 cases of schools and 5000 cases of elderly homes.

The output of this research shall support designers and policy makers identifying the potentials of building ranges that can benefit from this technology.

PRELIMINARY RESULTS AND CONCLUSIONS

The first step is to transform the linear building stock data into building energy models using a geometrical fitting process. A tool has been developed in python to transform the fitting geometries into multizone building energy models, while incorporating the building physical variables. An automatization process was implemented to simulate thousands of cases of the different typologies.

The Second step is analysing the estimated heating and cooling demand output simulations. By implementing a baseload algorithm developed by [4], this algorithm is able to define the shares of the GEOTABS (primary system) and secondary heating and cooling system.

The preliminary results show that 50% of cases have more than 50% of GEOTABS share while achieving balance in the ground (where heating and cooling demand covered by primary system are equal).

The paper will further identify the building design properties that leads to the highest environmental performance with hybridGEOTABS for different EU climates.

MAIN REFERENCES

This work is performed under the EU-Horizon 2020 hybrid GEOTABS project (MPC-GT) with project number 723649. Project: website: http://www.hybridgeotabs.eu

[1] Improved system design and control of GEOTABS buildings Design and operation of GEOTABS systems, Guidebook no.20, REHVA, 2013.

[2] R. Mahmoud, M. Sharifi, E. Himpe, M. Delghust and J. Laverge, "Estimation of load duration curves from general building data in the building stock using dynamic BES-models," in E3S Web of Conferences, vol. 111., Bucharest, 2019.

[3] R. Mahmoud, E. Himpe, M. Delghust and J. Laverge, "A modelling approach to reduce the simulation time of building stock models," in Proceedings of the 16th IBPSA Conference, Rome, 2019.

[4] M. Sharifi, R. Mahmoud, E. Himpe and J. Laverge, "Integrated sizing methodology for a hybridGEOTABS building," in ASHRAE TRANSACTIONS, Kansas City, Kansas, USA, 2019.

Keywords: hybridGEOTABS, Building stock dynamic simulations, Primary energy reductions, Schools typology

Paper ID 30228: BIM2SIM - Development of semi-automated methods for the generation of simulation models using Building Information Modeling

David Jansen* (1), Philipp Mehrfeld (1), Dirk Müller (1), Eric Fichter (2), Veronika Richter (2), Jérôme Frisch (2), Christoph van Treeck (2), Christian Warnecke (3), Andre Barz (3), Mike Dahnke (3), Jesse Brunkhorst (3), Pooyan Jahangiri (3), Bruno Lüdemann (3)

1: RWTH Aachen University, E.ON Energy Research Center, Institute for Energy Efficient Buildings and Indoor Climate; 2: RWTH Aachen University, Institute of Energy Efficiency and Sustainable Building (E3D); 3: Rud. Otto Meyer GmbH & Co. KG

AIM AND APPROACH

High complexity of buildings and their energy systems in today's construction projects requires advanced tools for predicting their behavior even before the planning phase. Dynamic simulations can ensure the optimum functionality of such complex systems in advance by analyzing various variants and identifying their optimization potentials. However, such simulations and the creation of the models are highly complex, time consuming and costly. Automated model generation based on already available building information modeling (BIM) models can significantly reduce the time and effort of dynamic simulation while at the same time increasing their precision.

The aim of this paper is to highlight the development of the methods and tools to derive descriptions to gain ready-to-simulate models for the improved design and optimization of building energy systems from BIM. This approach should significantly reduce the manual effort for model creation and thus support the dissemination of modern simulation applications in the building sector.

Three different topics are addressed, namely, Building Performance Simulations (BPS), Heating, Ventilation and Air Conditioning (HVAC) simulations and Computational Fluid Dynamics (CFD) simulations. The basis for all simulation types is the open exchange format IFC. The information hold in an IFC file are analyzed by topological and semantic algorithms.

SCIENTIFIC INNOVATION AND RELEVANCE

The BPS simulation is addressed by two different simulation approaches: using the open-source Modelica library AixLib and the open-source Python-based tool TEASER, as well as EnergyPlus. The goal of both tools is the calculation of dynamic heating and cooling demands, which can then be used in HVAC simulations to calculate the final energy requirements of a building and to compare different system variations. This information can then be coupled with a CFD simulation for simulating air flow behaviors, comfort, or smoke extraction.

The HVAC simulation models are created in Modelica using two different libraries: an inhouse library of the industry partner and the open-source library AixLib.

For analysis purposes, the energy system including the hydraulics, coming from the IFC file, is transferred into a graph network. This way the semantic studies can be extended by topological analyses.

The tool also serves as an interface program for CFD (Computational Fluid Dynamics) simulation. For this purpose, it offers functions for opening, viewing, filtering, simplifying and exporting IFC building models using a graphical user interface.

PRELIMINARY RESULTS AND CONCLUSIONS

The analysis and transfer of the data from the IFC file is performed in Python. For this purpose, in the scope of this work a comprehensive tool has been developed that serves all the simulation areas. The BIM2SIM tool is implemented as a console-based program but comes with a web application to make it more user-friendly. The tool allows simulations to be carried out quickly and simplifies the use of dynamic simulations in all phases of a construction process, from planning phase with only abstract data to the operation phase with complete data sheets and components.

Besides the automated creation of simulation models, the project deals with the identification of gaps in the IFC standard by close collaboration with IFC developers and software vendors. Especially, prescribed entities and property sets of HVAC systems are still missing in the IFC standard, which are not only necessary for the parameterization of simulation models, but also are requirements to achieve the goal of gathering all the building information in a standardized model.

MAIN REFERENCES

No papers are cited in this abstract. They will be part of the paper.

Keywords: BIM, HVAC, CFD, BEPS, Modelica

Session T4.4: Improving indoor environmental quality

Paper ID 30439: High fidelity multi-objective optimization of VAV system design

Kevin Cant*, Ralph Evins

University of Victoria, Canada

AIM AND APPROACH

Variable air volume (VAV) air handling units are the predominant HVAC configuration in North American commercial buildings. These systems have been proven to be able to operate very efficiently if well-designed, however, minimum damper position has the potential to require vast amounts of energy. It is time-consuming for HVAC designers to select unique minimum box positions for each space. This study couples zone-specific minimum damper position selection and central VAV control algorithms within energyplus simulations to a genetic algorithm to optimize between the competing objectives of energy consumption and adequate ventilation.

SCIENTIFIC INNOVATION AND RELEVANCE

The latest thrust in academic research oriented towards VAV systems is centered around high-performance control, such as model-predictive control, or artificial intelligence. There is a large building stock that is not equipped with building automation systems that have the level of fidelity required to perform real-time sub-hourly minimum damper position controls in conjunction with central system setpoints. These buildings present an opportunity for retuning or recommissioning of their VAV terminal box setpoints in order to improve ventilation and decrease energy consumption, however, the time required for manual damper selection is a barrier.

PRELIMINARY RESULTS AND CONCLUSIONS

The results of this study quantify the impact of default VAV box reheat position settings compared to a detailed design with individual setpoints selected for each space based on expected occupancy and sub-hourly cooling and heating loads. The tradeoff between ensuring adequate ventilation and energy consumption is presented for a range of different design regimes in an office building archetype. Additionally, it provides a rapid method for selecting minimum box positions using a genetic algorithm that can be applied to new and existing buildings.

MAIN REFERENCES

Brambley, M.R., and S. Katipamula. 2009. "Commercial Building Re-Tuning: A Low-Cost Approach to Improved Performance and Energy Efficiency." ASHRAE Journal, 51(10):12-23.

Energy Design Resources. 2009. Advanced Variable Air Volume VAV System Design Guide.Accessed June 5, 2012, from http://www.energydesignresources.com/media/2651/EDR_DesignGuidelines_VAV.pdf

Cho, Y. H., & Liu, M. (2009). Minimum airflow reset of single duct VAV terminal boxes. Building and Environment, 44(9), 1876-1885.

Murphy, J. (2011). High-performance VAV systems. Ashrae Journal, 53(10), 18.

Nassif, N., Kajl, S., & Sabourin, R. (2005). Optimization of HVAC control system strategy using two-objective genetic algorithm. HVAC&R Research, 11(3), 459-486.

Okochi, G. S., & Yao, Y. (2016). A review of recent developments and technological advancements of variable-air-volume (VAV) air-conditioning systems. Renewable and Sustainable Energy Reviews, 59, 784-817.

Wright, J., & Zhang, Y. (2008). The minimum capacity of HVAC secondary systems (with capacity reduction by interzonal airflow)(RP-1049). HVAC&R Research, 14(3), 397-415.

Kaam, S., Raftery, P., Cheng, H., & Paliaga, G. (2017). Time-averaged ventilation for optimized control of variable-air-volume systems. Energy and Buildings, 139, 465-475.

Cho, Y. H., & Liu, M. (2008, January). Optimal Minimum Airflow for Air Circulation of Single-Duct VAV Terminal Boxes. In Energy Sustainability (Vol. 43208, pp. 117-122).

Keywords: VAV, optimization, ventilation, energy, retuning

Paper ID 30861: "Assessing energy poverty and potential retrofit scenarios in residential stock in Albania".

Alessia Tafani*, Gliti Mazniku, Sokol Dervishi

Epoka University, Albania

AIM AND APPROACH

The paper aims to assess the prevalence of energy poverty in residential stock in Albania, focusing on the cities of Tirana and Pogradec, and its eradication. The study aims to provide the most adequate retrofit scenario, climate and typology wise, by exploring with different passive envelope retrofit scenarios, in order to achieve the proper indoor thermal comfort, minimize the energy consumption and thereafter reduce its cost per household.

The research is based on both, qualitative and quantitative approach, developed as follows:

i. A representative group of housing typologies for both cities is selected. To increase the impact and value of the study, the selection criteria is the representativity on city scale.

ii. A detailed questionnaire, designed based on reliable references, is conducted to the inhabitants of the selected buildings. It provides information on indoor temperature comfort, the impact of households' economic situation in shaping the heating and cooling regime, type of heating and cooling used, electricity and biomass bills etc.

iii. Calibrated simulation models are generated for each selected building typology, to increase relevance and credibility on the study.

iv. Several retrofit scenarios are explored, to provide the optimal solution in terms of sustainability and cost are explored.

SCIENTIFIC INNOVATION AND RELEVANCE

This paper explores a comprehensive research on prevalence of energy poverty in housing stock in Tirana and Pogradec, highlighting the vulnerability of households towards it and providing potential passive retrofitting measures for its eradication.

Tirana is selected as the representative city of Mediterranean climate in Albania, including a great territory in south and central regions of it, and Pogradec as the representative of temperate oceanic climate, including south-eastern and north-eastern regions of Albania, thus increasing the impact of the study in nation-scale.

Considering that 83.7% of the residential stock is constructed during 1945-1990 and doesn't respond to the contemporary building criteria, and the fact that 22% of houses have insufficient heating, and 13% of them suffer from humidity, the quality of life results to be very low. The greatest issue remains energy performance, considering residential buildings consume 60% of the national energy consumption and are responsible for 26% of the final energy consumption in Albania.

For this purpose, the paper investigates variations on energy consumption, the rate of comfortability related to housing conditions, climate and economical capability of the household, in order to increase life quality, thermal comfort, energy efficiency and reduce overall cost, through passive retrofit scenario implementation.

PRELIMINARY RESULTS AND CONCLUSIONS

Results from the questionnaire show that Tirana has a higher economic capacity, compared to Pogradec. Tirana results to rely more on electricity usage, while Pogradec relies more on biomass. Despite differences, results on indoor thermal comfort, housing conditions and heating/cooling regime set according to affordability are conceptually similar.

Results on energy performance show that different scenarios have different impact within different typologies and climates.

The best scenario, in terms of energy performance optimization and cost of implementation, for Tirana results to be a combination of a 5cm wall, roof insulation and double glazing, while for Pogradec is the same combination, but with a 7.5cm wall. Implementing thesse scenarios, reduces the energy consumption from the base case by an average of 71.4% for Tirana and 64.1% for Pogradec.

The results suggest that Pogradec requires a higher amount of energy consumption compared to Tirana in winter, while that relation is reversed in summer season.

Considering the orientation and position within the building, results suggest that generally the last floor apartments, require more energy to achieve the same level of comfort, compared to other apartments. Apartments oriented towards north or north-west require higher energy consumption, compared to south or south-east oriented ones.

MAIN REFERENCES

i. F. Ardente et al, Energy and environmental benefits in public buildings as a result of retrofit actions, Renew. Sustain. Energy Rev. 15 (1) (2011) 460–470

ii. JRHT, Temple Avenue Project: Energy Efficiency Refurbishment Homes for the 21st Century, Retrieved 1 17, 2017, from https://www.jrf.org.uk/sites/default/files/jrf/ migrated/files/energy-efficient-refurbished-homes-report.pdf

iii. B. Basarir, et al., Energy Efficient Retrofit Methods at the Building Envelopes of the School Buildings, <www.academia.edu/Ellipsis/Energy_efficient_retrofit_methods_at_the_building_envelopes>, 2012

iv. D. Farmer et al, Off-the-shelf solutions to the retrofit challenge: thermal performance, in: M. Dastbaz et al. (Eds.), Sustainable Ecological Engineering Design, Springer, Leeds, 2016, pp. 73– 93.

v. S. Attia et al, Impact and potentials of community scale low- energy retrofit: case study in Cairo, in: Proceedings of the 3rd CIB International Conference on Smart and Sustainable Built Environment, TU Delft, Delft, 2009.

vi. H. Erhorn-Kluttig, et al., Energy-Efficient Renovation of Educational Buildings,

vii. <www. oecd-ilibrary.org/Ellipsis/energy-efficient-renovation-of- educational-buildings_5622Ellipsis>, 2005.

viii. O. Osama et al, Zero energy university buildings energy performance evaluation of faculty of architectural engineering, Arch. Plan. J. 23 (2015) 13–21

Keywords: Energy performance, Energy poverty, Retrofit scenarios, Simulation, Residential stock

Paper ID 30372: Investigation of the resiliency of passive and natural cooling solutions through uncertainty analysis in a NZEB residential building in Denmark

Attila Kopányi*, Karolina Poczobutt, Lajos Ádám Pallagi

ABUD, Hungary

AIM AND APPROACH

The purpose of this research paper is to examine resiliency of natural and passive cooling technologies against increased cooling loads in the future. A resilient cooling technology is defined as a low-energy, robust solution that can maintain good thermal comfort of the occupants. The investigated solutions included shading, natural ventilation, and phase change materials (PCM). The technologies were investigated separately and in combinations, in the context of a NZEB single family house located in Denmark. Using local sensitivity analysis, parameters that have the largest impact on the operative temperature in the building were identified for each solution. The robustness of the cooling techniques against expected climate change until 2100, extreme warm weather events, changing occupant loads and technology-related uncertainties were examined with Monte Carlo uncertainty analysis, using the EnergyPlus calculation engine. Comfort evaluation was based on the adaptive comfort model. The performance of the technologies was assessed with Key Performance Indicators (KPIs): Percentage Outside Range (POR), number of hours when the operative temperature exceeds 27 and 28 °C and robustness.

SCIENTIFIC INNOVATION AND RELEVANCE

Overheating becomes a growing problem in Danish low-energy houses, especially during summer periods. The phenomenon is driven by multiple factors, including climate change, oversimplified design process, tight energy codes and more electrical equipment introduced in the living spaces. Excessive heating affects wellbeing, health and productivity of the occupants. Despite the availability of various cooling options on today´s market, only natural and passive methods provide the possibility to limit carbon emissions. Natural and passive methods also present the highest potential for resilient cooling in the northern European climate, but their real performance still needs to be assessed and performance gaps need to be addressed in relation to thermal comfort improvement. When it comes to resiliency assessment of the cooling solutions, proper methods have not yet been developed. Therefore, this research paper aims at bridging this gap in knowledge by providing examples of methods to determine robustness against climate change, increasing occupant load and technology related uncertainties.

PRELIMINARY RESULTS AND CONCLUSIONS

The results revealed that all the technologies show relatively low POR values, below 5%. The maximum acceptable number of hours above 27 and 28 °C was never exceeded, if the technologies were combined. PCM and natural ventilation showed low robustness against climate change, below 14% for both cases. When it comes to occupant loads, a combination of technologies indicated the highest robustness. All technologies reached approximately 85% of robustness when combining technology related uncertainties, weather, and occupational time. In terms of the individual technologies, shading provided the best results, but in the total assessment, the combination of all three technologies showed the highest resiliency against the examined uncertainties.

MAIN REFERENCES

The definition of resilient cooling was based on IEA Annex 80 on Resilient Cooling for Small and Non-Residential Buildings. The KPI-s were defined based on The Danish Building Regulations 2018, standard DS/EN 15251 2007 and the Ventilative Cooling Design Guide (which was based on IEA-EBC Annex 62). The predicted weather data was developed based on scenarios taken from the 4th IPCC report. For the local sensitivity analysis, the method proposed by Spitz et al. (2012) was used. A similar approach to account for uncertainties in the cooling technologies' performance was developed previously by Parys et al. (2012).

References:

G., BREESCH, H., BELLERI, A. and FLOURENTZOU, F., 2018. Ventilative Cooling Design Guide. Aalborg, Denmark: Department of Civil Engineering, Aalborg University.

PARYS, W., BREESCH, H., HENS, H. and SAELENS, D., 2012. Feasibility assessment of passive cooling for office buildings in a temperate climate through uncertainty analysis. Building and Environment, 56, pp. 95-107.

SPITZ, C., MORA, L., WURTZ, E. and JAY, A., 2012. Practical application of uncertainty analysis and sensitivity analysis on an experimental house. Energy & Buildings, 55, pp. 459-470.

Keywords: Resilient cooling, robustness, building simulation, uncertainty analysis, overheating

Session T4.6: The role of occupants

Paper ID 30529: A stochastic occupancy modeling approach to enhance the energy efficiency of residential heating and cooling through occupancy sensing technology

Tarek Rakha* (1), Tarek Sherif (1), Senem Velipasalar (2)

1: School of Architecture, Georgia Institute of Technology, United States of America; 2: Department of Electrical Enginereing and Computer Science, Syracuse University, United States of America

AIM AND APPROACH

People consume energy in buildings either by their passive presence indoors, which increases mechanical loads to attain thermal comfort, or by actively engaging with the building through control of said mechanical systems, as well as lighting and equipment. It is, therefore, important for individuals to engage efficiently with their building to reduce energy consumption through effective behavior and control of systems. Consequently, a high performing building aims to also engage with its users by employing human sensing technologies as a means to automate occupancy detection to reduce energy consumption and turn systems off in their absence. Occupancy schedules are used in Building Performance Simulation (BPS) to act as proxies for human presence. However, they were not previously used to explore the potentials and limitations of human sensing systems. In this paper, we develop a simulation-based approach to specifically examine the impact of human presence system sensing errors, like false positives, using occupancy schedules to quantify potential residential building heating and cooling energy loss. The aim is to examine varying effects of human detection system configurations on thermal energy consumption in false sensing scenarios, and to introduce occupancy schedules as a means to inform processes of such sensing systems.

SCIENTIFIC INNOVATION AND RELEVANCE

Occupancy schedules in BPS play a key role in representing human regulation of building energy consumption. Critical developments in sensing technologies over the last decade make such schedules a candidate for simulation to aid in testing and calibrating human detection systems in buildings. However, such emerging technologies may only have disruptive effects in residential building markets, specifically, if the system is significantly low in cost, easily self-commissioned, and requires minimal maintenance. These stringent constraints make them susceptible to inaccuracies. Therefore, the scientific innovation in this paper is the development of a stochastic approach to examine the role of low-cost sensing configurations, like scanning frequency and seasonal performance, in an integrated way for a symbiotic relationship between sensing and simulation schedules. A Markov-Chain analysis of the 2018 American Time Use Survey (ATUS) is used to extrapolate stochastic transition matrices and generate probabilistically driven occupancy schedules. We evaluate the impact of false positives in binary occupancy modelling scenarios using Honeybee as a front-end interface in Rhino/Grasshopper, and EnergyPlus as a backend engine. Relevance to residential building sectors is significant if such sensing technologies are to be deployed on a large scale, since BPS can now evaluate false sensing's negative energy impacts.

PRELIMINARY RESULTS AND CONCLUSIONS

Preliminary results showcase the role of scanning frequency. The percentage of weekly energy loss per false positive detection increased from 0.51% to 1.49% in correspondence with 10 to 60-minute scanning frequencies. Error timing also influenced the amount of energy lost, which was showcased in a standard deviation ranging between 0.53 – 0.74. Simulations were further expanded at the annual scale across 6 United States climate zones. The distribution of the annual investigation passed the Kolmogorov–Smirnov test for normality. The output was fitted to a normal distribution to measure energy loss ranges. Initial results indicate an approximate mean of 0.5% weekly energy loss per-hour error across tested climates. An observed large error range can be attributed to daily and seasonal climatic variations, as well as occupancy patterns prior to the false detection. An investigation of the average error duration on the probabilistic driven occupancy schedules identified critical system scanning points in a (low cost) limited battery power scenario. These were 08:30 AM and 07:00 PM for weekdays, and 09:30 AM and 06:00 PM for weekends. Overall, the paper's aim is to recommend guidelines for various system configurations in which the use of low-cost sensing is justified for heating and cooling regulation.

MAIN REFERENCES

Annex 66 Final Report. 2018. International Energy Agency.

Benezeth, Y., H. Laurent, B. Emile, and C. Rosenberger. 2011. "Towards a sensor for detecting human presence and characterizing activity." Energy and Buildings 43 (2-3): 305-314.

D'Oca, Simona, and Tianzhen Hong. 2015. "Occupancy schedules learning process through a data mining framework." Energy and Buildings 88: 395-408.

Duarte, Carlos, Kevin Van Den Wymelenberg, and Craig Rieger. 2013. "Revealing occupancy patterns in an office building through the use of occupancy sensor data." Energy and Buildings 67: 587-595.

Gunay, H. Burak, William O'Brien, Ian Beausoleil-Morrison, Rhys Goldstein, Simon Breslav, and Azam Khan. 2014. "Coupling stochastic occupant models to building performance simulation using the discrete event system specification formalism." JBPS 7 (6): 457-478.

Mahdavi, Ardeshir, and Farhang Tahmasebi. 2015. "Predicting people's presence in buildings: An empirically based model performance analysis." Energy and Buildings 86: 349-355.

Page, J., D. Robinson, N. Morel, and J. L. Scartezzini. 2008. "A generalised stochastic model for the simulation of occupant presence." Energy and Buildings 40 (2): 83-98.

Yan, Da, William O'Brien, Tianzhen Hong, Xiaohang Feng, H. Burak Gunay, Farhang Tahmasebi, and Ardeshir Mahdavi. 2015. "Occupant behavior modeling for building performance simulation: Current state and future challenges." Energy and Buildings 107: 264-278.

Keywords: Occupancy, Sensing, Stochastic, Simulation, Residential

Paper ID 30525: Occupant satisfaction with automation: experimental data vs. simulation

Alessandra Luna-Navarro* (2), Isabella Gaetani (1), Alistair Law (3), Francesco Anselmo (1), Mauro Overend (4)

1: Arup, Smart Buildings, London, United Kingdom; 2: Department of Engineering, University of Cambridge; 3: Arup, Facade Engineering, London, United Kingdom; 4: Structural Design and Mechanics, Faculty of Architecture and the Built Environment, TU Delft, The Netherlands

AIM AND APPROACH

The aim of this study is to compare actual occupant interaction with and response to a number of automated blind strategies to occupant behavioral and comfort models in a test facility in Cambridge (UK). Then, alternative automated strategies are compared in terms of energy performance and occupant satisfaction.

To achieve the aforementioned aim, this study adopts the following approach: firstly, alternative degrees of automation to control shading devices and increasing levels of occupant interaction with the automation in an office environment are investigated to understand their effect on occupant comfort and energy performance; Secondly, building performance simulation models are used to represent the control strategies and the appropriateness of such models is evaluated, both in terms of realistic modelling and capabilities to assess occupant comfort in a holistic manner.

Data on occupant response is collected in a novel test facility (MATELab) in Cambridge (UK) and building performance simulations are used to assess the impact of occupant interaction on the predicted energy performance.

SCIENTIFIC INNOVATION AND RELEVANCE

This study contributes to existing knowledge by:

- extending the knowledge on holistic and transient occupant response to increasing levels of automation;

- evaluating the importance of user overrides and control interfaces on holistic and transient occupant satisfaction; results from previous studies are extended by adopting a holistic approach;

- providing a better understanding of when automation can increase occupant satisfaction and energy efficiency of office spaces;

- presenting a fit-for-purpose aspect of automation in office environments. This includes the quantification of energy savings from shading devices control optimization, different control strategies based on occupant presence and further insights on the conflicts between energy-efficiency and comfort;

- highlighting the limitations of simulation tools for realistically predicting the extent and impact of occupant interaction in smart buildings;

- demonstrating the importance of an occupant-centered approach when designing control strategies to ensure occupants satisfaction with their workspace and acceptance of automation

PRELIMINARY RESULTS AND CONCLUSIONS

The experimental scenarios were characterized by an increasing level of automated control of shading devices, from fully manual to fully-automated control for glare minimization and daylight maximization, as well as energy-efficiency (comfort over glare). The option to automatically lower the blinds when sun in the field of view (FOV) was evaluated and user overrides were permitted throughout the study.

Additionally, during these scenarios, different control interfaces were available to occupants. Occupant satisfaction was measured in terms of discomfort votes, glare, daylight glare probability (DGP), useful daylight index (UDI), daylight autonomy, thermal comfort (operative temperature), user overrides and times shades were adjusted (manually and automatic).

Simulations included standard IF-THEN blind models, available stochastic blind models and actual interaction from data.

The results indicated that occupants are more inclined to accept automation when informed about the rationale and when the control strategy restored views. Also, EnergyPlus alone was not adequate to represent holistic comfort, but a number of key areas could be realistically evaluated.

MAIN REFERENCES

L. G. Bakker, E. C. M. Hoes-van Oeffelen, R. C. G. M. Loonen, and J. L. M. Hensen, "User satisfaction and interaction with automated dynamic facades: A pilot study," Build. Environ., vol. 78, pp. 44–52, 2014.

S. Stevens, "Intelligent facades : occupant control and satisfaction," Int. J. Sol. Energy, vol. 21, no. 2–3, pp. 147–160, 2001.

W. O'Brien and H. B. Gunay, "The contextual factors contributing to occupants' adaptive comfort behaviors in offices - A review and proposed modeling framework," Build. Environ , vol. 77. pp. 77–88, 2014.

A. Luna-Navarro et al., "Occupant-Adaptive Façade Interaction: relationships and conflicts," in FAÇADE 2018 Final conference of COST TU1403 "Adaptive Facades Network" Lucerne, November 26/27 2018, 2018.

I. Gaetani, R. N.P. van Woensel, P. Hoes, J. L.M. Hensen, "Occupant behavior modelling to support the development of adaptive facade control strategies" in Proceedings of the 16th International IBPSA Conference (Building Simulation 2019), Rome, Italy

Keywords: Occupant behavior, Automation, Holistic comfort, Satisfaction, Blinds

Paper ID 30590: People who? Why occupant behaviour modelling is not (yet) included in the design workflow

Isabella Gaetani (1), Alessandra Luna-Navarro* (2), Francesco Anselmo (1)

1: Arup, United Kingdom; 2: Department of Engineering, University of Cambridge

AIM AND APPROACH

People can be modelled as movement (pedestrian modelling), presence and behaviours. The modelling of people is particularly important during the building design phase for building safety design, building architecture and engineering (AE) design and for building performance analysis. While pedestrian modelling is part of the typical design workflow, presence and behaviour modelling seldom find applications in practice.

This paper uses surveys, first-person interviews and literature review to shed some light on why people modelling and its design implications are still poorly understood, what are the main barriers to overcome, and what scientists and researchers can do to ensure a better penetration of models in practice.

The research method consists of a 30-question survey and first-person interviews directed uniquely towards practitioners from leading engineering consultancies and design agencies. The survey results and interviews are complemented by a literature review regarding the existing design workflows and the integration of people modelling.

SCIENTIFIC INNOVATION AND RELEVANCE

This work is extremely in line with the aim of the conference 'Creating impact' as it bridges the work of academics and practitioners towards a higher degree of implementation of occupant behaviour research.

The relevance is clear when it comes to the implications that the successful adoption of occupant behaviour modelling would have on building performance simulation. A more accurate representation and prediction of the energy and comfort performance has obvious benefits such as optimization for actual in-use operation, higher degree of personalization, and reaching desired performance targets. However, studies such as this are needed to better understand the final target audience of our efforts and to tailor the methods we propose in academia towards practitioners more effectively.

PRELIMINARY RESULTS AND CONCLUSIONS

Generally, architects and designers consider people primarily in terms of the recipients of the experience they envision through their design. Pedestrian modelling is a recognized tool to inform flow analysis and layout design, as well as to comply with safety regulations (e.g. fire-safety). In contrast, the modelling of presence and behaviour is perceived as a specialist domain and it hence not part of the typical design workflow.

The most commonly used architecture tools do not encourage a further consideration of people within the design workflow: people modelling is troublesome and inefficient, and its benefits are poorly understood. Among the most important barriers to an appropriate consideration of people modelling within the design workflow are:

- Stakeholders, intended as lack of skills, understanding, and vision;

- Software, both in terms of robust models and methodologies, and their integration in architecture tools; and

- Feasibility factors such as the lack of support from standards and regulations, the unclear benefits and feedback, and the increased expenditure in terms of time and effort that appropriate modelling would imply.

The key solution to improve the current status is increasing the level of communication and collaboration among design team, modelling team (or domain specialists), and clients.

MAIN REFERENCES

Roetzel, "Occupant behaviour simulation for cellular offices in early design stages—Architectural and modelling considerations," Build. Simul., 2015

T. Abuimara, W. O'Brien, B. Gunay, and J. S. Carrizo, "Towards occupant-centric simulation-aided building design: a case study," Build. Res. Inf., 2019

I. Gaetani, "A strategy for fit-for-purpose occupant behavior modelling in building energy and comfort performance simulation," Technische Universiteit Eindhoven, 2019

A. Heydarian, J. P. Carneiro, D. Gerber, B. Becerik-Gerber, T. Hayes, and W. Wood, "Immersive virtual environments: Experiments on impacting design and human building interaction," in CAADRIA 2014, 2014

S. Altomonte, P. Rutherford, and R. Wilson, "Human factors in the design of sustainable built environments," Intelligent Buildings International. 2015

S. Attia, L. Beltrán, A. De Herde, and J. Hensen, "'Architect friendly': A comparison of ten different building performance simulation tools," in IBPSA 2009 - International Building Performance Simulation Association 2009, 2009

Keywords: Occupant behavior, Practitioners, Building Performance Simulation

Session T4.7 (Online Track): Practice and industry related case studies

Paper ID 30341: Simulation of historical buildings: bedford cottages, devon, uk

Jack Morewood, Pieter de Wilde*

University of Plymouth, United Kingdom

AIM AND APPROACH

This paper employs building performance simulation to assess the application of energy saving measures to historical buildings. The work focusses on the historic Duke of Bedford Cottages in West Devon, United Kingdom. These Bedford Cottages are both listed and are a UNESCO World Heritage Site in the Cornwall and West Devon Mining Landscape and generally have poor thermal properties, yet any interventions need to be sympathetic to the cottages and within constraints of cost, heritage and occupancy. A row of four cottages has been modelled in Revit and then simulated using Design Builder for assessment of thermal performance, whereas WUFI has been used to assess damp risk. Uncertainty analysis is carried out to study the behaviour of the cottages under predicted climate change conditions.

SCIENTIFIC INNOVATION AND RELEVANCE

This work presents a deep building performance study based on a whole building approach, considering retrofits in their full context and exploring how these may interfere with the entire cottage. The study explores the use of Heritage BIM (H-BIM), a specific stream of BIM developed to manage building information from historic buildings. The work highlights some of the challenges in applying building simulation to historic buildings.

PRELIMINARY RESULTS AND CONCLUSIONS

Initial result confirm that the Bedford Cottages are energy intensive. The uninsulated stone and brick walls, particularly for the end-of-terrace cottages, cause large energy losses in the winter, with heating demand was as high as 151kWhm-2 annually. Thermal comfort is also problematic: using ASHRAE-55 adaptive comfort, some cottages fail to meet the 90% threshold for three hours per day. This is largely down to a low air temperature within the buildings. Using a CIBSE TM59 assessment in DesignBuilder, overheating was not currently an issue for the cottages, however using future weather files, all cottages are expected to overheat in the long-term.

MAIN REFERENCES

[1] ASHRAE Standard 55 (2017). Thermal Environmental Conditions for Human Occupancy.

[2] CIBSE TM59 (2017) Design methodology for the assessment of overheating risk in homes

Keywords: historical buildings, thermal performance, damp risk, energy saving measure

Paper ID 30818: Multi-criteria robustness assessment of a sequential whole-building design optimization

Riccardo Talami*, Jonathan Wright, Bianca Howard

Loughborough University, United Kingdom

AIM AND APPROACH

This research aims to determine the additional reliability gained through the deployment of an integrated and simultaneous design process, as compared to a conventional sequential process, to determine the optimal selection of building geometry, fabric, energy system and its controls to minimize energy consumption and ensure thermal comfort. In this study, the reliability is defined as the ability of the search to achieve Pareto-optimal design solutions under scenario-based uncertainties related to occupant behaviour and changing climate. For both approaches, a full-factorial (exhaustive) search method was adopted to eliminate the uncertainty in the convergence behaviour that occurs with population-based metaheuristic search methods, such as evolutionary and particle swarm algorithms, guaranteeing to find all optima for a valid comparison. To ensure computational feasibility for the performance evaluation, a single-zone office located in the UK is formulated as a case study. The parametric modelling, building performance simulation, and optimization are implemented in Grasshopper for Rhino, allowing individual EnergyPlus simulation models to concurrently run. The resulting set of design options and scenarios, considering the results of the exhaustive search method, are postprocessed to identify robust Pareto-optimal designs.

SCIENTIFIC INNOVATION AND RELEVANCE

This work seeks to determine the magnitude of the benefits of a simultaneous whole-building design optimization approach when considering uncertainty. Research has shown how better solutions can be generated from simultaneous rather than sequential design optimization approaches (Ferrara et al., 2019; Waibel et al., 2019), although the better solutions come at the cost of additional computational burden (Talami et al., 2020) and the requirement to specify the entire design space at the initial design phase (Kanters and Horvat, 2012). However, the magnitude of the benefits is not often evaluated considering the additional assessment of uncertainty, as a robustness indicator for the Pareto-optimal solutions. This work seeks to fill this gap by evaluating the performance of an integrated and simultaneous design optimization approach with a sequential approach, mimicking the traditional design process, whilst considering uncertain scenarios. The work will determine the difference in the optimal designs, as defined by each search approach, to minimize energy consumption and maintain thermal comfort under various occupancy, usage, and climate scenarios. The insights from this work will enable for first comparison of a whole-building integrated and simultaneous design optimization approach with the conventional sequential design approach considering uncertainty.

PRELIMINARY RESULTS AND CONCLUSIONS

The preliminary results indicate that the simultaneous design approach is more frequently able to identify design solutions that are optimal across the range of uncertainty scenarios considered. The identification of these solutions however would lead to significantly more evaluations of the building simulation model. For the simultaneous design optimization approach, the deviation of optimal solutions is lower for climate scenarios compared to that of occupancy and usage scenarios, indicating that the designs are more sensitive to occupant behaviour rather than climate change. Additionally, the designs with lower insulation levels are found to be more robust for occupancy and usage scenarios whereas the designs with higher insulation levels are found to be more robust for climate change. The sequential design optimization approach however was more sensitive to climate as the internal loads due occupancy affect most acutely the design and operation of the mechanical systems and controls. This indicates that to be confident of robust design an integrated design solution should be considered.

MAIN REFERENCES

- Ferrara, M., Prunotto, F., Rolfo, A., Fabrizio, E. (2019). Energy Demand and Supply Simultaneous Optimization to Design a Nearly Zero-Energy House, Applied Sciences, 9, 2261.

- Waibel, C., Evins, R., Carmeliet, J. (2019). Co-simulation and optimization of building geometry and multi-energy systems: Interdependencies in energy supply, demand and solar potentials. Applied Energy, 242, 1661–1682.

- Kanters, J. and Horvat, M. (2012) The Design Process known as IDP: A Discussion. Energy Procedia, 30, 1153-1162.

- Talami, R., Wright, J., Howard, B. (2020). A comparison between sequential and simultaneous whole-building design optimization for building performance (ongoing research).

Keywords: Whole-building design, Uncertainty, Building performance, Optimization

Paper ID 30893: An integrated analysis of structural safety, embodied carbon, and construction cost in a prefabricated Chinese timber house within a BIM-based environment

Xi Zhang* (1), Zhantang Miao (2), Gang Feng (2), Jiangtao Du (1)

1: University of Liverpool, United Kingdom; 2: Tianjin University, China

AIM AND APPROACH

This study aims to assess performance of embodied carbon and construction cost in a newly built Chinese timber house, which was developed through an innovative way of prefabricated construction. This study first shows an optimization analysis of structural optimization using a BIM method. Then, a comparison analysis of embodied carbon and construction cost is presented between before and after the optimization.

Approaches:

1) Case study: a new timber terrace house located in Tianjin, China [1]; two floors; a total floor area of 143.56m^3. This house adopted a platform timber structure, which is currently the most common type of lightweight timber housing construction. Key timber components of this house were built using factory prefabrication.

2) Structural optimization: this was conducted to improve the construction design of this house using Revit-Dynamo (BIM tool) and Abaqus Unified FEA (Finite Element Analysis).

3) Analysis of construction cost: before and after the structural optimization, the material usage of this house was assessed using Revit-Dynamo (BIM tool) and relevant construction cost was calculated using a standard method in reference [2].

4) Assessment of embodied carbon: before and after the structural optimization, embodied carbon of key construction components was assessed using these methods in reference [3, 4].

SCIENTIFIC INNOVATION AND RELEVANCE

Scientific Innovation:

In recent studies of housing construction sector [5, 6], a superior characteristic of wooded structure has been identified in terms of embodied carbon and energy. It has been found that novel off site panellised modular timber frame systems can save up to 50% of embodied carbon and 35% embodied energy when compared to traditional residential building methods and materials. On the other hand, the prefabricated house currently receives the most attention in building sector [7]. One of the benefits of prefabricated homes is that they tend to be highly energy efficient at both stages of construction and operation [7].

This article reported an investigation into embodied carbon and construction cost in a house with the combined two characteristics: timber structure and prefabricated construction. More importantly, according to current situation in China, both prefabricated construction and application of timber structure in residential buildings are the pioneer areas, especially at northern China with a high energy demand in residential buildings.

Relevance: This study focused on the evaluation of embodied carbon and cost through the application of an advanced BIM tool and relevant theoretical methods. This study can well address two themes: Zero Energy Buildings (ZEB), Building Information Modelling (BIM).

PRELIMINARY RESULTS AND CONCLUSIONS

Key results: Three types of timber and non-timber materials were applied for wall, floor, and roof, including Kiln Dried Softwood (KDS), Oriented Strand Board (OSB), and Gypsum Board (GB).

After the optimization, results were achieved as:

1) Usage of construction materials (m3): A relative reduction of material usage was found as 19.3% (wall-KDS), 42.4% (wall-OSB), 25.8% (wall-GB), 7.3% (floor-KDS), 22.6% (floor-OSB), 7.0% (roof-KDS), and 26.4% (roof-OSB).

2) Construction cost (RMB): A relative reduction of construction cost was found as 30.7% (total wall construction), 12.5% (total floor construction), 21.9% (total roof construction).

3) Embodied carbon (kg CO2 eq.): A relative reduction of embodied carbon (from cradle to site; incl. distribution and construction only) can be found as 12.4% (total-KDS), 35.0% (total-OSB), 25.8% (total-GB).

In summary, key conclusion:

1) The performance of embodied carbon can significantly benefit from this BIM-based optimization of building structure in this prefabricated timber house. This is a new finding which might be different from the house built using traditional material (e.g. steel, concrete) and construction method.

2) The reduction of embodied carbon and cost was achieved through this structural optimization. This could expose the necessity to implement an integrated optimization when applying new construction methods in houses.

MAIN REFERENCES

[1] Zhang, X. 2020. Research on Optimization Design of Modern Light-Frame Timber Housing Based on BIM Technology. MSc Dissertation, Tianjin University, Tianjin, China.

[2] MHUD (Ministry of Housing and Urban-Rural Development). 2013. GB 50500: Code of bills of quantities and valuation for construction works. Beijing, China.

[3] BSI (British Standard Institute). 2011. BS&EN 15978:2011: Sustainability of construction works. Assessment of environmental performance of buildings. Calculation method. UK.

[4] Wood for Good: Lifecycle-database: https://woodforgood.com/lifecycle-database. Final access: 28/06/20202.

[5] Monahan J, Powell JC. 2011. An Embodied Carbon and Energy Analysis of Modern Methods of 5Construction in Housing: A Case Study Using a Lifecycle Assessment Framework. Energy and Buildings, 43: 179-188.

[6] Ramage M H, Burridge H, Busse-Wicher M, et al. 2017. The wood from the trees: The use of timber in construction. Renewable and Sustainable Energy Reviews. 68: 333-359.

[7] Tumminia G, Guarino F, Longo S, Ferraro M, Cellura M, Antonucci B. 2018. Life cycle energy performances and environmental impacts of a prefabricated building module. Renewable and Sustainable Energy Reviews, 92: 272-283.

Keywords: Embodied carbon, Construction cost, Timber structure, Structural safety, House

Paper ID 31003: Predicted energy savings by adopting novel radiant cooling systems in combination with natural ventilation

Denon Sheppard* (1), Adam Rysanek (1), Eric Teitelbaum (2), Kian Wee Chen (2), Dorit Aviv (3), Kipp Bradford (2), Forrest Meggers (2)

1: School of Architecture and Landscape Architecture, University of British Columbia, Vancouver, Canada; 2: Andlinger Center for Energy and Environment, Princeton University, Princeton USA; 3: Weitzman School of Design, University of Pennsylvania, Philadelphia, USA

AIM AND APPROACH

This paper presents a TRNSYS model developed to evaluate the whole-building performance of a novel membrane-assisted radiant cooling panel which can operate in hot and humid indoor environments without condensation. This type of cooling system has been developed to significantly improve the energy efficiency of buildings in hot and humid climates by allowing for sensible cooling to occur without dehumidification. A prior prototype of the radiant cooling system was developed and tested previously in Singapore. Empirical data from the prototype is used to calibrate a transient numerical model of the system's behaviour in Python. This Python code is incorporated into a TRNSYS simulation which allows the radiant cooling system to interact with a building energy model (Type 56) and weather data. A archetypal building model of a naturally-ventilated school classroom and office space is constructed. The simulated performance of the radiant cooling system is compared to a conventional mechanical ventilation-based air-conditioning system for the context of Singapore, Toronto (Canada), and Vancouver (Canada).

SCIENTIFIC INNOVATION AND RELEVANCE

In tropical climates, such as Singapore, air conditioning is typically responsible for 43-60% of the energy load of a commercial building. It's been previously proposed however, that natural ventilation could satisfy indoor thermal comfort requirements 73-77% of the time in Singapore, if buildings would be reconstructed to take better advantage of shading, thermal mass, and ventilation dynamics. For existing buildings to take advantage of the energy-savings potential of natural ventilation while maintaining thermal comfort for building occupants, a novel radiant cooling panel has been developed which separates actively-cooled surfaces from humid indoor air via an air gap and a semi-transparent membrane. The panel can provide sensible radiant cooling, driven by a low-lift chiller, without the need for a dehumidification system. This panel has been thoroughly investigated in prior experimental research. This paper completes the bridge between the experimental results of the panel's performance and building simulation by creating a TRNSYS framework that allows for the simulation of a building energy system that utilizes this technology. It is the first work to estimate the total primary energy savings potential of a realized whole-building space cooling system that makes use of the developed radiant cooling panels.

PRELIMINARY RESULTS AND CONCLUSIONS

Several building use cases, including office and classroom environments in both Singapore and Toronto, have been simulated to compare the cooling primary energy use of a membrane-assisted radiant cooling panel system with a conventional forced-air air-conditioning system. Initial results suggest that in Singapore, the radiant cooling system could provide annual energy savings of 30 to 50%, while Toronto summers could experience a reduction in air-conditioning energy use of 20 to 40%. It has been shown that this technology can significantly reduce primary energy requirements for cooling in a variety of climates, and it should be gradually implemented into the building stock in order to prove its effectiveness while simultaneously collecting data to improve the calibration model calibration.

MAIN REFERENCES

Hossain, Md Muntasir, and Min Gu. 2016. "Radiative Cooling: Principles, Progress, and Potentials." Advanced Science 3 (7): 1500360. https://doi.org/10.1002/advs.201500360.

Liping, Wang, and Wong Nyuk Hien. 2007. "Applying Natural Ventilation for Thermal Comfort in Residential Buildings in Singapore." Architectural Science Review 50 (3): 224–33. https://doi.org/10.3763/asre.2007.5028.

Teitelbaum, Eric, Kian Wee Chen, Forrest Meggers, Jovan Pantelic, Dorit Aviv, and Adam Rysanek. 2019. "The Cold Tube: Membrane Assisted Radiant Cooling for Condensation-Free Outdoor Comfort in the Tropics." Journal of Physics: Conference Series 1343 (November): 012080. https://doi.org/10.1088/1742-6596/1343/1/012080.

Teitelbaum, Eric, Adam Rysanek, Jovan Pantelic, Dorit Aviv, Lea Ruefenacht, Megan Teitelbaum, Forrest Meggers, Kipp Bradford, Kian Wee Chen, and Denon Sheppard. 2020. "Cooling without Air Conditioning: Membrane-Assisted Radiant Cooling for Expanding Thermal Comfort Zones Globally," March. https://doi.org/10.36227/techrxiv.12034971.v1.

Keywords: Radiant Cooling, Mixed mode ventilation, Tropical Climate, Energy Efficiency, Radiative heat transfer

Session T4.8 (Online Track): Climate change and bioclimatic design

Paper ID 30730: Application of solar energy in the development of university buildings energetic sustainability

Eusebio Conceição* (1), João Gomes (2), Mª Inês Conceição (3), Mª Manuela Lúcio (1), Hazim Awbi (4)

1: University of Algarve, Portugal; 2: CINTAL, Portugal; 3: Instituto Superior Técnico, University of Lisbon, Portugal; 4: School of Built Environment, University of Reading, United Kingdom

AIM AND APPROACH

In this numerical work, the application of solar energy in the development of university buildings energetic sustainability is made.

The numerical model is based on energy and mass balance integral equations system and works under transient conditions. These equations system, solved by the Runge-Kutta-Felberg method with error control, is developed for indoor spaces, transparent surfaces, opaque surfaces, interior bodies and ducts systems. The energy balance integral equations consider the convective, conductive, evaporative and radiative phenomena. In the radiative phenomena are considered the internal and external solar radiations and the heat exchange by radiation. In the radiation phenomena, all internal and external shading devices are considered. The mass balance integral equations consider mass convective and diffusive phenomena. In the simulation, solar radiation, grass radiative proprieties, convective coefficients, and other parameters are calculated. All equations are built, by the numerical software, using the building geometry developed using CAD (Computing Aid Design).

This study is done in winter and summer conditions. In winter condition, are used greenhouse spaces, turned south, to heat the air in spaces turned to East, West and North. In summer conditions are used underground spaces to cool spaces turned South, East and West.

The thermal comfort and indoor air are evaluated.

SCIENTIFIC INNOVATION AND RELEVANCE

In this work the numerical simulation is made using a building dynamic software, developed in the last decades by the authors.

The software simulates a group of buildings with complex topology.

The software considers the human thermo-physiology to evaluate the human thermal comfort.

The software developed a mass and energy integral equations using the building geometry developed using CAD (Computing Aid Design).

The software evaluated the indoor air quality using the evolution of contaminants released by the occupants and buildings surfaces.

The software considers a virtual building, using the grid generation and the shading devices techniques, in the external and internal solar radiation.

The software considers internal greenhouses and underground spaces.

The software considers a virtual internal occupation and internal ventilation system.

The software considers the HVAC control based in PMV and other indexes.

PRELIMINARY RESULTS AND CONCLUSIONS

The presence of an internal greenhouse increases the air temperature in spaces subjected to air from the greenhouse. The air temperature, inside an occupied space subjected to air coming from a greenhouse, increases 1 to 3 ºC and the air temperature in the greenhouse decreases 4 to 6 ºC. The thermal comfort level in spaces subjected to the influence of air from coming from the greenhouse increases the thermal comfort levels. The indoor air quality is acceptable for all occupied spaces. The carbon dioxide concentration evolution is in accordance with the standards. The greenhouse presents an important contribution to increase the air temperature in winter conditions.

A numerical simulation without and with underground thermal energy storage is made. In the numerical simulation the occupation and the internal ventilation are considered. In accordance with the obtained results the internal airflow rate used and the external air temperature during the day and the night, guarantee the underground thermal energy storage to cool the more uncomfortable spaces. The air quality is acceptable and the thermal comfort level, considering the adaptive concepts, guarantee levels near the suggested by the standards.

MAIN REFERENCES

E. Z. E. Conceição and M. M. J. R. Lúcio, “Thermal study of school buildings in winter conditions,” Build. Environ., vol. 43, no. 5, 2008, doi: 10.1016/j.buildenv.2007.01.029.

E. Z. E. Conceição and M. M. J. R. Lúcio, “Numerical study of the influence of opaque external trees with pyramidal shape on the thermal behaviour of a school building in summer conditions,” Indoor Built Environ., vol. 19, no. 6, 2010, doi: 10.1177/1420326X10377546.

E. Z. E. Conceição and M. M. J. R. Lúcio, “Numerical study of the thermal efficiency of a school building with complex topology for different orientations,” Indoor Built Environ., vol. 18, no. 1, 2009, doi: 10.1177/1420326X08099550.

E. Conceição, J. Gomes, and H. Awbi, “Influence of the airflow in a solar passive building on the indoor air quality and thermal comfort levels,” Atmosphere (Basel)., vol. 10, no. 12, 2019, doi: 10.3390/ATMOS10120766.

E. Z. E. Conceição, M. C. Lopes, and M. M. J. do R. Lúcio, “Energy and thermal comfort management in a kindergarten school building in the South of Portugal in Winter conditions,” New Asp. energy, Environ. Ecosyst. Sustain. Dev. PT 1, pp. 90–95, 2008.

Keywords: Building dynamic software, internal greenhouse, underground spaces, thermal comfort, indoor air quality.

Paper ID 30959: A comparison of performance of three variance-based sensitivity analysis methods on an urban-scale building energy model

Pamela Jane Fennell*, Ivan Korolija, Paul Ruyssevelt

UCL, United Kingdom

AIM AND APPROACH

The vast number of different inputs required to model a complex urban environment makes it impossible to precisely quantify all inputs and complex energy flows within models must be simplified to achieve tractable solutions. As a result, the outputs of these models inevitably have a significant range of variation. Without understanding these limits of inference resulting policy advice is inherently defective. Uncertainty Analysis (UA) and Sensitivity Analysis (SA) offer essential tools to determine the limits of inference of a model and explore the factors which have the most effect on the model outputs. Despite a well-established body of work applying UA and SA to models of individual buildings, very limited work has been done to apply these tools to urban scale models.

This study presents a systematic comparison of a range of different SA methods to a high resolution, dynamic thermal simulation of a mixed-use neighbourhood in North London. Accuracy, processing time and complexity of application of each SA method is evaluated to provide guidance which can inform the application of these methods to other urban and large-scale building energy models.

SCIENTIFIC INNOVATION AND RELEVANCE

UBEMs are a class of model defined by their outputs rather than their structure or inputs, consequently a very large variety of approachs exists. UA and SA have been applied at the urban scale to reduced order models such as resistor-capacitor or quasi steady-state models (Schiefelbein et al., 2015; Zhao et al., 2016). However, where UA and SA are applied to dynamic thermal simulations this has either been at the level of the individual building archetype (e.g. Mauro et al., 2015) prior to an urban-scale analysis, rather than undertaking SA at the level of the urban model itself, or for a handful of parameters. There is an urgent need for guidance on the relative performance of SA methods for large scale models with fine resolution and large numbers of uncertain parameters. To address this need, this study evaluates 4 different SA variance based-methods (One-at-a-Time, Sobol', Derivative-based Global Sensitivity Measure and Elementary Effects) applied to a dynamic thermal simulation of a neighbourhood model. The simulation engine used is EnergyPlus and 51 uncertain parameters are considered in the analysis.

PRELIMINARY RESULTS AND CONCLUSIONS

Three metrics are used to compare the results from each method: Kendall's tau is used to evaluate the differences in parameter sensitivity ranking, the fraction of false positives and false negatives in the identification of sensitive and unsensitive parameters (using a threshold of 95% of the total sensitivity index) and computational burden is calculated as total CPU hours required. As no analytical solution exists, Sobol' (Saltelli et al., 2010) analysis of 2,120 model evaluations is taken as the baseline result for comparison.

Preliminary results suggest that model form has a strong influence on the results from different methods with the DGSM method in particular being ill-suited for a model with a rough response surface. The range of different scales of input parameters is also important suggesting that the best performing method is the Elementary Effects method (Campolongo et al., 2007) which incorporates input perturbations in the calculation of the sensitivity index.

MAIN REFERENCES

Campolongo, F., Cariboni, J., & Saltelli, A. (2007). An effective screening design for sensitivity analysis of large models. Environmental Modelling & Software, 22(10), 1509–1518.

Mauro, G. M., Hamdy, M., Vanoli, G. P., Bianco, N., & Hensen, J. L. M. (2015). A new methodology for investigating the cost-optimality of energy retrofitting a building category. Energy and Buildings, 107, 456–478. https://doi.org/10.1016/j.enbuild.2015.08.044

Saltelli, A., Annoni, P., Azzini, I., Campolongo, F., Ratto, M., & Tarantola, S. (2010). Variance based sensitivity analysis of model output. Design and estimator for the total sensitivity index. Computer Physics Communications, 181(2), 259–270. https://doi.org/10.1016/j.cpc.2009.09.018

Schiefelbein, J., Amir, P. J., Diekerhof, M., & Bode, G. (2015, December). Development of a Tool Chain for Complex City District Energy System Modeling and Simulation. Building Simulation 2015, Hyderabad, India.

Zhao, F., Lee, S. H., & Augenbroe, G. (2016). Reconstructing building stock to replicate energy consumption data. Energy and Buildings, 117, 301–312. https://doi.org/10.1016/j.enbuild.2015.10.001

Keywords: UBEM, Global Sensitivity Analysis, EnergyPlus

Paper ID 30790: Extreme weather data in building performance simulation

Andrea Gasparella (1), Drury B. Crawley* (2), Giovanni Pernigotto (1), Alessandro Prada (3), Linda K. Lawrie (4)

1: Free University of Bozen - Bolzano, Italy; 2: Bentley Systems Inc., Washington, DC, USA; 3: University of Trento - Trento, Italy; 4: DHL Consulting LLC, Pagosa Springs, CO, USA

AIM AND APPROACH

In this research, we began with a long-term weather data series collected by NOAA for 10 worldwide locations, selected to be representative of the most important climate classes. For these 10 locations, we prepared both typical and extreme weather files for building performance simulation, with typical years created according to ISO Standard 15927-4:2005. We implemented a series of different methods, using those proposed by the authors in previous research on eXtreme Meteorological Years XMY and Extreme Reference Years ERY. These weather files were used in a series of annual simulations of selected ASHRAE reference buildings, using TRNSYS and EnergyPlus as simulation models.

The analysis focused both on the features of the generated weather files and on the effects on simulated energy needs and final uses for space heating and cooling. Moreover, statistical tests were exploited to confirm the significance of the building performance deviations detected by using the different weather files, as well as the properties of generated reference years.

SCIENTIFIC INNOVATION AND RELEVANCE

Since multi-year building simulations are computationally demanding, typical or reference years are usually adopted, rather than using individual years of weather data. The first definitions of typical meteorological years TMY and test reference years TRY were proposed during the 1970s, followed by several variants available in the literature. Those are generally fictitious years, made of twelve representative months of hourly series of actual weather data recordings (i.e., dry bulb temperature, relative humidity, solar horizontal irradiance, wind speed and direction), selected through statistical tests to maximize the representativeness with respect to multi-year series. The Finkelstein-Schafer statistics is the most frequently adopted for the preparation of a reference year.

In the last decade, some authors have proposed to integrate the information included in TMY/TRY with weather files describing extreme weather conditions. Their use allows sensitivity analysis on building performance and designs to assess and enhance building resilience in the typical range of weather conditions for a specific location. While some researchers proposed to generate extreme years by means of morphing techniques applied to the typical ones. Others modified the very TMY/TRY procedure and employ actual weather data series.

This research aims at extensively comparing the available approaches to extreme years.

PRELIMINARY RESULTS AND CONCLUSIONS

Preliminary findings of this research show the potential of extreme weather data when it comes to analyse the performance of buildings aimed to be robust to scenarios of climate change and extreme meteorological events. In the analysed climates, generated extreme files were found significantly different from the typical and reference years, suggesting the suitability of the proposed methodologies to catch the effect of the entire weather variability in each location. Impacts of selected weather files were clearly appreciated in terms of building energy performance, indicating not only the potential arising by the combined use of typical and extreme weather data in design or redesign stages but also for long-term energy performance assessment.

MAIN REFERENCES

CEN (European Committee for Standardization) (2005). EN ISO 15927-4:2005 - Hygrothermal Performance of Buildings - Calculation and Presentation of Climatic Data– Part 4: Hourly Data for Assessing the Annual Energy Use for Heating and Cooling. Brussels: European Committee for Standardization.

Crawley, D. B. (1998). Which Weather Data Should You Use for Energy Simulations of Commercial Buildings?. ASHRAE Transactions 104(2): 498–515.

Crawley, D. B., and L. K. Lawrie (2015). Rethinking the TMY: Is the 'Typical' Meteorological Year Best for Building Performance Simulation? Proceedings of the 14th international IBPSA conference, Hyderabad, India, 7–9 December.

Crawley, D. B., and L. K. Lawrie (2019). Should We Be Using Just 'Typical' Weather Data in Building Performance Simulation? Proceedings of the 15th international IBPSA conference, Rome, Italy, 2–4 September.

Herrera, M., S. Natarajan, D. A. Coley, T. Kershaw, A. P. Ramallo- González, M. Eames, D. Fosas, and M. Wood (2017). A Review of Current and Future Weather Data for Building Simulation. Building Services Engineering Research and Technology 38(5): 602–627.

Lawrie, L. K, D. B. Crawley. 2019. Development of Global Typical Meteorological Years (TMYx). http://climate.onebuilding.org

Pernigotto, G., A. Prada and A. Gasparella (2020). Extreme reference years for building energy performance simulation. Journal of Building Performance Simulation 13(2), 152-166.

Keywords: Extreme Weather Data, Typical Weather Data, Performance Based Comparison

Paper ID 30796: Predicting solar radiation with Artificial Neural Network based on urban geometrical classification

Anas Lila* (1), Wassim Jabi (2), Simon Lannon (2)

1: Department of Architecture and Built Environment, faculty of environment and technology, university of west england, United Kingdom; 2: Welsh School of Architecture, Cardiff University

AIM AND APPROACH

This research introduces the adaptation and development of an open source Artificial Neural Network (ANN) with the aim of predicting solar radiation falling on buildings for newly generated neighbourhoods in Aswan, Egypt as an example of a hot arid zone. The outcomes are a result of training the ANN on a database of classified urban geometries and their solar radiation simulation results for local weather conditions. The classification of this database was first introduced and discussed in (Lila and Lannon 2019). This paper discusses the different stages of developing the ANN code and its final version capabilities.

The ANN code is developed to differentiate the training process from the prediction code to allow for the reuse of the trained ANN in multiple tests. The ANN code was tested for different database sizes to predict individual buildings' solar radiation and was also used to predict solar radiation for urban configurations that were not part of the training process. The results of these ANN predictions were compared to conventional solar radiation simulation results to establish the accuracy and time saved.

SCIENTIFIC INNOVATION AND RELEVANCE

The focus on solar radiation is a part of a multi-stage proof of concept framework that produces a novel method to optimize performance based neighbourhood geometry. This study highlights the potential of applying ANN methods to predict solar radiation at the urban scale and the role it will play in allowing this prediction to be conducted for other simulation aspects at the neighbourhood scale i.e, energy performance, daylight availability and outdoor thermal comfort. The paper also discusses the limitation of this approach and future improvements to achieve higher efficiency.

This process also utilizes an urban geometry classification method to build the training datasets for solar radiation simulation results. This research saves computational time to allow for performance based design decisions to be included in the early stages of urban design. These time savings will also pave the way for generating optimized neighbourhood geometrical options with solar radiation as a fitness function for this optimization process. Building an ANN that is based on novel geometrical classification of urban solar radiation is a new method to save simulation time and reach acceptable accuracy.

PRELIMINARY RESULTS AND CONCLUSIONS

The study used the Python programming language to build and develop the ANN code and a visual programming language platform (Grasshopper) to host the process of generating parametric models then simulating and predicting their performance. The solar radiation simulation was conducted using the Grasshopper plug-in Ladybug Tools. The differentiation between the training and prediction codes used the Pickle library which allowed for reusing the trained ANN in further generation and optimization studies of urban geometry.

The final version of the developed ANN has shown a time saving of more than 90% when compared to the time consumed by traditional simulation for the same number of models. It achieved a coefficient of determination of 0.8 R2 value for unseen urban configurations and 0.9 R2 individual buildings prediction results correlation to simulation results These results have shown the capabilities of this method which allows for optimization between different neighbourhood geometrical iterations. This led to finalizing a proof of concept framework that can optimize neighbourhood geometry based on its solar radiation performance in the early stages of design with significantly less time than consumed for conventional brute force simulation methods.

MAIN REFERENCES

Arlot, S. and Celisse, A. 2010. A survey of cross-validation procedures for model selection. Statistics Surveys 4, pp. 40–79. doi: 10.1214/09-SS054.

Henseler, J., Ringle, C.M. and Sinkovics, R.R. 2009. The use of partial least squares path modeling in international marketing. Advances in International Marketing 20, pp. 277–319. doi: 10.1108/S1474-7979(2009)0000020014.

Jain, A.K., Mao, J. and Mohiuddin, K.M. 1996. Artificial neural networks: A tutorial. Computer 29(3), pp. 31–44. doi: 10.1109/2.485891.

Lila, A.M.H. and Lannon, S. 2017. A parametric sensitivity analysis of the impact of built environment geometrical variables on building energy consumption. In: PLEA. Edinburgh, UK. Available at: https://orca.cf.ac.uk/102918/1/Lila PLEA 2017.pdf [Accessed: 20 September 2017].

Lila, A.M.H. and Lannon, S. 2019. Classifying urban geometry impact on solar radiation. In: Corrado, V., Fabrizio, E., Gasparella, A., and Patuzz, F. eds. the International Building Performance Simulation Association (IBPSA) 2019 16th Conference. Rome, Italy

Lin, J., Wan, H. and Cui, Y. 2020. Analyzing the spatial factors related to the distributions of building heights in urban areas: A comparative case study in Guangzhou and Shenzhen. Sustainable Cities and Society 52, p. 101854. doi: 10.1016/j.scs.2019.101854.

Paliwal, M. and Kumar, U.A. 2009. Neural networks and statistical techniques: A review of applications. Expert Systems with Applications 36(1), pp. 2–17. doi: 10.1016/j.eswa.2007.10.005.

Keywords: urban geometry, neural network, solar analysis, parametric modelling

Paper ID 30872: Evaluating and improving outdoor thermal comfort using surrogate modeling and optimization techniques

Elizabeth Young*, Caitlin Mueller, Christoph F. Reinhart

Massachusetts Institute of Technology, Cambridge, Massachusetts, USA

AIM AND APPROACH

With more than half of the world's population dwelling in cities, the carbon footprint of city dwellers plays a significant role in meeting carbon emission reduction goals. Given rising global surface temperatures, people are more likely to choose modes of travel that are shielded from the outdoors, such as driving or other motorized vehicles. These modes of transport are often carbon-intensive. To reduce their environmental footprint, cities are increasingly focused on encouraging carbon-free transportation, such as walking and biking. The quality of outdoor thermal comfort contributes to dwelling and travel patterns (Reinhart, Dhariwal, and Gero 2017) and improving the thermal comfort experience for commuters through urban interventions is key to encouraging walking and biking.

We propose a methodology to design urban interventions along bicycle pathways that would optimize thermal comfort for bikers and pedestrians. By simulating Universal Thermal Climate Index (UTCI) on an urban scale, we can evaluate existing outdoor thermal comfort conditions and quantitatively evaluate the improvements of urban interventions to outdoor thermal comfort. We then test a variety of optimization strategies to design urban interventions that would optimize improvements to outdoor thermal comfort, such as maximizing the number of comfortable hours in a year along bicycle pathways.

SCIENTIFIC INNOVATION AND RELEVANCE

Simulating UTCI and thus evaluating outdoor thermal comfort for a specific time and place is prevalent in many urban planning tools from ENVI-met to Urbano and the Ladybug toolset. Annual outdoor comfort can be simulated using the existing Eddy3D tool. However, we have yet to see the implementation of these tools to improve urban interventions such that outdoor thermal comfort is optimized for commuters.

This paper proposes a methodology that enables urban planners and designers to consider outdoor thermal comfort in real-time at an early-design stage. Designers will be able to create shading structures or place trees, among other possible interventions, in ways that attract cyclists and pedestrians. This work will draw on and integrate methods such as Kastner and Dogan (2019)'s high-resolution annual outdoor comfort simulation mapping tool, in the form of the Eddy3D tool for Grasshopper, Chokhachian and Hiller (2020)'s PANDO tool for modeling trees and canopies, and Tserandidis, Brown, and Mueller (2016)'s data-driven approximation algorithms for the optimization of civil structures, in the form of the Design Space Exploration tool for Grasshopper.

PRELIMINARY RESULTS AND CONCLUSIONS

Outdoor thermal comfort conditions along an urban sidewalk pathway in San Francisco were simulated at 9am for an entire year and results show that the pathway was comfortable just 51% of the time due to wind chills and solar radiation. An urban intervention was designed to block wind from predominant wind directions and provide shade along the pathway. This increased the proportion of comfortable days to 70%. This means that the urban intervention that was designed to improve thermal comfort along the route made the route walkable for 54 more days in a typical year. This result shows the significant impact that comfort-optimized urban intervention could have on commuter choices throughout a given year.

This method of designing urban spaces for the comfort of pedestrians and cyclists is intended to improve outdoor thermal comfort and ultimately encourage a shift in travel mode choices and decrease travel-related carbon emissions.

MAIN REFERENCES

Chokhachian, Ata, and Marion Hiller. 2020. "PANDO : Parametric Tool for Simulating Soil-Plant- Atmosphere of Tree Canopies in Grasshopper." In SimAUD2020, 495–502.

Gehl, Jan. 2011. Life between Buildings : Using Public Space. Island Press. https://books.google.com/books/about/Life_Between_Buildings.html?id=X707aiCq6T8C.

Kastner, Patrick, and Timur Dogan. 2019. "Towards High-Resolution Annual Outdoor Thermal Comfort Mapping In Urban Design." In Building Simulation 2019. Rome, Italy.

Kastner, Patrick, and Timur Dogan. 2020. "Predicting Space Usage by Multi-Objective Assessment of Outdoor Thermal Comfort around a University Campus." In SimAUD2020. http://simaud.org/2020/preprints/89.pdf.

Reinhart, Christoph F., Jay Dhariwal, and Katy Gero. 2017. "Biometeorological Indices Explain Outside Dwelling Patterns Based on Wi-Fi Data in Support of Sustainable Urban Planning." Building and Environment 126 (December). Pergamon: 422–430. doi:10.1016/J.BUILDENV.2017.10.026.

Tseranidis, Stavros, Nathan C. Brown, and Caitlin T. Mueller. 2016. "Data-Driven Approximation Algorithms for Rapid Performance Evaluation and Optimization of Civil Structures." Automation in Construction 72. doi:10.1016/j.autcon.2016.02.002.

Keywords: Universal Thermal Climate Index (UTCI), Outdoor Thermal Comfort, Optimization, Urban Interventions, Urban Commute

Session T4.9 (Online Track): Ensuring high quality building simulations

Paper ID 30435: ThRend: a ray tracing module for infrared rendering of urban scenes

Jose Pedro Aguerre*, Eduardo Fernández

UdelaR, Uruguay

AIM AND APPROACH

This paper presents ThRend, a ray tracing software that allows for accurate and physically-plausible infrared rendering of urban environments. ThRend is designed to rapidly generate simulated thermograms based on few input data. It can be considered as a post-processing tool that takes the output of thermal simulation software and simulates the behavior of long-wave radiation reaching a virtual infrared sensor. The software provides a simple interface that gives the user the possibility to try different emissivity and reflectivity configurations to render thermal images. The results can be used to compare thermal simulations against real measurements performed with thermal cameras. This approach enables the use of thermography as a more reliable tool to assess the energy efficiency and performance of buildings in their corresponding urban context. Combining computational simulation of classic thermal solvers with infrared rendering appears as an alternative to better understand the results captured in a measurement campaign.

SCIENTIFIC INNOVATION AND RELEVANCE

Thermography is used for building diagnostics and energy efficiency evaluation [EGGA18]. It allows to obtain spatialized information about the thermal exchanges within the city [BG19]. Nevertheless, the temperature results captured in a thermography campaign are not always directly reliable [DICM05], mainly because the camera produces a temperature estimation based on the incoming radiative flux [VM17]. This radiative flux depends on many aspects, such as the emissivity of the materials, the reflectivity behavior of surfaces, the atmosphere between the camera and the lens, and the temperature of reflected objects. At the urban scale, such complex phenomena produces significant bias in the apparent surface temperature of the buildings [ANG*19].

The state-of-the-art thermal solvers used by the building simulation community (e.g. EnergyPlus, CitySim, SOLENE) produce temperature distribution results that need to be post-processed for performing a reliable validation with thermography measurements [HMLV12,WZCH15]. In this paper, a post-processing ray tracing module named ThRend is presented for this purpose. A Monte Carlo approach is used to estimate the incident radiation on a virtual thermal camera considering directional emissivity models [AG*20] and microfacet BRDF theory [WMLT07]. This approach enables dealing with the multiplicity of material that are characteristic of urban environments.

PRELIMINARY RESULTS AND CONCLUSIONS

A preliminary version of ThRend is already available as a free software solution [Agu20]. It has already been validated with a real urban measurement campaign, using it as a post-processing tool for the results of a finite element urban thermal analysis. The software is able to read the output of the simulation in UCD format [AVS20], and the infrared properties are set through a very simple material definition txt file. The camera and image settings can be set through another txt file. Moreover, an interactive alternative GPU accelerated version [AW20] is available for real time inspection of the infrared scene.

The present article aims to present the flexibility of the tool by showing the results under different material configurations and several temperature conditions of the same urban scene. Careful attention is put on the algorithms that are used in the current implementation, as well as on the computational performance of the solution.

MAIN REFERENCES

[EGGA18]EVANGELISTI et al."Assessment of equivalent thermal properties of multilayer building walls coupling simulations and experimental measurements". Building and Environment 127(2018)

[BG19]BECKERS and GARCIA-NEVADO."Urban Planning Enriched by Its Representations, from Perspective to Thermography". Sustainable Vernacular Architecture. Springer International Publishing(2019)

[DICM05]DATCU et al."Improvement of building wall surface temperature measurements by infrared thermography". Infrared physics & technology 46.6(2005)

[VM17]VOLLMER and MÖLLMANN. Infrared thermal imaging: fundamentals, research and applications. Wiley&Sons(2017)

[ANG*19]AGUERRE et al."A street in perspective: Thermography simulated by the finite element method". Building and Environment 148(2019)

[HMLV12]HÉNON et al."An urban neighborhood temperature and energy study from the CAPITOUL experiment with the SOLENE model". Theor.and app.climatology 110(2012)

[WZCH15]WU et al."Real-time mid-wavelength infrared scene rendering with a feasible BRDF model". Infrared Physics & Technology 68(2015)

[AG*20]AGUERRE et al."Physically Based Simulation and Rendering of Urban Thermography". Computer Graphics Forum 39-3(2020)

[WMLT07]WALTER et al."Microfacet Models for Refraction through Rough Surfaces." Proceedings of the 18th Eurographics conference on Rendering Techniques(2007).

[Agu20]AGUERRE."ThRend: Infrared rendering for thermography simulation". Acces. July2020. URL:https://github.com/jpaguerre/ThRend.

[AVS20]Advanced Visual Systems (AVS). "Unstructured Cell Data file format". Acces. July2020. URL:https://people.sc.fsu.edu/~jburkardt/data/ucd/ucd.html.

[AW20]AGUERRE and WALD."ThRend ported to GPU with Optix Wrapper Library". Acces.July2020. URL:https://gitlab.com/ingowald/ThRend/-/tree/iw/owl

Keywords: Infrared rendering, urban computational physics, ray tracing

Paper ID 30495: CityFFD/CityBEM: Fast and high accurate urban microclimate model

Mohammad Mortezazadeh* (1), Senwen Yang (1), Jiwei Zou (1), Ali Katal (1), Sylvie Leroyer (2), Leon Wang (1)

1: Concordia University, Canada; 2: Environment and Climate Change, Dorval, Canada

AIM AND APPROACH

Urban microclimate and building energy models have been developed to address the increasing concerns over thermal and wind comfort, and building energy consumption due to rapid urbanization and building resilience as a result of climate change. Modeling urban scale problems is mostly computationally expensive. Currently, researchers need to consider some simplifications, such as uniform microclimate, for modeling building energy performance or simulate urban microclimate for small regions. Otherwise, using powerful hardware is unavoidable. In this study, we developed an integrated two-way platform by combining CityFFD (City Fast Fluid Dynamics), an urban-scale fast fluid dynamics model for microclimate modeling, and CityBEM (City Building Energy Model), a new urban building energy model with a library of 1700 building archetypes for facilitating urban model creation. Local aerodynamics and heat transfer information are exchanged between both models at each time step. Graphics processing unit computing is also applied to CityFFD for simulation speedup. This model has been validated by several numerical and experimental results.

SCIENTIFIC INNOVATION AND RELEVANCE

CityFFD/CityBEM is an in-house model for modeling urban microclimate and building energy model. This model has been written based on CUDA-C++ and includes a group of novel numerical schemes to improve the accuracy and speed of the simulation. CityFFD is a 4th order numerical scheme that is suitable for modeling large scale problems.

PRELIMINARY RESULTS AND CONCLUSIONS

In the present work, a group of urban regions in Canada have been modeled. Main focus of this research paper is to show the performance of the model for modeling various urban problems, including wind and thermal comfort, building energy performance, mitigation strategies, etc. Our research study shows significant impact of building energy consumption and urban microcliamte on local climate distribution.

MAIN REFERENCES

Mortezazadeh, M. and Wang, L.L., 2017. A high-order backward forward sweep interpolating algorithm for semi-Lagrangian method. International Journal for Numerical Methods in Fluids, 84(10), pp.584-597.

Mortezazadeh, M. and Wang, L.L., 2019. SLAC–a semi-Lagrangian artificial compressibility solver for steady-state incompressible flows. International Journal of Numerical Methods for Heat & Fluid Flow.

Mortezazadeh, M. and Wang, L.L., 2020. Solving city and building microclimates by fast fluid dynamics with large timesteps and coarse meshes. Building and Environment, p.106955.

Katal, A., Mortezazadeh, M. and Wang, L.L., 2019. Modeling building resilience against extreme weather by integrated CityFFD and CityBEM simulations. Applied Energy, 250, pp.1402-1417.

Keywords: CityFFD, CityBEM, Urban microclimate, Building energy performance, fast simulation

Paper ID 30844: A parametric analysis of the impact of thermophysical, geometry and urban context features on the energy demand of a simplified building shoebox model

Federico Battini*, Giovanni Pernigotto, Andrea Gasparella

Free University of Bozen-Bolzano, Italy

AIM AND APPROACH

In Urban Building Energy Modeling, one of the main issues is to obtain reliable results with limited computational effort [1]. To achieve such goal, one of the options is to reduce the complexity of the modeling domain. In this respect, reducing the building into shoeboxes showed to be particularly effective [2]. Nevertheless, no extensive step of sensitivity analysis and uncertainty screening were carried out at urban scale so far [3] to identify the most relevant features affecting the shoebox thermal demand [4]. In this work, a parametric analysis was performed with the aim to improve the shoebox definition from the real building characteristics. Besides conventional thermophysical and geometry properties, also context-related features were considered. After selecting the most suitable sampling technique to perform a multi-variate sensitivity analysis, several configurations of shoeboxes were simulated in the climate of Denver, Colorado, USA, using EnergyPlus as building simulation code. As in the ANSI/ASHRAE 140, Denver is chosen as reference climate due to its cold winters, hot summers and large daily temperature variations. Annual energy needs and peak loads for space heating and cooling were then analyzed. Finally, in order to identify the most significant features, the Spearman correlation coefficient was calculated.

SCIENTIFIC INNOVATION AND RELEVANCE

Other sensitivity analyses in the literature already focused on the impact of the urban context on the thermal load. However, the considered features mainly pertained the building and urban geometry [5] or focused on tropical cities [6]. Moreover, such sensitivity analyses did not fully consider the single building characteristics and the urban ones at the same time. Only characteristics describing the urban environment were considered, or just few geometric features of the building were implemented as well. To develop reliable and representative models, it is necessary to understand which are the most important variables and how they interact with each other, impacting on the building performance. Identifying the features influencing a shoebox's thermal behavior at multiple spatial scales permits to draw some guidelines on how to set shoeboxes' properties, while developing shoebox-simplified urban building energy models.

PRELIMINARY RESULTS AND CONCLUSIONS

As a whole, the variables showing the highest Spearman correlation coefficient with the heating and cooling demand outputs were ventilation and cooling setpoint, respectively. Among the urban variables considered, the context ended up being the most correlated to the output with moderate correlation with the cooling demand and very weak correlation with the heating demand. Instead, the albedo of the context surfaces showed very low correlation to the output, differently to what obtained for tropical climates [6]. For what regards the remaining single building features, the obtained results were in agreement with previous studies performed on the same kind shoebox [4]. According to the most influencing variables determined in this study, it will be possible to consistently choose the features to simplify and represent more effectively a building into a shoebox in urban simulation.

MAIN REFERENCES

[1] C. F. Reinhart and C. C. Davila, "Urban building energy modeling - A review of a nascent field", Building and Environment, vol. 97, pp. 196-202, 2 December 2015.

[2] T. Dogan and C. Reinhart, "Shoeboxer: An algorithm for abstracted rapid multi-zone building energy model generation and simulation", Energy and Buildings, vol. 140, pp. 140-153, 2017.

[3] C. Cerezo, J. Sokol, C. Reinhart and A. Al-Mumim, "Three methods for characterizing building archetypes in urban energy simulation - a case study in Kuwait city", 14th Conference of International Building Performance Simulation Association, Hyderabad, 2015.

[4] G. Pernigotto, A. Prada, A. Gasparella and J. L. M. Hensen, "Development Of Sets Of Simplified Building Models For Building Simulation", International High Performance Buildings Conference, Purdue, 2014.

[5] A. Vartholomaios, "A parametric sensitivity analysis of the influence of urban form on domestic energy consumption for heating and cooling in a Mediterranean city", Sustainable Cities and Society, vol. 28, pp. 135-145, 2017.

[6] T. A. d. L. Martins, L. Adolphe, L. E. G. Bastos and M. A. d. L. Martins, "Sensitivity analysis of urban morphology factors regarding solar energy potential of buildings in a Brazilian tropical context", Solar Energy, vol. 137, pp. 11-24, 2016.

Keywords: simplified building models, building energy simulation, shoebox, urban building energy modeling

Paper ID 30943: Modeling district heating and cooling systems with URBANopt, GeoJSON to Modelica Translator, and the Modelica Buildings Library

Nicholas Long* (1), Antoine Gautier (2), Hagar Elarga (1), Amy Allen (1), Ted Summer (3), Lauren Klun (1), Nathan Moore (1), Michael Wetter (2)

1: National Renewable Energy Laboratory, United States of America; 2: Lawrence Berkeley National Laboratory, United States of America; 3: Devetry

AIM AND APPROACH

The URBANopt project has been successfully leveraging OpenStudio/EnergyPlus, REopt, OpenDSS, and Reference Network Models to bridge the gap between buildings, distributed energy resources, and electric grids; however, URBANopt has been lacking the ability to model district thermal energy systems until recently. Conventional simulation tools such as EnergyPlus, have difficulties modeling modern district heating and cooling (DHC) systems due to the lack of pressure-driven analysis and their limited capabilities to represent tailored control logics and coupling between plants, distribution network and substations. This paper will present the modeling infrastructure that was developed specifically for the analysis of district heating and cooling systems, and how it is integrated into the existing URBANopt framework.

SCIENTIFIC INNOVATION AND RELEVANCE

The paper will discuss the development of a new extension to URBANopt to model various district energy system components. It will describe the new models that were added to the Modelica Buildings Library to represent the central plants, the distribution networks, the energy transfer stations (ETS), and ultimately the loads on the DES. The supported modeling approaches and system configurations will be discussed. For instance, the loads can be provided as time series (CSV), TEASER models, or Spawn of EnergyPlus models, and the ETS models can represent direct or indirect connections, or even distributed heat recovery chillers for ambient loop systems. URBANopt allows the user to switch between the various configurations. It uses a translator from an URBANopt GeoJSON file describing the various buildings and district systems to scaffold an entire Modelica system model ready for running analysis.

PRELIMINARY RESULTS AND CONCLUSIONS

The paper will conclude with a comparative analysis between a 1st generation district system and a 4/5th generation district system using load profiles generated from OpenStudio/EnergyPlus.

MAIN REFERENCES

URBANopt: https://www.nrel.gov/buildings/urbanopt.html

GMT: https://github.com/urbanopt/geojson-modelica-translator

MBL: https://github.com/lbl-srg/modelica-buildings

TEASER: https://github.com/RWTH-EBC/TEASER

Spawn: https://github.com/NREL/Spawn

Keywords: district heating and cooling, modelica buildings library, urban analysis

Paper ID 30842: Evolving co-heating tests to deliver additional performance metrics and support model calibration

Jon William Hand* (1), Lori Barbara McElroy (1), Colin Sinclair (2)

1: University of Strathclyde, United Kingdom; 2: Building Research Establishment, United Kingdom

AIM AND APPROACH

The paper compares using an innovative, short duration 'Pulse Test' with a standard 'Co-heating Test', which typically requires a number of weeks to establish an overall facade heat transfer metric. The method relies on monitoring the heating/cooling and temperatures in each room during a typical one week period in order to:

a) capture the response to a step-change in heating or cooling, to reach a 25-30oC delta T against ambient;

b) identify the energy inputs required to maintain the elevated/depressed conditions; and

c) profile the response of the building during its return to ambient.

This is used to calibrate a numerical model of the building, thus creating a digital twin of the monitored building and physical test. Having carried out both co-heating and pulse tests the paper compares and contrasts working procedures, experimental design, performance assessment model design, model calibration steps as well as the building performance appraisal facilities required.

SCIENTIFIC INNOVATION AND RELEVANCE

The innovation is in making use of room-by-room data from short term testing and monitoring to calibrate a simulation model. This minimises disruption to occupants in the case of existing buildings and avoids unnecessary delay in occupation of new buildings while significantly improving model accuracy, thus closing the performance gap between the virtual and real worlds.

The three phases of the Pulse Test are key to capturing the impact of internal mass as well as local variances in façade performance within both the building and its digital twin. Although the method does not identify faults in specific portions of a rooms, it is at a higher spacial resolution than many other approaches. Combined with real and virtual twin blower door tests and thermographic surveys the resulting model is a better platform on which to explore broader performance assessments.

PRELIMINARY RESULTS AND CONCLUSIONS

The application of the method in a number of major Government funded new build and retrofit projects in the UK has resulted in both an overall reduction in the time required to undertake the experimental setup and monitoring and the creation of numerical models which reflect the actual response characteristics of the building.

For example, in one project differences between the physical and virtual tests indicated that on-site floor build-ups differed from the construction details provided as well as highlighting the impact of façade air leakage faults when the tests were undertaken with and without formal pressure tests. In one case a second round of building and digital twin tests were carried out to confirm improvements after on-site corrections were carried out.

After calibration, the digital twin was used in performance appraisals with local weather data to establish longer term performance characteristics. Initial indications from performance patterns observed in the digital twin were also noted in POE.

From such well-calibrated models, further virtual experiments have been conducted, exploring issues such as matching heating system characteristics with lifestyle/ occupancy factors.

MAIN REFERENCES

HIT2GAP - Highly Innovative Building Control Tools - Grant Agreement number: 680708 - H2020 ENERGY-EFFICIENT BUILDINGS -2016-2017 - www.hit2gap.eu

Wall-ACE – Novel Wall Insulation Systems - Grant Agreement number: 723574 - H2020 ENERGY-EFFICIENT BUILDINGS -2016-2017 – www.wall-ace.eu

ESP-r www.strath.ac.uk/research/energysystemsresearchunit/

Keywords: co-heating evolution, model calibration, digital twins, simulation vs reality, monitoring

Session T5.1: Buildings paving the way for the energy transition

Paper ID 30423: A collective heat and cold distribution system with decentralized booster heat pumps: a sizing study.

Stef Jacobs*, Freek Van Riet, Ivan Verhaert

University of Antwerp, EMIB research group, Antwerp, Belgium

AIM AND APPROACH

The development and use of more energy-efficient heating systems is required to secure a sustainable future. In this respect a myriad of heating systems and concepts are being developed.

The combined distribution circuit (CDC) of centralized heating at lower temperatures and decentralized booster heat pumps (BHP) is such concept, notably one of a collective heating and cooling system for apartment buildings

[1], [2]. Although certain studies [2], [3] have demonstrated that the use of BHP result in potential energy savings, it seems that the concept has not yet been put into practice, nor that detailed studies are available. As a result no in depth knowledge was available on the sizing, controlling and evaluating of such system, and notably on determining the optimal balance between decentralized and centralized heat production by varying the size of the BHPs, the decentralized storage of DHW and the temperature in the CDC.

As the Belgian EPBD standardization rules have a major impact on system selection, but, for this kind of heating systems, do not provide any appropriate evaluation framework, such framework needs to be further developed. It should provide an insight on the impact of various design choices, and an objective comparison with other systems.

SCIENTIFIC INNOVATION AND RELEVANCE

In order to evaluate the CDC with BHP in apartment buildings a simulation environment has been developed in Matlab, so that all concept components, individually and collectively, can be analyzed. In this respect all design elements can be varied for the purpose to acquire in depth knowledge. Notably the following sizing parameters have been examined. Firstly, the nominal power of the BHP, and the corresponding size of the storage vessel for the DHW, were varied, as these two parameters may have a direct impact on the performances of the BHP. Secondly, the impact of the volume of the central storage vessel on the stability of the supply water temperature has been examined. Finally, the central system design temperatures were varied, as they have an impact on the SPF of the BHP and of the central HP, and thus on the efficiency of the entire system.

Furthermore the developed simulation environment allows to take into account a dynamic and realistic DHW-demand [4].

Finally the developed simulation environment is complementary to other studies performed, that have analyzed other heating systems, or complex centralized hybrid heating production [5].

PRELIMINARY RESULTS AND CONCLUSIONS

Firstly, the concept study has shown that a CDC at lower temperatures with BHP can achieve a system efficiency of 3,34 on a yearly basis (SPF). Moreover, the SPF of the central HP and also of the entire system increases when the central system design temperatures are reduced, even though the SPF of the BHP then decreases, as a result of a higher electricity consumption.

In other words, the analyzes have shown that, for central system design temperatures ranging between 29/22 and 40/33, the performance of the central HP is decisive for the total system efficiency. Furthermore, the analyzes have demonstrated that the nominal power of the BHP has an important impact on the performance of the BHP. Finally, the concept study that the annual equivalent CO_2-emission of the CDC with BHP is substantially lower, even up to 60%, than the one of a CDC with boilers.

The concept study proofs that the concept of a CDC with BHP can contribute to decarbonizing the heat (and cooling) production in apartment buildings. Moreover, when the electricity demand of the system would be met by renewable energy sources (e.g. windmills or solar panels), CO_2-emissions could even be further reduced.

MAIN REFERENCES

[1] Ivan Verhaert, “Conceptfiche : Combilus en afleversets,” no. September, p. 16, 2018.

[2] P. A. Østergaard and A. N. Andersen, “Booster heat pumps and central heat pumps in district heating,” Appl. Energy, vol. 184, pp. 1374–1388, 2016.

[3] O. Kleefkens, J. Van Berkel, C. Geelen, and M. Bos, “Booster Heat Pump, development of test procedure and calculation methodology in order to estimate the energy performance in various domestic applications,” p. 12, 2017.

[4] J. De Schutter, I. Verhaert, and M. De Pauw, “A methodology to generate realistic random behavior profiles for space heating and domestic hot water simulations,” no. April, pp. 1–8, 2018.

[5] F. Van Riet, "Hydronic design of hybrid thermal production systems in buildings," 2019.

Keywords: heat pumps, combined distribution circuit, apartment building, domestic hot water, cooling

Paper ID 30117: Digital Twin Design with on-line calibration for HVAC systems in buildings

Christian Vering* (1), Sebastian Borges (1), Daniel Coakley (2), Hannah Kruetzfeldt (1), Philipp Mehrfeld (1), Dirk Müller (1)

1: RWTH Aachen University, Germany; 2: MERCE Mitsubishi, Scotland

AIM AND APPROACH

With regard to climate policy, the Paris Convention 2015 aims to limit global warming to below 2 °C [1]. In order to achieve this goal, it is necessary to obtain energy from renewable energy sources and to increase energy efficiency of supply systems. With about half of the final energy consumption serving the building sector, HVAC systems play a crucial role in achieving this goal [2].

Digital Twins of HVAC systems show great potential to increase efficiency through operational optimization and predictive maintenance of real processes. By connecting the real and virtual worlds using information technology, it is possible to make predictions about plant behavior and incorporate them into the operation and maintenance strategy [3, 4]. With its five phases, Digital Twin Concept, Digital Twin Prototype, Digital Twin Instance, Digital Twin Aggregates and Digital Twin Knowledge, Digital Twin Design is a promising method for integrating Digital Twins into the product life cycle [5].

The goal is to develop a method for the implementation of a scalable Digital Twin Design for HVAC systems and make it accessible for the industry. This paper will outline the framework for model development and calibration, incorporating scalable processes for on-line semi-automated parameter tuning.

SCIENTIFIC INNOVATION AND RELEVANCE

HVAC systems are very complex due to their nonlinear behavior through stochastic interaction with their environment. Their efficiency is highly dependent on the component design and control [5] which makes the implementation of a Digital Twin concept essential to increase the efficiency. To achieve this goal in the context of a framework for model development and calibration there are several scientific challenges and questions that will be addressed. One central research question in the context of implementing a Digital Twin concept is the integration of a continuous calibration of on-line simulation models. This requires a framework that reads all necessary data, performs an automated calibration and evaluates the results using metrics.

Within the scope of this paper, we suggest an approach for a sensitivity analysis, in which we determine adjustable model parameters for tuning when using the method for on-line calibration. This also addresses the question of which tuning methods are appropriate and how often the calibration should check and re-tune the model. According to the application of the Digital Twin Concept the requirement for model scalability and accuracy as well as the metrics for assessment (CVRMSE, NMBE, etc.) are clarified.

PRELIMINARY RESULTS AND CONCLUSIONS

The Digital Twin Design has already been used for an energy recovery ventilation system. It shows that the concept of Digital Twins can be used to achieve energy and operational cost savings through predictive maintenance [5]. Next steps for a successful implementation and application in industry are a more detailed investigation of machine communication and software technology [5] as well as the implementation of the above mentioned calibration framework to increase the model accuracy and the investigation of optimized operation strategies regarding energy and cost savings.

In this contribution, we introduce an IoT-based framework that allows the communication between HVAC system and its Digital Twin as well as further clients like a time series data base and a supervisor dashboard. The obtained data is used to on-line calibrate the Digital Twin robustly. We show that it is possible to use this framework to optimize the operation using on-line model predictive control.

MAIN REFERENCES

1. Liobikienė G, Butkus M (2017) The European Union possibilities to achieve targets of Europe 2020 and Paris agreement climate policy. Renewable Energy 106, 298–309. 10.1016/j.renene.2017.01.036.

2. Perez-Lombard L, Ortiz J, Maestre IR (2011) The map of energy flow in HVAC systems. Applied Energy 88(12), 5020–5031. 10.1016/j.apenergy.2011.07.003.

3. Klostermeier R, Haag S, Benlian A (2018) Digitale Zwillinge–Eine explorative Fallstudie zur Untersuchung von Geschäftsmodellen. HMD Praxis der Wirtschaftsinformatik 55(2), 297–311.

4. Qi Q, Tao F (2018) Digital twin and big data towards smart manufacturing and industry 4.0: 360 degree comparison. Ieee Access 6, 3585–3593.

5. Vering C, Mehrfeld P, Nürenberg M, Coakly D, Lauster M, Müller D Unlocking Potentials of Building Energy Systems' Operational Efficiency: Application of Digital Twin Design for HVAC systems.

Keywords: Application of digital twins, cloud control, modeling and calibration

Paper ID 30637: A virtual test bed for evaluating advanced building automation algorithms

Thomas Storek* (1), Fabian Wüllhorst (1), Silas Koßler (1), Marc Baranski (2), Alexander Kümpel (1), Dirk Müller (1)

1: RWTH Aachen University, E.ON Energy Research Center, Institute for Energy Efficient Buildings and Indoor Climate, Aachen, Germany; 2: Aedifion GmbH, Cologne, Germany

AIM AND APPROACH

Reducing the primary energy demand and CO2 emissions in the building sector is a major part of today's energy policies. Besides retrofitting and the integration of renewable energy sources into building energy systems (BES), operation optimization is in the focus of these efforts. A key challenge is to develop intelligent algorithms that are capable of controlling the augmented complexity of modern BES efficiently. Control strategies such as model predictive control but also machine learning-based algorithms hold great potential for significant savings in energy and operating costs. Due to the increased digital cross-linking of technical equipment, the occasionally high demand for computing power of some of the algorithms, and the availability of advanced maintainability tools, many future automation programs will be operated as cloud services. An important step during the implementation of controllers is a comprehensive test of the automation functions. In buildings, these tests usually do not consider energy efficiency. Hence, energy efficiency is often evaluated during operation and the programs need to be adapted on-site. This process is not only complex and lengthy but also cost-intensive. Therefore, we propose a virtual testbed to support the development and a-priori testing and evaluation of cloud-based control algorithms for BES.

SCIENTIFIC INNOVATION AND RELEVANCE

In this paper, we present a virtual testbed, which does not only allow to check the plausibility of control concepts during the development process but also to evaluate the expected energy efficiency a-priori. This goal is achieved by coupling the algorithms with a 1-D simulation of the controlled system model. This approach is known from other industry sectors as rapid control prototyping but has not yet been established in practical applications in the building sector. The possible reasons for this may be that building automation programs are usually implemented individually for a specific building and that the procurement for construction sites leads to a fragmented market and high pressure on pricing. By using identical cloud-based communication infrastructure in the test bench and in the real plant, this infrastructure is also included in the tests in advance. As a result, the communication infrastructure only needs to be configured to an insignificant extent when it is later integrated into the real building, which greatly reduces the number of potential sources of error.

PRELIMINARY RESULTS AND CONCLUSIONS

As a containerized Python web application, the virtual testbed is extendable in its functionality and connectivity to different cloud providers, as well as deployable as a cloud service itself. The user can access it via a web frontend, which allows the upload of simulation models of the controlled systems as either Functional-Mock-Up (FMU) or Modelica file and the result evaluation. In a first testing scenario, we uploaded a model with two decoupled thermal zones with heating supply from a heat pump. The controller under test controls the zone temperatures by manipulating two admix circuits. Connecting the testbed to the open-source cloud platform FIWARE, we deploy two remote PID controllers on the cloud platform and successfully run several simulations varying the PID parameters and sample rates. Afterwards, we superimpose the trajectories of the individual runs to compare the effect of parameter variation and to identify the effect of transmission latency when using high sample rates or comparing to locally deployed PIDs. Overall, we state that the use of a virtual testbed can complement and support the development, commissioning, and maintenance of building automation systems in a cost-effective way.

MAIN REFERENCES

Michael Wetter. Co-Simulation of Building Energy and Control Systems with the Building Controls Virtual Test Bed. Journal of Building Performance Simulation, 4(3):185–203, 2011. doi: 10.1080/19401493.2010.518631. URL https://simulationresearch.lbl.gov/sites/all/files/wetter-paper-9711.pdf.

Raymond Sterling, Jesús Febres, Andrea Costa, Adeleh Mohammadi, Rafael E. Carrillo, Baptiste Schubnel, Yves Stauffer, Pietro de Cinque, Krzysztof Klobut, and Marcus M. Keane. A virtual test-bed for building Model Predictive Control developments. In V. Corrado, E. Fabrizio, A. Gasparella, and F. Patuzzi, editors, Proceedings of Building Simulation 2019: 16th Conference of IBPSA, pages 17–24, 2019. ISBN 978-1-7750520-1-2. doi:10.3384/ecp1915717. URL https://ep.liu.se/ecp/157/001/ecp19157001.pdf.

Joanna Sendorek, Tomasz Szydlo, and Robert Brzoza-Woch. Software-Defined Virtual Testbed for IoT Systems. Wireless Communications and Mobile Computing, 2018. doi: 10.1155/2018/1068261. URL https://www.hindawi.com/journals/wcmc/2018/1068261/.

David Blum, Filip Jorissen, Sen Huang, Yan Chen, Javier Arroyo, Kyle Benne, Yanfei Li, Valentin Gavan, Lisa Rivalin, Lieve Helsen, Draguna Vrabie, Michael Wetter, and Marina Sofos. Prototyping the BOPTEST Framework for Simulation-Based Testing of Advanced Control Strategies in Buildings. In V. Corrado, E. Fabrizio, A. Gasparella, and F. Patuzzi, editors, Proceedings of Building Simulation 2019: 16th Conference of IBPSA, pages 2737–2744, 2019. ISBN 978-1-7750520-1-2. doi: 10.26868/25222708.2019.211276.URL http://www.ibpsa.org/proceedings/BS2019/BS2019_211276.pdf.

Keywords: Building Automation, Control, HVACR, Operation Optimization, Testing

Paper ID 30505: Strengths of non-linear white-box MPC for building HVAC control

Filip Jorissen* (1), Damien Picard (1), Lieve Helsen (1,2)

1: KU Leuven, Department of Mechanical Engineering, Leuven, Belgium; 2: EnergyVille, Thor Park, Waterschei, Belgium

AIM AND APPROACH

Building heating and cooling account for 15 % of the world final energy use [1]. While the share of renewable energy sources increases and building design standards become stricter leading to better insulated buildings but also to more complex and hybrid HVAC systems, building control and exploitation of flexibility play an important role in the primary energy saving potential. However, HVAC systems are often oversized due to conservative selection and sizing, and by neglecting control in the early design phase. This leads to increased costs and decreased performance.

The Thermal Systems Simulation (The SySi) research group of KU Leuven aims to overcome these problems by developing automated workflows for design and control of buildings and their HVAC systems using high fidelity Modelica models. The open-source nature of the Modelica language and of some Modelica tools allows flexible integration in a larger software environment, while the equation-based nature and translation engines of Modelica allow efficient numerical algorithms to solve the underlying model equations in a fully automated way. Furthermore, high fidelity models allow capturing important non-linear effects such as fan power as a function of flow rates and the coefficient of performance of heat pumps.

SCIENTIFIC INNOVATION AND RELEVANCE

This presentation gives an overview of the authors' research on non-linear optimal control using Model Predictive Control and optimal design of buildings, both using high fidelity Modelica models, which builds upon the work of [2-7]. While research often focusses on linear or simplified modeling to overcome long computation times, we show examples where efficient numerical implementations and algorithm choice allow directly using detailed Modelica models for on-line optimization. Furthermore, a workflow is presented where the Modelica models are automatically generated from user input in a way that avoids modelling errors and that reduces the modelling effort. The end goal of this research is to develop tools and workflows for deploying MPC at scale, in a way that is economically feasible.

PRELIMINARY RESULTS AND CONCLUSIONS

Preliminary results indicate that an MPC for a complex 10 000 m2 office building consisting of 100 zones and 300 control variables can be solved in a few minutes. Furthermore, an optimal design framework is presented where the HVAC components for a similar building using a 1-year optimization horizon are sized. At the time of writing we project that the computation time will be a few hours on a modern workstation. Preliminary results indicate that the building envelope for this building can be modelled within a few hours using our Modelica building envelope model generator. The HVAC model generator is currently under development.

While these computation times are not small, computation requirements for smaller projects are of course smaller and the listed computation times are considered acceptable for this relatively large and complex project.

MAIN REFERENCES

[1] International Energy Agency (2015). World energy outlook 2015. Technical report.

[2] Jorissen, F. (2018). Toolchain for Optimal Control and Design of Energy Systems in Buildings. PhD thesis, Arenberg Doctoral School, KU Leuven.

[3] Jorissen, F., Boydens, W., Helsen, L. (2019). Model implementation and verification of the envelope, HVAC and controller of an office building in Modelica. Journal of Building Performance Simulation, 12 (4), 445-464. doi: 10.1080/19401493.2018.1544277

[4] Jorissen, F., Boydens, W., Helsen, L. (2018). TACO, an automated toolchain for model predictive control of building systems: implementation and verification. Journal of Building Performance Simulation, 12 (2), 180-192. doi: 10.1080/19401493.2018.1498537

[5] Jorissen, F., Reynders, G., Baetens, R., Picard, D., Saelens, D., Helsen, L. (2018). Implementation and Verification of the IDEAS Building Energy Simulation Library. Journal of Building Performance Simulation, Art.No. DOI: 10.1080/19401493.2018.1428361.

[6] Picard, D., Drgoňa, J., Kvasnica, M., Helsen, L. (2017). Impact of the controller model complexity on model predictive control performance for buildings. Energy and Buildings, 152, 739-751.

[7] Drgona, J., Picard, D., Helsen, L. with Drgona, J. (corresp. author) (2020). Cloud-based implementation of white-box model predictive control for a GEOTABS office building: a field test demonstration. Journal of Process Control.

Keywords: Model predictive control, optimization, optimal design, Modelica

Session T5.2: Ensuring high quality building simulations

Paper ID 31104: A parametric combined design and simulation tool for the optimisation of domestic hot water systems in residential care centres

Martijn Holvoet*, Elisa Van Kenhove, Klaas De Jonge, Lien De Backer, Wim Boydens, Jelle Laverge

UGent / Archipelago

AIM AND APPROACH

In this master dissertation research is conducted into the proper design and dimensioning of Domestic Hot Water (DHW) systems in public building

typologies. A basis for the optimisation of the design is created through the development of a flexible and parametric simulation tool in Modelica that can be used in the design phase. An extensive design study and research

into parameters and characteristics concerning the building, system and demand leads to assumptions of average values and data to fill in the information that is still unknown at this stage. Together with the integration of the general methods of dimensioning, the simulation tool can function with only a limited number

of input data and can therefore be used in the design

phase of DHW systems in public buildings.

SCIENTIFIC INNOVATION AND RELEVANCE

The tightening of energy performance requirements effectuates the ever-improving energy efficiency in buildings. Among others, the insulation level and the airtightness of the building envelope have evolved considerably. Only in the field of DHW, there is comparatively little innovation. The energy demand for DHW remained nearly unaltered over the years and consequently starts to represent an important share of the total energy demand of buildings. It is a domain with little research and innovation, certainly concerning public building typologies. Regarding the building design, more attention should be paid to the optimisation of DHW systems. As there are many degrees of freedom concerning the configuration of these systems, tools that allow to make an informed choice for the specific configuration of the DHW system at an early stage can be of great importance. In addition, nowadays, also indicators such as energy use play a crucial role, where previously only comfort related aspects were considered as important. Concretizing performance factors of certain design proposals in the early phase can be useful.

Normally, dimensioning is only done at a later stage of the design, at a time when modifications to the design already entail major implications and are therefore no longer possible.

PRELIMINARY RESULTS AND CONCLUSIONS

A parametric simulation tool for DHW systems that can be effortlessly used in the design phase and requires only a few input parameters, is developed and it can have the potential to be of great importance in the design and optimisation of DHW systems in public building typologies.

However, this requires a complex and extensive integration of algorithms and a research into parameters and characteristics concerning the building, the DHW system and the hot water demand. On the basis of a virtual model, the operation of the DHW system can be assessed. The tool can allow concretizing of the performance factors of different proposals for the configuration of DHW systems. DHW system designers will be able to estimate the impact of design decisions and to reduce energy demand for DHW, while keeping an equilibrium between healthy, comfortable and energy efficient buildings.

In addition, the tool can also be used in the optimisation of various parameters, from thickness of insulation to the care and washing procedure. It also has opportunities in the field of Legionella decontamination.

It can have the potential to be of great importance in the optimisation of DHW systems and in the definition of energy-saving methodologies.

MAIN REFERENCES

*Van den Abeele L., Dinne K., De Cuyper K. and Bleys B., 'Best Beschikbare Technieken (BBT) voor Legionella-beheersing in Nieuw Sanitaire Systemen'. VITO & WTCB, 2017.

*Van Kenhove E., 'Coupled thermohydraulic and biologic modelling of legionella Pneumophila proliferation in domestic hot water systems', 2018.

*Deutsches Institut für Normung, 'DIN 1988-300: Technische Regeln für Trinkwasser-Installationen - Teil 300: Ermittlung der Rohrdurchmesser'. Berlin, 2012.

*A. Bertrand, A. Mastrucci, N. Schüler, R. Aggoune, en F. Maréchal, "Characterisation of domestic hot water end-uses for

integrated urban thermal energy assessment and optimisation", Applied Energy, vol. 186, pp. 152–166, jan. 2017.

Keywords: domestic hot water, parametric simulation tool, Modelica, design study, energy use

Paper ID 30985: Simulation of Legionella risks when restarting a sanitary sport facility installation after a period of inactivity during COVID-19

Elisa Van Kenhove*, Lien De Backer, Jelle Laverge

Ghent University, Belgium

AIM AND APPROACH

During the COVID-19 lockdown it was repeatedly reported in the media that the prolonged closure of buildings (e.g. hotels, sport facilities, student homes) is not without danger as stagnant water is an important risk factor for Legionella growth. LoWatter, a service of Ghent University, carries out scientific research and consultancy into Legionella in sanitary hot water systems [1]. Based on an in-house developed simulation tool, the Legionella concentration can be predicted dynamically in the entire hot water circuit [2]. The aim of the presented case study is to demonstrate what the consequences are of a temporary shutting down of a system with regard to the risk of a Legionella pneumophila contamination of the system. To do so, the simulation tool is applied on a fictive, but representative case of a sport facility. Parameters, such as number of changing rooms and showers were derived from a comparative study. The distribution system consists of a circulation loop, located in the basement, with vertical distribution to each changing room. Sizing of pipe diameters, insulation thickness, circulation pump and nominal power of the production unit was done within the simulation tool according to DIN1988-300 [3], [4].

SCIENTIFIC INNOVATION AND RELEVANCE

The simulation tool applied for the described case study, is written in the Modelica language. This in-house developed tool is a parametric tool that allows to build up a simulation model of the hydraulic scheme quickly by defining building characteristics: in case of a sport facility such parameters are the number of changing rooms and number of showers. The sizing of all subcomponents present in the hot water system (production system, pump, pipes, valves,) is done automatically. Although this tool was used in this paper to study the impact of temporary shutting down, the tool can also be used in other situations that may cause Legionella contamination of in the water system. Examples are an improper balance of the sanitary hot water system and an undersized production system. This case-specific simulation model can then be used to first "virtually" test the effectiveness of possible solutions on a contaminated system (modifications on the system design, decontamination techniques), before applying this in practice. Based on the results, it is possible to provide advice on which case-specific measure(s) will ensure that the contamination disappears and does not reoccur. Furthermore, it allows a building owner to get a better understanding of the system.

PRELIMINARY RESULTS AND CONCLUSIONS

The case study of the sports facility shows the effect of a prolonged stagnation of the water in a domestic hot water system. As expected, the combination of stagnation and the relatively high ambient temperatures that were present during April till August resulted in legionella growth in the system. The model also showed the importance of the correct application of proposed measures at the start-up of the system: flushing each tap point at a high temperature for a sufficiently long time.

Furthermore, the case study also shows the potential of the developed tool. The sizing that was calculated in the background of the tool was compared to a fully manual calculation for the same system and results in the same sizing of the system. There are still a few work points in the parametric tool such as the integration of the iterative calculation process in case the pressure drop in the pipes requires to an adjustment of the pipe diameter. This will be further improved.

MAIN REFERENCES

[1] "Lowatter - Controlling Legionella in tapwater," 2020. [Online]. Available: www.lowatter.com.

[2] E. Van Kenhove, L. De Backer, M. Delghust, and J. Laverge, "Coupling of modelica domestic hot water simulation model with controller," in Proceedings of Building Simulation 2019 : 16th Conference of IBPSA, 2019, pp. 924–931.

[3] S. Kreps, K. De Cuyper, S. Vanassche, and K. Vrancken, "Best Beschikbare Technieken (BBT) voor Legionella-beheersing in Nieuwe Sanitaire Systemen," 2017.

[4] DIN Deutsches Institut für Normung e.V., "DIN 1988-300 Technische Regeln für Trinkwasser-Installationen - Ermittlung der Rohrdurchmesser; Technische Regel des DVGW," 2012.

Keywords: domestic hot water, Modelica, Legionella, sport facility, COVID-19 lockdown

Paper ID 30398: Towards the integration of energy performance certificates (EPC) and simplified building performance simulations using machine learning: initial findings

Roberto Boghetti (1), Roberto Rugani* (1), Marco Picco (2), Giacomo Salvadori (1), Marco Marengo (2), Fabio Fantozzi (1)

1: University of Pisa, Pisa, Italy; 2: University of Brighton, Brighton, United Kingdom

AIM AND APPROACH

A broad range of policies and supportive measures aimed at improving the performance of existing and new buildings has been introduced in the recent years. Among these, energy performance certificates provide a rating scheme to assess the energy efficiency of buildings based on quasi-steady state methods. The adoption of these certificates, however, is currently facing some critical issues, such as the lack of a quality control, a limited access to data for practitioners, and regional differences in the way the certificates are managed. Furthermore, the quasi-steady state calculation method used, while less onerous than a dynamic simulation, has a lower precision and still carries numerous overheads [1]. The most common risk is to use energy certification methods to predict the real consumption of buildings. Data-driven approaches may solve some of these problems and provide a faster yet powerful alternative to quasi-steady state methods which could also be easily integrated with existing building cadasters. The goal of this study is to examine the reliability of existing energy performance certificates and to investigate the possibility of using a data-driven approach, namely an artificial neural network (ANN), as an alternative to the current quasi-steady state method.

SCIENTIFIC INNOVATION AND RELEVANCE

This study is a preliminary step towards a broader research intent: developing a simplified simulation tool based on machine learning to predict both the energy consumption of buildings and their associated energy performance score. Machine learning has been successfully used on a research level to predict energy consumption and carbon emissions of buildings[2-3-4], using both measured and simulated data. Given the complexity of the problem and the difficulty of gathering the needed data, however, the resulting models were always limited to the respective case studies and lacked the generality to be applied in practical situations. Focusing on energy performance certificates partially solves this issue as several public repositories of these certificates were made available by different european countries and institutions. To cope with the fact that these databases are not standardized, in particular for what concerns the descriptive parameters of the buildings, an intermediate step will be made to generate a surrogate model of each considered building using the available information, so that it will be possible to create a single predictive model despite having heterogeneous input data. In the view of the authors, this approach could represent a valuable and novel addition to the ongoing scientific discussion.

PRELIMINARY RESULTS AND CONCLUSIONS

Preliminary results were obtained using both the old and new version of the Lombardy's energy performance certificates database (CENED). The two databases, created according to two subsequent versions of the Italian norm, differ for the provided input data and for the calculation method. In both cases, the available certificates were vastly incorrect [5], with many entries incorrectly reported. The proposed data-driven model, which calculates the energy performance index (EPH) in the heating season, demonstrated high precision on the old database, with a mean error of around 10%. On the newer database, however, where some characteristics of the buildings were not available anymore, the resulting error was unacceptable, ranging around 40%. To demonstrate that the problem was due to the lack of important information, a third attempt to calculate the updated EPH value using inputs from the old database was made. Even if the number of available buildings in this case was smaller than in the previous attempts, the error, concentrated in highly efficient buildings, was reduced by 35%. The positive results on this initial step indicate that a data-driven approach to EPC calculation could yield reliable estimations and therefore represents a promising support to the current methodology.

MAIN REFERENCES

[1] Ballarini, I., Primo, E. and Corrado, V., 2018. On the limits of the quasi-steady-state method to predict the energy performance of low-energy buildings. Thermal Science, 22(Suppl. 4), pp.1117-1127.

[2] Ahmad, M.W., Mourshed, M. and Rezgui, Y., 2017. Trees vs Neurons: Comparison between random forest and ANN for high-resolution prediction of building energy consumption. Energy and Buildings, 147, pp.77-89.

[3] Paterakis, N.G., Mocanu, E., Gibescu, M., Stappers, B. and van Alst, W., 2017, September. Deep learning versus traditional machine learning methods for aggregated energy demand prediction. In 2017 IEEE PES Innovative Smart Grid Technologies Conference Europe (ISGT-Europe) (pp. 1-6). IEEE.

[4] Boghetti, R., Fantozzi, F., Kämpf, J.H., Salvadori, G., 2019, Understanding the performance gap: a machine learning approach on residential buildings in Turin, Italy. Journal of Physics: Conference Series (1343), CISBAT 2019

[5] Khayatian, F. and Sarto, L., 2016. Application of neural networks for evaluating energy performance certificates of residential buildings. Energy and Buildings, 125, pp.45-54.

Keywords: Gradient boosting, Machine learning, Energy Performance Certificates, Energy policy, Simplified simulation tool

Paper ID 30925: Interactions of building- / urban-, and multi-energy systems design variables

Adam Bufacchi* (1), Georgios Mavromatidis (2), Arno Schlüter (1), Christoph Waibel (1)

1: Architecture and Building Systems, ETH Zurich, Switzerland; 2: Sustainability and Technology, ETH Zurich, Switzerland

AIM AND APPROACH

Significant synergies between the energy demand and supply side have been shown to exist which should be exploited in the design of low-carbon buildings and neighbourhoods by considering building design parameters and energy systems design parameters simultaneously, rather than addressing them sequentially (Waibel et al., 2019) (Ferrara et al., 2019). However, existing literature lacks an explicit quantification of such synergies, particularly at the building geometry scale. Therefore, this work aims to investigate the degree of parameter interdependence between parameters influencing energy demand (i.e. building construction, use, and geometry) and parameters defining the energy supply (i.e. the design of a neighbourhood's decentralised multi-energy system).

We apply a Morris Screening on a wide range of parameters, which informs us through a qualitative ranking of these parameters' importance. Subsequently, we compute first, second and total order Sobol indices: Global Sensitivity Analysis indicators, quantifying the parameters' influence on the design output individually, when interacting with one-, or all other parameters respectively. Hereby we seek to quantify the degree of parameter interdependence. The workflow includes sampling real building geometries (LoD1 data of Zürich as case study), energy demand simulations with Energyplus/Honeybee, and multi-energy systems design optimisation using a Mixed Integer Linear Programming model.

SCIENTIFIC INNOVATION AND RELEVANCE

Related work has investigated the significance of coupling the demand side with the supply side on larger district or urban scales, and with simplified, or archetypal geometrical parameters (Shi et al., 2020)(Mavromatidis et al., 2018). Our study, however, uses more detailed geometrical properties of individual buildings derived from existing building geometry data in its sensitivity analysis. The footprint and building height of randomly selected buildings are used to generate geometries of multi-story buildings with varying roof types and windows. Derived geometrical characteristics (such as compactness and significant faces) are calculated to understand the importance of different architectural features.

Furthermore, the sensitivities to the building- and energy parameters of not only the overall costs and emissions are analysed, as is typically the case, but also the sensitivities of the selection and sizing of the technologies in the energy hub are investigated.

Finally, due to the large number of dependent and independent variables considered in this study, we cluster several parameters into topical groups in the analysis (e.g. by overall performance of individual technologies, or by individual attributes such as size, efficiency or lifetime of the energy systems), which helps in understanding possible interactions between the energy demand and supply.

PRELIMINARY RESULTS AND CONCLUSIONS

The preliminary results of the Morris screening are consistent with the existing literature: parameters directly affecting the energy demand, such as the overall geometry, ventilation, the climate, temperature setpoints and wall insulation appear to have a larger impact on the overall cost and emissions than technical energy system parameters such as investment costs or efficiencies of various technologies. In the subsequent Sobol Analysis, we will obtain quantitative measures on (first and second order) interaction effects between architectural design parameters and energy system design parameters.

MAIN REFERENCES

Ferrara, M., Prunotto, F., Rolfo, A., & Fabrizio, E. (2019). Energy Demand and Supply Simultaneous Optimization to Design a Nearly Zero-Energy House. Applied Sciences. https://doi.org/10.3390/app9112261

Waibel, C., Evins, R., & Carmeliet, J. (2019). Co-simulation and optimization of building geometry and multi-energy systems: Interdependencies in energy supply, energy demand and solar potentials. Applied Energy, 1661–1682. https://doi.org/10.1016/j.apenergy.2019.03.177

Mavromatidis, G., Orehounig, K., & Carmeliet, J. (2018a). Uncertainty and global sensitivity analysis for the optimal design of distributed energy systems. Applied Energy, 214, 219–238. https://doi.org/10.1016/j.apenergy.2018.01.062

Shi, Z. ;, Hsieh, S. ;, Fonseca, J. A. ;, Schlueter, A., Shi, Z., Hsieh, S., & Fonseca, J. A. (2020). Interdependencies between the design of street grids and the cost-effectiveness of district cooling systems ETH Library 1 Interdependencies between the design of street grids and the cost-effectiveness of district cooling systems. https://doi.org/10.3929/ethz-b-000391818

Keywords: global sensitivity analysis, architecture, multi-energy system

Paper ID 30635: Using a surrogate model to analyze the impact of geometry on energy efficiency of buildings.

Bhumika Bhatta* (1), Ralph Evins (2), Paul Westermann (3)

1: University of Victoria, Canada; 2: University of Victoria, Canada; 3: University of Victoria, Canada

AIM AND APPROACH

Parametric exploration and optimization of building geometry is a powerful tool for designing energy efficient buildings. However, in practice this process is computationally expensive and time-consuming. In this research, we explore the use of surrogate models, i.e. efficient statistical approximations of ex-pensive physics-based building simulation models, to lower the computational burden of large-scale build-ng geometry analysis. For this purpose, we developed a novel dataset of 38,000 residential building models derived from real world floor plans from (Wu et al. (2019))[5] and train a surrogate model to emulate their simulated annual energy performance. We extract up to 20 parameters as surrogate model inputs to represent the building geometry and show that the trained surrogate model reaches a high accuracy (R2score = 0.999, MSE =0.007 and RMSE = 0.022) on test data. The current setup forms the basis for further research where the complexity of the building models will be increased.

SCIENTIFIC INNOVATION AND RELEVANCE

Surrogates are the statistical models which can be used to provide rapid approximations of more expensive models. Surrogate models can be taken as an alternative approach to replace the detailed simulations with simplified approximate simulations, thereby sacrificing accuracy for reduced computational time. In this research work we are going to develop a surrogate model with two different approaches (physics-based and machine learning-based) which will approximate the impact of geometry on energy demand. Previous research work has been done on studying the impact of building geometry on its energy demand. For example, AlAnzi et al. [1] take the geometry of building into account for the energy performance analysis but consider the shape of façade only rather than the overall footprint. Most research studies sought to identify the relationship between the relative compactness or building volume and building energy loads [2],[3],[4] On the other hand, many research studies have considered hypothetical building shapes with constant floor area and height or constant floor area with varying heights among different shapes. The work presented in this paper will use a model, based on Building Technology Assessment Platform (BTAP) small and medium office building for Victoria, BC, Canada.

PRELIMINARY RESULTS AND CONCLUSIONS

In this paper we provide a framework to develop neural network based surrogate models that predicts the EUI over different geometries. The frame work consists of conversion of 2D raster images of floorplans into 3D building simulation models, extraction of 20-building features for geometry, surrogate model training, and performance analysis of the impact of each feature. We developed the dataset consisting of 38,000 building simulation IDF files derived from real-world floor plans (Wu et al. (2019)). Our results for the training of the surrogate models show that off-the shelve neural network surrogate models pro-vided with manually engineered features are capable of emulating the simulation outcomes really well (R2score = 0.999, MSE = 0.007 and RMSE= 0.022) on test data. Overall, this paper shows that a simple machine learning model can perform very well in predicting energy use for various geometries, which provides architects with a great tool to get building performance estimates in real time while they are exploring various designs without the burden of simulating building designs individually.

MAIN REFERENCES

1. Adnan AlAnzi, DonghyunSeo, and MoncefKrarti. Impact of building shape on thermalperformance of office buildings in kuwait. Energy Conversion and Management, 50(3):822–828, 2009

2. JosifasParasonis, Andrius Keizikas, Audron ̇eEndriukaityt ̇e, and Diana Kalibatien ̇e. Architec-tural solutions to increase the energy efficiency of buildings. Journal of civil engineering and management, 18(1):71–80, 2012

3. Werner Pessenlehner and Ardeshir Mahdavi. Building morphology, transparence, and energy performance. na, 2003

4. Carlo Ratti, Dana Raydan, and Koen Steemers. Building form and environmental performance: archetypes, analysis and an arid climate. Energy and buildings, 35(1):49–59, 2003.

5. Wu, W., X.-M. Fu, R. Tang, Y. Wang, Y.-H. Qi, and

L. Liu (2019). Data-driven interior plan generation for residential buildings. ACM Transactions

on Graphics (SIGGRAPH Asia) 38 (6)

Keywords: Energy efficiency, Building geometry, Heating and Cooling demand, Surrogate Model, Machine learning.

Session T5.3: Buildings paving the way for the energy transition

Paper ID 30219: Co-simulation approach to evaluate MPC strategies for all-air systems: case study

Bart Merema*, Dirk Saelens, Hilde Breesch

KU Leuven, Belgium

AIM AND APPROACH

A predictive control framework is implemented in an educational building for an all-air system that controls both the indoor temperature and room CO2 concentration. The predictive control framework optimizes and actuates set points for the supply air temperature and supply airflow rate. The case study demonstrates the concept of predictive control and the method used for implementation. For the predictive control a data-driven model is used that is identified using measurement data of spring 2019. Energy use for the heating and fans as well as thermal comfort and indoor air quality are assessed. The aim of this study is to assess the performance of the implemented model predictive control (MPC) using a simulation-based comparison of MPC to a rule based control (RBC). A previously validated Modelica model of the case study building is used for the simulations. The simulated performance of MPC is verified with measured results of the implemented MPC to ensure a fair comparison with the baseline RBC.

SCIENTIFIC INNOVATION AND RELEVANCE

HVAC is reported to use 50 % of the energy use in buildings (Pérez-Lombard, Ortiz, & Pout, 2008). In order to reduce this energy use in buildings, the control of the HVAC system can be optimized. However, HVAC systems are challenging to control due to time varying dynamics and varying internal/external disturbances (Afram & Janabi-shari, 2014; Killian & Kozek, 2016). A model predictive control (MPC) could be a solution to control an HVAC system more energy efficiently since it takes into account the current measurement for room temperature and CO2 concentration and the future demand. In buildings with hydronic systems, the reported energy reductions after implementation of a predictive control are significant (De Coninck & Helsen, 2016; Sturzenegger, Gyalistras, Morari, & Smith, 2016). However, the question remains what the energy saving potential is of an MPC for all-air systems, since the dynamics of the system are different compared to hydronic systems. For example, the time constant is smaller compared to hydronic systems. In literature, only a few examples of implementation in practice (Bengea, Kelman, Borrelli, Taylor, & Narayanan, 2014) exist for an MPC of an all-air ventilation system controlling both the room temperature and CO2 concentration.

PRELIMINARY RESULTS AND CONCLUSIONS

The measurement results for the operation of the ventilation system indicates that the predictive control framework is able to control the all-air system while maintain the occupant comfort. Using a simple data driven model, the room temperature and CO2 concentration can be controlled under uncertainty of the forecasts for weather and occupancy. The VAV's controlled by the predictive control framework react well to the heating and ventilation demand in order to control the indoor environmental quality (IEQ). Using a cost function that reduces the room temperature and CO2 set point violations while at the same time minimizes the energy use, significant energy savings can be achieved. For the implemented control, the savings compared to a RBC are 26% for the fan electrical energy and 35% for the space heating energy. Comfort constraints, involving occupancy based room temperature set points and CO2 concentration set points are respected and discomfort is minimized compared to the RBC.

MAIN REFERENCES

Afram, A., & Janabi-shari, F. (2014). Theory and applications of HVAC control systems e A review of model predictive control (MPC). Building and Environment, 72, 343–355. https://doi.org/10.1016/j.buildenv.2013.11.016

Bengea, S. C., Kelman, A. D., Borrelli, F., Taylor, R., & Narayanan, S. (2014). Implementation of model predictive control for an HVAC system in a mid-size commercial building HVAC&R Research, 20, 121–135. https://doi.org/10.1080/10789669.2013.834781

De Coninck, R., & Helsen, L. (2016). Practical implementation and evaluation of model predictive control for an office building in Brussels. Energy and Buildings, 111, 290–298. https://doi.org/10.1016/j.enbuild.2015.11.014

Killian, M., & Kozek, M. (2016). Ten questions concerning model predictive control for energy efficient buildings. Building and Environment, 105, 403–412. https://doi.org/10.1016/j.buildenv.2016.05.034

Pérez-Lombard, L., Ortiz, J., & Pout, C. (2008). A review on buildings energy consumption information. Energy and Buildings, 40(3), 394–398. https://doi.org/10.1016/j.enbuild.2007.03.007

Sturzenegger, D., Gyalistras, D., Morari, M., & Smith, R. S. (2016). Model Predictive Climate Control of a Swiss Office Building : Implementation , Results , and Cost – Benefit Analysis. IEEE Transactions on Control Systems Technology, 24(1), 1–12.

Keywords: ventilation, MPC, building energy

Paper ID 30820: Experimental tests on the optimal management of all-electric dwellings

Jacopo Vivian* (1), Angelo Zarrella (1), Giorgio Besagni (2), Lorenzo Croci (2)

1: University of Padua, Italy; 2: Ricerca Sistema Energetico S.p.A.

AIM AND APPROACH

It is widely accepted that the large-scale deployment of heat pumps is a promising way toward the decarbonisation of the heating sector. In this framework, it is worth mentioning that the deployment of heat pumps will severely affect the power consumption patterns (viz., the load profiles) in the electrical distribution systems [1]. At the same time, increasing penetration of small-scale renewable energy sources such as PV/PVT systems is pushing the energy system towards a decentralised structure. Building Energy Management Systems (BEMS) play a key role in the energy transition as they can exploit the thermal inertia offered by building structures and thermal energy storage (TES) systems, to decouple the supply-side and the demand-side, for increasing the self-consumption of electricity produced by PV systems, while at the same time maintaining comfortable conditions in the indoor environment [2]. This paper contributes to the ongoing discussion and presents a laboratory developed by Ricerca Sistema Energetico (RSE SpA) in Piacenza (Italy) where these smart control strategies are being tested. In particular, the very first lab trials are presented and discussed.

SCIENTIFIC INNOVATION AND RELEVANCE

The laboratory is equipped to simulate the performance of an efficient "all-electric" building. To this end, the laboratory has been equipped with (i) an air-to-water heat pump air conditioning system connected to four fan coils with a 300-liter heat storage tank; (ii) a heat pump water heater with 200-liter storage tank for domestic hot water (DHW) production; (iii) an air/air conditioning system consisting of an outdoor condensing unit and four indoor units with a nominal cooling capacity of 6.8 kW; (iv) four air extractors; (v) a set of electrical appliances (e.g., washing machine, dryer, dishwasher and combined refrigerator). In the next months, the laboratory will be also equipped with a PV system with inverter and storage battery. A monitoring and control system realized by RSE will allow to read data logged by the environmental sensors such as temperatures, mass flow rates, heat and power flows. The present paper builds upon previously research activities carried out by the authors have carried in the field of optimal integration of renewables [2], energy flexibility of buildings [3] and demand response [4]. Hence, the present paper provides experimental evidence and data to support previous and future findings on these topics.

PRELIMINARY RESULTS AND CONCLUSIONS

This article presents the Laboratory and the first tests concerning calibration of the energy management system and heat pump control. The first tests will be carried out during late summer and early winter 2020.

MAIN REFERENCES

[1] Barton J, Huang S, Infield D, Leach M, Ogunkunle D, Torriti J, Thomson M. The evolution of electricity demand and the role for demand side participation, in buildings and transport. Energy Policy 2013; 52: 85-102.

[2] Vivian J, Mazzi N. An algorithm for the optimal management of air-source heat pumps and PV systems. Journal of Physics: Conference Series. Proc. of CISBAT 2019 – Climate Resilient Cities – Energy Efficiency & Renewables in the Digital Era. Lausanne (Switzerland) 4-6 Sep. 2019.

[3] Vivian J, Chiodarelli U, Emmi G, Zarrella A. (2020). A sensitivity analysis on the heating and cooling energy flexibility of residential buildings. Sustainable Cities and Society, vol. 52, 101815, ISSN: 2210-6707, doi: 10.1016/j.scs.2019.101815

[4] Pau M, Vivian J, Cunsolo F, Ponci F, Monti A. Optimal scheduling of electric heat pumps combined with thermal storage for power peak shaving. Proc. of the 18th Intern. Conf. On Environment and Electrical Engineering. IEEE –EEEIC 2018, Palermo (Italy).

Keywords: HVAC optimization, Laboratory, electrification

Paper ID 31111: Model predictive control framework for operation of smart sustainable buildings

Himashu Nagpal* (1), Florin Capitanescu (1), Per Heiselberg (2)

1: Luxembourg Institute of Science and Technology; 2: Aalborg University

AIM AND APPROACH

Buildings are a major contributor to global primary energy consumption and total greenhouse gas emissions. However, buildings have immense potential to green and improve the efficiency of global energy sector. This paper investigates the use of model predictive control (MPC) for operation of a large variety of energy devices in a building. The optimization problem is formulated as a mixed-integer-linear-program (MILP). Apart from cost minimization, an original feature of this works is that a sustainability criterion is implemented to maintain the net-zero-energy (nZE) target of the building. The proposed framework is simulated with two different electricity price tariff structures. The simulation results showed a 12\% reduction in energy cost for the proposed approach in comparison to the benchmark model when simulated for one month period.

SCIENTIFIC INNOVATION AND RELEVANCE

The overall contribution of the present work to the scientific literature can be summarized as follows –

Modelling framework for energy devices – The modelling framework presented in the paper includes energy devices which are relevant to a generic domestic building such as heating ventilation and air condition system, hot water system, battery storage system, photovoltaic panel, electric vehicle and other shiftable loads (washing machine, dryer, etc.)

Introduction of sustainability criterion – In addition to achieving the objective of minimization of operating cost, the building energy management system (BEMS) aims to maintain sustainability of the building by enforcing an adaptive sustainability criterion. This is generally a less explored idea in the context of optimization in buildings.

\end{itemize}

PRELIMINARY RESULTS AND CONCLUSIONS

At current stage, the work has only been carried out without considering sustainability criterion and the uncertainties in solar radiation forecasts. Therefore, the optimization problem is formulated as a deterministic mixed integer linear program.

The main preliminary result observed from the current study is that the proposed EMS exploits the demand response flexibility provided by the energy devices hosted in the building and ensures that during the entire simulation period, power is bought from the grid during low-cost period and fed back into the grid during higher cost period thus leading to minimized energy cost.

Further results from the study showed that in addition to minimizing the energy cost EMS ensures that the following constraints were satisfied –

- Thermal comfort of the occupant was kept between the desired temperature range.

- State-of-charge of battery storage was maintained between prescribed values.

- Appliances finished their task before the assigned deadlines.

- Sequential operation of appliances was ensured (e.g. clothes dryer runs after washing machine finished its operation).

- SOC of EV battery was kept between the desired range.

MAIN REFERENCES

[1] Serale, G., M. Fiorentini, A. Capozzoli, D. Bernardini, and A. Bemporad (2018). Model predictive control (mpc) for enhancing building and hvac system energy efficiency: Problem formulation, applications and opportunities. Energies 11(3), 631.

[2] Shareef, H., M. S. Ahmed, A. Mohamed, and E. Al Hassan (2018). Review on home energy management system considering demand responses, smart technologies, and intelligent controllers. Ieee Access 6, 24498-24509.

[3] Sou, K. C., J. Weimer, H. Sandberg, and K. H. Johansson (2011). Scheduling smart home appliances using mixed integer linear programming. In 2011 50th IEEE Conference on Decision and Control and European Control Conference, pp. 5144-5149. IEEE.

Keywords: Model predictive control, mixed integer linear program, demand response, Smart sustainable building

Session T5.4: Improving indoor environmental quality

Paper ID 30779: DL Abacus- A simplified simulation-based tool for daylighting performance assessment in Brazilian dwellings

Natalia Giraldo Vasquez* (1,2), Pedro O. Pizzetti Mariano (1), Raphaela Walger da Fonseca (1), Fernando O.R Pereira (1)

1: PósArq, Department of Architecture and Urbanism, Federal University of Santa Catarina, Florianópolis, Brazil; 2: ICIEE, DTU Civil Engineering, Technical University of Denmark, Kongens Lyngby, Denmark

AIM AND APPROACH

In this paper we present the development of a simplified simulation-based tool –DL Abacus-, to assess the performance of natural light in dwellings since the early design stages.The database used for the abacuses' development was generated through parametric climate-based simulations in Rhinoceros 6 by using Grasshopper component and Diva plug-in. For the cases definition, we considered different width and depth, visual transmittances, urban densities, presence of balconies and horizontal shading elements; while fixing the window area, surfaces reflectances' and ceiling and lintel heights, following the main characteristics of Brazilian dwellings. Four orientations and three latitudes (covering Brazilian territory) were studied by using TMYx weather files. The annual basis simulations considered an analysis time period of ten hours per day. As output, we obtained DA50% values from a horizontal grid placed at 0.75cm high. We use the most restrictive result cases to build the AbacusDA, by defining two sDA requirements that must be reached, simultaneously: i) a minimum requirement of sDA60/50% in at least 75% of the space and ii) the performance, with three different levels, would be reached when a sDA200/50% is guaranteed in 70% (Superior), 55% (Intermediate) or at least in 40% of the space (Minimum).

SCIENTIFIC INNOVATION AND RELEVANCE

With around 50% of the population living in a few big cities, urban centres in Latin America are characterized by high density and to provide adequate indoor environmental quality could be challenging. The current worldwide health crisis due to Covid-19 has highlighted the importance of the dwellings' indoor environment quality. This scenario has brought the daylight as an important matter for occupants' health, as stated earlier and not only under an energy savings perspective. In Brazil, the National Standard for the assessment of residential buildings' performance is mandatory. The compulsory application of the standard, previous construction's approval, might represent an increase in buildings' costs. For the last years, the current Standard has raised some complaints from the industry and consultancy offices due to conceptual inconsistencies in the assessment methods, leading to a revision and foreseen an update of it. Regarding the Standard's Daylighting Performance Section, the methods and criteria are being updated from the point in time to the climate-based Daylighting modelling analysis. Aiming to ease and make affordable the evaluation of daylighting performance, mostly in projects with low budgets, we propose a graphic tool that enables any professional verifying the compulsory daylighting requirements.

PRELIMINARY RESULTS AND CONCLUSIONS

We used data from 36000 simulated cases for the development of the abacuses. From the decision-making process for selecting the targets, the addition of the balcony to the indoors area showed a significant effect on the reached performance level. Balconies improve the quality of indoor spaces, not only in terms of daylight. Since such addition has been a common practice in residential buildings, the assessment of the daylighting provision should take into consideration such benefits. Our analyses showed that variations of 50lux and 100lux, in Etarget, had a significant impact on the number of cases meeting the proposed criteria. Also, a reduction of 5% in the values of Ftarget used for the definition of the level of performance had a significant impact on this mater.

Since we adopted the idea of a restrictive condition for the development of the abacuses, the assessments made by it can be conservative. It is, therefore, a reliable tool. However, the daylighting provision estimation can be lower compared to the results obtained through simulation.

Finally, the tool can be replicated to any location with available climate file.

MAIN REFERENCES

Brazilian Association of Technical Standards (2013). Residential Buildings – Performance. Part 1: General requirements. (ABNT NBR 15575-1:2013). In Portuguese.

Brembilla, E. and Mardaljevic, J.(2019). Climate-Based Daylight Modelling for Compliance Verification: Benchmarking Multiple State-of-the-Art Methods. Building and Environment 158, 151–64.

European Committee for Standardization (2018). Daylight in buildings (CEN 17037:2018).

Illuminating Engineering Society of North America (2012). Approved Method: IES Spatial Daylight Autonomy and Annual Sunlight Exposure (IES LM-83-12).

Lo Verso, V.R. M., Mihaylov. G., Pellegrino, A. and Pellerey, F. (2017). Estimation of the Daylight Amount and the Energy Demand for Lighting for the Early Design Stages: Definition of a Set of Mathematical Models. Energy and Buildings 155,151–65.

Mardaljevic, J. and Christoffersen, J. (2013). A Roadmap for Upgrading National/EU Standards for Daylight in Buildings. CIE Midterm Conference: Towards a New Century of Light. Paris (France),178-187.

Oxman, R., Hammer,R., and Ari, S. (2007). Performative Design in Architecture: Employment of Virtual Prototyping as a Simulation Environment in Design Generation. Proceedings eCAADe. Frankfurt (Germany), 227–234.

Pereira, F.O.R., Schmitt, M. and Moraes, L.N.(2015). Luminous zoning of the Brazilian territory based on external diffuse horizontal illuminances (Zoneamento Luminoso Do Território Brasileiro Através De Iluminâncias Horizontais Difusas Externas). Proceedings ENCAC/ELACAC. Campinas (Brazil), 1-10. In portuguese.

Keywords: Simplified tool, simulation-based, dwellings, daylighting, standard

Paper ID 30444: Heat balance analysis of an office building with fully glazed façade in tropical and temperate climates in Brazil

Mônica Martins Pinto*, Fernando Simon Westphal

Federal University of Santa Catarina, Brazil

AIM AND APPROACH

When focusing on the heat transfer through the envelope, the concern with glasses is emphasized. The thermal performance of a window depends mainly on its thermal transmittance, solar transmittance, and the tightness of the system, since the exchanges are more direct and intense in those elements due to their transparency to solar radiation and high thermal transmittance. Hence, it is required a special attention for local climate and building feature when selecting glazing systems as it can significantly impact on the building energy consumption.

The introduction of an international architecture model in Brazil occurs not only in the façade design, but also in the constructive system. The use of Insulated Glass Unit (IGU) is an example. This system is widely used in Europe and North America as a component for improving the energy performance of buildings, while providing thermal comfort for building occupants. However, it has been introduced in the Brazilian market without a proper weather assessment. Suitable glass choice is essential to reduce energy consumption, as the most intense heat exchange occurs through the windows.

This research aims to analyzes the effects of IGU use on the thermal balance of conditionate office buildings in the Brazilian climate context.

SCIENTIFIC INNOVATION AND RELEVANCE

A previous study analyzed the impact of the IGU use on the thermal energy performance of artificially conditioned office buildings in Brazil. Based on the results of cooling consumption, the authors state that the use of IGU is not justified in southern Brazil (temperate climate). On the other hand, the results for tropical climates showed similar performance among IGU and SGUL, with a slight advantage for IGU models.

These results disagree with the common sense that the IGU is a strategy to improve energy efficiency, especially in extreme cold or hot climates. For this reason, thermal balance analysis was carried out to understand the effects of IGU, in place of SGUL, on the internal heat flux of an office building, mainly focus on the heat flux through windows.

Thermal balance analysis of the zones was performed based on heat flux from building simulation on Energyplus software. Two Brazilian cities with different climates conditions were selected. While Curitiba (temperate climate) was chosen due to its lower temperatures and higher variation of solar radiation incidence throughout the year, Boa Vista (tropical climate) presents a high average annual temperature and constant intensity of solar radiation.

PRELIMINARY RESULTS AND CONCLUSIONS

Even if in hot climates, SGUL presents a better potential to internal load heat dissipation during occupation if steeping daily outdoor temperature range, and it significantly decreases during the night.

In Curitiba, the IGU shows up to 3.5 times higher heat gain through the windows, even in summer. During winter, the SGUL model presents from 17.7% to 37.7% lower cooling annual demand. This glass also increases the heat loss during the occupied period when the hourly balance is analyzed. This difference reaches 0,73kWh each hour. The dissipation becomes even more efficient during the night, when this gap achieves 26,80kWh (94% higher).

In Boa Vista, both glass types showed similar results in the annual analysis. In terms of hourly heat exchanges, the results also do not present meaningful differences between IGU and SGUL, being up to 0,50kWh.

During the occupied period, the IGU mitigates the heat gain through the windows when the outdoor temperature tends to be higher than the indoors. Nevertheless, the thermal insulation created by the air gap hinders the heat loss during the night. This heat loss would be advantageous as it would decrease the thermal load for HVAC at the beginning of the next day or week.

MAIN REFERENCES

ATZERI, Anna Maria et al. Comfort metrics for an integrated evaluation of buildings performance. Energy and Buildings, v.127, p.411–424, 2016. Available in: https://doi.org/10.1016/J.ENBUILD.2016.06.007.

GASPARELLA, Andrea et al. Analysis and modelling of window and glazing systems energy performance for a well insulated residential building. Energy and Buildings, v.43, n.4, p.1030–1037, 2011. Available in: https://doi.org/10.1016/j.enbuild.2010.12.032

HUANG, Yu; NIU, Jian-lei; CHUNG, Tse-ming. Comprehensive analysis on thermal and daylighting performance of glazing and shading designs on office building envelope in cooling-dominant climates. Applied Energy, v.134, p.215–228, 2014. Available in: https://ideas.repec.org/a/eee/appene/v134y2014icp215-228.html.

JABER, Samar; AJIB, Salman. Thermal and economic windows design for different climate zones. Energy and Buildings, v.43, n.11, p.3208–3215, 2011. Available in: https://doi.org/10.1016/j.enbuild.2011.08.019.

LEE, J. W. et al. Optimization of building window system in Asian regions by analyzing solar heat gain and daylighting elements. Renewable Energy, v.50, p.522–531, 2013. Available in: https://doi.org/10.1016/j.renene.2012.07.029.

PINTO, Mônica Martins; WESTPHAL, Fernando Simon. Energy Performance of Offices Buildings in Brazil using Insulated Glass Units. In: Building Simulation 2019. p.4944–4951. Rome, 2019.

POIRAZIS, Harris; BLOMSTERBERG, Åke; WALL, Maria. Energy simulations for glazed office buildings in Sweden. Energy and Buildings, v.40, n.7, p.1161–1170, 2008. Available in: https://doi.org/10.1016/j.enbuild.2007.10.011.

Keywords: Window, Energyplus, Thermal Performance, Insulated Glass Unit

Paper ID 30393: Defining the life-cycle energy implications of informal settlement redevelopment

Alex Nutkiewicz*, Rishee K Jain

Urban Informatics Lab, Stanford University, United States of America

AIM AND APPROACH

By 2030, over 2 billion people worldwide will be living in informal settlements, colloquially known as “slums.” Characterized by poor quality of life with regard to sufficient living area, thermal comfort, and energy access, there has been a massive push for improvement and redevelopment efforts to transform existing slums into dense, high-rise social housing [SRA, 1995]. However, previous work [Nutkiewicz et al., 2018] demonstrates that redesigned high-rise, low-income housing are potential sources of thermal discomfort – showing that a lack of well-informed designs or policies can result in severe human health consequences for some of the world’s most vulnerable communities. And with the demand for active cooling (e.g., air conditioning) expected to sharply increase in tropical countries [Mastrucci et al., 2019], these well-intentioned redevelopment efforts may result in increased energy poverty and urban building energy consumption. We use a validated informal settlement model for Dharavi, Mumbai, India [Debnath et al., 2016; Nutkiewicz et al., 2018] to evaluate the embodied and operational energy requirements to provide sufficient thermal comfort under various design decision scenarios. In doing so, we quantify the life cycle energy implications of informal settlement redevelopment and highlight the need to consider thermal comfort in its proposed plans.

SCIENTIFIC INNOVATION AND RELEVANCE

A significant amount of research has utilized energy simulation to assess the influence of various building design decisions (e.g., window-to-wall ratio, ventilation, building facades) on thermal comfort and energy performance. However, we lack an understanding for how combinations of these design choices affect the life cycle energy consumption of buildings in the Global South – a largely underrepresented sector in energy research literature, despite it characterizing the housing of nearly 1 in 8 people worldwide.

We utilize a previously validated model representing a typical urban informal settlement in Dharavi, Mumbai, India. We simulate this model under combinations of both existing design parameters and morphologies as well as ones described in proposed slum redevelopment plans. These simulation outputs are then used to predict the embodied and operational energy requirements needed achieve acceptable indoor thermal comfort conditions. Through this approach, our study will quantify the life cycle energy impacts of informal settlement redevelopment and provide an understanding for how various design decisions influence these results.

PRELIMINARY RESULTS AND CONCLUSIONS

This work leverages EnergyPlus and parametric simulation tool jEPlus to estimate potential exposure to indoor heat stressed conditions and quantify the subsequent operational and embodied energy needs for adequate space cooling under local material and construction scenarios. Our preliminary results indicate that while low-cost building materials used in existing informal settlements greatly contribute to indoor heat stress, shifting the morphology of these urban neighborhoods to a high-rise urban form may also worsen indoor thermal comfort if considerations are not made for including passive cooling design strategies. A full informal settlement redevelopment will inevitably require additional energy to produce more durable buildings, but if designed appropriately, it may reduce the overall life cycle energy needs and subsequent energy burden for its future occupants. As a result, this work has the potential to help inform policymaking for informal settlement redevelopment to improve the energy and indoor well-being outcomes for a vulnerable community of residents.

MAIN REFERENCES

Debnath, R., Bardhan, R., & Jain, R. K. (2017). A data-driven and simulation approach for understanding thermal performance of slum redevelopment in Mumbai, India. Proceedings of the 15th IBPSA Conference, 2745-2752. http://www.ibpsa.org/?page_id=962

Nutkiewicz, A., Jain, R. K., & Bardhan, R. (2018). Energy modeling of urban informal settlement redevelopment: Exploring design parameters for optimal thermal comfort in Dharavi, Mumbai, India. Applied Energy, 231, 433-445.

Mastrucci, A., Byers, E., Pachauri, S., & Rao, N. D. (2019). Improving the SDG energy poverty targets: Residential cooling needs in the Global South. Energy and Buildings, 186, 405-415.

SRA (1995). Government of Maharastra, Slum Rehabilitation Scheme – SRP/1095/CR37/Housing.

Keywords: informal settlements, life cycle energy, heat stress, space cooling, building energy

Paper ID 30193: Comparison of different clustering approaches on different databases of smart meter data

Martina Ferrando* (1,2), Debora Nozza (3), Tianzhen Hong (2), Francesco Causone (1)

1: Politecnico di Milano, Italy; 2: Lawrence Berkeley National Laboratory, California; 3: Università Bocconi, Italy

AIM AND APPROACH

This paper aims to compare the results from different clustering algorithms, and different performance indexes, applied on three databases, segmented from the Building Data Genome Project 2 based on three cities with different climate conditions [1]. Particularly, the goal is to find an algorithm that gives fair results on different databases for pre-analyses of data. The data exploration uses the main clustering methods from the literature (e.g., k-means, k-medoids, hierarchical) and a deep clustering algorithm to understand how different algorithms and indexes change the results within the same database and understand which of the option perform better among databases. In particular, the clustering is conducted by representing each building with 205 normalized electricity features including the annual consumption (1 value), the monthly consumption (12 values), and 24 hourly values for 8 days (an average weekend day and weekday for each of the four seasons). The implemented algorithms require to define the number of final clusters. For this purpose, different indexes are exploited and compared. The original database supplement information for each building, like its typology (e.g., residential, office), building area, and construction year. These data are compared to the final clustering results to understand possible correlations and the clusters' quality.

SCIENTIFIC INNOVATION AND RELEVANCE

The smart meters differ from traditional ones because of their ability to store high-frequency data for different resources (e.g., electricity, gas, water). Useful information is derived by this raw data and the derived applications are numerous, including benchmarking [2], occupant behavior detection [3], costumers classification [4], anomalies detection [5], end-use disaggregation [6], buildings changes/retrofit. These applications mainly exploit clustering approaches to derive groups of buildings or customers with similar resource usage. Currently, the main clustering techniques used on smart meter readings include k-means, k-medoids, Self-Organizing Map (SOM) coupled with k-means, and hierarchical. The number of final clusters is assessed via indexes, among them: Davies–Bouldin, Silhouette, Dunn, Calinski-Harabasz, Bayesian Information. However, the main publications make use of just one or a few of these methods and indexes and tune the algorithm on a specific database. This hinders a clear comparison of the performance of the different implemented algorithms and indexes. Moreover, the implementation just on a single database is helpful to find the best option for that specific case but not a general approach that shows fair results in most cases for pre-analyses of data. This paper aims to clarify the performance of various algorithms and indexes, comparing them directly.

PRELIMINARY RESULTS AND CONCLUSIONS

Different clustering methodologies are implemented to determine a few representative groups of buildings with similar energy use. The number of clusters and their characteristics change on the base of the used clustering algorithms and performance indexes. Some clustering methodologies, however, bring to very similar results (i.e., k-means, k-medoids, and SOM coupled with k-means) and for these algorithms also the different indexes bring to similar results. The original databases mix different building typologies (e.g., school, office, residential) which, in most cases, are identified by the clustering algorithms. Due to the normalization of the data, the building area does not result in strong correlations with the resulting clusters. While some of them identify the specific year of construction, however, this result could be misinterpreted due to the relatively small size of the databases. The comparison of the different algorithms gives a clear overview of the main clustering approaches existing and their strength and limitations. The resulting final cluster centroids could be exploited to better understand the energy patterns of different buildings and building typologies, especially, but not limited, to the final aims of benchmarking and costumers' classifications.

MAIN REFERENCES

[1] C. Miller, A. Kathirgamanathan, B. Picchetti, P. Arjunan, J.Y. Park, Z. Nagy, P. Raftery, B.W. Hobson, Z. Shi, F. Meggers, The Building Data Genome Project 2: Hourly energy meter data from the ASHRAE Great Energy Predictor III competition, (2020). http://arxiv.org/abs/2006.02273.

[2] X. Luo, T. Hong, Y. Chen, M.A. Piette, Electric load shape benchmarking for small- and medium-sized commercial buildings, Appl. Energy. 204 (2017) 715–725. doi:10.1016/j.apenergy.2017.07.108.

[3] F. Causone, S. Carlucci, M. Ferrando, A. Marchenko, S. Erba, A data-driven procedure to model occupancy and occupant-related electric load profiles in residential buildings for energy simulation, Energy Build. 202 (2019) 109342. doi:10.1016/J.ENBUILD.2019.109342.

[4] G. Chicco, R. Napoli, F. Piglione, P. Postolache, M. Scutariu, C. Toader, Load pattern-based classification of electricity customers, IEEE Trans. Power Syst. 19 (2004) 1232–1239. doi:10.1109/TPWRS.2004.826810.

[5] M. Devlin, B.P. Hayes, Non-Intrusive Load Monitoring using Electricity Smart Meter Data: A Deep Learning Approach, in: IEEE Power Energy Soc. Gen. Meet., IEEE Computer Society, 2019. doi:10.1109/PESGM40551.2019.8973732.

[6] K. Khalid, A. Mohamed, R. Mohamed, H. Shareef, Performance comparison of artificial intelligence techniques for non-intrusive electrical load monitoring, Bull. Electr. Eng. Informatics. 7 (2018) 143–152. doi:10.11591/eei.v7i2.1190.

Keywords: Smart Meter Data, Clustering, Machine learning, Classification, Deep learning

Paper ID 30780: Evaluation of different refurbishment or improvement strategies to reduce the environmental impact of University campuses

Iñigo León (1), Xabat Oregi (1), Cristina Marieta (1), Alba Juncal Arias* (1), Lara Mabe (2)

1: Basque Country University (UPV/EHU), Architecture Department, Donostia-San Sebastián, Spain; 2: Tecnalia Research & Innovation, Energy and Environment Division, Azpeitia, Spain

AIM AND APPROACH

Over the past few years, urban planners and architects have searched for solutions to improve the performance of urban projects in terms of environmental impact, life quality and socio-economic problems. NEST is an agile and fast tool developed to assess this environmental impact (baseline scenarios and improvements hypotheses) at the district level by a life cycle analysis. The authors of this publication have used NEST in different research projects and they have studied different baseline scenarios and evaluated improvement cases for several campuses and distrits.

The article presents a summary of the evolution of the research carried out, explaining each case of study and the general conclusions obtained. It also presents the current research project that aims to quantify the environmental impact of three University campuses with different characteristics that serve as representative examples for others. It is focused on the utilisation of renewable energy sources in the campuses in order to reduce their environmental impact. The results can be extrapolate to other campuses with similar characteristics.

SCIENTIFIC INNOVATION AND RELEVANCE

One of the goals of NEST is that it is an agile and fast tool for carrying out analysis of the baseline scenario and the improvement hypotheses. In contrast to other tools, in NEST, the characteristics of the buildings that constitute the district or area to be studied have an important weight. In addition, it does not require very detailed baseline data (no need to monitor real data), but allows to compare different scenarios and choose the most sustainable one in a very short period of time.It can be applied effectively to the university campus, cities or distrits scale in order to assess its environmental impact.

The relevance of this study is to suggest new sustainable solutions by a Life Cycle Analyze to decrease the environmental impact of distrits and campuses. Ultimately, it could be used to search for solutions that promote the renewable energy, circular energy economy and local production.

In addition, the research includes several case studies of different size and nature (campuses and districts from different regions, size, policies,...), and the results have been contrasted with monitoring, allowing to have a clear idea of the potential scope of the tool.

PRELIMINARY RESULTS AND CONCLUSIONS

The preliminary results present a significant difference in the environmental impact produced by each campus and district studied. This difference is due to the specific characteristics of each one of the buildings (i.e. energy rate, use, heating, cooling, and domestic hot water system), their locations, mobility systems, size, green spaces, and renewable energy production. Furthermore, it can be observed how, with a minimum increase on the impact of the building's production stage, the impact produced by the campus by the operational energy and water use stage, is considerable lower. It also shows that monitoring results obtained higher values than the simulation with NEST, between 1–17% less.

The research concludes that University campuses, as an independent entity from the city, have limited improvement values. Whereas, if it is considered as part of the city, the synergies and combined efforts between the city and the campus allow to achieve greater environmental performance using fewer resources. The use of renewable energies should be encouraged to reduce the environmental impact on operational energy use. NEST makes possible the simulation of these new scenarios and the evaluation and selection of the best solutions.

MAIN REFERENCES

Leon, I., Oregi, X. and Marieta, C., 2018. Environmental assessment of four Basque University campuses using the NEST tool. Sustainable Cities and Society, 42, pp 396-406.

Burgess, R., 2003. Ciudad y sostenibilidad: Desarrollo urbano sostenible. Cuadernos de la CEPAL, 88, pp. 193–213.

Oregi, X., Pousse, M., Mabe, L., Escudero, A. and Mardaras, I., 2016, November. Sustainability assessment of three districts in the city of Donostia through the NEST simulation tool. Natural Resources Forum, 40, pp. 156-168.

Leon, I., Oregi, X. and Marieta, C., 2020. Contribution of University to Environmental Energy Sustainability in the City. Sustainability, 12(3), pp. 774.

Keywords: Energy transition, Renewable energy, Decision support tool, Life cycle assessment, University environmental impact

Paper ID 30389: A methodology for applying energy renovation on public buildings

Mohamed Amer*, Mehdi Roukny, Salma Lasri

Renowatt, Belgium

AIM AND APPROACH

As a part of the EU strategy to reduce energy demands of the building sector to achieve climate-neutral levels by 2050, energy renovation is required with a rate of 3% of the total floor area of occupied heated and/or cooled buildings. The EPBD requires that buildings should undergo major renovation, in which a building's final energy demand for heating could be reduces by 50% to 80%.

Accordingly, this study aims to present a novel methodology that facilitates the process of creating detailed simulation models and the application of different Energy Conservation Measures (ECM) to achieve the EU strategy for energy renovations. This methodology constitutes 5 mains steps, from the acquisition of existing building information, to site visits, calibration, application of ECM and finally the calculation of the Net Present Value (NPV) of the taken measures.

SCIENTIFIC INNOVATION AND RELEVANCE

Compared to the classic approach of using static normative energy calculation models, the novelty of this methodology returns back to the usage of thermal dynamic simulation to realize energy gains from different renovation strategies, while overstepping the challenges related to the complexity and the time needed to construction simulation models. This approach has several benefits, such as the use of 3D models, visual verifications, and higher accuracy.

The usability of the methodology has been tested and developed throughout a continuous feedback loop between the developer and users with different backgrounds, such as engineers, architects, and technicians. This methodology showed its efficiency in communicating the results to best suit the requirements of the client / municipality, for example, studies that requires hourly simulation such as indoor thermal comfort or district heating sizing.

PRELIMINARY RESULTS AND CONCLUSIONS

The methodology has been successfully applied on several projects, from which 16 are presented in this study. Those projects present a wide variety of buildings types, such as sportive, educational, nursery, and administrative.

A total of 75% of energy gains, and 62% of CO_2 reductions have been achieved, while maintaining indoor thermal comfort.

The acquisition of building data is critical for the model's accuracy and reliability. Building inspection process and site visits preparations are the most time-consuming task and requires an adequate level of experience. Moreover, placing data loggers to monitor indoor air temperatures is necessary to identify set point temperatures and help monitoring occupant's behavior and the efficiency of energy systems. In addition, monthly energy consumption should be monitored for energy accounting measures.

MAIN REFERENCES

Economidou, M., Todeschi, V., Bertoldi, P. 2019 "Accelerating energy renovation investments in buildings – Financial & fiscal instruments across the EU" Publications Office of the European Union, Luxembourg. ISBN 978-92-76-12195-4, doi:10.2760/086805, JRC117816.

Poel, B. 2007 "EPA-NR Energy Performance Assessment for Existing Non-Residential Buildings" Intelligent Energy Europe (EIE)

Lee, S. Fei, Z., Augenbroe, A. 2011 "The Use of Normative Energy Calculation Beyond Building Performance Rating Systems" 12th Conference of International Building Performance Simulation Association, Sydney

Keywords: Carbon emissions, EPBD, Thermal simulation

Session T5.6: Buildings paving the way for the energy transition

Paper ID 30531: Development of life-cycle based strategies towards buildings with a positive ecological footprint using thermal building simulations

Michael Vollmer*, Hannes Harter, Kathrin Theilig, Daniel Kierdorf, Werner Lang

Technical University of Munich, Germany

AIM AND APPROACH

The goal of this study is to develop life cycle based implementation strategies for the realization of buildings with a positive ecological footprint using thermal building simulations in combination with a generic optimization algorithm and Life Cycle Assessment (LCA) methods. Since there is no methodological approach existing which covers this topic, first a method was developed to calculate the ecological footprint of buildings. With the developed method it is possible to calculate the ecological land demand to compensate the life-cycle based building emissions. Challenges in the planning stages of a building are the influences of different parameters on the building's life-cycle based energy demand and emissions. To develop implementation strategies and to quantify the influences an office building, built in 2016 in Regensburg, Germany was used as a case study. For the case study 176 possible combinations of the technical building services and the building construction type were defined. Regarding the different parameters, a sensitivity analysis was used to identify the ten most influential parameters on energy demand. In order to develop implementation strategies, the optimization algorithm GenOpt with IDA ICE was used to derive the best possible combination of the ten parameters to reduce the total ecological land demand.

SCIENTIFIC INNOVATION AND RELEVANCE

It is known that the building sector plays a significant role in regards to climate change. Especially, since the building sector uses a lot of mineral based, non-recyclable materials, creates a lot of waste and uses fossil fuels. In recent years the regulation of buildings lead to a decrease of the energy demand. But with the reduction of the energy demand for the building operation the relative share of the embedded/grey energy (energy needed to manufacture, refurbish and dispose materials) rises. Therefore, a life-cycle approach gets necessary to fully consider all stages of a building over the whole life cycle and to develop climate neutral buildings. An important task which gets neglected in the process of planning and building stages of a building are actual effects a building has on the environment and its sensible habitats. Since it is mandatory to maintain the ecological system services and to ensure biodiversity the need to quantify those effects rises and implementation strategies have to be developed. Since the complexity rises and one can't solve the challenging task the possibility of optimization algorithms has to be applied in the planning stages of buildings.

PRELIMINARY RESULTS AND CONCLUSIONS

Regarding the total ecological land demand, it can be seen that the most suitable implementation strategies are to reduce the energy demand and ensure a renewable energy supply, generate renewable energy on the building site, use renewable materials and ensure recyclable building materials in order to reduce waste. In general, it can be seen that a life-cycle approach is mandatory since the relative share of the embedded energy vs. the energy demand for building operation varies between approx. 30:70% and 65:35%. Furthermore, a reduction of the total ecological land demand which is needed to compensate building emissions from 2,1ha/a to 0,5ha/a was achieved. In summary a life cycle based calculation and evaluation approach leads to more precise results, a better understanding of the total environmental impacts and hence has to be implemented in the planning process of buildings.

MAIN REFERENCES

Operational vs. embodied emissions in buildings—A review of current trends," Energy and Buildings, vol. 66, pp. 232–245, 2013.

M. Wackernagel and W. E. Rees, Unser ökologischer Fussabdruck: Wie der Mensch Einfluss auf die Umwelt nimmt. Basel: Birkhäuser Verl., op. 1997.

S. K. Pal, A. Takano, K. Alanne, and K. Siren, "A life cycle approach to optimizing carbon footprint and costs of a residential building," Building and Environment, vol. 123, pp. 146–162, 2017.

Council-DGNB eV,. CLIMATE POSITIVE: NOW! How every building can make a contribution to climate action, 2020.

Keywords: Optimization, Simulation, Life Cycle Assessment, Environmental Impacts

Paper ID 30347: Analysis of step-by-step individual roadmaps – what can we learn about the practice?

Iná Maia*, Lukas Kranzl

TUW, Austria

AIM AND APPROACH

The recast of the Energy Performance of Buildings Directive (EPBD) 2018/844/EU introduced in the Article 19a the building renovation passport [1], which is an instrument to provide a long-term and step-by-step deep renovation roadmap for individual buildings. In Europe, there are already some demonstration projects, which focus on the key concept of building passports and step-by-step renovation as the EU-funded H2020 iBRoad project [2], [3]. Its main objective is to help eliminating barriers of single stage retrofitting measures, especially by planning step-wise, long-term retrofitting of owner-occupied single-family houses.

SCIENTIFIC INNOVATION AND RELEVANCE

During the project, a software tool was implemented to assist energy auditors by the development of individual step-by-step renovation roadmaps. The roadmaps contain technical information per step as for example: package of measures, energy needs, useful and primary energy demand and investment costs. This tool was tested by energy auditors on 55 real-life buildings, in three different EU-countries: Bulgaria, Portugal and Poland.

This papers aims at analysing the developed individual roadmaps, in other to provide deeper understanding on how the step-by-step concept is understood and interpreted in the practices. Therefore, this paper contributes to the overarching question, if the activities of building renovation advice are compatible with the building stock decarbonisation targets set by the EU for 2050. Following questions will be researched: 1) which recommended energy savings development for the cases studies can be observed, 2) how much energy savings for each building would need to be achieved according to the roadmaps, 3) are the developed roadmaps in line with the long-term national renovation strategies, 4) are there country-specific tendencies of the results?

PRELIMINARY RESULTS AND CONCLUSIONS

First insights showed that many of the developed individual renovation roadmaps did not include decarbonisation measures, for example: replacing the coal boiler by a more efficient (more efficient technology, but remaining the same energy carrier). This alerts to the fact that, in some cases in the practices the opportunity to decarbonise the building by a retrofitting project might be missed. Therefore, supporting energy auditors with trainings and education programs and the development of correspondingly clear advice methodologies and tools is a key aspect to achieve decarbonisation targets. Outlook from this analysis is to apply these studied cases in a step-by-step optimisation model, which calculates the optimum timing, when each step should be performed [4].

MAIN REFERENCES

[1] EU-Parliament, „Directive 2010/31/EU of the European Parliament and of the Council of 19 May 2010 on the energy performance of buildings", 2010. https://eur-lex.europa.eu/eli/dir/2010/31/oj (zugegriffen Okt. 04, 2019).

[2] M. Fabbri, J. Volt, und M. de Groote, „iBRoad _D.2.2 The Concept of the Individual Building Renovation Roadmap.pdf". Buildings Perfomance Institute Europe, Jan. 2018.

[3] „iBRoad Project – Individual Building Renovation Roadmaps". https://ibroad-project.eu/ (zugegriffen Juli 13, 2020).

[4] I. Maia und L. Kranzl, „Defining a framework to apply retrofitting optimisation models for long-term and step-by-step renovation approaches", IOP Conference Series: Earth and Environmental Science, Bd. 323, S. 012175, Sep. 2019, doi: 10.1088/1755-1315/323/1/012175.

Keywords: case study, building stock decarbonisation, energy efficiency, individual building roadmap

Paper ID 30194: Energy performance of ETICS integrated with flexible photovoltaic panel

Dariusz Heim* (1), Anna Wieprzkowicz (1), Dominika Knera (1), Simo Ilomets (2), Targo Kalamees (2), Zdeno Spitalsky (3)

1: Lodz University of Technology, Poland; 2: Tallinn University of Technology, Estonia; 3: Polymer Institute of the Slovak Academy of Sciences, Slovakia

AIM AND APPROACH

The opaque elements of building envelope (walls and roofs) is often considered as place where PV panels can be applied. Among many others well known technologies, the proposed one is an original solution combining features of traditionally insulated wall with building integrated photovoltaic. The general aim of the study is an integrated power and heat flow analysis of Energy Activated External Thermal Insulation Composite System (En-ActivETICS). The unique and original feature of En-ActivETICS is that the traditional external plaster was substituted by flexible photovoltaic panels (references hidden for the review process). The model was developed in ESP-r software using special material approach [1]. The power flow modelling approach was based on [2], using WATSUN-PV model to converting of solar radiation to electrical power [3]. The analysis was done for 20 selected locations in Europe (5 north, 10 central and 5 south). It was found out that the propose En-ActivETICS suffer from overheating of PV panel not only during summer but in other period of the year as well. The statistical analysis of the results shows the critical temperature cannot be exceeded and for which the system should be thermally stabilized.

SCIENTIFIC INNOVATION AND RELEVANCE

Traditional BIPV systems are well described and analysed in simulation and monitoring context [4,5]. However, this paper propose a new type of BIPV using thin and flexible PV integrated directly with ETICS. In En-ActivETICS the risk of overheating, thermal stresses and mechanical damages of interstitial joints is very high. One of the problems identified for the proposed solution is possible overheating of the outer layers (PV panel, glue and reinforcement net) as well as the unpredictable temperature difference between the PV covered area and the rest part of the wall. On the other hand, it is generally known, that the productivity of PV decreases with the increase with the increase of panel's temperature [6]. To take into account that phenomena the proposed model should consider the coupled heat and power flows in the scale of building elements. The analysis under weather conditions at different locations provided more information about the effect of overheating on overall performance of the system as well as how to protect it during extreme weather conditions. The analysis was conducted for an experimental room with external wall partly covered by En-ActivETICS and partly covered by traditional ETICS.

PRELIMINARY RESULTS AND CONCLUSIONS

The analysis of three, commercially available types of PV panels was done for 20 locations in Europe (north, central and south). The simulation results displayed temperature changes in characteristic parts of the wall as well as electricity generated by PV panel. In summer months, for Central European location the temperature of En-ActivETICS varies between 9 and 62 degC with the maximum power flow generated by PV around 60 W/m2. For the whole period of simulation (one year) the number of overheating hours was also determined. It was calculated for different temperature limit at 20, 25, 30, 35 and 40 degC respectively and for the period of time when solar radiation occurs. The overheating hours were additionally recalculated taking into account the weighting factor of solar radiation.

The conclusions regarding the thermal behaviour of En-ActivETICS leads to the general recommendations about increasing the thermal inertia of outer layer (between flexible PV and thermal insulation). One of the reasonable solutions considered by authors is an additional PCM layer placed behind the PV. The recommended kind of PCM including phase change temperature and latent heat is the general conclusion from the analysis.

MAIN REFERENCES

[1] Clarke J. Integrative modelling methods. Response 2001;90:22–63. https://doi.org/10.1016/B978-075065082-3/50002-4.

[2] Clarke JA, Kelly NJ. Integrating power flow modelling with building simulation. Energy Build 2001;33:333–40. https://doi.org/https://doi.org/10.1016/S0378-7788(00)00115-8.

[3] Mottillo M, Beausoleil-Morrison I, Couture L, Poissant Y. A comparison and validation of two photovoltaic models. Can. Sol. Build. Conf., 2006.

[4] Tripathy M, Sadhu PK, Panda SK. A critical review on building integrated photovoltaic products and their applications. Renew Sustain Energy Rev 2016;61:451–65. https://doi.org/10.1016/j.rser.2016.04.008.

[5] Shukla AK, Sudhakar K, Baredar P. Recent advancement in BIPV product technologies: A review. Energy Build 2017;140:188–95. https://doi.org/10.1016/j.enbuild.2017.02.015.

[6] Machniewicz A, Knera D, Heim D. Effect of transition temperature on efficiency of PV/PCM panels. Energy Procedia, vol. 78, 2015, p. 1684–9. https://doi.org/10.1016/j.egypro.2015.11.257.

Keywords: building facade, solar energy, thermal inertia, electrical power, climate classification

Paper ID 30619: LCA optimal sizing of smart buildings' energy system components considering flexible LCA-grid-footprints

Tobias Blanke* (1), Markus Hagenkamp (2), Joachim Göttsche (1), Bernd Prof. Dr. Döring (2), Vitali Reger (3), Markus Prof. Dr. Kuhnhenne (3)

1: FH Aachen, Solar Institut Jülich, Germany; 2: FH Aachen, Germany; 3: RWTH Aachen, Chair for Sustainable Metal Building Envelopes, Germany

AIM AND APPROACH

The aim of this paper is finding an optimal size of building energy system components (BESC) considering their product-based LCA and the grid interaction LCA footprint. The BESCs are cold and warm tanks, a heat pump, a photovoltaic system, a battery system, a solar-thermal system and the heating / cooling exchange area. The optimization is done for an exemplary industrial building [4]. The product-based LCA data for the considered BESCs [2] as well as the hourly carbon footprint data of the German electricity sector of 2018 [3] are taken from the literature. The optimization is based on a global particle swarm optimizer for the general size of the BESCs and a MILP solver for the yearly operation based LCA emissions. The MILP solver should find the optimal schedule of the activation of the different the BESCs for the complete year using time steps of one hour [1]. The objective of the optimizer is to find a minimum for the emissions using the LCA-approach.

SCIENTIFIC INNOVATION AND RELEVANCE

Usually just literature values or cost-optimized solutions are considered for sizing BESCs . In this research an LCA emission-based minimum for different LCA emissions like carbon or ecotoxicity as well as nonrenewable primary energy was calculated and evaluated. Therefore, hourly based emission data over the year are considered to estimate the operation-based emissions due to the grid interaction. Furthermore, this is done for three different insulation thicknesses of the exemplary industrial building.

PRELIMINARY RESULTS AND CONCLUSIONS

The preliminary results indicate different results for the carbon LCA and the nonrenewable primary energy based LCA optimization. For example, the battery is at an advantage preferred regarding the carbon emissions but not by regarding the nonrenewable primary energy demand. In general, most of the BESCs sizes are larger for the carbon emission-based optimization. The conclusion is that the ratio of investment-based emissions to operation-based emissions is higher for the nonrenewable primary energy demand than for the carbon emissions.

MAIN REFERENCES

[1] Cost optimal sizing of smart buildings' energy system components considering changing end-consumer electricity markets; Henryk Wolisz; Thomas Schütz, Tobias Blanke, Markus Hagenkamp, Markus Kohrn, Mark Wesseling, Dirk Müller; 2017; https://doi.org/10.1016/j.energy.2017.06.025

[2] Ökobaudat; Bundesministerium des Innern, für Bau und Heimat; 2020; https://www.oekobaudat.de/

[3] Agorameter; Agora Energiewende; 2019; https://www.agora-energiewende.de/service/agorameter/

[4] Plusenergiegebäude 2.0 in Stahlleichtbauweise; Vitali Reger Dipl.-Ing., Univ.-Prof. Dr.-Ing. Markus Kuhnhenne, Prof. Dr.-Ing. Helmut Hachul, Prof. Dr.-Ing. Bernd Döring, Dr.-Ing. Thiemo Ebbert, Tobias Blanke, Dr. rer. nat. Joachim Göttsche; 2019; https://doi.org/10.1002/stab.201900034

Keywords: Optimization, LCA, building energy system components

Paper ID 30836: Multi-response optimization of a hybrid solar gas heating system based on the response surface method

Mohammed Dhiya-eddine Sarmouk (1), Abdelatif Merabtine* (2), Hachimi Fellouah (3), Arezki Smaili (1)

1: 1Laboratoire de Génie Mécanique et Développement, Ecole Nationale Polytechnique, Algiers, Algeria.; 2: EPF School of Engineering; 3: Department of Mechanical Engineering, Université de Sherbrooke

AIM AND APPROACH

Solar heating systems play an important role in the energy transition policies and development of renewable energies across the world with its different technologies. Hybrid solar heating systems are becoming a major centre of interest making the most of the synergy between different energy sources.

In the current study, we present a hybrid solar/gas heating system. The project is established at the Ecole Nationale Polytechnique of Algiers (Algeria) for space heating needs. The system consists of a solar heating system coupled to a gas boiler. We analyse the solar fraction and the overall efficiency of the system as performance metrics for a better understanding and eventual improvement of the operating conditions of such system.

SCIENTIFIC INNOVATION AND RELEVANCE

Based on a validated numerical model of the solar heating system, developed on Trnsys software, a new model is developed considering the hybridization of the system to investigate the main performance indicators. The numerical simulations are then compared to the measurements. Subsequently, the model is used to conduct a sensitivity analysis by varying different design parameters to study their effects on the systems performance.

Using the surface response method, we conduct a series of simulations on the validated numerical model to develop an empirical model of the response. The fitted modes is then used to achieve the optimum of the response factor by determining the best operating settings in a specific range.

PRELIMINARY RESULTS AND CONCLUSIONS

The influence of the different design parameters on the solar fraction and the overall efficiency, the conducted sensitivity study showed the importance of the interaction between the different factors and their impact of the responses.

Metamodels were derived describing the the relationship between the chosen reponses and the design parameters under study, these metamodels might be used for the prediction of the system performance as well as for its optimisation.

Optimum design parameters were defined considering the solar fraction and the overall efficiency and the metamodels results were compared to those of the numerical model.

MAIN REFERENCES

E. Bellos et C. Tzivanidis, « Multi-objective optimization of a solar assisted heat pump-driven by hybrid PV », Applied Thermal Engineering, vol. 149, p. 528-535, févr. 2019, doi: 10.1016/j.applthermaleng.2018.12.059.

R. Li, Y. Dai, et G. Cui, « Multi-objective optimization of solar powered adsorption chiller combined with river water heat pump system for air conditioning and space heating application », Energy, vol. 189, p. 116141, déc. 2019, doi: 10.1016/j.energy.2019.116141.

A. R. Starke, J. M. Cardemil, et S. Colle, « Multi-objective optimization of a solar-assisted heat pump for swimming pool heating using genetic algorithm », Applied Thermal Engineering, vol. 142, p. 118-126, sept. 2018, doi: 10.1016/j.applthermaleng.2018.06.067.

D. C. Montgomery, Design and analysis of experiments, Ninth edition. Hoboken, NJ: John Wiley & Sons, Inc, 2017.

Keywords: hybrid solar system, Optimization, Numerical modelling, Response surface method, desirability function

Session T5.7 (Online Track): Ensuring high quality building simulations

Paper ID 30174: Simulating thermal resilience of buildings and their influence by urban microclimate using EnergyPlus

Xuan Luo*, Tianzhen Hong, Kaiyu Sun

Lawrence Berkeley National Laboratory, Berkeley, CA, United States of America

AIM AND APPROACH

With increasing frequency and severity of extreme weather events (e.g., heat waves), it is crucial to ensure urban buildings are resilient during those events to provide critical services to preserve occupant health [1]. Hence, there is a need to measure the building resilience, regarding the likelihood and consequence for each risk to facilitate better decision making in design and operation, and to prioritize risk mitigation measures. Building performance simulation is a powerful tool to quantify the building's resilience and vulnerability under extreme weather events. To provide a consistent and convenient path for users of resilience analysis through modeling and simulation, we identified and implemented a list of thermal resilience-related reports in EnergyPlus with Heat Index [2,3], Humidex [4,5], and Standard Effective Temperature (SET) [6] as metrics. The reports include the standard thresholds of these metrics and summarize the space level accumulated hours and occupant·hours within each range. Applying these new metrics, we conducted a case study with a medium-sized office building and a single-family house in San Francisco and ran simulations under different scenarios to measure the passive survivability of the building under a heatwave in September 2017. The results infer how microclimate plays an essential role in influencing thermal resilience.

SCIENTIFIC INNOVATION AND RELEVANCE

Studies have revealed the need for standard metrics that are a repeatable, reproducible, and true reflection of building resilience and sustainability [7]. To fill the gap, the paper intends to introduce how the resilience metrics and their levels are defined and reported in EnergyPlus and to provide modeling guidance on assessing a building's thermal resilience under extreme weather events. The metrics target occupants' health and wellbeing and involve the aggregations across space and time. For the spatial resolution, the reporting considers both the entire building for aggregated results and individual spaces for detailed and worst-case results. For the temporal resolution, the reporting considers the aggregated, average, and worst-case results and the duration of the worst-case during the risk event occurrence. The selected resilience metrics (Heat Index, Humidex, and SET) are well defined, calculable, and have been adopted by government agencies and the industry. The formation of standard and reportable metrics can further facilitate the development of rating systems and tools for building resilience.

PRELIMINARY RESULTS AND CONCLUSIONS

Applying the new resilience metrics reporting feature in EnergyPlus, we simulated and analyzed the thermal resilience of a single-family house under an urban heat wave event. On September 1st, 2017, the City of San Francisco set its heat record in downtown at 106 °F since 2000, and the heat wave lasted three days. The outdoor Heat Index reached above 100 °F, falling into the Extreme Caution zone indicating heat cramps and heat exhaustion are possible for occupants, and continuing activity could result in heatstroke. As a preliminary case study, we used San Francisco downtown weather records collected from local weather stations and ran a five-day simulation (from August 30th to September 3rd) assuming a power-off scenario during the last three days. Results show that for the core living room thermal zone, 2.2 out of 120 hours fall into the Extreme Caution range and 62.2 hours fall into the Heat Index Caution range (Fatigue possible with prolonged exposure and/or physical activity).

The total SET-hour is 3.1 °C·hours for SET > 30 °C, and the longest SET-unmet duration is 3.5 hours. Passive cooling techniques should be applied to mitigate the threats to human health and survivability.

MAIN REFERENCES

[1] K. Sun, M. Specian, T. Hong, Nexus of thermal resilience and energy efficiency in buildings: A case study of a nursing home, Build. Environ. 177 (2020) 106842. doi:10.1016/j.buildenv.2020.106842.

[2] G.B. Anderson, M.L. Bell, R.D. Peng, Methods to Calculate the Heat Index as an Exposure Metric in Environmental Health Research, Environ. Heal. Perspect. 121 (2013). doi:10.1289/ehp.1206273.

[3] R.G. Steadman, The assessment of sultriness. Part I. A temperature-humidity index based on human physiology and clothing science., J. Appl. Meteorol. 18 (1979) 861–873. doi:10.1175/1520-0450(1979)018<0861:TAOSPI>2.0.CO;2.

[4] R. Rana, B. Kusy, R. Jurdak, J. Wall, W. Hu, Feasibility analysis of using humidex as an indoor thermal comfort predictor, Energy Build. 64 (2013) 17–25. doi:10.1016/j.enbuild.2013.04.019.

[5] J M Masterton; F A Richardson, Humidex: a method of quantifying human discomfort due to excessive heat and humidity, Downsview, Ont. : Environment Canada, Atmospheric Environment, Downsview Ont., 1979.

[6] L.G. Gagge, A. P., Fobelets, A. P. and Berglund, A standard predictive Index of human reponse to thermal enviroment, Am. Soc. Heating, Refrig. Air-Conditioning Eng. (1986) 709–731. https://www.aivc.org/sites/default/files/airbase_2522.pdf%0Ahttp://oceanrep.geomar.de/42985/.

[7] G.E. Marjaba, S.E. Chidiac, Sustainability and resiliency metrics for buildings - Critical review, Build. Environ. 101 (2016) 116–125. doi:10.1016/j.buildenv.2016.03.002.

Keywords: thermal resilience, performance metrics, heat wave, EnergyPlus, urban microclimate

Paper ID 30243: Passive envelope design optimization of residential buildings using NSGA-II in different Algerian climatic zones

Samir Semahi* (1,2,4), Noureddine Zemmouri (2), Mohamed Hamdy (3), Shady Attia (4)

1: University of Blida-1,Blida, Algeria; 2: Laboratory of Design and Modeling of Architectural and Urban Forms and Ambiances (LACOMOFA), Dept. of Architecture, University of Biskra, Algeria; 3: NTNU Norwegian University of Science and Technology, Dept. of Civil and Transport Engineering, Trondheim, Norway; 4: Sustainable Building Design Lab, Dept. UEE, Applied Sciences, Universite de Liege, Belgium

AIM AND APPROACH

Worldwide, residential buildings are significant drivers of energy consumption and greenhouse gas emissions. The most sustainable building operation is that presents a minimum primary energy consumption (PE), which minimizes the environmental impact. Therefore, the objective of this paper is to optimize the passive envelope design of residential buildings for reducing heating and cooling primary energy consumption (PEC). Twenty (20) design variables are selected to be optimized. The selected design variables include several early-design phase decisions, such as the building orientation, the opaque building envelope proprieties (block thermal conductivity, Block thickness, insulation type, insulation thickness, solar absorptance), and the transparent building envelope proprieties (WWR, window type, overhang depth). A multi-objective non-dominated sorting genetic algorithm (NSGA-II) is coupled with EnergyPlus building energy simulation software to obtain the optimal solution that leads to reduce the primary energy consumption. The Pareto solution was used to obtain a set of optimal solutions for building design optimization. This approach is applied to a calibrated representative apartment building model, and twelve (12) locations are selected to investigate all the Algerian climate zones.

SCIENTIFIC INNOVATION AND RELEVANCE

In this research, an advanced simulation approach involving automated optimization (NSGA-II) was used and applied for the first time, in the Algerian context. There are several previous studies that conducted a parametric analysis to improve the energy efficiency of buildings in cities in the arid or temperate climate of Algeria and North Africa. However, most of those studies explore improvement opportunities using manual approached resulting in limited solution spaces and sub-optimal results. Thus, this study is one of the first studies that apply automated optimization of passive design measures for a mainstream building typology, in Algeria. The extensive evaluation of all the possible combinations of passive and energy efficiency through the NSGA-II achieved remarkable results in an acceptable computation time budget. The application of optimization techniques allowed us to perform an accurate analysis of the entire dimension of the energy efficiency problem while condensing considerably the sets of recommendations with evidence-based content.

Our findings contribute to the literature on bioclimatic and energy-efficient building design in the residential building sector. In particular, we consider this research results beneficial and transferable to other Middle East and North Africa (MENA) region countries that have similar demographic growth and climatic conditions.

PRELIMINARY RESULTS AND CONCLUSIONS

More than 28000 simulation iterations tool place for more than 700 hours of computation time, resulting into a solution space with 1.19 E+18 solutions. The optimization results of passive and energy measures applied for the base case, of a typical multi-family social residential building, in Algeria, show significant improvements of the energy performance. The current base case that represents the most commonly constructed architectural type in Algeria is far away from the optimal design recommendations provided, in this study.

Compared to the reference design, significant reductions of primary energy consumption (PEC) were achieved. Our optimization approach achieved energy saving ranging from around 24% to 53%. For heating dominated cities, the energy saving rate is between 44% and 53%. For cooling dominated cities, the energy saving rate is between 24% and 28%. The optimization results prove that the solution sets are effective within each city.

Our findings can inform and guide the decision making of architects and building engineers during the early phases of building design in order to enhance its energy efficiency. The findings can help to advance the building energy efficiency policies of North Africa region and more importantly help in the transition towards low-carbon energy-built environment.

MAIN REFERENCES

S. Semahi, N. Zemmouri, M. K. Singh, et S. Attia, « Comparative bioclimatic approach for comfort and passive heating and cooling strategies in Algeria », Build. Environ., vol. 161, p. 106271, août 2019, doi: 10.1016/j.buildenv.2019.106271.

Z. Tian, X. Zhang, X. Jin, X. Zhou, B. Si, et X. Shi, « Towards adoption of building energy simulation and optimization for passive building design: A survey and a review », Energy Build., vol. 158, p. 1306–1316, 2018.

M. Hamdy et G. Mauro, « Multi-Objective Optimization of Building Energy Design to Reconcile Collective and Private Perspectives: CO2-eq vs. Discounted Payback Time », Energies, vol. 10, no 7, p. 1016, juill. 2017, doi: 10.3390/en10071016.

F. Ascione, N. Bianco, G. M. Mauro, et D. F. Napolitano, « Building envelope design: Multi-objective optimization to minimize energy consumption, global cost and thermal discomfort. Application to different Italian climatic zones », Energy, vol. 174, p. 359–374, 2019.

F. Ali-Toudert et J. Weidhaus, « Numerical assessment and optimization of a low-energy residential building for Mediterranean and Saharan climates using a pilot project in Algeria », Renew. Energy, vol. 101, p. 327–346, 2017.

Keywords: Multi-objective optimization, dynamic simulation, genetic algorithm, primary energy consumption, jEPlus.

Paper ID 30280: ASHRAE Standard 205P: Progress towards representation of performance data for HVAC&R equipment

Neal Kruis* (1), Timothy McDowell (2), Charles S. Barnaby (3), Tanaya Mankad (1)

1: Big Ladder Software, Denver, CO, USA; 2: Thermal Energy System Specialists, Madison, WI, USA; 3: Independent Consultant, Moultonborough, NH, USA

AIM AND APPROACH

ASHRAE Standard 205P defines representations of HVAC&R equipment data in Concise Binary Object Representation (CBOR) files conforming to a JSON schema defined by a representation specification – a human-readable (text) document that defines data models and formats for an equipment type.

ASHRAE Standard 205P is intended to support the following use cases –

• Data Publication. Data publishers use representation specifications to guide implementation of data writing and testing software that produces correctly formed representation files.

• Application Development. Application software developers use representation specifications to guide implementation of software that correctly reads representation data. Such implementations may include verification tests, and developers may use representation specification example data for testing purposes.

• Data Application. Application software users use representation specifications to understand and check representation data. Data exchange will generally be automated, but the availability of representation specifications facilitates additional data review when needed.

SCIENTIFIC INNOVATION AND RELEVANCE

Simulation models allow prediction of the fuel consumption and carbon emissions of mechanical equipment given information about how the equipment performs under all conditions. The lack of detailed equipment performance data in readily-usable forms has hampered the application of building simulation models since their invention in the 1960s and 1970s. Data are often not available or published in inconsistent formats, forcing simulation users to resort to default information and/or time consuming (and error-prone) reformatting.

ASHRAE Standard 205P addresses this deficiency by defining common data models and serialization formats for facility equipment performance data, allowing automated exchange among data sources (manufacturers), simulation models, and other engineering applications. The formats and procedures specified by Standard 205P are developed under ASHRAE and ANSI consensus processes with participation of equipment manufacturers, application software developers, and engineering practitioners. The intent is that all data publishers write common-format data files and all application software will include suitable procedures to read files using that format.

PRELIMINARY RESULTS AND CONCLUSIONS

HVAC&R equipment can successfully be represented in a uniform data model for conveying product information from data publishers (e.g., manufacturers) to building performance simulation software applications. A software toolkit framework demonstrates the ability to author representations specifications, generate the JSON schema and documentation for the representation specifications, generate a C programmatic interface for reading/writing compliant representations, validate representations against the generated JSON schema and translate representations into other formats of varying degrees of human and machine readability/edit-ability. The purpose of this toolkit is to facilitate adoption of the standard in three primary use cases:

1. Authoring future representation specifications

2. Publication of compliant representations by manufacturers

3. Processing of compliant representations by simulation software developers

With ASHRAE Standard 205P, manufactures and practicing building performance modelers will have greater confidence that the equipment proposed for a building project is better represented in the models used to inform building design and operation.

MAIN REFERENCES

BSR/ASHRAE (2020). "BSR/ASHRAE Standard 205P: Second Public Review Draft - Representation of Performance Data for HVAC&R and Other Facility Equipment".

Kruis, N., Barnaby, C. (2018). "Reading and writing standardized HVAC performance data: An early implementation of ASHRAE Standard 205P".

Keywords: HVAC performance, Simulation data, Standardization

Paper ID 30363: New Standard 140 test suite: check the weather before you simulate

Timothy P McDowell* (1), Ralph T Muehleisen (2)

1: Thermal Energy System Specialists, LLC, United States of America; 2: Argonne National Laboratory, United States of America

AIM AND APPROACH

ASHRAE Standard 140: Standard Method Of Test for the Evaluation of Building Energy Analysis Computer Programs provides test suites that allow for software developers and users to evaluate the capabilities of a BPS software program versus analytical standards and/or other software programs. The Standard has proven through the years to be a good tool for identifying weaknesses in BPS software both in individual programs and in the software programs in general. Up to this point, it was assumed in the test cases that the programs could properly read standard weather files and process this information for the simulations. A new test suite is being developed to test this assumption. In this test suite, BPS software programs are run with different weather files and the outputs are compared to the values in the TMY3 files and against the output from the other programs. Statistical analyses are performed on the results to determine the level of agreement between the programs and the data file and between each other.

SCIENTIFIC INNOVATION AND RELEVANCE

Previous test suites in ASHRAE Standard 140 have assumed that the programs can correctly read in and interpret the weather information included in the standard weather files. These weather conditions are important driving forces in the simulation of building performance. This new test suite evaluates this assumptions and identifies whether there are weak points in specific software tools as well as the BPS software programs in general. Identifying these weak points will assist in determining potential research areas for improvements in BPS software. Introducing statistical analysis to the Standard 140 results will start to provide better feedback to the BPS software developers as to areas of their software that may not be as accurate as desired.

PRELIMINARY RESULTS AND CONCLUSIONS

The results from the initial field trial of the weather drivers test suite included the outputs from 8 different BPS software programs. The analysis of these results showed that the programs in general did well recreating the data from the weather files. However, greater disagreement was found when the properties derived from the data from the weather file were compared across the programs. These properties include sky temperature and solar radiation on tilted surfaces. More detailed outputs are being collecting at this point and automated analysis scripts written to enhance the analysis of the results.

MAIN REFERENCES

ASHRAE Standard 140-2017: Standard Method of Test for the Evaluation of Building Energy Analysis Computer Programs

Keywords: Standard 140, Weather Files, Software testing

Paper ID 30922: Derivation of high-resolution local thermal conditions from satellite-based data

Pelin Firat Ors*, Ardeshir Mahdavi

TU Wien, Austria

AIM AND APPROACH

The present contribution pursues a systematic approach to obtaining high-resolution thermal information for computational assessments of the urban microclimate. Thereby, we consider location-specific issues such as variations in the built environment and the influence of geographic location. To this end, we exploit land surface temperature (LST) data acquired at four temporal intervals within the diurnal cycle and at 1km resolution through satellite-based remote sensing. We first examine relationship of the LST data with the terrestrially observed near-surface air temperatures at weather stations in (or close to the) city of Vienna, Austria. This relationship is further analyzed in the light of existing classification systems of urban areas (morphology, surface properties, etc.) [1, 2]. The idea is to derive ambient microclimatic conditions at different urban locations directly from satellite-based LST data. The motivation behind this inquiry is to establish an effective method to attain high-resolution information on external boundary conditions suitable for deployment in pertinent simulation applications. Whereas the development of the proposed procedure relies on Vienna data, its performance will be tested in a different city (Izmir, Turkey).

SCIENTIFIC INNOVATION AND RELEVANCE

Existing challenges related with "weather data" utilized in building energy inquiries and other relevant assessments is widely discussed in literature (see, for instance, [3]). The proposed approach targets a more fine-grained customization of generally available data to specific local (urban) settings. Hence the role of urban domain in the emergency of building-level boundary conditions could be captured in a more detailed manner [1, 2]. As such, the present contribution further pursues the objectives set in our previous work [4], which addressed relationship of daytime thermal remote sensing data and conventional terrestrial air temperature observations. The present contribution extends the diurnal temporal range of the analyzed LST dataset via inclusion nighttime observations and through the analysis of a substantially larger collection of thermal remote sensing observations. Moreover, we investigate in more detail the influence of built environment parameters. Last but not least, the present contribution entails an independent assessment of proposed method's performance using data from a different geographic location.

PRELIMINARY RESULTS AND CONCLUSIONS

This study focuses on obtaining reliable thermally relevant high-resolution local information in urban settings. The initial results of the respective inquiry point to a significant relationship between LSTs and observed air temperatures within the main case study's domain (Vienna and its surroundings). According to the preliminary findings, the relationship is particularly strong when considering nocturnal observations. Moreover, data obtained from individual terrestrial weather station locations suggest that the relationship between LST and air temperature datasets remains in a good agreement regardless of the investigated station. These functions are currently under investigation in terms of their reducibility potential to a unified formalism. Thereby, coefficients associated with the relevant features of specific urban locations act as candidate vehicles for the unification. The underlying hypothesis is that the differences in the functions for different locations imply the presence of influencing factors related to local urban features. The proposed approach facilitates thus the exploitation of these influence factors (as captured by local climate zone classification systems) toward a general formalism for obtaining local thermally relevant microclimatic boundary conditions as contextual input for building performance simulation applications.

MAIN REFERENCES

1. Mahdavi A., Kiesel K., & Vuckovic M. (2013). A framework for the evaluation of urban heat island mitigation measures. Proceedings from SB13 Munich: Implementing Sustainability – Barriers and Chances Conference. Munich (Germany), 24-26 April 2013.

2. Stewart, I. D. and Oke, T. R. (2012). Local climate zones for urban temperature studies. Bulletin of the American Meteorological Society, 93(12), 1879-1900.

3. Crawley, D.B. and Barnaby, C. S. (2019). Weather and Climate in Building Performance Simulation. In Hensen, J.L.M., Lamberts R. Building Performance Simulation for Design and Operation. Routledge. London (UK).

4. Firat Ors, P., Vuckovic, M., & Mahdavi, A. (2019). Utilization of satellite-based remote-sensing for the representation of external boundary conditions in building energy modelling. Proceedings from 16th IBPSA Conference. Rome, (Italy), 2-4 September 2019.

Keywords: urban microclimate, built environment, remote sensing, land surface temperature, air temperature

Session T5.8 (Online Track): The role of occupants

Paper ID 30434: Indoor air quality during lockdown: A monitoring-based simulation-assisted study in London

Farhang Tahmasebi*, Elizabeth Cooper, Yan Wang, Daniel Godoy Shimizu, Samuel Stamp, Dejan Mumovic

UCL Institute for Environmental Design and Engineering, University College London, United Kingdom

AIM AND APPROACH

The Covid-19 outbreak has resulted in extraordinary patterns of home occupancy, whose implications for indoor air quality (IAQ) and energy use are unknown. In this context, the present study benefits from monitored IAQ and window operation data obtained from 8 London flats between 2019 and 2020 to investigate if during the Covid-19 lockdown (a) IAQ in the monitored flats has deteriorated, (b) the patterns of window operation by occupants have changed, and (c) a more effective approach to natural ventilation could enhance IAQ and reduce energy demand.

While basic information on occupancy and ventilation patterns is gathered through semi-structured interviews, a one-year long monitored dataset (including indoor and outdoor air temperature, relative humidity, CO2, TVOCs, NO2, PM10 and PM2.5 along with operation of windows by occupants) serves to analyse the impact of Covid-19 lockdown on IAQ and develop models of window operation tailored for these unusual circumstances. Furthermore, a thermal performance model of one of the flats incorporating the monitored window operation data has been developed and calibrated. This makes it possible to investigate the implications of alternative ventilation strategies for IAQ and energy demand.

SCIENTIFIC INNOVATION AND RELEVANCE

A large number of studies underline the significant impact of occupant behaviour on the energy and environmental performance of buildings [1]. Given the complex nature of occupant adaptive behaviour, numerous campaigns of building monitoring and data mining [2,3], a variety of data-driven occupant models [4,5] and novel workflows for integration of occupants into simulation tools [6,7] have advanced the representation of occupants in building performance simulation. However, as existing occupant behaviour models are predominantly based on limited sets of observational data (especially in the residential sector), to achieve more reliable assessments of building energy use and IAQ, further empirical studies are needed to widen the applicability of the models [8,9]. This has become especially critical as the extraordinary circumstances associated with Covid-19 outbreak has resulted in unprecedented patterns of household occupancy. As people spend much more time at home following the 2020's global pandemic, it is more critical than ever to ensure that IAQ in houses meets the recommended standards without excessive energy use. To this end, the present study combines monitored data and building simulation to analyse the changing IAQ and window operation patterns in eight flats, and investigates enhanced natural ventilation strategies commensurate with these unprecedented circumstances.

PRELIMINARY RESULTS AND CONCLUSIONS

An initial analysis of the collected indoor environmental data demonstrates significantly higher CO2 and PM10 concentration values during the lockdown as compared with the pre-lockdown period. Notably, the monitored data suggests that, despite the higher occupied hours, occupants have relied less on natural ventilation across the studied flats. The different pattern of window opening and closing is particularly evident in terms of the average duration of windows open state, which is noticeably shorter during the lockdown period.

In the next stages of the study, the collected data will be further analysed to obtain a better picture of IAQ and windows operation patterns throughout the Covid-19 timeline in comparison with the prior "normal" circumstances. This will, in turn, provide the basis for simulation-based investigation of enhanced ventilation strategies to maintain the household IAQ in the light of the changing home occupancy patterns.

MAIN REFERENCES

[1] Yan D., et al. Occupant behavior modeling for building performance simulation, Energy and Buildings 107(2015).

[2] Gunay H.B., et al. A critical review of observation studies, modeling, and simulation of adaptive occupant behaviors in offices, Building and Environment 70(2013).

[3] D'Oca S., Hong T., A data-mining approach to discover patterns of window opening and closing behavior in offices, Building and Environment 82(2014).

[4] Rijal H.B, et al. Using results from field surveys to predict the effect of open windows on thermal comfort and energy use in buildings. Energy and Buildings 39(2007).

[5] Haldi F., Robinson D., Interactions with window openings by office occupants. Building and Environment 44 (2009).

[6] Hong T., et al. An ontology to represent energy-related occupant behavior in buildings. Part I: Introduction to the DNAs framework, Building and Environment 92(2015).

[7] Gunay H.B., et al. Implementation and comparison of existing occupant behaviour models in EnergyPlus, Journal of Building Performance Simulation 9(2016).

[8] Mahdavi A., Tahmasebi F., On the quality evaluation of behavioural models for building performance applications, Journal of Building Performance Simulation 10(2017).

[9] Schweiker M., et al. Verification of stochastic models of window opening behaviour for residential buildings, Journal of Building Performance Simulation 5(2012).

Keywords: Indoor air quality, Covid-19 lockdown, Window operation, Calibrated simulation

Paper ID 30378: Energy Flexibility of Clusters of Buildings: development of an assessment tool

Alice Mugnini* (1), Fabio Polonara (1,2), Alessia Arteconi (3,1)

1: Dipartimento di Ingegneria Industriale e Scienze Matematiche, Università Politecnica delle Marche, Via Brecce Bianche 12, 60131 Ancona, Italy; 2: Consiglio Nazionale delle Ricerche, Istituto per le Tecnologie della Costruzione, Viale Lombardia 49, 20098 San Giuliano Milanese (MI), Italy; 3: Department of Mechanical Engineering, KU Leuven, B-3000 Leuven, Belgium

AIM AND APPROACH

In a context of increasing use of renewable energy sources, the flexible management of the energy demand in buildings [1] represents a potential solution to balance demand and supply. In order to optimally activate the energy flexibility of buildings, an important aspect concerns the planning of the cluster of buildings to be managed [2]. Purpose of this work is to develop a tool to easily quantify the effects of different demand management strategies on user-defined clusters of buildings. Different archetypal buildings, which differs for geometry, age class and HVAC system, can be selected to compose the cluster. Each building is modelled as a detailed RC-network in Python where also the cluster is defined. Through the comparison with a reference operation the energy flexibility of the cluster is evaluated in terms of energy and power variations. The tool is applied to selected case studies in order to provide a demonstration of its working principles.

SCIENTIFIC INNOVATION AND RELEVANCE

In presence of clusters of buildings subject to demand management strategies, it is not always easy to identify the right set of users to obtain certain demand management targets. The innovative aspect of the tool proposed in this work is to provide an instrument that allow to simulate different scenarios of load management. The idea is to extend the evaluation of the energy flexibility characterization from the single building level [3] to the cluster dimension. In this way the tool can be useful both for district planning or at the design stage of clusters of new buildings to be involved in demand side management programs.

PRELIMINARY RESULTS AND CONCLUSIONS

At present, the tool has been implemented in a preliminary version. It allows the evaluation of the behavior of individual buildings combined in a cluster. As an example of the tool features, a demand side management strategy is implemented on a specific cluster. In particular, the case analyzed refers to the quantification of the influence of thermostatically controlled loads variations on the overall heat demand curve. Results show how the involvement of some users rather than others has a significant impact on the outcome of the strategy and especially buildings of an old construction age can affect significantly the heat demand modification by limited temperature set-point variations.

MAIN REFERENCES

[1] Jensen SØ, Marszal-Pomianowska A, Lollini R, Pasut W, Knotzer A, Engelmann P, et al. IEA EBC Annex 67 Energy Flexible Buildings. Energy Build 2017. https://doi.org/10.1016/j.enbuild.2017.08.044.

[2] Vigna I, Pernetti R, Pasut W, Lollini R. New domain for promoting energy efficiency: Energy Flexible Building Cluster. Sustain Cities Soc 2018. https://doi.org/10.1016/j.scs.2018.01.038.

[3] Arteconi A, Mugnini A, Polonara F. Energy flexible buildings: A methodology for rating the flexibility performance of buildings with electric heating and cooling systems. Appl Energy 2019. https://doi.org/10.1016/j.apenergy.2019.113387.

Keywords: Load management, energy flexibility, cluster of buildings

Paper ID 30965: Impact of occupant behavior on performance optimized building retrofits

Julian Donges* (1), Alessandro Prada (2), Francesca Cappelletti (3), Andrea Gasparella (1)

1: Faculty of Science and Technology, Free University of Bozen-Bolzano, Italy; 2: Department of Civil, Environmental and Mechanical Engineering, University of Trento, Italy; 3: Department of Design and Planning in Complex Environments, University IUAV of Venice, Italy

AIM AND APPROACH

Since 2010 the European Committee has promoted the cost-optimal methodology to guide the selection of the best retrofit solutions for existing buildings. The cost-optimal approach, described for the first time by the Commission Delegated Regulation EU (European Commission, 2012), pursues a balance of energy and economic targets. Recently Directive 844/2018 establishes that each European Member State shall implement a long-term renovation strategy to support the renovation of the national stock of residential and non-residential buildings, both public and private, into a highly energy efficient and decarbonised building stock by 2050, facilitating the cost-effective transformation of existing buildings into nearly zero-energy buildings. In this framework practitioner more and more often rely on Building Energy Simulation (BES) combined with Multi-Objective Optimization (MOO) to find the optimal energy saving measurements, but they usually refer to technical standards procedures mainly developed for energy certification concerning occupancy schedules, shading management and setpoints, without taking into account realistic actions to actively manage thermal discomfort. This normal practice can lead to a significant gap between the performance of the optimal designed solution and its actual performance. This paper wants to investigate to which extent the actual behavior of the occupants is affecting the estimated optimized performance.

SCIENTIFIC INNOVATION AND RELEVANCE

In this study, we investigate how detailed user-behavior models for the daily building operation - e.g. opening/closing windows or blinds – impacts on the performance of the optimal retrofit design configuration of a residential building simulated considering a standard user-behavior model and three optimization objectives (i.e. energy, cost and comfort). In particular a rule-based adaptive model and a more complex probabilistic model have been considered to recalculate the energy demand, the total cost and the weighted discomfort time of the optimal solution obtained with the normal simulation practice. Moreover, the addition of a comfort objective function in a multi-objective optimization is a quite innovative approach that allows to highlight the consequences that different occupant behavior can determine.

PRELIMINARY RESULTS AND CONCLUSIONS

The results demonstrate that the buildings performance strongly depends on the occupant profile used for performance simulation. Consequently, the building performance of an optimal retrofit design solution can be affected by the actual user behavior if a standard behavioral model is applied. The importance of realistic user behavior modeling is highlighted to prevent misleading conclusions on optimal solutions in the assessment of energy efficiency measures for building retrofits.

MAIN REFERENCES

Penna, P., Prada, A., Cappelletti, F. and A. Gasparella. 2015. Multi-objectives optimization of Energy Efficiency Measures in existing buildings. Energy and Buildings 95: 57-69.

Fabi, V., Andersen, R. V., Corgnati, S. P., & Olesen, B. W. (2013). A methodology for modelling energy-related human behaviour: Application to window opening behaviour in residential buildings. Building Simulation, 6(4), 415–427. https://doi.org/10.1007/s12273-013-0119-6

Haldi, F., Calì, D., Andersen, R. K., Wesseling, M., & Müller, D. (2017). Modelling diversity in building occupant behaviour: a novel statistical approach. Journal of Building Performance Simulation, 10(5–6), 527–544. https://doi.org/10.1080/19401493.2016.1269245

European Parliament. 2012a. EU 244/20122012. Commission Delegated Regulation No244/2012 of 16 January 2012. Supplementing Directive 2010/31/EU of the European Parliament and of the Council on the energy Performance of Buildings by Establishing a Comparative Methodology Framework for Calculating Cost-optimal Levels of Minimum Energy Performance Requirements for Buildings and Building Elements. Off. J. Eur. Union (2012).

EU Directive L156/2018. Directive (EU) 2018/844 of the European Parliament and of the Coucil of 30 May 2018 amending Directive 2010/31/EU on the energy performance of buildings and Directive 2012/27/EU on energy. Off. J. Eur. Union (2018), pp. 75-91

Keywords: performance gap, multi-objective optimization, adaptive occupant behavior, probabilistic occupant behavior, user resilience

Paper ID 30999: How the occupancy affects the reverberation of large university halls

Giulia Fratoni*, Domenico De Salvio, Dario D'Orazio, Massimo Garai

University of Bologna, Italy

AIM AND APPROACH

The aim of the present work is to propose a method to accurately design the acoustic comfort of large university classrooms in their typical condition of use, i.e. with the presence of the students. Geometrical Acoustic (GA) principles may be employed as a reliable tool, even in the preliminary phase of the acoustic design, to simulate the effect of the occupancy on the sound field. University lecture halls are environments almost entirely made up of reflective and hard surfaces, i.e. floor, walls, ceiling. Therefore, the presence of listeners significantly affects the acoustic condition, especially because the most common seats usually are wooden or light plastic chairs. The prediction method has been applied to simulate the reverberation time of a sample of university lecture halls.

SCIENTIFIC INNOVATION AND RELEVANCE

The novelty of the study is in exploiting a valid tool in the acoustic design process of large classrooms, in compliance with the recent trend of European standards (e.g. the German DIN 18041:2016 and the Italian UNI11532:2018). Geometrical acoustic simulations are able not only to quantify the contribution of the presence of the students in terms of absorption but also to calculate and evaluate all the diffraction phenomena due to the occupancy. The same effects are instead generally neglected by the formulas provided by the standards. Since a consequence of diffraction effects generally is a decrease of the reverberation time, considering them in a preliminary phase may facilitate the design process and lead to more accurate results.

PRELIMINARY RESULTS AND CONCLUSIONS

The acoustic design of a sample of university classrooms has been developed both through prediction formulas and geometrical acoustic simulation. Values obtained with GA software have been compared both with measurements results – successively performed - and analytical predictions. In detail, a measurements campaign, in compliance with ISO 3382, confirmed the reliability of the method based on the numerical prediction.

MAIN REFERENCES

-ISO 3382-2:2008, Acoustics - Measurement of room acoustic parameters - Part 2: Reverberation time in ordinary rooms. Geneva: International Organization for Standardization.

- Building Bulletin 93, Acoustic design of schools - performance standards.

- DIN 18041:2016, Acoustic quality in rooms - Specifications and instructionsfor the room acoustic design, DIN-Normenausschuss Bauwesen.

- UNI 11532:2018., Acustica in edilizia - Caratterizzazione acustiche interne di ambienti confinati. Milano: Ente Nazionale Italiano di Unificazione.

- Choi, Y. J. (2016). Effect of occupancy on acoustical conditions in university classrooms. Applied Acoustics, 114, 36-43.

- Hodgson, M., & Nosal, E. M. (2002). Effect of noise and occupancy on optimal reverberation times for speech intelligibility in classrooms. The Journal of the Acoustical Society of America, 111(2), 931-939.

Keywords: Speech intelligibility, Acoustic simulation, Large classrooms, Occupancy

Paper ID 30660: Representations of people in urban building energy models

Kexin Zhao, Pamela Jane Fennell*, Julia Tomei

UCL, United Kingdom

AIM AND APPROACH

Although, the fundamental driver of energy consumption in the urban environment are people and their need for the services energy can provide, such as thermal comfort, entertainment, cooking, economic activity etc., representations of people in UBEMs are typically limited to a set of rules for interaction with buildings and systems. Salim et al. (2020)'s review found that these sets of rules could either be deterministic or applied stochastically. In both cases these engineering approaches are part of the physical – technical – economic model of energy consumption which has been criticised for failing to incorporate behavioural responses to technological improvements (Keirstead, 2006).

In response to Keirstead's (2006) call for a more holistic approach to urban energy consumption, combining cross-disciplinary approaches, this study uses a fine-grained literature review to evaluate the detailed schedules and rules used to model occupant behaviour in UBEMs to identify the underpinning assumptions and archetypes of behaviour. In making these assumptions explicit, this study provides the tools which allow inequalities in representation between different communities to be made visible.

SCIENTIFIC INNOVATION AND RELEVANCE

This study goes beyond the identification of modelling approaches which is found in Salim et al. and draws on insights from social science in order to create a framework which can be used to compare the representations of people in UBEMs with the diverse groups of people and behaviours which are found in large cities. The complex interplay between spatial planning and social constructs such as race and class is highlighted by Neely and Samura (2011) and the potential for similar injustices to be perpetuated through absence or under-representation in energy planning is clear. The absence of resource-constrained households whose energy consumption is dictated by economic resources rather than desirable temperatures and hours of operation in schedule-based approaches is a key example. This framework allows situations of under and over representation to be identified, their potential impact assessed and adjustments made accordingly.

PRELIMINARY RESULTS AND CONCLUSIONS

Preliminary results indicate that representations of people in UBEMs are typically based on the judgement of the modeller, likely reflecting their particularly biases and expectations of behaviour. Little evidence was found for consideration being given to the validation of the representation of people contained in the model, for example, by referring to relevant demographic details, nor is there evidence of consideration being given to the potential impacts of policy decisions being made based on models which do not adequately represent the diversity of the urban population, who are included spatially at least, within the model.

MAIN REFERENCES

Keirstead, J. (2006). Evaluating the applicability of integrated domestic energy consumption frameworks in the UK. Energy Policy, 13.

Neely, B., & Samura, M. (2011). Social geographies of race: Connecting race and space. Ethnic and Racial Studies, 34(11), 1933–1952.

Salim, F. D., Dong, B., Ouf, M., Wang, Q., Pigliautile, I., Kang, X., Hong, T., Wu, W., Liu, Y., Khan Rumi, S., Saiedur, R., An, J., Deng, H., Shao, W., Dziedzic, J., Sangogboye, F. C., Kjærgaard, M. B., Kong, M., Fabiani, C., … Yan, D. (2020). Modelling urban-scale occupant behaviour, mobility, and energy in buildings: A survey. Journal of Building and Environment, 49. https://doi.org/10.1016/j.buildenv.2020.106964

Keywords: UBEMs, occupant modelling, archetypes of behaviour, informal settlements, slums

Session T5.9 (Online Track): Ensuring high quality building simulations

Paper ID 30374: Peak extraction in daylight simulations using BSDF data

David Geisler-Moroder* (1), Gregory J. Ward (2), Taoning Wang (3), Eleanor S. Lee (3)

1: Bartenbach GmbH, Aldrans, Austria; 2: Anyhere Software, Berkeley, California, USA; 3: Lawrence Berkeley National Laboratory, Berkeley, California, USA

AIM AND APPROACH

Over the past years, tools and processes have been developed to accurately characterize and simulate the performance of optically complex daylighting and solar control systems in buildings by using bidirectional scattering distribution function (BSDF) data. High resolution data are needed to accurately represent the direct solar component, especially for systems with relevant direct transmission (e.g., fabrics with openness, blinds or louvers, ...). This in turn challenges the Monte Carlo sampling in a backward raytracing approach to adequately resolve the peaks in the distribution. An accurate representation of the direct solar contribution is very important whenever it comes to luminance-based output. This includes evaluation of visual comfort (e.g., using the daylight glare probability (DGP) metric) as well as realistic appearance of physically-based renderings (e.g., sun in the field of view or sharp shadow patterns).

For small but bright light sources (e.g., the sun), RADIANCE complements the stochastic backwards raytracing algorithm with a deterministic component. To be able to apply this approach also for the direct component passing through a daylighting system, a peak extraction (PE) algorithm has been developed that analyzes the BSDF data during the raytracing process. Thereby it is decided if a direct component is simulated or not.

SCIENTIFIC INNOVATION AND RELEVANCE

In theory, a "direct through" component is a delta peak in a BSDF. Every representation with finite resolution averages the peak (e.g. direct solar contribution) over a defined solid angle. While the transmitted flux is correct, the resulting peak luminance can be reduced by orders of magnitude, and the light flux is spread over a bigger area, where the degree of scattering depends on the resolution of the BSDF. This is of high relevance for luminance-based calculations, such as glare evaluations, and for realistic renderings where correct shadow patterns are desired.

The developed PE algorithm analyzes the BSDF data for every ray during Monte Carlo sampling that hits the respective surface and determines whether the underlying distribution has a peak in the tested direction. By checking the surrounding directions, the algorithm discovers a local strong peak in the distribution. If so, the peak is replaced with a direct specular component where the transmission is calculated from the local BSDF value. While relatively simple in its approach, resolving the details of assigning flux during PE are challenging, requiring careful tuning to satisfy energy conservation and rendering criteria.

PRELIMINARY RESULTS AND CONCLUSIONS

In the RADIANCE software the new method has been implemented in the "aBSDF" material. By assigning this material the user tells the software to look for a possible peak in direct transmission.

First tests with various BSDF data sets show the desired result of the PE algorithm. For a typical application with a fabric shade with openness factor the simulations show the difference in both – the visual appearance of the shadow pattern in the room and the size and luminance of the direct sun in the field of view. The maximum luminance in this example changes from 171 040 cd/m2 to 3 243 122 cd/m2 with PE in this example, and the DGP changes from 0.246 to 0.351.

In summary, the new PE algorithm improves accuracy and quality of daylight simulations with BSDF data. As it is difficult and costly to rectify problems associated with occupant discomfort after purchase of capital equipment such as windows and shades have been made, accurate annual simulations help to improve informed product selection and ultimately indoor environmental quality.

MAIN REFERENCES

- Geisler-Moroder, David, Eleanor S. Lee, Gregory J. Ward; Validation of the Five-Phase Method for Simulating Complex Fenestration Systems with Radiance against Field Measurements; Proceedings of Building Simulation 2017, International Building Performance Simulation Association, San Francisco, August 7-9, 2017.

- Lee, E.S., Geisler-Moroder, D. and Ward, G., 2018. Modeling the direct sun contribution in buildings using matrix algebraic approaches: Methods and validation. Solar Energy 160: 380-395.

- Ward, Greg, Rob A. Shakespeare. 1998. "Rendering with RADIANCE." The Art and Science of Lighting Visualization. Morgan Kaufmann Publishers

Keywords: Daylight simulation, BSDF, daylight system, rendering, direct solar component

Paper ID 30486: Modelling of transparent PV windows with CFS

Joaquim Romaní* (1), Marion Hiller (2), Jaume Salom (1)

1: Catalonia Institute for Energy Research (IREC), Spain; 2: Transsolar Energietechnik GmbH

AIM AND APPROACH

The aim of the article is to present the modification of fenestration models as integrated in building simulation software that allows modelling the performance of semi-transparent photovoltaic windows. The goal is to model the impact of the new developments on semi-transparent (coloured) and transparent photovoltaic thin films, accurately modelling the effect on the thermal demands, electric output, and daylighting.

The approach is to modify the current TRNSYS complex fenestration system (CFS) (Hiller & Schöttl 2014) to introduce the PV effect. The CFS already provides detailed modelling of the optical and thermal performance of the window layers, including glazing and shaders. Hence, it is a good base for introducing the PV effect, as it is able to calculate the temperature of each of the layers, such that PV efficiency dependence on temperature can be accounted. The power output of the PV cell is introduced into the energy balance of the window layers..

The model is empirically validated against IEA exercises (Loutzenhiser et al. 2006) for empirical validations of shadings/daylighting/load interactions in building energy simulation tools. The impact of the PV effect is evaluated, and the model is compared to results in the literature.

SCIENTIFIC INNOVATION AND RELEVANCE

Building Integrated Photovoltaics (BIPV) systems are currently used as technological solution for the contribution of buildings to the energy transition (Biyik et al. 2017). Semi-transparent, and more specifically, transparent building integrated photovoltaic in windows are growing in interest for its advantages in solar gains control, daylighting, renewable energy production, and aesthetical integration in buildings (Tech4win). However, the evaluation of these systems require complex modelling of the optical, thermal, and electrical performance, which needs to be linked to the behaviour of other building elements. The present article presents an improvement of the current CFS in TRNSYS, hence, the optical performance is modelled with bidirectional scattering distribution function and the energy balance is performed according ISO15099, but introducing the PV effect on the energy balance of the glazing. The thermal mass is also introduced to better assess the temperature and efficiency of the PV layers (Hiller et al. 2019). This improvement is a progress in the state of the art in modelling transparent PV windows, as it allows to evaluate the thermal, electrical, and daylighting impact in a single tool. Moreover, it allows to evaluate complex fenestrations integrating shadings and PV integrating the state-of-the-art optical and thermal models.

PRELIMINARY RESULTS AND CONCLUSIONS

The model was implemented successfully, showing robustness in computing the PV efficiency temperature dependence. Hence the model may be used to evaluate the impact of the glazing thermal mass into the PV output and overall PV glazing energy balance..

The impact of the PV of the PV window will be evaluated on the basis of IEA exercises (Loutzenhiser et al. 2006), performing a sensitivity analysis of the PV efficiency, temperature dependence constant, and inclusion of the thermal mass of the glazing in the energy balance. The impact of the PV window will be also evaluated on the validated model of real office building.

The improvement of the model and the conclusions of the research are expected to help increase to implementation of transparent PV in buildings, hence increasing the share of renewable and reducing the environmental impact. Finally, the limitations of the models on selective glazing and/or cell configuration will also be evaluated.

MAIN REFERENCES

E. Biyik, M. Araz, A. Hepbasli, M. Shahrestani, R. Yao, L. Shao, E. Essah, A.C. Oliveira, T. del Caño, E. Rico, J.L. Lechón, L. Andrade, A. Mendes, Y.B. & Atlı. A key review of building integrated photovoltaic (BIPV) systems. Engineering Science and Technology, an International Journal 20(3) (2017) 833–858.

P.G. Loutzenhiser, H. Manz, P.A. Strachan, C. Felsmann, T. Frank, G.M. Maxwell, P. Oelhafen. An empirical validation of modelling solar gains through glazing unit using building energy simulation programs. HVAC & Research 12(4)(2006) 1097-1116

M. Hiller, P. Schöttl. Modellierung Komplexer Verglasungssysteme in TRNSYS. Proceedings from BauSIM2014: German Building Performance Simulation Conference. Karlsruhe (Germany), 26-28 September 2014 (English translation available upon request hotline@transsolar.com).

M. Hiller, M. Gut, S. Holst. CFS model improvement based on measured data of a 1:1 scale test mock-up. Proceedings of the 16th IBPSA conference. Rome (Italy), 2-4 September 2019.

Tech4win. Disruptive sustainable technologies for next generation pv windows. EU H2020 Framework Programme under Grant Agreement no. 826002.

Keywords: BIPV, CFS, Window, TRNSYS

Paper ID 30641: The influence of envelope moisture in the building energy simulation

Piero Bevilacqua (1), Roberto Bruno (1), Natalia Shushunova (2), Stefania Perrella* (1), Daniela Cirone (1), Natale Arcuri (1)

1: Department of Mechanical, Energy and Management Engineering, University of Calabria, Italy; 2: Department of Integrated Safety in Construction, National Research Moscow State University of Civil Engineering

AIM AND APPROACH

Energy simulation of buildings plays a fundamental role in the communitarian and international framework as a useful tool to help in the design of new performant and efficient buildings that can minimize energy consumptions and reach the targets of nZEB and ZEB. Several legislations and regulations impose strict requirements in terms of water condensation in the building vertical and horizontal partitions. It is important, however, not only to verify the presence or absence of surface and interstitial condensation but also to evaluate how the thermo-hygrometric conditions, in different climatic zones, affect the overall energy performance of the envelope itself. For such evaluations, dynamic simulations are particularly suitable. They are based on a non-stationary calculation mechanism that considers the transport and accumulation of heat in building materials, user behavior, variable climatic conditions, including rain and solar radiation. Nevertheless, in the majority of cases, the presence of moisture in the building envelope materials is neglected and the simulations do not account for the variability of thermal properties dictated from the variable water content.

SCIENTIFIC INNOVATION AND RELEVANCE

Currently, most of the building simulations is performed with dynamic tools and software that, even though being accurate in determining the heat exchanges through the building envelope, rely in some basic assumptions. Among them, usually, the thermal properties of the building envelope constituents are set constant and are not allowed to vary during the simulation. With the aim to overcome this limitation, in this paper, a novel approach is proposed where the combined use of hygrothermal and energy simulation allowed to determine the influence and the effect of the envelope moisture content in the energy requirements of a building located in two different climates.

The influence of the water content on the thermal transmittance of the wall was assessed and it was highlighted how the presence of building site humidity, absence of latent heat in the solid/liquid phase and thermal conductivity increase the transmittance values for each individual construction. Finally, a dynamic energy simulation was performed with reference to a case study building located in two different climates, properly modelling the thermal properties of the construction components to account for moisture variability.

PRELIMINARY RESULTS AND CONCLUSIONS

The results clearly highlighted the variation of the water content in the building envelope constituents during the year, indicating that neglecting this aspect can lead to deviances in the building energy simulation. After setting a variability of the thermal properties of the building walls the dynamic energy simulation showed the effect of variable moisture content on the energy requirements of a building located in two different climates and produced overestimation of the building energy requirements with variations that overcame 10 % when compared to the reference approach in terms of thermal properties modelling.

MAIN REFERENCES

1) Wenqiang Dong, Youming Chen, Yang Bao, Aimin Fang,A validation of dynamic hygrothermal model with coupled heat and moisture transfer in porous building materials and envelopes, Journal of Building Engineering, Volume 32, 2020, 101484. ISSN 2352-7102. https://doi.org/10.1016/j.jobe.2020.101484.

2) Steven Kordziel, Samuel V. Glass, Charles R. Boardman, Robert A. Munson, Samuel L. Zelinka, Shiling Pei, Paulo Cesar Tabares-Velasco, Hygrothermal characterization and modeling of cross-laminated timber in the building envelope, Building and Environment, Volume 177, 2020, 106866. ISSN 0360-1323, https://doi.org/10.1016/j.buildenv.2020.106866.

3) Marián Vertaľ, Marek Zozulák, Anna Vašková, Azra Korjenic. Hygrothermal initial condition for simulation process of green building construction. Energy and Buildings, Volume 167, 2018, Pages 166-176. ISSN 0378-7788. https://doi.org/10.1016/j.enbuild.2018.02.004.

4) Guilherme B.A. Coelho, Hugo Entradas Silva, Fernando M.A. Henriques. Calibrated hygrothermal simulation models for historical buildings. Building and Environment, Volume 142, 2018, Pages 439-450. ISSN 0360-1323. https://doi.org/10.1016/j.buildenv.2018.06.034.

5) Marián Vertaľ, Marek Zozulák, Anna Vašková, Azra Korjenic. Hygrothermal initial condition for simulation process of green building construction. Energy and Buildings, Volume 167, 2018, Pages 166-176. ISSN 0378-7788. https://doi.org/10.1016/j.enbuild.2018.02.004.

Keywords: Building energy consumption, Moisture content, Hygrothermal simulation

Paper ID 30841: Deteriorate: a model to simulate envelope r-value fluctuations in natural porous insulation

Yasser El Masri*, Tarek Sherif, Tarek Rakha

Georgia Institute of Technology, United States of America

AIM AND APPROACH

As the threat of climate change exacerbates due to increasing energy demand and the carbon tax that is linked to it, driving down energy consumption is a goal that building professionals must seek to prioritize due to the key role that buildings play in global energy consumption. The use of insulation to drive down building energy consumption is ubiquitous. While inorganic types of insulation may offer high resistivity (R Values) per square unit, their embodied energy might make their use pointless from a climate change perspective. Natural based insulations, which are typically porous, offer an insulation solution with low to negligible embodied energy. The porosity of these materials, while offering better air tightness for buildings, can decrease thermal performance due to moisture build-up. Simulating these effects generally requires Building Energy Modelling (BEM) software that utilizes advanced algorithms to factor this variability. Due to the need to employ these materials more commonly and to anticipate their behavior for more informed decision making, we propose a computationally efficient model that discretizes porous insulation into three distinct layers: Solid, Air, and Moisture. The model associates moisture build up in the pores using material-specific Sorption Isotherms and relative humidity values only.

SCIENTIFIC INNOVATION AND RELEVANCE

Existing BEM measures in tools such as EnergyPlus that simulate decreased thermal conductivity efficiency differently depending on the solver selected. The default Conduction Transfer Function uses a static thermal conductivity value. As natural porous insulation materials have variable thermal conductivity due to variability in moisture content occupying the pores utilizing a static thermal conductivity value would lessen the simulation accuracy. While EnergyPlus's Conduction Finite Difference Solution Algorithm (ConFD) requires the user to input the thermal conductivity and the temperature which relates to it This solution, however, does not take into account the effects of moisture on thermal conductivity. On the other hand, the HAMT Model does take into account the effects of moisture dependent thermal properties. It does require the user to input the following properties: Porosity, Initial Water Content, Sorption Isotherm, Suction, Redistribution, Diffusion, and Thermal Conductivity. The availability of information on these properties would pose a challenge for users to obtain it. Our discretized model aims to factor in this variability only through the use of sorption curves for the given material and in the process make simulating thermal R-Value fluctuations easier to factor in without a computational penalty or extensive information.

PRELIMINARY RESULTS AND CONCLUSIONS

We run a simulation using our model on a building utilizing cellulose insulation. Results show that the model is validated by matching experimental data reported by other authors in terms of thermal conductivity and R-Value changes over time. Comparing the outcome to experimental values shows promise in mimicking thermal conductivity fluctuations in porous materials.

In conclusion, this paper presented a computationally inexpensive modeling approach that can mimic thermal conductivity fluctuation across porous materials. By discretizing the elements of porous materials, it allows the user to mimic their fluctuating transient thermal behavior and simulate their use more accurately in relation to real-life conditions. Further verification of results must be conducted in comparison to experimental data with different materials for further refinement of the model. However, these preliminary results indicate that the proposed approach shows promise in terms of presenting a simplified solution to simulate the behavior of variable thermal conductivity in porous materials.

MAIN REFERENCES

-Aldawi, Fayez, and Firoz Alam. 2016. "Residential Building Wall Systems: Energy Efficiency and Carbon Footprint." In Thermofluid Modeling for Energy Efficiency Applications, 169–96. Elsevier Inc. https://doi.org/10.1016/B978-0-12-802397-6.00008-7.

- Berardi, Umberto, and Matteo Naldi. 2017. "The Impact of the Temperature Dependent Thermal Conductivity of Insulating Materials on the Effective Building Envelope Performance." Energy and Buildings 144 (June): 262–75. https://doi.org/10.1016/j.enbuild.2017.03.052.

- Brzyski, Przemysław, Piotr Kosiński, Aneta Skoratko, and Wojciech Motacki. 2019. "Thermal Properties of Cellulose Fiber as Insulation Material in a Loose State." AIP Conference Proceedings 2133 (July). https://doi.org/10.1063/1.5120136.

- Hansen, Kurt Kielsgaard, Carsten Rode, Ernst Jan de Place Hansen, Tim Padfield, and Finn Harken Kristiansen. 2001. "Experimental Investigation of the Hygrothermal Performance of Insulation Materials." In Proc. Performance of Exterior Envelopes of Whole Buildings VIII, Dec 2-7, 2001, Clearwater Beach, Florida.

- Knöll, Sabine, and Uwe Welteke. 1991. "THE ZERO–ENERGY–HOUSE: THE RANKING OF THE SUN IN THE HEAT–REQUIREMENT MARKET." In Energy Conservation in Buildings, 66–71. Elsevier. https://doi.org/10.1016/b978-0-08-037215-0.50018-3.

- Kunič, Roman. 2017. "Carbon Footprint of Thermal Insulation Materials in Building Envelopes." Energy Efficiency 10 (6): 1511–28. https://doi.org/10.1007/s12053-017-9536-1.

- Petter Jelle, Bjørn. 2016. "Nano-Based Thermal Insulation for Energy-Efficient Buildings." In Start-Up Creation: The Smart Eco-Efficient Built Environment, 129–81. Elsevier Inc. https://doi.org/10.1016/B978-0-08-100546-0.00008-X.

Keywords: Building Envelope, Variable Thermal Conductivity, Sorption Curve, R-Value

Paper ID 31054: Distributed sensing for indoor environments: comparing measured and simulated indoor airflow data for complex commercial building interiors

Nick Novelli (1), Mohamed Aly Etman* (1), Justin Shultz (2), Anna Dyson (1)

1: Yale University, United States of America; 2: EYP, United States of America

AIM AND APPROACH

Indoor air quality (IAQ) sensing could illuminate critical information on building performance and the health of occupants who typically reside over 90% indoors. However, substantial research and development is required towards usefully measuring and anticipating exposures, particularly within commercial buildings comprising complex spaces and movements. These airflow patterns are of vital importance to designers, managers, and occupants, because they are dynamic and challenging to control; with the current pandemic, improving their evaluation and visualization has become a critical public health concern.

Despite the proliferation of sensing technologies, distributed air speed detection is not yet common. However, air speeds can yield easy-to-parse, actionable data when collected alongside other environmental factors such as temperature, humidity and carbon dioxide (CO2) concentrations, and are critical towards the characterization of biological exposures in the aérobiome.

To evaluate airflow patterns in a large adapted mixed-use factory space, serviced by air-side HVAC, a radiant slab, and ceiling-mounted circulation fans, sensors were distributed to occupied locations, measuring air speed and other factors including temperature, humidity, CO2, and total volatile organic compounds (TVOCs). In parallel, a simplified computational fluid dynamics (CFD) model of the measured space was constructed and run for multiple boundary condition sets, and results were compared.

SCIENTIFIC INNOVATION AND RELEVANCE

Although investigations into built-environment phenomena are traditionally rooted in reductive methods of isolating controllable variables and observing specific outcomes, more integrated methods of ecosystemic analysis, involving a greater number of interrelated variables, could be necessary to assess public health exposures, as built environments are increasingly understood to exhibit deep systems-level complexity.

For example, to characterize the real-time impact of IAQ and adaptive indoor air flow regimes to complex patterns of occupation, these factors must be integrated through multiple frameworks, such as envelope thermal performance, occupant behaviors, HVAC control strategies, and the indoor microbiome.

A finer-grained understanding is required of coupling variable occupant comfort requirements to distributed controls for adaptable building performance, in order to design and verify indoor environments, vis vis our emerging knowledge of the health impacts from time spent indoors, energy use, sustainability, and resiliency.

Improved design and verification of airflow in different indoor building types has long been critical but is now a crucial component to the global health crisis response. Moreover, reliable verification of building air flow behaviors is critical for assessment of various IAQ management technologies such as next-generation HVAC strategies, or alternatives such as smart fans and various devices for air 'purification'.

PRELIMINARY RESULTS AND CONCLUSIONS

Multi-factor sensors were positioned within occupied locations within a large factory bay on the ground and mezzanine floors, logging data over four months, through different seasonal patterns. CFD simulations were constructed and run for key instances. Preliminary conclusions are drawn on the relative efficacy of the integrated air flow and thermal comfort management strategy, comparing logged data to simulations. Measured air speeds were typically lower than optimal for thermal comfort, represented by Predicted Mean Vote with elevated air speed. This was true despite the location of some of these positions directly in the downwash of ceiling-mount fans. Moments were rare when increased air flow would benefit some positions, but reduce comfort in others. Inasmuch, dynamic fan adjustments could result in significant improvement on comfort levels. Simulated results (in a sensitivity analysis of fan speeds) indicated much higher possible air speeds at sensor positions without significant disruption. Given the encouraging results achieved with relatively coarse data (measured and simulated), integrating this sensors and simulation workflow is anticipated to contribute towards a developing holistic understanding of the built environment, and ubiquitous application in low-cost building management strategies, through the integration into visualization methodologies.

MAIN REFERENCES

"ASHRAE Standard 55-2017 - Thermal Environmental Conditions for Human Occupancy (ANSI/ASHRAE Approved)," 2017.

Dong, Bing, Vishnu Prakash, Fan Feng, and Zheng O'Neill. "A Review of Smart Building Sensing System for Better Indoor Environment Control." Energy and Buildings 199 (September 15, 2019): 29–46. https://doi.org/10.1016/j.enbuild.2019.06.025.

Kiatreungwattana, Kosol, Michael Deru, and Jason DeGraw. "GSA Green Proving Ground Smart Ceiling Fan - White Paper," 2016, 30.

Li, Jerry. "Comparing Modes of Operation for Residential Ceiling Fans to Achieve Thermal Destratification," 2014. https://www.academia.edu/13522636/COMPARING_MODES_OF_OPERATION_FOR_RESIDENTIAL_CEILING_FANS_TO_ACHIEVE_THERMAL_DESTRATIFICATION. Accessed May 9, 2019.

Loftness, Vivian, Volker Hartkopf, Azizan Aziz, Joon-Ho Choi, and Jihyun Park. "Critical Frameworks for Building Evaluation: User Satisfaction, Environmental Measurements and the Technical Attributes of Building Systems (POE + M)." In Building Performance Evaluation: From Delivery Process to Life Cycle Phases, edited by Wolfgang F.E. Preiser, Andrea E. Hardy, and Ulrich Schramm, 29–48. Cham: Springer International Publishing, 2018. https://doi.org/10.1007/978-3-319-56862-1_3.

Parkinson, T., A. Parkinson, and R. de Dear. "Continuous IEQ Monitoring System: Context and Development." Building and Environment 149 (2019): 15–25. https://doi.org/10.1016/j.buildenv.2018.12.010.

Keywords: Distributed sensing, Post-occupancy analysis, Thermal comfort, Environmental performance, Airflow

Session F1.1: Ensuring high quality building simulations

Paper ID 30365: Update of ASHRAE Standard 140/BESTEST thermal fabric test cases

Joel Neymark* (1), Ronald Judkoff (2), Michael Kummert (3), Ralph Muehleisen (4), John Allison (5), Jingjing An (6), Jason Glazer (7), Robert Henninger (7), Marion Hiller (8), Yue Jiang (9), Alec Johannsen (10), Neal Kruis (11), Timothy McDowell (12), Eikichi Ono (13), Paul Strachan (14), Michael Witte (7), Da Yan (15), Harunori Yoshida (16), Xin Zhou (9)

1: J. Neymark & Associates, United States of America; 2: National Renewable Energy Laboratory, United States of America; 3: Polytechnique Montréal, Canada; 4: Argonne National Laboratory, United States of America; 5: arbnco Ltd, United Kingdom; 6: Beijing University of Civil Engineering and Architecture, China; 7: GARD Analytics, Inc., United States; 8: Transsolar Energietechnik GmbH, Germany; 9: Southeast University, China; 10: Alec Johannsen Consulting Engineers, South Africa; 11: Big Ladder Software, United States; 12: Thermal Energy System Specialists, United States; 13: National University of Singapore, Singapore; 14: University of Strathclyde, United Kingdom; 15: Tsinghua University, China; 16: Kyoto University (professor emeritus), Japan

AIM AND APPROACH

The BESTEST (Building Energy Simulation Test) thermal fabric tests, originally published in 1995, comprise a portion of ANSI/ASHRAE Standard 140 Method of Test for the Evaluation of Building Energy Analysis Computer Programs. These are program-to-program comparative tests, where to assess model accuracy a tested program's results are compared with the range of agreement of the results of other programs that are typically a set of vetted example results provided with Standard 140.

The aim of this work is to: 1) adapt the building thermal fabric test cases to the current modeling state of the art, and 2) improve (narrow) the range of agreement among the test suite example simulation results, where some disagreement may be attributable to original test specification ambiguities. The approach of the work is to: 1) extensively revise and update the test specifications such that they would be unambiguous for the input structures of most whole-building energy simulations programs with time steps of one hour or less, 2) field-test the specifications with a variety of internationally developed simulation programs, and 3) assess the resulting improvement to (narrowing of) the range of agreement of the vetted simulation results that will be provided with Standard 140.

SCIENTIFIC INNOVATION AND RELEVANCE

A fundamental aspect of the BESTEST methodology is to provide test specifications with equivalent inputs that satisfy a variety of simulation program input schemes, thereby reducing disagreement among program results attributable to interpretive issues. The primary scientific innovation is to improve diagnostic capability by eliminating previous ambiguities for some of these equivalent inputs in the test specification, including those related to simulation inputs for: surface heat transfer, window construction properties, floor details, site altitude and terrain details, internal mass, simulation initialization and pre-conditioning, thermostat control, and others. The weather data was also updated from the original TMY data to new TMY3 data for a location within 30 km of the original location.

The relevance of the work is that BESTEST/Standard 140 is widely used for diagnostic testing by software developers and for qualifying software for various applications. Citations of Standard 140 include: ASHRAE building energy efficiency Standards 90.1 and 189.1, the U.S. tax code for certifying software used to evaluate building energy efficiency tax credits, the International Energy Conservation Code, many state and local agencies in the U.S., and many countries and international agencies.

PRELIMINARY RESULTS AND CONCLUSIONS

The updated building thermal fabric modeling test cases have been vetted via field trials with seven whole-building energy simulation programs with timesteps of one hour or less, run by the developers of the programs. These include, in alphabetical order (country in parenthesis): BSIMAC (South Africa), California Simulation Engine (U.S.), DeST (China), EnergyPlus (U.S.), ESP-r (U.K.), NewHASP (Japan), and TRNSYS (Germany). The field-trial process was iterative in that executing the simulations led to refinement of the test cases, and the results of the tests led to improving and debugging the models.

Final results indicate substantially improved agreement among simulations for the updated test specification versus results for the original test specification of 1995. This improvement was observed for annual heating and cooling loads and peak heating loads, where in many of the test cases the range of difference is reduced by half or more. For peak cooling loads, some test cases exhibit improved ranges of agreement while others do not. In general, peak cooling loads are considered more difficult to assess—because of timing of solar gains, along with mass and surface heat transfer interactions, so there may still be room for improvement in the modeling state of the art.

MAIN REFERENCES

ANSI/ASHRAE. (2017). ANSI/ASHRAE Standard 140-2017. Standard Method of Test for the Evaluation of Building Energy Analysis Computer Programs. Atlanta, Georgia, U.S.A.: ASHRAE.

ANSI/ASHRAE/IES. (2019). ANSI/ASHRAE/IES Standard 90.1-2019, Energy Standard for Buildings Except Low-Rise Residential Buildings. Atlanta, Georgia, U.S.A.: ASHRAE.

IECC. (2018). 2018 International Energy Conservation Code. Country Club Hills, IL: International Code Council.

Judkoff, R.; Neymark, J. (1995). International Energy Agency Building Energy Simulation Test (BESTEST) and Diagnostic Method, NREL/TP-472-6231 (300 pp.).

Neymark, J.; Judkoff, R.; Kummert, M.; Muehleisen, R.; et al.* (2020). Update of Standard 140-2017 Section 5.2 and Related Sections (BESTEST Building Thermal Fabric Test Cases). Lemont, Illinois: Argonne National Laboratory. May 2020. DRAFT. Publication planned for Jul/Aug 2020. *15 additional co-authors.

Neymark, J., Kummert, M., and Judkoff, R. (2019). "Developing Equivalent Surface Heat Transfer Input Parameters for Updated Standard 140/BESTEST Thermal Fabric Test Cases." Proceedings of Building Simulation 2019. Rome, Italy. Sept. 2-4, 2019. International Building Performance Simulation Association.

U.S. DOE. (2020). Qualified Software for Calculating Commercial Building Tax Deductions. Washington, DC: U.S. Department of Energy, Office of Energy Efficiency and Renewable Energy, Building Technologies Program. Accessed 08 July 2020.

Keywords: BESTEST, validation, building thermal fabric

Paper ID 30766: SimQuality - A novel test suite for dynamic building energy simulation tools

Andreas Nicolai (1), Stephan Hirth* (1), Madjid Madjidi (2)

1: TU Dresden, Germany; 2: University of Applied Sciences, Munich, Germany

AIM AND APPROACH

Based on existing test suites like the BESTest and DIN 13791, a new test suite for building energy simulation tools has been developed. The test cases investigate physical models and implementations individually and in combined tests, hereby identifying sources of deviations between tested tools. The methodology to built test cases incrementally and evaluate changes resulting from new features added to the tests, allows a much better isolation of error sources. The test suite is particularly focused on providing reference solutions for simulation tool developers. Another novel aspect of the test suite is a set of rules that ensures publication of models and simulation input data, such that results can be reproduced and verified easily by other users. Any software certified with the test suite can be considered to fulfil minimum requirements for the anticipated use cases. The article describes the test cases and variants formulated so far, and gives examplary result for the tested tools NANDRAD, IDA-ICE, TRNSYS, TAS, Modelica/AixLib, THERAKLES, EnergyPlus and ETU-Sim.

SCIENTIFIC INNOVATION AND RELEVANCE

New and detailed test methodology and certification procedure. Highly relevant in context of increasing use of building energy simulation tools. Provides a software-developer/producer independent testing and evaluation procedure.

PRELIMINARY RESULTS AND CONCLUSIONS

Procedure and methodoly describes has been tested with 8 tools, and 6 contributing parties over a period of 12 months. The test case descriptions have been improved and results collected for the tested tools. The simpler test cases (radiation loads, thermal transport + storage, ventilation) showed fairly good agreement between all tools, yet more advanced tests showed somewhat alarming deviations even in tools considered as industry-standard.

MAIN REFERENCES

ASHRAE 140 BESTest, DIN EN ISO 13791, VDI 6020/6002/2078, DIN EN ISO 15255

Keywords: Validation, SimQuality, Building Energy Simulation, BESTest

Paper ID 30130: Consistency in early simulation inputs with statistical analyses of historical models: Reducing the impact of mental models and simulation settings

Nanna Dyrup Svane* (1,2), Torben Østergård (1,2), Rasmus Lund Jensen (1)

1: Aalborg University; 2: MOE A/S

AIM AND APPROACH

Early Building Performance Simulations (BPS) are a helpful means for estimations of necessary airflow, window areas, HVAC systems, etc. Already in the planning phase of the design, before the drawing work has started, these estimations are of high value for assessing the necessary size of technical rooms and areas, building heights, etc. Due to unavailable drawing materiel, early guidelines are today based on rules-of-thumb and experience. But often these methods turn out to be inadequate and outdated for predictions of modern buildings because of the rapid increase in building performance requirements and high variability in early design.

In this study, we wish to develop a methodology for guidance in early building design phases. The guidance is based on thousands of prerun simulations of generic rooms. The method is planned to cover the most prevalent rooms in industrial and commercial buildings, such as offices, meeting rooms, and teaching rooms. With the use of the illustrative parallel coordinate plot, it will be possible to filter the simulations and adjust to the information level of the current design stage.

SCIENTIFIC INNOVATION AND RELEVANCE

Due to the great uncertainty in early design phases, the design process today is characterized by many redesigns and necessary savings rounds as the project progresses. We believe that a higher certainty in the early stages will require detailed software, many calculations to explore the design room, and a paradigm shift concerning the early decision-making process. This is possible with the use of generic geometry and a high level of predefined and locked parameters.

As the design process evolves, drawing material is developed and thereby enables more detailed calculations. If progressing from a simple rule-of-thumb to a detailed dynamic simulation, many aspects and parameters will be unconsidered or even misjudged. By basing the early guidance tool on several thousand detailed dynamic simulations, enabling the exploration possibility, but at the same time requiring a low level of project details, we ensure that all parameters at some point are considered by specialists.

Furthermore, we intend that the filtering in the many prerun simulations should be carried out at a collaboration meeting between the developer, architect, contractor, engineer, and other relevant parties to explore the consequences of potential design decisions and achieve a consensus of possibilities in technical layout and preliminary façade design.

PRELIMINARY RESULTS AND CONCLUSIONS

The results of a preliminary version of the method have been compared to three real case BPS to identify the sensitivity of the geometry. The three cases vary in complexity from simple to very complex geometry. The analysis shows comparable results with deviations in the magnitude of 1-12% for the outputs concerning the consumption of heating, cooling, and electricity as well as the atmospheric air quality. However, the output Risk of overheating is causing bigger deviations around 30-60% and reveals the need for further analysis regarding the necessary detail level, of both the number of inputs and the variability of inputs, compared to acceptable deviation.

The preliminary results show that simulations based on generic rooms can be a helpful means of exploring the consequences of potential design decisions and making high-quality early estimations.

MAIN REFERENCES

T. Østergård, "Thermal Comfort in Residential Buildings by the Millions - Early Design Support from Stochastic Simulations," in CLIMA 2016 - proceedings of the 12th REHVA World Congress volume 6, 2016, vol. 6.

C. Struck, "Uncertainty propagation and sensitivity analysis techniques in building performance simulation to support conceptual building and system design," Eindhoven Technical University, 2012.

S. Attia, E. Gratia, A. De Herde, and J. L. M. Hensen, "Simulation-based decision support tool for early stages of zero-energy building design," Energy Build., vol. 49, pp. 2–15, 2012.

F. Ritter, P. Geyer, and A. Borrmann, "Simulation-based Decision-making in Early Design Stages," in Proc. of the 32nd CIB W78 Conference 2015, 2015.

P. B. Purup and S. Petersen, "Requirement analysis for building performance simulation tools conformed to fit design practice," Autom. Constr., vol. 116, no. November 2019, p. 103226, 2020.

T. Østergård, R. Lund, and F. Søndergaard, "The best way to perform building simulations? One-at-a-time optimization vs . Monte Carlo sampling," vol. 208, pp. 1–13, 2020.

Keywords: Early Building Performance Simulations, Modeler dependency, statistical analyses, guidance and streamlining

Paper ID 30410: Evaluating different metrics for inter-model comparison of urban-scale building energy simulation time series

Hicham Johra* (1), Michael Mans (2), Konstantin Filonenko (3), Ina De Jaeger (4), Dirk Saelens (4), Torben Tvedebrink (1)

1: Aalborg University, Denmark; 2: RWTH Aachen University, Germany; 3: University of Southern Denmark, Denmark; 4: KU Leuven, Belgium

AIM AND APPROACH

IBPSA Project 1 aims to develop an open-source modelling framework for clusters of buildings connected to urban-scale energy networks [1]. This framework is based on sub-models' libraries developed in the Modelica language extensively applied for the design and control of smart energy grids at the district and city level. The project focusses in large part on district heating [2].

To develop such a framework, various sub-models have to be tested for different network configurations and sizes. The common bugs, issues and differences in modelling approaches from the Modelica users also have to be accounted for. Appropriate guidelines can thus be established for accurate simulation of urban-scale energy systems.

For that matter, several groups of modellers have simulated the same cases (common exercises). The challenge then results in choosing the adequate key performance indicators (KPIs) and metrics to perform the comparison of this pool of simulation results without reference case.

In this article, different KPIs and metrics found in the literature concerning the validation and comparison of building energy models are reviewed and discussed. After testing them on the pool of results from the common exercises, a selection is made for the current project and integrated in a Python-based tool.

SCIENTIFIC INNOVATION AND RELEVANCE

The quality of building energy models is usually defined by its capacity to accurately predict indoor environment state variables and power demand in the form of time series with a given time resolution. Most of the validation studies of such models are performed by comparison against a real system that has been monitored and described in detail.

Different graphical representations can be used to qualitatively assess the difference between the model and the reference measurements. Most of the time, the deviation is calculated formally with the normalized mean bias error (NMBE), the root mean square error (RMSE), and the coefficient of variation of root mean square error (CVRMSE), as suggested by the ASHRAE guidelines [3, 4]. The R2 is also a common comparison metric for numerical models, although its use is often misunderstood.

If deviations between time series can easily be assessed by simple statistical metrics like minimum, maximum, average, standard deviation, NMBE, or CVRMSE, other distance and shape comparison methods have been developed for classification purposes such as clustering analysis.

This study tests those different comparison methods and suggests a simple way to form reference time series for a pool of simulations when no measured reference case is available.

PRELIMINARY RESULTS AND CONCLUSIONS

To estimate the validity of simulations and detect potential mistakes from the different modellers, it is necessary to have a reference case that sets a benchmark for comparison of each model. If no real case measurement is available, it is suggested to form reference time series as the point-by-point mean average of all results from the pool of simulations. Those reference time series should be updated as the pool of simulations increases.

Once the reference time series have been generated, the deviation between the reference and each numerical model can be calculated. For the sake of simplicity (in terms of implementation in calculation tools), it is suggested to calculate 3 comparison metrics: NMBE of all data points (maximum time resolution) which represents the overall bias of the model; CVRMSE of the hourly-average data points which represents the fitting of the models but does not penalize too much small time lags; CVRMSE of the daily amplitude which how well building dynamics are modelled. Those comparison metrics are calculated for key parameters such as indoor temperature and power demand. An overall accuracy grade is then calculated from the ranking of each model in the simulation pool for each metric.

MAIN REFERENCES

[1] IBPSA Project 1: https://ibpsa.github.io/project1/

[2] Dirk Saelens, Ina De Jaeger, Felix Bünning, Michael Mans, Alessandro Maccarini, Enora Garreau, Øystein Rønneseth, Igor Sartori, Annelies Vandermeulen, Bram van der Heijde, Lieve Helsen. Towards a DESTEST: a District Energy Simulation Test Developed in IBPSA Project 1. pp. 3569-3577. 2019.

[3] ANSI/ASHRAE Standard 14-2011. Standard method of test for the evaluation of building energy analysis computer programs. American Society of Heating, Refrigerating and Air-Conditioning Engineers. 2011.

[4] ASHRAE Guideline 14-2002. Measurement of energy and demand savings. American Society of Heating, Refrigerating and Air-Conditioning Engineers. 2002.

Keywords: Numerical model validation, numerical model comparison, inter-model comparison, comparison metrics

Session F1.2: Ensuring high quality building simulations

Paper ID 30954: Effect of reduced air change rates on indoor air quality and air conditioning energy consumption in retail buildings

Konstantin Finkbeiner*, Martin Kremer, Martin Rätz, Xuchao Ying, Paul Mathis, Dirk Müller

RWTH Aachen University, Germany

AIM AND APPROACH

In the studies of [Mathis] and [Finkbeiner] the energy saving potential was investigated by measures such as reducing the air exchange rate in sales rooms and simultaneously increasing the water temperature of the refrigeration circuit. Savings of up to 35 % were proven

in the overall system.

Although the measures examined appear to be very promising, the condensation of water from moist air cannot be expected to be unproblematic. The reduction in fresh air volumes implies a reduced removal of moisture caused in shopping centers, especially by people. This can lead to an uncomfortable relative humidity level, even if the supply air humidity is controlled by an AHU.

This paper examines the effect of reduced air exchange rates on indoor humidity, assuming that only persons are the source of moisture. The minimum limit of the air exchange rate determined in [Mathis] is examined with regard to the aspect of comfortable indoor air humidity and, if necessary, further restricted.

For this purpose, the building model already presented in [Finkbeiner] is extended to include balancing of room air humidity. Moisture loads by people are implemented. In addition, the setting of the supply air humidity of the AHU is included in the analysis.

SCIENTIFIC INNOVATION AND RELEVANCE

The low-order building model of [Lauster] is validated and is used in the context of simulative investigations of the energy demand of buildings. The model was primarily developed for the investigation of thermal energy demand of city districts and shows comparatively high simulation speed. Nevertheless, the accuracy of the simulation results is satisfactory. In the context of the consideration of city districts, the heat demand is the focus of the investigations. Thus, the building model of[Lauster] is also primarily intended for the sensitive change of state of the indoor air. Condensation effects, which mainly occur when using air water systems for indoor room

conditioning, are not considered. In this paper, the building model of[Lauster] is extended to include the consideration of moist indoor air in the energy demand calculation. For this purpose, the room air node is extended by the balancing of the water content in the moist air. On the other hand, the model of the air water systems is supplemented by balancing the water content of the humid air, so that the additional energy consumption that results from condensation during indoor air cooling can be taken into account.

PRELIMINARY RESULTS AND CONCLUSIONS

The model of large retail buildings, especially shopping centres, presented by[Finkbeiner] was further calibrated on the basis of more detailed monitoring data, so that the accuracy of results could be further increased, as the simulation results shown in the attachment show. The simulated electrical demand of the monitored shopping center was primarily determined by converting the thermal demand. Looking at a period of six months, a deviation of 2% can be observed. A comparison of the simulated interior temperature with the measured one also shows a comparatively small deviation, see attachment. The water content of moist air indoors is largely determined by room air temperature. Thus, the calibrated model offers a good prerequisite for extending the air node model to include water content balancing of the moist indoor air. In addition, the model presented by[Finkbeiner] already takes into account the water content of the

outside air, the control of the supply air humidity via the AHU and the surface temperature of the air water systems.

MAIN REFERENCES

[Finkbeiner] K. Finkbeiner, P. Mathis, F. Hintz, E. Bykhovskaya, L. Engelmeyer, D. Bohne and D. Müller, "Modeling a Building Energy System for Development of Energy Efficient Systems of Shopping Centers," in Proceedings of the 15th IBPSA Conference, San Francisco, USA, 2017.

[Mathis] P. Mathis, H. Freitag, D. Hegemann, M. Schmidt and D. Müller, "New Concepts for the Ventilation of Shopping Centers: Reducing Air Change Rate, Applying Active Chilled Beams and Elevating Cold Water Supply Temperature," Indoor and Built Environment, vol. 26, no. 2, pp. 08-225, 2017.

[Lauster] M. Lauster, J. Teichmann, M. Fuchs, R. Streblow and D. Müller, "Low order thermal network models for dynamic simulations of buildings on city district scale," Building and Environment, p. 223–231, 2014.

Keywords: shopping center, energy efficiency, indoor air quality, moist air

Paper ID 30709: Metrics for evaluating envelope performance in next generation energy codes

Rasoul Asaee* (1), Adam Wills (2), Alex Ferguson (1)

1: Natural Resources Canada, Canada; 2: National Research Council, Canada

AIM AND APPROACH

A performance path, which is a common approach for compliance in building codes, requires the whole building to perform up to a minimum limit. The performance path provides an opportunity for builders to use innovative design and technologies. This study aims to create criteria to capture the contribution of the building fabric and airtightness to the energy-efficient operation of the building while excluding the effects of mechanical equipment (e.g., space heating system type and ventilation system) and associated gains (e.g., standby losses of hot water tank). The metric should find a balance between the complexity of calculation and accuracy of output. Energy codes should not be dependant on building performance simulation software, and the building designers should be able to calculate the metric using a wide range of software options. Auhtors developed multiple metrics to assess the performance of envelope design and evaluated the impact of those metrics on building design in several climate zones.

SCIENTIFIC INNOVATION AND RELEVANCE

Criteria for envelope improvement are considered desirable because these elements of a building are less frequently replaced and upgraded during the life of the building and can actively reduce the heat demand, whatever equipment is subsequently installed. Energy efficiency savings that must be achieved through envelope improvements alone (airtightness, insulation, glazing, solar design), and demonstrated through building performance modeling. The addition of a building envelope metric to the energy codes prevent compliance with the performance target by relying solely on a heat pump technology.

PRELIMINARY RESULTS AND CONCLUSIONS

This paper presents the results obtained from the impact analysis of the proposed performance target in the new building code. The authors determined whether the new requirements set forward for the envelope metric can be achieved with technologies and approaches that builders have previously used in low-energy homes. We evaluated the energy savings and capital cost impact for achieving the envelope targets in different housing types and climate zones. Results indicate that the envelope metrics prioritizes investments in insulation, air-sealing, and energy-efficient windows. This philosophy presumes that cost savings can be found by reducing heating loads, thereby permitting the use of smaller and simpler heating equipment.

MAIN REFERENCES

Government of British Columbia, "Energy Step Code - Building Beyond the Standard," 2019. [Online]. Available: https://energystepcode.ca/.

Natural Resources Canada, "Lessons learned and key findings from the ecoEII Net-Zero Demonstration and the R-2000 Net-Zero Energy Pilot," 2019.

BC Housing, "Energy Step Code 2017 Metrics Research," 2017. [Online]. Available: https://www.bchousing.org/research-centre/library/residential-design-construction/energy-step-code-2017-full-report&sortType=sortByDate.

Keywords: building envelope, energy code, performance path, energy efficiency, simulation

Paper ID 30259: Developments in simulation and evaluation of large-scale hot water tanks and pits in renewables-based district heating

Abdulrahman Dahash, Fabian Ochs*, Alice Tosatto

University of Innsbruck, Austria

AIM AND APPROACH

The efforts to phase-out the conventional fuels and achieve the decarbonization of the heating sector are notably increasing. Thus, the integration of renewable energy resources (e.g. solar energy, geothermal) found its place favorably in district heating (DH) whereby the share of renewables is gradually increasing in R-DH systems [1]. Yet, renewables integration can somehow alter the energy scheme and the operation leading to shortcomings (e.g. security of supply not met) [2]. Consequently, large-scale thermal energy storage (TES) is advantageous to bridge the gap between the renewables' availability and heat demand eliminating the mismatching.

Accordingly, it is held that a large-scale and long-term TES enables a more flexible integration of renewables in DH systems and, as a result, has benefits in terms of lower fossils consumption, higher primary energy savings and lower CO2 emissions. Yet, the high capital cost associated to STES is frequently a major downside. Together with the space availability, complex planning layout and the presence of groundwater tables, these are the set of major challenges in STES domain to be tackled among others [2]. Thus, simulation-based analyses found its place favorably in the planning phase of such large-scale systems since its planning is considered a complex process.

SCIENTIFIC INNOVATION AND RELEVANCE

To select properly the design, geometry and construction type for a large-scale TES, a number of inputs (hydro-geological factors, system characteristics, thermal losses, investment cost, etc.) are repetitively evaluated until a compromise between the technical performance and the economic feasibility is found [2]. Therefore, modeling is ideally suited to tackle the different challenges during planning to produce the optimal layout for the chosen TES.

Therefore, this work reports the development of large-scale TES numerical models in the framework of an international project entitled "Giga-Scale Thermal Energy Storage for Renewable Districts" [3] addressing the limitations, downsides and advantages based on experience in order to transfer the knowledge for other researchers and planners. Further, the work provides the major modeling parameters and their influence on TES operation and answers relevant questions such as to what extent and accuracy it is required to include these parameters. Moreover, the work compares the functionality of different techno-economic measures and addresses its applicability for large-scale TES. Such measures are energy efficiency, TES capacity efficiency, stratification and MIX numbers, thermocline thickness and rate of entropy generation.

PRELIMINARY RESULTS AND CONCLUSIONS

In the modeling process hierarchy for the TES modeling, a pre-design tool is crucial for the early TES planning to examine TES potential. Next, the details are systematically included in the model for a detailed design. Later, it is necessary to examine the functionality of the integrated TES into a specific energy system and, herein, technology integration tools are required. As a result, the technology is thoroughly evaluated and the outputs are post-processed in the evaluation level. If the planners are not yet satisfied with the results, then TES planning reaches to the optimization level whereby different players (e.g. construction type, insulation thickness) are examined.

Within "Giga-Scale Thermal Energy Storage for Renewable Districts", also co-simulations scenarios were proposed for large-scale TES. Thereby, a co-simulation scheme between a detailed design tool (i.e. COMSOL Multiphysics) and technology integration tool (i.e. Modelica/Dymola) was also tested and the results demonstrated the applicability of this approach. Yet, the feasibility of this method was limited due to the required computation efforts.

MAIN REFERENCES

[1] Ochs et al., "Techno-economic planning and construction of cost-effective large-scale hot water thermal energy storage for Renewable District heating systems," Renewable Energy, vol. 150, pp. 1165-1177, 2019.

[2] Dahash et al., "Advances in seasonal thermal energy storage for solar district heating applications: A critical review on large-scale hot-water tank and pit thermal energy storage systems," Applied Energy, vol. 239, pp. 296-315, 2019.

[3] Giga_TES, "Giga-Scale Thermal Energy Storage for Renewable Districts", https://www.gigates.at/, Flagship project (FFG, Energieforschung), 2019.

Keywords: Hot water tank, buried pit, numerical modeling and simulation, seasonal thermal energy storage, co-simulation approaches.

Paper ID 30468: Energy flexibility quantification with integrated heat pump, floor heating system, and thermal storage in a school building

Navid Morovat* (1), José Agustín Candanedo (1,2), Andreas K Athienitis (1)

1: Concordia University, Montreal, Canada; 2: CanmetENERGY, Montreal, Canada

AIM AND APPROACH

School buildings are an important part of the building stock; they also represent a sizeable portion of the total energy use in the building sector. However, HVAC systems in schools are often non-optimized in terms of energy consumption and efficiency, and therefore, the energy cost of educational buildings is considerable. Thus, satisfying the increasing requirements of environmental performance standards –along with the need to enhance energy flexibility– is a challenging task[1-4].

While simulations have often been used to investigate energy consumption patterns in schools [5], there are relatively few publications on measured energy use in schools in general [6,7] and even fewer publications on new/low-energy schools. Considering that many schools need to be built in the near future, it is imperative to address this gap in data-based research on the energy performance and flexibility of new schools.

In this paper, an iterative calibration process is performed to evaluate the quality of grey-box models and identified parameters in a thermal zone with a floor heating system. The predictions from the model are then compared to experimental data and validated models are used to further study energy performance and energy flexibility of the thermal zone with floor heating in a school.

SCIENTIFIC INNOVATION AND RELEVANCE

The case study school has been in operation since 2017, and it is a 100 % electric building. Features of the school include a high-temperature thermal storage device, local water-to-air heat pumps, and a water-to-water heat pump used for radiant floor heating. There are several sources of high-resolution data, including the BAS and dedicated electrical sub-meters.

This paper focuses on the evaluation of the different control strategies aimed at reducing peak demand and quantifying the energy flexibility potential of the thermal zone. The developed models are based on a data-driven grey-box model formulation and calibrated with measured data which is suitable for application in the context of demand-side management in Smart Grids.

Energy flexibility is an essential part of the solution to address the electrical grid's challenges such as providing a contingency reserve to be used for emergencies. To address this need, real-time energy flexibility is needed to be predicted ahead of time or calculated continuously and available at short notice (e.g. 10 minutes) over a time duration of a few hours. Using the calibrated model, this paper investigates the application of a Building Energy Flexibility Index (BEFI) to quantifying real-time energy flexibility relative to a reference as-usual energy consumption profile.

PRELIMINARY RESULTS AND CONCLUSIONS

The developed grey-box model shows acceptable accuracy for both the total and instantaneous heat demand in the building, making it suitable for the development of demand-side management control strategies in a Smart Grid context. This study also investigates the calibration of data-driven grey-box models for simulation of energy demand and energy flexibility of a thermal zone with floor heating in the school building.

This paper develops a methodology consisting of four steps: (1) modeling and calibration with experimental data, (2) flexibility characterization, (3) scenario modeling to assess the impact on electricity demand, and (4) evaluating the application of Building Energy Flexibility Index (BEFI) to quantifying real-time energy flexibility and enable interaction between building, aggregators and utilities.

MAIN REFERENCES

1. Finck, C., R. Li, and W. Zeiler, Optimal control of demand flexibility under real-time pricing for heating systems in buildings: A real-life demonstration. Applied Energy, 2020. 263: p. 114671.

2. Hu, M., et al., Price-responsive model predictive control of floor heating systems for demand response using building thermal mass. Applied Thermal Engineering, 2019. 153: p. 316-329.

3. Jensen, S.Ø., et al., IEA EBC annex 67 energy flexible buildings. Energy and Buildings, 2017. 155: p. 25-34.

4. Reynders, G., et al., Energy flexible buildings: An evaluation of definitions and quantification methodologies applied to thermal storage. Energy and Buildings, 2018. 166: p. 372-390.

5. Dasgupta, A., A. Prodromou, and D. Mumovic, Operational versus designed performance of low carbon schools in England: Bridging a credibility gap. HVAC&R Research, 2012. 18(1-2): p. 37-50.

6. Van Dronkelaar, C., et al., A review of the energy performance gap and its underlying causes in non-domestic buildings. Frontiers in Mechanical Engineering, 2016. 1: p. 17.

7. Golshan, M., H. Thoen, and W. Zeiler, Dutch sustainable schools towards energy positive. Journal of Building Engineering, 2018. 19: p. 161-171.

Keywords: School buildings, Floor heating system, Grey-box model, Energy Flexibility

Paper ID 30526: Integrating energy systems into building design with Hive: Features, user survey and comparison with Ladybug and Honeybee tools

Christoph Waibel*, Daren Thomas, Amr Elesawy, Illias Hischier, Linus Walker, Arno Schlüter

Chair of Architecture and Building Systems, ETH Zurich, Switzerland

AIM AND APPROACH

Synergistic effects between the energy demand and supply side suggest that building energy systems should be treated as integral parts in the design of buildings (Ferrara et al. 2019, Waibel et al. 2019). In architectural practice and education, however, building systems are commonly regarded as to be specified by (HVAC) engineers to fit a given architectural design. To close this gap, we present a decision support tool that makes the most relevant aspects related to energy systems easily accessible to architects at a conceptual design stage. We discuss results from qualitative studies and semi-structured interviews with architecture students on their experience with performance-driven design and on their experience with the tool. From these surveys, we can identify common challenges that arise in an increasingly complex and technology driven architectural design process.

SCIENTIFIC INNOVATION AND RELEVANCE

This study includes contributions in model development (decision support tool), as well as in empirical research (interviews and questionnaires). The decision support tool is implemented as a plug-in in Rhinoceros Grasshopper. The major innovation of the tool is its flexible software architecture that allows coupling to various third party and in-house model libraries. This framework is realized by establishing clean input-output relations with the introduction of a "distributor" component that processes all inputs into appropriate data types for the calculation models used. All results are then streamed back together and collided into a central results viewer. Thus, it leads to an easily accessible toolset with low entrance barriers, allowing an intuitive integration of energy systems into architectural design - a major difference to other existing workflows and tools. The empirical research with the architecture students is also adding novel findings to the building simulation research community, as it identifies real-world challenges that arise when integrating energy systems into architectural design processes.

PRELIMINARY RESULTS AND CONCLUSIONS

First user tests of the new energy systems integrated design tool with architecture students have shown that the accessibility of novel decision support tools significantly impact design decisions made by architects. To foster the uptake of such tools, we found that (i) providing default values (such as for simulation model settings or technology parameters) but with the option to set custom values, (ii) visual representations (e.g. showing energy systems in 3D as part of the building design), (iii) as well as immediate performance feedback (e.g. implications of architectural design elements such as form, orientation or construction on energy systems design, cost and carbon emissions) are most vital. Finally, due to the growing complexity of architectural design processes with an increasing degree of specialized domain expertise becoming relevant, it is indispensable to provide a design framework that allows flexible compilation of third party model libraries packaged into one coherent architecture-focused design workflow.

MAIN REFERENCES

M. Ferrara, F. Prunotto, A. Rolfo, E. Fabrizio, Energy demand and supply simultaneous optimization to design a nearly zero-energy house, Appl. Sci. 9 (2019) 2261, https://doi.org/10.3390/app9112261.

C. Waibel, R. Evins, J. Carmeliet, Co-simulation and optimization of building geometry and multi-energy systems: interdependencies in energy supply, energy demand and solar potentials, Appl. Energy 242 (2019) 1661–1682, https://doi.org/10.1016/j.apenergy.2019.03.177.

Keywords: Integrated Building Systems, Co-Simulation, Decision Support, Design, Grasshopper

Session F1.3: Buildings paving the way for the energy transition

Paper ID 30792: Cooling energy use reduction in private residential buildings in Egypt accounting for global warming effects.

Mohammad Abdollah Fadel Abdollah*, Rossano Scoccia, Giulia Filippini, Mario Motta

Politecnico di Milano, Italy

AIM AND APPROACH

Residential and commercial buildings are responsible for almost 50% of the total electricity consumption in Egypt. Most of the electricity production is from coal-based power plants which have significant environmental impact. This percentage is expected to exponentially increase due to the global warming effect. This work deals with the cooling energy reduction strategies, thermal comfort analysis and the usage of solar water heating system compatible with the Egyptian market and accounting for the global warming effects for a privately owned residential building in Cairo as a part of the Mediterranean Investment Facility project in Egypt (MIF) which is an initiative under UN Environment with the overall objective of scaling up clean energy technologies. A study of the Egyptian market was done to explore the best available envelope technologies. Series of dynamic simulations were executed in TRNSYS 18 and a multi criteria decision analysis was conducted to choose the best alternatives. The envelope alternatives were then tested again against the predicted future weather which was generated using statistical downscaling methods to test the continuous validity of the chosen alternatives.

SCIENTIFIC INNOVATION AND RELEVANCE

After assessing common technologies and practices in the Egyptian market. Parametric dynamic simulation campaign was launched for the envelope alternatives for the specific building including different thermal masses and insulation thicknesses and window types. Financial, Energetic, and environmental factors were taken into consideration and comparative analysis was done to assess the best alternatives. Moreover, simple measures to further reduce the cooling energy need was explored such as the usage of more efficient lighting and night ventilation. Additionally, the efficiency and the potential financial and energy savings from the solar water heating system was assessed. Using statistical downscaling methods on the global climate models, future weather files adapted to climate change were generated and the selected passive strategies were tested to assess the validity of such strategies in the future keeping the same level of occupant's thermal comfort. The size sufficiency and the variation in performance of the HVAC systems were checked to ensure the system adequacy in the future with accordance to the generated adapted weather files.

PRELIMINARY RESULTS AND CONCLUSIONS

This work led to a reduction of 41% in the cooling energy needs and CO_2 emissions, with a discounted payback period of 6 years. The cooling energy needs are expected to increase by 39% while the peak cooling loads are also expected to increase by 23% by 2080 rendering the current installed HVAC system undersized. Although the alternative chosen according to the current external conditions were still the best out of the other alternatives, the favourability of the strategies is declining with respect to the alternatives with higher insulation thickness up to 8cm of insulation thickness.

MAIN REFERENCES

1. A. A. Saleem, A. K. Abel-Rahman, A. H. H. Ali, and S. Ookawara, "An Analysis of Thermal Comfort and Energy Consumption within Public Primary Schools in Egypt," IAFOR J. Sustain. Energy Environ., vol. 3, no. 1, 2016.

2. J. F. Nebojsa Nakicenovic, Joseph Alcamo, Gerald Davis, Bert de Vries, T. K. Stuart Gaffin, Kermeth Gregory, Amulf Griibler, Tae Yong Jung, T. M. Emilio Lebre La Rovere, Laurie Michaelis, Shunsuke Mori, A. R. William Pepper, Hugh Pitcher, Lynn Price, Keywan Riahi, P. S. Hans-Holger Rogner, Alexei Sankovski, Michael Schlesinger, and Z. D. Steven Smith, Robert Swart, Sascha van Rooijen, Nadejda Victor, "Special Report on Emissions Scenarios," The Edinburgh Building, Cambridge, 2000.

3. R. A. Cox, M. Drews, C. Rode, and S. B. Nielsen, "Simple future weather files for estimating heating and cooling demand," Build. Environ., vol. 83, pp. 104–114, 2015.

4. S. E. Belcher, J. N. Hacker, and D. S. Powell, "Constructing design weather data for future climates," Build. Serv. Eng. Res. Technol., vol. 26, no. 1, pp. 49–61, 2005.

5. L. Abdallah and D. El-shennawy, "Evaluation of CO2 emissions from electricity generation in Egypt: Present Status and Projections to 2030." .

Keywords: building envelope optimization, Global warming adapted weather files, TRNSYS

Paper ID 30722: Multi-criteria optimization of an earth-air heat exchanger based system for several climates of European capital cities.

Arnaud Lapertot*, Benjamin Kadoch, Olivier Le Metayer

Aix Marseille Université, CNRS, IUSTI UMR 7343, 13453, Marseille, France

AIM AND APPROACH

In Europe, the energy consumed from the heating, air conditioning and ventilation system represents 68 % in the residential sector [1]. The European Union's strategies planned to reduce energy consumption in the building in order to fight global warming. One possible solution is to use Earth-Air Heat Exchangers (EAHE) based on renewable energies. The EAHE is a heating, ventilating and air-conditioning system using outside air flowing through underground tubes to recover energy from the soil [2]. This energy is used to preheat or cool the air in the building by ventilation. Moreover, the addition of a Heat Recovery Ventilation (HRV) [3] improves the performance of the system by recovering the energy lost by the exhaust air. However, the system requires an auxiliary system to satisfy the energy needs [4]. The energy system can also be coupled with a heat pump and photovoltaic collectors in order to satisfy the heating needs in winter and the cooling demands in summer [5]. The main objective is to determine the optimal sizing and regulation of the energy system using an optimization procedure. This one is composed of a sensitivity analysis, a multi-criteria optimization and a decision-making.

SCIENTIFIC INNOVATION AND RELEVANCE

An innovative multi-criteria optimization methodology applied to this EAHE energy system is achieved. The energy system is first dynamically modelled over a full year using several models such as the EAHE, the HRV, the bypass, the building, the heat pump and photovoltaic collectors. The bypass corresponds to the EAHE regulation which allows to have a building temperature as close as possible to the comfort temperature. The regulation operates with four modes: the insufflated air comes from the outside air, the EAHE, the HRV or the coupling EAHE/HRV. Furthermore, a sensitivity analysis, using the FAST method [6], is performed to select the most significant parameters. Then, a multi-criteria optimization study, based on the genetic algorithm NSGA-II [7], is carried out to determine the best compromises. The multiple-criteria decision making TOPSIS method [8] selects the optimal sizing and regulation. Finally, this optimization procedure can be used for different climates of European capital cities which are different in nature due to the diversity of climates in Europe (Mediterranean, oceanic, continental, ...).

PRELIMINARY RESULTS AND CONCLUSIONS

A preliminary study involving different French climates has been performed using the electrical grid to complete the heating and cooling demands. The associated results show that the system operates more frequently with the EAHE in summer and with the coupling EAHE/HRV in winter. The sensitivity analysis reveals that tube radius, tube length, burial depth, air renewal and regulation temperature are the most significant parameters. A multi-criteria optimization study illustrates that when the cost of the energy recovered and the coefficient of performance decrease, the fraction of renewable energy increases. A multiple-criteria decision making method exhibits the optimal parameter values are obtained for a large tube length and air renewal, an average tube radius and a burial depth and a small regulation temperature. The results show even though the system depends on weather conditions, it can achieve high energy performance for several climates. In addition, the energy system combined with a heat pump and photovoltaic collectors allows to avoid the use of auxiliary energy resources from the electrical network. This system satisfies the energy needs because the fraction of renewable energy is close to unity. Finally, the optimized system improves performance compared to the reference case.

MAIN REFERENCES

[1] L. Pérez-Lombard, J. Ortiz, C. Pout, A review on buildings energy consumption information, Energy and Buildings 40(2008)394–398.

[2] S. Thiers, B. Peuportier, Thermal and environmental assessment of a passive building equipped with an earth-to-air heat exchanger in France, Solar Energy 82(2008)820–831.

[3] H. Li, L. Ni, G. Liu, Z. Zhao, Y. Yao, Feasibility study on applications of an earth-air heat exchanger (EAHE) for preheating fresh air in severe cold regions, Renewable Energy 133(2018)1268–1284.

[4] M. Cuny, A. Lapertot, B. Kadoch, O. Le Métayer, Multi-criteria optimization of an earth-air heat exchanger for different french climates, Renewable Energy 157(2020)42–352.

[5] P. Poulet. R. Outbib. Energy production for dwellings by using hybrid systems based on heat pump variable input power. Applied Energy 147(2015)413-429.

[6] A. Saltelli, M. Ratto, T. Andres, F. Campolongo, J. Cariboni, D. Gatelli, M. Saisana,S. Tarantola, Global Sensitivity Analysis. The Primer, Vol. 304, John Wiley & Sons, 2008.

[7] K. Deb, A. Pratap, S. Agarwal, T. Meyarivan, A fast and elitist multi-objective genetic algorithm: NSGA-II, Evolutionary Computation 6(2002)182–197.

[8] G.H. Tzeng, J.J. Huang, Multiple attribute decision making: Methods and applications, CRC Press, Taylor and Francis Group, A Chapman & Hall Book, Boca Raton, 2011.

Keywords: Earth-air heat exchangers, Heat pump, Multi-criteria optimization, Sensitivity analysis.

Paper ID 30666: Control of a state-wide pool of hybrid heat pumps to decrease the GHG emissions of heating

Marianne Biéron* (1,2), Jérôme Le Dréau (1), Benjamin Haas (2)

1: LaSIE (UMR CNRS 7356) - La Rochelle University, La Rochelle, France; 2: Engie Lab Future Buildings and Cities, CRIGEN, ENGIE, France

AIM AND APPROACH

Decarbonizing space heating and domestic hot water production is a key to cope with climate change as these are mostly provided through fossil fuels and are energy intensive. In our work we focus on the case of France where these usages represent around 25% of the final energy consumption . Since these consumptions are spread over different energy vectors and space heating is highly seasonal, choosing the right vector at the right time is crucial to decarbonize them. Demand side management in buildings is a way to improve the integration of the fatal renewable energies into the power grid and for the interruption of demand for power when the marginal technologies for electricity production are fossil.

According to the French Low Carbon Strategy and its associated Multi Annual Energy Plan, the share of electricity for heating in the used primary energy source should grow and aerothermal heat pumps should be widely spread on the territory. In this article we will therefore consider the control of a pool of hybrid heat pumps supplied with electricity or with green gases (mixed or not with natural gas). The selection of the right energy will be driven by the GHG emissions of the electricity grid.

SCIENTIFIC INNOVATION AND RELEVANCE

In the proposed demand side management strategy the share of gas boilers to activate at each time is optimized in order to minimize the carbon intensity of both electricity and gas consumed during the entire operating time of the hybrid heat-pumps. As the reduction of the consumption of electricity during the activation of the gas boilers influences the electricity grid and thus the level of the GHG emissions, coupling the supply and the demand models is essential. In order to do this, we modeled the French electricity system using an approach inspired by the priority list methods. Due to their low complexity these methods are suitable for real time control of systems. The influence of the change in the electricity production is strongly related to the amplitude, the duration and the time of the variation of the demand, parameters that are not always considered in the marginal emission factors calculation. In this system the marginal GHG emission is the difference between the emission of the overall electric system without gas activation and with activation of a share of gas boilers.

PRELIMINARY RESULTS AND CONCLUSIONS

The developed model for simulation of a national electric system was calibrated and validated for France. The same methodology can be used for other countries. The influence of the heating demand electrification in France and the optimal share of hybrid heat pump in the country has been estimated using this model. The pattern (duration, frequency and amplitude) of the activation of the gas boilers has been estimated in order to minimize the GHG emissions of the energy systems. In our study a first scenario of 50,000 hybrid heat pumps was considered and the aggregated load of the overall pool was simulated. At a later stage the control of one heat pump at a time, at the scale of a single building and the optimal aggregation of these individual heat-pumps will be studied. Our results show a significant positive impact on the current electrical mix, this impact being sensible to the considered GHG intensity of the gas in future and to the number of installed heat-pumps.

MAIN REFERENCES

Delarue, Erik. (2009). Modeling electricity generation systems. Development and application of electricity generation optimization and simulation models, with particular focus on CO2 emissions.

Delarue, E, Cattrysse, D., & D'Haeseleer, W. (2013). Enhanced priority list unit commitment method for power systems with a high share of renewables.

J. Clauß, C. Finck, P. Vogler-Finck, et P. Beagon, "Control strategies for building energy systems to unlock demand side flexibility – A review, Building Simulation 2017", San Francisco, 7-9 August 2017.

T. Péan, J. Salom, J. Ortiz," Environmental and Economic Impact of Demand Response Strategies for Energy Flexible Buildings", 2018

C. Roux, P. Schalbart, et B. Peuportier, "Development of an electricity system model allowing dynamic and marginal approaches in LCA—tested in the French context of heating in buildings", Int. J. Life Cycle Assess., vol. 22, no 8, p. 1177-1190, August 2017.

S. Heinen et M. O'Malley, "Power system planning benefits of hybrid heating technologies", in 2015 IEEE Eindhoven PowerTech, 2015, p. 1-6.

Z. Zheng, F. Han, F. Li, J. Zhu, "Assessment of Marginal Emissions Factor in Power Systems Under Ramp-Rate Constraints". 1(4), 37–49., December 2015

Keywords: Electric system model, hybrid heat pump, Marginal emission factor, heating, green gas

Paper ID 30222: Simulation based evaluation of building integrated solar envelope systems on building level

Fabian Ochs* (1), Mara Magni (1), Martin Hauer (2), Samuel De Vries (3), Paolo Bonato (4)

1: University of Innsbruck, Austria; 2: Bartenbach, Austria; 3: TUe, The Netherlands; 4: Eurac Research, Italy

AIM AND APPROACH

"Building Integrated Solar Envelope Systems for HVAC and Lighting" were investigated within IEA SHC Task 56 by means of literature review, workshops, laboratory tests, onsite monitoring as well as dynamic building and system simulation. A solar envelope consists of elements that use and/or control solar energy and deliver renewable thermal and/or electric energy to the building HVAC system providing heating, cooling and ventilation, and/or daylighting control. For the detailed technical and economic analysis of these solar façade systems on building level, building and HVAC simulations were performed using reference buildings. Furthermore, experience from demo projects were included in the analysis.

An integral and systemic approach is required to promote solar envelopes integrated into the building's HVAC and lighting systems. Both, residential and tertiary buildings (office, school, library) as well as new-built and retrofitted ones were considered. The energy performance, primary energy savings, indoor air quality, thermal and visual comfort as well as architectural integration were investigated on component and on building level. Virtual case studies were examined using several cross-validated calculation and simulation platforms and for three European climates (cold: Stockholm, moderate: Stuttgart, warm: Rome), representing heating and cooling dominated climates as well as different solar potential.

SCIENTIFIC INNOVATION AND RELEVANCE

The European Commission requests that member states implement nZEB (EPBD) requirements applying from 2021 on, enabling buildings with very low energy demand due to efficient building envelopes and HVAC systems as well as RE integration on a cost-optimal level.

This contribution focuses on simulation-assisted evaluation of the cost-optimal level. Techno-economic analysis is performed using different building and HVAC models that were previously cross-validated. (Pre-) Design calculation tools as well as dedicated simulation platforms (TRNSYS, E+, Modellica, Matlab) were used for the virtual case studies. A sensitivity analysis as well as multi-objective optimization was performed.

The evaluation was extended to monthly primary energy factors accounting for different scenarios for the future electricity mix, i.e. with different RE shares (hydro, wind, PV). The cost-intensity (€/saved kWh) was used as a performance indicator to compare cost-optimal solution integrating high shares of RE into buildings. The comparison was extended to different European countries allowing to compare possible PE savings and additional costs to achieve these primary energy savings. For certification of buildings, simplified, fast and easy to use methods are required, contrariwise, only dynamic simulation models allow a realistic prediction of e.g. the PV self-consumption considering thermal mass and (thermal/electric) storage.

PRELIMINARY RESULTS AND CONCLUSIONS

For all virtual case studies, primary energy savings and the capitalized total annual costs were evaluated. A techno-economic analysis of different technologies including passive components (envelope, mechanical ventilation with heat recovery (MHVR), shower drain water recovery (SDWR)), active components (heat pump) and renewables (BI-ST, -PV, -PVT) was performed. Cost-optimal solutions based on suitable combinations of passive and active technologies can be identified depending on the climate and type of building i.e. residential buildings and non-residential buildings as well as application (heating, cooling, DHW, lighting, appliances). The cost efficiency of different technologies was evaluated vs. their primary energy savings. The investigated virtual cases of solar façade systems show the potential of integrating passive and active solar technology. A methodology was developed to analyse and compare different solutions. The capitalized life cycle costs per saved kWh of PE of different passive and active technologies are plotted vs. the primary energy, or primary energy savings, respectively. Based on the presented results design guidelines can be elaborated. Furthermore, the investigated numerical models can be used in future work to further foster and optimize solar façade systems.

MAIN REFERENCES

Magni, M.; Ochs, F.; Bonato, P.; D'Antoni, M.; Geisler-Moroder, D.; de Vries, S.; Loonen, R-; Maccarini, A.; Afshari, A.; Calabrese, T. (2019): Comparison of simulation results for an office building between different BES tools: the challenge of getting rid of modeller Influence and Identifying Reasons for deviations. IBPSA., Rome, 2019.

Calabrese T., Ochs F., Siegele D., Dermentzis G., Potential of covering electricity needs of a flat of a MFH with decentral compact heat pumps with PV – Simulation study for different DHW profiles and PV field sizes, EuroSun 2018, Rapperswil

D'Antoni M., Geisler-Moroder D., Bonato P., Ochs F., Magni M., de Vries S., Loonen R., Fedrizzi R., Definition of a reference office building for simulation based evaluation of solar envelope systems, EuroSun 2018, Rapperswil

Ochs, F.; Dermentzis, G. (2018): Evaluation of Efficiency and Renewable Energy Measures Considering the Future Energy Mix. In: 7th IBPC, Syracuse, 2018.

Taveres-Cachat E., Goia F. Grynning S., „Solar efficiency index of building envelopes and load matching in low energy buildings," in 7th IBPC, Syracuse, 2018.

Dermentzis G., Ochs F., Passive House with PV façade and electric heating, Low costs and minimal installation effort vs. Performance, PV potential and evaluation of PE consumption, ABS 2018, Bern.

Keywords: Techno Economic Analysis, Primary Energy, cost-optimality, HVAC, BISES

Paper ID 30420: A machine learning-based methodology for harnessing the energy flexibility potential of residential buildings

Adamantios Bampoulas* (1,2), Fabiano Pallonetto (1), Eleni Mangina (1,3), Donal P. Finn (1,2)

1: UCD Energy Institute, University College Dublin, Dublin, Ireland; 2: UCD School of Mechanical and Materials Engineering, University College Dublin, Dublin, Ireland; 3: UCD School of Computer Science, University College Dublin, Dublin, Ireland

AIM AND APPROACH

The current paper aims to develop a data-driven machine learning model to assess the energy flexibility potential of residential buildings based on day-ahead load prediction. The most informative predictors are selected based on random forests and the autocorrelation and Pearson coefficients to forecast the space heating electrical demand of the ground source heat pump (GSHP). The candidate features include weather variables (ambient temperature, solar radiation, etc.), temporal variables (minute of day, day of week, etc.), time-shifted lag values of the GSHP operation (electric load, water tank temperature, water mass flow rate), and the zone thermostat setpoint. The forecasts are produced daily by using the next-day prediction dataset as the testing set, which is based on random forests and a series cross-validation method. This approach emulates practical scenarios for day-ahead energy flexibility prediction. In the full paper, energy flexibility will be assessed from an integrated systems perspective by considering various types of machine learning models and DR events.

The synthetic datasets are generated by utilising a calibrated white-box model of a residential building. The model is developed with EnergyPlus by using a 15-min simulation time-step and comprises inter alia, a 12-kW GSHP coupled with a 0.8 m3 hot water storage tank.

SCIENTIFIC INNOVATION AND RELEVANCE

A key issue in energy flexibility assessment is the lack of a robust and practicable approach to assess the flexibility of residential buildings on a one-by-one basis [1]. The accurate quantification and characterisation of residential building flexibility potential are likely to allow electricity aggregators to evaluate a portfolio of buildings. White-box physical models developed using building simulation tools have been shown to accurately predict baseline and modulated energy consumption [2]. However, energy flexibility depends both on the building thermal dynamics and on time-variant attributes such as occupancy, weather, electrical component mix, and any active storage systems. In contrast to white-box models, data-driven models can be attuned to incoming smart meter data and adapt to these changes without manual intervention. This feature may not only enable customer-tailored model development but also ensure scalability.

Previous studies on machine learning approaches to building load forecasting have been developed for building performance evaluation [3,4], and energy management systems [5]; however, few works develop data-driven models to assess the flexibility potential of residential buildings. Preliminary results show that a priori knowledge of zone thermostat setpoints, as well as the use of the GSHP operation and zone temperature time-shifted values, shape the day-ahead power profile forecast.

PRELIMINARY RESULTS AND CONCLUSIONS

The adopted thermostatic setpoint is based on a daily average occupancy profile resulting from a UK Time Use Survey [6]. 35 candidate variables were selected, based on data that an energy management system could measure, including variables related to weather, occupancy, and the GSHP operation. By using the Pearson coefficient for one heating season it is shown that the water tank temperature (WT(t)), the water mass flow rate (MFR(t)), the zone thermostat setpoint (TH_SET(t)), and the zone temperature (ZT(t)) exhibit the highest correlation; nevertheless, only the TH_SET(t) can be used directly as a predictor. The lag terms, from the GSHP power and the WT(t), MFR(t), ZT(t) exhibiting the highest autocorrelation, are selected as candidate features and random forests are used both to assess feature importance and train the model.

Based on a parametric analysis and the daily root mean square error (RMSE), the optimal feature number is 11. A weekly simulation has shown that the average daily RMSE is 164 W and the model generalises better for weekdays (RMSE=141W) compared to weekend days (RMSE=221W). The TH_SET(t) is the most meaningful feature while the optimal feature set also comprises a series of the GSHP operation lag terms and the ambient temperature.

MAIN REFERENCES

1. G. Krishnadas and A. Kiprakis. A Machine Learning Pipeline for Demand Response Capacity Scheduling. Energies, 2020, doi: 10.3390/en13071848

2. G. Reynders, J. Diriken, D. Saelens. Generic characterization method for energy flexibility: Applied to structural thermal storage in residential buildings. Applied Energy, 2017 URL: https://doi.org/10.1016/j.apenergy.2017.04.061

3. A. Attanasio, M. Savino Piscitelli, S. Chiusano, A. Capozzoli, T. Cerquitelli, Towards an Automated, Fast and Interpretable Estimation Model of Heating Energy Demand: A Data-Driven Approach Exploiting Building Energy Certificates. Energies 2019, URL: https://doi.org/10.3390/en12071273

4. Y. Guo, J. Wang, H. Chen, G. Li, J. Liu, C. Xu, R. Huang, Y. Huang, Machine learning-based thermal response time ahead energy demand prediction for building heating systems. Applied Energy, 2018, URL: https://doi.org/10.1016/j.apenergy.2018.03.125

5. D. Zhang, S. Li, M. Sun, and Zheng O'Neill, "An Optimal and Learning-Based Demand Response and Home Energy Management System," IEEE Transactions on Smart Grid, vol. 7, no. 4, pp. 1790-1801, 2016, doi: 10.1109/TSG.2016.2552169.

6. G. Buttitta, D. P. Finn, “A high-temporal resolution residential building occupancy model to generate high-temporal resolution heating load profiles of occupancy-integrated archetypes,” Energy & Buildings, 2020, https://doi.org/10.1016/j.enbuild.2019.109577

Keywords: energy flexibility, machine learning, data-driven, residential building, smart grid

Session F1.4: Practice and industry related case studies

Paper ID 30678: Hydrogen explosion hazards mitigation in industrial lead-acid battery rooms under different ventilation conditions

Dorota Brzezińska*

Lodz University of Technology, Poland

AIM AND APPROACH

Hydrogen gas is emitted when charging most types of industrial lead-acid batteries. In the absence of an adequate ventilation system ot the battery room an explosion hazard may be created.

This paper presents full scale tests results of hydrogen emission and dispersion phenomena completed for worst case scenarios indicated on the basis of event tree analysis, which demonstrate conditions that can occur in a battery room in the event of a ventilation system breakdown. On the base of the experiments a CFD model of Fire Dynamic Simulator (NIST) was validated for such phenomena modelling and then FDS was used for confirmation that the lack of ventilation in a battery room can be the cause of an explosive atmosphere developing, and leading to a potential huge explosive hazard. It is also demonstrated that different ventilation systems provide battery rooms with varying efficiencies of hydrogen removal.

SCIENTIFIC INNOVATION AND RELEVANCE

The research presents three significant scientific innovations:

1. Provides unique results from real-scale experiments of hydrogen emission and dispersion phenomena;

2. Demonstrates that hydrogen dispersion in battery rooms is uniform in the entire room instead of its cumulation below the ceiling what was expected previously;

3. Demonstrates the most effective ventilation system which can eliminate explosive hazard in the battery rooms.

PRELIMINARY RESULTS AND CONCLUSIONS

The full-scale experiments of hydrogen dispersion in the battery room confirmed that detection and ventilation systems are the key to eliminating fire or explosive risk. It was found that distribution of hydrogen gas in battery rooms is uniform in the entire room and hydrogen wouldn't cumulate below the ceiling what was expected previously. In consequence, in the absence of efficient ventilation system the lower explosive limit would be reached in the whole room, leading to a very high explosive hazard. Different ventilation solutions were verified on the base of CFD simulations. The CFD analyses confirmed that, in non-ventilated battery rooms, a high explosive hazard could be expected, and each ventilation system can give a different level of efficiency in hydrogen removal. The most effective solution appeared to be the natural ventilation system with an exhaust opening in the ceiling of the battery room. The observations could be taken into account during preparation a new version of British standard BS EN 62485-2014.

MAIN REFERENCES

1. BS EN 62485:2014, Safety Requirements for Secondary Batteries and Battery Installations;

2. FM Global Property Loss Prevention Data Sheets 7-91, January 2012

3. Brzezinska D., „Emergency ventilation of buildings", Publishing House of the Lodz University of Technology, Lodz 2019, Poland, ISBN 978-83-7283-978-7;

4. Brzezinska D., „Ventilation System Influence on Hydrogen Explosion Hazards in Industrial Lead-Acid Battery Rooms", Energies 11 (2018) (DOI: 10.3390/en11082086);

5. Hesse H.C., Schimpe M., Kucevic, D., Jossen A., Lithium-Ion Battery Storage for the Grid - A Review of Stationary Battery Storage System Design Tailored for Applications in Modern Power Grids. Energies 2017, 10, 2107, https://doi.org/10.3390/en10122107

6. Ubertini S., Facci A. L., Andreassi L, Hybrid Hydrogen and Mechanical Distributed Energy Storage, Energies 2017, 10, 2035; doi:10.3390/en10122035

Keywords: Hydrogen, battery, ventilation, CFD modelling, explosion

Paper ID 30183: On the reliability of room acoustics simulation models: A case study of multi-purpose performance halls

Edita Vinca, Helene Teufl, Josef Lechleitner, Ardeshir Mahdavi*

Department of Building Physics and Building Ecology, TU Wien, Vienna, Austria

AIM AND APPROACH

In the last few decades, the capability of room acoustic simulation and auralization techniques has significantly increased [1]. Nonetheless, as with other areas of building performance simulation, various sources of uncertainty must be addressed. Specifically, the definition of the pertinent properties of materials and elements in the simulation model requires detail information that is frequently unavailable in the early stages of the design process. In this context, investigations that compare predicted and measured values of performance variables could provide useful information concerning the potential and limitation of room acoustics simulation tools [2, 3, 4]. The present contribution presents an instance of such an investigation. Thereby, a number of room acoustic variables were obtained via empirical measurements in four existing multi-purpose performance halls in the city of Vienna, Austria. The measured variables pertained to both transient processes in the reverberant field and the stead-state spatial distribution of sound pressure field. Independently, these halls were modelled using an existing advance room acoustics simulation tool. Simulation-based values of the aforementioned variables were compared with their empirically obtained counterparts within an iterative assessment process.

SCIENTIFIC INNOVATION AND RELEVANCE

As alluded to in the previous paragraph, the reliability of results obtained from room acoustics simulation depends on multiple factors, including the fidelity of input information concerning not only the simulated room's geometry, but also the acoustically relevant attributes of the room's constitutive physical components (e.g., enclosure elements, surface material characteristics, furnishing). Corresponding technical information is frequently unavailable in the early stages of design process, requiring the users thus to make educated assumptions. The associated uncertainty cannot be quantified in general, but give a sufficiently large number of studies covering rooms of different size, shape, layout, and materials, general tendencies such as recurrent sources of errors could be identified, thus gradually improving the fidelity of the simulation process [5, 6]. The present contribution represents a specific yet significant instance of such studies, as it pertains to a class of rooms for acoustical performance known as multi-purpose halls.

PRELIMINARY RESULTS AND CONCLUSIONS

The initial simulation results (first iteration) were obtained without the knowledge of and influence from the measurement results. Given the age of the selected existing (partly historical) buildings, no technical information was available regarding acoustically relevant material-related simulation input data. Therefore, the comparison of simulated and measured values in this case resembles the simulation deployment scenario in early stages of design. The second set of simulation runs were informed and influenced by the knowledge of the measurements. Comparison of the simulation results with measurements resembles thus in this case the simulation deployment scenario in building retrofit cases. The paper illustrates the scope of prediction errors as relevant to design support use cases based on the results of the first iteration. Moreover, the potential for measurement-supported calibration of room acoustics models (relevant to retrofit use cases) is demonstrated based on the results of the second iteration.

MAIN REFERENCES

1. Mahdavi, A. (2019): Computational modeling in architectural acoustics. Building Performance Simulation for Design and Operation - Expanded Second Edition. Hensen, J., Lamberts, R. (Ed.); Routledge, NY, ISBN: 978-1-138-39219-9; pp. 315 - 340.

2. Mahdavi, A., Kainrath, B., Orehounig, K., Lechleitner, J. (2008): Measurement and Simulation of room Acoustics Parameters in Traditional and Modern Bath Buildings. Building Simulation, 1, 3; pp. 223 - 233.

3. Mahdavi, A., Orehounig, K., Kainrath, B. (2008): Assessment and evaluation of the thermal and acoustical conditions in traditional bath buildings. Conference on Passive and Low Energy Architecture. J.O. Lewis (Ed.); Paper-Nr. 187.

4. Mahdavi, A., Lechleitner, J., JinGuk, P. (2008): Measurement and predictions of room acoustics in atria. Journal of Building Performance Simulation, 1, 2; pp. 67 - 74.

5. Jakic, N., Kiesel, K., Mahdavi, A. (2019): Exploring the Potential of Simulation Model Calibration: An Acoustical Retrofit Case Study. Applied Mechanics and Materials, 12th Envibuild - Buildings and Environment, 887; pp. 561 - 568.

6. Tugrul, A., Lechleitner, J., Schuss, M., Mahdavi, A. (2013): Simulation-Supported Room Acoustics Retrofit of Office Spaces. Building Simulation 2013 - 13th International Conference of the International Building Performance Simulation Association, Chambery, France. IBPSA (Ed.); ISBN: 978-2-7466-6294-0; pp. 339 - 344.

Keywords: room acoustics, multi-purpose halls, measurements, simulation

Paper ID 30938: Case study: the influence of model size on local room acoustic parameters in Odeon

Yannick Sluyts*, Leopold Kritly, Arnon Vandenberghe, Christ Glorieux, Monika Rychtarikova

KULeuven, Belgium

AIM AND APPROACH

The paper deals with the room acoustic modelling of a shopping centre situated in Liège, Belgium, which is approximately 400 meters long. The main hall has a long and curved shape, which is 8 metres wide at its narrowest and 18 meters at its widest. It is covered by a complex steel structure holding ETFE cushions.

The aim is to compare the simulation results for a few room acoustic parameters of the full-sized model to a medium-sized and small-sized model. The results were taken from the same area regardless of size.

SCIENTIFIC INNOVATION AND RELEVANCE

Simulating the room acoustic properties of large spaces with complex geometries is challenging, due to the additional required modelling time and heavy computational load. Understanding the relationship between the virtual model size of functional spaces and their acoustic modelling accuracy allows for efficient modelling and accurate simulation results.

PRELIMINARY RESULTS AND CONCLUSIONS

The room acoustic parameters of the space were simulated by ODEON, a program which starts from the geometry of indoor surfaces and their absorption properties and evaluates the propagation, reflection, scattering and absorption generated by a source and detected by a receiver, using the principles of ray tracing and the image source method.

A parametric study showed that for the modelling of a specific part of the long main hall of the building, reducing the model length to a relatively small size near the area of interest (and therefore reducing computation time) has a negligible impact on the result.

MAIN REFERENCES

ISO 3382-1, Acoustics – Measurement of Room Acoustic Parameters – Part 1: Performance Spaces. (2009).

ISO14257:2001, Measurement and parametric description of spatial sound distribution curves in workrooms for evaluation of their acoustical performance

Blevins, M. G., T.Buck, A., Peng, Z., & M.Wang, L. (2013). ISRA Quantifying the just noticeable difference of reverberation time with band-limited noise centered around 1000 Hz using a transformed up-down adaptive method. International Symposium on Room Acoustics, ISRA, (June 9-11).

Lamnatou, C., Moreno, A., Chemisana, D., Reitsma, F., & Clariá, F. (2018). Ethylene tetrafluoroethylene (ETFE) material: Critical issues and applications with emphasis on buildings. Renewable and Sustainable Energy Reviews. https://doi.org/10.1016/j.rser.2017.08.072

Lisa, M., Holger Rindel J., Christensen C. L., J. (2004). PREDICTING THE ACOUSTICS OF ANCIENT OPEN-AIR THEATRES: THE IMPORTANCE OF CALCULATION METHODS AND GEOMETRICAL DETAILS. Retrieved from https://www.researchgate.net/publication/237511143

Rychtáriková, M., Šimek, R., Húsenicová, J., & Chmelík, V. (2020). Prediction of noise levels in large shopping streets covered by glass and ETFE. Architectural Engineering and Design Management, 1–8.

Savioja, L., & Svensson, U. P. (2015). Overview of geometrical room acoustic modelling techniques. The Journal of the Acoustical Society of America, 138(2), 708–730.

Keywords: room acoustic simulation, size, long models, 3d models, odeon

Paper ID 30900: A new building thermal model to a better evaluation of the thermal comfort during heat waves

Adrien Toesca* (1), Damien David (1), André Kuster (1), Michel Lussault (2), Kévyn Johannes (1)

1: CETHIL, France; 2: EVS, France

AIM AND APPROACH

In the context of global warming, heat waves will be longer, more intense and more frequent. In order to study the comfort of inhabitants in buildings during heat waves, we must build appropriate thermal models. An appropriate model consists in an appropriate simulation tool, and appropriate settings of the tool parameters. The present paper describes a methodology to properly parametrize a building simulation tool to perform simulations during heatwave. The methodology is based on in-situ measurements.

The simulation tool in this paper is a chain tool with Urban Weather Generator for the simulation of the urban heat island, UrbaWind for the determination of pressure coefficients, and EnergyPlus for the simulation of the building thermal behavior.

In summer, heat transfers through openings, the behaviour of the occupants and the implantation of the building in the urban fabric, which allows us to know the solar masks as well as the effects of the urban heat island are more important than heat transfers through the wall.

The paper describes how the in-situ measurements setup was designed to provide sufficient information to enhance the model settings. It also demonstrate the adjustment of the model settings based on the measured data.

SCIENTIFIC INNOVATION AND RELEVANCE

In summer periods, the actions of the occupant have a determining influence on the thermal behavior of the building and on comfort [2]. After conducting a literature review, we found a lack in modelling the behavior of occupants in summer. So we classified their actions which are related to their thermal comfort management in three types by decreasing order of impact on the thermal behavior of the building.

Thus, our metrology includes a continuous recording of the type 1 occupants' actions, which directly modify the thermal characteristics of the building: manipulation of shutters, doors and windows [1,3,4,5]. A questionnaire is proposed to the occupants to monitor their type 2 (physical activity, usage of electrical device [1,3,4,5]) and type 3 actions (drinking, changing clothes, resting) [1,3].

The metrology also includes measurement data on each node of the simulation chain too to evaluate the relevance of the prediction of the different tool separately. Air temperature is measured outside the apartment to determine the relevance of the Urban Heat Island predictions. And output quantities of building thermal simulation software are measured [5,6]: air temperatures and humidity in different living rooms, as well as surface temperatures over high thermal mass walls.

PRELIMINARY RESULTS AND CONCLUSIONS

We focus on four cases studies, which are four apartments, representative of four distinct construction periods of Lyon (France): the Canut apartments (1850-1900), the industrial period (1900-1950), the reconstruction period (1950-1980), the contemporary period (1980-2020). The geometry and the materials of the apartment's buildings have been roughly surveyed. The surrounding urban morphology has been extracted from a 3D urban database.

With the instrumentation of the occupants' behavior, the outside temperature as well as the information on the case studies (materials, orientation...) we have a good knowledge of the input data of the tool chain which allows us to build a solid model. And with the measurement of air temperature and humidity in different living rooms, and the surface temperatures over high thermal mass walls [5,6], we have data of the real thermal behavior of building that we will use to compare with the output quantities of our model.

The measurement campaign will be held during 2020th summer. Simulation results, fed with the appropriate building occupants' behavior, will be confronted to the measurement results in order to identify and correct discrepancies.

MAIN REFERENCES

[1] Bui, R., Labat, M., Lorente, S., 2019. Impact of the occupancy scenario on the hygrothermal performance of a room. Building and Environment 160, 106178. https://doi.org/10.1016/j.buildenv.2019.106178

[2] Fabi, V., Andersen, R.V., Corgnati, S., Olesen, B.W., 2012. Occupants' window opening behaviour: A literature review of factors influencing occupant behaviour and models. Building and Environment 58, 188–198. https://doi.org/10.1016/j.buildenv.2012.07.009

[3] Hong, T., Yan, D., D'Oca, S., Chen, C., 2017. Ten questions concerning occupant behavior in buildings: The big picture. Building and Environment 114, 518–530. https://doi.org/10.1016/j.buildenv.2016.12.006

[4] Mavrogianni, A., Davies, M., Taylor, J., Chalabi, Z., Biddulph, P., Oikonomou, E., Das, P., Jones, B., 2014. The impact of occupancy patterns, occupant-controlled ventilation and shading on indoor overheating risk in domestic environments. Building and Environment 78, 183–198. https://doi.org/10.1016/j.buildenv.2014.04.008

[5] Piselli, C., Pisello, A.L., 2019. Occupant behavior long-term continuous monitoring integrated to prediction models: Impact on office building energy performance. Energy 176, 667–681. https://doi.org/10.1016/j.energy.2019.04.005

[6] Rodler, A., Guernouti, S., Musy, M., Bouyer, J., 2018. Thermal behaviour of a building in its environment: Modelling, experimentation, and comparison. Energy and Buildings 168, 19–34. https://doi.org/10.1016/j.enbuild.2018.03.008

Keywords: In-situ measurement, Thermal Comfort, Heat waves, New building thermal mode

Paper ID 30178: Assessing the effect of network order on epistemic uncertainty quantification for reduced-order grey-box energy models

Mohammad Haris Shamsi* (1,4), Usman Ali (1,3), Eleni Mangina (2), James O'Donnell (1)

1: School of Mechanical & Materials Engineering, Energy Institute, University College Dublin (UCD), Ireland; 2: School of Computer Science and Informatics, University College Dublin (UCD), Ireland; 3: Sustainable Energy Authority Of Ireland, Dublin, Ireland; 4: University of Victoria, British Columbia, Canada

AIM AND APPROACH

Building stakeholders usually rely on building energy simulation strategies to design energy-efficient buildings or retrofit existing energy-intensive buildings. As such, the need for reliable reduced-order building energy models has surfaced over past few decades. Energy simulation inputs and model parameters in such models could introduce uncertainty and hence, highly affect the accuracy and reliability of energy simulation results. Furthermore, it is often required to identify an appropriate balance between model complexity and accuracy. Increasing the reduced-order model complexity would eventually increase the epistemic uncertainty (lack of knowledge) in energy simulation results due to an associated increase in number of model parameters. Therefore, a systematic framework to create and analyze reduced-order building energy models could provide confidence in model accuracy. The objective of this study is to devise a framework to create reduced-order grey-box energy models and henceforth, quantify and analyze the effect of epistemic uncertainties through variation of network order.

SCIENTIFIC INNOVATION AND RELEVANCE

This research introduces novelty through the generalized and integrated use of model identification and quantification procedures. The identification process generalizes, simplifies and automates the generation of grey-box models; uncertainties can then be quantified in these models through the quantification procedure. Furthermore, the implemented 2D fuzzy Monte Carlo approach segregates the quantification and propagation of different uncertainty types, which is crucial as different uncertainty types can trigger different responses and the possible amalgamation of these might lead to erroneous inferences. The devised framework further enables the identification of a balance between network order, complexity and accuracy.

PRELIMINARY RESULTS AND CONCLUSIONS

This study lays out a framework to create reduced-order grey-box models. Furthermore, the framework introduces a process to quantify epistemic uncertainties and assess the effect of variation of network order on these uncertainties. The methodology implements two processes, namely, (1) a model identification process and (2) an uncertainty quantification process. Model identification uses easily available building metrics, for instance, building floor area to identify the grey -box network order. Uncertainty quantification employs a fuzzy Monte Carlo approach to quantify uncertainty types, namely, aleatory and epistemic. Following this procedure, the network order is then varied to determine whether there is an increase in epistemic uncertainty. The methodology is tested for an administrative building located on the campus of University College Dublin. A strong relationship exists between network order and model parameter uncertainty. Although increasing the network order enhances the accuracy of the model, associated epistemic uncertainty increases at the same time. This necessitates the identification of a network order that will optimize the balance between complexity and uncertainty.

MAIN REFERENCES

Tian, W., Heo, Y., de Wilde, P., Li, Z., Yan, D., Park, C. S., Feng, X., & Augenbroe, G. (2018). A review of uncertainty analysis in building energy assessment. Renewable and Sustainable Energy Reviews, 93, 285–301. https://doi.org/10.1016/J.RSER.2018.05.029

Fu, X., Sun, H., Guo, Q., Pan, Z., Xiong, W., & Wang, L. (2017). Uncertainty analysis of an integrated energy system based on information theory. Energy, 122, 649–662. https://doi.org/10.1016/j.energy.2017.01.111

de Wilde, P., & Tian, W. (2009). Identification of key factors for uncertainty in the prediction of the thermal performance of an office building under climate change. Building Simulation, 2(3), 157–174. https://doi.org/10.1007/s12273-009-9116-1

Heo, Y., Choudhary, R., & Augenbroe, G. A. (2012). Calibration of building energy models for retrofit analysis under uncertainty. Energy and Buildings, 47, 550–560. https://doi.org/10.1016/j.enbuild.2011.12.029

Rivalin, L., Stabat, P., Marchio, D., Caciolo, M., & Hopquin, F. (2018). A comparison of methods for uncertainty and sensitivity analysis applied to the energy performance of new commercial buildings. Energy and Buildings, 166, 489–504. https://doi.org/10.1016/j.enbuild.2018.02.021

Keywords: uncertainty, building energy modeling, simulation, grey-box modeling

Paper ID 30960: Applying sensitivity analysis to urban building energy models

Pamela Jane Fennell (1), Matthias Van Hove* (2), Lia Weinberg (3), George J. Bennett (1), Marc Delghust (2), Sebastian Forthuber (4), Martin Jakob (3), Lukas Kranzl (4), Érika Mata (5), Claudio Nägeli (6), Janet L. Reyna (7), Giamomo Catenazzi (3)

1: UCL; 2: UGent; 3: TEP Energy; 4: TU Wien; 5: IVL; 6: Chalmers University; 7: NREL

AIM AND APPROACH

Uncertainty Analysis (UA) and Sensitivity Analysis (SA) offer essential tools to determine the limits of inference of a model and explore the factors which have the most effect on the model outputs. However, despite a well-established body of work applying UA and SA to models of individual buildings, a review of the literature relating to energy models for larger groups of buildings undertaken by Fennell et al. (2019) highlighted very limited application at larger scales. This contribution describes the efforts undertaken by a group of research teams working with a diverse set of Urban Building Energy Models (UBEMs) to apply global sensitivity analysis methods and compare their results. Since UBEMs are a class of model defined by their output and coverage rather than their structure and inputs, they represent a diverse set of modelling approaches. Key challenges for the application of SA are identified and explored, including the influence of model form, input data types and model outputs. This study combines results from 8 different modelling teams, each using different models across a range of urban areas to explore these challenges and begin the process of developing standardised workflows for SA of UBEMs.

SCIENTIFIC INNOVATION AND RELEVANCE

UA and SA have been applied at the urban scale to reduced order models such as resistor-capacitor or quasi steady-state models (Schiefelbein et al., 2015; Zhao et al., 2016). However, where UA and SA are applied to dynamic thermal simulations this has either been at the level of the individual building archetype (e.g. Mauro et al., 2015) prior to an urban-scale analysis. Applying SA at the urban scale presents many challenges: (i) models are often multi-scale amalgams of sub-models at different scales (ii) model inputs are ill-defined, encompassing uncertainty due to natural variation, lack of knowledge and variance from typical values for a particular class of age or type making defining input distributions challenging, (iii) model inputs are highly diverse, from categorial record data, material properties to geometric data from surveys making characterising input uncertainties a complex task and (iv) temporal dynamics add complexity to many models with inputs changing over time from an initial state. Together with the computational burden associated with SA for complex models these challenges mean that applying SA to UBEMs is not simply a question of repurposing existing techniques for a new class of model.

PRELIMINARY RESULTS AND CONCLUSIONS

Initial results suggest a need for a much more detailed analysis of the forms of uncertainty encompassed in the analysis and greater clarity about how this reflects the structure of the model itself. A key example is the question of whether uncertainty in an input parameter is uncertainty related to a particular instance of a parameter or to that parameter's variation across the building stock. The selection of uncertain parameter is also a key choice which requires greater consideration, parameter which are treated as variable in one model may be embedded within the model code in another and treated as fixed as a result. The fixing such parameters may owe more to choices made early in the development of the model rather than to the nature of the parameter itself. A template for reporting how SA has been undertaken is proposed with the aim of enabling greater transparency of results and improving the ability to compare results between models and cases.

MAIN REFERENCES

Fennell, P. J., Ruyssevelt, P. A., Mata, É., & Jakob, M. (2019). A Review of the Status of Uncertainty and Sensitivity Analysis in Building-stock Energy Models. In V. Corrado & A. Gasparella (Eds.), Proceedings of Building Simulation 2019 (p. 9).

Mauro, G. M., Hamdy, M., Vanoli, G. P., Bianco, N., & Hensen, J. L. M. (2015). A new methodology for investigating the cost-optimality of energy retrofitting a building category. Energy and Buildings, 107, 456–478. https://doi.org/10.1016/j.enbuild.2015.08.044

Saltelli, A., Annoni, P., Azzini, I., Campolongo, F., Ratto, M., & Tarantola, S. (2010). Variance based sensitivity analysis of model output. Design and estimator for the total sensitivity index. Computer Physics Communications, 181(2), 259–270. https://doi.org/10.1016/j.cpc.2009.09.018

Schiefelbein, J., Amir, P. J., Diekerhof, M., & Bode, G. (2015, December). Development of a Tool Chain for Complex City District Energy System Modeling and Simulation. Building Simulation 2015, Hyderabad, India.

Zhao, F., Lee, S. H., & Augenbroe, G. (2016). Reconstructing building stock to replicate energy consumption data. Energy and Buildings, 117, 301–312. https://doi.org/10.1016/j.enbuild.2015.10.001

Keywords: UBEM, Global Sensitivity Analysis, model amalgams, input uncertainty, structural uncertainty

Paper ID 30245: Data-driven statistical modelling of real energy use in modern Flemish single-family houses: a feasibility study

Matthias Yvan C. Van Hove* (1), Mieke Deurinck (2), Wim Lameire (2), Jelle Laverge (1), Arnold Janssens (1), Marc Delghust (1)

1: Ghent University, Department of Architecture and Urban Planning, Building Physics Research Group, Belgium; 2: Flemish Energy and Climate Agency (VEKA), Brussels, Belgium

AIM AND APPROACH

How accurately do regulatory energy calculations predict the real energy use in modern Flemish single-family houses and how does their performance relate to comprehensive data-driven statistical machine learning models built from common dwelling parameters? This question was investigated by analysing 122680 single-family houses. Data on building characteristics and performance from the Flemish Energy Performance of Buildings (EPB) open data portal were complemented with extra datasets from the Flemish Energy and Climate Agency and data from Flemish energy utilities on real gas and electricity consumption and installed PV power.

SCIENTIFIC INNOVATION AND RELEVANCE

A building stock simulation study on this scale (e.g., 200000 Flemish dwellings) has never been performed before in any other research so far in Flanders. Most grey and black box modelling and calibration studies are always on individual building whereas this study is on building stock level. Therefore, all conclusions will be new, relevant and innovative.

PRELIMINARY RESULTS AND CONCLUSIONS

As the EPB and EPC datasets are being analysed at the moment, the only results and conclusions there are so far are on that part. This study is mainly executed on behalf of the Flemish Energy Agency, therefore we cannot share any definite results in this stadium, but we will of course at the upcoming paper deadline with their consent. Grey and black box modelling is about to begin in the upcoming months (after dataset analysis completion).

MAIN REFERENCES

OpenData: https://opendata.vlaanderen.be/dataset?q=aangifte&organization=vlaams-energieagentschap-vea

https://www.energiesparen.be/

Keywords: statistical modelling, linear regression, EPB data, Flanders

Session F2.2: Ensuring high quality building simulations

Paper ID 30250: CNN-based quick energy prediction model using image analysis for shape information

Manav Mahan Singh* (1), Philipp Geyer (1,2)

1: KU Leuven; 2: TU Berlin

AIM AND APPROACH

An early stage of design requires a few hundreds of simulations to support informed decision-making for energy-efficient building design (Clarke & Hensen, 2015; Lee et al., 2016). However, high computational time and re-modelling efforts required by dynamic energy simulation tools limit their use at the early design stage (Geyer & Schlüter, 2014). It requires quick energy prediction approach integrated into the design process. Previously, other researchers have proposed use of data-driven approaches such as deep learning (DL) model and their integration into building information modelling (BIM) tool (Amasyali et al., 2018). A designer experiments with building shape, envelope and technical specifications at an early design stage (Picco et al., 2014). Thus, we aim to develop a DL model which predicts the energy demand for various combinations of building shapes and other design parameters. Since there is a lot of information which is uncertain at the early design stage, we aim to develop BIM-integrated tool to make probabilistic energy prediction based on the range of undecided design parameters (Singh & Geyer, 2019). The tool should be able to evaluate several design options for energy performance within a few seconds to make use of energy prediction results at the early design stage.

SCIENTIFIC INNOVATION AND RELEVANCE

The building shape evolves continuously at early stages and representing it with a fixed set of parameters is cumbersome. However, we can represent a building having the same floor plan on each floor, by a two-dimensional image. We are using a convolutional neural network (CNN) to capture the information related to the building shape. Besides the building shape information, there is a set of technical specifications and other parameters which are adequately represented by numeric values. Thus, we need a model architecture which utilises both the information of building shape and other design parameters. We propose a CNN-DL model architecture, in which CNN captures the information from image (floor plan) and DL captures the information about technical parameters. Both models will be merged to get CNN-DL model, which will be trained to make quick energy predictions. The developed model will be useful for predicting energy demand of different building configurations within a few seconds. This reduction in time will make the energy prediction information available quickly as expected to make an informed decision at the early design stage.

PRELIMINARY RESULTS AND CONCLUSIONS

The approach is tested on a medium-size office building in Munich. We developed the EnergyPlus model and calibrated against the real energy consumption of a three-storey office building. Then, it is extended to predict the energy demand of various shapes and combinations of design parameters. We generated the training dataset of 5000 samples and test dataset of 1000 samples. Building samples in both training and test dataset have random rectilinear floor plans. We trained CNN-DL model for several combination of hyper-paramters and the best model based on the least validation loss is selected. The model shows an accuracy of 0.983 in terms of R2 and root mean square error (RMSE) of 7 MWh/a.

We integrated these energy prediction model into a BIM tool to make energy prediction under uncertainty. The developed Approach makes probabilistic energy prediction five design options with 500 samples within a minute with the accuracy. Thus, in this research, we developed an approach to make use of energy prediction results to steer the design decisions at the early design stage.

MAIN REFERENCES

Amasyali, K., Gohary, N., & El-Gohary, N. M. (2018). A review of data-driven building energy consumption prediction studies. Renewable and Sustainable Energy Reviews, 81, 1192–1205. https://doi.org/10.1016/j.rser.2017.04.095

Clarke, J., & Hensen, J. (2015). Integrated building performance simulation: Progress, prospects and requirements. Building and Environment, 91, 294–306. https://doi.org/10.1016/j.buildenv.2015.04.002

Geyer, P., & Schlüter, A. (2014). Automated metamodel generation for Design Space Exploration and decision-making – A novel method supporting performance-oriented building design and retrofitting. Applied Energy, 119, 537–556. https://doi.org/10.1016/j.apenergy.2013.12.064

Lee, B., Pourmousavian, N., & Hensen, J. L. M. (2016). Full-factorial design space exploration approach for multi-criteria decision making of the design of industrial halls. Energy and Buildings, 117, 352–361. https://doi.org/10.1016/j.enbuild.2015.09.028

Picco, M., Lollini, R., & Marengo, M. (2014). Towards energy performance evaluation in early stage building design: A simplification methodology for commercial building models. Energy and Buildings, 76, 497–505. https://doi.org/10.1016/j.enbuild.2014.03.016

Singh, M. M., & Geyer, P. (2019). Statistical decision assistance for determining energy-efficient options in building design under uncertainty. In P. Geyer, K. Allacker, M. Schevenels, F. De Troyer, & Pieter Pauwels (Eds.), 26th International Workshop on Intelligent Computing in Engineering. http://ceur-ws.org/Vol-2394/paper08.pdf

Keywords: Deep Learning, Early Design Stage, Energy Simulation, Convolutional Neural Network, Image Analysis

Paper ID 30153: Application of MyBEM, a BIM to BEM platform, to a building renovation concept with solar harvesting technologies

Mathias Bouquerel* (1), Kevin Ruben Deutz (1), Benoît Charrier (1), Thierry Duforestel (1), Mickael Rousset (1), Bart Erich (2), Gerrit-Jan van Riessen (3), Thomas Braun (4)

1: EDF R&D, Moret-Loing-et-Orvanne, France; 2: TNO, Eindhoven, Netherlands; 3: Emergo, Netherlands; 4: Pilkington Germany, Gelsenkirchen, Germany

AIM AND APPROACH

MyBEM is an innovative and modular platform designed to generate and simulate a Building Energy Model (BEM) from a Building Information Model (BIM), through an automated workflow. The platform is partly the result of the MERUBBI ANR project [1]. Building energy modelling is based on the Modelica library BuildSysPro [2].

The platform has been used to assess the energy performance of a case study within the ENVISION H2020 project [3][4]. The Envision project aims to develop and demonstrate a building renovation concept, which integrates solar harvesting technologies on the whole building envelope, including opaque and glazed parts of the vertical façades.

This case study required the development of specific models for solar harvesting elements (opaque solar collectors and PV windows), and their integration in the building energy model. The simulation of both the energy demand and the energy production can be simulated, to assess how local production can meet the energy needs.

SCIENTIFIC INNOVATION AND RELEVANCE

The design of MyBEM is modular, in order to allow as much flexibility as possible [5]. The platform functionalities are handled by several independent modules, which communicate through file exchange. Open standards are used as much as possible for the purpose of interoperability.

A pre-processing module is able to import geometrical and construction data from various sources, and to produce a gbXML file – an extraction of BIM files dedicated to building energy modeling. This module is also able to run annual solar calculations through efficient ray tracing algorithms, so that for each façade element, solar masking and solar reflections are accurately taken into account [6].

A second module generates a building energy model in Modelica from the previously exported gbXML file. The high versatility of the Modelica library BuildSysPro is used to adapt the model to the specific needs of the Envision project study case. Indeed, Modelica models for solar harvesting technologies have been developed and integrated in the model generation algorithm.

The Modelica model is then simulated under Dymola. Post-processing routines are used to extract energy and bioclimatic indicators from the simulation.

PRELIMINARY RESULTS AND CONCLUSIONS

The MyBEM platform has been applied to a demonstration building of the Envision project. It is a 3-storey residential building with 24 dwellings, located in the Netherlands, and built in the 1950s. The renovation configuration integrates:

- Insulation of the façades, roof and lower floor

- Change of windows

- Closure of balconies

- Integration of solar collectors on the vertical façades

- Integration of PV windows on the staircase

From a SketchUp geometric file and additional technical data, a gbXML file has been automatically generated within the pre-processing module. From this gbXML, the Modelica model has also been automatically generated.

Then simulations with Dymola have been carried out to calculate the energy needs and the energy production through solar harvesting. The coupling of both the space heating system and the solar harvesting system (based on solar collectors) is used to assess the self-consumption potential, the remaining energy needs that require an additional energy source and the unused harvested energy that can be injected in the energy networks (electrical grid or district heating network).

MAIN REFERENCES

[1] Mathieu Schumann et al. Interdisciplinarity around design tools for new buildings and districts: the ANR MERUBBI project. Proceedings of 33rd PLEA International Conference, pages 2148-2155, Edinburgh, UK, July 2017.

[2] Gilles Plessis, Aurélie Kaemmerlen, Amy Linday. BuildSysPro: a Modelica library for modelling buildings and energy systems. Proceedings of Modelica Conference 2014, Lund, Sweden, March 2014

[3] ENVISION Project description : https://cordis.europa.eu/project/id/767180/fr

[4] Bart Erich et al. Energy harvesting by invisible solar façade collector. 14th Conference on Advanced Building Skins, Bern, Switzerland, Oct. 2019.

[5] Mathias Bouquerel, Sébastien Bermes, Adrien Brun, Hassan Bouia, Régis Lecussan, Benoît Charrier. Building Energy Modeling at District Scale through BIM Based Automatic Model Generation – Towards Building Envelope Optimization. Proceedings of Building Simulation 2019 Conference, Roma, Italy, Sept. 2019.

[6] Clément Ribault, Mathias Bouquerel, Adrien Brun, Mathieu Schumann, Gilles Rusaouen, Etienne Wurtz. Assessing tools relevance for energy simulation at the urban scale: towards decision-support tools for urban design and densification. Energy Procedia 122:871-876, September 2017

Keywords: BIM, BEM, renovation, sola

Paper ID 30668: Optimal energy system sizing with independent load and weather time series.

François Lédée* (1,2), Gaëlle Faure (1,2), Curran Crawford (1,3), Ralph Evins (1,2)

1: Institute for Integrated Energy Systems, University of Victoria, British Columbia, Canada; 2: Energy in Cities group, Department of Civil Engineering, University of Victoria, British Columbia, Canada; 3: Sustainable Systems Design Lab, Department of Mechanical Engineering, University of Victoria, British Columbia

AIM AND APPROACH

To size an energy system with stochastic programing, hourly energy demand and weather variables over multiple years are required. Unlike weather data, hourly building's energy demand over multiple years is usually not available. Therefore, some authors [1,2] highlight the possible use of stochastic generators. These algorithms do not jointly generate load and weather data [1,2,3], resulting in a loss of their natural correlation, possibly affecting further applications.

This study aims at assessing if an energy system sizing ran with an energy hub, an optimization framework for multi-source energy systems [4], can be conducted with independently stochastically generated time series. In this aim, we investigate if the proposed sizing for a reference simulation, where the building electric load and the solar data used are issued from the same year, differs significantly from other simulations, where the years of employed data differ. If so, we would conclude that jointly generated time series are required for multi-source energy system sizing.

We focus on the sizing of energy systems comprising a PV installation, for residential buildings in different climate zones. For each building, one year of hourly electrical load is used along with 20 different years of hourly solar data of the same location.

SCIENTIFIC INNOVATION AND RELEVANCE

Optimization-based energy models is a growing research area aiming at supporting strategic energy planing at various scales and for different time horizons. These models require the use of time series, commonly with an hourly granularity, for example for the energy demand or the resource's availability.

Most of the commonly used models do not consider uncertainty and rely whether on past data or forecasts, often inaccurate [5]. To deal with optimization under uncertainty, two main approaches are considered: the use of robust optimization or stochastic programming. While the first "aims to find a solution with the best worst performance" [6], the second "optimizes the expected value of the objective over all possible realizations" [5].

In a multi-source energy model, the time series represent the most important parameters to consider. In an energy system, the demand and the weather conditions are often related. The current work proposes to investigate the importance of this relation in the optimization of an energy system, to see if its optimal sizing through stochastic programming can be based on the use of independent weather and building's load stochastic generators.

PRELIMINARY RESULTS AND CONCLUSIONS

Our first results regarding a residential building in Austin (Texas, USA) reveal a undersizing of the solar facilities of 10% in average with the use of random years of sun availability, compared to the reference simulation. The sizing of other energy facilities varies between -6% and +10% and expected costs are slightly overestimated (+1.5% in average).

To assess if these differences are significantly acceptable, two analysis approaches are adopted. The first one is based on statistical tests. For a p-value of 5%, the results for the first simulated case reveals no significant difference between the reference case and other simulations for the cost and all facility size, except for the installed capacity of photovoltaic modules.

A second analysis is based on geometric approaches. Both a principal component analysis and a clustering-based analysis fail at rejecting the hypothesis of a significant overall difference between the reference simulation and other optimization sizing processes.

We will consolidate these preliminary results by multiplying the cases of study in various climate zones and for buildings with different characteristics. This would help drawing appropriates conclusions regarding the influence of the use of independent time series on the sizing of multi-source energy systems.

MAIN REFERENCES

[1] Sun et al. (2019), Using Synthetic Traces for Robust Energy System Sizing, e-Energy 19, pp. 251-262.

[2] Sharafi M, ElMekkawy TY. Stochastic optimization of hybrid renewable energy systems using sampling average method. Renewable and Sustainable Energy Reviews. 1 déc 2015;52:1668-79.

[3] Patidar S, Jenkins DP, Peacock A, McCallum P. Time Series decomposition for simulating electricity demand profile. In: Proceedings of Building Simulation 2019: 16th Conference of IBPSA. Rome (Italy); 2019.

[4] Evins et al. (2014), New formulations of the 'energy hub' model to address operational constraints. Energy 73, pp. 387–398.

[5] Moret et al (2020), Decision support for strategic energy planning: A robust optimization framework. European Journal of Operational Research 280, pp. 539-554.

[6] Moazeni et al. (2019), A Risk-Averse Stochastic Dynamic Programming Approach to Energy Hub Optimal Dispatch. IEEE Transactions on power systems 34 (3), pp. 2169-2178.

Keywords: Energy system sizing, Stochasticity, Solar, Energy hub

Paper ID 30215: Evaluating the Performance of different Window Opening Styles for single-sided buoyancy-driven natural Ventilation using CFD Simulations

Akshit Gupta*, Annamaria Belleri, Francesco Babich

Institute for Renewable Energy, Eurac Research, 39100 Bolzano, Italy

AIM AND APPROACH

The aim of this research is to investigate the effectiveness of different types of windows such side-hung, top-hung, and horizontal pivot for natural and mixed-mode ventilation. This paper will focus on the CFD analysis that was performed as a preliminary step to better define the windows' prototypes before being tested in full-scale experimental facilities.

In this study, both wind- and buoyancy-driven ventilation was considered. For both cases, the reference room included in the CFD model was one of the chambers (4m x 8m x 3m, width x depth x height) of the lab that will be later used for the tests. However, different levels of simplifications and different ways to model the boundary conditions were tests.

Firstly, the use of an external air domain was investigated to evaluate whether this was required or not, what would be its optimal size and its most appropriate boundary conditions. Secondly, different levels of geometrical simplification of the opening were compared. Thirdly, transient and steady-state simulations were performed.

For this CFD analysis, the preliminary validation of the models was based on similar previously published studies. A further and more comprehensive validation will be made after the completion of the tests with the prototypes.

SCIENTIFIC INNOVATION AND RELEVANCE

Natural and mixed-mode ventilation (i.e. combination of natural and mechanical ventilation) are effective means to provide comfortable indoor environments (such as thermal comfort and good indoor air quality) while minimizing the energy consumption [1,2,3,4].

However, the use of different types of windows and control strategies usually lead to different indoor thermal conditions [1,5]. CFD is a powerful modelling technique to compare the air distribution within a room for varying scenarios. The boundary conditions and modelling assumptions should be carefully evaluated to ensure that the differences among the simulated scenarios are due to variations in the actual systems, and not due to other aspects such as oversimplification of the geometry or driving forces [2,4]. For instance, wind pressure coefficients are often given on the plane of the wall, but the same wind is likely to have a different indoor effect depending on the type of opening and its position (e.g. outer edge, inner edge or in the middle of a wall) [1,2,4,6].

Hence, this work aims to make a step forward by showing the effect of using different modelling assumptions and opening geometrical simplifications, in terms of results' accuracy and required computational power.

PRELIMINARY RESULTS AND CONCLUSIONS

In all cases, initial results indicate that, the use of an external air domain provide better results when comparing the CFD predictions with published similar studies [1,3,4].

For buoyancy-driven flows, the size of the external domain may be limited, while its boundary-conditions play an essential role. The most realistic results were achieved with an external domain as big as the testing chamber. The boundary conditions of the external domain with the top surface as opening and all the others as No-slip were most reasonable.

For wind-driven flows, the complexity increases depending on the wind direction and speed value, and on the position of the simulated room within the buildings (e.g. ground floor, mid floor, on an edge, etc.) [3,7].

Once the initial conditions are correctly defined, transient simulations usually provide the most accurate results as they can better capture the unsteady and cyclic nature of these flows, which is especially true for buoyancy-driven flows.

Lastly, the oversimplification of the geometry of walls and windows (e.g. overlooking the wall thickness, or the way in which the opened and closed sections of a window are modelled) is likely to cause unreliable results, and therefore to question the validity of the CFD analysis.

MAIN REFERENCES

1. Wang, J., Wang, S., Zhang, T., & Battaglia, F. (2017). Assessment of single-sided natural ventilation driven by buoyancy forces through variable window configurations. Energy and buildings, 139, 762-779.

2. Cook, M. J., Ji, Y., & Hunt, G. R. (2003). CFD modelling of natural ventilation: combined wind and buoyancy forces. International Journal of Ventilation, 1(3), 169-179.

3. Gan, G. (2010). Simulation of buoyancy-driven natural ventilation of buildings—Impact of computational domain. Energy and Buildings, 42(8), 1290-1300.

4. Allocca, C., Chen, Q., & Glicksman, L. R. (2003). Design analysis of single-sided natural ventilation. Energy and buildings, 35(8), 785-795.

5. von Grabe, J., Svoboda, P., & Bäumler, A. (2014). Window ventilation efficiency in the case of buoyancy ventilation. Energy and Buildings, 72, 203-211.

6. Favarolo, P. A., & Manz, H. (2005). Temperature-driven single-sided ventilation through a large rectangular opening. Building and Environment, 40(5), 689-699.

7. Bangalee, M. Z. I., Lin, S. Y., & Miau, J. J. (2012). Wind driven natural ventilation through multiple windows of a building: A computational approach. Energy and Buildings, 45, 317-325.

Keywords: windows, CFD modelling, natural ventilation, geometrical simplification, external domain

Paper ID 30333: Beyond normal: Guidelines on how to identify suitable model input distributions for building performance analysis

Giorgos Petrou* (1), Anna Mavrogianni (2), Phil Symonds (2), Mike Davies (2)

1: UCL Energy Institute, United Kingdom; 2: UCL Institute for Environmental Design and Engineering, United Kingdom

AIM AND APPROACH

This work presents a step-by-step guide on how to identify the probability distribution function that best describes a given dataset of building-related parameters. We demonstrate the process for a set of wall U-value measurements. Firstly, histograms and cumulative distribution plots are used to visualise the data to establish whether the data appears to be normally distributed. Second, data cleaning is performed through observation of histogram extremes and removal of outliers. This is preferred over automated procedures based on the data's interquartile range or standard deviation when the data does not appear to be normally distributed. Third, a set of candidate distributions are selected using the data's empirical distribution and the 'Cullen and Frey' graph of kurtosis and square of skewness. Next, the candidate distributions are then fitted to the data using Maximum Likelihood Estimation – this is easily achieved with the R package fitdistrplus [1]. Finally, drawing from Information Theory, the Akaike Information Criterion (AIC) and its derivates are used to identify the best fitting distribution [2]. Density plots, Q-Q plots, P-P plots and Cumulative Distribution Function plots provide a supplementary measure of goodness-of-fit and inform the modeller whether the best-fitting distribution is satisfactory.

SCIENTIFIC INNOVATION AND RELEVANCE

The importance of uncertainty propagation and model calibration in the built environment is widely recognised and distributions are an integral part of this process [3]. The normal distribution is commonly assumed in building performance simulation due its convenience and familiarity. Similarly, the uniform distribution is often used to express lack of knowledge about the possible value or distributional form of a model input. However, with data availability on the rise, distributions used for uncertainty quantification could in some cases be based on empirical evidence. If the modeller identifies the distribution that best describes the observed data relating to a model input, they can capture its expected value and shape more accurately. Alternative, the use of inappropriate distributions could contribute to the 'performance gap'. Examples of using non-normal and non-uniform distributions exist within the field of building modelling [4, 5], however, no clear guidance on how to identify the most suitable distributions for a given dataset could be found. Therefore, by providing a detailed, step-by-step guidance of this process using code snippets in R, this paper aims to enable building performance modellers to make the best use of available data, potentially improving their modelling workflow and the accuracy of their predictions.

PRELIMINARY RESULTS AND CONCLUSIONS

We demonstrate the step-by-step process using open source wall U-value data from English homes [6]. The data cleaning process did not reveal any outliers. Based on the empirical distribution and the Cullen and Frey graph, the candidate distributions were chosen to be the normal, Weibull, gamma and lognormal. The AIC was lowest for the gamma distribution with a value of -16.1. The difference in AIC between the best fitting distribution and the normal was 8, suggesting fairly weak support for the normal distribution to be a plausible alternative to the gamma [2]. By estimating the Akaike Weights, the probability that the gamma best describes the data amongst the candidate distributions was 0.75 while the normal had a probability of 0.01. As AIC is a relative goodness-of-fit measure, Q-Q plots, P-P plots and Cumulative Distribution Function plots were used to confirm that the gamma provides both the best fit amongst the candidate distributions and a sufficiently good fit for the given data and its intended use. To conclude, this paper shows that an alternative distribution better represents wall U-values in English homes than the more popular normal distribution. The steps taken could be followed by others to improve their modelling assumptions.

MAIN REFERENCES

[1] Delignette-Muller ML, Dutang C. fitdistrplus: An R Package for Fitting Distributions. J Stat Softw 2015; 64: 1–34.

[2] Burnham KP, Anderson DR. Multimodel Inference: Understanding AIC and BIC in Model Selection. Sociol Methods Res 2004; 33: 261–304.

[3] Tian W, Heo Y, de Wilde P, et al. A review of uncertainty analysis in building energy assessment. Renew Sustain Energy Rev 2018; 93: 285–301.

[4] Kristensen MH, Choudhary R, Pedersen RH, et al. Bayesian Calibration Of Residential Building Clusters Using A Single Geometric Building Representation. 2017; 10.

[5] Booth AT, Choudhary R, Spiegelhalter DJ. Handling uncertainty in housing stock models. Build Environ 2012; 48: 35–47.

[6] Hulme J, Doran S. In-situ measurements of wall U-values in English housing. 290–102, Building Research Establishment (BRE) on behalf of the Department of Energy and Climate Change (DECC), 2014.

Keywords: Uncertainty, Distribution Fitting, Building Performance Analysis, Building Simulation, Akaike Information Criterion

Session F2.3: Buildings paving the way for the energy transition

Paper ID 30142: Modeling and simulation of the Ukrainian residential building stock : application to the evaluation of gas demand

Tarik Bouhal (1), Valentin Gavan* (2)

1: Energeo - LR Technologies Groupe, 35 rue des chantiers, 78000 Versailles, France; 2: ENGIE Lab CRIGEN, 4, Street Josephine Baker, 93240 Stains, France

AIM AND APPROACH

The present study focuses on the modeling and simulation of the Ukrainian residential building stock. The aim of this research is to evaluate the gas demand for different end-uses such as space heating, domestic hot water and cooking, at a local, regional and national scale. The building stock model is developed using a Modelica based low-order building model. A bottom-up approach is then applied to obtain and parameterize the building stock model by using data from several Ukrainian organisms, such as the State Statistics Service of Ukraine.

The building stock model is able to run for a whole year, by taking into account when to operate in heating or non-heating modes, and with the possibility of combining space as space heating, domestic hot water and cooking. Various occupancy profiles were defined and different building archetypes were considered (individual and collective buildings) with several thermal performance levels. Four different regions and climate zones were considered, based on the Ukrainian climate, and the results are presented for all cases with regards to the different local characteristics.

SCIENTIFIC INNOVATION AND RELEVANCE

The major scientific innovation of the study is represented by the bottom-up model development of the Ukrainian building stock. Using census data and local architectural information, building archetypes were identified for different construction periods. From these archetypes, energy models with physical parameters such as building envelope performance and glazing ratio, or utilization parameters such as heating setpoints or domestic hot water draw function of the occupancy, were developed. Thus, hourly gas demand (space heating, domestic hot water and cooking) was obtained by dynamically simulating the building stock. The developed modes were validated at building level using field data from meters from 200 end-users.

Therefore, the current work aims to derive a knowledgeable data base for the energy performance parameters concerning the use of gas in Ukraine. It provides useful guidelines about the residential gas demand and the consumption under Ukrainian conditions (technical and economic). The combined effects of building archetype, consumer profile and climatic conditions on the evolution of gas demand were carried out. The novelty of the present work consists of conducting not only an energy assessment but also providing insights for the development of a roadmap intended to improve the national gas consumption metering and pricing system.

PRELIMINARY RESULTS AND CONCLUSIONS

Thanks to these studies, we draw the gas demand pattern for heating, domestic hot water and cooking. Findings show that the gas demand depends mainly on the building type, consumer behaviour and climate. The developed numerical models developed in this work can serve as a tool to assess and improve the performance of the buildings and thus evaluate structural reforms from a technical and economic point of view, destinated to decrease the gas demand of the Ukrainian building stock. The applicability of model remains valid for other countries for improving the performance of potential buildings and their renovation because, in addition to technical parameters related to building archetypes, other aspects are taken into account in the model such as building envelope, energy systems and ambient conditions.

MAIN REFERENCES

[1] Bayramov A, Marusyk Y. Ukraine's unfinished natural gas and electricity reforms: one step forward, two steps back. Eurasian Geography and Economics 2019;60(1):73-96.

[2] Energy efficiency requirements in building codes, energy efficiency policies for new buildings. IEA information paper.

[3] Gavan V., M. A. (7-9 August 2017.). Development of a district modeling approach for buildings using 3DEXPERIENCE and Dymola/Modelica environments. Building Simulation 2017. San Francisco.

[4] Bulych Y. Spatial planning and urban development. Country Profiles on Housing and Land Management: Ukraine, United Nations, New York and Geneva 2013;:35-48.

[5] Spoladore A, Borelli D, Devia F, Mora F, Schenone C. Model for forecasting residential heat demand based on natural gas consumption and energy performance indicators. Applied Energy 2016;182:488-99.

[6] Bruckner T, Blavdzevitch A, Groscurth HM, Kummel R. Rational use of energy in the Ukraine: a pilot study for desna. Energy 1994;19(4):489-97.

[7] Cousin J., Leo J., Duchayne C., Gavan V. Development and validation of a low-order thermal model for building behavior in Modelica. Building Simulation 2019, Rome, 2-4 September 2019.

Keywords: Simulation, Residential gas consumption, Consumer profiling, Prediction, Ukraine.

Paper ID 30616: Application of optimisation, building energy simulation and life cycle assessment to the design of an urban building construction project.

Aurore Wurtz*, Patrick Schalbart, Bruno Peuportier

Mines ParisTech, France

AIM AND APPROACH

Building is the E.U. economic sector with highest final energy consumption. In response to concerns about climate change, energy security and social equity, countries around the world plan to substantially reduce energy demand and greenhouse gas emissions. Moreover, 55% of the world's population lives in urban areas, a proportion that is expected to rise to 68% by 2050. For these reasons, neighbourhood environmental impacts have to be minimised with reliable tools offering economically viable solutions.

SCIENTIFIC INNOVATION AND RELEVANCE

This paper presents a multicriteria optimisation methodology developed in order to reduce both investment and environmental costs of buildings. The life cycle assessment methodology used in this study allows to evaluate environmental impacts over the whole life cycle of buildings. Individual building optimisation and simultaneous optimisation of several buildings were carried out considering various criteria such as environmental impacts and construction cost, in the case of a 5 hectare urban project.

A genetic algorithm was implemented in order to identify Pareto-optimal solutions associating a life cycle assessment tool and building energy simulation. NSGA-II algorithm was implemented in a first step and a comparison with NSGA-III algorithm is considered in perspective.

PRELIMINARY RESULTS AND CONCLUSIONS

Results present optimal Pareto fronts minimising investment costs and CO_2 emissions of buildings. In this way, a decision-making aid is proposed to urban projects planners.

MAIN REFERENCES

BRANKE, J.,DEB, K.,MIETTINEN, K. et SLOWINSKI, R., 2008. Multiobjective optimization: Interactive and evolutionary approaches. In : Springer Science & Business Media.

COSTA-CARRAPIÇO, I., RASLAN, R. et GONZÁLEZ, J. N., 2020. A systematic review of genetic algorithm-based multi-objective optimisation for building retrofitting strategies towards energy efficiency. In : Energy and Buildings. 2020. Vol. 210.

DEB, K., 2001. Multi-Objective Optimization using Evolutionary Algorithms. S.l. : John Wiley & Sons.

DEB, K.,PRATAP, A.,AGARWAL, S. et MEYARIVAN, T., 2000. A fast and elitist multiobjective genetic algorithm: NSGA-II. In : IEEE Transactions on Evolutionary Computation. 2000. Vol. 6, n° 2, p. 182-197.

KISS, B. et SZALAY, Z., 2020. Modular approach to multi-objective environmental optimization of buildings. In : Automation in Construction. March 1, 2020. Vol. 111.

SCHWARTZ, Y., RASLAN, R. et MUMOVIC, D., 2016. Implementing multi objective genetic algorithm for life cycle carbon footprint and life cycle cost minimisation: A building refurbishment case study. In : Energy. 2016. Vol. 97, p. 58-68.

ZITZLER, E., 1999. Evolutionary algorithms for multiobjective optimization: Methods and applications. Doctoral thesis. Swiss Federal Institute of Technology of Zurich (ETH).

Keywords: Optimisation, Life cycle assessment, Building energy simulation, urban project

Paper ID 30603: Renovating Flanders: a building classification approach to assess large-scale renovation costs

Guillermo Borragán*, Dorien Aerts, Glenn Reynders, Yixiao Ma, Lukas Engelen, Stijn Verbeke

VITO-Energy Ville, Belgium

AIM AND APPROACH

As part of the strategy to improve energy efficiency and decarbonize the building stock, the Flemish government has set the target to renovate the residential stock 1 by 2050. However, an old housing stock integrated by a large number of detached buildings suggest that finding a good cost-efficiency balance is neither an easy nor an inexpensive task. Having accurate figures about the costs and benefits of renovation is essential not only to anticipate public aids but also to boost private investment. Earlier studies trying to describe building typologies (e.g. IEE-Tabula2) focused on energy use profiles rather than on renovation potential. The recent availability of new (big)(open) data (e.g. GIS, consumption data, heat maps...) enables the development of machine learning classification techniques to create more accurate building set representations. The purpose of the present study is to develop a massive classification approach to identify the type of renovation plan and the associated costs for the different building typologies in Flanders to comply with the 2050 target.

SCIENTIFIC INNOVATION AND RELEVANCE

With more than 9 out of 10 buildings in Flanders requiring substantial renovations during the next three decades3, there is a growing need to develop automatic cost-effective approaches that provide home owners and tenants with comprehensive renovation advice. The approach undertaken in this study goes a step forward and generates specific renovation plans for all the building typologies identified at the level of Flanders. Our bottom-up classification algorithm based on K-prototypes uses both numerical and categorical publicly available statistical and geometrical information to categorize the different buildings according to the type of intervention needed. Then, the renovation costs are estimated using the EBECS tool, an in-use software developed at Energy Ville that uses real energy consumption to calculate renovation needs and costs. In contrast to previous classification schemes, the typologies generated in this work are not subjected to any predefined limitation, providing thus a more accurate profiling. The methodology is also very interesting from a timing point of view as it is coincident with the European renovation wave and the new directives on the Flemish Renovation Pact. Besides, it proposes a new approach to update the first classification of reference building types in Europe (IEE-Tabula) developed already 9 years ago.

PRELIMINARY RESULTS AND CONCLUSIONS

Although still in a preliminary phase, our firsts simulations confirmed that the selected variables (including construction year, number of occupants, building type and building physics parameters such as u-values and WWR) were sufficient to provide decent classification results. Besides, results disclosed the existence of comparable renovation needs for buildings located closely. This fact makes sense given that the construction year is often a relevant variable in our models. Nevertheless, it also advises the relevancy of focusing the renovation efforts on buildings clusters4 rather than individual buildings. This is interesting since large/medium-scale renovation plans might bring reduced renovation costs. On top of this, we find also significant differences in renovation potential which suggest that our classification algorithm presents a good sensitive to detect the building heterogeneity of the Flemish building stock. This last fact could contribute to trigger new renovation strategies in Flanders.

MAIN REFERENCES

1. Pact R. The energy targets for Flemish homes by 2050. https://www.energiesparen.be/energiedoelstellingen-tegen-2050.

2. Ballarini I, Corgnati SP, Corrado V. Use of reference buildings to assess the energy saving potentials of the residential building stock: The experience of TABULA project. Energy Policy. 2014. doi:10.1016/j.enpol.2014.01.027

3. Vlaamse Regering. Long-term strategy for the renovation of Flemish Buildings. Performance, Energy Dir Build. 2020;(May).

4. Ghiassi N, Mahdavi A. Reductive bottom-up urban energy computing supported by multivariate cluster analysis. Energy Build. 2017;144:372-386. doi:10.1016/j.enbuild.2017.03.004

Keywords: Classification, Clustering, Building renovation costs, Flanders

Paper ID 30903: Can Archive Film Stores in the global South achieve net zero and reliable Energy?

Elli Nikolaidou*, Harrison King, David Coley

University of Bath, United Kingdom

AIM AND APPROACH

Archive film stores aim to preserve the cultural heritage of countries by protecting their photographic films from degrading. This is however challenging for many countries in the global south due to the associated construction cost. Even if this is surpassed, there is commonly a lack of funding for running costs, which can be very high due to the need to store films at low temperatures. Such countries also often need to cope with highly aggressive climates and intermittent electricity grids, while acting on the global need for "affordable, reliable, sustainable and modern energy services" (UN 2015).

Various service-side and construction-side solutions are tested, in an effort to investigate the potential of archive film stores to achieve net zero and reliable energy and hence protect their film stocks and the environment. An exemplar film store is modelled in different locations in the global south, with its performance being predicted by a popular thermal simulation engine.

SCIENTIFIC INNOVATION AND RELEVANCE

The importance of achieving universal access to sustainable and reliable energy is recognised in the seventh Sustainable Development Goal set by the United Nations (UN 2015). Nevertheless, according to surveys conducted between 2010 and 2020 by the World Bank, 75.5% of manufacturing firms in sub-Saharan African countries still suffer from frequent energy outages, with a monthly average of 8.3 outages (The World Bank 2020). In addition to disrupting the smooth operation of buildings, such outages detrimentally affect the environment due to the energy inefficient backup generators that need to be used, these being typically powered by diesel fuel (Farquharson et al. 2018).

Focusing on film stores, sustainable and reliable energy is critical, as building operation significantly affects the lifespan of film stocks. Despite the increasing popularity of low/net zero energy buildings, there is a lack of low/net zero archive film stores in the global south, which may jeopardise the goal for "sustainable and reliable energy for all". Hence, this study focuses on identifying design solutions that can help the film stores located in such countries achieve net zero and reliable energy.

PRELIMINARY RESULTS AND CONCLUSIONS

Despite their high space cooling energy demand, archive film stores demonstrate a great potential for net zero and reliable energy in the majority of the simulated locations. Renewable energy production with the help of photovoltaic panels was vital for securing the energy surplus that is required for the energy independence of the store. The best-performing solutions are presented, these being found to vary from one location to another due to the gamut of climates that are observed in the global south. Meeting the requirements of building regulations and/or standards (such as the Passivhaus standard) calls for an increase in the cooling setpoint, this however making film stocks more prone to degrading. This trade-off is illustrated and discussed to provide guidance on the design of archive film stores in the global south.

MAIN REFERENCES

Farquharson, D., Jaramillo, P., and Samaras, C., 2018. Sustainability implications of electricity outages in sub-Saharan Africa. Nature Sustainability, 1 (10), 589–597.

The World Bank, 2020. Enterprise Surveys [online]. Available from: http://www.enterprisesurveys.org/ [Accessed 25 Apr 2020].

UN, 2015. Transforming our world: the 2030 Agenda for Sustainable Development.

Keywords: Film stores, net zero energy, energy reliability

Paper ID 30350: Methodology for comparison of nZEB ambition level in different countries by building and system simulation

Carsten Wemhoener* (1), Fabian Ochs (2), Mara Magni (2), Christina Betzold (3)

1: Institute of Energy Technology, University of Applied Sciences Eastern Switzerland, Switzerland; 2: Unit for Energy Efficient Building, UIBK, University of Innsbruck, Austria; 3: Technische Hochschule Nuremberg Georg Simon Ohm, Germany

AIM AND APPROACH

By the beginning of January 2021 all new buildings in the EU member states have to comply with nearly Zero Energy Building (nZEB) requirements according to the implementation of the Energy Performance of Buildings Directive (EPBD recast, 2018). Despite different harmonisation initiatives, e.g. ISO (2017) the implementation of nZEB requirements in the single EU member states differs in criteria, metrics and limits. On top of that, national calculation methods and boundary conditions are used for the nZEB compliance rating in the different countries. By the different implementation among the EU member states, it can be presumed that some member states set higher requirements to the building performance than other member states. However, it is hard to compare this ambition level among the EU member states, i.e. the member state's ambition to reach highly performant buildings as future new built standard. In the framework of the IEA research project HPT Annex 49, a methodology has been developed and tested by building and system simulation of a reference building according to the reference framework (Dott et al., 2013) for the nZEB requirements and boundary conditions in the participating countries.

SCIENTIFIC INNOVATION AND RELEVANCE

In order to further develop the requirements for future building standards across Europe, it is important to characterise the state-of-the-art in EU member states after the introduction of the EPBD recast by the beginning of 2021. Therefore, a consistent methodology is needed to assess the performance level of building requirements among the EU member states. A more harmonised nZEB definition among countries is of high interest both for the building technology industry in order to develop performant system solution sets, which comply with requirements across European countries and for policy makers to set adequate targets in the EU member states to promote the energy transition required to achieve climate protection targets.

Furthermore, benchmarks for different ambition levels have been defined. The Net Zero Energy building where only the self-consumed on-site electricity is taken into account in the balancing as upper performance limit, the building physics requirements in terms of avoidance of surface moisture condensation and comfort temperatures as lowest possible limit and the cost optimal building as minimum requirement according to EU regulations.

PRELIMINARY RESULTS AND CONCLUSIONS

The proposed methodology outlined in Wemhoener et al. (2019) has been further developed to include measures for the building envelope, HVAC technology and renewable production on-site. The method consists of the implementation of the national nZEB requirements for a reference building, which is then transferred to common boundary conditions by simulation and compared to references. Three approaches have been tested:

(i) Transfer of the reference building to a common site,

(ii) comparison of the reference building to a high performance building of a passive house in local climate conditions and

(iii) comparison of the reference building with the cost optimal building, which is considered as minimum ambition level for the national nZEB implementation.

From these comparisons a graphical depiction of the ambition level has been deduced in the colour code of energy performance certificates, where further ambition level, e.g. the EU benchmarks given in EC (2016) can be included. The depiction takes into account trade-offs between primary energy and cost.

Furthermore, a tabled version can be set-up to list relative one-to-one comparison between buildings of different ambition levels, e.g. the national implementation compared to the EU benchmarks or compared to the best NZEB as mentioned above.

MAIN REFERENCES

Dott, R., Haller, M.Y., Ruschenburg, J., Ochs, F., Bony, J. 2013. The Reference Framework for System simulations of the IEA SHC Task 44 / HPP Annex 38, Part B: Buildings and space heat load, Report C1 Part B.

EU. 2018.DIRECTIVE (EU) 2018/844 OF THE EUROPEAN PARLIAMENT AND OF THE COUNCIL of 30 May 2018 amending Directive 2010/31/EU on the energy performance of buildings and Directive 2012/27/EU on energy efficiency, Official Journal of the European Union, L/156/75-91

EC. 2016. Commission recommendation (EU) 2016/1318 of 29 July 2016 on guidelines for the promotion of nearly zero-energy buildings and best practices to ensure that, by 2020, all new buildings are nearly zero-energy buildings, Official Journal of the European Union, L/208/46-57

International Organisation for Standardisation (2017). Energy performance of buildings-Overarching EPB assessment -- Part 1: General framework and procedures (ISO 52000-1:2017).

Wemhoener et al. (2019). Simulation-based methodology for comparison of nZEB require-ments in different countries including results of model calibration tests, Proceedings of the 16th IBPSA Conference Rome, Italy, Sept. 2-4, 2019, pp. 5060-5067

Keywords: nZEB, ambition level, EPBD implementation

Session F2.4: The role of occupants

Paper ID 31130: Demonstration of operation of anti-Legionella controller in test setup

Elisa Van Kenhove*, Lien De Backer, Klaas De Jonge, Jelle Laverge

UGent, Belgium

AIM AND APPROACH

Domestic Hot Water (DHW) is produced, stored and distributed at temperatures above 60°C to mitigate the risk of contaminating DHW systems with Legionella pneumophila. This elevated temperature however, is not necessary for most DHW applications, such as showering or washing hands, which requires a maximum temperature of about 40°C. The disparity between 60°C and 40°C has a negative effect on the heat losses and on the efficiency of DHW production units such as heat pumps.

Reducing the production temperature could therefore reduce the energy use. This paper describes the steps towards a sustainable alternative, which involves lowering the DHW temperature in combination with simulation based heat shocks.

Previously, a theoretical proof of concept has been reached, and a DHW controller prototype has been developed (Van Kenhove et al., 2019) and tested in controlling the total viable microbial cells in drinking water (De Backer et al., 2020). In this current paper the tests are extended to L. pneumophila sg. 1 specifically. The purpose of this paper is to implement a stand-alone controller prototype in a laboratory test setup to demonstrate its operation.

SCIENTIFIC INNOVATION AND RELEVANCE

Alternative measures exist in case the standard 60°C measure cannot be applied. For public buildings in Flanders (Belgium), approved alternative measures rely on chemical or electrochemical decontamination. However, concerns are raised about these measures' harmful by-products and other health risks (Pūle, 2017).

This research focuses on a new alternative technique that relies on thermal decontamination. Instead of producing water continuously at 60°C, the water temperature will alternate between the comfort temperature required at the tap (± 40 à 45°C) and a high decontamination temperature to mitigate L. pneumophila in water and biofilm (> 60°C). The heat shocks are initiated by the controller as soon as the concentration limit is reached. To apply this technique, it is mandatory to know the L. pneumophila concentration in the system. As it is inconvenient and time-consuming to measure L. pneumophila concentrations, the concentration throughout the system is predicted using a simulation model that runs in the background of the controller. Running a case specific model is necessary because it is impossible to define general guidelines for the heat shock time and duration, as this is system dependent. General guidelines may either lead to dangerous concentrations in certain building systems or undermine the energy savings.

PRELIMINARY RESULTS AND CONCLUSIONS

A detailed simulation study of such a controller showed savings of 35% (distribution energy) and 14-64% (production energy) without increasing contamination risk.

In this paper the operation of the anti-Legionella controller is demonstrated with L. pneumophila sg. 1. The laboratory tests show that the anti-Legionella controller is able to predict the L. pneumophila sg. 1 concentrations and to initiate heat shocks accordingly and correctly.

The conclusions of this paper are opening the path towards controlling L. pneumophila in an alternative way (instead of the constant 60°C), making it possible to save energy without endangering the health of the occupants.

MAIN REFERENCES

Van Kenhove, E., De Backer, L., Delghust, M. et al. (2019). Coupling of Modelica Domestic Hot Water Simulation Model with Controller. Proceedings of Building Simulation 2019: 16th IBPSA Conference, 24-931.

Van Kenhove, E., De Backer, L., Janssens, A. and Laverge, J. (2019). Simulation of Legionella concentration in domestic hot water: comparison of pipe and boiler models. J. Build. Perform. Simul., 12(5).

De Backer, L., Van Kenhove, E., Delghust, M., Laverge, J. (2020). From Theory to Practice – Development of a Model Predictive Anti-Legionella Controller. Proceedings of Indoor Air 2020: the 16th Conference of the International Society of Indoor Air Quality & Climate.

Pūle, D. (2017). Conventional and Alternative Disinfection Methods of Legionella in Water Distribution Systems – Review. Construction Science, 19(1), 21-26.

Keywords: Legionella pneumophila, health, energy savings, controller

Paper ID 30588: Contribution of building energy simulations to the assessment of overheating health risks in urban dwellings

Flavia Barone* (1), Lucie Merlier (1), Virginie Chasles (2), Frédéric Kuznik (1)

1: Univ Lyon, CNRS, INSA-Lyon, Université Claude Bernard Lyon 1, CETHIL UMR5008, F-69621, Villeurbanne, France; 2: Université Jean Moulin Lyon 3, UMR 5600 Environnement Ville Société (EVS)

AIM AND APPROACH

This exploratory study introduces a methodology and indicators aiming to characterize individual indoor exposure for health risk assessment related to overheating using building energy simulation (BES).

The intensification in peak and frequency of heat waves raises a public health issue, which challenges the way of designing buildings, especially in urban areas due to the urban heat island phenomenon. Current health risk indexes are mostly based on large scale statistical data and on outdoor weather data [1]. Yet, people spend more than 80% of their time indoors. Indicators at the dwelling scale appear thus necessary to better assess health risk, which requires adapted methods and tools. Hence, this contribution evaluates the possibilities of current building energy models to tackle this issue as suggested in [1].

Focussing on the evaluation of overheating for health risk assessment, this contribution goes beyond more common thermal comfort studies [2]. In particular, a set of exposure indicators and thresholds [3,4] are suggested based on an extensive literature review. Building energy simulations are also carried out for different types of urban dwellings and passive cooling strategies are modelled to highlight effects of the relative adaptive capacity of people and buildings [5].

SCIENTIFIC INNOVATION AND RELEVANCE

This contribution highlights critical physical outputs and timescales of BES to be processed to evaluate overheating exposure for health risk assessment taking into account acclimatization. Night-time and day-time exposures are especially analysed with respect to the suggested suitable equivalent temperature and related thresholds. Application cases are studied considering different types of dwellings and building-users interaction (individually implemented passive cooling strategy) [5]. The impact of local urban weather conditions and user's behaviours is also studied.

The contribution also highlights the limits of current building simulation approaches to deal with such a problem. Identified issues especially deal with the definition of boundary conditions, the reproduction of users' behaviours and the modelling traditional passive cooling strategies, as well as the coupling with other health-related parameters such as air quality.

PRELIMINARY RESULTS AND CONCLUSIONS

Preliminary results show that :

- Even though outside temperatures are acceptable, indoor overheating can occur when no specific cooling action is implemented.

- Efficiency of passive cooling actions depends on building insulation properties.

- Inhabitants' behaviour (ex: appropriate natural ventilation and occultation) can significantly reduce overheating during the night and even cancel it during days.

- The dynamics and intensity of the exposure to indoor overheating can strongly vary depending on the building type and considered period

Current developments focus now on a better integration of urban effects and inhabitant's behaviour, and on the model validation. A test survey relating people's perceived health, dwelling properties and people and buildings' adaptive capacities will also take place during the 2020 summer. BES results will be criticized based on the results of this test survey to challenge the developed approach, although a direct correlation between BES and survey's results can obviously not be relevantly drawn.

MAIN REFERENCES

[1] A. Amengual and others, 'Projections of Heat Waves with High Impact on Human Health in Europe', Global and Planetary Change, 119 (2014)

[2]Leen Peeters and others, 'Thermal Comfort in Residential Buildings: Comfort Values and Scales for Building Energy Simulation', Applied Energy, 86.5 (2009).

[3] 'O Cassandra R. O'Lenick and others, 'Urban Heat and Air Pollution: A Framework for Integrating Population Vulnerability and Indoor Exposure in Health Risk Analyses', Science of The Total Environment, 660 (2019).

[4] Khaiwal Ravindra and others, 'Generalized Additive Models: Building Evidence of Air Pollution, Climate Change and Human Health', Environment International, 132 (2019).

[5] T. van Hooff and others, 'Reprint of: On the Predicted Effectiveness of Climate Adaptation Measures for Residential Buildings', Building and Environment, 83 (2015).

Keywords: Building energy simulation, exposure to overheating, passive cooling, health risk, users' behaviour

Session F2.5 (Online Track): Practice and industry related case studies

Paper ID 30249: A simulation approach for the optimization of distributed energy supply systems based on multiple energy indicators in commercial districts.

Takahiro Ueno*

Building Research Institute / Japan, Japan

AIM AND APPROACH

Urban energy supply systems are changing to distributed energy supply systems (DESS) due to the spread of renewable energy. We aim to construct a method to optimize DESS on a city scale by using geographical information system (GIS) data.

This paper discusses a simulation study for the optimal layout of DESS in commercial districts based on three energy indicators. The first indicator aims to minimize the annual primary energy consumption (PEC) with consideration of the dynamic coefficient of primary energy consumption in order to reduce thermal power generation in the power system of the target district. The second indicator evaluates the annual value of life cycle CO2 emissions (LCE) in just the target district by taking into account all power sources including renewable energy such as solar and wind power in the grid. The third was an indicator of the standard deviation of power supply and demand (SDP) which aimed to level the power supply and demand between the target area and the power grid in order to stabilize the power system. The simulation study calculates energy consumption at five-minute intervals in target commercial districts, and then evaluates the calculation results of each case according to three indicators.

SCIENTIFIC INNOVATION AND RELEVANCE

This paper optimized DESS based on power and heat demands at five-minute intervals. Some studies have evaluated the DESS performance in commercial districts by one-hour demands. Other studies have examined energy supply systems in a small scale such as one building. But few studies have optimized the same complex energy supply systems as our research based on detailed calculations such as five-minute intervals. Compared to these previous studies, we have sufficient scientific innovation from the following six points.

(1) examine both electricity and heat.

(2) calculate demands and supplies at five-minute intervals.

(3) use many static equipment models such as absorption chiller, air heat-source heat pump, combined heat and power (CHP), storage battery, water heat storage tank.

(4) change the value of energy demand in each building according to the type of buildings (office, hospital, hotel, store, restaurant and school), total floor area, outdoor air temperature and so on.

(5) incorporate pump power, restrictions and loss on heat transport.

(6) evaluate DESS performance based on three energy indicators.

PRELIMINARY RESULTS AND CONCLUSIONS

We evaluated the DESS performance in four commercial districts with different characteristics of located buildings. We analyzed thousands of calculation results of each district in a case study and confirmed the energy saving effect, CO2 saving effect, and power system operation stability improvement effect by DESS in Japanese commercial districts. We also set the case of the smallest value in each indicator in the district as a best case for each indicator. All best cases in all districts reduced PEC by more than 10% and SDP by more than 50%. On the other hand, some best cases for the indicators increased LCE by several percent to several tens of percent. In order to clarify the reason, we examined breakdowns of the CO2 emissions in all best cases. They showed that gas from CHP accounted for more than half of the CO2 emissions in DESS. Further investigations also indicated that CHP in fine weather during the cooling season could not reduce LCE unless the power generation and exhaust heat matched exactly the power and heat demands of the target buildings.

MAIN REFERENCES

Georgios Dermentzis, et al, Primary-energy Based Optimization of a New Building District through Simulations on Flat, Building, Block and District Level, Proceedings of the 16th IBPSA Conference, pp.3628-3635, 2019.9

Anja Willmann, et al, A Comparison of Bottom-up and Top-down Modelling Approaches in Urban Energy Simulation for the Assessment of City District Data Models, Proceedings of the 16th IBPSA Conference, pp.3303-3310, 2019.9

Martina Ferrando, Francesco Causone, An Overview Of Urban Building Energy Modelling (UBEM) Tools, Proceedings of the 16th IBPSA Conference, pp.3452-3459, 2019.9

Eui-Jong KIM, et al, Simulation of Hourly Energy Needs at Urban Scale: A Methodology of Adapted Modeling, Proceedings of the 16th IBPSA Conference, pp.3484-3490, 2019.9

Patrick Beguery, et al, MicroGrid Energy Management Optimization – A Common Platform For Research, Development And Design Tools, Proceedings of the 16th IBPSA Conference, pp.3163-3170, 2019.9

Usman Ali, et al, Application Of Intelligent Algorithms For Residential Building Energy Performance Rating Prediction, Proceedings of the 16th IBPSA Conference, pp.3177-3184, 2019.9

Alex Nutkiewicz, Rishee K. Jain, Exploring the Integration of Simulation and Deep Learning Models for Urban Building Energy Modeling and Retrofit Analysis, Proceedings of the 16th IBPSA Conference, pp.3209-3216, 2019.9

Keywords: Distributed energy, Power demand, Heat demand, Life cycle CO2 emissions, Peak demand

Paper ID 30478: A Comparative analysis on the impact of different weather datasets used in Australia on building energy simulation results

Samuel Udom*, Saeed Banihashemi, Charles Lemckert

University of Canberra, Australia

AIM AND APPROACH

This paper presents a comparative analysis of 3 different EnergyPlus weather data sets from 3 different sources commonly used in Australia against actual weather data measured at a building site to determine the weather file most suited for dynamic building energy simulations in remote Australia. Weather characteristics of monthly average dry bulb temperature and relative humidity were statistically compared against similar actual data collected from the building site along with data from the Australian Bureau of Meteorology (BOM) for the same location. The statistical calculations are based on common metrics found in literature such as mean, median, standard deviation mean bias error (MBE), mean absolute percentage of error (MAPE), root mean square error (RMSE), and coefficient of variance RMSE (CV-RMSE) from the hourly data. In addition, annual energy simulations were then performed on a residential building using the different weather files for comparison of energy performance. The results show that the difference in annual energy consumption, heating and cooling loads can vary by ±6%, ±7% and ±10% respectively as a function of the source of the weather data used for the same location.

SCIENTIFIC INNOVATION AND RELEVANCE

The commonly used weather file for residential energy simulation in Australia is the Typical Meteorological Year (TMY) from Nationwide House Energy Rating Scheme (NatHERS) and the Australia Representative Meteorological Year (RMY) from EnergyPlus. Among many other sources such as Meteonorm and climateOne, it is increasingly difficult selecting an appropriate weather file for energy simulation that accurately predicts expected/ current energy use in comparison to actual energy use in buildings. The study presents a simplified and efficient approach for the selection of appropriate EPW files based on available data that closely represents existing climatic conditions for a specific location of interest in Australia. It is expected that the findings will provide greater understanding and insight on the degree of uncertainty associated with each of the 3 weather files commonly used in Australia especially in relation to the outcome of energy simulation results. This will have practical implications for building practitioners and stakeholders with respect to predicting energy use in new and existing buildings, especially in the remote parts of Australia.

PRELIMINARY RESULTS AND CONCLUSIONS

Preliminary findings reveal the following:

For dry bulb temperature comparisons with actual data, statistical analysis of key metrics reveal:

• MBE: Meteonorm records the lowest at 2.38, Nathers records highest at 2.75.

• MAPE: Meteonorm records the lowest at 9.87, Nathers records the highest at 11.88

• RMSE: Meteonorm records the lowest at 8.12, Nathers records the highest at 9.52

• CV-RMSE: Meteonorm records the lowest at 0.32, Nathers records the highest at 0.38

All compared dry bulb temperature data record a strong correlation with the actual data with Meteonorm recording the highest at 0.96 while Nathers is lowest at 0.94.

For relative humidity,

• MBE: RMY records the lowest at -6.52, BOM records highest at -0.67

• MAPE: BOM records the lowest at 22.61, RMY records the highest at 59.21

• RMSE: BOM records the lowest at 2.31, RMY records the highest at 22.59

• CV-RMSE: BOM records the lowest at 0.08, RMY records the highest at 0.77

All compared relative humidity data did not show any significant correlation with the actual data. However, BOM recorded the strongest at 0.46 while Meteonorm had the weakest at 0.008.

Preliminary findings suggests notable differences between variables that may impact simulation results.

MAIN REFERENCES

Hong, T., W.-K. Chang, and H.-W. Lin, A fresh look at weather impact on peak electricity demand and energy use of buildings using 30-year actual weather data. Applied energy, 2013. 111: p. 333-350.

2. Bhandari, M., S. Shrestha, and J. New, Evaluation of weather datasets for building energy simulation. Energy and Buildings, 2012. 49: p. 109-118.

3. Ohlsson, M.B., et al., Predicting System Loads with Artificial Neural Networks--Methods and Results from" The Great Energy Predictor Shootout". ASHRAE Transactions-American Society of Heating Refrigerating Airconditioning Engin, 1994. 100(2): p. 1063-1074.

4. Dodier, R.H. and G.P. Henze, Statistical analysis of neural networks as applied to building energy prediction. J. Sol. Energy Eng., 2004. 126(1): p. 592-600.

5. Crawley, D.B. and C.S. Barnaby, Weather and climate in building performance simulation, in Building Performance Simulation for Design and Operation. 2019, Routledge. p. 191-220.

6. Kreider, J.F. and J.S. Haberl, Predicting hourly building energy usage. ASHRAE Journal (American Society of Heating, Refrigerating and Air-Conditioning Engineers);(United States), 1994. 36(6).

7. Hensen, J.L. and R. Lamberts, Building performance simulation for design and operation. 2012: Routledge.

8. Scott West, P., et al., Energy Simulation Aided Design for Buildings. ASHRAE Journal, 2019. 61(12): p. 20-26.

Keywords: Building Energy Simulations, Weather data, EnergyPlus, Climate, Remote Australia

Paper ID 30781: Spatio-temporal data analysis for development of microclimate prediction models

Tageui Hong*, Yeonsook Heo

Korea University, Korea, Republic of (South Korea)

AIM AND APPROACH

This study aims to reveal underlying relationships between urban characteristics and urban microclimate in dense cities. Seoul is a studied area in which hourly outdoor air temperature, wind speed, and wind direction measurements were obtained from about 270 weather stations. GIS and satellite data were used to provide urban characteristics. This study is conducted in three steps. First, urban heat island (UHI) phenomena were analyzed based on a set of UHI indicators for identifying key differences in UHI patterns. Second, we developed a comprehensive set of urban indicators to represent potential heat fluxes in different locations across the city. The set includes urban morphological, anthropogenic heat and vegetation indicators. The relative importance of these indicators on the UHI will be examined, and key factors will be used in the next step. Finally, a regression model with the selected variables and regional climatic variables will be developed to investigate the combined effects of urban and weather characteristics. The effect of spatial resolution will be also considered to determine the appropriate buffer size for urban microclimate predictions. Practical discussions for developing microclimate prediction models will be provided on the basis of analysis results.

SCIENTIFIC INNOVATION AND RELEVANCE

Owing to increasing urbanization, the weather conditions around a building can vary substantially depending on the urban context. The urban surroundings of a site directly affect the energy consumption of buildings by influencing local air temperatures, wind speeds, and shading effects on buildings. A simulation study of a typical office building with local air temperature measurements suggested that the annual cooling load in an extremely high-density urban area increases by 25% and that the heating load is reduced by 22% solely due to the raised ambient temperatures (Kolokotroni et al., 2007) .

Microclimate studies have been conducted through CFD-based or measurements-based modelling. Previous studies based on CFD tools has focused on the effect of a few urban morphological factors on the UHI by creating various urban scenarios. However, a significant amount of heat emitted by human activities is typically neglected in this approach. In contrast, urban microclimate data are the result of various elements that compose the city. However, a comprehensive study of urban microclimate in relation to various urban factors is currently lacking due to the shortage of extensive urban measurement data. This study performs data analysis based on extensive spatio-temporal data to provide a comprehensive overview of microclimate phenomena.

PRELIMINARY RESULTS AND CONCLUSIONS

Hourly microclimate data of 270 locations in Seoul were analyzed. When hourly outdoor temperature data were aggregated into daily average hourly profiles, the profiles showed a similar trend within certain temperature ranges. However, when hourly data were used instead of the average, more distinct differences were observed. The differences in the peak temperature ranged up to 5.26°C in December and 6.67°C in July, and showed maximum 11.7°C differences between the weather stations at the same day. Also, there is substantial variation observed in the peak timing, with a time-lag of 1 to 3 hours. This implies the value of using extensive spatio-temporal microclimate data to correctly investigate UHI phenomena.

In our previous study, the GIS data of Seoul was obtained and processed to calculate ten common spatial indicators related to buildings and roads, and additional information about subway stations explain some anthropogenic heat. Additional data sources are in process to represent other anthropogenic heat sources and vegetation. This dataset will be used to identify the urban characteristics of locations corresponding to microclimate data. Regression analysis based on the dataset will be performed to develop a microclimate prediction model, and issues relevant to model development will be discussed.

MAIN REFERENCES

Kolokotroni, M., Zhang, Y., & Watkins, R. (2007). The London heat island and building cooling design. Solar Energy, 81(1), 102-110.

Oleson, K. W., Bonan, G. B., Feddema, J., & Vertenstein, M. (2008). An urban parameterization for a global climate model. Part II: Sensitivity to input parameters and the simulated urban heat island in offline simulations. Journal of Applied Meteorology and Climatology, 47(4), 1061-1076.

Liao, W., & Heo, Y. (2018). The Effect of Urban Spatial Characteristics on Microclimate. In Building Simulation and Optimization Conference.

Keywords: Urban microclimate, Urban characteristics, Urban heat island, Spatio-temporal data

Paper ID 30884: Design and analysis of a renewable energy network based on fuel cells in residential communities

Jiwook Byun*, Yeonsook Heo

Korea University, Korea, Republic of (South Korea)

AIM AND APPROACH

This paper aims to elicit guidelines from simulation studies for designing the optimal renewable energy network targeting residential communities. Key design-related questions are the effect of integrating a heat network, integrating the mix of various renewable energy systems, and adding a hydrogen storage system on the energy supply performance. A residential community with 12 households was modelled and simulated in TRNSYS to evaluate the performance of different power/heat network scenarios under dynamic power and heat loads. The first scenario includes gas-based fuel cell systems integrated to the heat network providing domestic hot water and heating to the residential community. In addition, renewable energy systems such as PVs, batteries, electrolyzers, and geothermal systems are incrementally added to create a set of scenarios. The last scenario includes a hydrogen storage system to investigate its effect on balancing seasonal mismatch and accordingly reducing supply system capacities. The energy system capacities will be optimized for each design scenario, and the energy network performance will be compared in terms of the cost, unmet demands, and dumped energy.

SCIENTIFIC INNOVATION AND RELEVANCE

As the transition from the existing centralized grid to the decentralized grid, the concept of decentralized grids has been studied and realized by installing renewable energy systems and providing energy for different scales of communities. The energy hub concept is a scalable approach that forms a local energy-sharing community on the basis of renewable energy technologies at the neighborhood scale and has the potential to expand to larger areas (Orehounig, 2015). Decentralized energy systems have been investigated through simulation-based case studies in which a specific energy network is given and the main design interest is optimizing system capacities. In these studies, the overall performance of the whole system network is sufficiently investigated, but there is lack of knowledge to understand the effect of individual supply systems on the overall performance. In order to create knowledge base for optimizing the local-scale energy supply network, this study creates a set of design scenarios by incrementally adding a different type of supply systems and systematically evaluates the role of individual systems on the overall system. performance.

PRELIMINARY RESULTS AND CONCLUSIONS

Preliminary analysis of building energy demands showed that the hourly electricity demand profile is consistent throughout the year, whereas the heating demand showed a large seasonal variation: 0.4% of the total heating demand in summer and 67% in winter. However, the domestic hot water demand is relatively constant; 27% in winter and 19% in summer. Four scenarios with a different mix of energy supply systems were modelled and compared. In Scenario 1, fuel cells are suitable to replace boilers because they meet both the total heating demand and the total electricity demand by consuming the gas by 72% more than boilers. When fuel cells operate to meet the heating demand, their electricity yield is minimal during the summer. PVs and batteries were added in Scenario 2 to make up the unmet power demand in summer because the electricity yield during the summer was 36% higher than during the winter. Geothermal systems were included in Scenario 3 to meet the dynamic heating demand and potentially assist in the stable operation of fuel cells. In Scenario 4 with the sufficient capacity of PVs, the usability of a hydrogen storage system to balance the seasonal mismatch will be examined.

MAIN REFERENCES

Orehounig, K., Evins, R., & Dorer, V. (2015). Integration of decentralized energy systems in neighbourhoods using the energy hub approach. Applied Energy, 154, 277-289.

Al-Falahi, M. D., Jayasinghe, S. D. G., & Enshaei, H. J. E. C. (2017). A review on recent size optimization methodologies for standalone solar and wind hybrid renewable energy system. Energy conversion and management, 143, 252-274.

Oh, S. D., Kim, K. Y., Oh, S. B., & Kwak, H. Y. (2012). Optimal operation of a 1-kW PEMFC-based CHP system for residential applications. Applied Energy, 95, 93-101.

Keywords: fuel cell, renewable energy network, distributed energy resources

Paper ID 30891: The design of distributed control system for a hybrid energy supply network in a residential community

Jaehyun Go*, Yeonsook Heo

Korea University, Korea, Republic of (South Korea)

AIM AND APPROACH

The objective of this paper is to identify the minimal level of information that ensures the performance of distributed controls for distributed energy resources (DER) in residential communities. A local energy-sharing community with 12 households was modelled and simulated in TRNSYS to evaluate the performance of distributed control with varying information. The studied energy supply network is composed of PVs, batteries, ground source heat pumps, thermal storage tanks, electrolyzers, fuel cells, and hydrogen tanks. The effect of information level on the control performance depends on the level of distribution in the system network. Hence, four scenarios were created to represent different system partitioning. In each scenario, different levels of information will be provided to distributed controllers by incrementally removing possible information (e.g., control variables of other controller(s), state variables of subsystems). The control performance will be evaluated in terms of the operating cost (average operating time, number of starts), unmet demands, and dumped energy. In addition, the performance of distributed controllers will be compared with the centralized controller in order to examine the relevance of distributed control in optimizing the DER operation. On the basis of simulation results, the minimal level of information required for distributed controls will be discussed.

SCIENTIFIC INNOVATION AND RELEVANCE

For control of distributed energy resources (DER), there is the growing interest in developing distributed control systems. The traditional centralized control approach is suitable for small-scale energy systems on the basis of communication networks connecting all components in the system. However, as the energy supply network becomes larger and more complex, the applicability of the centralized control is quite limited due to the high computational cost and practical difficulties of constructing communication networks. Recently, the distributed control approach has received attention for the DER operation to overcome the major limitations of the centralized control. In order to ensure the performance of the distributed control comparable to the centralized control, system partitioning and information exchange level between distributed controllers need to be carefully designed. Various control methods such as multi agent systems and model predictive control has been applied for distributed control systems. Many of existing studies have developed and demonstrated distributed controllers based on selected control methods. However, the effect of system partitioning and information exchange level on the control performance have not been sufficiently considered. This study will quantify the effect of information level under different system partitioning to examine the minimal level of information required for distributed controls.

PRELIMINARY RESULTS AND CONCLUSIONS

In the case study, the energy supply network was partitioned into a different number of groups controlled by individual distributed controllers. The model predictive control (MPC) approach was implemented in TRNSYS as distributed controllers. As a preliminary study, the performance of decentralized controllers (i.e., one for the power network, including fuel cells and the other for the heat network without information exchange between them) was compared against the centralized controller for the entire energy supply network using all accessible information. In case of the centralized controller, the water temperature in the thermal storage tank exceeded the temperature range (40-45℃) only 2 hours. In contrast, the decentralized controllers without information exchange exceeded for 67 hours during the 72-hour test period. This difference is mainly due to the lack of coordination between the ground source heat pumps and the fuel cells. This comparison highlights the potential significance of information exchange between the controllers in the control performance. The performance of distributed controllers under three different system partitioning will be further investigated with varying levels of information exchange.

MAIN REFERENCES

[1] Scattolini, R. (2009). Architectures for distributed and hierarchical model predictive control–a review. Journal of process control, 19(5), 723-731.

[2] Ulleberg, Ø. (2004). The importance of control strategies in PV–hydrogen systems. Solar Energy, 76(1-3), 323-329.

[3] Khan, M. R. B., Jidin, R., & Pasupuleti, J. (2016). Multi-agent based distributed control architecture for microgrid energy management and optimization. Energy Conversion and Management, 112, 288-307.

[4] Roche, R., Blunier, B., Miraoui, A., Hilaire, V., & Koukam, A. (2010, November). Multi-agent systems for grid energy management: A short review. In IECON 2010-36th Annual Conference on IEEE Industrial Electronics Society (pp. 3341-3346). IEEE.

[5] Shabani, M. J., & Moghaddas-Tafreshi, S. M. (2020). Fully-decentralized coordination for simultaneous hydrogen, power, and heat interaction in a multi-carrier-energy system considering private ownership. Electric Power Systems Research, 180, 106099.

[6] Bilodeau, A., & Agbossou, K. (2006). Control analysis of renewable energy system with hydrogen storage for residential applications. Journal of power sources, 162(2), 757-764.

[7] Milo, A., Gaztañaga, H., Etxeberria-Otadui, I., Bacha, S., & Rodríguez, P. (2011). Optimal economic exploitation of hydrogen based grid-friendly zero energy buildings. Renewable Energy, 36(1), 197-205.

[8] Fontenot, H., & Dong, B. (2019). Modeling and control of building-integrated microgrids for optimal energy management–A review. Applied Energy, 254, 113689.

Keywords: Distributed energy resources, Distributed control, Model predictive control, System partitioning, Information level

Session F2.6 (Online Track): Improving indoor environmental quality

Paper ID 30123: Performance evaluation of passive hygrothermal control for houses using a thermodynamic HAM model

Haksung Lee* (1), Akihito Ozaki (2)

1: Department of Architecture, Chungbuk National University, Cheongju, Republic of Korea; 2: Kyushu University, Japan

AIM AND APPROACH

With the aim of this study is to design a passive house which have constant temperature and humidity performance and functions of cooling/dehumidification in hot and humid summer and heat collection in cold and dry winter using renewable energy, we have developed a high-performance envelope system that can also be applied to houses of dry construction method which are mainly used for industrial building materials. The proposed system is a variable thermal performance intelligent passive system (PDSC: Passive Dehumidification and Solar Collection), which utilizes a thermodynamic potential [1] difference between indoors and outdoors to cool and dehumidify naturally in the summer and to collect the solar heat in the winter by air circulation.

We have already reported on fundamental research by conducting numerical simulations based on laboratory experiments using roof models [2] and a housing model [3]. Based on the previous research, we designed and constructed a full-scale house to verify the applicability and effectiveness of this system to actual houses. The temperature and humidity control performance and the energy-saving effect of this system have been studied through outdoor experiments using a full-scale demonstration house and simulation of building temperature, humidity, heat load that considers coupled heat, moisture, and air transfer.

SCIENTIFIC INNOVATION AND RELEVANCE

It is estimated that the energy consumption would be slightly reduced even if the heat insulation and airtight performance of houses are further strengthened. Therefore, the introduction of new technology is indispensable for further energy saving aiming at a zero-energy house in the future. In particular, because there is no effective reduction method for the latent heat load during cooling of humid summer, innovative energy-saving technology is required as a countermeasure. Also, to reduce the heating load, which occupies most of the heat load, passive technology using solar thermal energy can be expected.

The PDSC system uses the thermodynamic potential difference between indoor and outdoor, which is related to various factors, as the driving force for heat and moisture transfer. Therefore, various physical quantities related to heat and water vapor diffusion such as temperature, concentration, pressure, external force, and adsorption force are unified and expressed in energy of the same dimension (potential defined by thermodynamic function). The fundamental concept of this system can be theoretically explained by expressing the coupled phenomenon as a flow of energy according to nonequilibrium thermodynamics.

PRELIMINARY RESULTS AND CONCLUSIONS

The analysis code of the PDSC envelope system is incorporated into the software "THERB for HAM" [4] for temperature/humidity/heat load calculation of an entire building that considers the coupled heat, moisture, and air transfer. The temperature and humidity control performance and energy-saving performance of this system is examined by the numerical simulation. First, we confirmed the high calculation accuracy of the developed software by performing a numerical simulation on a demonstration house and comparing the calculated and measured values. Next, by numerical experiments using standard weather data (expanded AMeDAS (Automated Meteorological Data Acquisition System) weather data), the influence of differences in weather conditions and building specifications (regions, presence/absence of PDSC system, differences in thermal insulation materials (moisture capacity), presence/absence of moisture-proof sheets (moisture adsorption/desorption)) are clarified.

As a result, the house with PDSC system can reduce the sensible heat and latent heat load in summer by about 5% and about 20-41%, respectively, compared to conventional housing, and the sensible heat load in winter can be reduced by 10% or more.

MAIN REFERENCES

[1] A. Ozaki, T. Watanabe, T. Hayashi, Y. Ryu, Systematic analysis on combined heat and water transfer through porous materials based on thermodynamic energy, Energy Build. 33 (2001) 341–350. https://doi.org/10.1016/S0378-7788(00)00116-X.

[2] H. Lee, A. Ozaki, M. Lee, W. Cho, A fundamental study of intelligent building envelope systems capable of passive dehumidification and solar heat collection utilizing renewable energy, Energy Build. 195 (2019) 139–148. https://doi.org/10.1016/j.enbuild.2019.04.039.

[3] H. Lee, A. Ozaki, W. Cho, M. Lee, Smart passive system for dehumidification , cooling , and heating utilizing renewable energy in detached house, Proceedings of the 16th IBPSA Conference, (2019) 2442–2449.

[4] A. Ozaki, T. Tsujimaru, Prediction of hygrothermal environment of buildings based upon combined simulation of heat and moisture transfer and airflow, Proceedings of the Ninth International IBPSA Conference (2005) 899–906.

Keywords: Passive dehumidification, Radiative cooling, Humidity control, Solar heat collection, Latent heat reduction

Paper ID 30212: Airflow optimization for thermal comfort and energy efficiency for room air conditioners

Ryuta Tanaka* (1), Saleh Nabi (2), Mio Nonaka (1)

1: Mitsubishi Electric Corporation Advanced Technology R&D Center, Japan; 2: Mitsubishi Electric Research Laboratories, USA

AIM AND APPROACH

The purpose of this paper is to increase the energy efficiency of room air conditioners while achieving a thermally comfortable indoor space. We investigate the solution to this problem as an optimization problem in which the control variables are inlet temperature, air speed, and angle. We formulated such an optimization as a minimization problem for the objective function representing the temperature uniformity of the room and the energy consumption of the room air conditioner. The temperature uniformity of the room was formulated as the squared error between the temperature in a given region of the interest within room, which includes the occupants and the set temperature. The energy consumption is formulated as a function of the control variables using the coefficient of performance (COP). The optimization problem was optimized using sensitivity and the gradient descent method. Such implementations have been performed using a coupling between the CFD solver of OpenFOAM and Matlab for optimization. To validate our optimization results, laboratory model was built and experiments have been carried out. We also compared and analyzed the effects of different models of room air conditioners (i.e., different degrees of freedom in the control variables) on energy savings and comfort.

SCIENTIFIC INNOVATION AND RELEVANCE

Improving indoor comfort and reducing energy consumption are of primary concern in the control of room air conditioners. In the past, we have used trial-and-error CFD simulations or experiments to determine the optimal control parameters to achieve these goals. However, such method is intractable when the number of control variables are large and also there is no mathematical guarantee that such solution is in fact optimal. Moreover, previous studies mostly focused on either the thermal comfort or power consumption, but less attention is given to both tasks considering the multi-physics of the dynamics of airflow as well as the refrigerant cycle. We propose a gradient-based framework for CFD optimization to circumvent such difficulties that includes a multi-objective optimization for both temperature uniformity of the room energy consumption as a function of the control variables. This makes it possible to optimize not only the indoor comfort but also the energy consumption of the air conditioner. Furthermore, our full-scale experimental model helps to validate the efficacy of the CFD-based optimization for room ventilation.

PRELIMINARY RESULTS AND CONCLUSIONS

In this paper, we employed a CFD-based optimization framework to analyze the impact of room air-conditioner control variables on the air velocity and indoor temperature in a typical room for the heating mode. The control variables are the inlet volume flux, temperature and the vane's angle. The cost function is a multi-objective function of both thermal comfort and the energy consumption. The results show that for the optimal values of control variables, the air-conditioning system can undertake indoor heat load to provide thermally comfortable environment while maintaining a minimal energy budget for power consumption. We also demonstrate that more degrees of freedom on the air-conditioning system, e.g. having additional yaw angles, results in further improvement of thermal comfort for relatively large heat loads while the energy consumption is decreased. Moreover, our results show that, the optimal air velocity in the region of interest is low to moderate such that the wind velocity cannot be felt by the residents. We corroborate our results by comparing them with experimental data. Our results show that CFD-based optimization will have a reference value to create thermally comfort indoor environments while maintaining energy consumption to a minimum.

MAIN REFERENCES

Nabi, S., Grover, P., Caulfield, C.C., "Adjoint-Based Optimization of Displacement Ventilation Flow", Building and Environment, 2017

Nabi, S., Grover, P., Caulfield, C.C., "Nonlinear optimal control strategies for buoyancy-driven flows in the built environment", Computers & Fluids, 2019

Keywords: airflow optimization, thermal comfort, energy efficiency, room air conditioners

Paper ID 30750: Designing a data-driven model predictive control framework for residential buildings

Hyeong Seok Lee*, Yeonsook Heo

Urban Energy&Environment Lab, Dept. of Civil, Environmental and Architectural Engineering, Korea University

AIM AND APPROACH

This study aims to optimize the data-driven model predictive control (MPC) framework for residential buildings. The data-driven MPC framework is largely divided into three parts: prediction model(s), an optimization scheme, and MPC settings. This study investigates the effect of key factors in the data-driven MPC framework on the MPC performance through a simulation study of a case residential building based on TRNSYS. First, the essential features of the prediction model (i.e., non-linearity, time-correlation, model order) are determined by comparing the performance of linear regression, autoregressive with exogenous input model(ARX), and nonlinear autoregressive with exogenous input model(NARX) models with varying sets of predictors in the MPC application. Second, the two commonly used optimization approaches, linear programming (LP) and quadratic programming (QP), were investigated, and each approach requires certain considerations in designing the optimization scheme (e.g., objective function, constraints) to ensure both the building energy efficiency and the thermal comfort. Different optimization designs under the two approaches are evaluated in terms of their resulting MPC performance. Last, the effect of three MPC settings (i.e., a control time step and prediction and control horizons) is examined.

SCIENTIFIC INNOVATION AND RELEVANCE

Model Predictive Control (MPC) allows for optimally controlling system operation in a proactive manner with considering future disturbances. A key component in the MPC is a prediction model that evaluates the effect of testing control actions. Although simulation-based MPC has been sufficiently demonstrated to show the energy-saving potentials in building applications, it requires manual creation of a simulation model for every targeting building, which substantially hinders the applicability of this approach. As an alternative to enhance the applicability, MPC based on data-driven statistical models has been developed, and commonly used statistical models include artificial neural networks, fuzzy logic models, etc. These complex models have been tested to provide accurate predictions at a fine time resolution. However, whether such sophisticated models are necessary for MPC applications has not been tested yet. These models are typically stored in a central computing cloud due to the high computational demand. Reliance on the computing cloud is not suitable for residential buildings where privacy is an important issue. Therefore, it is necessary to develop a distributable MPC framework based on lean, computationally efficient methods.

PRELIMINARY RESULTS AND CONCLUSIONS

The virtual building of a residential unit was created in TRNSYS, and statistical models were created on the basis of simulation results. To reflect actual variation in the occupancy pattern and associated electricity usage, the ECO data set, collected by ETH Zurich for actual occupancy and load profiles were used in this study. Two statistical models were developed as part of the MPC framework to predict the indoor temperature and heat supplied by the boiler and radiant underfloor. The effect of different features in the statistical model on the prediction accuracy was analyzed. It was confirmed that a modeling feature to represent time-correlation (i.e., the ARX model in comparison to the linear regression model) shows a more significant effect on the prediction accuracy than adding more predictors or capturing nonlinearity (i.e., comparing the ARX model with the NARX model). As the next step, the effect of different modelling features on the MPC performance will be investigated. In addition, different optimization schemes and MPC settings will be investigated in terms of their effect on the MPC performance and computational time.

MAIN REFERENCES

Killian, M. and M. Kozek (2016). "Ten questions concerning model predictive control for energy efficient buildings." Building and Environment 105: 403-412.

Mirakhorli, A. and B. Dong (2016). "Occupancy behavior based model predictive control for building indoor climate—A critical review." Energy and Buildings 129(Review papaer to MPC): 499-513.

Beckel, C., W. Kleiminger, R. Cicchetti, T. Staake and S. Santini (2014). The ECO data set and the performance of non-intrusive load monitoring algorithms. Proceedings of the 1st ACM Conference on Embedded Systems for Energy-Efficient Buildings. Memphis, Tennessee, Association for Computing Machinery: 80–89.

Keywords: Data-driven model predictive control, Autoregressive with exogenous input model, Linear Programming, Residential buildings

Paper ID 30945: Predicting occupant thermal comfort for multiple air-side systems and seasonal scenarios using Autonomous HVAC CFD

Sandip Jadhav*, Praveen Kumar, Rohit Chavan, Avinash Goen

Centre for Computational Technologies Pvt Ltd, India

AIM AND APPROACH

Achieving the desired occupant thermal comfort in indoor spaces is a challenge that the HVAC industry is trying to address in recent times. It's seen that a thermally comfortable working environment enhances productivity and pleasant mood amongst the employees. A popular method to analyze thermal comfort is to perform CFD simulations for the indoor spaces with the selected air-side systems for seasonal scenarios and calculate the evaluation parameters such as PMV, PPD, Mean age of air etc. The complete and sound analysis of the occupant thermal comfort is possible when one has simulated the given indoor space for multiple air-side systems, varying seasonal and occupant density scenarios throughout the year. Such a complete analysis will require a large number of CFD simulations along with the CFD expertise and high-performance computing capabilities with the HVAC consultant. With the advent of powerful cloud computing, a new 'Autonomous HVAC CFD' application has been developed by simulationHub, that performs the CFD simulations on the cloud for multiple design configurations and scenarios and provides essential parameters to analyse the occupant thermal comfort. Through introduction and case studies, this paper illustrates the 'Autonomous HVAC CFD' app to assess the occupant thermal comfort in indoor spaces.

SCIENTIFIC INNOVATION AND RELEVANCE

Our innovation resides in the automation of the complete process of performing the CFD simulation to evaluate occupant thermal comfort in indoor spaces. Using the approach developed in the app, any user can perform an array of CFD simulations involving multiple air-side systems for multiple scenarios simultaneously. All the inputs related to CFD including external domain size, mesh density, turbulence model, numerical schemes are intelligently selected by the algorithm. The complete CFD workflow right from the fluid domain creation, meshing, solving governing equations, monitoring convergence, post-processing, and report generation is automated, thus removing the key barrier of the requirement of CFD expertise while performing such simulations. Being a cloud-based app, Autonomous HVAC CFD does not demand in-house high-performance computing resources or hard-wired servers. The app 3D CAD model of the indoor space as input along with the details of the air side systems and scenarios and predicts quantitative results such as PMV, PPD, EDT, DR, Mean age of air etc along with a detailed CFD report automatically.

PRELIMINARY RESULTS AND CONCLUSIONS

This paper includes the case study of office space with 7 rooms. The rooms serve different purposes such as workspace, pantry, conference room, lobby etc. A total of 5 air-side systems which include Constant Air Volume, Variable Air Volume and Underfloor Air Distribution Systems are chosen. A total of 6 different scenarios based on seasons and occupant density are simulated for this case study which combines to a total of 210 CFD simulations. Based on the results generated by these simulations, PMV, PPD, EDT, DR, Mean age of air and qualitative results such as contour plots and flowlines are generated to evaluate the occupant thermal comfort of the complete indoor office space. A detailed observation of these predicted results is discussed and the optimum air-side system for a year-round application is suggested. This paper shall try to demonstrate the usefulness and accessibility of the app to the HVAC consultants while determining the occupant thermal comfort in indoor spaces.

MAIN REFERENCES

1) ISO 7730:2005

Ergonomics of the thermal environment — Analytical determination and interpretation of thermal comfort using calculation of the PMV and PPD indices and local thermal comfort criteria

2) ANSI/ASHRAE Standard 62.1-2019 - Ventilation for Acceptable Indoor Air Quality

3) ASHRAE Standard 55, Thermal Environmental Conditions for Human Occupancy

4) P.V.Nielsen, S.Murakami, S.Kato, et al. "Benchmark test for Computer Simulated Person" Aalborg University, 2003

5) N. Martinho,A.Lopes,et al. "CFD modelling of benchmark tests for flow around a detailed computer simulated person", 7th International Thermal Manikin and Modelling Meeting - University of Coimbra, September 2008

Keywords: hvac, thermal comfort, cfd, simulations, BIM

Paper ID 30990: A simulation study to analyze the impact of Integrated Passive Strategy on air-ventilation and daylighting in a slum house of Mumbai

Abdul Moeed Chaudhary* (1), Sahil Priyadarshi (2), Shailee Goswami (3)

1: Essential India, Mumbai; 2: International Finance corporation (IFC); 3: Centre for Environmental Planning and Technology, Ahmedabad

AIM AND APPROACH

In India, an estimated 44 to 105 million people live in urban slums. Census data also indicates that a large section of the metropolitan city population lives in these informal settlements. For instance, around 41% of Mumbai's population lives in slums or 'Chawls', and it is projected that this statistic will further increase. The comfort conditions in these settlements are very poor and the occupants live in substandard daylighting and air ventilation levels.

This study aims at analyzing the capability of the proposed strategy to improve air ventilation and daylighting levels in a slum dwelling. The compact and adjoined spaces in Chawls don't allow open planning and restrict the construction of external fenestrations which results in poor airflow. Also, the presence of utilities and toilets within the living spaces makes the air quality worse. The daylight penetration in slum houses is also limited due to the dense planning and small windows that hinder the sky view. Thus, the main objective of this simulation study is to analyze the impact of Integrated Passive Strategy (IPS) in the form of a chimney, which is enhancing the daylighting and air ventilation levels.

SCIENTIFIC INNOVATION AND RELEVANCE

The compact and adjoined spaces in Chawls don't allow open planning and restrict the construction of external fenestrations which results in poor airflow. Also, the presence of utilities and toilets within the living spaces makes the air quality worse. The daylight penetration in slum houses is also limited due to the dense planning and small windows that hinder the sky view. The constricted space only allows minimum daylight inside the rooms and results in substandard illuminance levels.

By considering these design aspects, the simulation model is created for a ground floor slum-dwelling unit in CFD and daylighting simulation software. The feasibility of different combination of chimney designs are analyzed in this model. The impact of the solar-induced stack effect due to chimney is analyzed on the overall air exchange rate (ACH) in the room. The proposed chimney used in the model is a customized low-cost prototype that is manufactured locally. The effect of its integration in the space is quantified in terms of improvement in illuminance levels (lux).

PRELIMINARY RESULTS AND CONCLUSIONS

The proposed solution is exhibiting significant improvement in indoor environmental quality. The proposed solar chimney is inducing a stack effect that aids in providing adequate air exchange in the rooms. The results are showing 2.5 to 7 times increment in the air changes per hour (ACH) of proposed case comparison to base case and 2°C to 3°C reduction in temperature. The ACH rate is in compliance with NBC code along with the indoor operative temperature, which is meeting the standards for naturally ventilated spaces.

From daylighting perspective, the integration of proposed component is enabling effective daylight penetration within the spaces. This proposed solution is resulting in an average increase of almost 300 (lux) illuminance level in the living spaces. Uniformity ratio (min/avg) has also improved from 0.004 to 0.13. The Illuminance levels have also met the NBC standards requirement in the proposed case. The increase in illuminance levels has resulted in significant improvement in visual comfort conditions.

In conclusion, the integration of chimney component is significantly improving the room comfort conditions for the occupants living in slum houses. The existing compact planning calls for an effective solution that delivers adequate air quality and daylighting levels in the spaces.

MAIN REFERENCES

Bardhan, Ronita, et al. “Low-Income Housing Layouts under Socio-Architectural Complexities: A Parametric Study for Sustainable Slum Rehabilitation.” Sustainable Cities and Society, vol. 41, Elsevier B.V., 2018, pp. 126–38, doi:10.1016/j.scs.2018.04.038.

Debnath, Ramit, et al. A Data-Driven and Simulation Approach for Understanding Thermal Performance of Slum Redevelopment in Mumbai , India Department of Civil and Environmental Engineering , Stanford University, USA Centre for Urban Science and Engineering , Indian Institute O. 2016.

Karandikar, Priyanka, and Nadia Anderson. Chawls: Analysis of a Middle Class Housing Type in Mumbai, India. 2010, p. 96, http://libproxy1.nus.edu.sg/login?url=http://search.proquest.com/docview/848940263?accountid=13876.

O'Hare, Greg, et al. “A Review of Slum Housing Policies in Mumbai.” Cities, vol. 15, no. 4, 1998, pp. 269–83, doi:10.1016/S0264-2751(98)00018-3.

Sarkar, Ahana, and Ronita Bardhan. “Improved Indoor Environment through Optimised Ventilator and Furniture Positioning: A Case of Slum Rehabilitation Housing, Mumbai, India.” Frontiers of Architectural Research, no. xxxx, Elsevier Ltd, 2020, doi:10.1016/j.foar.2019.12.001.

Trani, Marco L., et al. “Template Customization for Construction Site Information Models.” Procedia Engineering, vol. 164, 2016, pp. 495–502, doi:10.1016/j.proeng.2016.11.650.

Keywords: Urban slums, Ventilation, Daylighting, Solar chimney, CFD

Session F2.7 (Online Track): Buildings paving the way for the energy transition / optimisation /control

Paper ID 30319: Prediction of HVAC loads at different spatial resolutions and buildings using deep learning models

Antonio Liguori* (1), Shiying Yang (1), Romana Markovic (2), Thi Thu Ha Dam (1), Andreas Wagner (2), Jérôme Frisch (1), Christoph van Treeck (1)

1: RWTH Aachen, Germany; 2: KIT, Germany

AIM AND APPROACH

This paper explores the applicability of deep learning-driven heating ventilation and air conditioning (HVAC) energy consumption models for agnostic commercial buildings. For that purpose, the modeling is conducted using two data sets. The first data set consisted of the energy consumption profiles from a 12-floor office building located in Seattle, USA. The second data set is collected on a significantly different 3-floor office building located in Aachen, Germany.

The modeling approach relies on recurrent neural networks (RNNs), while the input consists of physical data streams such as indoor air temperatures and data obtained from the central HVAC. The research steps include the implementation of an existing RNN-based model for energy consumption and further model optimization using training and validation set. Eventually, the final model was evaluated using the data from two data sets and the evaluation performance was evaluated in case of the varied spatial and system granularities. For that purpose, the energy consumption boundaries were defined as a) HVAC consumption in single office b) multiple thermal zones (floor-wise consumption) as well as c) building-wise HVAC consumption.

SCIENTIFIC INNOVATION AND RELEVANCE

Significant proportion of the worldwide energy consumption is caused by buildings. In particular, HVAC systems are responsible for 50 % of the total building energy consumption in United States (Pérez-Lombard et al. (2008)), while the similar trend is confirmed across different locations (Al Amoodi and Azar, (2018); Knight, (2012)). Due to its' amplitude and autoregressive properties, the energy consumption required for HVAC represents a major optimization potential towards a better use of renewable energy and more energy efficient buildings.

One of the approaches to achieve the latter two goals is to predict the energy improve the prediction of the required energy loads. In past, the energy loads were commonly predicted by applying analytical simulation-based approaches. However, such physical simulation models are not well scalable to a large number of buildings and they require extensive model calibration. As an alternative, data-driven methods present a promising approach that is more applicable to agnostic buildings or energy prediction system boundaries (such as thermal zone, building or district level). In that regard, the presented deep RNN model is identified as a suitable modeling approach that leads to satisfying predictive accuracy.

PRELIMINARY RESULTS AND CONCLUSIONS

The results showed, that the existing RNN-based model proposed by Chen et al. (2017) is well applicable to different buildings, given repeated model training using data from the target domain (i.e. data set from Germany). The accuracy for building-wise HVAC energy consumption ranged between 3 % and 8 % in terms of normalized root mean squared error, while the spatial granularity had a direct impact on the predictive performance. Furthermore, the data analysis showed that the energy consumption at more coarse spatial resolution is less stochastic and it shows higher autocorrelation. According to the conducted analytics, these properties could be accounted for simplier model formulation and improved accuracy in case of the building-wise predictive modeling. Further results as well as conclusions will be presented in the main paper.

MAIN REFERENCES

Al Amoodi, A., & Azar, E. (2018). Impact of Human Actions on Building Energy Performance: A Case Study in the United Arab Emirates (UAE). Sustainability, 10(5), 1404.

Chen, Y., Shi, Y., & Zhang, B. (2017). Modeling and optimization of complex building energy systems with deep neural networks. In 2017 51st Asilomar Conference on Signals, Systems, and Computers (pp. 1368-1373). IEEE.

Knight, I. P. (2012). Assessing electrical energy use in HVAC systems. REHVA Journal (European Journal of Heating, Ventilating and Air Conditioning Technology), 49(1), 6-11.

Kusiak, A., Li, M., & Tang, F. (2010). Modeling and optimization of HVAC energy consumption. Applied Energy, 87(10), 3092-3102.

Pérez-Lombard, L., Ortiz, J., & Pout, C. (2008). A review on buildings energy consumption information. Energy and buildings, 40(3), 394-398.

Keywords: HVAC, recurrent neural networks, generic modeling, time-series

Paper ID 30371: Data-driven calibration of joint building and HVAC dynamic models using scalable Bayesian optimization

Ankush Chakrabarty* (1), Emilio Maddalena (2), Hongtao Qiao (1), Christopher Laughman (1)

1: Mitsubishi Electric Research Labs, United States of America; 2: École polytechnique fédérale de Lausanne, Switzerland

AIM AND APPROACH

Physics-informed simulation models of heating, ventilation, and cooling (HVAC) systems play a critical role in predicting system dynamics and enabling analysis, control, and optimization of buildings and equipment. The predictive performance of these simulation models is strongly linked to calibration mechanisms: algorithms that systematically select parameter values that optimize a given calibration-cost map (e.g., L-2 error). Poorly selected parameter values typically result in large deviations between measured building data and simulated data, limiting the utility of the simulation model in subsequent design.

State-of-the-art calibration methods explore the parameter space by computing numerical gradients that are susceptible to measurement noise or employ population-based search mechanisms that require exorbitant data. To improve robustness and curtail data requirements, one can 'learn' or approximate the calibration-cost map and subsequently leverage the topology of the approximated function to find good search directions despite noisy measurements.

Concretely, we employ machine learning to construct a calibration-cost map to direct model calibration for systems with joint dynamics of buildings and HVAC equipment. The learner explores subregions of the parameter space with high uncertainty and queries the model only where collecting simulation data yields useful information. This leads to lower simulation data-requirements compared to widely used calibration mechanisms.

SCIENTIFIC INNOVATION AND RELEVANCE

Calibration-cost maps are not always differentiable/convex due to coupled interactions in joint building/equipment dynamics. Furthermore, measured data is corrupted by environmental and process noise, limiting the effectiveness of gradient-based methods. Population-based, gradient-free searches are effective, but incur high computational expenditure as they require an exorbitant number of simulations. Dynamical estimators such as Kalman filters also underperform due to multi-rate dynamics and limited generalizability of state-space models used to design these estimators.

Our contributions are as follows.

- We study the problem of model calibration for joint building/equipment dynamics.

- We employ data-driven Gaussian processes (GPs) for learning a parameter-to-calibration-cost function that contains the true cost with high probability. The GP also generates confidence bounds around predicted function values that quantify the prediction uncertainty at various regions in the parameter space.

- We utilize confidence bounds to explore the parameter space without requiring exorbitant simulations by collecting simulation data for parameters in regions with large uncertainty bounds and high likelihood of containing global optima, both of which are estimated by designing an appropriate acquisition function (as per standard Bayesian optimization).

- After sufficient exploration, we exploit the topology of the learned function to obtain optimal parameters without requiring additional simulation data.

PRELIMINARY RESULTS AND CONCLUSIONS

We employ Modelica to design a physics-informed model of a building system with closed-loop controllers embedded at the equipment level. We calibrate solar and infrared emissivity coefficients for the roof along with an infiltration/volumetric flow rate of the building. The calibration is performed using measured temperature data (space and ambient) obtained at 15-minute intervals over 7 days. A challenging aspect of using temperature data to calibrate building parameters is that the calibration-cost map is confounded by environmental and climactic factors. Despite this difficulty, our proposed Bayesian optimization method estimates these 3 parameters to >95% accuracy within 100 simulation runs. In contrast, gradient-based methods report <80% accuracy with 2 parameters and diverge when searching for all 3. Population-based methods with a modest population size (e.g., 20 agents) require >100 iterations (thus, >2000 simulation runs) to find good estimate, which is far more computationally demanding than our proposed methodology.

Encouraged by preliminary results using Bayesian optimization for selecting building parameters, our ongoing efforts are devoted to developing and testing scalable Bayesian optimization frameworks in joint models of buildings and equipment. Evolution of system dynamics at multiple timescales, along with coupled nonlinear interactions between these systems, makes this joint problem particularly challenging.

MAIN REFERENCES

[1] D. Coakley, P. Raftery, M. Keane, A review of methods to match building energy simulation models to measured data, Renewable Sustainable Energy Rev., Vol. 37, pp. 123–141, 2017.

[2] A.-T. Nguyen, S. Reiter, P. Rigo, A review on simulation-based optimization methods applied to building performance analysis, Appl. Energy, Vol. 113, pp. 1043–1058, 2014.

[3] S. Asadi et al., Building energy model calibration using automated optimization-based algorithm, Energy and Buildings, Vol. 198, pp. 106–114, 2019.

[4] J. Snoek, H. Larochelle, and R. P. Adams. Practical Bayesian optimization of machine learning algorithms. Adv. in Neural Information Processing Systems, pp. 2951-2959. 2012.

[5] A. Chakrabarty, M. Benosman. Safe Learning-based Observers for Unknown Nonlinear Systems using Bayesian Optimization. arXiv preprint arXiv:2005.05888 (2020).

[6] C. R. Laughman, C. Mackey, S. A. Bortoff, and H. Qiao. Modeling and Control of Radiant, Convective, and Ventilation Systems for Multizone Residences. Proc. 16th IBPSA Conference, 2019.

Keywords: Bayesian optimization, machine learning, Modelica, parameter calibration, Gaussian processes

Paper ID 30636: A performance-driven simulation workflow for PV integration into the design process: application on an innovative building project in Switzerland

Sergi Aguacil Moreno*

Building2050 group, Ecole Polytechnique Fédérale de Lausanne (EPFL), Fribourg, Switzerland

AIM AND APPROACH

In the construction sector, the integration of active elements functioning both as building envelope material and on-site electricity generator is identified as a key measure to achieve the 2050 targets and carbon neutrality [1,2]. Despite an increasing confluence of the photovoltaic (PV) industry and the building glass manufacturers, which offers high design freedom in relation to size, colour and texture, architects and engineers continue to address the issue in a traditional way, by limiting themselves to the application of "standard" catalogue products that particularly constrain their designs [3,4]. This way of working, which makes it difficult to integrate PV systems into the building envelope and, therefore, into the overall building design, often leads to a method of sizing that largely makes abstraction of the building's electricity needs. This article presents the application – in an ongoing design process – of an innovative active-surface selection method adapted to the design phases using 3D-modeling and hourly-step simulations. Through an integrated-design approach, the proposed workflow allows quickly obtaining visual and quantitative results on the building envelope, and supports a design decision-making process that proposes an alternative approach to usual practice.

SCIENTIFIC INNOVATION AND RELEVANCE

The literature review shows a lack of reliable design-driven methods to support the sizing and implementation of building-integrated (BI)PV installations in projects [5,6]. Most existing methods are highly time-consuming and based on complex optimisation algorithms not tailored to the workflow of architects/designers [7].

We propose the application of a novel method that relies on BIM-integrated parametric tools to allow architects to get instant feedback on the impact of their decisions on building energy performance along the design process. The method consists in a streamlined and automated workflow supporting BIPV design (sizing and positioning) in coherence with the building's context, architectural features, and electricity needs. It uses the Rhino.Inside Revit add-on [8], an open source Rhino WIP project that allows Rhino [9] and Grasshopper [10] to run inside other 64-bit Windows applications such as Revit, by which we are able to apply, to any Revit informed-3D model, our Grasshopper definition for the calculation and optimisation of PV systems upon building envelopes. This in-depth analysis from the early design stages aims to reduce the costs due to design changes [11]. The method's added-value is here tested in a real project and compared to more common PV-sizing approaches.

PRELIMINARY RESULTS AND CONCLUSIONS

Our method was applied in the ongoing design of the future Smart Living Lab building in Fribourg, Switzerland. Prior to this test, we observed that the design of the PV installation for the building tended to follow a common approach, i.e. adjust standard, building-applied (BA)PV panels on the least visible surfaces (roof), while aiming to produce the maximum amount of energy possible. This approach is partly explained by the desire to achieve the Minergie-A label [12], which requires the production of more energy than the building's needs in absolute annual value. Moreover, the most economical solution was to be favoured, to the detriment of architectural integration. The method was then applied, focusing on reaching a self-sufficiency and self-consumption adapted to the needs of the building, regardless of the annual amount of energy produced. This resulted in a better-adapted installation that not only produces the needed energy, but also reduces the building's environmental impact since the integrated panels substitute part of the materials that would have been otherwise used for the building envelope. Results show that although the building is compact and located in a dense urban context, high-level of performance such as the 2000-Watt society targets for 2050 are reachable.

MAIN REFERENCES

[1] S. Aguacil, S. Lufkin, E. Rey, Active surfaces selection method for building-integrated photovoltaics (BIPV) in renovation projects based on self-consumption and self-sufficiency, Energy Build. 193 (2019).

[2] S. Aguacil, Architectural Design Strategies for Building Integrated Photovoltaics (BIPV) in Residential Building Renovation (Thesis N°9332), EPFL Lausanne, (2019).

[3] P. Bonomo et al., BIPV: building envelope solutions in a multi-criteria approach. A method for assessing life-cycle costs in the early design phase, Adv. Build. Energy Res. 11 (2017) 104–129.

[4] I. Zanetti et al., BIPV- Product overview for solar building skins- Status Report, (2017).

[5] C. Ballif et al., Integrated thinking for photovoltaics in buildings, Nat. Energy. 3 (2018) 438–442.

[6] P. Heinstein et al., Building Integrated Photovoltaics (BIPV): Review, Potentials, Barriers and Myths, Green. 3 (2013).

[7] T. Østergård et al., Building simulations supporting decision making in early design - A review, Renew. Sustain. Energy Rev. 61 (2016) 187–201.

[8] Robert McNeel&Associates, Rhino.Inside, (2020). https://www.rhino3d.com/inside.

[9] Robert McNeel&Associates, Rhinoceros software, (2020). https://www.rhino3d.com/.

[10] S. Davidson, Grasshopper, Algorithmic modelling for Rhino, (2020). http://www.grasshopper3d.com/.

[11] A. Hollberg, A parametric method for building design optimization based on Life Cycle Assessment, Bauhaus-Universität Weimar, (2016).

[12] Minergie, Minergie standard certification label, (2020)

Keywords: active surfaces, building-integrated photovoltaics, architectural design, carbon emission, building information modelling

Paper ID 30658: An inquiry into the accuracy of the energy model calibration process

Ipek Yilmaz (1), Burak Gunay* (1), Guy Newsham (2), Adam Wills (2)

1: Carleton University, Canada; 2: National Research Council, Canada

AIM AND APPROACH

This paper aims at quantifying the uncertainty inherent in the building energy model calibration process. The following research questions are investigated: (a) how accurately can the unknown parameters of a calibrated energy model be estimated through the model calibration process?; (b) how do the imperfections in the model calibration process affect the calibrated model's ability to make operational decisions? To answer these questions, we needed a building with all thermophysical properties accurately characterized. To this end, a three-storey office building in Ottawa, Canada is simulated with different occupancy and envelope property scenarios to generate metered energy data for calibration. Subsequently, several input parameters (e.g., air infiltration, thermal conductances) are then assumed unknown. The model is coupled to a custom optimization script which searches for these unknown parameters by minimizing the deviation to the metered energy data subject to practical and physical constraints. The process is repeated with monthly and hourly metered energy data while introducing varying levels of systematic error to the meter data. The performance of the model calibration process is assessed based on the optimization algorithm's ability to estimate the unknown parameters. Further, the calibrated energy models are used to make operational decisions related to air handling units.

SCIENTIFIC INNOVATION AND RELEVANCE

Energy models for existing buildings are needed to make operation and retrofit decisions. However, pure data-driven modelling approaches may not be suitable to many existing buildings due to issues in obtaining metadata and limitations in the sensing and data collection infrastructure. Thus, physics-based energy models calibrated with metered energy use data have gained popularity among the building simulation community. In practice, a calibrated physics-based energy model is largely assumed to be an adequate representation of the actual building when it complies with a measurement and verification standard's criteria for fitness (e.g., ASHRAE Guideline 14). To this end, the analysis presented in this paper advances our understanding about the uncertainty in the energy model calibration process. It demonstrates the imperfections in the model calibration process and highlights how these imperfections translate into the building performance simulation-based operational decision-making process.

PRELIMINARY RESULTS AND CONCLUSIONS

Preliminary results of this investigation point out that calibration with monthly energy use data does not generate accurate estimates of the unknown parameters – even though the goal was to estimate only seven parameters. Due to the multicollinearity amongst many of the parameters, it was challenging to acquire correct estimates, particularly for the window and wall thermal conductances, and the air-infiltration rate. The unknown parameters were accurately estimated through calibration with hourly energy use data, albeit only when the systemic error in the metered energy data was less than ±5%. Imperfections in the model calibration process appeared to affect only some of the operational decisions. Thus, a calibrated model, despite not being able to adequately represent the thermophysical properties of a building, may support the operational decision-making process.

MAIN REFERENCES

Gunay, H. B., Ouf, M., Newsham, G., & O'Brien, W. (2019). Sensitivity analysis and optimization of building operations. Energy and Buildings, 199, 164-175.

O'Neill, Zheng, and Bryan Eisenhower. "Leveraging the analysis of parametric uncertainty for building energy model calibration." Building simulation. Vol. 6. No. 4. Springer Berlin Heidelberg, 2013.

Sun, Kaiyu, et al. "A pattern-based automated approach to building energy model calibration." Applied Energy 165 (2016): 214-224.

Lim, Hyunwoo, and Zhiqiang John Zhai. "Influences of energy data on Bayesian calibration of building energy model." Applied Energy 231 (2018): 686-698.

Guyot, Dimitri, et al. "Building energy model calibration: a detailed case study using sub-hourly measured data." Energy and Buildings (2020): 110189.

Chong, Adrian, et al. "Bayesian calibration of building energy models with large datasets." Energy and Buildings 154 (2017): 343-355.

Mihai, Andreea, and Radu Zmeureanu. "Bottom-up evidence-based calibration of the HVAC air-side loop of a building energy model." Journal of Building Performance Simulation 10.1 (2017): 105-123.

Keywords: Calibration, Optimization, Uncertainty quantification, Operational decision-making

Paper ID 30879: Deep reinforcement learning-based optimal building energy management strategies with photovoltaic systems

Minjeong Sim, Geonkyo Hong*, Dongjun Suh

Kyungpook National University, Korea, Republic of (South Korea)

AIM AND APPROACH

<Research abstract>

Owing to the spread of solar photovoltaic (PV) systems, a significant amount of research has been conducted on the development of efficient energy management methods. Significantly, the energy operation strategies are essential for residential buildings due to the difference between peak demand and solar power generation time.

Therefore, we proposed a novel deep reinforcement learning-based model considering both, direct use of the generated energy to the buildings and selling to utilities to minimize the building's total energy operating cost in a residential building with PV-energy storage system (ESS) installed.

To verify the performance of the proposed model, case studies such as rule-based, selling-only case, and consumption-only case were conducted, showing that the proposed model minimized energy operating costs.

SCIENTIFIC INNOVATION AND RELEVANCE

-Practical Implications

This study supports a deep reinforcement learning-based optimal PV-ESS management system for residential facilities considering economic efficiency based on several energy management strategies.

Furthermore, this study can provide proper guidelines for deriving the optimal energy management methods of residential buildings equipped with PV-ESS systems, according to various operation strategies, energy policies, and building types.

- Research framework

PV power generation is considered for the purpose of reducing the total operating cost and aims to minimize the operating cost of buildings by lowering the peak of energy consumption of buildings.

The energy management system is then modeled using the Deep Reinforcement Learning(DRL)-based method. The energy stored in the ESS can be determined by discharging used to lower the peak load of the building and selling to grid. Based on this behavior, the operating cost of the building can be optimized.

In summary, PV/ESS scheduling method is presented through time series-based building energy flow analysis. In addition, we derive an energy management algorithm based on deep reinforcement learning through the exchange of PV/ESS and external grid energy.

PRELIMINARY RESULTS AND CONCLUSIONS

This study proposes an optimal operational management model of PV-ESS system based on DRL to minimize the building's total operating costs in residential buildings where a difference between peak demand and solar power generation time is observed. The proposed model minimizes the operating cost of the building through the charging and discharging scheduling of ESS. In the proposed DQN-based model, the ESS agent learns actions until it maximizes the total cumulative reward while preventing overcharging and undercharging.

A simulation-based case study was conducted to verify the performance of the proposed model. The experimental results show that the proposed model reduced the energy operation cost by $21,609 compared to the rule-based approach, confirming the monetary benefits. Thus, the proposed model can provide a guideline for optimal energy management methods of residential buildings equipped with PV-ESS systems.

Further research is needed to improve the PV and ESS optimally linked energy management methods, considering several energy policies in the development of energy management techniques of the general-purpose PV-ESS system.

MAIN REFERENCES

IEA (2017) Market Report Series Renewables 2017

Gul, M., Kotak, Y. and Muneer, T. (2016) Review on recent trend of solar photovoltaic technology, Energy Exploration and Exploitation

Kim, B. and Suh, D. (2020) 'A hybrid spatio-temporal prediction model for solar photovoltaic generation using numerical weather data and satellite images', Remote Sensing, 12(22), pp. 1–21.

Teleke, S. et al. (2010) 'Rule-based control of battery energy storage for dispatqianng intermittent renewable sources', IEEE Transactions on Sustainable Energy. IEEE, 1(3), pp. 117–124.

Hong, T. et al. (2014) 'An economic and environmental assessment for selecting the optimum new renewable energy system for educational facility', Renewable and Sustainable Energy Reviews. Elsevier, 29, pp. 286–300.

Keywords: Optimization, Building Energy Simulation, PV, ESS, Deep Reinforcement Learning

Session F3.1: Improving indoor environmental quality

Paper ID 31010: Improving indoor environmental quality through smart operable windows; a machine learning approach

Farimah Moezzi* (1), Amirhossein Fathi (2)

1: Shahid beheshti university, Iran, Islamic Republic of Iran; 2: Shiraz University, Iran, Islamic Republic of Iran

AIM AND APPROACH

Global warming has drawn public attention to the rise in fossil fuels' price, air pollution and climate change throughout the world. In today's smart buildings, thermal comfort is directly related to indoor air quality (IAQ) and indoor environmental quality (IEQ), which in turn depends on weather conditions such as air velocity, solar heat gain and relative humidity (RH). The notion of IEQ improvement is obtained by different ways of controlling the parameters of internal condition in buildings. The effect of indoor environmental quality on thermal comfort and occupants' function is undeniably entwined with one another. In this research the aim is to find the relations between IEQ parameters and occupant behaviors and preferences using a Building Energy Management System (BEMS) framework. The study attempts to realize the factors and roles of IEQ parameters impacted by occupants autonomy via availing of two of the most effective machine learning and deep learning techniques, namely Support Vector Regression (SVR) and Recurrent Neural Network (RNN), respectively. This study is conducted on an educational building and the two aforementioned machine learning techniques- for better accuracy and speed in prediction- are utilized to explore the further possible IEQ improvements.

SCIENTIFIC INNOVATION AND RELEVANCE

The IEQ is measured through multiple parameters such as indoor/outdoor temperatures, RH, CO2 concentration and air velocity both inside and outside of a conditioned space. One of the main challenges to the buildings is stochastic and non-controllable behavior of active occupants like opening the windows inside a smart building as the main factor of dissatisfaction. Occupants unawareness of outside conditions like adverse RH and high CO2 concentration leads to sporadic and irrational decision such as opening window when they are not supposed to be open. Owing to the aforementioned, this study aims to provide an online questionnaire through a Building Energy Management System, which simultaneously adapts with the collected data from the sensors to improve the independency and practicability of the BEMS framework by operating window openings considering conditions air pollution (CO2 concentration). Moreover this haphazard occupants behavior is prone to have intensified impacts due to the occupants limited responsibility in tertiary sector buildings (e.g. educational buildings), for instance leaving a window open for a while. The online questionnaire will be updated every 4 hours over the period of an academic semester through keeping the record of occupant's presence, which will be limited to the working hours of the building.

PRELIMINARY RESULTS AND CONCLUSIONS

The outcomes of this study on the one hand will facilitate the acceptance of BEMS technology for elderly people and people with disabilities by assisting to taking the control of windows operation; on the other hand, for tertiary buildings with educational and office functions the energy use will be drastically ameliorated. The acceptance and compatibility of such frameworks which depend significantly on the buildings function, occupants' behavior, pollution level and the climatic conditions, will be enhanced by deploying SVR and RNN machine learning techniques. The collected data of temperature, relative humidity and CO2 sensors will be gathered from classrooms located in different thermal zones of the building. Exploiting machine learning techniques has proven to be instrumental in achieving a better IEQ, alongside reducing the energy consumption of the building, while taking into account the occupants preferences and behavior, which in most of the previous studies have been neglected.

MAIN REFERENCES

Peng, Y., Rysanek, A., Nagy, Z., & Schlüter, A. (2018). Using machine learning techniques for occupancy-prediction-based cooling control in office buildings. Applied Energy, 211, 1343–1358.

Sattarpour, T., Nazarpour, D., & Golshannavaz, S. (2018). A multi-objective HEM strategy for smart home energy scheduling: A collaborative approach to support microgrid operation. Sustainable Cities and Society, 37, 26–33.

Schieweck, A., Uhde, E., Salthammer, T., Salthammer, L. C., Morawska, L., Mazaheri, M., & Kumar, P.

(2018). Smart homes and the control of indoor air quality. Renewable and Sustainable Energy Reviews, 94, 705–718.

Zou, Z., Yu, X., & Ergan, S. (2020). Towards optimal control of air handling units using deep reinforcement learning and recurrent neural network. Building and Environment, 168, 106535.

Keywords: IEQ, machine learning, operable window, occupants behavior

Paper ID 30164: Hygrothermal modeling of green roof made with substrate and drainage layers of coarse recycled materials

Mostafa Kazemi*, Luc Courard, Shady Attia

University of Liège, Belgium

AIM AND APPROACH

The thermal performance of an extensive green roof can be influenced by the initial hygrothermal conditions of substrate and drainage layers. Moreover, coarse recycled materials can affect the thermal resistance of green roof layers, while there is a demand for optimizing their thickness. Therefore, the main objective of this study is to optimize the thickness of green roof layers, once coarse recycled materials were used for substrate and drainage layers: WUFI software has been used for such application, which was suitable for modeling the initial hygrothermal conditions (heat and moisture properties) of green roof layers. The thermal resistance of green roof with substrate of coarse recycled material and the drainage layer of recycled coarse aggregate was tested and evaluated concerning ISO 9869-1 (2014). After that, the heat and moisture properties of green roof layers were introduced into WUFI software and then the modeling outputs were compared and validated with experimental results. Later on, the temperature distribution through the depth of green roof layers was assessed once the the thickness of substrate and drainage layers was numerically changed. Finally, the optimum thicknesses of green roof models with a reliable thermal performance were introduced.

SCIENTIFIC INNOVATION AND RELEVANCE

Although some researches have been carried out on green roof systems' insulation performance, few studies have assessed and simulated the hygrothermal conditions of green roof layers, mainly including coarse recycled materials. Also, the optimization of substrate and drainage layers with recycled coarse aggregate has rarely been assessed so far. There are no European standards to propose the optimum thickness for green roof layers concerning their thermal performance. Therefore, there is a demand for evaluating the thermal resistance of substrate and drainage layers with coarse recycled materials based on their hygrothermal properties.

PRELIMINARY RESULTS AND CONCLUSIONS

The results showed that the insulation performance of green roof is dependent on the type of materials used for the substrate and drainage layers. Meanwhile, the fluctuation of temperature and humidity decreases through the depth of the green roof. Furthermore, the insulation performance of green roof increases by increasing the thickness of substrate and drainage layers.

MAIN REFERENCES

1. L. Rincón, J. Coma, G. Pérez, A. Castell, D. Boer, L.F. Cabeza, Environmental performance of recycled rubber as drainage layer in extensive green roofs. A comparative life cycle assessment. Building and Environment, 74 (2014) 22–30.

2. J. Coma, G. Pérez, C. Solé, A. Castell, L.F. Cabeza, Thermal assessment of extensive green roofs as passive tool for energy savings in buildings. Renewable Energy, 85(11) (2016) 1106–1115.

3. P.C. Tabares-Velasco, M. Zhao, N. Peterson, J. Srebric, R. Berghage, Validation of predictive heat and mass transfer green roof model with extensive green roof field data. Ecological Engineering, 47 (2012)165–173.

4. G. Pérez G, A. Vila, L. Rincón, C. Solé, L.F. Cabeza, Use of rubber crumbs as drainage layer in green roofs as potential energy improvement material, Applied Energy, 67(2012) 347-354.

5. M. Wanielista, M. Kelly, M. Hardin, A comparative analysis of green roof designs including depth of media, drainage layer materials, and pollution control media; Florida Department of Environmental Protection: Tallahassee, FL, USA, (2008).

6. S. Parizotto, R. Lamberts, Investigation of green roof thermal performance in temperate climate: A case study of an experimental building in Florianópolis city, Southern Brazil, Energy and Buildings, 43 (2011) 1712–1722.

Keywords: Thermal behaviour, relative humidity, extensive green roof, drainage layer, substrate

Paper ID 30353: A computational approach for checking compliance with European view and sunlight exposure criteria

Eleonora Brembilla*, Shervin Azadi, Pirouz Nourian

Delft University of Technology, The Netherlands

AIM AND APPROACH

In 2018 a new European standard on daylighting, EN 17037 [1] was introduced and ratified in all European Union countries, largely superseding existing national daylighting codes. The standard includes recommendations for indoor daylight provision, view out, exposure to sunlight and protection from glare. The proposed paper focuses on the 'view out' and 'exposure to sunlight' sections, introducing a computational approach to assess them. Presently, for these two performance parameters, the standard suggests evaluation methods that are solely based on building geometrical characteristics and on 2D layout representations. The approach of this work is twofold: establish that the compliance check can be systematically performed on a 3D model by a novel simulation tool developed by the authors, and quantify the uncertainty of the building performance targets set by the standard. Three scenarios characterised by different levels of density and complexity are analysed. The work aims at testing that the analysed sections of the new standard can be successfully integrated in contemporary design practice – characterised by the use of 3D models and parametric/procedural workflows – and that the recommended targets are achievable in realistic urban and rural scenarios.

SCIENTIFIC INNOVATION AND RELEVANCE

The EN 17037 standard is still relatively new and its implementation into commercial software is at a preliminary stage [2]. The sections of the standard on 'daylight provision' and 'protection from glare' are going through a more extensive scrutiny [3-4], as tools to systematically simulate and assess them already exist, or require little modification to be used for these purposes. Conversely, the sections on 'view out' and 'exposure to sunlight' stemmed from more traditional, manual approaches to assess windows and apertures, not suitable for systematic analyses, typical of building performance simulation. This paper introduces what is likely to be the first computational tool for simulation and verification of compliance with EN17037 for 'view out' and 'sunlight exposure' recommendations. A simulation tool that well integrates with computational design workflows capable of processing 3D CAD or BIM models is fundamental to guarantee a widespread adoption of this new standard. Moreover, such a tool can provide a basis for generative design methods aimed at systematically seeking out optimal building shapes. Lastly, quantifying the uncertainty of the metrics prescribed by the standard will shed light on its strengths and limitations, and provide stronger arguments pro or against future modifications of the current standard document.

PRELIMINARY RESULTS AND CONCLUSIONS

In its current form, the EN 17037 standard is presented as a series of non-normative recommendations to improve indoor visual quality, hence leaving great freedom of interpretation and relying on the designer's best judgement. This results in high uncertainties related to user decisions, with variations of up to 80% in the case of the view out horizontal sight angle parameter. Design uncertainties are lower but nevertheless significant, with variations mostly related to the level of detail of the model under analysis and of the surrounding buildings. The development of a computational tool for compliance verification on 3D building models allows the authors to draw conclusions on whether the 'simple', two-dimensional methods recommended by the standard can be generalised and applied in assessing buildings within more complex settings. Wherever the standard does not provide sufficient information, several assumptions – listed in the paper – are introduced by the authors. This list will serve other developers and building performance simulation users in implementing the standard, as well as policymakers in potential revisions of the current standard version.

MAIN REFERENCES

[1] European Committee for Standardization, 2018. "Daylight in Buildings – EN 17037:2018".

[2] Paule, B, J Boutillier, S Pantet, and Y Sutter. 2018. "A Lighting Simulation Tool for the New European Daylighting Standard." In Proceedings of BSO 2018: 4th Building Simulation and Optimization Conference. Cambridge, UK.

[3] Jones, Nathaniel L. 2019. "Fast Climate-Based Glare Analysis and Spatial Mapping." In Proceedings of the 16th Building Simulation Conference, 982–89. Rome, Italy. https://doi.org/10.26868/25222708.2019.210267.

[4] Šprah, Nataša, and Mitja Košir. 2019. "Daylight Provision Requirements According to EN 17037 as a Restriction for Sustainable Urban Planning of Residential Developments." Sustainability 12 (1): 315. https://doi.org/10.3390/su12010315.

Keywords: daylighting, window design, computational tool, layout planning, building quality standardisation

Paper ID 30528: Building performance simulation supporting typical design activities: the case of 'Organising 2D layout'

Steffen Petersen*, Mike Steen Kingo Danielsen, Pernille Klavsen

Aarhus University, Denmark

AIM AND APPROACH

The research presented in this paper contributes to the development of building performance simulation (BPS) tools that are conformed to fit into common design practices in the early design stage. Purup and Petersen (2020) have through semi-structured interviews with practicing architects identified various archetypes of design activities undertaken by designers in the early stages of the building design process. One of them is where the designer is manipulating the plan solution inside an already defined outer building boundary - i.e. organising the 2D layout of the building. The aim of this study was to enable building designers to conduct building performance simulations (BPS) for energy and thermal indoor climate code compliance during this specific design activity. The approach was to enable an automated generation of BPS model in ICEbear (Purup and Petersen, 2017) and DIVA from arbitrary 2D CAD drawings and a few other user-defined inputs, and the present the performance results to the designer directly in the CAD environment. This way the designer can manipulate the 2D layout on apartment level and room level while getting fast and reliable simulations of the consequence on energy performance and overheating risk.

SCIENTIFIC INNOVATION AND RELEVANCE

There are research-based efforts focusing on proposing procedures for enhancing efficient use of building simulation tools for proactive performance prediction in the early design stage, e.g. Petersen (2011), Attia et al (2012), and Gerber and Lin (2014), to cite but a few. However, to our knowledge, there is no research-based evidence on whether any of such procedures are actually being adopted by designers in professional design practice; our notion is that it is rare. A reason might be that current proposals for procedures and methodologies are only assumptions of, or they interprets wrong, what architects actually need (de Souza 2011). Purup and Petersen (2020) found that the conceptual building design process is different from project to project, and therefore difficult to generalize. However, the design process can be regarded as a project-specific or personalized sequence of generalizable design activities; BPS tools should be integrated in these common design activities to inform the decision-making process with regard to e.g. indoor climate, daylight and energy performance in an easy, non-disruptive, and non-intrusive way. This paper follows this realisation by proposing how BPS can be integrated in the common design activity of "Organising 2D planning".

PRELIMINARY RESULTS AND CONCLUSIONS

The support of the design activity "2D layout" is currently possible in the CAD environment of Rhino/Grasshopper. Here, the building designer can manipulate their own 2D planning sketches, and then with a few additional inputs get a thermal and daylight simulation of their own design suggestions. The results are visualised directly in the Rhino model. The integration of additional inputs and the output visualisation was developed in a co-creation process involving practicing architects who ended up adopting the integration as a standard in their design practice.

The preliminary conclusion is that BPS tools can be integrated in 2D layout activities in such a way that the designer find it neither intrusive, or disruptive but rather informative and constructive; it is suggested that the co-creation process is key to this success.

MAIN REFERENCES

Attia, S., Gratia E., De Herde, A. and Hensen J.L.M. Simulation-based decision support tool for early stages of zero-energy buildings. Energy and Building 49 (2012): 2-15.

de Souza, C. B. (2011) Contrasting paradigms of design thinking: The building thermal simulation tool user vs. the building designer. Automation in Construction, Vol. 22, pp. 112–122

Gerber, D. J. and Lin S.E. 2013. Designing in complexity: Simulation, integration, and multidisciplinary design optimization for architecture. Simulation 0 (0): 1-24.

Lin, S-H and Gerber D.J. Evolutionary energy performance feedback for design: Multidiciplinary design optimization and performance boundaries for design decision support. Energy and Buildings 84 (2014): 426-441.

Petersen, S. 2011. Simulation-based support for integrated design of new low-energy office buildings. PhD diss., Danish Technical University (DTU), Kgs. Lyngy, Denmark.

Purup, P. B. and Petersen S. 2020. Requirement analysis for building performance simulation tools conformed to fit design practice. Automation in Construction 116, 103226

Purup P.B. and Petersen S. Rapid Simulation of Various Types of HVAC Systems in the Early Design Stage. Energy Procedia 122 (2017) 469-474

Keywords: Thermal performance, Daylight performance, Parametric modelling, Design Practice

Paper ID 30683: Natural ventilation predictions for a slum house in Dhaka using large-eddy simulations within a multi-fidelity simulation framework with uncertainty quantification

Yunjae Hwang*, Catherine Gorle

Stanford University, United States of America

AIM AND APPROACH

According to UNICEF, approximately one million children die every year from pneumonia, which is the leading cause of death in children under five[1]. A previous study in Dhaka, Bangladesh, indicates that there might be an association between the occurrence of pneumonia and the presence of cross-ventilation in slum housing [2]. The objective of this research is to establish a validated computational framework to accurately estimate household ventilation rates and support further investigation of this association.

To achieve this objective, we perform high-fidelity computational fluid dynamics (CFD) simulations that resolve the turbulent flow field. The results of these computationally demanding large-eddy simulations (LES) are used to develop relationships for the air change per hour (ACH) as a function of the indoor-outdoor temperature difference and wind conditions; these relationships are then implemented into a computationally efficient building thermal model with uncertainty quantification (UQ). The resulting multi-fidelity model results are validated against field measurements, considering both the air change per hour (ACH) measured using a tracer concentration decay technique, and time-series of air and wall temperatures.

SCIENTIFIC INNOVATION AND RELEVANCE

Efficient use of CFD for predicting natural ventilation system performance is challenging because of inherent variability in boundary and operating conditions. In this study, we aim to address this challenge by leveraging high-fidelity LES results within a multi-fidelity computational framework with UQ. The low-fidelity model is a computationally efficient building thermal model that uses an empirical model to calculate ACH and solve for the time-evolution of the volume-averaged indoor air temperature. This model has inherent limitations in representing the complex turbulent flow field and its dependency on spatial variability in the temperature field, which can introduce significant uncertainty in the results. High-fidelity LES can resolve these effects, but the computational cost imposes restrictions on the number of simulations that can be performed to investigate variability in boundary and operating conditions. Using a carefully selected small number of LES, we develop a more accurate functional relationship for ACH that represents uncertainty introduced by the effects not resolved in the building thermal model. Employing this relationship within the thermal model then enables us to explore the full range of conditions, such that the framework can provide predictions for the mean and 95% confidence interval (CI) of both volume-averaged temperature and ACH.

PRELIMINARY RESULTS AND CONCLUSIONS

Initial results obtained using a UQ analysis of the low-fidelity building thermal model with an empirical ACH model provide ACH predictions that correspond to the trends observed in the field measurements. Analysis of the importance of the different uncertain parameters revealed that the assumption regarding the occurrence of cross- versus single-sided ventilation in the empirical ACH model is a dominant uncertainty. In addition, the empirical models are highly sensitive to the specification of the pressure coefficients on the openings, and the single-sided model tends to overestimate the effect of turbulence. These findings indicate the potential of using high-fidelity LES to reduce uncertainty in the predictions.

High-fidelity LES have first been validated on a canonical test case for wind-driven ventilation. Comparison of the flow pattern and ACH indicates good agreement with wind tunnel measurements. In ongoing work, LES of the full-scale test house in Dhaka is being performed to obtain more accurate relationships for ACH and employ them in the building thermal model with UQ. The final predictions of the ACH will be compared to the field measurements, and the results will be analyzed to identify robust ventilation strategies that will work under a variety of weather and housing conditions.

MAIN REFERENCES

[1] Unicef, "One is too many: Ending child deaths from pneumonia and diarrohea." (2016): 1-74.

[2] Ram, Pavani K., et al. "Household air quality risk factors associated with childhood pneumonia in urban Dhaka, Bangladesh." The American journal of tropical medicine and hygiene 90.5 (2014): 968-975.

Keywords: Natural ventilation, Large eddy simulation (LES), Uncertainty quantification (UQ)

Session F3.2: Ensuring high quality building simulations

Paper ID 30527: Analysis of the thermal inertia in historic buildings and related effects for retrofit strategies applied to urban energy modeling

Laura Carnieletto*, Enrico Prataviera, Giuseppe Emmi, Michele De Carli, Angelo Zarrella

University of Padova, Italy

AIM AND APPROACH

The retrofit of historic buildings is one of the major challenges to reduce energy use in urban city centers. In particular, Italian building stock accounts of historic buildings (i.e. older than 50 years) for more than 55%. Energy retrofit is usually complicated, due to many limitations by national laws that protect their cultural heritage or the urban assessment of the city. Therefore, few actions can improve the non-insulated high-mass envelope and the HVAC system, limiting the possibilities to reduce their energy use.

An historic building located in the city center of Padua (northern Italy) and built up in four different historic period has been investigated. The first construction belongs to XII century and the other blocks were built during the centuries, until the last part that was built during the Fascist regime. Energy Plus has been used for the detailed dynamic simulation and was compared to a Resistance-Capacitance (RC) model developed for urban building energy simulations. The RC model was applied firstly to the single parts of the building and later to the whole construction, to study the influence of thermal inertia with multiple approaches and to approach the modelling of historical buildings in urban building energy simulations.

SCIENTIFIC INNOVATION AND RELEVANCE

Old constructions were made of massive walls and structures with high thermal inertia that can be exploited to optimize the operating condition of the system. To evaluate this effect, a Resistance-Capacitance (RC) model has been applied to each part of a historic building in the city center of Padua, which was built in 4 different periods with four different envelope structures joint together during the years. The building is mainly used for office and teaching activities, with rooms dedicated to meetings and academic ceremonies, therefore with different operating and boundary conditions during weekdays and weekends.

Results have been compared to a detailed dynamic model of the whole building implemented with the software Energy Plus, analyzing the hourly profiles. A sensitivity analysis will be performed to evaluate the effects of the input parameters, including the occupants' presence and the operating conditions related to the different end uses. The influence of these parameters and the thermal inertia on the energy need is analysed to define more efficient strategies for historic building retrofit, taking advantage of high thermal mass and related inertia to flatten the temperature trends and optimizing systems' operation exploiting the heat that can be stock and released by the structure.

PRELIMINARY RESULTS AND CONCLUSIONS

The comparison between the single RC model for each construction and the overall model of the building with Energy Plus aim at evaluating the influence of massive walls and the related thermal inertia on buildings' energy performance. This approach will be useful to model historic buildings in an urban context. Focusing on the heating and cooling profiles obtained from the simulations, optimized control strategies for heating and cooling systems can be developed, possibly reducing the operating hours, and thus the building energy use. Moreover, temperature flatten can be investigated to guarantee indoor thermal comfort of the occupants.

Furthermore, the evaluation of the thermal inertia increases the possibility of defining retrofit strategies that take advantage of this property, for example using proper insulating materials avoiding the risk of overheating due to the high thermal mass. More performing windows have been considered as building envelope refurbishment, while operating schedule modifications and preliminary sizing of a cogeneration system have been investigated for the system improvement.

This analysis can be useful to apply new strategies to groups of buildings in historic urban city centers, supporting the reduction of energy use and related CO2 emissions at wider level while respecting the existing limitations.

MAIN REFERENCES

[1] Italian Institute of Statistic (ISTAT), Census of the Italian building stock (available at: http://dati-censimentopopolazione.istat.it, last seen 14/07/2020)

[2] Martinez-Molina A., Tort-Ausina I., Cho S., Vivancos J. L., Energy efficiency and thermal comfort in historic buildings: A review, Renewable and Sustainable Energy Reviews 61, 2016

[3] Webb A. L., Energy retrofits in historic and traditional buildings: A review of problems and methods, Renewable and Sustainable Energy Reviews 77, 2017

[4] Stazi F., Vegliò A., Di Perna C., Munafò P., Experimental comparison between 3 different traditional wall constructions and dynamic simulations to identify optimal thermal insulation strategies, Energy and Buildings 60, 2013

[5] Mancini F., Cecconi M., De Sanctis F., Beltotto A., Energy retrofit of a historic building using simplified dynamic energy modelling, 71st conference of the Italian Thermal Machine Engineering Association, ATI2016, 14-16 September 2016, Turin, Italy

[6] Caro R., Sendra J. J., Evaluation of indoor environment and energy performance of dwellings in heritage buildings. The case of hot summers in historic cities in Mediterranean Europe, Sustainable Cities and Society 52, 2020

Keywords: Historic building, Resistance-Capacitance model, energy efficiency, retrofit and control strategies

Paper ID 30227: Simulation of outdoor thermal comfort: a tweak with energyplus

Edouard Walther* (1), Carla Delmarre (2), Séverine Huet (1)

1: AREP L'hypercube, France; 2: INSA Lyon, Département Génie Civil et Urbanisme, France

AIM AND APPROACH

The aim of the present work is to provide for a simplified evaluation of surface temperatures in the outdoor built environment using an open source, state-of-the-art building energy simulation (BES) software.

A limited number of tools allow for the detailed determination of the urban micro-climate, such as the recognized ENVI-met (Bruse & Fleer, 1998) and SOLENE (Gros et al. 2014). These approaches are based on the coupling of heat transfer on built surfaces with non-isothermal fluid dynamics, their computational expense reaching days of CPU time.

At the early stage of construction projects, the available level of detail is low and iterations are often required in the design process. The aforementioned tools hence become practically unaffordable.

The work proposed here is a simplified, de-coupled approach using the open source softwares EnergyPlus and openFOAM respectively for BES and CFD. The hourly air velocity field is reconstructed from a limited number of isothermal CFD simulation similarly to (Delpech et al. 2005) and (Kastner & Dogan, 2020). The method relies on state-of-the art models for the simulation of heat transfer and fluid flow and allows for the outdoor comfort evaluation of design variants with an accessible computational effort.

SCIENTIFIC INNOVATION AND RELEVANCE

Building energy simulation softwares are obviously able to determine the surface temperatures of external walls depending on outdoor boundary conditions, amongst which solar flux, air velocity or sky vault temperature. In order to simulate outdoor comfort, the approach used here relies in constructing thermal zones that encompass the urban environment considered, forming one "building".

This method advantageously allows for the usage of the available features in EnergyPlus , such as the addition of shadings or the simulation of low vegetation via the green roof object. Materials of the environment are chosen depending on their constitution: building wall, roof, pavement or road surface. The computational expense is reduced as a few hours only are required for a yearly simulation of the surface temperature for the BES part. The isothermal CFD computations may run independently overnight.

Latent heat transfer to the ambiance from ponds, lawns or trees is not accounted for. However, as far as vegetation is concerned, the shading effect was proved to be the leading contributor to cooling rather than water evaporation (Morakinyo et al. 2020), (Tsoka et al. 2018). Trees are hence integrated as mere shading objects.

PRELIMINARY RESULTS AND CONCLUSIONS

The results obtained exhibit a satisfactory behaviour compared to the experimental and numerical literature:

- The phenomenon is well reproduced in terms of dynamics and amplitude of the surface temperatures compared to (Musy et al. 2014)

- Reflexion is taken into account and visible between buildings and ground, influenced by the albedo of each surface, similarly to (Musy et al. 2014). Specular reflection of the solar flux on glazed facades is also represented.

- The shading from the buildings and the trees is correctly represented. The buildings are completely opaque whereas the trees transmit 20-30% of the solar flux to the soil (Toudert & Mayer 2005).

- The radiative flux decreases with the sky-view factor . Therefore, the radiant flux is inhibited in presence of trees, in narrow streets or close to buildings. Similar results can be observed in (Musy et al. 2014).

- The heat storage in environment's surfaces (walls, roofs, ground) is modelled with EnergyPlus' integrated wall model .

As a perspective, the calculation of local heat transfer coefficients using the air velocity fields is currently implemented. The integration of detailed view factors, following the recent evolution of energyplus is also examined (Luo et al. 2020).

MAIN REFERENCES

Bruse M, Fleer H. Simulating surface–plant–air interactions inside urban environments with a three dimensional numerical model. Environmental modelling & software. 1998.

Kastner P, Dogan T. A cylindrical meshing methodology for annual urban computational fluid dynamics simulations. Journal of Building Performance Simulation. 2020.

Morakinyo TE, Ouyang W, Lau KK, Ren C, Ng E. Right tree, right place (urban canyon): Tree species selection approach for optimum urban heat mitigation-development and evaluation. Science of The Total Environment. 2020.

Delpech P., et al. 2005. Pedestrian wind comfort assessment criteria: A comparative Study. EACWE4. Prague: J. Naprestek & C. Fischer.

Luo X, Hong T, Tang YH. Modeling Thermal Interactions between Buildings in an Urban Context. Energies. 2020.

DoE US. EnergyPlus engineering reference. LBNL. 2009.

Musy M et al. Impact of vegetation on urban climate, thermal comfort and building energy consumption–Overview of VegDUD project results. InIC2UHI 2014 .

Toudert A, Mayer H. Thermal comfort in urban streets with trees under hot summer conditions. PLEA 2005–Passive and Low Energy Architecture, 2005, Beirut. Proceedings. 2005.

Tsoka S, Tsikaloudaki A, Theodosiou T. Analyzing the ENVI-met microclimate model's performance and assessing cool materials and urban vegetation applications–A review. Sustainable cities and society. 2018.

Keywords: outdoor comfort, urban environment, energyplus

Paper ID 30634: Lumped-capacitance models to evaluate the urban cooling energy consumptions: analysis on a case-study district in the University of Padua.

Enrico Prataviera*, Pierdonato Romano, Laura Carnieletto, Jacopo Vivian, Angelo Zarrella

University of Padua, Italy

AIM AND APPROACH

In the last decade, the research in buildings physics has focused on the extension from standard building simulation tools to Urban Building Energy Modelling (UBEM) which is a wider and novel research topic aiming at the modelling of buildings in cities and urban environments. In the next future, cities' population will grow up exponentially, increasing primary energy consumptions and global warming emissions connected to urban areas. In this perspective, several new software and platforms have been developed by research groups worldwide, trying to propose modelling solutions to the challenging issues that arise in urban environments.

Energy consumption in cooling season is becoming more and more relevant due to the increasing user requests and solar heat gains, causing temperature rise and affecting chillers' efficiencies itself. The present paper proposes a detailed analysis on the cooling energy consumptions of a district of 9 buildings within the University campus located in Padua, northern Italy, utilizing a new UBEM platform (named EUReCA) developed by the Authors. Particular attention was also given to possible actions to reduce energy consumption and gas emission for the cooling season in a district scale perspective.

SCIENTIFIC INNOVATION AND RELEVANCE

The proposed model, EUReCA, Energy Urban Resistance Capacitance Approach, consists of Python based tool to predict energy demand and consumption in district and city scale applications. Semantic georeferenced data and archetypes' databases are used to derive buildings Resistance-Capacitance thermal networks (based on both one and two thermal capacitances), thus solving the thermal zone balance. Moreover, latent thermal zone balance and plants components' models are included.

The present work is divided into two phases. Firstly a detailed analysis of the model results is proposed, comparing hourly temperature and humidity profiles to the detailed simulations carried out in EnergyPlus, used as benchmark, in all 9 buildings. Hourly and seasonal cooling demand are verified as well, and attention has been paid to the peak load difference. Secondly, the effect of several retrofit actions is discussed and compared to the actual system, which consists of a district cooling network with centralized chillers coupled to cooling towers. Envelope actions (e.g. windows substitution, low absorptance roofs and external walls) and plant actions (latent and sensible heat air handling unit recovery, night free cooling and new reversible heat pump systems) are considered and compared in a consumption reduction perspective.

PRELIMINARY RESULTS AND CONCLUSIONS

Several simulation parameters and results were compared to the single building detailed simulation in EnergyPlus. The testing procedure has been carried out with different calculation methods, one or two thermal capacitances model considering also the effect of the climate. Results display a good prediction on the monthly and seasonal demand for both thermal networks. Hourly temperature is in good agreement for the entire district of buildings, but the zone relative humidity prediction can be inaccurate because of EUReCA's simplified thermal zoning. The model based on two thermal capacitances presents a better performance when power peak load evaluation is considered.

Several improvements have been then modelled using EUReCA. Reducing the windows solar gain appears to be a valuable choice to reduce cooling demand, while, on the plant side, consumptions benefits from a night free cooling solution. The substitution of the cooling towers with a ground source heat pump could strongly decrease energy consumptions, although their installation involves a high initial cost and technical issues related to the available space.

MAIN REFERENCES

T. Hong, Y. Chen, X. Luo, N. Luo, and S. H. Lee, "Ten questions on urban building energy modeling," Build. Environ., vol. 168, p. 106508, Jan. 2020.

United Nations Department of Economic and Social Affairs, "Revision of World Urbanization Prospects," 2018.

A. Sola, C. Corchero, J. Salom, and M. Sanmarti, "Multi-domain urban-scale energy modelling tools: A review," Sustain. Cities Soc., p. 101872, Oct. 2019.

International Organization for Standardization, "ISO 13790:2008 Energy performance of buildings — Calculation of energy use for space heating and cooling," 2008.

German Association of Engineers, "Calculation of transient thermal response of rooms and buildings - modelling of rooms (VDI 6007-1)," 2012.

Keywords: Urban Building Energy Modelling, Resistance-Capacitance model, Cooling demand, District retrofits

Paper ID 30177: GIS-based multi-scale residential building energy modeling using a data-driven approach

Usman Ali* (1), Mohammad Haris Shamsi (1), Mark Bohacek (3), Karl Purcell (3), Cathal Hoare (1), Eleni Mangina (2), James O'Donnell (1)

1: School of Mechanical & Materials Engineering, Energy Institute University College Dublin (UCD), Ireland; 2: School of Computer Science and Informatics, University College Dublin (UCD), Ireland; 3: Sustainable Energy Authority Of Ireland, Dublin, Ireland

AIM AND APPROACH

In order to adapt to climate change, urban planning and development strategies are undergoing a transformation from conventional design to more innovative approaches. As such, city planners often develop strategic sustainable energy plans to minimize overall energy consumption and CO2 emissions. Planning at such scales could be informed by the spatial analysis of building stock using Geographic Information Systems (GIS) based mapping. Generally, the building stock is available in the form of building Energy Performance Certificates (EPC) database that provides an indication of buildings' energy use and carbon emission predictions and can play an important role in energy and climate policymaking. The creation of an EPC for any individual building requires extensive information surveys. Hence these ratings typically only exist for a subset of the entire building stock. A data-driven methodology could aid in the identification of building energy performance using existing available building data.

SCIENTIFIC INNOVATION AND RELEVANCE

This research proposes a novel generalizable methodology for data-driven GIS-based residential building energy modeling at multiple scales. Different machine learning algorithms are examined to predict building energy ratings for GIS-based multi-scale mapping.

PRELIMINARY RESULTS AND CONCLUSIONS

This paper develops a methodology for GIS-based residential building energy modeling at a multi-scale using a data-driven approach. The machine-learning algorithm predicts building energy ratings from local to national scale using the bottom-up approach. The predictive modeling results are coupled with GIS for multi-scale mapping. The methodology is applied to Irish residential building stock and the energy rating is examined at multiple scales. The modeling results indicate priority geographical areas that have the greatest potential for energy savings.

MAIN REFERENCES

Wei, Y., X. Zhang, Y. Shi, L. Xia, S. Pan, J. Wu,M. Han, and X. Zhao (2018). A review of data-driven approaches for prediction and classificationof building energy consumption.Renewable andSustainable Energy Reviews 82, 1027–1047.

Amasyali, K. and N. M. El-Gohary (2018). A re-view of data-driven building energy consumptionprediction studies.Renewable and Sustainable En-ergy Reviews 81, 1192–1205.

Y. Sun, F. Haghighat, B. C. Fung, A review of the-state-of-the-art in data-driven approaches forbuilding energy prediction, Energy and Buildings (2020) 110022.

Keywords: data-driven, machine learning, urban energy modeling, GIS, energy performance certificate

Session F3.3: Buildings paving the way for the energy transition

Paper ID 31127: Data-driven estimation of parametric uncertainty of reduced order RC models for building climate control

Anke Uytterhoeven* (1,2), Ina De Jaeger (1,3), Kenneth Bruninx (1,2), Dirk Saelens (1,3), Lieve Helsen (1,2)

1: EnergyVille, Genk, Belgium; 2: KU Leuven, Mechanical Engineering Department, Leuven, Belgium; 3: KU Leuven, Civil Engineering Department, Leuven, Belgium

AIM AND APPROACH

Current model predictive control (MPC) applications for residential space heating typically rely upon accurate building models, obtained via extensive data acquisition and/or experts' knowledge. Here, the feedback mechanism of the receding horizon implementation of MPC offers sufficient robustness against the small remaining uncertainties. In contrast, in the context of older, existing residential buildings, one needs to rely on sparse, publicly available data to construct controller models. In that case, the parametric uncertainty of the controller model can become non-negligible, and additional measures might be necessary (Bengea et al. (2011); Oldewurtel et al. (2010,2012); Ioannou and Itard (2015)). However, precisely because of the lack of information, a scientifically sound characterization of the building model and associated parametric uncertainty, as well as an assessment of the impact of the uncertainty, is a challenging task. Therefore, the aim of this paper is to come up with an estimate of the parametric uncertainty of a building controller model in case neither detailed information about the building thermal properties nor experts' knowledge is available. In addition, the impact of this uncertainty on the resulting optimal space heating strategy is investigated.

SCIENTIFIC INNOVATION AND RELEVANCE

This paper determines a substantiated estimate of the parametric uncertainty for building simulation or controller models generated with the help of a statistical, data-driven building characterization method leveraging publicly available data (De Jaeger et al. (2021)). Since neither detailed information about the building thermal properties nor experts' knowledge is incorporated, this can be seen as a worst-case assessment. In addition, this paper investigates the impact of the obtained variation in model parameters on the energy demand profile determined by optimal control, and on the resulting yearly energy use, via a Monte Carlo analysis. This is a first step in assessing the importance of building model parametric uncertainty regarding MPC performance.

PRELIMINARY RESULTS AND CONCLUSIONS

The results show that the considered approach gives rise to a rather large uncertainty. The obtained variation in model parameters is shown to markedly affect the optimal space heating control, both in terms of dynamic effects and yearly energy use, thereby indicating the need for improved data acquisition and/or dedicated control strategies that operate robustly under uncertainty.

MAIN REFERENCES

Bengea, S., V. Adetola, K. Kang, M. J. Liba, D. Vrabie, R. Bitmead, and S. Narayanan (2011). Parameter estimation of a building system model and impact of estimation error on closed-loop performance. In 50th IEEE Conference on Decision and Control and European Control Conference, 2011, Orlando, FL, USA, pp. 5137-5143. IEEE.

Oldewurtel, F., D. Gyalistras, M. Gwerder, C. Jones, A. Parisio, V. Stauch, B. Lehmann, and M. Morari (2010). Increasing energy efficiency in building climate control using weather forecasts and model predictive control. In Clima RHEVA World Congress, 2010, Antalya, Turkey.

Oldewurtel, F., A. Parisio, C. N. Jones, D. Gyalistras, M. Gwerder, V. Stauch, B. Lehmann, and M. Morari (2012). Use of model predictive control and weather forecasts for energy efficient building climate control. Energy and Buildings 45, 15-27.

Ioannou, A. and L. C. Itard (2015). Energy performance and comfort in residential buildings: Sensitivity for building parameters and occupancy. Energy and Buildings 92, 216-233.

De Jaeger, I., J. Lago, and D. Saelens (2021). A probabilistic building characterization method for district energy simulations. Energy and Buildings 230, 110566.

Keywords: Building climate control, model predictive control (MPC), building model, RC model, parametric uncertainty

Paper ID 30574: Formulation and implementation of a model predictive control (MPC) strategy for a PCM-driven building ventilation cooling system

Tao Yang* (1), Konstantin Filonenko (1), Jonathan Dallaire (1), Viktor Bue Ljungdahl (1), Muhyiddine Jradi (1), Esther Kieseritzky (2), Felix Pawelz (2), Christian Veje (1)

1: Center for Energy Informatics, University of Southern Denmark, Denmark; 2: Rubitherm Technologies GmbH, Berlin, Germany

AIM AND APPROACH

This paper aims to integrate phase change material (PCM) based ventilation cooling system with a building model and studies feasibility and overall performance of MPC controllers in different context and conditions.

To achieve this, a surrogate model of a PCM ventilation unit is built in Dymola, the model is validated based on results from an actual experimental setup. The performance of the PCM based energy storage system is simulated and its cooling capacity is assessed. Subsequently, the PCM model is integrated with a calibrated gray-box office zone model. The simulation results of the whole system are performed under real occupancy and climate data.

An MPC controller controlling supply airflow rate of ventilation cooling is implemented by running a sequential quadratic programming optimization algorithm on the system functional mock-up unit (FMU). Performance of the entire system is compared with a conventional system employing rule-based control approach in terms of thermal comfort violation and electrical power consumption. The surrogate PCM model serve both as emulation model and control model in this setup.

Sensitivity with respect to control and zone model parameters affecting the MPC performance is inspected including model complexity, optimization horizon, objective function and weather conditions.

SCIENTIFIC INNOVATION AND RELEVANCE

Phase change material, as a storage medium in latent thermal energy storage systems, has received considerable attention in recent research studies and investigations as it is able to reduce overall energy consumption and shift peak loads [1]. Meanwhile, advanced control strategies such as MPC have been proven to facilitate better performance in buildings as compared to conventional control strategies. There is a trend of applying PCM-based energy storage systems and MPC in buildings aiming for improving the performance. Previous research studies of building MPC mainly focus on traditional heating, ventilation, and air conditioning (HVAC) systems such as controlling radiators, cooling and heating coils in ventilation. There are some studies reporting MPC on PCM-based energy storage system, but they either focus on PCM for heating or PCM embedded in building envelope [2-4]. Few studies have been found in literature regarding PCM units for separate ventilation cooling in summer, which forms the basis of this paper's novelty.

PRELIMINARY RESULTS AND CONCLUSIONS

The surrogate model of the PCM energy storage system yields considerable accuracy of predicting the thermal dynamics as compared with experimental setup measurements. It enables passive cooling and shows capability of regulating indoor thermal comfort when integrated in an office room in summer. Results show that implementing the MPC approach enables enhanced cooling capability of a PCM energy storage system as compared to conventional control. The complexity of the control model has a direct impact on the overall MPC performance in terms of results accuracy and simulation time, while optimization horizon, objective function and weather conditions allow for an appropriate tuning depending on a specific archetype data.

MAIN REFERENCES

[1] E. Osterman, V. Butala, and U. Stritih, "PCM thermal storage system for 'free' heating and cooling of buildings," Energy Build., vol. 106, pp. 125–133, Nov. 2015

[2] A. C. Papachristou, C. A. Vallianos, V. Dermardiros, A. K. Athienitis, and J. A. Candanedo, "A numerical and experimental study of a simple model-based predictive control strategy in a perimeter zone with phase change material," Sci. Technol. Built Environ., vol. 24, no. 9, pp. 933–944, Oct. 2018.

[3] G. Serale, M. Fiorentini, A. Capozzoli, P. Cooper, and M. Perino, "Formulation of a model predictive control algorithm to enhance the performance of a latent heat solar thermal system," Energy Convers. Manag., vol. 173, pp. 438–449, Oct. 2018.

[4] M. Fiorentini, P. Cooper, Z. Ma, and D. A. Robinson, "Hybrid model predictive control of a residential HVAC system with PVT energy generation and PCM thermal storage," in Energy Procedia, 2015, vol. 83, pp. 21–30.

Keywords: Phase change material, HVAC, Calibration and validation, Model predictive control

Paper ID 30907: Data-driven prediction of flow parameters in a ventilated cavity using high-fidelity CFD simulations

Nina Morozova*, Fransesc Xavier Trias Miquel, Roser Capdevila Paramio, Asensio Oliva Llena

Heat and Mass Transfer Technological Center (CTTC), Universitat Politècnica de Catalunya (UPC), Spain

AIM AND APPROACH

In this work, we elaborate data-driven models for predicting the flow parameters in a three-dimensional ventilated cavity with a heated floor. It was first studied experimentally by Blay et al. [1]. Cold air enters the cavity through the inlet at the top of the left wall. The air is discharged at the bottom of the right wall. The bottom wall is hot, while other walls are cold. This configuration is typical for many indoor environmental applications. This study aims to develop an affordable data-driven model, which accurately predicts the airflow parameters. The predictions of the model are based on the results of high-fidelity CFD simulations. The elaborated model could be considered a cheaper alternative for CFD in indoor environmental design and control. The input parameters for the model are the results of CFD simulations with different geometrical aspect ratios, boundary, and initial conditions. The CFD simulations are conducted using an in-house code [2] with symmetry-preserving finite volume discretization on staggered grids. This numerical configuration has provided the best trade-off between the computational cost and accuracy in our previous studies [3, 4]. We compare artificial neural networks (ANN) with support vector regression (SVR) and gradient boosting regression (GBR) using open-source libraries.

SCIENTIFIC INNOVATION AND RELEVANCE

CFD is a reliable tool for indoor environmental applications. However, accurate CFD simulations require large computational resources, whereas significant cost reduction can lead to unreliable results. The high cost prevents CFD from becoming the primary tool for indoor environmental simulations. Our previous findings [3, 4] suggest that the growth of computational resources in the near future would not be enough to make CFD available for routine use in building applications. This means more work is required on developing better models and numerical methods, to reduce the computational cost of the simulations while maintaining accuracy. In our study, we develop machine learning algorithms based on data from previous CFD simulations. The algorithms aim to accurately predict the airflow parameters while having lower than CFD computational cost. The main focus of our research is on investigating the capabilities and limitations of machine learning algorithms as a cheaper alternative to CFD simulations. We compare the computational cost and accuracy of different machine learning algorithms for the prediction of flow parameters in a ventilated cavity.

PRELIMINARY RESULTS AND CONCLUSIONS

We consider a characteristic test case of turbulent($Ra= 2.4\times109$)mixed convection in a ventilated square cavity. In the previous work [3, 4], we analyzed this test case using different turbulence models, discretization techniques, and mesh resolutions. This test case is difficult to be solved accurately due to the small aspect ratios of the inlet and outlet openings. The LES simulation with staggered symmetry-preserving discretization provided the best trade-off between computational cost and accuracy, while the accuracy of the RANS turbulence model appeared to be insufficient. Reliable, high-fidelity CFD simulations require significant computational resources, thus alternative methods of obtaining high-fidelity simulation results with lower computational cost should be explored. We use the obtained results as the guidance for the turbulence model and the discretization method choice for the CFD simulations, which form the basis of our data-driven model. Moreover, the performed analysis allows investigating the requirements for the minimal set of simulations to build up a reliable data-driven model, such as the quality of CFD (high-fidelity or coarse-grid), the number of datasets, and the principal data components, which affect the prediction the most.

MAIN REFERENCES

[1] D. Blay, S. Mergui, J. L. Tuhault, and F. Penot. Experimental turbulent mixed convection created by confined buoyant wall jets. In Proceedings of the First European Heat Transfer Conference, UK, pages 821–828, 1992.[2] A. Gorobets, F. X. Trias, M. Soria, and A. Oliva. A scalable parallel Poisson solver for three-dimensional problems with one periodic direction. Computers & Fluids, 39(3):525–538, 2010.

[3] N. Morozova, F. X. Trias, R. Capdevila, C. D. Pérez-Segarra, and A. Oliva. On the feasibility of affordable high-fidelity CFD simulations for indoor environment design and control. Building and Environment,(accepted for publication), 2020.

[4] N. Morozova, R. Capdevila, F. X. Trias, and A. Oliva. On the feasibility of CFD for transient airflow simulations in buildings. In Proceedings of Building Simulation 2019: 16th Conference of IBPSA, September2-4, 2019.

Keywords: Computational Fluid Dynamics, Large Eddy Simulation, Data Driven Models, Artificial Neural Network, Mixed Convection

Paper ID 30569: Evaluating data-driven building stock heat demand forecasting models for energy optimization

Tohid Jafarinejad*, Ina De Jaeger, Arash Erfani, Dirk Saelens

Katholieke Universiteit Leuven, Belgium

AIM AND APPROACH

The building sector is one of the most energy intensive sectors. In order to decarbonize society, the focus on reducing the energy use of dwellings is important. This reduction can be achieved by looking at individual building level, but many researchers also made assessments on the district and city scale by developing innovative techniques to improve district energy planning, develop optimal controllers for network operation, and demand side management. For these analyses researchers use district models to simulate the energy demand under varying weather conditions and end use decisions. In many cases, the energy management in district level involves non-convex optimization schemes. Predicting these energy needs requires accurate and fast responsive models. In this study, various data-driven models, which forecast the aggregate energy demand of the building stock, are developed based on a residential district data generated by the Modelica simulation environment. To see how well and fast they predict the Modelica generated results, their suitability for the aforementioned applications is assessed based on several Key Performance Indicators (KPIs), namely the coefficient of determination (R2), Normalized Root Mean Square Error (NRMSE), and run time, which ascertains the agility of the developed data-driven model for optimization over a specific time horizon.

SCIENTIFIC INNOVATION AND RELEVANCE

Conventional and prevalent high order white-box models that simulate and forecast the district energy demand are difficult to be deployed for optimization purposes since they are time intensive. In many cases, using surrogate simplified models is proved to be an interesting alternative. Nowadays with the development of building energy management tools and having a huge load of data associated with buildings in a neighborhood at hand, deriving surrogate models for districts is facilitated. These data-driven ersatz models, however a bit less accurate than the conventional white-box models, are precise enough to represent a building stock, while taking a much shorter time to simulate the district demand at each run, and carry out the optimization task. The two main criteria that evaluate the appropriateness of a black-box model for district level optimization applications are the forecast accuracy and the response time (computational load) of the model. In this study, the competence of most frequent black-box models including ANN (Artificial Neural Network) models, SVM (Support Vector Machine) models and etc. are investigated. In contrast to previous studies, this study also takes into account the influence of the district size on the developed models in terms of the aforementioned criteria.

PRELIMINARY RESULTS AND CONCLUSIONS

The black-box techniques for modeling a building stock are divided into two main categories; statistic based and machine learning based models. Although on the one hand machine learning based models are more accurate than statistic based models, they are prone to be over-fit and unable to generalize on new datasets. In this study also the generalization of black-box models is evaluated on test datasets as well. The training dataset is 4380 hourly data points (half a year) of simulation data in a year, and the test dataset is 4380 hourly data points (half a year) of simulation data. Preliminary results demonstrate that through opting a suitable regularization mechanism, lack of generalization could be avoided while maintaining the expedient precision. The outcomes signify that machine learning based models have more edge as compared to statistic based ones. Moreover, they perform more rapidly in comparison to the white-box models, which makes them adept for the intended optimization purposes. Last but not least, in districts with a large number of buildings, Support Vector Machine identification technique shows a remarkable performance in terms of agility (response time) and accuracy both combined together.

MAIN REFERENCES

Idowu, S., Saguna, S., Åhlund, C., & Schelén, O. (2016). Applied machine learning: Forecasting heat load in district heating system. Energy and Buildings, 133, 478–488. https://doi.org/10.1016/j.enbuild.2016.09.068

Swan, L. G., & Ugursal, V. I. (2009). Modeling of end-use energy consumption in the residential sector: A review of modeling techniques. Renewable and Sustainable Energy Reviews, 13(8), 1819–1835. https://doi.org/10.1016/j.rser.2008.09.033

Xie, J., Li, H., Ma, Z., Sun, Q., Wallin, F., Si, Z., & Guo, J. (2017). Analysis of Key Factors in Heat Demand Prediction with Neural Networks. Energy Procedia, 105, 2965–2970. https://doi.org/10.1016/j.egypro.2017.03.704

Zhao, F., Lee, S. H., & Augenbroe, G. (2016). Reconstructing building stock to replicate energy consumption data. Energy and Buildings, 117, 301–312. https://doi.org/10.1016/j.enbuild.2015.10.001

Zygmunt, M., & Gawin, D. (2019). Application of ANN for analysing a neighbourhood of single-family houses constituting an Energy Cluster. MATEC Web of Conferences, 282(2019), 02072. https://doi.org/10.1051/matecconf/201928202072

Keywords: Building stock, Data driven model, Black box model, Identification

Session F3.4: Climate change and bioclimatic design

Paper ID 30395: The effect of climate change on the renovation of social housing

Els Van de moortel*, Delphine Ramon, Karen Allacker, Frank De Troyer

KU Leuven, Belgium

AIM AND APPROACH

To achieve the 2050 energy goals set by Europe, deep renovations should be stimulated. Buildings, moreover, have to adapt to climate change as the temperature rise might affect their indoor climate and energy demand for heating and cooling. Recent research showed a decrease of the HDD with 30% for Belgium, while an increase of the CDD with a factor 2.4 was noticed for a high-end climate change scenario by the end of the 21st century. This paper investigates the effect of climate change on the selection of renovation strategies for a poorly insulated residential case study building. The energy demand for heating is estimated by using the equivalent heating degree day (eHDD) method, while the energy demand for cooling is calculated by using the (adapted) EPBD method. Three locations (urban versus rural, seaside versus inland) across Belgium are considered to evaluate regional effects on the selection of renovation strategies.

SCIENTIFIC INNOVATION AND RELEVANCE

To date, only current climate conditions are taken into account to evaluate renovation strategies in the Belgian context. Further, these strategies mostly consider the energy need for heating as currently there is no need for active cooling in residential buildings. Climate change will cause an important decrease in the need for heating and an increase in the need for cooling. To select the most appropriate renovation measure, it is hence important to investigate if considering climate change leads to different decision taking. To estimate the energy use for cooling two parameters are adapted in the current Flemish EPBD method when the outdoor temperature is higher than the indoor temperature: 1) the flow of transmission is reversed and added to the solar and internal gains, 2) no utility rate for heat losses is assumed.

Further, average climate data based on past observations for Uccle are used to estimate the energy use of buildings in Flanders, regardless of where the building is situated. Recent studies have, however, shown important differences caused by regional effects such as elevation differences or the urban heat island effect. This study investigates if optimal renovation strategies might differ over the country because of these regional effects.

PRELIMINARY RESULTS AND CONCLUSIONS

The first results reveal that the number of EHDD and the heating demand have decreased by 28-39% when future climate data are used, depending on the region and building characteristics. The cooling demand is 8-22 times higher in the future climate compared with the current climate.

The NZEB renovation of the case study building leads to a 74-76% decrease in energy use for heating in the current climate. A similar effect is found for an NZEB renovation in the future climate. In the current climate no cooling is needed, while in the future climate, cooling is only needed for the existing building in the inland region, and increases from 50 - 175 kWh for the NZEB renovation, depending on the region.

As the differences are significant it seems interesting to include climate change and regional climate data in further dynamic assessments of renovations. It will be studied whether active cooling is needed or passive shading systems will be sufficient in the future to guarantee indoor comfort.

MAIN REFERENCES

Ürge-Vorsatz et al. (2012). GEA: Global Energy Assessment - Toward a Sustainable Future. Cambridge, UK and New York, USA and the International Institute for Applied Systems Analysis, Luxenburg, Austria.: Cambridge University Press.

Keywords: climate change, renovation, LCA, energy simulation

Paper ID 30811: The feasibility of ventilative cooling in an office building in a Flemish urban context and the impact of future climate scenario's

Joost Declercq* (1), Delphine Ramon (2), Fabrice Derny (3), Karen Allacker (2)

1: archipelago architects, Belgium; 2: KU Leuven, Belgium; 3: MATRIciel, Belgium

AIM AND APPROACH

A Flemish architectural office is designing its new offices for 60 collaborators in the city center of Leuven. To emphasize their ambition to design high-performant low-tech buildings the aim is to mitigate the dependence on complex AHU's and as such to reduce the operational energy use and environmental impact of the building. The optimization of the natural ventilation potential has been considered from the first architectural sketches. A systematic approach has been developed. The conceptual design was guided by rules of thumb, supported by quantitative estimations with the natural ventilation module of the Swiss DIAL+ software. The near-final design has been assessed with TRNSYS. Not only a typical year has been considered, but also custom-made climate data, based on extreme winters and summers of the last 10 years. An additional assessment will be done with a high-resolution climate model data for an RCP 8.5 climate change scenario. Similarly, a typical, an extreme warm, and an extreme cold year will be considered. The paper focuses on the workflow and the impact of the future climate data on the feasibility to realize a medium-sized office in a dense urban context using ventilative cooling for its summer comfort.

SCIENTIFIC INNOVATION AND RELEVANCE

Climate change and hygienic concerns are putting high demands on the ventilation and cooling concepts of buildings. To date, technical installations represent more than 35% of the building budget and more than 30% of the initial environmental impact of a typical office building in Europe. The growing space demand of the technical installations is at cost of the usable and architectural space for the building occupants and the building program. To provide a way-out to this growing problem, methodologies and software-tools are available to assess the feasibility of concepts as ventilative cooling but are not sufficiently exploited or used by architects. The challenge is to make architectural considerations meet with complex engineering knowledge. In this case study, the architectural office is also the final client which gives the opportunity to develop a new workflow. The workflow and toolbox are now ready for upscaling to other contexts and clients. The fact that the building is situated in a dense urban context and that future climate is considered, including the urban heat island effect, makes it a robust case for referencing for other projects.

PRELIMINARY RESULTS AND CONCLUSIONS

The current simulations show a big potential for the ventilative cooling concept. Full summer comfort according to EN 15251 is reached in all the building zones under a standard weather scenario. Under the custom-made extreme climate year data, slight discomfort is visible in some zones which only have single-sided ventilation. In further simulations, the results under the RCP 8.5 climate change scenario for the city center of Leuven will be investigated. The robustness of the concept under changing internal loads will also be analyzed. The current concept has a backup cooling system under the form of a radiant ceiling, connected to an air-water heat pump. This is primarily used as a heating device, but can also be used as a cooling source if it would be needed. The need for this backup cooling will be assessed.

MAIN REFERENCES

Breesch, H., (2006) Natural Night Ventilation in Office Buildings, doctoral thesis UGent, 2006

Florentzou, F., Paule, B., Pantet, S., (2012). Natural ventilation and passive cooling simulation is not any more a privilege for experts, AIVC conference, Copenhagen

Flourentzou F., Pantet S., Ritz K., (2015). Automatic natural ventilation in large spaces: a passive ventilation technology for passive buildings,

Ramon, D., Allacker, K., De Troyer, F., Wouters, H., & van Lipzig, N. P. (2020). Future heating and cooling degree days for Belgium under a high-end climate change scenario. Energy and Buildings, 109935

Ramon, D., Allacker, K., Van Lipzig, N., De Troyer, F., Wouters, H. (2019). Future Weather Data for Dynamic Building Energy Simulations: Overview of Available Data and Presentation of Newly Derived Data for Belgium. In: E. Motoasca, A.K. Agarwal, H. Breesch (Eds.), Energy Sustainability in Built and Urban Environments, Chapt. 6, (111-138). (Energy, Environment, and Sustainability). Springer, Singapore. ISBN: 978-981-13-3283-8.

Keywords: ventilative cooling, office building, belgium, climate change scenario, urban

Paper ID 30112: Performance simulation of a geothermal rain water tank coupled to building ventilation

Jean-Baptiste Bouvenot*

INSA Strasbourg/ICube Laboratory, France

AIM AND APPROACH

Large (several m3) buried rain water recovery tanks are similar to single but large superficial geothermal probes that can be "thermally activated" by immersion of a heat exchanger. This publication presents the heat exchanger sizing and the energy relevance of a rainwater tank used as geothermal probe in a heat recovery double-flow mechanical ventilation unit to cool down fresh air in summer and to preheat cold air in winter to avoid operating the defrosting electrical heating resistance. The system will be stressed in actual and future climates taking into account mainly the variation of rainfalls along the season and the variation of temperature. A model of variable ground thermal conductivity has been applied to take into account the influence of summer drought on ground thermal properties. A parametric study on the control strategy and on the safety water volume has been led. This study shows the good performance of this passive and low tech system to provide sufficient passive cooling to ensure smmer thermal comfort.

SCIENTIFIC INNOVATION AND RELEVANCE

This system is innovative because it combines two systems in a unique one. The literrature review shows this hybrid system has been poorly investigated while it is an interesting alternative in the current climate change, energy efficiency and low technology contexts.

The relevance is linked to the hybrid and passive nature of the system, which makes it possible to limit the use of raw materials while ensuring the production of renewable thermal energy locally. Finally, the hybrd nature of the system makes it cheap in comparison wih expensive geothermal systems. This system is used in parallel as buffer rain storm basin and as cold water provider for some applications.

PRELIMINARY RESULTS AND CONCLUSIONS

Preliminay results show this system can provide significant cooling powers (about 1000 W for 5 m3 of rain water tank volume) and avoid using the defrosting heating electrical resistance during a whole winter in european climates. The results show the water never freeze in winter and that the heat exttraction is sustainable in the time (stabilization of the temperature in the ground after 3 years). The results also show that the systems keeps good performances in future climates.However, the results show the cold loading in winter provides low impacts on summer performances.

MAIN REFERENCES

[1] A. Cardona-Maestro, Etude sur la géothermie très basse énergie afin de redynamiser la filière, Etude ADEME, 9 Juin 2016

[2] A. Choorapulakkal, M. Nogughi, Performance Analysis Of A Proposed 'Water Tube Heat Exchanger' Space Cooling System in Kerala, PLEA2013 - 29th Conference, Sustainable Architecture for a Renewable Future, Munich, Germany 10-12 September 2013

[3] Gan, G., Riffat, S.B., Chong, C.S.A., 2007. A novel rainwater–ground source heat pump – Measurement and simulation. Applied Thermal Engineering 27, 430–441.

[4] Gao, Yan, Fan, R., Li, H., Liu, R., Lin, X., Guo, H., Gao, YuTing, 2016. Thermal performance improvement of a horizontal ground-coupled heat exchanger by rainwater harvest. Energy and Buildings 110, 302–313.

[5] Irei, S., Asakura, T., 2001. Air conditioning system for refreshment utilizing rainwater. US6227000B1.

[6] Kalz, D.E., Wienold, J., Fischer, M., Cali, D., 2010. Novel heating and cooling concept employing rainwater cisterns and thermo-active building systems for a residential building. Applied Energy 87, 650–660.

[7] Yumrutaş, R., Ünsal, M., 2005. Modeling of a space cooling system with underground storage. Applied Thermal Engineering 25, 227–239.

[8] Yumrutaş, R., Ünsal, M., 2000. A computational model of a heat pump system with a hemispherical surface tank as the ground heat source. Energy 25, 371–388

Keywords: Rain water harvesting, Surface geothermal energy, Geocooling, Summer comfort, ventilation

Paper ID 30279: Development of an urban microclimate model to assess the air temperature field during the summer period

Julie Soriano* (1,2), Lucie Merlier (1), Damien David (1), Frédéric Kuznik (1), Lionel Soulhac (2)

1: Univ Lyon, CNRS, INSA-Lyon, Université Claude Bernard Lyon 1, CETHIL UMR5008, F-69621, Villeurbanne, France; 2: Laboratoire de Mécanique des Fluides et d'Acoustique UMR CNRS 5509, École Centrale de Lyon, Université Claude Bernard Lyon 1, INSA Lyon, 69134 Écully Cedex, France

AIM AND APPROACH

This contribution presents a micrometeorological model designed to assess the urban heat island effect (UHI) distribution over a summer period. Indeed, the UHI significantly impacts on the building energy behaviour and citizens' thermal comfort and health. Its effect becomes dramatic when combined with the more frequent and intense heat waves.

There is nowadays a gap of precision in the urban microclimate models. On the one hand, microclimatic models such as ENVI-met [4] explicitly represent the urban area geometry and use CFD to simulate airflows. Their computational costs limit the simulation period to a few days. On the other hand, urban canopy models [5] require few computational resources and can be run over a year, but they generally represent urban areas with equivalent street canyons. There are very few models able to represent an air temperature field in an urban district over a whole summer, considering a realistic 3D urban geometry. It would yet be the suited tool to evaluate the impact of an urban configuration on the UHI effect, and urban overheating problems. Hence, the proposed model aims to make a step toward filling this gap and improve assessment of the UHI effect.

SCIENTIFIC INNOVATION AND RELEVANCE

To compute the field of air temperature over a city during a summer period with limited computational resources, the model adapts the methodology used in the pollutant dispersion software SIRANE [1].

In particular, the urban geometry is modelled using two types of elements: (1) a realistic network of streets with exchanges at the intersections and with the above atmosphere, and (2) courtyards with exchanges only with the above atmosphere. A parametric airflow description is used in order to limit the computational costs, but the parameterization of convective fluxes will be improved thanks to the use and analysis of large eddy simulations (LES) based on the Lattice Boltzmann Method (LBM [2][3]).

This approach will further ease cross analyses of urban thermal and air quality fields, which is particularly interesting for health risk studies.

PRELIMINARY RESULTS AND CONCLUSIONS

The developed model already allows to compute the surfaces heat fluxes and temperatures as well as the air temperature in a street canyon. It has been confronted with reference measurements [6], showing consistent results.

Current developments aim to handle district scale simulations, including realistic street networks and courtyards. In parallel, the convective parametrisation will be improved based on detailed CFD simulations.

MAIN REFERENCES

[1] L. Soulhac, P. Salizzoni, F. X. Cierco, and R. Perkins, "The model SIRANE for atmospheric urban pollutant dispersion; part I, presentation of the model," Atmos. Environ., vol. 45, no. 39, pp. 7379–7395, 2011.

[2] L. Merlier, J. Jacob, and P. Sagaut, "Lattice-Boltzmann Large-Eddy Simulation of pollutant dispersion in street canyons including tree planting effects," Atmos. Environ., vol. 195, no. August 2019, pp. 89–103, 2018.

[3] J. Jacob and P. Sagaut, "Wind comfort assessment by means of large eddy simulation with lattice Boltzmann method in full scale city area," Build. Environ., vol. 139, pp. 110–124, 2018.

[4] M. Bruse, "Simulating microscale climate interactions in complex terrain with a high-resolution numerical model: A case study for the Sydney CBD Area (Model Description)," Int. Conf. Urban Climatol. Int. Congr. Biometeorol. Sydney, no. Microscale models, p. 6, 1999.

[5] C. S. B. Grimmond et al., "Initial results from Phase 2 of the international urban energy balance model comparison," Int. J. Climatol., vol. 31, no. 2, pp. 244–272, 2011.

[6] M. Nunez and T. R. Oke, "Energy Balance of an Urban Canyon.," J. Appl. Meteorol., vol. 16, no. 1, pp. 11–19, 1977.

Keywords: Urban microclimate model, Urban heat island, Street canyon

Paper ID 30305: Simulation of outdoor thermal conditions of sub-urban neighbourhood typology in Ha Tinh, Viet Nam

Dung Ngo Hoang Ngoc (1,3), Emilia Motoasca (2), Alexis Versele (1), Cuong Pham Hung (3), Hilde Breesch* (1)

1: Faculty of Engineering Technology, KU Leuven, Ghent, Belgium; 2: VITO NV, Mol, Belgium; 3: Faculty of Architecture and Planning, National University of Civil Engineering, Ha Noi, Viet Nam

AIM AND APPROACH

The city of Ha Tinh is a medium-size city of the North Central Vietnam with relatively high annual growth rate at 3% [1], and it will be expanded by 2030 [2] with new roughly 874 ha of surrounding countryside acquired for urban development. Due to this plan, changes on outdoor thermal condition are foreseen over these areas. The aim of this paper is to investigate the correlation between neighbourhood typologies and microclimate in outskirts of Ha Tinh.

In order to archive the targets, simulations of existing and future microclimate conditions are performed with ENVI-Met [3] at Thach Dai, a southwest suburban neighbourhood of Ha Tinh city. This suburban area is now landscaped by sparsely-located low-rise rural houses and agricultural fields but will be more urbanized due to urbanization. Simulation area is of 500-500m dimension. Simulation results are verified with monitoring data of ambient temperature, wind speed and direction, solar radiation, collected by a grid of mini-weather stations of Renkforce WH2600 installed in the selected neighbourhood.

SCIENTIFIC INNOVATION AND RELEVANCE

In the context that Vietnamese cities are in evolution of rapid urbanization and become vulnerable to urban heat island and heat stress, there is a need of resilience- and adaptation-specific advices to the practices of planning and design at urban and neighbourhood scales. This study combines on-site monitoring and simulation to investigate the effects of neighbourhood typology on outdoor microclimate, which provides further a concrete basis for the design or improvement of new climate resilient neighbourhood typologies. The simulation with ENVI-Met verified with measurements in the study is novel in Vietnam and relevant in other climates [4,5] including tropical conditions [6].

Although studies on thermal comfort and energy efficiency on building level are being valued in Vietnam, the reshaping of original urban and suburban landscape due to urbanization requires studies on thermal comfort of community level to ensure the effectiveness of resilient approaches for buildings such as passive cooling, design of building envelope and greenery.

Simulations in this study allow a comprehensive analysis of individual parameters like built density, materials, aspect ratio, green and blue network that form a neighbourhood typology and are associated with outdoor thermal environment, which complements the evaluation of outdoor thermal condition through only monitoring.

PRELIMINARY RESULTS AND CONCLUSIONS

Gathered measurement data in 2019 showed that outdoor thermal comfort level in Ha Tinh city is different in each of the studied neighbourhood typologies and it provides clear evidence on the positive effects of low built density and greenery on improving outdoor thermal condition. However, it is still necessary to deepen the study further i.e to quantify the impact of a neighbourhood typology on the outdoor thermal condition, including the above factors, the used materials and the blue/green network typology as well. This analysis can be realized with on-site monitoring during summer and simulation of a selected neighbourhood as explained in the aim and approach section.

The simulation results of current and future outdoor thermal condition in chosen neighbourhood, it is possible to show that the neighbourhood typologies and the urban development affect outdoor thermal conditions in summer in a manner that can be implemented rigorously in simulation models. These results can be used for designing new neighbourhood typologies. The analysis allows drawing conclusions regarding the most influent parameters of a neighbourhood typology on the outdoor thermal comfort of Ha Tinh suburban neighbourhoods, providing further an important basis for the development of climatic resilient neighbourhood typologies of Ha Tinh.

MAIN REFERENCES

1. Annual report of Ha Tinh city (2019), Government of Ha Tinh city

2. Master plan of Ha Tinh city and the vicinity (2015), Ha Tinh institute of planning and architecture in collaboration with consultants of AREP

3. ENVI-Met [Computer software] (2020). Retrieved from http://www.envi-met.com

4. Elmira Jamei & Priyadarsini Rajagopalan (2018): Effect of street design on

pedestrian thermal comfort, Architectural Science Review, DOI: 10.1080/00038628.2018.1537236

5. Mohammad Taleghani, Laura Kleerekoper, Martin Tenpierik, Andy van den Dobbelsteen (2015): Outdoor thermal comfort within five different urban forms in the

Netherlands, Building and Environment 83, p65-78

6. Amirhosein Ghaffarianhoseini, Umberto Berardi, Ali Ghaffarianhoseini (2015): Thermal performance characteristics of unshaded courtyards in hot and humid climates, Building and Environment 87, pp154-168

Keywords: neighbourhood typology, ENVI-Met, urban microclimate, field measurement, outdoor thermal conditions

Session F3.5 (Online Track): Practice and industry related case studies

Paper ID 30342: Case studies for feasibility and potential estimation of demand response control of district heating and cooling system in Japan

Hisataka Kitora* (1), Shohei Miyata (2), Yasunori Akashi (2)

1: The Kansai Electric Power Company,Incorporated, Japan; 2: The University of Tokyo

AIM AND APPROACH

In recent years, with the increase in power generation by renewable energy, "demand response (DR)", which regulates the electric power demand not only on the supply side but also on the consumer side, has been attracting attention. A large-scale thermal storage system is often used for district heating and cooling (DHC) and DHC is expected as a DR resource as HVAC system in buildings [1,2,3]. In addition, DHC often has sub-plants, whose potential DR resource can be larger than the sum of resource of each plant, by linking plants with different chiller and thermal storage tank capacity balance. Since DR is commanded immediately just before the start of DR, it is necessary to secure the thermal storage amount at the start time of DR in order to implement DR to the maximum. However, it is required to be compatible with high-efficiency operation at normal times too, because in Japan the days to be announced DR are a few days a year. In this study, using the simulation, we aimed to investigate control and operating method to achieve both the high efficiency operation of during normal operation and the maximum amount of DR.

SCIENTIFIC INNOVATION AND RELEVANCE

In Japan, Incentive-type DR market to reduce electric power started in 2017 when the weather is harsh. There are some reports on demonstration test, and using heating demand [4], however, in this paper, approaching from the actual DR regulation clarifies what operation is more efficient from the customer's perspective. To utilize the thermal storage system as a DR resource without fail, it is necessary to anticipate the DR announcement and perform the thermal storage / dissipation operation. In this study, we examined the appropriate control of DHC based on the forecast by usage rate of electric grid, announced at 18:30 on the previous day and 9:30 on the day, and the DR announcement issued 3 hours before the start. The significance of this research is to clarify the effect of a simpler control rule in to ensure reliable control without trouble in actual operation, without using an optimization algorithm that searches for a stricter optimal solution.

Therefore, it is novel that we constructed a highly reproducible simulation model that can calculate not only the behavior of equipment but also the pressure of piping [5], and examined the DR operation method of the existing DHC while evaluating the control state.

PRELIMINARY RESULTS AND CONCLUSIONS

The results from our simulation model reasonably agree with measurement data for existing DHC plant composed of two sub-plants. In this study, this simulation model was conducted from the viewpoint of (1) whether to make a preparatory operation (PO) to secure the thermal storage amount for DR beforehand, (2) conditions to start preparation, and (3) expected time to issue DR. Control logic was aimed at the optimal DR implementation method in terms of DR amount, energy saving, and increase of peak power consumption.

In calculations, there were assumption that there was one DR announcement to reduce power consumption once in July to August. If the PO started from the morning of the day, the DR amount becomes 1129kW, which is larger than the 923kW without PO. The peak electric power did not increase when performing PO, but without PO it increased by 371kW, which would cancel the economic effect of the DR amount. The power consumption was almost the same whether PO was performed or not. From these, it is necessary to make proper PO to obtain a larger DR effect. The results of each calculations confirmed that the system was properly controlled from the viewpoint of heat supply.

MAIN REFERENCES

[1] E. Guelpa, V. Verda, Thermal energy storage in district heating and cooling systems: A review, Applied Energy. 252 (2019)

[2] Omagari Y, Sugihara H, Tsuji K, Funaki T, An economic effect of demand response with Thermal Storage Air-conditioning systems in electricity markets: 2009 IEEE PES/IAS Conference on Sustainable Alternative Energy (SAE), 28-30 Sept. (2009)

[3] Watanabe Y, Fukuta A, Ozawa H, Mizutani K, Takada O, Kosai Y, Report on Demand Response Demonstration Utilizing Thermal Storage System, Study to Build a Virtual Power Plant Utilizing Various Energy Resources; Proceedings from Shase 2017. Kochi (Japan), 13-15 Sep.(2017

[4] A. Khabdullin, Z. Khabdullina, G. Khabdullina, D. Lauka, D. Blumberga, Demand response analysis methodology in district heating system, Energy Procedia. 128 (2017) 539-543.

[5] Kitora H, Akashi Y, Lim J, Study on Efficient Heat Interchange Control in District Heating and Cooling System with Multiple Sub-plants: Proceedings from BS2019. Rome(Italy), 2-4 Sep.2019

Keywords: District heating and cooling, Demand response, thermal storage

Paper ID 30357: Improving the training efficiency of automated fault detection and diagnosis for central chilled water plants by transfer learning

Shohei Miyata* (1), Yasunori Akashi (1), Yasuhiro Kuwahara (2), Katsuhiko Tanaka (3)

1: Department of Architecture, School of Engineering, The University of Tokyo, Japan; 2: MTD Co.,Ltd., Japan; 3: Tokyo Electric Power Company Holdings, Inc., Japan

AIM AND APPROACH

HVAC systems are designed aiming to achieve both high-quality indoor environment and energy efficiency. However, during the operation phase, faults, such as deterioration in equipment performance and inappropriate settings, often occur. It has been reported that efficiency losses due to faults can be 15%-30% [1]; therefore, automatic fault detection and diagnosis (AFDD) is essential for the appropriate operation.

Various methods have been proposed for AFDD, ranging from rule-based to machine learning [2,3]. The authors have proposed an AFDD method that firstly calculates dozens of types of faulty behaviors by system simulation and secondly trains convolutional neural networks (CNN) as classifiers by the faulty behavior data [4]. This method has performed high accuracy in learning the faulty behavior data and performed adequately in diagnosing measured data in Building Energy Management System. However, it requires very high human and calculation cost in constructing simulations, calculating fault states, and training the networks. Therefore, in this paper, to reduce one of these costs, the cost of training the networks, we applied transfer learning that utilizes trained networks and investigated its validity. In this paper, the objective is central chilled water plants which is water side of HVAC system.

SCIENTIFIC INNOVATION AND RELEVANCE

AFDD methods that utilize machine learning have advantages in less human cost in training the algorithm, however, they have disadvantages in more computational cost. Regarding equipment faults such as chillers and air handling units, some AFDD methods are applicable to various systems as long as the product model is the same. On the contrary, regarding control faults such as inappropriate setpoints whose features varies among systems, it is necessary to customize the AFDD method for each system. This is one of the barriers to the widespread adoption of AFDD for HVAC systems.

Transfer learning (TL) utilizes networks that is originally trained for different objectives aiming less computational cost and higher performance. The validity of TL has been shown in the fields of image recognition [5, 6]. The proposed AFDD method utilizes CNN which is originally developed for image recognition; therefor, in this paper, TL is applied to AFDD in central chilled water plants. Specifically, whether the CNN trained on one plant is applicable to a similar plant was investigated. Although this study consists of case studies, it is scientifically innovative in that it shows that TL is effective for AFDD in central chilled water plants.

PRELIMINARY RESULTS AND CONCLUSIONS

Firstly, system simulation with fault conditions were conducted for an existing plant and a similar virtual plant. Secondly, CNN was trained by faulty behavior data of an existing plant (CNN_1). Then, faulty behavior data of the similar virtual plant was diagnosed by following methods:

a) directly using CNN_1

b) using CNN_1 whose output layer were trained by faulty behavior data of the similar virtual plant

c) using CNN whose all layer were trained by faulty behavior data of the similar virtual plant

The computational cost increases in the order of the above. As a result, diagnosis accuracy in each case was 27.6%, 97.9%, 98.4%, respectively. Using the same computer, the training time was 112 and 245 minutes for cases b) and c), respectively. Regarding case a), no additional training was required except for CNN_1. This result implies that it can be difficult to use the trained CNN for one system directly to other systems. Chilled water plants are designed with a balance of heat load, equipment characteristics, and control logic; therefore, the characteristics of the system behavior with faults will change when the systems are different. In terms of computational cost, TL was showed to be effective without performance loss.

MAIN REFERENCES

[1] Katipamula Srinivas, Brambley Michael R., Methods for fault detection, diagnostics, and prognostics for building systems-a review, part I, HVAC&R Research 11. 1 (2005) 3-25.

[2] Schein Jeffrey, Bushby Steven T., Castro Natascha S., House John M., A rule-based fault detection method for air handling units, Energy and Buildings 38.12 (2006) 1485-1492.

[3] Du Zhimin, Fan Bo, Jin Xinqiao, Chi Jinlei, Fault detection and diagnosis for buildings and HVAC systems using combined neural networks and subtractive clustering analysis, Building and Environment. 73 (2014) 1-11.

[4] S. Miyata, J. Lim, Y. Akashi, Y. Kuwahara, K. Tanaka, Fault detection and diagnosis for heat source system using convolutional neural network with imaged faulty behavior data, Science and Technology for the Built Environment. 26 (2020) 52-60.

[5] M. Oquab, L. Bottou, I. Laptev, J. Sivic, Learning and transferring mid-level image representations using convolutional neural networks, Proceedings of the IEEE Conference on Computer Vision and Pattern Recognition (CVPR) (2014) 1717-1724.

[6] Shin, Hoo-Chang, et al. "Deep convolutional neural networks for computer-aided detection: CNN architectures, dataset characteristics and transfer learning." IEEE transactions on medical imaging 35.5 (2016) 1285-1298.

Keywords: Automated fault detection and diagnosis, central chilled water plant, transfer learning, convolutional neural network, HVAC system

Paper ID 30421: Evaluating stack effect impact of thermal comfort in high rise office towers

Gary Chang*, Karim El Mokhtari, Russell Richman, J.J. McArthur

Ryerson University, Canada

AIM AND APPROACH

Stack effect adversely affects energy consumption and thermal comfort in high-rise buildings due to the excess infiltration caused by the resultant pressure differentials. Mitigating the effects of this phenomenon is crucial to achieve ambitious energy targets such as Canada's commitment to COP21. A novel data-driven modelling approach is leveraged using return air temperature as a proxy for thermal comfort issues caused by stack effect. A case study investigated the prediction algorithms, where it was determined that random forest was the most suited for stack effect modelling. The research demonstrates the utility of return air temperature data to identify situations where stack effect is most detrimental to occupant comfort.

SCIENTIFIC INNOVATION AND RELEVANCE

Current literature on stack effect focuses on approaching the problem through the multi-zonal method [2], computation fluid dynamics [4], or the grey box method [5]. The lack of literature investigating a purely data driven model suggest there is opportunity to develop a novel modeling method to address the weakness of the traditional techniques. The case study building is located in Toronto, Canada and represents a typical high-rise office building in the city, allowing the model to be leveraged towards expedited analysis of similar towers. This research has significant potential to contribute to the Canada's COP21 commitment while providing economical savings for building owners.

PRELIMINARY RESULTS AND CONCLUSIONS

This research provided a framework for future data-driven stack models. Using return air temperature as a proxy to predict thermal comfort issues related to stack effect was proven feasible as the data from the upper and lower parts of the fans indicated influence due to stack effect. This was reinforced by the learning weights of the prediction model, since the core zone parameters were a pivot factor in classification, but pressure differential sensors are the final component required to validate this method. The study concluded that random forest is the most suited algorithm for predicting thermal comfort issues related to stack effect as it achieved 93% accuracy while maintaining good generalization. Future research will leverage targeted testing in order to acquire data during various conditions in order to expand the captured conditions to further improve the generalizability of the predictions. The end goal of this research is to create a recommender algorithm to predict instances where stack effect is detrimental to thermal comfort, and energy consumption in order to output corrective actions to mitigate the impact. The recommender algorithm will be visualized via a digital twin to aid facilities with an effective tool for stack effect management.

MAIN REFERENCES

[1] A. Wilson and G. Tamura, "Stack effect in buildings," Canadian Building Digest, vol. 104, 1968.

[2] Y. Jung-yeon, K.-d. Song and C. Dong-woo, "Resolving Stack Effect Problems in a High-Rise Office Building by Mechanical Pressurization," Sustainability, vol. 9, no. 10, p. 1731, 2017.

[3] S. J. Emmerich, A. K. Persily and T. P. McDowell, "Impact of Infiltration on Heating and Cooling Loads in U.S. Office Buildings," in 26th IEA Conference of the Air Infiltration and Ventilation Center, 2005.

[4] N. H. Wong and H. Heryanto, "The study of active stack effect to enhance natural ventilation using wind tunnel and computational fluid dynamics (CFD) simulations," Energy and Buildings, vol. 36, no. 7, pp. 668-678, 2004.

[5] S. Yoon, J. Seo, W. Cho and D. Song, "A calibration method for whole-building airflow simulation in high-rise residential buildings," Building and Environment, vol. 85, pp. 253-262, 2015.

[6] Lim, H., Seo, J., Song, D., Yoon, S., & Kim, J. (2020). Interaction analysis of countermeasures for the stack effect in a high-rise office building. Building and Environment, 168, 106530.

Keywords: Data Analysis, Stack Effect, Machine Learning, Offices

Paper ID 30503: End-to-end Model for Estimating Heat loss of Building envelope using Deep learning with Infrared image

Heesung Park* (1), Sangho Son (1), Ji Young Kang (1), Haemin Jung (1), Seungeon Lee (2), Deuk-Woo Kim (2), Wooju Kim (1)

1: Department of Industrial Engineering, Yonsei University, Seoul, Republic of Korea; 2: Korea Institute of Civil Engineering and Building Technology, Goyang-Si, Republic of Korea

AIM AND APPROACH

Estimating heat loss on the envelope of buildings is important in efficient management of energy within buildings. To determine the degree of heat loss, various methods which use infrared images as input data have been suggested. However, the studies have limitations; they require consideration of environmental conditions and lots of additional metadata inputs to describe them. This paper presents a novel method structured as an end-to-end process using a convolutional neural network (CNN) to estimate heat loss, only with infrared images of the building envelope. The process is divided into three steps: CNN-based object detection to identify building components from infrared images, edge detection to calculate actual area, and estimation of relative heat loss among objects with identical material. The results of our process are estimated heat loss types of the objects with their classes and locations. And we experimented with our method for building windows and showed its usability. In the future, we will generalize our method by applying it to other parts of the building envelope.

SCIENTIFIC INNOVATION AND RELEVANCE

The proposed methodology is divided into three stages, and the required input is only infrared images. The final output comes in two ways; first, information about the type, location, and proportion of objects presents in infrared images, and second, information about the area where heat losses are expected. Based on the aforementioned information, it is possible to determine the heat energy losses present on the exterior of a building. In this study, in heat loss estimation, comparisons are required for the same area. Therefore, comparison and analysis were conducted around the window among the components of the building's exterior walls. The main innovations and contributions of this research are:

- We propose a relative heat loss evaluation and a temperature numericalization methodology for infrared images with minimal user intervention.

- It presents a CNN based architecture for heat loss evaluation. and this model is organized to make it simple to secure training datasets and has scalability.

- We construct a dataset for object detection of building images and propose a method for object detection of infrared images using real building images.

PRELIMINARY RESULTS AND CONCLUSIONS

We proposed an end-to-end process using convolution neural networks that can estimate heat loss. This work consists of a model for detecting objects in infrared images and estimating heat loss.

Summarizing the results of this study, when object detection is applied to the building domain, our datasets can be used as a benchmark. In model, by changing this general image to grayscale, we create a model that detects infrared images well. And we accurately obtain split area information using edge detection. In estimation, based on comparison of relative temperature distributions, we determine estimation indicators for heat loss.

In addition, from the user's perspective, there is an advantage that there is no input of additional information and estimation can be obtained only with infrared images. In the process of extracting heat loss estimating results, the degree of desired conditions can be applied, enabling contextual analysis.

Efforts in this study are focused on developing end-to-end framework based on CNN to estimate heat loss. Furthermore, Since the proposed method can be utilized in real time, it intended to develop a framework for building energy efficiency management in future.

MAIN REFERENCES

Bochkovskiy, A., Wang, C.Y., and Liao, H.Y.M. (2020). YOLOv4: Optimal speed and accuracy of object detection. arXiv:2004.10934.

Canny, J. (1986). A computation approach to edge detection. IEEE Transactions on Pattern Analysis and Machine Intelligence PAMI-8(6), 679-698.

Matas, J., Galambos, C., Kittler, J. (2000). Robust Detection of Lines Using the Progressive Probabilistic Hough Transform. Computer Vision and Image Understanding 78(1), 119-137.

Garrido, I., Lagüela, S., Sfarra, S. et al. (2019). Automatic detection of moistures in different construction materials from thermographic images. Journal of Thermal Analysis and Calorimetry 138, 1649–1668.

Park, G., Lee, M., Jang, H., Kim, C. (2021). Thermal anomaly detection in walls via CNN-based segmentation. Automation in Construction 125, 103627.

Bauer E., Pavón, E., Barreira, E., Castro, E.K.D. (2016). Analysis of building facade defects using infrared thermography: Laboratory studies. Journal of Building Engineering 6, 93-104.

Keywords: Infrared image object detection, Deep learning, Neural Network, Heat loss estimation, End-to-end model

Paper ID 30589: Life cycle metrics of certified green school buildings' envelope shading scenarios

Paula Brumer Franceschini*, Lizzie Pulgrossi, Vanessa Gomes, Doris Kowaltowski, Leticia Oliveira Neves

University of Campinas (UNICAMP), Brazil

AIM AND APPROACH

This study aims to analyse the life-cycle energy performance of four building envelope scenarios of AQUA-certified school buildings located in the state of Sao Paulo, Brazil, by comparing the operational and the embodied energy consumption. Two representative cases studies were selected within FDE's (Foundation for the Development of Education) 19-certified building database: one with single-sided ventilated classrooms and the other with cross ventilated classrooms. Classrooms' annual operational energy consumption was simulated through EnergyPlus version 9.3 software for a reference case, without window shading elements, and for three scenarios with variations of shading elements - (i) galvanized steel perforated screen, (ii) horizontal aluminium device, (iii) ceramic cobogós, a widely used solar protection element in Brazil. The embodied energy of the different shading device configurations was calculated considering a 50-year life-cycle with maintenance (painting steel elements every 5 years), using the Cumulative Energy Demand (CED v1.11) and the CML-IA baseline v3.05 impact assessment methods. SimaPro 9.0 life-cycle assessment (LCA) platform and Ecoinvent database v.3.4/3.5 were used to perform the cradle-to-grave modeling (modules: A1-5, B4, C1-2). Embodied-to-operational energy ratio and Global Warming Potential (GWP) results identified the optimal configuration within the considered scenarios.

SCIENTIFIC INNOVATION AND RELEVANCE

The operational energy of buildings is usually assumed to be more significant than the embodied energy to analyse energy consumption over the long term. However, the former frequently is decreased by increasing the latter (e.g. incorporating materials that provide energy savings) and this balance should be explicitly considered in decision-making. Few studies analysed the trade-off between embodied and operational energy. Most of them were published in the last five years, confirming that this is a relatively new subject powered by increased research interest. Yet, even fewer studies cross checked decisions made on certified buildings, particularly schools, and were rarely based away from temperate climates. Therefore, our research contributes to these growing field, adding data and information related to certified school buildings in the humid subtropical conditions of the state of Sao Paulo, Brazil.

PRELIMINARY RESULTS AND CONCLUSIONS

The operational energy was simulated for a reference case (classroom without shading elements) and three envelope scenarios of two cases studies (single-sided and cross-ventilation). The reference case presented higher operational energy consumption in both cases studies, followed by the scenario with ceramic cobogós. The operational energy consumption of the cross-ventilated classroom was approximately 60% lower than the single-sided ventilated classroom. CED and GWP were calculated for the three shading elements' scenarios, considering a 50-year life-cycle with maintenance. The scenario with ceramic cobogós presented 47% to 60% less impact than the galvanized steel perforated screen and the horizontal aluminium device. The maintenance presented a major contribution, especially for the perforated screen, in which it accounts for 35% of the total primary energy and 42% of the GWP. The embodied-to-operational ratio analysis showed a lower impact of the ceramic cobogó scenario, besides its higher operational energy consumption. In addition, the embodied energy accounted for 19.6% to 40% of the total energy demand, confirming previous results.

MAIN REFERENCES

The trade-off between operational and embodied energy has been studied in the last five years mainly in residential context (Koezjakov et al., 2018; Kovacic et al., 2018; Najjar et al., 2019; Resalati et al., 2020; Shadram et al., 2020; Shadram & Mukkavaara, 2019; Su et al., 2020; Tagliabue et al., 2018) and office context (Giordano et al., 2017; Ma et al., 2017). These studies were performed in different countries, such as Italy, Germany, China, USA and UK, by defining scenarios of building envelope for one case study or selecting a few case studies to compare energy consumption and CO2 emissions. Ma et al. (2017) found out that the climate may affect the operational energy to a large extent, since the operational energy accounts for a higher proportion of the life-cycle of buildings that require both heating and cooling. Su et al. (2020) compared passive and conventional buildings and discovered that the former had higher values of embodied energy and CO2 emissions (+37.6% and +16.2%, respectively), but lower values of operational energy (-26.7%). These previous studies showed a variation range of the embodied energy from 12.5% to 46% of the total energy demand, depending on the climate and the design solutions.

Keywords: energy performance, embodied energy, operational energy, building envelope, school building

Paper ID 30629: A simplified open-loop control strategy for integrated shading and lighting systems using machine learning

Jiarong Xie*, Azadeh Omidfar Sawyer

Carnegie Mellon University, United States of America

AIM AND APPROACH

Access to daylight has significant benefits for workers' health, performance, satisfaction, and circadian rhythm. However, excessive and uncontrolled daylight through windows can cause glare, a common problem in offices. Appropriate automatic blind control systems can be a promising approach to effectively prevent glare and maximize daylight and view access. However, most existing glare-based control strategies rely on sensor readings as input, which can entail excessive cost and labor for the installation, calibration, maintenance and replacement of sensors, especially in an open-plan office requiring multiple sensors. This burden is one of the main reasons that daylight-linked shading control is not widespread. Hence, this study aims to develop a simplified, simulation-assisted shading control strategy that can be implemented in complex real-world settings, especially in an open-plan office. The strategy relies on machine learning models developed from pre-simulated data to predict glare in real conditions. More specifically, the model will be trained using Typical Meteorological Year (TMY) weather data and implemented in a simulated open-plan office. The accuracy and effectiveness of the control framework will be validated with actual historical weather data.

SCIENTIFIC INNOVATION AND RELEVANCE

Advancements of simulation and modelling tools have directed more attention to simulation-assisted glare control. The significant advantage of this method is that it can eliminate the use of multiple sensors while obtaining sufficient data to control shading systems. Existing studies usually use real-time daylight simulation for shading system control in private offices or at one workstation in an open-plan office. However, a system providing real-time glare simulation for multiple workstations at the same time would require intensive computation, which could be problematic for the system's local controller. The method proposed in this study avoids heavy simulation in the controller by using pre-simulated data. Using a data-driven method, a model can be developed to predict and prevent glare in real conditions. In addition, current simulation-based control usually assumes a fixed sitting position and view direction, which may not be sufficient to fully characterize glare conditions in practice. In contrast, this method uses pre-simulated data and pre-computation to allow evaluation of glare at multiple workstations with different orientations and varied positions and view directions.

PRELIMINARY RESULTS AND CONCLUSIONS

The proposed control strategy has been tested to control the slat angle of venetian blinds in a simulation of an existing open-plan office. The predictive model was developed using hourly glare data (represented by Daylight Glare Probability, DGP) obtained from daylight simulation using TMY weather data, accounting for the date and time, solar radiation intensity, sun position, and blind slat angle. Different models were trained with blind slat angles of 0º, 30º, 45º, 60º and 90º. The model was then used to predict the glare profile using actual weather data from 2018. The predictions were compared with the results from daylight simulation using the same weather data. The resulting prediction accuracy of each model was 92–96% using common classification methods, such as Support Vector Machine (SVM), k-nearest neighbours (KNN) and Random Forest. The high-glare hours during occupied periods also decreased from 14.8% to a maximum of 2.8%. This preliminary result suggests the feasibility and potential of applying the proposed method in real-world settings.

MAIN REFERENCES

Shen, H., and Tzempelikos, A. (2017). Daylight-linked synchronized shading operation using simplified model-based control. Energy and Buildings, 145, 200–212. https://doi.org/10.1016/j.enbuild.2017.04.021

Hu, J., & Olbina, S. (2011). Illuminance-based slat angle selection model for automated control of split blinds. Building and Environment, 46(3), 786–796. https://doi.org/10.1016/j.buildenv.2010.10.013

Yun, S. I., Kim, H. R., Park, D. Y., & Jeong, J. W. (2020). Sensor minimization method for integrated daylighting control by a mathematical approach. Energy and Buildings, 214. https://doi.org/10.1016/j.enbuild.2020.109891

Charpentier, V., Meggers, F., Adriaenssens, S., & Baverel, O. (2020). Occupant-centered optimization framework to evaluate and design new dynamic shading typologies. PLoS ONE, 15(4), 1–30. https://doi.org/10.1371/journal.pone.0231554

Katsanou, V. N., Alexiadis, M. C., & Labridis, D. P. (2019). An ANN-based model for the prediction of internal lighting conditions and user actions in non-residential buildings. Journal of Building Performance Simulation, 12(5), 700–718. https://doi.org/10.1080/19401493.2019.1610067

Chan, Y. C., and Tzempelikos, A. (2013). Efficient venetian blind control strategies considering daylight utilization and glare protection. Solar Energy, 98(PC), 241–254.

Katsifaraki, A., Bueno, B., & Kuhn, T. E. (2017). A daylight optimized simulation-based shading controller for venetian blinds. Building and Environment, 126(September), 207–220. https://doi.org/10.1016/j.buildenv.2017.10.003

Keywords: Glare control, automated shading system, data-driven method, daylight simulation

Paper ID 30745: Exterior insulation thickness and its effect on energy usage and thermal comfort on a high-performance multi-unit residential building

Edward Chun Ming Lau*, Fitsum Tariku

British Columbia Institute of Technology, Canada

AIM AND APPROACH

Orion is a high-performance three storey low-rise wood-frame construction multi-unit residential building currently under construction in Pemberton, British Columbia, Canada. The purpose of this project is to use DesignBuilder, a dynamic energy modelling simulation tool, to identify the relationship between the exterior insulation's thickness and the heating and cooling energy uses of high-performance multi-unit residential buildings. The exterior wall is made up of exterior EPS insulation, sheathing, wood studs spaced at 400mm o.c. with R-24 batt insulation inside the cavity and gypsum board on the interior. Thicknesses of 152mm, 101mm, and 51mm exterior EPS insulation were simulated and results were recorded. Typical thermal bridges were taken into account by using the equivalent U-value method and THERM 7.4. The main HVAC system for Orion is designed with a centralized air-to-air heat pump system with an energy recovery ventilator. All other simulation parameters followed the City of Vancouver's Modelling Guidelines and Orion's architectural drawings.

SCIENTIFIC INNOVATION AND RELEVANCE

Orion, a high-performance building, is designed with 152mm of exterior EPS insulation. The authors of this paper wanted to explore the effects on the total energy usage of the building and thermal comfort of the occupants by varying the insulation's depth on a high-performance multi-unit residential building. Because the total energy percentage to condition a high-performance building is low relative to other energy end uses, adding 152mm of exterior EPS insulation may have had diminishing returns. The findings of this paper would potentially be helpful for future design considerations of similar building types in Climate Zone 6 and may potentially be used to extrapolate cost and life cycle analysis of adding additional exterior insulation.

PRELIMINARY RESULTS AND CONCLUSIONS

The change in total energy usage for all three models are only slightly different because the heating and cooling electricity usage only accounts for about 13.8% of the total energy. Any improvements in the heating and cooling energy usage would be very minor when reflected in the total energy usage. However in terms of heating and cooling energy alone, the heating energy used increased by 6.15% for the 101mm EPS model and by 16.06% for the 51mm EPS model. The cooling energy used decreased by 0.77% for the 101mm EPS model and decreased by 2.57% for the 51mm EPS model. In terms of thermal comfort, the average number of hours that the cooling setpoint is not met and the number of units overheated increases as the thickness of the insulation decreases. Intuitively, heat energy can disperse out to the exterior more easily if the insulation is thinner, so a decrease in cooling energy should occur. However, in Climate Zone 6 the benefit of having more insulation to maintain the heat inside the building outweighs the benefit of having less insulation to allow the heat to disperse to the exterior by a factor of 20 in terms of kilowatt-hour.

MAIN REFERENCES

1) NECB, National Energy Code for Buildings 2017, National Research Council, Canada.

2) City of Vancouver Modelling Guidelines Version 2.0, Vancouver, Canada.

3) Martin K., A. Rekoreka, I. Flores, M. Odriozola, J.M. Sala, 2011. Problems in the calculation of thermal bridges in dynamic conditions. Energy and Buildings (43): 529-535.

4) Ge H., F. Baba, 2015. Dynamic effect of thermal bridges on the energy performance of a low-rise residential building. Energy and Buildings (105): 106-118.

5) McLeod R.S., C.J. Hopfe, A. Kwan , 2013. An investigation into future performance and overheating risks in Passivhaus dwellings. Building and Environment (70): 189-209.

6) Mitchell R., S. Natarajan, 2019. Overheating risk in Passivhaus dwellings. Building Services Engineering Research and Technology (40): 446-469.

7) Blight T.S., D. A. Coley, 2013. Sensitivity analysis of the effect of occupant behaviour on the energy consumption of passive house dwellings. Energy and Buildings (66): 183-192.

8) Rouleau J., L. Gosselin, P. Blanchet, 2018. Understanding energy consumption in high performance social housing buildings: A case study from Canada. Energy (145): 677-690.

9) Rouleau J., L. Gosselin, P. Blanchet, 2019. Robustness of energy consumption and comfort in high performance residential building with respect to occupant behavior. Energy (188): 155978

Keywords: insulation, high-performance building, energy usage

Paper ID 30995: A GIS-based regional scale analysis of urban water balance with an aquifer-based water supply

Shirley Tandoh, Robert Ries*

University of Florida, United States of America

AIM AND APPROACH

One of the biggest challenges in urban development is balancing water demand and supply given a growing population. In an aquifer-based water supply system, addressing water scarcity must consider urban hydrology, including maintaining groundwater levels and the interactions among elements of the water cycle.

A GIS-based Python simulation of urban water balance was created to estimate the components of the water cycle in a region over time. The analysis includes changing land cover over time and the resulting variable aquifer extraction and waste water treatment and disposition volumes over time. The simulation models the basic elements of the water cycle, e.g., evapotranspiration, infiltration, and runoff, which are used to estimate impact as well as evaluate alternative water specific strategies.

The simulation uses readily available water supply and treatment volumes, land cover and weather data, and pervious surface analysis to calculate outcome variables. The simulation spatial unit of analysis is a parcel, typically over a scale of a county, and the time step is daily, over multiple years. Parcel level data used includes metered water use and impervious surface area, weather input data includes precipitation and variables needed to calculate potential evapotranspiration, and regional-level data includes waste water treatment and disposition.

SCIENTIFIC INNOVATION AND RELEVANCE

The simulation tool analyzes the impact of urbanism on the hydrologic cycle by evaluating the components of the water balance on both developed and undeveloped land cover over time. The study focuses on residential water use unlike many land use and land cover change studies. The study does not only evaluate potable water use at the parcel level but also considers wastewater distribution and importantly, the disposition of waste water. The analysis includes both natural and artificial aquifer recharge such as treated wastewater discharged to wetlands and direct injection of treated wastewater into the aquifer, and the use of reclaimed water for irrigation.

Parcel-level aquifer withdrawal volume is metered by the utility. The analysis evaluates groundwater recharge from infiltration by estimating parcel-level water flows as precipitation and irrigation and parcel-level water outflows as evapotranspiration, runoff, and wastewater.

The use of a Python programming interface with GIS for the simulation helps incorporate spatial data from various sources such as land cover, parcels, hydrologic soil and meteorological zone classifications, impervious surfaces, and metered water use. Python integrated with GIS automates the analysis and makes it more versatile such that the spatial scale and time step do not become a challenge.

PRELIMINARY RESULTS AND CONCLUSIONS

The GIS platform effectively incorporates all input datasets and the Python code using the application programming interface generates parcel-level daily results for the defined study period. Daily infiltration, evapotranspiration, run-off and waste water generation outputs and precipitation, potable and reclaimed water inputs are calculated for each parcel. Basic and spatial statistical analyses are conducted for either the entire study area or any portion thereof. Output formats are tabular data, maps, and charts.

Simulation studies were performed on four communities in Alachua County, Florida. Two communities are in the west and two are in the east of the county. The communities differ in age, i.e., year built, average parcel and floor area, density, volume of potable and irrigation water use per parcel, and income of the residents.

The comparison of communities showed different water use patterns including water for irrigation. Differences in the type of development pattern partially explains these differences. A comparison between the water cycle in the developed state and the undeveloped state showed more recharge on pristine land. However, aquifer recharge after waste water treatment, including injection wells, wetlands, and use of reclaimed water for irrigation in the developed case increases recharge of extracted potable water.

MAIN REFERENCES

Drummond MA, Stier MP, Auch RF, Taylor JL, Griffith GE, Riegle JL, Hester DJ, Soulard CE, McBeth JL, Assessing Landscape Change and Processes of Recurrence, Replacement, and Recovery in the Southeastern Coastal Plains, USA, Environmental Management (2015) 56:1252–1271

Manzardoa A, Lossa A, Fialkiewiczb W, Rauch W, Scipionia A, Methodological proposal to assess the water footprint accounting of direct water use at an urban level: A case study of the Municipality of Vicenza, Ecological Indicators 69 (2016) 165–175

McMahon, T. A., Peel, M. C., Lowe, L., Srikanthan, R., & McVicar, T. R. (2013). Estimating actual, potential, reference crop and pan evaporation using standard meteorological data: a pragmatic synthesis. Hydrol. Earth Syst. Sci, 17(4), 1331-1363.

Mejía, A., Daly, E., Rossel, F., Jovanovic, T., & Gironás, J. (2014). A stochastic model of streamflow for urbanized basins. Water Resources Research, 50(3), 1984-2001.

Eslinger, A., Pendleton, M., & Burkhalter, S. (2013). Impervious Surface Analysis Tool (ISAT) for ArcMap 10.x. National Oceanic and Atmospheric Administration (NOAA) Coastal Services Center.

Keywords: Water balance, hydrologic cycle, urban hydrology, regional impacts of urban water use

Paper ID 31049: Reducing reliance on deterministic asset ratings to improve the operational energy performance of buildings

Minu Agarwal (1), Parag Rastogi* (2), George Mavromatidis (3)

1: CEPT, Ahmedabad, India; 2: arbnco Ltd., Glasgow, UK; 3: ETH Zurich, Switzerland

AIM AND APPROACH

Climate change is already occurring, and recent years have shown that the rate of change is accelerating. This means that the future climate is unknown, introducing uncertainty into an important 'boundary condition' for planning and design. Buildings designed today reflect a cost-effective response to historical weather information and occupant needs. However, these buildings will have to perform under increasingly stringent performance benchmarks in an unknown future climate. Uncertainty in occupant behavior can further impact compliance with future benchmarks. We assess if an architect's prudent choice between two contrasting design alternatives of residential neighborhood designs can ensure compliance with the prevailing energy regulations under weather and occupant uncertainty. Two neighborhood scale design schemes, with different building layouts and exterior exposed surface area, were modeled. Future weather time series scenarios were generated using a stochastic weather generator [1] that modifies the distribution of historical data by combining it with the outputs of climate models. We expanded the scope of occupant behavior modelling at the neighborhood scale by sampling household types based on demographic data and modelling behavior based on household type [2] to present a holistic estimate of occupants' impact on building energy use at the neighborhood scale.

SCIENTIFIC INNOVATION AND RELEVANCE

This case-study-based paper demonstrates the significance of design decisions made by architects when limiting the energy demand of buildings under important aleatory forms of uncertainty. Given the high degree of control occupants have on their respective energy needs, we assess if knowledge of, or degree of control over future occupant types should be considered when making design decisions. We also propose a new performance metric for decision making based on the probability of continued compliance of a given design proposal.

PRELIMINARY RESULTS AND CONCLUSIONS

Both design proposals in the case study were first simulated with only one stochastic input at a time, and then both, against a baseline with deterministic inputs for weather and occupancy. A deterministic input for occupancy is a nominal fixed profile while for weather it is the measured historical data over 37 years. At the building scale, the mean and standard deviation of annual heating energy use over historical years were 21 and 3.5 kWh/m2 respectively for the design proposal with less exposed surface areas. Simulating with stochastic occupancy-only changed those values to 41 and 7.4 kWh/m2, with stochastic weather-only to 23 and 5.4 kWh/m2, and finally with both stochastic inputs to 54 and 7.7 kWh/m2. Comparison of such results for the two competing design proposals indicate that if occupant behavior characteristics align with low energy needs, then both design proposals can achieve compliance with energy code under the harshest weather conditions. However under large diversity in occupant behavior, design choices become important.

MAIN REFERENCES

1. P. Rastogi, 'On the sensitivity of buildings to climate: the interaction of weather and building envelopes in determining future building energy consumption', PhD, Ecole polytechnique fédérale de Lausanne, Lausanne, Switzerland, 2016.

2.Flett, Graeme, and Nick Kelly. "An occupant-differentiated, higher-order Markov Chain method for prediction of domestic occupancy." Energy and Buildings 125 (2016): 219-230.

Keywords: Design Decisions, Uncertainty, Probabilistic Compliance Metrics

Paper ID 31131: Modeling of VOC removal in a commercial airplane hangar by multiple exhaust hoods

Fei Liu*, Tengfei Zhang

Tianjin University, China, People's Republic of

AIM AND APPROACH

To improve the VOC removal efficiency, this study proposed a novel upside-supply and multiple air-exhaust (UM) system. CFD modeling and measurement were conducted to evaluate the system's performance.

SCIENTIFIC INNOVATION AND RELEVANCE

• Portable, multiple exhaust hood ventilation was proposed for hangars.

• The CFD simulation showed that the proposed ventilation forms a push-pull flow pattern to exhaust pollutants.

• The VOC concentrations in the occupied zone are greatly reduced as compared with existing systems.

PRELIMINARY RESULTS AND CONCLUSIONS

(1) The UM system formed a nearly push-pull flow pattern in the hangar, and short-distance direct flow from the pollutant sources to the exhaust hoods, while the conventional UF and UD systems generated strong mixing flow.

(2) During painting of the aircraft wings, the occupied-zone TVOC concentration under the UM system was 71.9% to 82% lower than under the conventional UF and UD systems. For painting of the fuselage or vertical tail, the UM system maintained a TVOC concentration in the occupied zone that was at least 90% lower than the UF or UD system.

MAIN REFERENCES

Bennett, J. S., Marlow, D. A., Nourian, F., Breay, J., & Hammond, D. (2016). Hexavalent chromium and isocyanate exposures during military aircraft painting under crossflow ventilation. Journal of Occupational and Environmental Hygiene 13(5), 356–371.

Li, J., Uttarwar, R. G., & Huang, Y. (2013). CFD-based modeling and design for energy-efficient VOC emission reduction in surface coating systems. Clean Technologies and Environmental Policy 15(6), 1023–1032

Keywords: Ventilation, Painting, VOCs, Hangar, CFD

Paper ID 30783: Virtual reality enabled building-data management through the combination of a fully integrated IFC-BIM model and an IoT-based building management system

Sergi Aguacil Moreno (1), Sebastian Duque Mahecha* (1), Alexandre Dennis Stoll (1), Tiago De Sousa Pereira (1), Laurent Deschamps (1), Jean-Phillippe Bacher (2)

1: Building2050 group, Ecole Polytechnique Fédérale de Lausanne (EPFL), Fribourg, Switzerland; 2: Energy Institute, University of Applied Science of Western Switzerland (HEIA-FR, HES-SO), Fribourg, Switzerland

AIM AND APPROACH

Professionals involved in building construction and monitoring increasingly implement new practices made available by Building Information Modelling (BIM), Virtual Reality (VR) and the Internet of Things (IoT). Acknowledging the need to integrate BIM and IoT technologies to effectively link the processes carried-out before and after construction, certain methodologies have been accurately defined and explored [1–7]. In turn, successful experiments with VR-enabled environments have shown how virtual and augmented reality technologies may make Facility Management (FM) interfaces more reactive and efficient [8–10]. Despite these efforts, most methods integrating BIM and IoT rely on Application Programming Interfaces (APIs) related to Closed Source Software (CSS), which practically exclude the basic BIM format IFC [2]. Within the fewer methods departing from an IFC model, crucial IFC information such as textures, IFC hierarchy, and most IFC parameters tend to get lost on the way to user VR-enhanced interfaces [11,12]. This article describes a workflow of a BIM-IFC to IoT VR-enabled data management definition, articulating a number of systems operating in a real building. That is Building Information Models (BIMs), a network of sensors and actuators, the Building Management Systems' (BMS) database (or, as an alternative, the IoT system), and the facility management interface.

SCIENTIFIC INNOVATION AND RELEVANCE

Since IFC is the basic BIM open international standard [13–15], Facility Management-enabled Building Information Models (FM-BIMs) should be 100% IFC compatible. Paradoxically, a review of current industry practice and scientific literature shows that most successful tests are based on closed-source software such as Autodesk Revit-Dynamo [7]. For other tests, including our own research, results show that IFC data loss is the common denominator for current FM-BIMs workflows. The principal reason for this is that most modelling and control software have not been certified for the latest updates and improvements made by the IFC platform [16]. Simply put, dealing with IFC appears as a headache to professionals who eventually prefer relying on closed BIM solutions. This article describes thus an IFC-based VR-enabled FM-BIMs workflow, which secures the transmission and automatic update of IFC information between the digital model and the FM user's interface, which in turn is linked to a BMS database. This experience is evidently relevant for the application case described above, but also may open new possibilities to enhance IFC usage in other workflows for professional and research practices.

PRELIMINARY RESULTS AND CONCLUSIONS

Three major parts constitute the protocol. (1) IFC-Model import into game-engine software: a script that allows IFC-data verification and the incorporation of extra information, outputs a DAE file compatible with a visualisation software, in this case Unity [17]. (2) Integration of user interface and virtual actuators: a code rendering certain model-elements interactive, such as light switches, as well as a graphic interface making this interaction explicit. (3) Link between user interface and BMS (or IoT) database: an API connecting model-elements, user interactions, and database values transmitted by the network of sensors and actuators installed in the building space. As a result, we are able to set an IoT network controllable by the user in VR. The API transmits information from and to the BMS database, managing a two-way control. When touching the light-switch in the virtual environment, the corresponding light in the real space is turned-on in real-time. The actuator state of light and switch automatically update to their new state, and the API transmits the information back to the virtual environment: virtual light shows switched-on and the switch, by a graphic convention, changes from red (off) to green (on). During this real-time synchronisation, all IFC information remains available.

MAIN REFERENCES

[1] S.Tang et al., Automation in Construction A review of building information modeling (BIM) and the internet of things (IoT) devices integration: Present status and future trends, Autom. Constr. 101 (2019)127–139.

[2] M.Jourdan et al., Towards an integrated approach of building-data management through the convergence of Building Information Modelling and Internet of Things, J. Phys. Conf. Ser. 1343(2019(.

[3] H.T.Nguyen, Integration of BIM and IoT to improve the building performance for occupants' perspective, (2016).

[4] C.Quinn et al., Automation in Construction Building automation system - BIM integration using a linked data structure, Autom. Constr. 118(2020)103257.

[5] K.M.Chang et al., An automated IoT visualization BIM platform for decision support in facilities management, Applied Sciences 8, 2018.

[6] D.Jo et al., AR enabled IoT for a smart and interactive environment: A survey and future directions, Sensors 19, 2019.

[7] P.Parsanezhad, Towards a BIM-enabled Facility Management: Promises, Obstacles and Requirements, 2019.

[8] A.Kiviniemi et al., Review of the Development and Implementation of IFC compatible BIM Executive Summary, ERA Build. (2008)1–2

[9] M.Poljanšek, Building Information Modelling (BIM) standardization, 2017.

[10] Olof Granlund et al., Series 12: Use of models in facility management, Common BIM Requir. (2012)23.

[11] UnityTechnologies, Unity platform, 2005. https://unity.com/.

Keywords: Building Information Modelling, Internet of Things, Immersive Virtual Reality, Buildings Automation Systems (BAS), Facility Management-enabled Building Information Models (FM-BIMs)

Paper ID 30386: Achieving carbon emission reduction goals through calibrated energy modeling and parametric analyses

Samuel Mason* (1), Andre Legault (2)

1: Encompass Energy, United States of America; 2: MRA, Quebec, Canada

AIM AND APPROACH

Calibrated energy models can be used to predict the effect on energy consumption and carbon emissions related to complex modifications to an existing building. This calculation methodology is appropriate when the nature of a building renovation includes energy measures that have interactive effects such as facade upgrades combined with interior load and HVAC modifications. With new legislation in New York City related to carbon emissions limits for existing buildings (Local Law 97), there is a need to better quantify energy usage and carbon emissions of current building operations and the energy-conservation measures needed to achieve the related carbon emission reductions.

This presentation will also illustrate the impact of tenant energy usage on the energy consumption in buildings using a large (1.6 million square foot, 32 story) commercial office building in New York City. Multiple strategies of tenant load reduction were modeled to determine the most cost-effective strategies. These strategies include lighting power allowances, lighting controls (occupancy and daylight), receptacle and plug load controls, and management plans, server room planning and cooling strategies, and overall space planning, The effectiveness of the lighting and plug load controls, which are dependent on occupant behavior, are analyzed in further detail through parametric analyses.

SCIENTIFIC INNOVATION AND RELEVANCE

Existing building renovations are becoming more prevalent as a method of delivering energy, utility cost, and carbon emission savings for building owners. Using a combination of sub-meter and interval data for the base building and every individual tenant, electric and steam load profiles can be generated that allow a detailed view of energy flows in a building. From this data set, approaches to reducing energy consumption are tailored for the base building and each tenant. From a list of over 100 measures, 32 tenant specific measures were selected for implementation after detailed parametric analysis and sensitivity studies.

The methodology and results will be useful to practioners, building owners, and research groups to understand the time commitment and results from an in-depth calibration study. The presentation is intended to be part of a larger conversation currently happening about tenant engagement in reducing energy consumption and carbon emissions in New York City and worldwide. It is also expected to generate discussion amongst conference attendees about their own experiences with the intent to foster knowledge sharing.

PRELIMINARY RESULTS AND CONCLUSIONS

The energy conservation measures are grouped into three packages based on current acceptability of the technology, first cost, and simple payback. The first package of measures, which includes currently available and applicable technologies, are estimated to reduce operational costs by $0.53/ft2/yr and annual carbon emissions by 14.6% with a simple payback of 3 years. The second package of measures, which carry a higher first cost and require better design integration and commissioning, are estimated to reduce operational costs by $0.14/ft2/yr and annual carbon emissions by 7.4% with a simple payback of 2.6 years The third package of measures, which require significant tenant engagement and/or are still not readily available technologies to all tenant types, are estimated to reduce operational costs by $0.05/ft2/yr and annual carbon emissions by 2.6% with a simple payback of 10.4 years. Combined, all measures are estimated to reduce operational costs by $0.72/ft2/yr and annual carbon emissions by 24.6% with a simple payback of 3.5 years.

A pilot office project is underway in the building to implement the entire package of measures and future work will include a presentation on the actual energy consumption of the space.

MAIN REFERENCES

ASHRAE 90.1-2013

IECC 2018

International Performance Measurement and Verification Protocol Volume 1

New York City Local Law 97

Keywords: calibration, tenant design and interaction, NYC carbon limits, carbon emissions, energy code

Paper ID 30155: Modeling and simulation of district heating and cooling systems using CityBES

Han Li*, Tianzhen Hong, Wanni Zhang

Lawrence Berkeley National Laboratory, United States of America

AIM AND APPROACH

District Energy Systems (DES) have advantages including space and expenses savings by avoiding distributed, individual equipment installations, more predictable demand due to load aggregations, reduced power generation and transmission loss over traditional individual in-building systems [1,2]. With the increasing environmental awareness and needs for flexibility and resiliency, DES have gained more interests worldwide. Methods and tools are needed to efficiently plan and design DES. This paper covers two parts around the new DES feature of CityBES [3] – a web-based urban scale simulation and data analytics platform. It first introduces the functionalities of the DES feature, which includes visualization of district load profiles and key statistics, and district systems selection, configuration and simulations. Then, it presents a case study conducted with the DES feature of CityBES to showcase how it can support preliminary comparison and selection among several common DES options. Two example districts with different sizes in two cities in the United States are selected for the case study. The energy consumption and simple economics metrics are compared among common district systems in both examples, including: (1) central boilers and chillers, (2) central boilers and chillers with ice-storage, (3) heat-recovery chillers with heat-pumps, and (4) ground-source heat-pumps (GSHP).

SCIENTIFIC INNOVATION AND RELEVANCE

In recent years, DES have gained growing interests as they bring benefits in engineering, economic, and social perspectives like reducing GHG emissions, providing resiliency and complementary services to grid [2]. Currently, DES in the U.S. serves over 5.5 billion sqft of heating floorspace and over 1.9 billion sqft of cooling floorspace [4]. The global growth of DES is anticipated to be 5.8% annually in the coming decade [5]. Despite the momentum, there are still challenges of DES applications, which include uncertainties in long-term economic viability, and lack of easily accessible system evaluations. Various methods were developed to model DES. Among them, physics-based models yield the most accurate demand load-profiles and allow comprehensive DES simulations[6]. However, this approach is not scalable because it's hard to model each building at district-level. The DES feature of CityBES is innovative as (1) it leverages demand load profiles from physics-based simulation of each individual building; (2) it provides quick analyses of the load profiles, and compares common DES options in terms of their energy, peak demand, and simple economic factors. This feature provides simple yet informative preliminary evaluations of DES at the early conceptual phase. The feature is free to use for public.

PRELIMINARY RESULTS AND CONCLUSIONS

A San Francisco district and a Chicago district are selected to demonstrate CityBES's DES feature. The time-series heating, cooling, electricity and natural gas load-profiles, which serve as the input of DES simulations, are aggregated from individual building simulation results provided in CityBES. The visualization of the load-profiles reveals the annual heating and cooling demands in the Chicago example are well-balanced, while they are significantly imbalanced in the San Francisco example. Therefore, all four DES system options are evaluated with the Chicago case, while only two system options are evaluated for the San Francisco case, as the other two options (the heat recovery chiller, and the GSHP) require well-balanced heating and cooling demands. The DES evaluation also allows simple economics parameter inputs, including fuel prices, and incremental costs. The DES simulations yield a set of energy and utility cost savings, and electric-load pattern results across different system types in both examples, which will be presented in the full paper. The case study demonstrated CityBES's capability of (1) aggregating individual building's load-profiles to district-level, (2) guiding DES selections from load-profile visualization, and (3) supporting preliminary comparison of different system types from energy and economic perspectives, which can inform DES planning and designing.

MAIN REFERENCES

[1] B. Talebi, P.A. Mirzaei, A. Bastani, F. Haghighat, A Review of District Heating Systems: Modeling and Optimization, Front. Built Environ. 2 (2016) 22. https://doi.org/10.3389/fbuil.2016.00022.

[2] B. Rezaie, M.A. Rosen, District heating and cooling: Review of technology and potential enhancements, Appl. Energy. 93 (2012) 2–10. https://doi.org/10.1016/j.apenergy.2011.04.020.

[3] T. Hong, Y. Chen, M.A. Piette, X. Luo, Modeling City Building Stock for Large-Scale Energy Efficiency Improvements using CityBES Overview of CityBES, (2016).

[4] L.T. Cooper, N.B. Rajkovich, An Evaluation of District Energy Systems in North America: Lessons Learned from Four Heating Dominated Cities in the U.S. and Canada, in: ACEEE Summer Study Energy Effic. Build., 2012.

[5] District Heating and Cooling Systems Distributed Energy, (n.d.). https://www.distributedenergy.com/home/article/13034512/district-heating-and-cooling-systems (accessed June 29, 2020).

[6] G. Schweiger, R. Heimrath, B. Falay, K. O'Donovan, P. Nageler, R. Pertschy, G. Engel, W. Streicher, I. Leusbrock, District energy systems: Modelling paradigms and general-purpose tools, Energy. 164 (2018) 1326–1340. https://doi.org/10.1016/j.energy.2018.08.193.

Keywords: District Energy System, Modeling, Tools

Paper ID 30392: Impact of multiple urban microclimates in building performance. A simulation approach to support climate-resilient building design.

Jesus Lizana, Eduardo Diz-Mellado, Victoria Patricia Lopez-Cabeza*, Carlos Rivera-Gomez, Carmen Galan-Marin

Universidad de Sevilla, Spain

AIM AND APPROACH

For decades, the main research attention has been paid to the energy efficiency of buildings, as existing buildings play an important role in energy consumption and associated carbon emissions. The rapid development of computer technology and building energy simulation tools have supported the methods to compare the cost-effectiveness of energy conservation measures, which has been studied and debated over year. However, current tools present several challenges aiming at facing future building needs under climate change projections. This research provides a first quantification of existing gaps in building simulation using a case study in Spain, and develop a holistic approach to support a climate-resilience design in buildings. The use of weather data not representative of urban heat island (UHI), the consideration of potential benefits of specific urban microclimate design, the implementation of changeable operating profiles on the road to smart buildings and the consideration of additional indicators to strong climate resilience have been analyzed. The results will quantify the importance of these issues and will provide a first holistic approach to implement these gaps in building energy simulation models.

SCIENTIFIC INNOVATION AND RELEVANCE

The scientific innovation of this research raises on the importance of a reliable and accurate building energy simulation to efficiently support a climate-resilient design of buildings under climate change scenarios. Current weather data for building simulation are from weather stations in rural or suburban areas, whose data are not representative of urban climate, not considering urban heat island (UIH). There is no consideration of the potential benefits of thermodynamic performance enhancement associated with specific urban microclimate design. Most of the building energy simulation tools are based on fixed operating profiles, related to fixed hourly occupancy and system operation, which make difficult the consideration of potential resilient solutions based on the smart building design. Moreover, most of the previous studies investigating retrofitting measures for climate change scenarios are strictly focus on energy demand, and a few include the effects on thermal comfort through adaptive models. Thus, as a step forwards in building simulation, this research evaluates a holistic approach to assess climate impacts in building retrofit analysis reliably and accurately.

PRELIMINARY RESULTS AND CONCLUSIONS

Previous studies have demonstrated that current Urban Heat Island effect in cities can increase the outdoor temperature in urban climate by more than 5ºC. Moreover, previous monitoring campaigns have shown that the courtyard's microclimate in different building typologies can reduce outdoor peak temperatures by 8 ºC. These issues involve an important gap, not considered in building performance assessment. Taking into account that the annual mean temperature for 2100 could increase up to 11ºC, ranging in Europe between 3ºC and 6ºC according to EURO-CORDEX, special attention should be considered in future building energy models to increase the mitigation potential of climate risks to improve the comfort and wellbeing of citizens in buildings.

MAIN REFERENCES

- Shen, P., Braham, W., & Yi, Y. (2019). The feasibility and importance of considering climate change impacts in building retrofit analysis. Applied Energy, 233–234(October 2018), 254–270.

- IEA/UN. (2019). 2019 Global Status Report for Buildings and construction: Towards a zero-emission, efficient and resilient buildings and construction sector.

- Coakley, D., Raftery, P., & Keane, M. (2014). A review of methods to match building energy simulation models to measured data. Renewable and Sustainable Energy Reviews, 37, 123–141.

- Peiró, M. N., Sánchez-Guevara, C., & González, F. J. N. (2017). Update of the Urban Heat Island of Madrid and Its Influence on the Building's Energy Simulation. In Sustainable Development and Renovation in Architecture, Urbanism and Engineering.

- Rivera Gómez, C., M. Diz Mellado, E., Galán Marín, C., & López Cabeza, V. (2019). Tempering potential-based evaluation of the courtyard microclimate as a combined function of aspect ratio and outdoor temperature. Sustainable Cities and Society, 51(August), 101740.

- Climate change projections for Europe provided by the EURO-CORDEX initiative. https://www.eea.europa.eu/soer/data-and-maps/data/external/euro-cordex.

Keywords: Climate-resilient design, Building energy simulation, Urban heat island, Climate Resilience, Courtyard

Paper ID 30437: Integration of raytracing and CFD simulations to assess the impact of UVGI technology on the control of aerosol transmission in an occupied room

Miaomiao Hou (2,1), Jihun Kim (1,4), Jovan Pantelic (3), Dorit Aviv* (1)

1: University of Pennsylvania; 2: Tongji University; 3: KU Leuven; 4: City University of New York

AIM AND APPROACH

A simulation technique is proposed as a tool for assessment of the efficacy of Ultraviolet Germicidal Irradiation (UVGI) devices as measures for controlling virus spread in occupied rooms' retrofits. UVGI devices disinfect the air and surfaces in their range through intense UV-C radiation which causes DNA or RNA damage to microorganisms. This disinfection method is proved to be highly effective for reducing aerosol transmission risk. For UVGI to be effective, maintaining a high-level UV-C fluence rate in the upper-zone of the room over sufficient time period is required in order to achieve the necessary radiation dose for the microorganism deactivation. At the same time, ensuring a safe lower zone with irradiation levels low enough to avoid any potential damage to human eyes and skin, is necessary in occupied rooms. We created a workflow between a CFD analysis of airflow patterns and virus concentration, and a raytracing simulation technique for determining the spatial distribution of the UV irradiation. We show how design variables can optimize the distribution for effective disinfection and what safety precautions must be taken to avoid overexposure of people to the UV radiation, when deploying these devices within existing buildings.

SCIENTIFIC INNOVATION AND RELEVANCE

Due to the rapid indoor spread of the current COVID-19 pandemic, UVGI technology in the upper zone of occupied hospital patient rooms has been discussed as an effective measure for disinfecting the air close to the infection source and thus reduce the risk of virus transmission from a patient. However, the spatial impact of such installation and the material and geometric variables that affect its performance and risk level, are not well understood or characterized. For this reason, we use the raytracing light simulation RADIANCE engine (Ward 1994) based on the radiosity equation, which governs the UV-C range as well as visible light. We adapt the surface response to UV-C reflectance and use lab-tested UVGI-specific ray distribution, accounting for multiple interreflections. Coupled with airflow analysis based on CFD simulation (Gilkeson and Noakes 2013), we achieve high-resolution mapping of UV-C irradiation intensity at the lower level of the room, with potential dangerous "hotspots" and check for effective ranges of radiation in the "kill zone" above human height. Checking for multiple variables, we determine which design parameters have the greatest impact on the resultant irradiation. We collaborate with industry partners and medical institutions to apply the conclusions to real-world scenarios.

PRELIMINARY RESULTS AND CONCLUSIONS

Preliminary results show a range of 0.21-3028.77 μW/cm2 in the upper kill zone and 0.01-1.43 μW/cm2 in the lower occupied zone. The effective radiation value in the upper zone is a minimum of 10 μW/cm2 for air disinfection, while in the lower zone, any value above 0.2 μW/cm2 of surface radiation is considered dangerous to human exposure over time and should be avoided (First et al 1999). A combination of design variables impact the distribution of UV-C radiation in space. The position of UVGI device is critical in determining the span of the effective kill zone. The room proportions and reflectance coefficients of materials have a major impact on the amount of UV interreflections in the room, creating potentially dangerous high irradiation value in the lower zone. The CFD simulation results show a variety of potential airflow paths based on air supply and return locations, and the free convection around people and equipment in the room. Specific airflow patterns are evaluated based on their alignment with the patient's need for fresh air supply and the zone of highest radiation intensity where the airflow should slow down for effective disinfection.

MAIN REFERENCES

ACGIH, TLVs. 2010. "Threshold Limit Values for Chemical Substances and Physical Agents and Biological Exposure Indices." In USA: American Conference of Governmental Industrial Hygienists, 200–219.

Boyce, P., S. Vasconez, and P. Brons. 2003. "Controlling Tuberculosis Transmission with Ultraviolet Irradiation." Delta. Troy (NY): Rensselaer Polytechnic Institute.

First, M. W., E. A. Nardell, W. Chaisson, and R. Riley. 1999. "Guidelines for the Application of Upper-Room Ultraviolet Germicidal Irradiation for Preventing Transmission of Airborne Contagion -- Part 2: Design and Operation Guidance." CONF-990102-. ASHRAE, Inc., Atlanta, GA (US).

Gilkeson, Carl A., and Catherine Noakes. 2013. "Application of CFD Simulation to Predicting Upper-Room UVGI Effectiveness." Photochemistry and Photobiology 89 (4): 799–810.

Rutala, William A, Maria F Gergen, Brian M Tande, and David J Weber. 2013. "Rapid Hospital Room Decontamination Using Ultraviolet (UV) Light with a Nanostructured UV-Reflective Wall Coating." Infect Control Hosp Epidemiol 34 (5): 527–29.

Ward, Gregory J. "The RADIANCE lighting simulation and rendering system." In Proceedings of the 21st annual conference on Computer graphics and interactive techniques, pp. 459-472. 1994.

Wengraitis, Stephen, and Nicholas G Reed. 2012. "Ultraviolet Spectral Reflectance of Ceiling Tiles, and Implications for the Safe Use of Upper-Room Ultraviolet Germicidal Irradiation." Photochemistry and Photobiology 88 (6): 1480–88.

Keywords: Ultraviolet Germicidal Irradiation, Raytracing, Simulation, Aerosol, CFD

Paper ID 30460: Numerical prediction of dew condensation in EAHE system in summer

Kazuma Hirata*, Kento Tomoda, Yasuyuki Shiraishi

The University of Kitakyushu, Japan

AIM AND APPROACH

In recent years, Earth-to-Air Heat Exchanger (EAHE) system that pre-cool/pre-heat introduced outside air has been brought to public attention as one of the passive systems. On the other hand, during operation of EAHE, there has been a concern about air pollution due to growth of microorganisms by dew condensation [1]. In order to suppress the dew condensation, firstly it is important to estimate the detailed dew condensation in underground air tunnels. In this study, we predicted and evaluated dew condensation in EAHE (underground air tunnel) in the actual building by numerical analysis. Besides, in this analysis, we realized the long-term prediction by the unsteady CFD analysis with reduced calculation load proposed in the previous study. This analysis method reduces the calculation load for CFD analysis by using flow fields simulated in advance, and linearizing the heat/ humidity transport equation by loading these flow fields. Moreover, we consider the moisture transfer and moisture absorption/desorption of coals with humidity control by coupling simultaneous heat and moisture transfer equation in coals to the proposed method. We verified the prediction accuracy through comparison between actual survey and analysis, and then conducted detailed evaluation of dew condensation in this underground air tunnel.

SCIENTIFIC INNOVATION AND RELEVANCE

Tomoda et al. [1] proposed the unsteady CFD analysis method with does not analyze flow fields to reduce the calculation load (uncoupled analysis) for EAHE. Results showed that the proposed method has a practical use in terms of calculation times for predicting annual performance. Muta et al. [2] predicted annual performance of a trench in a medical facility by uncoupled analysis, and verified the prediction accuracy in comparison with a long-term actual survey. Results showed that uncoupled analysis had high prediction accuracy in temperature. Tomoda et al. [3] carried out case studies on the introduction path of outside air and the technique of introducing outside air by uncoupled analysis for a pit to evaluate dew condensation and the countermeasure effect of it. Results showed that it can prevent to some extent occurrence of dew condensation by the dew point temperature control. Various studies on EAHE were carried out by uncoupled analysis. However, dew condensation is likely to occur due to the introduction of high humidity outside air in summer. In this study, we conducted detailed evaluation of dew condensation in an underground air tunnel in summer. Moreover, we verified the countermeasure effect of dew condensation by coals with humidity control.

PRELIMINARY RESULTS AND CONCLUSIONS

The RMSE were as follows. Outlet temperature : 0.32°C, Outlet humidity : 2.49%. Thus, it was confirmed that the proposed method had high prediction accuracy. Furthermore, for dew condensation, the analysis generally captured the tendency of it throughout the period. Dew condensation frequently occurred in summer. In July, there were many rainy days and the high humidity environment was frequently occurred in the pit. Therefore, in July, the condensation area ratio was high, and it was 25.4% at the maximum. On the other hand, it was likely that dew condensation occurred only in some areas, since the condensation area ratio was very small in August and September. Moreover, the condensation time ratio and the condensation amount were large in internal corner, so internal corner seems to be the area where dew condensation is most likely to occur in the pit. However, it was confirmed that dew condensation was suppressed, since dew condensation was not generated on the floor surface laying coals with humidity control. From the above results, dew condensation frequently occurred in the underground air tunnel (Especially in internal corner) in summer. Therefore, it was suggested that introducing the outdoor air should be limited taking into account the indoor air quality.

MAIN REFERENCES

1) Kento Tomoda, Yasuyuki Siraishi, "Annual Prediction Method for The Thermal Performance of Earth-to-Air Heat Exchanger By CFD Analysis where Calculation Loads were Reduced" Evaluation method of cooling and heating effect of an earth-to-air heat exchanger (Part 2), Journal of Environmental Engineering (Transactions of AIJ), Vol.81, No.722, pp. 393-401, 2016.4.(in Japanese).

2) Akitoshi Muta, Kento Tomoda, Yasuyuki Shiraishi, Eri Takaoka, Shiro Tsukami, "Annual Performance Prediction of Cool and Heat Trench in the Medical Facility (Part1)-Performance Verification Based on Kong-Term Actual Survey and Unsteady CFD Analysis That Reduced Calculation Load", Transactions of the Society of Heating, Air-conditioning and Sanitary Engineers of Japan, Vol.82, No.238, p.1-7, 2017.1. (in Japanese).

3) Kento Tomoda, Akitoshi Muta, Yasuyuki Shiraishi, "Prediction of Dew Condensation and Effectiveness Study of Countermeasure of Dew Condensation by Annual Simulation Evaluation method of cooling and heating effect of an earth-to-air heat exchanger (Part 3)", Journal of Environmental Engineering (Transactions of AIJ), Vol.81, No.736, pp. 579-587, 2017.6.(in Japanese).

4) Satomi Matsuu, Hanako Ueda, Hideyo Ninomiya, "The fungal flora in the geothermal utilization pit Part1. The fungal concentration actual situation in the building and identification technique", Summaries of technical papers of annual meeting at Architectural Institute of Japan, pp.729-730, 2016.8. (in Japanese).

Keywords: Earth-to-Air Heat Exchanger System, CFD analysis, Calculation load reduction, Dew condensation risk

Paper ID 30562: Perspectives of building simulation in Brazil: Brazilian building simulation market landscape and the applicability of ASHRAE standard 209

Alberto Hernandez Neto (1), Juliana Pellegrini Lemos Trigo (1), Mônica Pereira Marcondes Cavaleri (1), Ana Paula de Almeida Rocha (2), Maíra André* (3), Rodrigo Dias Machado (1), Thiago Montenegro Góes (4)

1: University of Sao Paulo, São Paulo, Brazil; 2: FAE University, Curitiba, Brazil; 3: Federal University of Santa Catarina, Florianópolis, Brazil; 4: University of Brasilia, Brasília, Brazil

AIM AND APPROACH

This paper proposes to investigate the Brazilian building simulation market landscape and analyzes the possible adherence of the ASHRAE Standard 209 guidelines to the national context, especially for the compliance of ASHRAE's standard to the most used certification, labeling and normative processes in Brazil.

In order to accomplish these goals, two surveys were carried among building simulation practitioners, as well as among companies that request building simulation services in their projects. The purpose of such surveys was to evaluate how the simulation tools were applied in the Brazilian market, especially what kind of analysis was employed, to what end, and in what design phase according to the ASHRAE's 209 guidelines. Furthermore, a survey's results evaluation was carried out to serve as a roadmap to the draft of a national standard for building simulation application like the ASHRAE 209.

SCIENTIFIC INNOVATION AND RELEVANCE

The use of building simulation has increased in the last 10 years in Brazil, mainly because of building certification, standards and labeling requirements. Also, several pieces of research show that the use of the simulation in the early stages of the design process can provide insight for the design team to improve the building performance in a more integrated approach.

National standards can be one of the main drivers in improving industry performance, however, there is a need to first comprehend its current state. In that sense, many countries have surveyed building simulation market landscape before drafting its industry standards. In this context, ASHARE 209 can provide minimum requirements for the building simulation process and guidelines of the role of the building simulation in each stage of the design process.

PRELIMINARY RESULTS AND CONCLUSIONS

The results show that building simulation is mainly used for thermal and energy evaluation in the schematic design of construction documentation. It could be argued that certification, labeling, and normative processes have driven this trend, as they are indicated to be the company's main concern. Additionally, there are signs of implementation of building performance simulation in earlier phases, however, they seem to be more focused in the evaluation of active systems and envelope materials. These aspects highlight the importance of a common guideline to standardize the practice and foster best practices to all phases, earlier and after executive stages, in the same way as the ASHRAE 209. The dissemination of this analysis and the proposed standard may also contribute to make companies aware of the advantages of using computer simulation during the design process and throughout the building's life cycle. Furthermore, the emphasis in thermal and energy analysis draws attention to the need for a more comprehensive approach that can include other sorts of analysis, such as daylight, air quality, microclimate, acoustics, passive bioclimatic solutions and others in the same framework and workflow.

MAIN REFERENCES

ANSI/ASHRAE. Standard 209 - Energy Simulation Aided Design for Buildings Except Low-Rise Residential Buildings. 2018

ATTIA, S, HENSEN, JLM., BELTRÁN, L., DE HERDE, A. Selection Criteria for Building Performance Simulation Tools: Contrasting Architects and Engineers Needs. Journal of Building Simulation Performance. V. 5 Issue 3. 2012

ØSTERGÅRD, T., JENSEN, RL., MAAGAARD, SE. Building simulations supporting decision making in early design – A review. Renewable and Sustainable Energy Reviews. V. 61, 187-201. 2016

SAMUELSON, HW., LANTZ, A., REINHART, CF. Non-technical barriers to energy model sharing and reuse. Building and Environment. V. 54, 71-76. 2012

SAWAF, S., BRANNON, B.; HASKELL, J. Energy Modelling Around the World: From a Practitioners View. Conference: Conference: BS2015: 15th Conference of International Building Performance Simulation Association, Session Title: Best Practices 04: Views from Around the World. San Francisco, United States. 2017

THE AMERICAN INSTITUTE OF ARCHITECTS. An Architect's Guide to Integrating Energy Modeling in the Design Process. USA, 2012. Disponível em: http://aiad8.prod.acquia-sites.com/sites/default/files/2016-04/Energy-Modeling-Design-Process-Guide.pdf

Keywords: Standard application, certification process, professional qualification

Paper ID 30661: "Methodology to Evaluate the Room ADPI through CFD Simulation"

Chirag Shah, Praveen Kumar*, Rohit Chavan, Sandip Jadhav

Center for Computational Technologies Pvt. Ltd. (CCTech), India

AIM AND APPROACH

Computational fluid dynamics (CFD) is a powerful tool to predict the thermally comfortable environment for occupants living in indoor spaces. There are several parameters like a predicted mean vote (PMV), percentage of people dissatisfied (PPD), air diffusion performance index (ADPI), draft rating (DR), etc. used to predict thermal comfort. This paper focuses on the methodology for evaluating the room ADPI through CFD analysis. ADPI is a single-number rating of the room air diffusion performance of a system of diffusers for a specified supply airflow rate and thermal load. As per ASHRAE standard, ADPI is based only on air speed and effective draft temperature (EDT) at the measuring locations. The experimental procedure for evaluating ADPI is limited to a few measuring locations in the room that may be misleading to evaluate the overall thermal comfort level in the room. Whereas from the CFD simulation one can get the data at every location in the room, so ADPI can be evaluated based on every location in the room.

SCIENTIFIC INNOVATION AND RELEVANCE

In the current study, two methodologies are compared to evaluate ADPI of the room. In the first approach, the measuring locations are at every cell point in the occupied zone of the room and in the second approach, the measuring locations are at selected points as per the ASHRAE standard. The methodology results are validated against other thermal comfort criteria such as PMV and DR.

PRELIMINARY RESULTS AND CONCLUSIONS

Both methodologies are evaluated in the case of a typical meeting room in the office with occupants seated around the table. The preliminary results show that evaluating the ADPI by taking all measuring locations in the room gives a better prediction than the selected few locations as per ASHRAE standard. After performing a detailed comparison study, the limitations and advantages of both methodologies are presented in the paper.

MAIN REFERENCES

1) ANSI/ASHRAE 55-2013, Thermal Environmental Conditions for Human Occupancy.

2) ANSI/ASHRAE 113-2013, Method of Testing for Room Air Diffusion.

3) AbouDief, T. M., and E. E. Khalil, "Effects of the Geometry of Air Outlets on Predicted Human Comfort inside Rooms: CFD vs. ADPI", Proceeding Zagazig University Journal OF Science &Technology, August 2014, vol. 23, issue 1, pp. 1-18, 2014.

Keywords: HVAC, Thermal Comfort, ADPI, CFD

Paper ID 30873: A sensitivity analysis to investigate UHI impact on neighborhood building energy consumption

Sedigheh Ghiasi (1), Ulrike Passe* (2), Yuyu Zhou (3), Janette Thompson (4)

1: Iowa state university, United States of America; 2: Iowa state university, United States of America; 3: Iowa state university, United States of America; 4: Iowa state university, United States of America

AIM AND APPROACH

Urban heat island (UHI) analysis is a complex problem from an energy consumption standpoint. The primary purpose of this study is to investigate UHI impact on building energy consumption and energy bills in a residential neighborhood in the Midwest of the U.S. The data outputs have been generated for two significant scenarios. Considering the UHI effect (scenario 1) and considering actual weather data in an airport measured near to the neighborhood (scenario 2). The sensitivity analysis includes parameters related to the building energy model such as their construction materials and landscape characteristic. The simulation is based on Ashrae 90.1 standard occupancy schedules.

The urban building energy model used has been developed in a platform that enables the evaluation of the environmental performance of neighborhoods with respect to operational and embodied energy use, walkability, and daylighting potential called Urban Modeling Interface (UMI). UMI is a design environment for Rhinoceros 3d. The neighborhood buildings were modeled based on GIS shapefile for Capitol East and Sherman Hill and Oakridge neighborhoods in Des Moines Iowa, which included building footprints and their height. Polk Country assessor data provided the second type of information such as number of building stories, construction materials, and type of occupancy.

SCIENTIFIC INNOVATION AND RELEVANCE

Many studies focus on the impact of UHI on building energy consumption. These studies vary in their simulation methods and the method of gathering weather data for UHI. The innovation of this study is presenting a sensitivity analysis instead of traditional simulations.

The sensitivity analysis provides an opportunity to include post-design characteristics, such as occupant habits, operational energy-use, and outdoor conditions such as the existence of trees. Another benefit of our method is considering the neighborhood instead of focusing on individual buildings.

We consider various building construction materials, occupancy schedules, and the existence of outdoor vegetation as our parameters. Our sensitivity analysis provides more accurate results while taking into account the effect of buildings on each other in a neighborhood.

Another challenge for investigating UHI is preparing weather data to measure the UHI impact. We developed a weather data file based on remote sensing observations and geospatial modeling

PRELIMINARY RESULTS AND CONCLUSIONS

We ran all the simulations based on two main weather data files. The first weather data file was prepared through a satellite-based method to consider urban heat island, and the second one was based on actual measured weather data on a station near the neighborhood in a non-urbanized location at the airport. Comparing two sets of simulations confirms an increase in the cooling load and a reduction in heating load. The increasing cooling load can be considered as a burden for the families and the government. Considering the residential comfort zone, our results will discuss how much the cooling load would be increased, and the heating load would be decreased in the mentioned neighborhood. Accordingly, we will discuss whether this effect is desirable for the residents who are mostly low-income families or not?

Aside from increasing expenditures, cooling loads can increase carbon dioxide emissions and create unhealthy indoor conditions if no air conditioning is available.

The residential of the neighborhood are low-income families. The result of this study can help policymakers seeking to reduce UHI impact in the USA, especially for the low-income part of society.

MAIN REFERENCES

•Koupaei, D. Malekpour, et al. "Development Of A Modeling Framework For Refined Residential Occupancy Schedules In An Urban Energy Model." Building Simulation 2019: 16th Conference of IBPSA. 2019.

•Li, X., Y. Zhou, G. R. Asrar and Z. Zhu (2018). Developing a 1 km resolution daily air temperature dataset for urban and surrounding areas in the conterminous United States. Remote Sensing of Environment 215: 74-84.

•Li, X., Y. Zhou, G. R. Asrar and Z. Zhu (2018). Creating a seamless 1 km resolution daily land surface temperature dataset for urban and surrounding areas in the conterminous United States. Remote Sensing of Environment 206: 84-97.

•Li, Xiaoma, et al. "Urban heat island impacts on building energy consumption: A review of approaches and findings." Energy 174 (2019): 407-419.

•Malekpour Koupaei, Diba, Manon Geraudin, and Ulrike Passe. "Lifetime Energy Performance of Residential Buildings: A Sensitivity Analysis of Energy Modeling Parameters." (2020): 585.

•Roxon, J, F-J. Ulm, and RJ-M. Pellenq. "Urban heat island impact on state residential energy cost and CO2 emissions in the United States." Urban Climate 31 (2020): 100546.

•Reinhart, Christoph, et al. "Umi-an urban simulation environment for building energy use, daylighting and walkability." 13th Conference of International Building Performance Simulation Association, France (2013)

Keywords: Urban heat island, UMI, building energy consumption, simulation, sensitivity analysis

Paper ID 30877: Evaluation of smart community energy performance using bottom-up energy simulation model

Ayumu Sawamura*, Monica I.C. Gondokusuma, Yuki Kitagawa, Yoshiyuki Shimoda

Osaka University, Japan

AIM AND APPROACH

In Japan, the recent development of "smart communities" which is a low-carbon community equipped with renewable energy sources and high energy efficient devices, are currently being developed by the private sector and local governments. However, the specific impact of these appliances and equipment on the community's energy performance has not be quantified sufficiently. Therefore, it is necessary to evaluate the smart community energy performance comprehensively by analyzing each characteristic of the smart community. The authors have developed a model that considered the characteristics of smart communities and evaluated their energy performance1). However, many of the actual smart community developed in Japan are equipped with a Three-Cell-System: solar cell, fuel cells, and battery, which system are not fully examined previously. Also, when evaluating a community's energy performance. In addition, it is important to evaluate the energy consumption for mobility in the community, especially dissemination of electric vehicle (EV) because EV consumes a large amount of electricity when charged at home. This paper aims to evaluate the community's overall energy performance by developing a Three-Cell-System model and an EV model in addition to the current model that considers the community characteristics.

SCIENTIFIC INNOVATION AND RELEVANCE

To make a highly accurate evaluation of the energy performance of a smart community consisting of hundreds of houses, it is necessary to consider the diversity of occupant's behavior, electricity consumption, and electricity generation in each household. The simulation model developed by the authors probabilistically determines each occupant's behavior, generates a behavior schedule for 5 minutes interval, then calculates the appliance energy consumption based on the behavior schedule in each household. Also, since this model is a bottom-up model, many estimation conditions such as family composition, house equipment, and weather condition can be considered.

Furthermore, this model can calculate both PV, fuel cell, battery for storage, and EV. Therefore, the effect and influence of PV, fuel cell, battery are accurately evaluated. Furthermore, using this simulation model, it is possible to evaluate various evaluation indexes such as life cycle performance (LCP) evaluation, ZEC (Zero Energy Community) achievement rate, and CO2 emissions.

PRELIMINARY RESULTS AND CONCLUSIONS

The simulation was conducted in the 125 detached houses with Three-Cell-System and energy-saving devices with 4 family composition and 12 types of houses based on real smart community developed in Japan.

From this study, without the SOFC fuel cell, about 33% of the electricity consumption of the house self-produced, and the remaining 67% was purchased from the grid. However, with the SOFC fuel cell, the self-sufficient rate of power consumption increased to 99%. Furthermore, when the surplus power generated can be sold to the grid, the SOFC fuel cell will be operated for 24 hours a day and the community becomes a positive energy community that creates surplus energy of 416GJ. Therefore, the environmental performance of the community with a SOFC fuel cell is relatively good, but a large amount of power is sold to the grid and the system load is large. Thus, the effective use of surplus power is needed.

MAIN REFERENCES

Evaluation of Energy Performance of Smart Community Considering Occupant's Behavior, Yuki Kitagawa, Gondokusuma Monica Irisa Clara, Yoshiyuki Shimoda, Proceedings of the 16th IBPSA Conference, 3555-3562, 2019.09, International Conference(Proceedings)

Keywords: smart community, energy performance, fuel cells, electric vehicle

Paper ID 30969: CFD modelling of contaminant control in designing ventilation system for an airborne infection isolation room (AIIR)

Aaditya Ruiker, Shadab Mohammed, Rohit Chavan, Praveen Kumar*, Sandip Jadhav

Center for Computational Technologies Pvt. Ltd. (CCTech)

AIM AND APPROACH

Current epidemic situation created a challenge for the healthcare industry in designing quick-to-build isolation rooms to prevent the spread of infection. Designing proper ventilation for the isolation room plays a crucial role in controlling contaminant behavior inside the room and protecting healthcare workers from the cross-infection. The diffusion of contaminant and airflow behavior mainly depends on supply air and exhaust air locations. The aim of this study is to examine airflow behavior and the distribution of contaminated air for three different ventilation strategies in an isolation room using computational fluid dynamics (CFD). The methodology uses the open-source CFD software package OpenFOAM with cloud-based technology to obtain results as quickly. In which, Navier Stokes and energy equations have been solved by the control volume method with transport equations to obtain the contamination diffusion field. The simulation uses a steady-state, k-ε turbulence model, and compressible fluid flow by considering the effect temperature.

SCIENTIFIC INNOVATION AND RELEVANCE

Various standards have been defined by ASHRAE in designing isolation rooms for airborne diseases for maintaining pressure differential with adjacent spaces and minimum air-exchange rates. But there is still uncertainty whether the room is refreshed uniformly or there are no stagnant, under-ventilated areas in the room where infectious contaminants might be concentrated. Our innovation resides in the development of CFD methodology in designing an effective ventilation system for the modular isolation rooms using an open-source software solution OpenFOAM. Using open-source CFD technology along with cloud-based computational resources can lead to reducing the cost and time required for the simulations and thus in analyzing the performance of ventilation design.

PRELIMINARY RESULTS AND CONCLUSIONS

The paper includes the case study of three different ventilation strategies for the modular isolation room using computational fluid dynamics (CFD). Strategy-I is a parallel airflow pattern induced using opposite directional wall-mounted supply and exhaust. Strategy-II will be ceiling mounted supply and low-level exhaust ventilation. Strategy -III will retain supply diffuser as in Strategy-II but exhaust will be relocated near the patient bed. The qualitative analysis of air-flow behavior and contamination distribution are presented using contour plots, flow lines from the simulation results. The quantitative results of average contaminant residue time and maximum concentration of the contaminant will be presented for simulated ventilation strategies. This paper tries to demonstrate the CFD methodology of analyzing the effectiveness of the ventilation system for contaminant control inside an isolation room.

MAIN REFERENCES

[1] Cho, Jinkyun. "Investigation on the contaminant distribution with improved ventilation system in hospital isolation rooms: Effect of supply and exhaust air diffuser configurations." Applied Thermal Engineering 148 (2019): 208-218.

[2] Kao, P. H., and Ruey-Jen Yang. "Virus diffusion in isolation rooms." Journal of Hospital infection 62.3 (2006): 338-345.

[3] Li, Xiangdong, et al. "Numerical investigation of particle transport and inhalation using standing thermal manikins." Building and Environment 60 (2013): 116-125.

[4] ASHRAE, HVAC Applications, American Society of Heating Refrigerating and AirConditioning Engineers (ASHRAE Handbook), Atlanta, GA, 2015.

Keywords: airborne infection, CFD simulation, HVAC, airborne infection isolation room

Paper ID 31097: A framework for integrating high-resolution trees in urban energy use models

Diba Malekpour Koupaei, Ulrike Passe*, Janette Thompson

Iowa State University, United States of America

AIM AND APPROACH

Urban vegetation has been widely accepted to be effective in mitigating Urban Heat Island (UHI) effect and reducing building energy use, specifically cooling loads. However, in order to clarify the urban trees cooling effect, the influence of tree characteristics on the cooling effectiveness and correspondingly building energy use in urban environments need to be quantitatively assessed and reflected in energy simulation efforts. The major impact factors of trees affecting the cooling demands of buildings have been identified to be tree height, position of trees relative to the built infrastructure, tree canopy shape and density, tree species, and foliation periods. In this study, a grasshopper plugin is developed to facilitate the integration of a higher-resolution geometry for trees in urban energy simulation models, which accounts for all the impact factors listed previously. The TREE plugin uses data from previous studies, on tree canopy shape categories, foliation periods, and canopy density values, and gives the user the ability to build the geometry of any given tree at any point in time during the year. The input information required in this plugin include tree species, foliation periods, trunk diameter, tree height, canopy shape/height, canopy width in two dimensions, and latitude/longitude coordinates.

SCIENTIFIC INNOVATION AND RELEVANCE

As mentioned in the previous section, many factors impact the shading properties of trees and thus, their effectiveness as UHI mitigation measures. However, it is often the case that simplified geometry of trees are used in building energy simulation models, which often tend to ignore the impact of foliation periods and tree shape or species on the shading properties of the given trees all together. The use of simplified tree geometries as an input to building energy simulation models is a source of uncertainty that results in undesired gaps between the predicted and actual energy use of buildings. This is specifically problematic when it comes to predicting cooling loads and the expected savings in terms of reduced need for cooling in adjacent buildings. Our plugin enables the user to easily build high-resolution geometric models of trees and assess their effectiveness in terms of reducing the cooling loads more realistically. Therefore, the TREE plugin is a step towards narrowing the gap between the predicted and actual cooling loads of buildings.

PRELIMINARY RESULTS AND CONCLUSIONS

A sample case study of our plugin's effectiveness is conducted by developing a 3-D urban energy model, consisting of 1142 trees and 340 buildings. In essence, 12 variations of this model are created in the Urban Modeling Interface (umi) environment to represent 12 consecutive months of a year and account for the impacts of tree shedding seasons on the modelled trees' shading geometries. The results of these set of simulations are compared against a baseline scenario that uses simplified geometries for trees and show a 2-7% change in monthly cooling loads, depending on the intensity of changes in canopy density values over that specific month. This suggests that any model using simplified geometry for trees is expected to predict results that are not necessarily accurately representative of the actual consumption rates. Therefore, our plugin has been successful in increasing the accuracy of energy simulations' predictions and will be a valuable tool for all users interested in studying the impacts of trees on energy consumption in buildings.

MAIN REFERENCES

• Chokhachian, A., & Hiller, M. PANDO: Parametric Tool for Simulating Soil-Plant-Atmosphere of Tree Canopies in Grasshopper.

• Hashemi, F., Marmur, B. L., Passe, U., & Thompson, J. R. (2018). Developing a workflow to integrate tree inventory data into urban energy models.

• Hsieh, C. M., Li, J. J., Zhang, L., & Schwegler, B. (2018). Effects of tree shading and transpiration on building cooling energy use. Energy and Buildings, 159, 382-397.

• McPherson, E. G. (1984). Planting design for solar control. Energy-conserving site design. Washington, DC: American Society of Landscape Architects, 141-164.

• McPherson, E. G. (1999). Carbon dioxide reduction through urban forestry: guidelines for professional and volunteer tree planters (Vol. 171). US Department of Agriculture, Forest Service, Pacific Southwest Research Station.

• McPherson, E. G., Xiao, Q., van Doorn, N. S., Johnson, N., Albers, S., & Peper, P. J. (2018). Shade factors for 149 taxa of in-leaf urban trees in the USA. Urban Forestry & Urban Greening, 31, 204-211.

• Simpson, J. R., & McPherson, E. G. (1998). Simulation of tree shade impacts on residential energy use for space conditioning in Sacramento. Atmospheric Environment, 32(1), 69-74.

Keywords: Urban building energy simulation, Tree shade, Urban forest

Paper ID 30298: Optimal control of TABS using internal heat load prediction

Fumiaki Deguchi*, Yasuyuki Shiraishi

The University of Kitakyushu, Japan

AIM AND APPROACH

In recent years, Thermally Activated Building System (TABS) has attracted attention, as a means of creating an indoor thermal environment that satisfy both energy-saving and thermal comfort in the office, and there have been many studies on TABS. In the previous study [1], we proposed optimal control method using model predictive control (MPC), which takes into consideration thermal response of TABS, in order to control the indoor thermal environment provided by TABS in an office. So far, we verified the effectiveness of this control method by performing a coupled analysis using MATLAB/Simulink and CFD. In actual operation, although it is effective to control TABS based on prediction and measurement of heat load generated indoors, we have not conducted a detailed study on load prediction. Therefore, in the present study, we propose an optimal control method that incorporates internal heat (office automation equipment, lighting equipment) load prediction into MPC. First, we created the prediction model of internal heat load based on Neural Network (NN) trained on measurement data of an actual office. Then, using the obtained predicted data, we verified the effectiveness of the proposed method by numerical simulation in a general office space where TABS is introduced.

SCIENTIFIC INNOVATION AND RELEVANCE

In general office buildings, air-conditioning accounts for a large proportion of the total energy consumption, and new technologies that contribute to energy saving are expected in recent years. Furthermore, it is also required to consider the intellectual productivity of workers in these buildings. Therefore, radiant heating and cooling has attracted attention, and TABS has been introduced in Japan in recent years. TABS utilizes the building frame to store and discharge heat, so TABS offers higher energy efficiency, a more comfortable environment for workers, and cost advantages compared with a conventional radiant system. However, since the thermal response of the ceiling surface temperature is slow due to the ceiling's large thermal mass in TABS, optimal control based on MPC that takes into account the load prediction results is desirable. On the other hand, there are many studies on load prediction, and Nigitz et al. [2] performed load prediction by regression analysis and stated that the developed method could be generally applied because the prediction accuracy was sufficient. Furthermore, in recent years, many studies on artificial intelligence-based energy prediction have been reported [3], and among them, Wang et al. [4] have performed heat load prediction by shallow machine learning and deep learning.

PRELIMINARY RESULTS AND CONCLUSIONS

As the result, this study suggests that it is possible to reduce the water flow rate of TABS while maintaining comfort by performing optimum control with load prediction incorporated in MPC, compared with MPC without load prediction. It was confirmed that the load prediction by NN has a high prediction accuracy in the prediction for one week. In addition, regarding indoor thermal environment, Predicted Mean Vote (PMV) is almost uniform in the room, and the vertical temperature difference in the work area is less than 3 degrees. Furthermore, it was suggested that the water flow rate of TABS on weekdays was greatly reduced, and it was possible to save about 52% in the cumulative water flow rate per week compared with the method without load prediction. It was also confirmed that energy saving can be realized by using this proposed method [5].

MAIN REFERENCES

[1] Yohei Ogawa and Yasuyuki Shiraishi, Study on optimal control of Thermo-Active Building System based on Model Predictive Control, Journal of Environmental Engineering (Transaction of AIJ), Vol.85, No.771, pp.379-387, 2020. 5 (in Japanese)

[2] Thomas Nigitz, Markus Gölles, A generally applicable, simple and adaptive forecasting method for the short-term heat load of consumers, Applied Energy, Vol. 241, pp.73–81

[3] Zeyu Wang, Ravi S.Srinivasan, A review of artificial intelligence based building energy use prediction: Contrasting the capabilities of single and ensemble prediction models, Renewable and Sustainable Energy Reviews, Vol. 75, pp.769-808, 2017

[4] Zhe Wang Tianzhen Hong, Mary Ann Piette, Building thermal load prediction through shallow machine learning and deep learning, Applied Energy, Vol. 263, 114683, 2020

[5] Fumiaki Deguchi and Yasuyuki Shiraishi (in press), Study on Optimal Control Method of TABS in the Office (Part1) Introduction of Load Prediction by Neural Network, SHASE (The Society of Heating, Air-Conditioning and Sanitary Engineers) of Japan (in Japanese)

Keywords: TABS, MPC, CFD, Load prediction, Neural Network

Paper ID 30628: Solar Load modelling in OpenFOAM for indoor HVAC applications

Shadab Mohammed, Sadashiv Khadilkar, Unnati Jadhav, Sandip Jadhav*

Center for Computational Technologies Pvt. Ltd., India

AIM AND APPROACH

The designing of thermal comfort inside buildings with large glass facades is a challenging task especially, concerning the total energy consumption required for maintaining the desired cooling. The solar load constitutes a major contributing factor in the total load of the system. The semi-transparent glazing material or coating is applied on glass walls to reduce energy consumption. CFD simulations come handy in evaluating the thermal comfort and best HVAC design for indoor spaces with windows or glass facades. In CFD, the solar load is modelled by combining the solar calculator with the ray-tracing algorithm and applying the load on the surfaces that directly exposed to the sun. The inbuilt solar load model that comes with OpenFOAM is not suitable for indoor HVAC simulations. Instead, it is designed for external flow applications to evaluate the Urban Heat Index (UHI) and outdoor thermal comfort parameters. Additionally, it has no support for semi-transparent glass walls, and it is also not compatible with convective heat flux boundary conditions which deviates a lot from reality and physics. Therefore, this paper focuses on overcoming the shortcoming of inbuilt solar load model by developing our in-house codes and methodology to model the solar load accurately for indoor HVAC applications.

SCIENTIFIC INNOVATION AND RELEVANCE

An in-house solar load model with the support for semi-transparent glass walls and compatible with convective heat flux boundary condition is developed based on OpenFOAM for the "Autonomous HVAC CFD" application. It is a web-based application, developed by simulationHub, specially designed for architects, HVAC engineers, and designers to evaluate their heating and cooling system design and perform the design iterations quickly.

PRELIMINARY RESULTS AND CONCLUSIONS

The code implementation and its results are validated with ASHRAE handbook and Solpos NREL code. To further evaluate our model qualitatively and quantitatively, a software comparison case studies are performed with ANSYS FLUENT software. The case studies involve thermal manikins and a semi-transparent glass wall facing the west direction. From the results, it is concluded that our implementation is correct and is in good agreement with the results produced with the commercial software package. In addition to this, CFD best practises guidelines are also proposed in the paper to achieve fast and accurate simulation results.

MAIN REFERENCES

1) ASHRAE Handbook: Fundamentals. Atlanta, GA: American Society of Heating, Refrigeration and Air-Conditioning Engineers, 2001.

2) Reda, Ibrahim & Andreas, Afshin. (2004). Solar position algorithm for solar radiation application. Solar Energy. 76. 577-589.

3) E. U. Finlayson, D. K. Arasteh, C. Huizenga, M. D. Rubin, and M. S. Reilly. WINDOW 4.0: Documentation of Calculation Procedures. Technical Report Publication LBL-33943/ TA-309, Lawrence Berkeley Laboratory, Energy and Environmental Division, Berkeley, CA 94720, 1993.

4) Nielsen, Peter V., S. Murakami, Shinsuke Kato, Claus Topp and Jeong Hyu Yang. Benchmark Test for a Computer Simulated Person. (2003).

5) Martinho, N. Gameiro Lopes, Antonio Gameiro da Silva Manuel. CFD Modeling of Benchmark Test for Flow Around a Detailed Computer Simulated Person. (2008).

Keywords: CFD, Solar Load, Indoor HVAC Simulations, Semi-transparent Glass Walls, OpenFOAM Development

Paper ID 30685: Comparative thermal performance analysis of a shaded and non-shaded Double Skin Façade in hot arid areas.

Yomna Saad Elghazi* (1,2), Neveen Hamza (1), Martyn Dade-Robertson (1)

1: Newcastle University, United Kingdom; 2: Arab Academy for Science, Technology & Maritime Transport, Cairo, Egypt.

AIM AND APPROACH

The demand for more energy-efficient and sustainable architecture is leading to a growing interest in Double-skin facades (DSFs). A reduction between 12-30% of indoor cooling loads can be achieved by applying DSFs. This paper studies the impact of DSF geometrical variables including cavity width, façade layering and domain size on the internal glazing surface temperature. The comparative study looks at:

• Four different cavity widths of 0.4, 0.6, 0.8, and 1m

• Two inner skin configurations; single and double inner glazing.

• A 3D vertical segment of 0.4m deep and full cavity of 3.2m.

The optimized DSF without shading is compared with an integrated solar-shading screen with 3 opening percentages (30%, 50% & 70%). All configurations are tested as Point-in-time simulations on a peak summer and winter day, at 9:00am, 12:00pm and 15:00pm, for a single-storey south-orientated DSF in a hot arid climate. The paper assesses DSF thermal performance using Computational Fluid Dynamics (CFD) to minimize internal heat gains. CFD allows for detailed numerical modelling of interactive physical phenomena inside DSF cavities by coupling RNG k-epsilon turbulence model with DO radiation method and solar ray-tracing model. This method is validated by the replication of Mei et al., (2007) experimental case.

SCIENTIFIC INNOVATION AND RELEVANCE

The literature review highlights the numerical simulation setup of DSFs has many challenges associated with turbulence modelling and conjugate heat transfer that include solar and radiation fields (Baldinelli, 2009; Coussirat et al., 2008; Pasut & De Carli, 2012). The paper compares laminar and turbulent air flow taking place in non-shaded and shaded DSF cavities. The paper discusses the validation process using coupled radiation and turbulence sub-models

This study tests a simplified CFD modelling method using a 3D vertical section to reduce computational cost against a full cavity model. The 3D CFD approach combines the solar load model and the turbulent mixing in the cavity and assesses complex forms.

Finally, the study focuses on the impact of geometrical parameters of non-shaded cavity in hot countries around peak solar radiation is expected. It helps in understanding how the non-shaded and shaded cavity perform in peak summer and winter.

PRELIMINARY RESULTS AND CONCLUSIONS

The results show that the 1m cavity resulted in a drop of 17.5°C between external single glazing temperature (47.8°C) and internal double glazing temperature (29.5°C). All the smaller cavities recorded higher internal glazing temperatures. The 0.8, 0.6 and 0.4m cavities resulted in 30°C, 30.3°C and 30.5°C internal double glazing surface temperature. By monitoring total heat flux of internal surface, the heat gains are reduced from 74.36 to 63.54 w/m2 by increasing the cavity width from 0.4m to 1m. The results indicate that increasing the cavity width can improve the thermal performance of the DSF.

The internal double glazing lowered the internal surface temperature by 24% from 39.4°C recorded by single glazing to 30.08°C recorded by the double glazing.

The 0.4m 3D vertical section allows for complex form assessment with reduced computational cost, but it has some limitations. The temperature profile at mid cavity of both methods are similar with less than 5% variation. Both methods are valid for summer middays but full cavity model is used in case of side components of sun angles.

The results of comparative analysis of non-shaded and shaded cavities could be encouraging for architects to be applied in hot countries to improve energy savings.

MAIN REFERENCES

- Baldinelli, G. (2009) 'Double skin façades for warm climate regions: Analysis of a solution with integrated movable shading system', Building and Environment, 44(6), pp.1107-1118.

- Hamza, N (2008) Double versus single skin facades in hot arid areas, Energy and Buildings, Vol(40) issue3, pp240-248

- Parra, J., Guardo, A., Egusquiza, E. and Alavedra, P. (2015) 'Thermal Performance of Ventilated Double Skin Façades with Venetian Blinds', Energies, 8(6), pp.4882-4898.

- Pasut, W. and De Carli, M. (2012) 'Evaluation of various CFD modelling strategies in predicting airflow and temperature in a naturally ventilated double skin façade', Applied Thermal Engineering, 37, pp. 267-274.

- Pourshab, N., Tehrani, M. D., Toghraie, D., & Rostami, S. (2020). Application of double glazed façades with horizontal and vertical louvers to increase natural air flow in office buildings. Energy, 200. doi:10.1016/j.energy.2020.117486

- Mei, L., Loveday, D., Infield, D., Hanby, V., Cook, M., Ji, Y., Bates, J. (2007). The influence of blinds on temperatures and airflows within ventilated double-skin façades. Paper presented at the Proceedings of Clima 2007 WellBeing Indoors.

- Velasco, A., Jiménez García, S., Guardo, A., Fontanals, A. and Egusquiza, M. (2017) 'Assessment of the Use of Venetian Blinds as Solar Thermal Collectors in Double Skin Facades in Mediterranean Climates', Energies, 10(11).

Keywords: CFD, DSF, Simulation, Cavity-integrated solar-shading

Session F3.7 (Online Track): The role of occupants

Paper ID 30187: The impact of using actual occupancy data on building performance simulation results for a multi-purpose university building

Mahsa Sayfikar*, David P. Jenkins

Heriot Watt University, United Kingdom

AIM AND APPROACH

Building performance simulation (BPS) software are powerful tools for studying energy use in buildings. BPS programs can be used in the design phase of the buildings and in refurbishments to predict energy demand, perform code compliance calculations, and for analyzing the energy performance of a building in operation. However, considerable gaps have been reported between simulation results and actual energy consumption levels. The model of occupants' presence used in building simulation is one of the contributing factors to the "performance gap". Typically, due to empirical occupancy data not being readily available, estimated occupancy schedules provided by building codes such as UK's National Calculation Method (NCM) are used as input in building simulations. However, it has been argued that such schedules are overly simplified and don't reflect real occupancy patterns which leads to significant errors in calculation results. This study investigates the implications of using NCM occupancy profiles compared to using monitored occupancy data in performance simulation of a multi-purpose university building. A framework is presented for using building-level occupancy data to produce zone-level occupancy profiles suitable for use in BPS tools. The results are compared to simulation results using NCM profiles as well as to actual energy consumption data.

SCIENTIFIC INNOVATION AND RELEVANCE

The impact of occupancy profiles on energy simulation results has been the focus of many studies during the past few years. Most of these studies are focused on residential and office buildings and are usually limited to smaller spaces such as single or multiple occupancy offices. These studies use motion sensors to monitor occupancy which only record the occupancy state of a space (occupied/unoccupied) and don't record the number of people in that space. Compared to such buildings, it is more difficult to collect long-term occupancy data in university buildings since the number of people inside the building has to be monitored as well as the occupancy state. High numbers of people, as well as multiple spaces to be monitored, are among the challenges that have prevented more studies on such buildings. This paper uses the occupancy data in a multi-purpose university building over an entire year to show the impacts on the energy simulation results.

PRELIMINARY RESULTS AND CONCLUSIONS

In this study, NCM profiles for equipment and lighting usage have been used in order to focus on the impact on only occupancy profiles. The studied building is not equipped with occupant centred controls, however, the results confirm that occupancy has a direct impact on the building's heating, cooling, ventilation and hot water demands.

Simulation results show when NCM occupancy profiles were used, annual heating demand was 31% lower than the results with measured occupancy data. Annual cooling demand and domestic hot water energy consumption were respectively 16.3% and 59% higher when using NCM profiles compared to measured occupancy data. Overall, simulated annual natural gas consumption was 17% lower with NCM occupancy profiles while electricity consumption was 0.6% higher compared to measured occupancy data. If actual equipment and lighting profiles are also incorporated in the study, the electricity results would be impacted by the occupancy as well. These results confirm the extent of the impact of occupancy inputs on building simulation results for a building with high numbers of occupants. Therefore, creating better representations of occupancy for such buildings is crucial for more accurate energy consumption predictions.

MAIN REFERENCES

D'Oca, S. & Hong, T. 2015. Occupancy schedules learning process through a data mining framework. Energy and Buildings, 88, 395-408.

Davis, J. A. & Nutter, D. W. 2010. Occupancy diversity factors for common university building types. Energy and Buildings, 42, 1543-1551.

Dong, B., Yan, D., Li, Z., Jin, Y., Feng, X. & Fontenot, H. 2018. Modeling occupancy and behavior for better building design and operation—A critical review. Building Simulation, 11, 899-921.

Duarte, C., Van Den Wymelenberg, K. & Rieger, C. 2013. Revealing occupancy patterns in an office building through the use of occupancy sensor data. Energy and Buildings, 67, 587-595.

Hong, T., Taylor-Lange, S. C., D'oca, S., Yan, D. & Corgnati, S. P. 2016. Advances in research and applications of energy-related occupant behavior in buildings. Energy and Buildings, 116, 694-702.

Parker, J., Hardy, A., Glew, D. & Gorse, C. 2017. A methodology for creating building energy model occupancy schedules using personal location metadata. Energy and Buildings, 150, 211-223.

Tahmasebi, F. & Mahdavi, A. 2017. The sensitivity of building performance simulation results to the choice of occupants' presence models: a case study. Journal of Building Performance Simulation, 10, 625-635.

Keywords: building energy, university buildings, occupancy, building performance simulation, data collection

Paper ID 30267: Impact of the UK's COVID-19 lockdown on the residential electricity consumption

Kumar Biswajit Debnath*, David P Jenkins, Sandhya Patidar, Andrew D Peacock, Peter McCallum

Heriot-Watt University, United Kingdom

AIM AND APPROACH

Extreme disruptive scenarios such as pandemic lockdown force people to alter regular daily routines, impacting their energy consumption pattern. The implication of such a disruptive scenario for a more extended period on energy consumption is uncertain. This study aimed to investigate the impact of COVID-19 lockdown on residential electricity consumption in 100 houses from the southwestern UK. For the study, we analysed highly granular (1-minutely) electricity consumption data for April-September 2020 compared to the same months in 2019 for the same houses. Our study showed statistically significant differences during the lockdown period (the analysed six months) in energy demand. The minutely average electricity demand was 1.4-10% lower during April-September 2020 than in 2019. Our analysis showed that not all houses had similar type of changes during the lockdown. Some houses demonstrated a 38% increase in electricity demand, whereas some houses showed a 54% reduction during the lockdown period compared to 2019. Some houses showed significantly higher electricity use during the morning and afternoon than in 2019, which might be due to working and schooling from homes during the lockdown.

SCIENTIFIC INNOVATION AND RELEVANCE

Key Innovations:

- Lockdown impacted the residential electricity demand.
- Electricity demand distribution shifted towards lower bins on weekdays and weekends.
- Mean electricity demand reduced during April-September 2020 than 2019.
- Some houses had higher electricity consumption, but some houses had lower demand than in 2019.
- Some houses had higher electricity consumption in the morning and afternoon due to working and schooling from home.

Practical Implications:

The study outcomes shed some light on the uncertainties of the long-term impact of the COVID-19 on the UK's energy demand and characteristics of electricity consumption in the residences due to working and schooling from home. Although some households encountered higher electricity demand in some parts of the day, some families did not. Also, the distribution of aggregated electricity consumption shifted towards lower amounts. Our study provided insight into the lockdown's implication on the (aggregated) electricity consumption of 100 houses and some selected houses.

PRELIMINARY RESULTS AND CONCLUSIONS

Our analysis showed a significant rise in aggregated electricity consumptions as compared to the same period of 2019. Moreover, working from home practice impacted the electricity consumption pattern even after lockdown as the practice was ongoing. Another factor of such rise during lockdown may induce from children being home (whole day) as many of the houses had children. Aggregated minutely electricity consumption per month for April-September 2020 showed a substantial shift in consumption data distribution from unimodal distribution to bimodal distribution, suggesting that two distinctive consumption patterns were happening in those 100 houses. The weekdays demonstrated a more significant shift towards lower and higher consumption bins than weekends where the distribution shift was towards lower consumptions.

Individual house electricity consumption analysis showed some houses had more significant rise during the afternoon and evening period which resembled the reported change in grid consumption by National Grid, possibly due to working from home and increasing evening family activities such as watching TV. However, some houses showed a very different shift in electricity consumption patterns such as between 12 and 4:59 am. The differences in consumption pattern and change in higher and lower consumption with bimodal distribution raised more questions requiring further investigation.

MAIN REFERENCES

CREDS. (2020, 04 29). Ending the daily work commute may not cut energy usage as much as one might hope. (Centre for Research into Energy Demand Solutions) Retrieved 07 02, 2020, from https://www.creds.ac.uk/ending-the-daily-work-commute-may-not-cut-energy-usage-as-much-as-one-might-hope/

EAC. (2018). Heatwaves: Adapting to climate change (Report of the Session No. 9). Environmental Audit Committee. London: House of Commons. Retrieved from https://publications.parliament.uk/pa/cm201719/cmselect/cmenvaud/826/826.pdf

Kovats, R., & Osborn, D. (2016). UK Climate Change Risk Assessment 2017: Evidence Report. Chapter 5: People & the built environment. Adaptation Sub-Committee of the Committee on Climate Change. Retrieved from https://www.theccc.org.uk/wp-content/uploads/2016/07/UK-CCRA-2017-Chapter-5-People-and-the-built-environment.pdf

National Grid. (2020, 07 21). 4 ways lockdown life affected UK electricity use. Retrieved from www.nationalgrid.com: https://www.nationalgrid.com/uk/stories/grid-at-work-stories/4-ways-lockdown-life-affected-uk-electricity-use

National Grid. (2020). Working through COVID-19: Challenging times bring out the best in the best. Gridline, p. 05. Retrieved from https://www.nationalgrid.com/uk/electricity-transmission/document/135051/download

Nichols, D., Kane, D., Parish, D., Cooper, D., Peacock, A., & McCallum, P. (2021). Centrica PLC: Cornwall LEM Residential Electricity Dataset with Solar Production and Battery Storage. (UK Data Service) Retrieved from https://beta.ukdataservice.ac.uk/datacatalogue/studies/study?id=854578

Keywords: Residential community, adaptation, electricity consumption, COVID-19 lockdown

Paper ID 30337: Spatiotemporal variation in occupant behavior: statistical analysis and application in American households

Yuanmeng Li*, Yohei Yamaguchi, Yoshiyuki Shimoda

Graduate school of engineering,Osaka University, Japan

AIM AND APPROACH

Occupancy at home and conducted activity have been recognized as one of the most important factors in energy demand simulation as they characterize the scale and temporal pattern of residential energy demand [1, 2]. The literature revealed that the occupancy and activity exhibit a great extent of diversity, which is called occupant diversity, depending on their demographic attributes and other factors. Clustering, regression and combined approaches have been proposed to address occupant diversity [3]. However, the spatial and temporal variation in occupant behavior have not been well understood. Ignoring these factors may involve significant error in simulation results [4]. Based on this background, we aim to investigate the spatial and temporal variation in occupant behavior (OB). For this aim, we conducted a statistical analysis on the influence of spatial and temporal factors, such as region, geo-climate zone, and season, based on the American Time Use Survey data. Then, simulation cases representing different ways considering the spatial and temporal conditions were developed to quantitatively evaluate the spatiotemporal variation due to the factors.

SCIENTIFIC INNOVATION AND RELEVANCE

Most of previous researches considered simple predictors to determine modeling parameters of OB, such as basic demographic and household related information [1, 3, 5]. The spatiotemporal variation and granularity have not been fully investigated [6]. In this study, we considered the following spatiotemporal factors as predictor:

1) Spatial factor: A) regional boundary, B) metropolitan status, and C) geo-climatic zones.

2) Temporal factor: D) time of day, E) weekdays and weekends, F) observation month and G) season.

In our model, the activity starting probability and activity duration were used as modeling parameters to generate activity sequence. Random forest which constructs a multitude of decision trees to realize the multi-classification was used to model the activity starting probability at every time interval of diary day for activities. Significant predictors were selected according to the mean accuracy and mean Gini indicators. For the activity duration, a distribution fitting method was conducted considering the relationship between the duration and start time. Based on the result, an OB model was developed and the resultant spatiotemporal variation in simulated OB was quantified.

PRELIMINARY RESULTS AND CONCLUSIONS

The result of random forest revealed that activity-related factors, i.e. previous activities and type of accompanied people, were recognized as the most important. The spatial (county, metropolitan status) and temporal (month, dairy day) factors were recognized as more influential compared to demographic factors. Moreover, some activities especially outdoor activities happened and ended later in warmer regions (West North Central, South Atlantic) compared to colder regions. As for the seasonal factor, some activities such as traveling were more likely to happen during the daytime in summer seasons (June to August) compared to the winter seasons (December to February), however activities such as leisure activities were more likely to happen at evening in the winter seasons.

Activity duration also showed spatiotemporal variation as well. The duration of some outdoor activities (e.g., work related activities, commuting and schooling) which conducted in the evening or night were shorter in the winter seasons and colder regions (New England, East North Central, West North Central and Mountain).

Through the results obtained so far, we concluded that there is a significant spatiotemporal variation in OB, which implies that the consideration of spatiotemporal factors improves the reliability of OB models.

MAIN REFERENCES

[1]Wilke, U. (2013). Probabilistic Bottom-up Modelling of Occupancy and Activities to Predict Electricity Demand in Residential Buildings. PhD Thesis, 5673, 172. https://doi.org/10.5075/epfl-thesis-5673

[2]Zhao, J., Lasternas, B., Lam, K. P., Yun, R., & Loftness, V. (2014). Occupant behavior and schedule modeling for building energy simulation through office appliance power consumption data mining. Energy and Buildings, 82, 341–355. https://doi.org/10.1016/j.enbuild.2014.07.033

[3] Wilke, U., Haldi, F., Scartezzini, J., & Robinson, D. (2013). A bottom-up stochastic model to predict building occupants ' time-dependent activities. Building and Environment, 60, 254–264. https://doi.org/10.1016/j.buildenv.2012.10.021

[4]Hong, T., Yan, D., D'Oca, S., & Chen, C. fei. (2017). Ten questions concerning occupant behavior in buildings: The big picture. Building and Environment, 114, 518–530. https://doi.org/10.1016/j.buildenv.2016.12.006

[5]Hong, T., Taylor-Lange, S. C., D'Oca, S., Yan, D., & Corgnati, S. P. (2016). Advances in research and applications of energy-related occupant behavior in buildings. Energy and Buildings, 116, 694–702. https://doi.org/10.1016/j.enbuild.2015.11.052

[6]Li, J., Yu, Z. (Jerry), Haghighat, F., & Zhang, G. (2019). Development and improvement of occupant behavior models towards realistic building performance simulation: A review. Sustainable Cities and Society, 50(February), 101685. https://doi.org/10.1016/j.scs.2019.101685

Keywords: occupant behavior, diversity, spatiotemporal variation

Paper ID 30383: Development of synthetic ageing occupant behaviour through generative adversarial network

Yuezhong Liu*

Nanyang Technological University, Singapore

AIM AND APPROACH

The energy performance of residential buildings is highly influenced by occupants' behaviour, especially for older adults. However, the collection of real occupant behaviour faces difficulties in privacy issues and various sensors with relatively high costs.

This study proposed a novel approach for generating synthetic ageing occupant behaviour of residential buildings through the Generative Adversarial Network (GAN), which is capable of predicting electricity consumption by simulation tool (EnergyPlus) during the design process. The proposed approach has three main steps: (1) normalizing the daily 24-hour occupant behaviour activities, (2) clustering the occupant behaviour activities with the k-means algorithm, and (3) using GAN to generate occupant behaviour activities for each cluster.

SCIENTIFIC INNOVATION AND RELEVANCE

Generative Adversarial Network is used to generate Synthetic Ageing Occupant Behaviour.

The generated electrical loads could capture both the general trend and stochastic dynamics.

The proposed approach was tested using real local survey data with two-generational cohorts (40–64 years and 65 years and above) .

Potential applications include energy-efficient buildings for an ageing population, anonymization.

PRELIMINARY RESULTS AND CONCLUSIONS

The approach was tested with local survey data. The initial real raw occupant behaviour data contains 600 participants with two-generational cohorts (40–64 years and 65 years and above) in Singapore. We validated the proposed method by comparing the mean, standard deviation, and distribution of key parameters of the generated electrical loads from EnergyPlus with those of the real ones. The Kullback–Leibler divergence of the generated and real ageing occupant behaviour are within 0.23 for majority of parameters and clusters. Additionally, results showed the ageing occupant behaviour generated by GAN can capture not only the general trend but also the random variations of the actual electrical loads in buildings. The proposed GAN approach can be used to generate synthetic ageing occupant behaviour, anonymize real older adults information for sharing, to support research and design of energy-efficient buildings for an ageing population.

MAIN REFERENCES

Z. Chen, C. Jiang, Building occupancy modeling using generative adversarial network, Energy Build., 174 (2018), pp. 372-379

X. Feng, D. Yan, T. Hong, Simulation of occupancy in buildings, Energy Build., 87 (2015), pp. 348-359

I. Richardson, M. Thomson, D. Infield A high-resolution domestic building occupancy model for energy demand simulations, Energy Build., 40 (8) (2008), pp. 1560-1566

C. Liao, Y. Lin, P. Barooah, Agent-based and graphical modelling of building occupancy, J. Build. Perform. Simul., 5 (1) (2012), pp. 5-25

I. Goodfellow, J. Pouget-Abadie, M. Mirza, B. Xu, D. Warde-Farley, S. Ozair, A. Courville, Y. Bengio, Generative adversarial nets, Adv. Neural Inf. Process. Syst. (2014), pp. 2672-2680

Keywords: Synthetic Ageing Occupant Behaviour, Generative Adversarial Network, Machine Learning, Energy Simulation

Paper ID 30508: The impact of occupancy patterns on artificial light energy demand - simulation and post-occupancy-evaluation

Sascha Hammes* (1), Martin Hauer (1), David Geisler-Moroder (1), Johannes Weninger (1), Rainer Pfluger (2), Wilfried Pohl (1)

1: Bartenbach GmbH, Austria; 2: University of Innsbruck, Unit of Energy Efficient Building, Austria

AIM AND APPROACH

Despite the use of energy-efficient LED technologies and automated lighting systems, workplace lighting accounts for 21% of energy costs in the EU-27 countries [1]. The increase in the contribution of artificial light to the total energy demand of the building results from to reduced daylight-incidence as a result of higher thermal insulation and increasing building density [2].

User-centred lighting control systems, which separate the normative planar room illumination into user-oriented, workplace-zoned lighting, promise great potential for reducing the use of artificial light and thus energy consumption. The principle of uniform lighting for all workplaces and employees of all age groups is no longer up-to-date. The individually appropriate illuminance level and color temperature depend on age and activity [3]. The neglect of specific user and application requirements results in low acceptance rates, which can have a negative impact on energy system performance and biological effectiveness [4].

Similar effects can also result from planning based on generalized requirement criteria. An increased use of artificial light compared to the calculated requirements is the consequence of insufficient consideration of the end applications. Therefore, user and application aspects must already be considered in the context of artificial lighting calculations and daylight simulation.

SCIENTIFIC INNOVATION AND RELEVANCE

While the mapping of user-centered effects in the control logic can be implemented by an increased use of sensors, artificial lighting calculations and daylight simulations have so far largely neglected user and application aspects, resulting in an inadequate energy system mapping. Against this background, this paper highlights the importance of user and application aspects in the simulation environment.

The consideration of user aspects in the simulation not only offers the chance to plan more resource-oriented and to calculate energy balances more accurate, but also to create a basis for further research in this area. In this context we also present that the consideration of the specific building or room use in the simulation can provide indications of the control logic to be used.

This topic can also be of further importance in the BIM workflow. Requirement criteria can be checked more efficiently and a system evaluation can be carried out on the basis of the information feedback. In addition, simulation-based control systems could gain more importance through more precise system mapping.

PRELIMINARY RESULTS AND CONCLUSIONS

The role of user-centricity in simulation is clarified by comparing common planning methods with generalized requirement criteria to a simulation considering user and application aspects. Preferences for illuminance are used in conjunction with industry-specific characteristics of the number of employees, both broken down by age, to show the influence on the use of artificial lighting. In addition, various attendance time models are compared to the generalized model in order to highlight the influence of the operating times of general traffic routes in the building. The results are compared with measured values from a specially initiated long-term study of an open-plan office with zonal individual workplace lighting. The measured values were collected under real working conditions to record the energetic effects of a zoned lighting control concept. In addition, the significance of the obtained results for the choice of the control system is listed.

If the LED is hailed as a great revolution in energy-efficient lighting, individual lighting offers the chance for another great leap forward in energy efficiency. However, the implementation of individual lighting concepts starts with simulation. In this way, it is not only possible to plan in a resource-oriented manner, but also to improve people's quality of life.

MAIN REFERENCES

[1] Bertoldi, P., et al., (2012), Electricity Consumption and Efficiency Trends in the EU-27, Energy Efficiency Status Report 2012, DOI: 10.2788/37564

[2] Kyba, C., et al., (2017). Artificially lit surface of Earth at night increasing in radiance and extent. Science Advances, 3(11), DOI: 10.1126/sciadv.1701528

[3] Leger, D., et al., (2011), Underexposure to light at work and it's association to insomnia and sleepiness, Journal of Psychosomatic Research, 70(1), pp. 29-36, DOI: 10.1016/j.jpsychores.2010.09.006

[4] Nagy, Z., Yong, F. Y., Schlueter, A., (2016), Occupant centered lighting control: a user study on balancing comfort, acceptance, and energy consumption. Energy and Buildings, 126, pp. 310-322, DOI: 10.1016/j.enbuild.2016.05.075

Keywords: user-centered, energy efficiency, day- and artificial light

Paper ID 30578: Numerical and experimental evaluation of a granular PCM-enhanced plaster for the energy refurbishment of historic buildings

Eleonora Baccega*, Michele Bottarelli

University of Ferrara, Italy

AIM AND APPROACH

The construction sector represents more than one-third of the global energy consumption and the energy demand on a global scale for air conditioning is estimated to be about 40% of the total requirement, thus forcing policymaker actions to manage energy consumptions. However, great part of the building stock consists in historic buildings that frequently have no air conditioning systems and poor performances in terms of consumptions and indoor comfort: in Europe 14% of the buildings were built before 1919, while in Italy the rate rises up to 20%. These buildings, many of which with relevant qualities, usually suffer of a lack of regulations and few are the improvements already made.

Among the possible interventions to reduce summer consumptions of historic buildings, especially of those in Mediterranean countries, the application of Phase Change Materials (PCMs) mixed within the plaster of external walls shows interesting performances. PCMs can be used to increase the thermal inertia of the building envelope thus obtaining a reduction of the incoming heat flux and of the peak temperature on the surface. More in deep, a focus was made on granular PCMs, conducting both numerical and experimental analyses with different percentages of material mixed within different plasters.

SCIENTIFIC INNOVATION AND RELEVANCE

PCMs have been deeply studied over the last decades and their characteristics and behaviour are well known. Many methods to incorporate them within the building envelope as a passive solution to reduce energy consumptions and improve thermal comfort have been developed. However, their application on the so-called Cultural Heritage is still to investigate.

Considering the construction methods typical of the past centuries, a focus of this research was on lime plaster, considered to be more suitable for the restoration of historic buildings instead of lime-cement one, more common nowadays but that risks of damaging the wall on which it is applied. Moreover, among the available technologies of PCMs, instead of the wide studied micro-encapsulation of the material, the granule form, consisting in paraffin absorbed in silica granules produced by PCM Products Ltd, was selected. This type of PCM, recently introduced through the European project TESSe2b, is of great interest for its expected advantages in terms of strength and reliability of the mixture.

PCM-enhanced plaster on the exterior walls can be a valid option for the improvement of energy performance of the so-called Cultural Heritage without affecting the value of the buildings or violating any legal restrictions.

PRELIMINARY RESULTS AND CONCLUSIONS

Small samples of plaster mixed with granular PCM were realized and tested at a lab scale under real conditions. The monitoring showed a reduction of the peak temperature on the surface and of the heat flux through the thickness of the plaster were shown.

At the same time, a numerical model of the same samples was implemented in COMSOL Multiphysics and calibrated through the experimental data acquired. This model was then used for further simulations, in case of application of granular PCM-enhanced plaster on a historic building wall as well as changing some characteristics of the sample: the thickness of the plaster, the amount of PCM mixed within the plaster, the melting point of the PCM or sun exposition of the sample. Results outlined relevant and useful considerations for the design of applications on real buildings, in terms of correlation between melting point and solar gain.

MAIN REFERENCES

Cabeza LF, Castell A, Barreneche C, de Gracia A, Fernández AI. Materials used as PCM in thermal energy storage in buildings: A review. Renewable and Sustainable Energy Reviews 2011; 15: 1675-1695.

Navarro L, de Gracia A, Niall D, Castell A, Browne M, McCormack S, Griffiths P, Cabeza LF. Thermal energy storage in building integrated thermal systems: A review. Part 2. Integration as passive system. Renewable Energy 2016; 85: 1334-1356.

Bernardi A, Becherini F, Romero-Sanchez MD, Lopez-Buendia A, Vivarelli A, Pockelé L, De Grandi S. Evaluation of the effect of phase change materials technology on the thermal stability of Cultural Heritage objects. Journal of Cultural Heritage 2015; 15:470-478.

Bianco L, Serra V, Fantucci S, Dutto M, Massolino M. Thermal insulating plaster as a solution for refurbishing historin building envelopes: First experimental results. Energy and Buildings 2015; 95: 86-91.

Ventolà L, Vendrell M, Giraldez P. Newly-designed traditional lime mortar with a phase change material as an additive. Construction and Building Materials 2013; 47: 1210-1216.

Rady M. Granular phase change materials for thermal energy storage: experiments and numerical simulations. Applied Thermal Engineering 2010; 29: 3149-3159.

Keywords: thermal energy storage (TES), granular phase change materials (PCMs), Cultural Heritage, energy efficiency

Paper ID 30601: Predicting thermal comfort for diverse populations

Nathaniel Jones* (1), Ingrid Chaires (1), Ian Mackenzie (2), Timothy Arioto (1), Alexej Goehring (1)

1: Arup, United States of America; 2: Arup, Canada

AIM AND APPROACH

Most thermoregulatory models used to understand human comfort are based on a standard human physiology — a 175 cm-tall man weighing 74.1 kg whose body composition is 15% fat. However, this body type is not representative of the world's population, and variation in body build may significantly affect human perception of temperature. We use diverse, rapidly generatable body type models to predict thermal sensation and comfort of populations. Using multi-nodal heat transfer models of the body and environment, we can simulate the physiological and psychological response to environmental conditions for a variety of individuals, including cases of non-uniform and time-varying conditions.

SCIENTIFIC INNOVATION AND RELEVANCE

We use parallel computing to simulate the multi-nodal human thermoregulatory models with a variety of physical characteristics. Our approach is implemented in the cloud and accessible through a web interface. The results offer increased specificity over a Predicted Percent Dissatisfied approach because they account for transient effects and do not simplify thermal sensation to a normal distribution. Furthermore, using our method, we can identify subgroups within the population with increased sensitivity to certain types of conditions and design for them accordingly.

PRELIMINARY RESULTS AND CONCLUSIONS

Our tool provides a fast and easy-to-use method for predicting thermal sensations felt by diverse populations. We will present validation cases that compare our predictions to human subject studies where body build was recorded. Our study also includes sensitivity analysis showing which factors in a population are most important to consider when designing the thermal environment that will be comfortable for people of all body builds.

MAIN REFERENCES

Stolwijk, Jan, 1971. A mathematical model of physiological temperature regulation in man, NASA CR-1855.

Tanabe, Shin-ichi et al., 2002. Evaluation of thermal comfort using combined multi-node thermoregulation (65MN) and radiation models and computational fluid dynamics (CFD). Energy and Buildings 34, 637-646.

Zhang, Hui et al., 2001. Considering individual physiological differences in a human thermal model, Journal of Thermal Biology.

Zhang, Hui et al., 2010. Thermal sensation and comfort models for non-uniform and transient environments, part III: Whole-body sensation and comfort, Building and Environment.

Zhao, Yin et al., 2014. Thermal sensation and comfort models for non-uniform and transient environments, part IV: Adaptive neutral setpoints and smoothed whole-body sensation model, Building and Environment.

Keywords: thermal comfort, occupant satisfaction, diversity

Paper ID 30744: Generating traffic-based building occupancy schedules in Chattanooga, Tennessee from a grid of traffic sensors

Anne S Berres (1), Brett Bass (2), Joshua New* (3), Piljae Im (3), Marie Urban (4), Jibonananda Sanyal (1)

1: Computational Science and Engineering Division, Oak Ridge National Laboratory, TN, USA; 2: Bredesen Center, University of Tennessee at Knoxville, Knoxville, TN, USA; 3: Electrification and Energy Infrastructures Division, Oak Ridge National Laboratory, TN, USA; 4: Geospatial Science and Human Security Division, Oak Ridge National Laboratory, TN, USA

AIM AND APPROACH

Occupancy significantly impacts energy use, timing for demand impacts, and is a significant source of uncertainty in building energy models. While an ASHRAE Multidisciplinary Task Group is attempting to extend occupancy information, there are relatively few sources (e.g, Building America protocols[1], ASHRAE Standard 90.1[2], and prototype building models[3]) that define building occupancy schedules and number of occupants per building type or space type. More importantly, these sources define traditional schedules that are likely not to reflect the true occupancy of a given building. In this research, we address this by using travel data to more realistically estimate occupant arrivals, departures, and counts in individual buildings[4].

About 25% of Chattanooga's signalized intersections are equipped with modern sensors which track objects as they traverse an intersection. This produces a time series of turn movements, size (distinguishing among buses, cars, pedestrians), and other characteristics. Downtown Chattanooga is particularly well-connected with densely packed traffic sensors, which allow us to infer arrival/departure times of people at city blocks. Based on proximity and building capacity, we construct a mapping of people to buildings, and derive detailed occupancy counts.

SCIENTIFIC INNOVATION AND RELEVANCE

Traditional building occupancy schedules have very harsh transitions from hour to hour, whereas actual occupancy changes more smoothly in large commercial buildings. Local commute differences, traffic impacts (e.g. construction), and quarantine work patterns lead to significant variations in occupancy. With a mobility-informed approach to building occupancy, it is possible to not only generate more realistic building occupancy schedules from historic traffic data, but also to update them in real-time as the traffic sensors provide continuous updates on the state of the traffic system. In the future of smart buildings, this could be used to project changes in arrival times, and preemptively activate or better manage energy-consuming appliances within a building to match the projected level of occupancy.

We empirically validate the occupancy schedules by comparing building model outputs using these schedules to actual energy use recorded by the local electricity provider. Furthermore, we visualize the differences in occupancy and energy use in the form of charts, and we evaluate the variation in differences by building type.

PRELIMINARY RESULTS AND CONCLUSIONS

In a previous study, we applied a similar method to simulated traffic data in the Chicago loop, which was based on generalized traffic volumes [4] as well as more detailed travel survey data [5] which provided household-level information on travel times, trip purpose, and destination [6] by day-of-week at a census block group level. This provided promising results, however s this approach is based on general travel information, it cannot model changes in travel times or behavior, including traffic jams, road closures, weather conditions, or the recent increase in telecommuting. Due to the general nature of the data, there was little variation between different simulated days. In this work, we bridge this gap by using measured traffic data for each day, which provides travel information with much higher fidelity, and at a much higher spatial resolution where intersection-level traffic sensing is available. The resulting occupancy schedules are smoother than the previously available schedules, and we see up to a 19% reduction in error in simulated energy use than we did with the previous schedules, when we compare with the measured energy consumption.

MAIN REFERENCES

[1] Robert Hendron and Cheryn Engebrecht (2010). "Building America House Simulation Protocols." NREL Report TP-550-49426. Available WWW: https://www.nrel.gov/docs/fy11osti/49246.pdf

[2] ASHRAE/IES (1989) ASHRAE STANDARD 90.1-1989 ENERGY EFFICIENT DESIGN OF NEW BUILDINGS EXCEPT LOW-RISE RESIDENTIAL BUILDINGS.

[3] DOE commercial prototype building models (https://www.energycodes.gov/development/commercial/prototype_models)

[4] Anne Berres, Piljae Im, Kuldeep Kurte, Melissa Allen, and Jibonananda Sanyal. "A Data-Driven Urban Scale Mobility Model for Estimating Building Occupancy for EnergyPlus.": ASHRAE Building Performance Analysis Conference. 2019

[5] Anne Berres, Piljae Im, Kuldeep Kurte, Melissa Allen-Dumas, Gautam Thakur, and Jibonananda Sanyal. "A Mobility-Driven Approach to Modeling Building Energy.": Dec. 2019, pp. 3887–3895. DOI:10.1109/BigData47090.2019.9006308.

[6] N McGuckin and A Fucci. Summary of travel trends: 2017 national household travel survey. Technical Report FHWA-PL-18-019, US Department of Transportation, Federal Highway Association, 2018.

Keywords: occupancy, data-driven, sensors, mobility, traffic

Paper ID 31077: Personas to understand occupant behavior and energy usage in buildings

Vinu Subashini Rajus* (1), William O'Brien (2), Gabriel Wainer (1)

1: Department of Systems and Computer Engineering, Carleton University, Canada; 2: Department of Civil and Environmental Engineering, Carleton University, Canada

AIM AND APPROACH

Occupants come with varying comfort preferences. Their interactive behavior often depends on contextual parameters like building type or available controls and personal parameters like clothing preferences, adaptive knowledge, and metabolic levels. Occupancy hours vary based on building context. For example, in an academic building context, occupancy hours vary for professors, administrators, young researchers, and students. Likewise, in an industrial office space, the occupancy may be regular 9 am to 5 pm, or it may vary for office that follows Results-Only Work Environment (ROWE) that gives flexibility in working hours. Currently, advanced simulation modeling for occupancy behavior allows us to evaluate the building performance for varying occupancy hours. But they do not consider various individual parameters —comfort preferences and interactive behavioral aspects in the modelling. Even if the simulation models consider varying occupancy hours, they are similar for all zones. Hence, we propose a novel occupancy modeling approach using Personas that considers contextual, personal, and behavioral aspects to simulate potential interactive behavior and energy usage in buildings. The designed tool generates random occupancy hours, comfort preferences, and interactive behavior based on comfort, building performance, and context.

SCIENTIFIC INNOVATION AND RELEVANCE

Advanced modelling of occupancy behavior has moved the research from static occupancy analysis to dynamic. These models enable us to understand different occupancy hours and energy usage in the buildings. But these advanced models still assume similar occupancy hours for all the spaces. It ignores variation in individual preferences and behavioral aspects contextually and spatially. Personas is a common practice in other fields but not in building simulation. The objective of personas is not necessarily to improve accuracy, but to make the process a bit more tangible and human. The novelty of our approach is the model considers not only contextual parameters but also personal parameters like comfort preferences (thermal, lighting, views), occupant characteristics (age, clothing, activities), and behavioral aspects (sustainable behavior, trade-offs between comfort and energy, adaptive knowledge). Considering these aspects at the design stage will enable designers to perceive various complex or unexpected situations for solutions. Personas could be a bridge between design and building automation. Once the building is occupied, building managers or occupants can define the comfort preferences, occupant characteristics, and behavioral aspects for better building performance. The goal of the research is to reduce the performance gap between the predicted energy and actual energy usage.

PRELIMINARY RESULTS AND CONCLUSIONS

The personas were generated based on monitored data collected from an academic office context. The data reveal several occupancy patterns and interactive behavior like frequent, moderate, and passive interactions. Analysis of the monitoring data shows that building automation overrides occupancy behavior. In this context, building automation took control, and actual behavioral pattern could not be derived. Specific parameters for the personas were defined from the analysis of monitored data and other occupancy parameters from a large body of literature on occupants' behavior. The advanced modelling simulation was performed for thermal comfort with various contextual scenarios. Comparative analysis of building zones with similar window sizes, varying window sizes with similar occupancy hours and comfort preferences reveal that the difference in energy usage was not significant unless the zones were in the corner of the building. During winter, if the occupancy hours were from 8 am to 5 pm or from 2 pm to 5 pm, the difference in energy usage was not significantly different. But during summer, varying occupancy hours and comfort preferences show significantly different energy usage. In conclusion, personas could help in understanding occupancy patterns, interactive behavior, and energy usage.

MAIN REFERENCES

De Wilde, P. (2014). The gap between predicted and measured energy performance of buildings: A framework for investigation. Automation in Construction 41, 40–49.

Cole, R. and Z. Brown (2009). Reconciling human and automated intelligence in the provision of occupant comfort. Intelligent buildings international 1 (1), 39–55.

Meinke, A., Hawighorst, M., Wagner, A., Trojan, J., & Schweiker, M. (2017). Comfort-related feedforward information: occupants' choice of cooling strategy and perceived comfort. Building Research & Information, 45(1-2), 222-238.

D'Oca, S., Gunay, H. B., Gilani, S., & O'Brien, W. (2019). Critical review and illustrative examples of office occupant modelling formalisms. Building Services Engineering Research and Technology, 40(6), 732-757.

Gunay, H. B., O'Brien, W., & Beausoleil-Morrison, I. (2016). Implementation and comparison of existing occupant behaviour models in EnergyPlus. Journal of Building Performance Simulation, 9(6), 567-588.

Keywords: Occupant Behavior, Building Performance, Personas, Energy Modelling, Adaptive Behavior

Paper ID 30303: Building energy performance analysis of the application of a vision-based deep learning approach for occupancy activity detection

Paige Wenbin Tien*, Shuangyu Wei, John Kaiser Calautit, Jo Darkwa, Christopher Wood

University of Nottingham, United Kingdom

AIM AND APPROACH

This study utilises a vision-based data-driven deep learning approach for the recognition of occupancy activities within building spaces to provide live detections and form occupancy profiles for building energy management systems. The study aims to analyse the feasibility of such approach for a demand-driven control strategies for heating, ventilation and air-conditioning (HVAC) systems.

To establish the vision-based deep learning approach, a convolutional neural network was configured and trained using image-based datasets. The trained deep learning model was deployed to an artificial intelligence-powered camera and was used to perform live detection to form real-time Deep Learning Influenced Profiles. To evaluate the performance of the trained model, a series of experimental test approaches were evaluated using a building simulation tool. Various case study buildings, including private office spaces, open-planned office spaces and university lecture theatres were used to perform these tests. Furthermore, unique scenarios such as days with large number of occupants present in the desired space were established to assist the analysis of the impacts of occupancy behaviour. Building energy performance were analysed based on the assignment of the general building operational conditions and various occupancy profiles which provided discussion on the building energy demands and the operations of HVAC systems.

SCIENTIFIC INNOVATION AND RELEVANCE

A significant element affecting the usage of HVAC is the behaviour of occupants [1]. For instance, rooms in offices or lecture theatres are not fully utilised or occupied during the day, and in some cases, some rooms are routinely unoccupied. While the HVAC systems operates based on a fixed set point schedules, it assumes maximum occupancy during the entire working week. The used of fixed set points in combination with various occupancy patterns could lead to rooms frequently being over or under-conditioned. This leads to significant waste in energy consumption [2, 3]. It indicates the need for the development of a demand-driven controls that adapts to occupancy patterns in real-time and optimise HVAC operations. Information on real-time occupancy patterns is central to the effective development and implementation of a demand-driven control strategy for HVAC [4]. Previous works have shown the capabilities of strategies [4-6] in sensing occupancy information through the count and location of occupants. However, there is limited research on sensing the actual activities performed by occupants. A potential solution is to use artificial intelligence based techniques to detect and recognise the activities performed by occupants to predict the heat emitted by occupants to estimate better building energy requirements.

PRELIMINARY RESULTS AND CONCLUSIONS

Based on the initial analysis, an experimental test was performed within an office space of a selected case study building. Average detection accuracy of 92.20% was achieved for all activities. The application of the deep learning detection approach provides real-time data to establish the generated Deep Learning Influenced Profile (DLIP). Compared with an Actual Observation Profile of the 'true' occupancy activities performed, it suggests a difference of 0.0362%. Furthermore, results suggest the use of static occupancy profiles currently used in most building HVAC systems operations and in building energy modelling and simulations presents an overestimation in occupancy gains by up 54.92% with the Actual Observation. Therefore, effective understanding and monitoring of occupancy behaviour within buildings can provide a better estimation of the occupancy internal gains, heating and cooling loads. Thus, it is important to perform a series of building performance analysis to provide more advantageous evaluations of the novel vision-based deep learning approach for it to become an effective building energy management solution.

MAIN REFERENCES

[1] Hong, T., Yan, D., D'Oca, S., Chen, C., Ten questions concerning occupant behavior in buildings: the big picture, Build Environ, Vol. 114, pp. 518-530, 2017, https://doi.org/10.1016/j.buildenv.2016.12.006

[2] Dong, J., Winstead, C., Nutaro, J., Kuruganti, T. Occupancy-Based HVAC Control with Short-Term Occupancy Prediction Algorithms for Energy-Efficient Buildings. Energies, Vol. 11, pp. 2427, 2018, https://doi.org/10.3390/en11092427

[3] Kwok, S.S., Lee, E.W., A study of the importance of occupancy to building cooling load in prediction by the intelligent approach, Energy Convers. Manag., Vol. 52, pp. 2555–64, 2011, https://doi.org/10.1016/j.enconman.2011.02.002

[4] Gunay, H.B., O'Brien, W., Beausoleil-Morrison, I., Development of an occupancy learning algorithm for terminal heating and cooling units, Building and Environment, Vol. 93, Part 2, pp. 71-85, 2015, https://doi.org/10.1016/j.buildenv.2015.06.009

[5] 18. Nagy, Z., Yong, F.Y., Frei, M., Schlueter, A., Occupant centered lighting control for comfort and energy efficient building operation Energy Build., Vol. 94, pp. 100-108, 2015, https://doi.org/10.1016/j.enbuild.2015.02.053

[6] Benezeth, Y., Laurent, H., Emile, B., Rosenberger, C., Towards a sensor for detecting human presence and characterizing activity. Energy and Buildings. Vol 43(2), pp. 305-314., https://doi.org/10.1016/j.enbuild.2010.09.014

Keywords: Artificial intelligence, deep learning, energy management, occupancy detection, HVAC system

Session F4.1: Improving indoor environmental quality

Paper ID 30761: The impact of glazing characteristics, orientation and wall reflectance on energy, daylighting, visual comfort and environmental performance in patient rooms

Nazanin Eisazadeh*, Frank De Troyer, Karen Allacker

KU Leuven, Belgium

AIM AND APPROACH

The aim of this study is to evaluate the impact of glazing characteristics, orientation and different wall reflectance on the energy use/cost, environmental performance, daylighting and visual comfort in patient rooms. In order to study the impact of glazing characteristics (U-value, g-value and Tvis) on the performance, six glazing types are considered. To analyse the role of internally reflected component of daylight on visual comfort and daylight levels, different wall reflectance is assumed while keeping the reflectance of ceiling and floor constant. A typical patient room from a hospital in Belgium is used as case study. The results of alternative design strategies are analysed in terms of the Energy Use Intensity (EUI, kWh/m2), Annual Utility Cost (AUC, €/m2), environmental cost (€/m2), sDA300/50% (%), UDI100-3000lux (%), UDI>3000lux (%) and DGP<0.4 (%). Based on the comparative analysis, potential design options are discussed that allow for energy efficient, daylit and comfortable patient rooms with the least life cycle environmental impact. The methodology combines dynamic energy and daylight simulations and integrates these in a life cycle assessment of the patient rooms.

SCIENTIFIC INNOVATION AND RELEVANCE

Literature review revealed that there is various research on window systems and different design parameters influence on the performance of office buildings (Goia 2016) (Grynning, Time, and Matusiak 2014) (Mangkuto, Rohmah, and Asri 2016) (Nielsen, Svendsen, and Jensen 2011) (Ochoa et al. 2012) (Shen and Tzempelikos 2012) (Teixeira et al. 2020), but only a limited number of studies focus on patient rooms and the existing literature primarily concentrates on hot arid climates (Sherif et al. 2014; 2015; 2016; Wagdy et al. 2017). As for the environmental performance, only a few studies address the LCA of window systems (Salazar and Sowlati 2008; Carlisle and Friedlander 2016) and few publications focus on the LCA of building façade and envelope (Kim 2011) (Azari 2014; Stazi, Mastrucci, and Munafò 2012). It can be concluded that there appears to be no comprehensive study focusing on combining dynamic energy simulations, dynamic visual comfort and daylight analysis in a building-integrated life cycle assessment study of window systems, covering a wide range of environmental indicators.

PRELIMINARY RESULTS AND CONCLUSIONS

The results revealed that the glazing characteristics (U-value, g-value and Tvis) affects the window system performance significantly and requires special attention during the early design stage. The results also show that the most significant environmental impact indicators in the design options are global warming, human toxicity (cancer and non-cancer effects), particulate matter formation and eutrophication. This study demonstrates the importance of multi-criteria analysis of window systems performance and assists with understanding and observing the interaction between the different design parameters and the various metrics. Hence, this approach allows architects to recognize the impact of these parameters on buildings overall performance, which could assist with more selective choices in the early design stage and lead to more sustainable and high performing buildings.

MAIN REFERENCES

Ochoa, Carlos E., Myriam B.C. Aries, Evert J. van Loenen, and Jan L.M. Hensen. 2012. "Considerations on Design Optimization Criteria for Windows Providing Low Energy Consumption and High Visual Comfort." Applied Energy.

Teixeira, Henriqueta, M. Glória Gomes, A. Moret Rodrigues, and Júlia Pereira. 2020. "Thermal and Visual Comfort, Energy Use and Environmental Performance of Glazing Systems with Solar Control Films." Building and Environment.

Wagdy, Ayman, Ahmed Sherif, Hanan Sabry, Rasha Arafa, and Islam Mashaly. 2017. "Daylighting Simulation for the Configuration of External Sun-Breakers on South Oriented Windows of Hospital Patient Rooms under a Clear Desert Sky." Solar Energy.

Keywords: Daylight, Energy efficiency, Life cycle assessment, Comfort

Paper ID 31101: Lighting for circadian well-being: a simulation-based workflow for the visual and non-visual lighting design of a nursing home

Donatienne Carmon*, Sergio Altomonte

Universite' catholique de Louvain, Belgium

AIM AND APPROACH

For almost a century, the luminous efficiency function V(λ) has been the sole means to characterise human responses to light based on its ability to excite the visual system. Recent research, however, provides evidence that light exposure, other than enabling "visual" comfort, offers a cue to entrain "non-visual" physiological, endocrine, and behavioural processes that are conducive to the circadian well-being of observers. Several simulation tools are currently available to support designers in estimating the daylighting and glare performance of a building. However, the development and application of software tools to robustly model the non-visual stimulation characterising a naturally or electrically-lit space are still at their infancy. This is due to uncertainties in the metrics and thresholds to be utilised, but also to the difficulty in characterising the effective needs - and their variation over time - of building users, particularly when they belong to different age groups. In response, this study is among the first to use a novel software tool – ALFA (Adaptive Lighting for Alertness) – to estimate the lighting performance of a nursing home, computing the levels of visual comfort and non-visual stimulation received at different times by a population of elderly patients, and proposing strategies for their improvement.

SCIENTIFIC INNOVATION AND RELEVANCE

Over recent years, significant research has focused on the characterisation of the non-visual (melanopic) effects of light based on its intensity, spectrum, duration, timing, and history of exposure. New metrics have been proposed to complement conventional visual (photopic) measures (e.g., vertical and horizontal illuminance), including the EML (Equivalent Melanopic Lux), Circadian Stimulus, and M/P ratio (i.e., the ratio between melanopic and photopic illuminance).The software ALFA, developed by Solemma and based on a Rhinoceros 3D model, is one of few existing tools able to comprehensively integrate visual and non-visual metrics to allow the point-in time and annual simulation of a built space based on different viewpoints and directions. The tool enables the simultaneous prediction of photopic and melanopic measures at different times of the day and year, which can be compared to established thresholds. Potential applications of this tool are particularly relevant for the comprehensive lighting design of buildings, especially when they are aimed at a population of users with specific and variable needs. This is the case of the patients of a nursing home, especially in light of recent international recommendations that highlighted the differences in visual and non-visual requirements characterising the elderly with respect to the general population.

PRELIMINARY RESULTS AND CONCLUSIONS

This study applied a simulation-based approach to characterise the lighting quantities and qualities of a nursing home in Brussels (Belgium), estimate the levels of visual and non-visual stimulation received by occupants, and propose a series of adapted strategies to respond to their multi-faceted needs of visual comfort and circadian well-being. The simulations conducted with ALFA highlighted that most spaces (e.g., in-patient rooms and dining halls) respond to photopic and melanopic requirements over the summer and in the early hours of the day. However, required thresholds are seldom achieved in winter, while for some spaces (e.g., western-oriented) an excessive melanopic stimulation in the late hours of the day during the summer might interfere with the proper circadian rhythmicity of building occupants, with potential detrimental consequences on their well-being (e.g., decreased sleep quality). The strategies proposed include modifications to the electric lighting, alterations in the arrangement of furniture, and the installation of protective blinds. Overall, the study provided evidence that robust methodologies and tools for the simulation of photopic and melanopic metrics can support the design of spaces in order to comprehensively respond at once to the needs of visual comfort, task performance and satisfaction, while enhancing the circadian well-being of occupants.

MAIN REFERENCES

Altomonte S. (2008). Daylight for energy savings and psycho-physiological well-being. Journal of Sustainable Development 1(3):1-16.

Andersen M., Mardaljevic J., Lockley S. (2012). A framework for predicting the non-visual effects of daylight. Lighting Research & Technology 44(1):37-53.

Berson D.M., Dunn F.A., Takao M. (2002). Phototransduction by retinal ganglion cell that set the circadian clock. Science 295:1070-1073.

Boyce, P. (2014) Human factors in lighting. New York: CRC Press.

Figueiro M.G., Gonzales K., Pedler D. (2016). Design with Circadian Stimulus. LD+A:31-34

International Commission on Illumination (2019). CIE Position Statement on Non-Visual Effects of Light. Recommending proper light at the proper time. Vienna: CIE.

Konis K. (2018) Field evaluation of the circadian stimulus potential of daylit and non-daylit spaces in dementia care facilities. Building and Environment, 135:112-123

Lockley S. (2009). Circadian rhythms: Influence of light on humans. In: L. Squire (Ed.) Encyclopedia of Neuroscience (Vol. 2). Amsterdam: Elsevier.

Lucas R.J., et al. (2014). Measuring and using light in the melanopsin age. Trends in Neuroscience 37(1):1-9.

Reinhart C.F., Jakubiec J.A., Ibarra D. (2013) Definition of a reference office for standardized evaluations of dynamic façade and lighting technologies. Proceedings of BS 2013, Chambery, France.

Solemma (2020). Adaptive Lighting for Alertness (ALFA). https://solemma.com/Alfa.html

Keywords: Light, Non-visual effects, Circadian, ALFA, Elderly

Paper ID 30886: A daily-based approach to study daylighting performance of openwork façades of classrooms in hot-humid climates: glare protection, daylight sufficiency and circadian potential.

Natalia Giraldo Vasquez* (1,2), Luisa Fernanda Callejas (3), Lucas Arango Diaz (4)

1: PósArq, Department of Architecture and Urbanism, Federal University of Santa Catarina,Florianópolis, Brazil; 2: ICIEE, DTU Civil Engineering, Technical University of Denmark, Kongens Lyngby, Denmark; 3: University of Bio-Bio, Bio-Bio, Chile; 4: Faculty of Integrated Arts, University of San Buenaventura, Medellin, Colombia

AIM AND APPROACH

Open-work facades are a common strategy used in hot-humid climates. In Colombia, these façades solutions have been widely incorporated in educational facilities due to its low cost and easy maintenance. In this paper, we study the influence that subtle changes made in the reflectance index of open-work façades could cause in the occupants' glare sensation, the illuminance levels and the non-visual effects of daylight. Our database was generated through climate-based simulations by using DIVA-for-Rhino, considering weather files from three Colombian cities. In a hypothetical school classroom, we located two observers, with three view directions each. We fixed the classroom internal surfaces reflectances and opening area while four façade solutions were defined by changing the reflectances of the open-work façade components. We evaluated two orientations for a period between 07h to 17h. As output, we obtained DPG values, work plane illuminance and vertical illuminance at eye level. The adequate reflectance index for open-work façades would be that one with DGP < 0.35, DA > 50% and EML > 4hrs/day. Finally, an adaptive approach, aiming to reduce the glare sensation, was incorporated in our analysis by mean of the time spent by the observers looking in each direction.

SCIENTIFIC INNOVATION AND RELEVANCE

The large amount of time spent by children inside the classrooms has led some studies searching for the influence of the characteristics of the classroom on the students' learning and academic outcomes. The façade design is a factor that mediated the interactions between the interior and the exterior environment. Regarding the visual environment of the classrooms, external views, visual comfort and non-visual effects of daylight should be contemplated in the façade design, to improve visual tasks and academic performance. In the last years, parametric and genetic algorithms have conducted to sophisticate façade solutions, resulting in optimized architectural responses to different aspects. However, these kinds of solutions are usually projected for commercial buildings, since their construction and maintenance can be high. Despite open-work façades are common in buildings located in hot-humid regions, due to their thermic benefits and low cost, the daylighting performance of these solutions has been barely studied. Aiming to highlight how small changes in open-work façades, that can be easily implemented and afford by educational buildings, we investigated the differences in the daylighting performance, in terms of glare, daylight sufficiency and non-visual effects, while introducing a temporal approach for the users' behaviour.

PRELIMINARY RESULTS AND CONCLUSIONS

The analysis of the results based on % of days in the year in which 1, 2 or 3 criteria associated with this triple daylight perspective were met, showed differences in the assessment of the daylighting performance of the spaces according to the type of facade, the city, the orientation and the observer´s position. Although the surface reflectance's had a small effect on the DGP and EML, the consideration of the target illuminance was decisive.

When considering a target illuminance according to people's perception of daylight sufficiency, the optical properties of the façade were decisive to achieve the 3 criteria in some % of days and it could be concluded that the use of openwork walls could more efficient in locations with cloudier skies. However, when a typical target illuminance of 300 lux was considered, the orientation turns out to be the variable with the greatest weight in the % of days that the criteria are reached. This paper showed that the design decisions on the type of facade that should be chosen to optimize the daylighting performance from the triple perspective should be very different if one considers outdoor daylight availability and people's perception of daylight sufficiency.

MAIN REFERENCES

Arango, L. et al. (2020). Perforated Blocks Efficiency from Natural Ventilation Perspective In Warm-Humid Climate. Proceedings PLEA. Coruna (Spain), 1698–1703.

Arango-Díaz, L. et al. (2020). Daylight Performance Comparison of OpenWork Walls in the Facades for Interior Spaces. Proceedings PLEA. Coruna (Spain), 944–949.

Cordeiro, A.C.A. (2018). Use of open-work as a second skin in office buildings: analysis of the lighting performance of different geometries. Masters Dissertation.

Vásquez, N.G. et al. (2016). Investigating a Method to Dynamic Assessment of Glare Using the HDR Technique. Proceedings PLEA. Los Angeles (USA), 1-7.

Goedert, G.S., Vásquez, N.G. and Pereira, F.O. (2020). Influence of Gaze Direction and Time on the Glare Assessment : A Case Study in an Office Space. Proceedings PLEA. Coruna (Spain), 4–9.

Salazar, T.J.H. (2015). Openwork Walls and Their Solar Applicability Range. Journal of Civil Engineering and Architecture 9:373–83.

Konis, K. (2019). A Circadian Design Assist Tool to Evaluate Daylight Access in Buildings for Human Biological Lighting Needs. Solar Energy, 449–58.

Lucas, R. J. et al. (2014). Measuring and Using Light in the Melanopsin Age. Trends in Neurosciences 37(1):1–9.

Reinhart, C.F., Mardaljevic, J., and Rogers, Z.(2006). Dynamic Daylight Performance Metrics for Sustainable Building Design. LEUKOS, 7–31.

Keywords: Openwork façade, glare, daylight sufficiency, equivalent melanopic lux

Paper ID 30514: Spatial evaluation of potential saturation and contrast effects of discomfort glare in an open-plan office

Geraldine Quek* (1), Stephen Wasilewski (2,1), Jan Wienold (1), Marilyne Andersen (1)

1: École Polytechnique Fédérale de Lausanne (EPFL), Lausanne, Switzerland; 2: Lucerne University of Applied Sciences and Arts, Horw, Switzerland

AIM AND APPROACH

This paper analyses the relative distribution of annual saturation and contrast effects of discomfort glare in a deep open-plan office. A previous comparative study has shown that while hybrid glare metrics were found to predict glare well in most scenarios occurring in the investigated datasets, contrast-driven glare metrics predict discomfort glare better than saturation-driven metrics in daylit conditions when vertical illuminance is lower than 3000 lux; and saturation-driven glare metrics outperform the contrast-driven metrics when vertical illuminance is above 3000 lux. The focus of this paper is to determine the potential effects of contrast-driven and saturation-driven glare across the floor plan, simulating annual hourly vertical illuminance (Ev) and contrast (log_gc) were simulated in 8 view directions with a grid spacing of 0.75m, at a typical eye level of 1.2m, for daylight hours in the context of Geneva, Switzerland. To identify viewpoints where saturation and contrast effects of glare dominate, a synthetic and versatile spatial visualization approach was established. An in-development simulation method, Raytraverse, allowed us to calculate detailed annual saturation and contrast metrics in a reasonable timeframe. Results show that high contrast lighting conditions occur far more frequently than high saturation conditions except for a narrow area near the façade.

SCIENTIFIC INNOVATION AND RELEVANCE

Discomfort glare prediction metrics usually account for contrast and saturation effects that cause discomfort glare. A study across a combined dataset of 6 laboratory studies emulating typical daylit offices established that Daylight Glare Probability (DGP), a hybrid metric that accounts for both saturation and contrast effects, is the best performing metric thus far in predicting glare (1). However, the thresholds proposed with DGP were developed for brightly lit office spaces. In reality, office workspaces can also have vertical illuminances in the lower range, especially in open-plan configurations (2). Researchers have investigated in post-occupancy evaluations and proposed metrics to target conditions prevalent in open-plan office buildings, Unified Glare Probability (UGP), built upon a purely contrast-based metric, Unified Glare Rating (UGR). (3) To this end, a recent comparative study conducted by the authors revealed that purely saturation-based metrics perform better in predicting glare when the adaptation level (measured by vertical illuminance) is more than 3000 lux, while purely contrast-based metrics seem to perform better in dimmer conditions below 3000 lux (4). Therefore, in this study, we aim to visualize the annual potential saturation and contrast effects of discomfort glare across a synthetic array of viewpoints in a deep open-plan office.

PRELIMINARY RESULTS AND CONCLUSIONS

Faster simulations using Raytraverse have allowed us to look into annual hourly contrast, across a large open office floor plan for many points and view directions, kickstarting the generation of highly granular plan based data visualizations of luminous conditions responsible for discomfort glare. The results show an observable variation of saturation and contrast effects of glare in a deep open-plan office, indicating where saturation-driven, contrast-driven and hybrid glare metrics could predict glare best in distinct zones. We found a significant zone with a lower frequency of saturation effects and a higher frequency of contrast effects annually. This highlights the importance of taking into account the risk of glare experienced in areas further away from the façade, typically subjected to contrast-driven discomfort glare. The size of the zones were found to differ based on urban surroundings and, to a lesser extent, orientation, but would also depend on other factors; overall office design and floorplan, materials, presence of electric lighting, dynamic shading devices, climates and latitudes, etc. Future user studies on visual comfort should consider studying dimmer scenes with higher contrast glare so as to expand the datasets presently available and improve discomfort glare metrics by increasing their applicability to such conditions.

MAIN REFERENCES

1. Wienold J, Iwata T, Sarey Khanie M, Erell E, Kaftan E, Rodriguez R, et al. Cross-validation and robustness of daylight glare metrics. Light Res Technol. 2019 Nov 1;51(7):983–1013.

2. Pierson C, Sarey Khanie M, Bodart M, Wienold J. DISCOMFORT GLARE CUT-OFF VALUES FROM FIELD AND LABORATORY STUDIES. In: PROCEEDINGS OF the 29th Quadrennial Session of the CIE [Internet]. Washington DC, USA: International Commission on Illumination, CIE; 2019 [cited 2019 Jul 8]. p. 295–305. Available from: http://files.cie.co.at/x046_2019/x046-OP41.pdf

3. Hirning MB, Isoardi GL, Cowling I. Discomfort glare in open plan green buildings. Energy Build. 2014 Feb 1;70:427–40.

4. Quek, G., Wienold, J., Andersen, M., Sarey Khanie, M., Erell, E., Kaftan, Tzempelikos, T., Konstantzos, I., Christoffersen, J., & Kuhn, T. (2020). Comparing performance of discomfort glare metrics under different ranges of vertical illuminance (ongoing publication)

Keywords: glare, discomfort, metrics, daylighting, simulation

Paper ID 31110: Carbon-cost efficient retrofit of passive and active systems in residential buildings using genetic algorithm

Negar Mohtashami* (1,2), Rita Streblow (1), Linda Hildebrand (2), Dirk Müller (1)

1: Institute for Energy Efficient Buildings and Indoor Climate, RWTH Aachen University, Germany; 2: Chair of Reuse in Architecture, Faculty of Architecture, RWTH Aachen University, Germany

AIM AND APPROACH

In the past two decades, improving the energy performance of existing buildings is a major trend to reduce the environmental impacts since existing buildings make up for the largest share compared to new buildings. The energy consumed in a building consists of both embodied and operational form. However, most current conducted energy optimizations only consider operational energy and remain reluctant to the embodied energy invested in building materials and services during the life cycle of a building. This study addresses both types of energy together and considers building as a whole in order to optimize properties of the building envelope and HVAC systems simultaneously.

SCIENTIFIC INNOVATION AND RELEVANCE

The current research uses MOGA (Multi Objective Genetic Algorithm) to minimize CO2 equivalent emissions, and costs in order to find optimal retrofit scenarios for a typical multi-family house (MFH06) according to TABULA building classification for Germany.

PRELIMINARY RESULTS AND CONCLUSIONS

Findings show four clusters of refurbishment scenarios that are mainly categorized based on the amount of insulation materials. it is also perceived that the most optimal carbon-cost retrofit options focus on increasing the thermal energy storage capacity and remain reluctant in insulating the envelope or changing the windows for a typical multi-family house of 70s mainly due to high embodied energy in the insulation materials and devices.

MAIN REFERENCES

Evins R. A review of computational optimisation methods applied to sustainable building design. Renewable and Sustainable Energy Reviews 2013; 22: 230–45.

Ramesh, T., Prakash, R., & Shukla, K. K. (2010). Life cycle energy analysis of buildings: An overview. Energy and buildings, 42(10), 1592-1600.

Verbeeck G, Hens H. Life Cycle Optimization of Extremely Low Energy Dwellings. Journal of Building Physics 2007; 31(2): 143–77.

Vilches, A., Garcia-Martinez, A., & Sanchez-Montanes, B. (2017). Life cycle assessment (LCA) of building refurbishment: A literature review. Energy and Buildings, 135, 286-301.

Keywords: Operational and Embodied energy, Carbon-Cost Efficient Retrofit, Life Cycle Carbon Footprint, Life Cycle Cost, Genetic Algorithm

Paper ID 30997: Application of a long-term MPC formulation to hybrid GEOTABS buildings

Iago Cupeiro Figueroa* (1), Lieve Helsen (1,2)

1: KU Leuven, Belgium; 2: Energyville, Belgium

AIM AND APPROACH

This research evaluates the benefits of applying a long-term model predictive control (MPC) formulation to hybrid GEOTABS buildings [1], which consist of a ground-source heat pump (GSHP) and a thermally activated building system (TABS) emission system to cover most of the building energy needs, and a secondary but fast-reacting system to assist the primary system during peak periods. To that end, the methodology presented at [2] is applied to a generic modular hybrid GEOTABS building simulation model for different building balance degrees (i.e., heating or cooling dominated), electricity-gas price ratios and borefield sizes. The possibility of active regeneration is also analyzed.

SCIENTIFIC INNOVATION AND RELEVANCE

To ensure an optimal coordination between the components of a hybrid GEOTABS system an MPC methodology can be applied, reducing the energy use and increasing the cost savings of the building while keeping or even improving the indoor comfort [3]. However, MPCs typically optimize the control actions for an horizon that foresights a few days, while the geothermal borefield dynamics last for several years [4]. The contribution of this research is to evaluate in which situations it is beneficial to apply a long-term formulation that foresights a full year of operation. For example, in the case of passive cooling, is it more beneficial to use it as early during the summer season as possible or to save its capacity towards the end of the season?

PRELIMINARY RESULTS AND CONCLUSIONS

A modular model that comprises a typical hybrid GEOTABS configuration is constructed using Modelica. An MPC optimization problem is constructed using TACO [5], which allows the translation and compilation of non-linear problems (NLPs) using Optimica and in a similar fashion as JModelica. The set of models is simulated for a period of 5 years.

MAIN REFERENCES

[1] Himpe, E., Vercautere, M., Boydens, W., Helsen, L., & Laverge, J. (2018). GEOTABS concept and design: state-of-the-art, challenges and solutions. In REHVA Annual Meeting Conference Low Carbon Technologies in HVAC.

[2]Cupeiro Figueroa, I., Cimmino M., Helsen L. A methodology for Long-term Model Predictive Control of hybrid geothermal systems: theshadow-cost formulation. Under review

[3]Cupeiro Figueroa, I., Cigler, J., & Helsen, L. (2018). Model Predictive control formulation: a review with focus on hybrid geotabs buildings. In https://www. rehvam2018atic. eu/images/workshops/9/Figuereroa. pdf (pp. 1-9). Proceedings of the REHVA Annual Meeting Conference.

[4]Cupeiro Figueroa, I., Picard, D., & Helsen, L. (2020). Short-term modeling of hybrid geothermal systems for Model Predictive Control. Energy and Buildings, 109884.

[5] Jorissen, F., Boydens, W., & Helsen, L. (2019). TACO, an automated toolchain for model predictive control of building systems: implementation and verification. Journal of Building Performance Simulation, 12(2), 180-192.

Keywords: hybridGEOTABS systems, Model Predictive Control, Geothermal HVAC, Long-term control, Shadow-cost

Paper ID 31069: Development and validation of radiation-coupled multimodal heat flux boundary condition for incompressible buoyant internal flows in OpenFOAM

Jake Haskell* (1), Mike Jäger (1), Pia Riedel (1), Jerell Gill (2), Paul Lynch (2), Talbot Kingsbury (2), Steven Downie (2)

1: Arup, Advanced Building Engineering, Germany; 2: Arup, Advanced Digital Engineering, UK

AIM AND APPROACH

The open-source CFD code OpenFOAM has been gaining relevance for the investigation of internal flow phenomena for building physics applications over the course of the past decade, but a major limitation to more widespread adoption has the necessity to use more computationally expensive compressible models to effectively calculate the effects of heat transfer. Furthermore radiation coupling with the default boundary conditions is limited at best and poorly documented to date. Most internal flows can be accurately modelled using the Boussinesq approach and more computationally efficient incompressible solvers, however the adequate thermal boundary conditions are not yet implemented for the OpenFOAM incompressible buoyant solvers and the existing compressible thermal boundary conditions are ill-suited for building physics applications. A new thermal boundary condition is developed to close this gap to allow more accurate CFD modelling of thermally driven buoyant internal flows using OpenFOAM.

SCIENTIFIC INNOVATION AND RELEVANCE

This study documents the development and validation of a new thermal boundary condition that offers the user the ability effectively calculate specific and absolute wall heat fluxes as well as the heat flux resulting from the definition of an external temperature and wall thermal resistance. This new boudary condition allows improved CFD approximation of thermally internal flow for building applications and has furthermore been coupled with the DOM and view factor radiation models to further improve the quality of heat transfer modelling using CFD.

PRELIMINARY RESULTS AND CONCLUSIONS

The new boundary condition is developed in C++ first without radiation coupling and tested to ensure adequate calcuation of a heat flux for the three different modes. Following successful implementation and error testing, the boundary condition is tested on a simple L-shaped room case and benchmarked for accuracy against ANSYS CFX. The radiation coupling is then integrated and the boundary condition is verified against the analog case calculated in CFD; the results show very good agreement for air, surface and mean radiant temperatures. Six different parameter configurations using the boundary condition are investigated and the results show maximum deviation of 0.2°C between the two simulation models. The DOM model shows slightly better agreement with regard to the radiation modelling compared the view factors model.

MAIN REFERENCES

Cid, K. 2016. A Comparative study between thermal radiation models P1 and discrete ordinates using CFD software OpenFOAM, Congresso Internacional de Fluidodinâmica Computacional

Frie, F. 2014. Calculation of radiative losses of solar receivers using viewfactors, OpenFOAM User Conference 2014

Junior et al. 2016. Numerical validation of viewFactor and FVDOM radiation models of OpenFOAM® and application. Revista chilena de ingeria 26, S. 546-556.

Haskell et al. 1994.Boundary Conditions for the Diffusion Equation in

Radiative Transfer. Journal of Optical Society of America 10. S. 2727-2740

Koncar, B. and Koncar L. 2018. Open access peer-reviewed chapter

Use of CFD Codes for Calculation of Radiation Heat Transfer: From Validation to Application. In: Heat Transfer: Models, Methods and Applications. Kingston University London.

Modest, M. 2013. Radiative Heat Transfer, 3rd Edition. Academic Press

Programmers Guide, OpenFOAM v1806, 2018

Vdovin, A. 2009. Radiative Heat Transfer in OpenFOAM, Chalmers University of Technology

Keywords: CFD, Heat Transfer, Radiation Modelling

Paper ID 30551: Optimal control of TABS by Sparse MPC

Yasuyuki Shiraishi* (1), Masaaki Nagahara (1), Dirk Saelens (2)

1: The University of Kitakyushu, Japan; 2: KU Leuven, Belgium

AIM AND APPROACH

In this study, continuous energy use of Thermally Activated Building System (TABS) is minimized by the sparse optimal control that combines MPC and sparse modeling in order to improve the unsteady actual operation performance of TABS. In other words, we propose a method to improve unsteady energy saving performance of TABS and durability of air conditioning equipment. Furthermore, the purpose of this study is to clarify the energy-saving performance and the control performance of indoor environment by introducing the proposed method to TABS based on numerical simulation.

In this analysis, the effectiveness of the proposed method is verified by using a coupled analysis tool [1] of unsteady CFD analysis and MTLAB/Simulink for a general office space with TABS. Specifically, the water flow rate of TABS has been optimized by the combined method of MPC and sparse modeling so that the ceiling surface temperature of TABS reaches the target value is given as an input condition for CFD analysis. The effectiveness of the proposed method is verified through evaluation of control performance and energy saving performance.

As a case study, we will analyze the results of two cases: normal MPC control (Case1) and combined control of MPC and sparse modeling (Case2).

SCIENTIFIC INNOVATION AND RELEVANCE

TABS, which utilizes the building structure to emit and store energy, is becoming increasingly popular. So far, the introduction of TABS has been progressing mainly in Europe [2], but in recent years, the introduction has also been attempted in Japan [3].

As shown in previous studies [2], TABS is expected to have various advantages such as peak shift, reduction of heat source capacity, and cost reduction by utilizing large heat capacity in addition to energy saving and comfort. On the other hand, since the thermal response is slow due to the large heat capacity, control for creating a comfortable indoor thermal environment is important.

For this reason, study on Model Predictive Control (MPC) [4] that determines the current manipulated variable while predicting the behavior of the controlled variable has attracted attention as a new control method that replaces classical control for the operation of TABS. Furthermore, by combining this MPC and sparse modeling [5], it is possible to realize further energy savings while maintaining control performance and improve durability by minimizing the start and stop frequency of air conditioning equipment. However, application of this method to engineering problems has not been reported.

PRELIMINARY RESULTS AND CONCLUSIONS

In both cases, by using MPC, the TABS is activated before the heat load occurs. As a result, the control error of the ceiling surface temperature was reduced throughout the analysis period, and the average control error on each day was less than 0.3℃.

Preliminary results showed that, in Case1, water was often delivered at a low flow rate of 0.5 L/min. On the other hand, in Case2, compared with Case1, the operating time for 5 days on weekdays was reduced by about 14 hours in total. Therefore, the sparseness of the water flow rate (expansion of the time zone when the water flow rate became zero) was confirmed by using both MPC and sparse modeling. Regarding cumulative water flow rate, in Case2, it was possible to reduce by approximately 39% over the entire analysis period compared with Case1.

By controlling the ceiling temperature, the PMV values in the workspace satisfy the comfort zone in both cases, and the distribution in the horizontal plane was extremely small.

From the above, it has been confirmed that by combining MPC and sparse modeling, it is possible to achieve sparseness of the operation amount while satisfying comfort in the workspace.

MAIN REFERENCES

[1] Y. Ogawa, Y. Shiraishi, Multi-Objective Optimization of Energy Saving and Thermal Comfort in Thermo Active Building System based on Model Predictive Control, 40th AIVC conference Ghent, Belgium, 2019

[2] D. Olsthoorn, F. Haghighat, A. Moreau, and G. Lacroix, Abilities and limitations of thermal mass activation for thermal comfort, peak shifting and shaving: A review, Building and Environment, VoL.118, pp.113-127, 2017

[3] E. Kataoka, Y. Shiraishi, et al., Radiation Cooling and Heating using Building Thermal Storage for Outside Insulation Building -Part 1. The Architectural Summary of the Building and Equipment System and the Report of Piece Model Experiment Result-, SHASE (The Society of Heating, Air-Conditioning and Sanitary Engineers) of Japan, J-30, pp.317-320, 2016(in Japanese)

[4] A. Afram, and F. Janabi-Sharif, Theory and applications of HVAC control systems – A review of model predictive control (MPC), Building and Environment, VoL.72, 2014

[5] M. Nagahara et al.: Maximum hands-off control - a paradigm of control effort minimization-, IEEE Trans. Automatic Control, 61(3), pp.735-747, 2016

Keywords: TABS, MPC, Sparse modeling, CFD

Paper ID 31009: Developing an archetype building stock model for new cities in Egypt

Fady Abdelaziz*, Rokia Raslan, Phil Symonds

UCL Institute for Environmental Design and Engineering

AIM AND APPROACH

Egypt's growing population in the last few decades has led to a significant rise in housing demands. This directed the government to initiate a national project with the goal of developing 50 new satellite cities in the arid desert by 2030 to mitigate the high population density in major cities such as Cairo and Giza. The Egyptian building stock comprises around 13.5 million residential buildings, with almost 37 million residential units. More than 50% of these buildings were built in the last two decades. This has resulted in the residential sector accounting for 47% of total electrical energy consumption in Egypt, as well as 5% of the total CO_2 emissions. Developing a residential building stock model, based on representative archetypes presents a strong potential for understanding and analyzing the overall energy performance of the stock. In addition, it allows the prediction of future energy demand and CO_2 emissions. Existing data in Egypt provides an overview of the most common archetypes of building stock in terms of building height, construction method, and year of construction, however, these databases lack crucial information regarding the building's physical data that can be used in evaluating energy performance.

SCIENTIFIC INNOVATION AND RELEVANCE

This paper presents the development of ENCEM (Egyptian New Cities Energy Model), which is based on a bottom-up model of building stock archetypes for the new cities within the Greater Cairo region. The study develops the framework of the archetypes using data from 201,440 domestic buildings sampled from 9 new city districts that were constructed in Cairo and Giza governorates over the last 30 years. The study is built on data from CAPMAS (Central Agency for Public Mobilization and Statistics) Egyptian Housing Survey and Census. The energy simulation software tool, EnergyPlus is employed in developing this model to generate an hourly energy demand profile for each archetype. This study will provide a more accurate representation of the overall building stock variability in terms of building type, geometric form, building envelope, and overall energy consumption.

PRELIMINARY RESULTS AND CONCLUSIONS

From a bottom-up approach, the resulting model can be generalized to other cities with similar climates and can be studied further to develop a national building stock for Egypt. Developing this model will act as a key tool for governmental decision-makers and stakeholders in Egypt to be informed with the energy use and the energy-saving potential of the building stock, and will also help policy developers and building scientists to identify replacement or retrofit measures for the various categories of Egypt's housing archetypes. Understanding these measures will not only reflect on existing building stock but will also provide guidelines for the government's building codes and legislation to be implemented in new constructions.

MAIN REFERENCES

- Kavgic, M., Mavrogianni, A., Mumovic, D., Summerfield, A., Stevanovic, Z., & Djurovic-Petrovic, M. (2010). A review of bottom-up building stock models for energy consumption in the residential sector. Building And Environment, 45(7), 1683-1697. doi: 10.1016/j.buildenv.2010.01.021

- Pasichnyi, O., Wallin, J., & Kordas, O. (2019). Data-driven building archetypes for urban building energy modelling. Energy, 181, 360-377. doi: 10.1016/j.energy.2019.04.197

- Famuyibo, A., Duffy, A., & Strachan, P. (2012). Developing archetypes for domestic dwellings—An Irish case study. Energy And Buildings, 50, 150-157. doi: 10.1016/j.enbuild.2012.03.033

- Krarti, M., Aldubyan, M., & Williams, E. (2020). Residential building stock model for evaluating energy retrofit programs in Saudi Arabia. Energy, 195, 116980. doi: 10.1016/j.energy.2020.116980

Keywords: Building Stock Model, Housing Archetype, Bottom-up Approach, Egypt

Session F4.3: Buildings paving the way for the energy transition

Paper ID 30701: Bayesian approach to accounting for uncertain heating system efficiency in co-heating tests

Sarah Juricic, Simon Rouchier*

UMR 5271 Univ. Savoie Mont-Blanc CNRS, France

AIM AND APPROACH

To provide relevant energy conservation measures, an accurate estimation of the actual overall Heat Transfer Coefficient of the building envelope is necessary. Integrated co-heating tests, i.e. using the dwelling's own heating system, and more generally any characterisation method that uses on-board monitoring faces the issue of accounting for the heating system efficiency.

Indeed, fixing its value to that of the manufacturer may induce a significant systematic error on the final estimation of the HTC, whereas measuring the system efficiency on-site is cumbersome. Conversely, considering the system's efficiency as an unknown in the model results in a structurally non identifiable model, making data exploitation impossible in most of the usual calibration methods, as they use a frequentist approach.

This paper proposes to tackle this uncertainty issue by including the heating system efficiency as an unknown in the model formulation. The inherent structural non identifiability is dealt with by calibrating the model in a Bayesian approach, which provides a sort of regularization of the problem thus defined. The method is applied on data from an integrated co-heating test and its relevance illustrated by a comparison to the usual frequentist approach.

SCIENTIFIC INNOVATION AND RELEVANCE

The novelty of this work is to tackle structurally non identifiable models and to propose a canvas to calibrate such models. Instead of using the usual least square or gradient-descent methods, a Bayesian approach that samples the posterior distribution is used. This approach successfully provides estimates of the model parameters while reflecting on the uncertainty of the heating system's efficiency.

Although the proposed method is only illustrated by exploitation of an integrated coheating test, it is expected to be equally applicable on other methods for on-board characterisation.

The proposed canvas therefore reflects in a straightforward manner on one of the major sources of uncertainty in the HTC estimation and contributes to trustworthy on-board characterisation methods.

PRELIMINARY RESULTS AND CONCLUSIONS

The methodology has so far been tested on a synthetic dataset mimicking an integrated coheating test where steady state indoor conditions are achieved by a heating system of known efficiency.

Four modelling and calibration options are tested to illustrate the benefits of the methodology. First, a frequentist least square method is used with a fixed heating system efficiency in the linear regression model, once with a lucky guess at the actual value and once with a significant error. Then, a second model is formulated in a Bayesian approach, once with a narrow prior normal distrbution and once with an inaccurate but wide prior normal distribution.

The preliminary results indicate that a wrong prior estimation of the heating system efficiency in a frequentist approach leads to a significant error on the HTC estimation with misleading narrow confidence intervals. Conversely, the proposed methodology in a Bayesian approach leads to estimations with concededly wider uncertainties but that include at least the target HTC value. The HTC estimations thus obtained actually better reflect on the actual state of uncertainty in this type of experiment.

MAIN REFERENCES

[1] D. Farmer, D. Johnston, and D. Miles-Shenton, "Obtaining the heat loss coefficient of a dwelling using its heating system (integrated coheating)," Energy Build., vol. 117, pp. 1–10, 2016.

[2] D. J. Poirier, "Revising Beliefs in Nonidentified Models," Econom. Theory, vol. 14, no. 4, pp. 483–509, 1998.

[3] A. Raue, C. Kreutz, F. J. Theis, and J. Timmer, "Joining forces of Bayesian and frequentist methodology: a study for inference in the presence of non-identifiability," Philos. Trans. R. Soc. A Math. Phys. Eng. Sci., vol. 371, no. 1984, p. 20110544, Feb. 2013.

[4] S. K. Sahu and A. E. Gelfand, "Identifiability, Improper Priors, and Gibbs Sampling for Generalized Linear Models.," J. Am. Stat. Assoc., vol. 94, no. 445, pp. 247–254, 1999.

[5] M. Senave, G. Reynders, P. Bacher, S. Roels, S. Verbeke, and D. Saelens, "Towards the Characterization of the Heat Loss Coefficient via On-Board Monitoring: Physical Interpretation of ARX Model Coefficients," Energy Build., May 2019.

Keywords: HTC estimation, heating system efficiency, integrated co-heating, Structural identifiability, Bayesian approach

Paper ID 30723: Robust optimal identification and scheduling of modernization measures for typical buildings

Jan Richarz*, Yuhan Hu, Marco Wirtz, Tanja Osterhage, Dirk Müller

RWTH Aachen University, E.ON Energy Research Center, Institute for Energy Efficient Buildings and Indoor Climate

AIM AND APPROACH

Existing buildings play a key role in achieving the worldwide emission reduction targets. In this regard realized modernization measures have long-term effects on the energetic performance of buildings.

Optimization approaches for the modernization of buildings mostly consider one-time investments for the design decision. As boundary conditions, especially technical and economic parameters, change in the future, also the optimal building energy system (BES) configuration will change over the life time of a building. Hence, taking into account multiple points in time for the design decision will result in a BES design that will lead to lower costs and emissions over the life time of a building compared to only considering an initial design.

Richarz et al. [1] therefore developed a mixed-integer linear program (MILP) that generates modernization schedules over the next 30-50 years. Therefore, boundary conditions from prognoses (e.g. energy prices) are set explicitly for each prospective year of the considered time horizon.

This work contains the handling of the prognoses' uncertainty. The aim is to determine the optimal set of measures under many possible prognoses instead of under one. For this purpose, robust optimization is a promising approach and we hereby present a robust extension of the mentioned MILP.

SCIENTIFIC INNOVATION AND RELEVANCE

Robust optimization is rarely used in the field of energy system planning. Recently, Moret el al. [2] and Hollermann [3] showed promising approaches for its application in energy models.

By calculating modernization schedules, the original MILP ([1]) deals with a mutable building energy system over a time horizon and uses many prognoses. Therefore, taking into account the robustness against significant boundary conditions seems promising to generate more realistic modernization schedules. All investment decisions of the considered time horizon are formulated as the optimization problem. Investments contain modernization measures for the envelope and the energy supply system. Finally, the program returns a modernization schedule i.e. investment path.

Three steps are part of our proposed method. First, based on the proposed approach by [4], all model parameters are categorized concerning uncertainty behavior, and their uncertainty ranges are determined. Second, a sensitivity analysis based on [5] leads to the most influencing parameters of the model. Finally, we use these most influencing parameters for the so-called robust counterparts, built up on the approach by [6]. The extended robust MILP executes a multi-criteria optimization by minimizing equivalent annual costs and emissions. In this study, it is applied on a typical office building.

PRELIMINARY RESULTS AND CONCLUSIONS

In the nominal (non-robust) as well as in the robust solutions, the constellation of the building energy system of the typical building underlies multiple changes over its life cycle. Therefore, we observe advantages in scheduling modernization measures over a buildings' lifetime compared to one-time investment approaches.

The conducted sensitivity analysis of the original MILP shows that parameters of user behavior, emission factors of energy sources, discount rate, energy prices, and investments in some plants are highly influencing the results.

The robust Pareto frontiers turn out to save emissions less cost-efficiently in comparison to the non-robust ones, which is due to the nature of robust optimization. Furthermore, their shape is different compared to the nominal ones. However, first results show acceptable deviations of the objective values and optimal modernization schedules compared to the nominal solutions. Despite to the robust formulation, still a significant share of carbon emissions can be avoided cost-efficiently.

Depending on the chosen degree of robustness, especially economic and ecological optimum differ over the robust solutions. Concerning the chosen BES in the modernization schedules, we see larger capacities and more gas-based technologies in robust solutions than in less robust solutions.

MAIN REFERENCES

[1] Richarz J., Eller P., Henn S., Osterhage T., Müller D., Multi-objective optimization for the systematic identification and scheduling of modernization measures for non-residential buildings, ECOS 2020: Proceedings of the 33rd International Conference on Efficiency, Cost, Optimization and Environmental Impact of Energy Systems (ECOS), Osaka, Japan, 2020

[2] Moret S., Babonneau F., Bierlaire M., Francois M., Decision support for strategic energy planning: a robust optimization framework, European Journal of Operational Research, pp. 539-54, 2020.

[3] Hollermann D., Reliable and Robust Optimal Design of Sustainable Energy Systems [dissertation], Wissenschaftsverlag Mainz GmbH, Aachen, 2020.

[4] Moret S., Codina Gironès V., Bierlaire M., Maréchal F., Characterization of input uncertainties in strategic energy planning models, Applied Energy, p. 597–17, 2017.

[5] Heiselberg P., Brohus H., Hesselholt A., Rasmussen H., Seinre E., Thomas S., Application of Sensitivity Analysis in Design of Sustainable Buildings, SDBE 2007: International Conference on Sustainable Development in Building and Environment, Chongqing, China, 2007.

[6] Bertsimas D., Sim M., The Price of Robustness, Operations Research, pp. 35-53, 2004.

Keywords: Building energy system, Investment pathways, Modernization schedule, Mixed-integer linear programming, Robust optimization

Paper ID 30789: Revealing the hidden dynamics of the energy signature model

Jaume Palmer Real*, Christoffer Rasmussen, Davide Cali, Henrik Madsen

DTU, Denmark

AIM AND APPROACH

The energy signature model is a well-known method to quantify the thermal characteristics of a building. This method uses daily averages of heat load and outdoor temperature to: i) find the base temperature that separates the weather-dependent regime from the weather-independent energy consumption, ii) quantify the heat-loss due to the indoor-outdoor temperature difference. However, by using daily averages, the classical energy signature model omits the dynamics of the thermal elements of the house. This work presents an assessment of the information lost by using the classical coarse model. Different measurement resolutions are studied, in order to study which provides the best picture of the house thermal characteristics. Moreover, the heat loss of a building is affected by weather variables, such as the solar irradiation and the wind speed. These conditions vary faster than the outdoor temperature, thus their effects on the heat loss are also lost during the daily average. The impact of these effects are non-linear, in order to incorporate them into the model, non-parametric methods, such as Kernel estimation, and spline curve fitting are used.

SCIENTIFIC INNOVATION AND RELEVANCE

The energy signature is used as a fix model that provides insight about the thermal characteristics of buildings. It is extensively used and it impacts the understanding that we have of the buildings' energy response. Revealing how this model evolve in time, and which variables affect it the most, will improve estimating the energy consumption for future buildings, thus narrowing the performance gap.

PRELIMINARY RESULTS AND CONCLUSIONS

The classical energy signature model can be improved by adding other variables, like weather variables. By doing so, the residuals of such model approach white noise. Nevertheless, studying the evolution of residuals over time shows a clear auto-correlation. This correlation is present with daily values, and will be analysed with higher resolutions in the presented work.

MAIN REFERENCES

This research departs from a submitted journal article by C. Rasmussen , P. Bacher, D. Cali, H. A. Nielsen and H. Madsen. This work was submitted late June and it is still waiting fo acceptance.

Other relevant work:

H. A. Nielsen, S. B. Mortensen, P. Bacher, H. MAdsen. Analysis of energy consumption in single family houses. DYNASTEE International Workshop on Dynamic Methods for Building Energy Assessment, 2010.

Keywords: Energy signature, dynamic modelling, performance gap; thermal characterisation

Session F4.4: Climate change and bioclimatic design

Paper ID 31008: Urban heat islands in future climate scenarios

Violeta Lialios-Bouwman, J. Alstan Jakubiec*

University of Toronto, Canada

AIM AND APPROACH

Climate change is a pressing global event [1], and the possibility of mitigating the impacts of anthropogenic effects on climate seems remote as shown by the fact that carbon emission trends are still growing [2] and at least until 2009 followed the worst case 'A1' scenario of the Intergovernmental Panel on Climate Change (IPCC). Anthropogenic changes are also evident on localized urban climates [3], often known as urban heat island (UHI). UHI can increase heat stress and operational energy use. Both climate change and UHI should be taken into account when designing for the future; however, typical climate files [4] are always based on the past. In the case of Canadian standard CWEC weather files, source weather data goes back as far as 1950. With this paper, we assess UHI under different future climate scenarios and under a variety of urban design strategies using simulations [5, 6] in order to understand the impact of future climate on design decision making to reduce UHI, the size of that effect, and how it varies over time. We simulated over 48,000 UHI scenarios under five climates and nine other urban and building design variables to build a robust predictive analysis.

SCIENTIFIC INNOVATION AND RELEVANCE

We assess nine different urban design parameters in terms of their impact on UHI: height, green coverage ratio, green roofs, site coverage ratio, tree canopy ratio, building material selection, façade-to-site ratio, building program types, and albedo. All parameters are assessed under five climate files for Toronto, ON, Canada. Two are based on measured data—TMYx [7] files for 1950-2018 and 2004-2018—while three others are simulated future climates [5] for 2020, 2050, and 2080. Overall 48,558 simulated climate files accounting for climate change and UHI were created. Each file is assessed for typical UHI metrics: maximum UHI, mean daily UHI, maximum daytime temperature, and maximum nighttime temperature. In addition, proxies for the frequency of heat and cold stress are included—the number of hours above 30°C and the number of hours below 5°C. This study is highly relevant to the Building Simulation community, because it is a novel application of simulation technology to assess how urban design may need to change over time and how urban masterplans and projects designed today can have variable impacts in the future.

PRELIMINARY RESULTS AND CONCLUSIONS

We found that as the Toronto climate warms, the maximum UHI decreases (TMYx 1950-2018 median 7.9°C; CCWWG 2080 median 6.1°C) as does the mean daily UHI (TMYx 1950-2018 median 2.0°C; CCWWG 2080 median 1.8°C). However, the range of possible heat stress outcomes rises dramatically under future climates, underscoring the importance of intelligent urban design to mitigate UHI (TMYx 1950-2018 range 41 hours; CCWWG 2080 range 180 hours). Using Pearson correlation analysis, we found that the influence of site coverage ratio and building height on mean UHI decreased slightly over time, but the influence of urban greenery becomes significantly more important in reducing UHI in future climate files.

MAIN REFERENCES

1. PCC Working Group III, Climate Change 2014-Mitigation of Climate Change: Summary for Policymakers. 2014: Intergovernmental Panel on Climate Change.

2. Friedlingstein, P., et al., Global carbon budget 2019. Earth System Science Data, 2019. 11(4): p. 1783-1838.

3. Oke, T.R., City size and the urban heat island. Atmospheric Environment (1967), 1973. 7(8): p. 769-779.

4. Hall, I.J., et al., Generation of a typical meteorological year. 1978, Sandia Labs., Albuquerque, NM (USA).

5. Jentsch, M.F., et al., Transforming existing weather data for worldwide locations to enable energy and building performance simulation under future climates. Renewable Energy, 2013. 55: p. 514-524.

6. Bueno, B., et al., The urban weather generator. Journal of Building Performance Simulation, 2013. 6(4): p. 269-281.

7. Crawley, D.B. and L.K. Lawrie, Should We Be Using Just 'Typical' Weather Data in Building Performance Simulation?

Keywords: urban heat island, future climate, heat stress

Session F4.5 (Online Track): Practice and industry related case studies

Paper ID 30176: Impact of changes in design/operation guidelines against the COVID-19 on the energy performance of an institutional building: a case study

Pedro Guaraldi*, Radu Zmeureanu

Centre for Zero Energy Building Studies, Department of Building, Civil, and Environmental Engineering, Gina Cody School of Engineering and Computer Science, Concordia University, Montreal, Canada

AIM AND APPROACH

This paper presents the evaluation of impact of changes in design and operation guidelines (ASHRAE and REHVA), for reducing the COVID-19 propagation, on the buildings energy performance. So far there are no proofs of the link between virus propagation and ventilation systems/ducts. However, theoretically it can happen. The knowledge/experience from COVID-19 changes every day. Scientists and medical personnel bring new evidence, sometimes conflicting evidence, every day. Given the recentness and emergency nature of these responses against COVID-19, there are no publications, to the best knowledge of authors, addressing the building energy performance impact of the new recommendations.

The evaluation of energy impact is performed by computer simulation using eQuest program. The base case model of an existing institutional building is modified to comply with National Energy Building Code of Canada (NECB-2017). Design and operation recommendations from ASHRAE and REHVA are implemented in the model. The impact of each recommendation on the base case performance is assessed using a series of energy Key Performance Indicators (KPIs) that include whole-building Energy Utilization Index (EUI), and indices of performance at system and equipment levels.

SCIENTIFIC INNOVATION AND RELEVANCE

The paper presents a summary of those recommendations such as: (1) the operation time at nominal air flow rate should be extended, (2) the continuous exhaust ventilation of toilets, (3) the de-activation or by-pass of heat recovery units, and stopping the operation of rotary heat exchangers, (4) the elimination of recirculation of indoor air, and (5) the increase of outdoor air floor rate. Some recommendations will likely increase the annual energy use and cost.

The findings of this study would present only the tip of the iceberg, which would encourage the consultants and researchers to explore and implement different methods to mitigate the increase of energy use and cost.

PRELIMINARY RESULTS AND CONCLUSIONS

Preliminary results show the increase of overall building energy consumption for heating and cooling. This increase can be explained by the increase of outdoor air ventilation rates without recirculation, especially in cold climate, keeping the systems running for longer hours, increasing the pressure drop using improved HVAC filters, and bypassing energy recovery ventilation systems that leak contaminated exhaust air back into the outdoor air supply. The system Coefficient of Performance (COP) is expected to increase, since the part load values of existing HVAC equipment will be higher due to additional space and ventilation heating and cooling loads. However, keeping the systems running for longer hours with lower occupation and increasing the pressure drop in the ventilation system yield the opposite effect. The overall change in the system COP is to be determined. A discussion specifically addressing changes in HVAC design guidelines and the limitation of changes for existing buildings will be provided.

MAIN REFERENCES

1. L.J. Schoen. Guidance for building operations during the COVID-19 pandemic. ASHRAE Journal, 72-74, May 2020.

2. J. Kurnitski. How to operate and use building services in order to prevent the spread of the coronavirus disease (COVID-10) virus (SARS-CoV-2) in workplaces. REHVA COVID-19 Task Force. Build Up Webinar, April 28, 2020.

3. L. Mazzarella. Showcasing the Italian COVID-19 guidance for reducing the diffusion risk of SARS-CoV-19. REHVA COVID-19 Task Force. Build Up Webinar, April 28, 2020.

Keywords: COVID-19 guidelines, case study, HVAC system, energy use and cost, building energy simulation

Paper ID 30425: HVAC performance evaluation and optimization algorithms development for large buildings

Mike Stock* (1), Mohamed Kandil (1), Brian Macleod (2), Brad Pilgrim (2), Jennifer McArthur (1)

1: Ryerson University, Toronto, Canada; 2: Parity Inc, Toronto, Canada

AIM AND APPROACH

Data-driven models can be coupled with optimization algorithms to predict optimal setpoints for heating and cooling systems in large buildings. Informed by equipment thermodynamics, these models ingest available building data to infer building performance under a range of weather and control conditions and use this to develop an optimization strategy. This paper presents the development of these models using a case study building in ASHRAE CZ5. Initial results indicate savings up to 45% for heating applications can be achieved beyond standard outdoor air reset control without any physical changes to equipment, offering a low-cost, rapidly deployable solution to reduce building energy consumption and GHG emissions.

SCIENTIFIC INNOVATION AND RELEVANCE

This research addresses two scientific and technical challenges to implement building-scale optimization of heating and cooling systems: (1) the difficulty in scaling model predictive control approaches - particularly empirically based - to large buildings; and (2) the lack of available as-built information and monitored points in legacy buildings. Model predictive control has traditionally shown ability to reduce energy expenditure in buildings, however physical modeling of complex buildings is time and labour-intensive and is difficult to generalize between different buildings effectively. We propose the use of a hybridized data driven model to allow rapid deployment of prediction algorithms in existing sites, limiting ourselves to the data that can be collected in most buildings.

This is achieved through a review of typical system configurations, controls, operational parameters, and captured sensor data; the development of an emulator to simulate the full range of normal and edge-case operating conditions; analysis, processing and feature engineering of existing datasets from available test facilities; training machine learning algorithms to classify HVAC equipment performance and determine optimal operational setpoints based on data available in common large buildings and cloud-based weather forecasting APIs.

PRELIMINARY RESULTS AND CONCLUSIONS

Various state-of-the-art machine learning and deep learning techniques have been considered. A great effort has been devoted to the pre-processing and feature engineering the data to maximize the accuracy of the embedded predictive models and improving optimization results. Simulations using data-driven models with Bayesian optimization suggest that there is opportunity for substantial savings in air conditioning systems in large buildings trained strictly on available BAS logs.

These results also show that the model describing heating system must be provided with sufficient data to describe all expected operating conditions. Energy savings can still be achieved with imbalanced data, however the magnitude of savings is significantly reduced. Future work will examine this implication and will explore training intervals or potential for on-line training to ensure the application provides consistent energy saving recommendations.

MAIN REFERENCES

[1] M. Molina-Solana, M. Ros, M. D. Ruiz, J. Gómez-Romero, and M. J. Martin-Bautista, "Data science for building energy management: A review," Renew. Sustain. Energy Rev., vol. 70, no. August 2015, pp. 598–609, 2017, doi: 10.1016/j.rser.2016.11.132.

[2] Y. Zhao, T. Li, X. Zhang, and C. Zhang, "Artificial intelligence-based fault detection and diagnosis methods for building energy systems: Advantages, challenges and the future," Renew. Sustain. Energy Rev., vol. 109, no. April, pp. 85–101, 2019, doi: 10.1016/j.rser.2019.04.021.

[3] Bengea, S. C., Kelman, A. D., Borrelli, F., Taylor, R., & Narayanan, S. "Implementation of model predictive control for an HVAC system in a mid-size commercial building." HVAC&R Research, Vol 20, Issue 1, pp. 121-135, 2014.

[4] A. Afram and F. Janabi-Sharifi, "Theory and applications of HVAC control systems - A review of model predictive control (MPC)," Build. Environ., vol. 72, pp. 343–355, 2014, doi: 10.1016/j.buildenv.2013.11.016.

[5] P. Li, H. Qiao, Y. Li, J. E. Seem, J. Winkler, and X. Li, "Recent advances in dynamic modeling of HVAC equipment. Part 1: Equipment modeling," HVAC R Res., vol. 20, no. 1, pp. 136–149, 2014, doi: 10.1080/10789669.2013.836877.

Keywords: Existing building performance, data analytics, HVAC system simulation, optimization, continuous commissioning

Paper ID 30655: AI-based building type assignment

Brett Czech Bass (1), Joshua Ryan New (2), Evan Ezell (1), Eric Garrison (1), Piljae Im* (2), William Copeland (3)

1: University of Tennessee, Knoxville, TN United States of America; 2: Oak Ridge National Laboratory, Oak Ridge, TN United States of America; 3: Electric Power Board, Chattanooga, TN, United States

AIM AND APPROACH

Building energy modeling allows researchers to simulate buildings to gain insights into causes of energy use, possible energy saving opportunities, as well as gauge the effects of the changing climate and environment. Many of the building features needed to create building energy models are readily available such as 2D footprints and building height. A critical feature that is not generally less obtainable is the type of building. The building type can have a significant impact on the energy profile of a building. In partnership with the Electric Power Board of Chattanooga, TN; a year's worth of real-world 15-minute electrical use data has been examined. The smart meter data for a sample of 100 Chattanooga buildings is compared to 97 different prototype building energy models to assign a building type using three different methods. The models are then simulated using EnergyPlus and compared to real-world data to assess simulation quality.

SCIENTIFIC INNOVATION AND RELEVANCE

Building type assignment is a critical piece of urban-scale energy modeling with huge impact on simulation and energy conservation measure efficacy. While building type can be easily assigned when working with a small number of buildings, assigning building type to larger number of buildings can be cumbersome or impossible. Little work has been done in the realm of assigning building type without publicly available (often tax assessor) data [1][2]. In this research, features are extracted from prototype simulations, with these features being used to train a machine learning model to assign building type to buildings by extracting the same features from real-world data. With this method, given one has access to a time series of electricity data for a building for a year, a building type can be assigned at a large scale. An additional element inherently included in a machine learning classifier is a prediction probability which provides a pseudo confidence in the prediction which should help determine if a building type assignment should be re-assessed.

PRELIMINARY RESULTS AND CONCLUSIONS

Training a machine learning algorithm with time-based statistics from prototype building model and vintage combinations was done. Several algorithms were evaluated with simulations for each building type assignment compared to real-world data. While real-world data will not be available; simulation metrics can be presented and compared. Variable importance is considered in the machine learning models to determine which features have the largest impact on building type. Confidence in the form of the distributions of prediction probabilities can also be reported.

MAIN REFERENCES

[1] Garrison, Eric, New, Joshua R., Adams, Mark (2019). "Accuracy of a Crude Approach to Urban Multi-Scale Building Energy Models Compared to 15-min Electricity Use" Conference: ASHRAE Winter Conference - Atlanta, Georgia, United States of America, January 12, 2019

[2] New, Joshua R., Adams, Mark, Garrison, Eric, Bass, Brett, and Guo, Tianjing (2020). "Scaling Beyond Tax Assessor Data." ASHRAE/IBPSA-USA 2020 Building Performance Analysis Conference & SimBuild (BPACS), Chicago, IL, Aug. 12-14, 2020.

Keywords: Building Energy Modeling, Urban Scale, Machine Learning, Artificial Intelligence

Paper ID 30597: Evaluation and test operation of different model predictive control approaches for an energy system

Christina Betzold*, Arno Dentel

Technische Hochschule Nürnberg Georg Simon Ohm, Germany

AIM AND APPROACH

An energy system of terraced houses, which is monitored since 2018, consists of two shared modulating heat pumps (MHPs), eight decentralized domestic hot water heat pumps (DHW-HPs) as well as shared battery and photovoltaics (PV). Beside a rule-based control, the energy system is used for testing control strategies in real application. A model predictive control (MPC) strategy was tested for several hours, controlling the MHPs and some DHW-HPs. However, different approaches regarding models and optimization algorithms lead to different operation results. Two different model approaches, simple on basis of an energy balance or detailed on basis of physical processes, are investigated. Another aspect is the investigation of optimization algorithms, which can be generic or mixed-integer linear programming (MILP). The influence of both, models and optimization algorithms, on the heat pump operation in simulation and especially in real application is not clear.

The aim of this research is evaluation and test operation of different MPC approaches for an energy system that consists of variable and constant speed heat pumps as well as a battery and PV system.

SCIENTIFIC INNOVATION AND RELEVANCE

In the last years, the installation of heat pumps in buildings increased [1]. Continuing the process, it will lead to a high number of "all electric buildings", which can burden the public grid. However, intelligent control strategies in "all electric buildings" enable flexibility for interacting with the grid and offer potential for cost savings due to flexible prices for electricity and service. In order to fully exhaust this effect, MPC strategies offer the possibility to operate the energy system according to clearly defined targets, e.g. minimal costs, limits to maximal electrical load peaks. However, there is a multitude of optimization algorithms and models available and the question arises which MPC approach is suitable for the respective system.

PRELIMINARY RESULTS AND CONCLUSIONS

A previous simulation study compared a PV optimized control (PVC) and MPC to heat controlled strategies. Results from the annual simulation show a potential of the MPC to increase PV self-consumption by 2.1 % and a cost reduction by 36.1 %. Furthermore, an efficient heat-controlled control has less operating costs compared to PVC, as it offers lower overall energy consumption with a better Seasonal Performance Factor [2]. A previous test of a MPC approach with simple modelling and a mixed-integer linear programming in the real energy system showed good implementation and application. Nevertheless, deviations for predictive loads as well as the ability of setting variable speed of heat pumps have high influence on the application [3].

As energy meters and temperature sensors monitor the energy system extensively, the operation by the MPC can be analyzed regarding behavior of the heat pumps, seasonal performance factor, PV self-consumption and grid interaction.

MAIN REFERENCES

[1] Nowak, T., 2018. Heat Pumps – Integrating technologies to decarbonize heating and cooling. European Copper Institute

[2] Betzold, C., Bordin S. und Dentel, A., 2021. Control strategies for modulating heat pumps in a plus energy building. Heat Pump Conference 2021. (Paper accepted for publication)

[3] Betzold, C., Dentel, A. and Bordin, S., 2020. Entwicklung und Implementierung von Betriebsführungsstrategien in einem Plusenergiegebäude. 8. Deutsch-Österreichische ibpsa Konferenz BauSIM 2020. (Paper accepted for publication)

Keywords: MPC, modelling approaches, optimization approaches

Paper ID 30984: Embodied facade retrofit design with augmented reality in the built environment

John Sermarini, Robert A. Michlowitz, Joseph J. LaViola Jr., Lori C. Walters, Roger Azevedo, Joseph T. Kider Jr.*

University of Central Florida, United States of America

AIM AND APPROACH

Building Performance Simulation is often conducted during later design stages by complex software, but integrating it into early-phase planning provides opportunities for changes. The architecture, engineering, and construction (AEC) industry is facing a shortage of tools and processes to properly conduct advanced energy retrofits in the built environment. Simultaneously, there is innovative growth of computer-based Building Information Modeling (BIM) tools and research focusing on enhancing communication and collaboration through the usage of 3D models. This research investigates how Augmented Reality (AR) and BIM technologies interactively integrate to display daylighting and energy changes from façade retrofits. When surrounded by the retrofit design in AR the designer is enabled to rapidly iterate and interactively understand alterations in the space. As the AEC industry has become more dependent upon 3D graphics to create or retrofit the built environment, there is a desire to visualize designs on-site and in context. We demonstrate that cognition of the design's energy impact emerges from the augmented embodiment in the space producing better and more innovative designs.

SCIENTIFIC INNOVATION AND RELEVANCE

The increased use of BIM tools is driving the collection of detailed information through a building's life cycle. AR combines the seen real-world with computer-generated simulated information which allows various users to visualize proposed changes on site. Our approach integrates these disruptive technologies into an early-phase design tool for facade retrofitting through to the final design. A system to display façade changes on-site prior to construction that utilizes AR has numerous benefits as alterations are graphically simulated permitting interactive feedback for various design elements. We demonstrate the ability of this prototype to address a range of design and retrofit scenarios to improve the daylighting and energy efficiency of a structure. Maintaining a building's functionality and habitability in mind while maximizing the cost/benefit trade-off is the primary goal of this project.

PRELIMINARY RESULTS AND CONCLUSIONS

Our system currently maintains a building's BIM data in Autodesk Revit allowing model exports to the Unreal Engine to be used in HoloLens 2 for augmented reality. Facade designers typically create their proposed plans in Rhino 3D, 3DS Max, or Revit permitting designs to easily integrate with our AR interactive simulation system. Providing a tool that enables designers to modify facades either directly in an AR environment or with desktop software allows flexibility. As they change the facade in AR, modifications are propagated back to the original BIM model. The designer has the ability to examine a real-time ray-traced output using the global illumination of the design to visualize the impact on the lighting of the space. Additionally, combining this with the data from ClimateStudio can be used to simulate daylighting, glare, and energy impact embodied within the augmented environment by an augmented view. Providing iterative, interactive feedback to designers permits the exploration of alternatives. Our system demonstrates the advantage of connecting AR to BIM for interactive interior and facade design on three spaces and numerous facade designs.

MAIN REFERENCES

Rockcastle, Siobhan, María L. Ámundadóttir, and Marilyne Andersen. "OCUVIS: a web-based visualizer for simulated daylight performance." In Proceedings of the Symposium on Simulation for Architecture and Urban Design, pp. 1-8. 2018.

Jones, Nathaniel L., and Christoph F. Reinhart. "Speedup potential of climate-based daylight modelling on GPUs." In Building Simulation Conference, pp. 1438-1447. 2017.

CimateStudio Software.https://www.solemma.com/ClimateStudio.html (accessed July 20, 2020)

Nasman, Joshua, and Barbara Cutler. "Evaluation of user interaction with daylighting simulation in a tangible user interface." Automation in construction 36 (2013): 117-127.

Turan, Irmak, Andrea Chegut, Daniel Fink, and Christoph Reinhart. "The value of daylight in office spaces." Building and Environment 168 (2020): 106503.

Rahimian, Farzad Pour, Rahinah Ibrahim, Jun Wang, Xiangyu Wang, Wenchi Shou, and Bo Xu. "Integrating BIM and augmented reality for interactive architectural visualisation." Construction Innovation (2014).

Keywords: BIM, daylighting, energy, augmented reality, on-site

Session F4.6 (Online Track): Ensuring high quality building simulations

Paper ID 30272: Tree shading factor as input for thermal model through hemispherical photography

Suyashi Srivastava* (1), Swati Puchalapalli (2), Dr. Marshal Maskarenj (3)

1: Building Energy Performance, CEPT University; 2: Terra Viridis, Hyderabad; 3: Universite Catholique de Louvain, Louvain-la-Neuve, Belgium

AIM AND APPROACH

Solar irradiation has a significant impact on building energy consumption due to solar heat gains through the building envelope. Considering this accurately by including exterior obstructions in initial simulations will help understand building performance better and optimize building design for energy efficiency. Trees are one of the parameters that account for the micro-climate of the place and shading factor is the metric that quantifies the effect of trees on a building. The existing methods to model trees in thermal models comprise of complex geometry and boundary conditions.

The aim of this study is to propose a simplified methodology, recognized by various simulation engines to reduce computational power. Actual measured data through hemispherical photography and simulated results from RHINO honeybee and grasshopper are therefore brought together.

The detailed procedure is divided into two phases where,

The first phase– involves capturing and collecting tree canopy void ratio from the field study, followed by simplifying its geometry for simulation to note the change in irradiation received over the building element for different tree species.

The second phase– focusses on researching how existing simulation engines consider tree shading along with finding the easiest pathway to add collected first phase tree data into simulation software.

SCIENTIFIC INNOVATION AND RELEVANCE

Presently, there is no study that considers native Indian trees as a passive measure in simulation engines. Even though several papers quantify ways by which tree shade can be integrated into dynamic energy modeling, research on individual effects of shading and tree characteristics from building design perspective is rather limited. Another challenge lies in accurately incorporating these complex 3D dynamic entities in simulation software. This requires computationally intensive geometry modeling which is both cumbersome and time-consuming. Together these challenges render the importance to form a simplified method for avoiding tree geometries in the thermal model, without compromising on data accuracy for detailed energy estimation.

'Shading factor' is a possible solution and is defined as a fractional reduction in irradiation received over a building element due to tree shading. It is determined by finding canopy 'void ratio' or 'transmissivity' based on field parameters like canopy area, canopy density, tree species, tree location and orientation with respect to the building element.

To obtain this factor, HDR (high-dynamic-range) tree image is captured with the help of Raspberry pi and fisheye lens. This is converted into workbook data with the help of python code followed by filtering surrounding data captured by the hemispherical lens.

PRELIMINARY RESULTS AND CONCLUSIONS

The reduction in solar irradiation through simulation is noted and calculated to obtain the shading factor of a given tree at a given distance from the building element. This resultant trendline is then added as a schedule based on how one can input site data into a simulation engine.

The study result consists of 2 distinct schedules as input for 'type1' & 'type 2' simulation engines where,

Type 1– requires geometrical input of tree/shading device in the form of coordinates apart from defining a transmissivity value schedule for simulation engines to consider a variable external source like- Energy plus (Design Builder and Open Studio), IESVE and e-QUEST.

Type 2– does not require geometrical input to consider external shading device and allows direct input of 'shading factor schedule'- Tas.

The methodology proposed can be extended to determine the shading factor of various tree species for different seasons throughout the year. Notably, this research data collected is specific to an individual tree in a climate zone and is not representative of their entire species. The generalized shading factor of native trees would aid energy analysts, consultants and modelers to inculcate the shading advantages offered by surrounding vegetation into a simulation model.

MAIN REFERENCES

Villalba, A., Pattini, A., & Correa, E. (2014). An approach to urban tree daylight permeability simulation using models based on louvres.

Balakrishnan, P. (2019). Measuring and Modelling Equatorial Light.

Al-Sallal, K. A., & Al-Rais, L. (2012). A novel method to model trees for building daylighting simulation using hemispherical photography.

Hashemi, F., Marmur, B., Passe, U., & Thompson, J. (2018). Developing a workflow to integrate tree inventory data into urban energy models.

Cascone, Y., Corrado, V., & Serra, V. (2011). Calculation procedure of the shading factor under complex boundary conditions.

Sattler, M. A., Sharples, S., & Page, J. K. (1987). The geometry of the shading of buildings by various tree shapes.

Hsieh, C. M., Li, J. J., Zhang, L., & Schwegler, B. (2018). Effects of tree shading and transpiration on building cooling energy use.

Pandit, R., & Laband, D. N. (2010). Energy savings from tree shade.

Bhalla, A., & Manu, S. (2014). Impact of native evergreen trees on the visual comfort in an office space in Ahmedabad.

McPherson, E. G., Xiao, Q., van Doorn, N. S., Johnson, N., Albers, S., & Peper, P. J. (2018). Shade factors for 149 taxa of in-leaf urban trees in the USA.

Keywords: Transmissivity, Tree shading factor, Indian tree species, Hemispherical photography

Paper ID 30289: Development of the City of London thermal comfort guidelines

Ryan Danks, Ruth Shilston*, Ozkan Ender, Jayyaratnam Krishan

RWDI, United Kingdom

AIM AND APPROACH

In August 2019, the City of London published Wind Microclimate guidelines to ensure that new developments proposed within the City do not create adverse wind impacts on surrounding public spaces. However, The City recognized that wind is only one aspect of urban comfort and wanted to build on these existing guidelines to ensure suitable thermal comfort as well. The authors were engaged by the City of London to develop a set of Thermal Comfort Guidelines which sets out a simulation methodology and acceptability criteria to ensure a robust and consistent assessment of the impact of new developments on thermal comfort within the urban realm.

Building on previously conducted simulations of the City's wind environment, the authors undertook detailed solar simulations which were combined with climate data to provide an assessment of thermal comfort in all existing public spaces within the City. This data allowed for the development a metric and criteria to be used to assess the thermal comfort impact of a proposed development on public spaces and associated reporting to support a planning submission.

SCIENTIFIC INNOVATION AND RELEVANCE

The requirement to undertake a thermal comfort assessment as a part of the planning process is completely new for the UK. Therefore, care needed to be taken when defining the methodology to ensure it is robust and consistent but also appropriate in scope for the planning permission stage, when many building parameters are not fully defined.

Through this lens, the authors conducted a review of existing thermal comfort metrics and their acceptability criteria, with a focus on their suitability for London's temperate climate, the level of detail required in their inputs and the ease at which they can be communicated to those not well versed in thermal comfort studies. Various sources of climatic data were also analysed to ensure that the key parameters for thermal comfort are statistically well-represented and acknowledged the impact of London's urban heat island. Multiple urban scale simulations of thermal comfort across the City were conducted to assess the influence of these factors on thermal comfort predictions and used to define acceptability criteria for the various types of public spaces found in the City which form the basis of the assessment for a proposed development.

PRELIMINARY RESULTS AND CONCLUSIONS

A simulation methodology based on the Universal Thermal Climate Index (UTCI) was developed for use within the guidelines. Spaces are categorized based on the frequency they are within an 'acceptable' range of UTCI values based on five recent years of historic climate data computed through mesoscale modelling localized to the City. A slightly wider range of UTCI values than the standard 'no thermal stress' range was found to better represent conditions within the City when using this data. The use of categories based on comfort frequency was found to be the most intuitive reporting mechanism for planners. This was due to its similarity to the current method for assessing wind, and because it avoids the use of equivalent temperatures, which was found to be challenging for lay-people to interpret correctly.

A document defining simulation procedures, criteria and reporting requirements was developed and in June 2020 provided to other UK-based consultants for review. A final draft was issued to the City in July 2020, and is expected to be published and in force by the end of 2020.

This presentation will provide details of the simulations undertaken to develop the guidelines and the evolution of the methodology and future work.

MAIN REFERENCES

City of London, Wind Microclimate Guidelines for Developments in the City of London, August 2019, https://www.cityoflondon.gov.uk/assets/Services-Environment/wind-microclimate-guidelines.pdf

Jayyaratnam, Krishan., Shilston, Ruth., Hackett, Daniel. "The use of CFD and wind tunnel testing in wind microclimate assessments." SIMAUD '17: Proceedings of the Symposium on Simulation for Architecture and Urban Design May 2017 Article No.: 35 Pages 1–8

Shilston, Ruth., Ozkan, Ender., Hackett, Daniel. "Development of the City of London Wind Assessment Guidelines." The 7th International Symposium on Computational Wind Engineering 2018. June 18-22, 2018 in Seoul, Republic of Korea

Jayyaratnam, Krishan., Shilston, Ruth., Hackett, Daniel. “New Wind Microclimate Assessment Methods, and the Implications for Urban Planners.” 7th European-African Conference on Wind Engineering EACWE 2017

Reinhart, Christoph F., Jay Dhariwal, and Katy Gero. "Biometeorological indices explain outside dwelling patterns based on Wi-Fi data in support of sustainable urban planning." Building and Environment 126 (2017): 422-430.

Keywords: Thermal Comfort, guidelines, computational fluid dynamics, UTCI

Paper ID 30361: Albedo impact on the microclimate of courtyards, potential implications on the UHI. A case study in Seville, Spain.

Sebastian Alzate-Gaviria, Victoria Patricia Lopez-Cabeza*, Eduardo Diz-Mellado, Carlos Rivera-Gomez, Carmen Galan-Marin

Universidad de Sevilla, Spain

AIM AND APPROACH

The Urban Heat Island (UHI) effect is being worsened by the rise in temperatures caused by climate change. One of the strategies that is being analyzed in many cities is the use of the appropriate albedo factor (the fraction of solar radiation that is reflected by a surface) mainly in the horizontal surfaces of the cities, roofs and pavements. In this research, this strategy is extended to smaller outdoor spaces such as courtyards. The courtyard is a passive strategy used in traditional architecture in many regions of the world to enhance indoor and outdoor comfort. The courtyard generates a microclimate that tempers the outdoor temperature thus it is not as hot or cold as the outdoor environment. However, comfort not only depends on air temperature, but also on other factors, such as the mean radiant temperature. As many studies conclude, this is the variable responsible for the highest impact on the human body and surface albedo is directly related to this. In this sense, the study analyses the effects of albedo in both, temperature and comfort in a courtyard in Seville, south of Spain, by means of monitoring and simulations.

SCIENTIFIC INNOVATION AND RELEVANCE

This research starts with the validation and calibration of ENVI-met v4 software, used for outdoor space simulation and using monitoring data. The monitoring campaign takes place in the inner courtyard of the UNED building in Seville in 2020 summer. Air temperature, surface temperature, relative humidity and estimation of different albedos are recorded using data loggers, a weather station, a pyranometer, a thermal camera and a surface temperature thermometer. Once the software is validated, it will be used to simulate different scenarios of surface albedo in the courtyard that cannot be reproduced through experiments, and to analyze human comfort. The study aims to determine the optimal albedo distribution for the pavement and wall surfaces of the selected courtyard and provide general guidelines for designers of this kind of space, considering the lack of studies on the effects of albedo on the thermal behavior of outdoor spaces, especially at a local scale.

PRELIMINARY RESULTS AND CONCLUSIONS

Previous monitoring campaigns developed in this case study (2017 – 2019) have shown that the courtyard's microclimate provides a tempering effect of the air temperature that reaches 8 ºC. ENVI-met simulated the trend of the thermal performance of the courtyard in a similar way but it was approximately 2 ºC above the monitored data. Although software accuracy predicting the temperature in the courtyard could be higher, it is a good approximation to estimate the thermal performance. Early simulations results show that scenarios with high albedo reduced the surface temperatures, however, it significantly increased mean radiant temperatures and comfort indexes (PMV - PET) at 12:00 and 16:00. Thus, this phenomenon caused discomfort due to the multiple reflections irradiated from surfaces to users at the center of the courtyard. Changing the albedo of wall surfaces instead of pavement seems to be more significant, specifically for the analyzed courtyard conditions.

MAIN REFERENCES

A. Ghaffarianhoseini, U. Berardi, y A. Ghaffarianhoseini, «Thermal performance characteristics of unshaded courtyards in hot and humid climates», Build. Environ., vol. 87, pp. 154-168, may 2015.

E. Erell, D. Pearlmutter, D. Boneh, y P. B. Kutiel, «Effect of high-albedo materials on pedestrian heat stress in urban street canyons», Urban Clim., vol. 10, n.o P2, pp. 367-386, dic. 2014.

F. Rosso et al., «On the impact of innovative materials on outdoor thermal comfort of pedestrians in historical urban canyons», Renew. Energy, vol. 118, pp. 825-839, abr. 2018.

M. Taleghani, «The impact of increasing urban surface albedo on outdoor summer thermal comfort within a university campus», Urban Clim., vol. 24, pp. 175-184, jun. 2018.

M. Taleghani, D. J. Sailor, M. Tenpierik, y A. van den Dobbelsteen, «Thermal assessment of heat mitigation strategies: The case of Portland State University, Oregon, USA», Build. Environ., vol. 73, pp. 138-150, mar. 2014.

Keywords: Heat mitigation, Outdoor thermal comfort, Courtyard, Albedo, ENVI-met

Paper ID 30546: Sensitivity analysis of the UK Standard Assessment Procedure for dwellings

Benjamin Halls*, Steven Firth, Kevin Lomas

Loughborough University, United Kingdom

AIM AND APPROACH

The aim of this study is to identify the sensitivity of the inputs within the Standard Assessment Procedure (SAP), the UK energy model for the building regulation compliance of dwellings. SAP is a steady-state energy model with hundreds of model inputs representing all facets of dwelling design in the UK. This study will: i) Develop an open and extensible SAP model in the Python programming language; ii) Construct several building models including archetypes of varying built type and energy efficiency levels; iii) Apply several sensitivity analysis techniques on the model inputs including One at A Time (OAT), Morris method and Monte Carlo Analysis; and iv) Analyse the results of the sensitivity analysis to review the various techniques and draw conclusions on the sensitivity of the SAP model to its inputs.

SCIENTIFIC INNOVATION AND RELEVANCE

Building simulation tools can be used in the design stage to predict the energy performance of a building. These tools contain a large number of input variables which incorporate information about the building form (construction materials, dimensions etc.), the behaviour of occupants (presence, use of services etc.) and the external conditions (weather, wind speed, solar gains etc.). Understanding the influence each of these input variables on the simulation output is crucial in identifying the most significant factors which drive the energy performance of a building and where improvements can have a substantial impact on improving energy efficiency.

PRELIMINARY RESULTS AND CONCLUSIONS

The work will conclude with a discussion of the potential impacts of uncertainties in the model inputs on the overall SAP results and subsequent design decisions for new-build UK dwellings.

MAIN REFERENCES

Kristensen, M. and Petersen, S., 2016. Choosing the appropriate sensitivity analysis method for building energy model-based investigations. Energy and Buildings, 130, pp.166-176.

Firth, S., Lomas, K. and Wright, A., 2010. Targeting household energy-efficiency measures using sensitivity analysis. Building Research & Information, 38(1), pp.25-41.

Hughes, M., Palmer, J., Cheng, V. and Shipworth, D., 2014. Global sensitivity analysis of England's housing energy model. Journal of Building Performance Simulation, 8(5), pp.283-294.

Keywords: Global sensitivity analysis, Local sensitivity analysis, Uncertainty, Building simulation

Paper ID 30621: Calibration of a multi-residential building energy model - Part I: Cluster-based sensitivity analysis

Charlotte Marguerite* (1), Van Long Lê (1), Charlotte Beauthier (1), Olivier Fontaine de Ghélin (1), Cécile Goffaux (1), Loïc De Moffarts (2)

1: Cenaero, Belgium; 2: Thomas&Piron, Belgium

AIM AND APPROACH

Designing and operating energy efficient buildings is a key solution to meet the European targets regarding the reduction of building impacts on GHG emissions. In the framework of the TPEE project, co-financed by

Thomas&Piron Group and the Wal-e-cities FEDER project, an energy toolbox is developed for multi-residential buildings which are both simulated with OpenStudio/EnergyPlus (opensource energy simulation engine) and IoT equipped, in order to forecast their energy needs and detect and diagnose faults in their energy systems.

To do so, the building energy model must be calibrated accordingly (the calibration work is presented in a separate paper: Part II - Calibration using surrogate-based optimisations). As the building energy model is based on more than 1200 parameters, a preliminary step consists in applying a sensitivity analysis, to identify the most influent parameters on which the calibration should focus.

The present paper describes the developed cluster-based sensitivity analysis method. First, the apartments are grouped into clusters defined by a combination between a Self Organising Maps clustering and an artificial intelligence method (Affinity Propagation method from Scikit-Learn machine learning package). Then two sensitivity analysis methods, ANOVA (Sobol based) and Morris methods, are carried out with different simulated periods and scales.

SCIENTIFIC INNOVATION AND RELEVANCE

The developed sensitivity analysis methodology constitutes a novel approach considering that publications on sensitivity analysis applied to multi-residential buildings can barely be found and none of them propose an analysis based on clustering methods and various levels (multi-period and multi-scale analyses). Indeed, several examples exist for tertiary or industrial buildings, but the case of residential buildings presents different challenges such as the many apartments involved, each one having its physical characteristics, its specific equipment and energy use habits or the assumptions regarding consumer behaviors that have to be made in the modelling phase because of data privacy issues.

PRELIMINARY RESULTS AND CONCLUSIONS

The developed methodology is applied on a multi-residential building of 40 apartments and a triplex. In the clustering step, a combination of clustering methods, applied with 4 architectural characteristics, identifies 8 clusters of apartments. The interest of clustering is double: 1) for time computing purpose (to avoid studying all the 40 apartments); 2) for this building, it is not trivial to scale up the results, because most apartments have their own physical characteristics (building architecture, height, orientation, etc.) that influence the internal thermal gains.

In the sensitivity analysis step, for individual apartments, results show that the infiltration rate, the temperature setpoint and the nominal power of electric appliances (cooking range, lights, plugs) have the most impact on the heating and electricity consumption. Further analyses show that the adjacent apartments have no meaningful impact on the energy needs of a specific apartment. At building scale, the boiler characteristics (efficiency, mass flow rate, maximum setpoint outlet temperature) the circulating pump efficiency (distribution side), the radiators inlet temperature (from the apartments side) and the parameters characterizing the physical elements (insulation and thermal parameters of the walls and windows) have the highest influence on the gas and electricity consumption in the building energy system.

MAIN REFERENCES

Enrico Fabrizio and Valentina Monetti, Methodologies and Advancements in the Calibration of Building Energy Models, Energies 8(4):2548-2574, 2015.

Kaiyu Sun et al., A Pattern-based automated approach to building energy model calibration. Applied Energy, 165:214-224, 2016.

Valentina Monetti et al., Optimization-based calibration for dynamic building simulation models: a case study. Building Simulation Conference 2015 Hyderabade, India.

David Goldwasser et al., Advances in calibration of building energy models to time series data. ASHRAE Building Performance Analysis Conference and SimBuild, Chicago, Illinois 2018.

Zhihong Pang, Zheng O'Neill, A comparison study of various sensitivity analysis methods in building applications. Building Simulation Conference 2019, Rome, Italy.

Yanjun Gan et al., A comprehensive evaluation of various sensitivity analysis methods: A case study with a hydrological model. Environmental Modelling and Software, 51: 269-285, 2014.

Juha Vesanto and Esa Alhoniemi, Clustering of the Self-Organizing Map. Ieee transactions on neural networks, 11(3), 2000.

Mad D. Morris: Factorial sampling plans for preliminary computational experiments. Technometrics, 33:161–174, 1991.

Andrea Saltelli et al., Sensitivity analysis. Wiley Series in Probability and Statistics. Wiley, 2000.

Julien Jacques, Université de Lille, Pratique de l'analyse de sensibilité : comment évaluer l'impact des entrées aléatoires sur la sortie d'un modèle mathématique, 2011.

Keywords: multi-residential building, clustering methods, sensitivity analysis, calibration

Paper ID 30122: Numerical study on the performance of the temperature-controlled airflow ventilation system during the movement of a patient from the operating room to the MRI room

Parastoo Sadeghian* (1), Henrik Skredsvik (2), Marco Baroni (2), Cong Wang (1), Sasan Sadrizadeh (1)

1: KTH Royal Institute of Technology, Stockholm, Sweden; 2: Avidicare, Lund, Sweden

AIM AND APPROACH

Ventilation systems have an important role in controlling infection risks during surgery in the operating room. Ventilation systems supposed to remove bacteria carrying particles from the clean zone that are released from the surgical staff members. Human movement is among the important factors that may break the stability of airflow field and reduce ventilation performance.

This study numerically investigated the performance of the temperature-controlled airflow (TcAF) ventilation systems while moving the patient from the operating room to the MRI room during a given brain surgery. Moreover, effect of the internal constellation of staff members on airborne particle distribution was also studied.

SCIENTIFIC INNOVATION AND RELEVANCE

Using temperature-controlled airflow (TcAF) ventilation system for controlling infection risks during the imagining in the MRI room and movement of the patient is a new approach.

PRELIMINARY RESULTS AND CONCLUSIONS

In this regards, the computational fluid dynamics method was used for simulating the airflow field and bacteria carrying particle distribution in the operating and MRI room. The simulated results showed that the TcAF ventilation system can reduce the infection risks to an acceptable level in both rooms. Moreover, the location of the surgical personnel had considerable effect on limiting the concentration of bacteria carrying particles in the MRI device, where the patient located.

MAIN REFERENCES

[1] M. Alsved, A. Civilis, P. Ekolind, A. Tammelin, A.E. Andersson, J. Jakobsson, T. Svensson, M. Ramstorp, S. Sadrizadeh, P.A. Larsson, M. Bohgard, T. Šantl-Temkiv, J. Löndahl, Temperature-controlled airflow ventilation in operating rooms compared with laminar airflow and turbulent mixed airflow, J. Hosp. Infect. 98 (2017).

[2] C. Wang, S. Holmberg, S. Sadrizadeh, Numerical study of temperature-controlled airflow in comparison with turbulent mixing and laminar airflow for operating room ventilation, Build. Environ.

[3] F. Memarzadeh, A.P. Manning, Comparison of operating room ventilation systems in the protection of the surgical site, ASHRAE Trans. 108 PART 2 (2002) 3–15.g room ventilation, Build. Environ. 144 (2018) 45–56.

Keywords: Temperature-controlled airflow ventilatin, Infection risks, Airflow field

Paper ID 30310: Interdisciplinary parametric modelling and modularization to improve air quality, acoustics and lighting in school buildings

Clara-Larissa Lorenz*, Tobias Burgholz, Maximilian Schildt, Jérôme Frisch, Christoph van Treeck

RWTH Aachen, Germany

AIM AND APPROACH

Poor indoor air quality and acoustics can affect well-being and productivity in classrooms[1][2]. . In naturally ventilated buildings, weather conditions and outside noise may prevent necessary window openings, necessitating technological solutions. The implementation of solutions to mechanically ventilate spaces may however be met with resistance, either due to the perceived effort and cost of renovation measures, or disruptions to school activities. The aim of this study is therefore to improve air quality and acoustics in classrooms and furthermore facilitate interventions by reducing complexity in planning and installation via integrated parametrization and modularization.

The study is done in three stages. First, a parametric model is developed to design any variant of a typical classroom configuration. Second, parametric rules are implemented to automate the generation of solutions for the layout and distribution of a ventilation unit, acoustic panels and lighting. These three cross-disciplinary components are then formulated as a new product unit and modularized to facilitate assembly and installation. Third, the model is implemented to assist with renovation measures for a classroom in three different schools. Parametric rules are adjusted and effectiveness in facilitating planning and installation are evaluated.

SCIENTIFIC INNOVATION AND RELEVANCE

Environmental qualities including air-quality, acoustic and visual aspects are known to improve occupants' attention and performance[2]. Mechanical ventilation can help ensure appropriate levels of CO2, which can fall below appropriate levels in manually aired classrooms[3]. Acoustic measures can improve speech communication and reduce noise from ventilation units and good light conditions can improve attention[4], highlighting the need to consider all three aspects when designing for occupant well-being in classrooms. This work provides a tool that considers the interdependencies of all three aspects in practical implementation.

The implementation divides design parameters into design rules and module parameters as a process of modularization[5]. This work is made unique by establishing building component related to ventilation, acoustic and lighting as a new product unit. By doing so, competition for space (e.g. ceiling space) is reduced, which keeps interactions outside modules at a minimum, as recommended by[6]. Another contribution can be found in the integration of parametric modeling and modularization, which allows for the design non-repetitive modules. This reduces restrictions to design in advance.

PRELIMINARY RESULTS AND CONCLUSIONS

Results are presented on i) the process of developing a highly customizable parametric model to replicate any given classroom configuration, ii) the dependencies and constraints implemented to generate an integrated design solution for ventilation, acoustic and lighting, iii) the modularization of product components to an assembly unit to speed-up preassembly and montage, and iv) model performance and limitations when applied to three case studies.

The parametric model was developed in Grasshopper with 20 required, and 12 optional input variables to reconstruct typical classroom configurations through simple slider adjustment, therefore ensuring ease-of-use for planners.

The rules imposed to generate solutions are based on availability of spaces, structural constraints, cost effectiveness (e.g. installing supply and exhaust air outlets over the window head, into the wall or in place of a top light), and compliance to regulations requirements DIN18041 for acoustics. The modularization of all components, including lighting, ensures that clashes between components and with building structure are avoided, thereby reducing planning efforts through interface reduction. Through modularization, renovations are expected to take one day, thus hardly interfering with school operation. Modularization is expected to reduce onsite resources and improve project quality.

MAIN REFERENCES

[1] J. Toftum, B. U. Kjeldsen, P. Wargocki, H. R. Menå, E. M. N. Hansen, and G. Clausen, "Association between classroom ventilation mode and learning outcome in Danish schools," Build. Environ., vol. 92, pp. 494–503, 2015.

[2] S. P. Corgnati, M. Filippi, and S. Viazzo, "Perception of the thermal environment in high school and university classrooms: Subjective preferences and thermal comfort," Build. Environ., vol. 42, no. 2, pp. 951–959, 2007.

[3] L. Stabile, M. Dell'Isola, A. Frattolillo, A. Massimo, and A. Russi, "Effect of natural ventilation and manual airing on indoor air quality in naturally ventilated Italian classrooms," Build. Environ., vol. 98, pp. 180–189, 2016.

[4] A. Astolfi and F. Pellerey, "Subjective and objective assessment of acoustical and overall environmental quality in secondary school classrooms," J. Acoust. Soc. Am., vol. 123, no. 1, pp. 163–173, 2008.

[5] S. Isaac, T. Bock, and Y. Stoliar, "A methodology for the optimal modularization of building design," Autom. Constr., vol. 65, pp. 116–124, 2016.

[6] A. Mohamad, G. Hickethier, V. Hovestadt, and F. Gehbauer, "Use of modularization in design as a strategy to reduce component variety in one-off projects," 21st Annu. Conf. Int. Gr. Lean Constr. 2013, IGLC 2013, pp. 285–294, 2013.

Keywords: Parametric modeling, modularisation, indoor air quality, acoustic performance, environmental design for school buildings

Paper ID 30866: A new method for comparing long-term thermal conditions in buildings

Shide Salimi* (1), Holly Samuelson (1), Esteban Estrella Guillén (2)

1: Harvard University, United States of America; 2: Universidad de Monterre, Mexico

AIM AND APPROACH

This study aims at developing a new discomfort index based on equivalent temperature sensation, such as Standard Effective Temperature (SET), which accounts for all six main personal and environmental factors affecting thermal comfort and measures the severity of discomfort experienced. Expressing discomfort in building-related terms (i.e., temperature) helps building engineers to easily relate thermal comfort to building performance while using it in building simulation and/or optimization analyses.

The proposed method can be applied to both air-conditioned and naturally-ventilated buildings. For the former, a comfort zone is plotted on a psychrometric chart per ASHRAE-55 to determine reference thermal zones. The proposed method then includes nuances of how variations in thermal comfort affecting factors, such as metabolic rate impact reference zones. For an uncomfortable point in question, the closest point on the comfort zone boundary to the point is determined. The difference between the SET of a point and that of its boundary line projection determines the severity of discomfort. When applied to naturally-ventilated buildings, the proposed method calculates the discomfort severity by determining the difference between the SET of the point in question and that of the corresponding upper or lower limit based on the Adaptive Comfort model.

SCIENTIFIC INNOVATION AND RELEVANCE

In current practice, thermal comfort is determined by expressing environmental conditions in terms of a given index and comparing the results to thermal comfort ranges suggested in predefined standards or models. Most of the proposed indices define the duration of discomfort by accumulating periods in which indoor conditions exceed comfort conditions. In these indices, the severity, if presented, is accounted for by using weighting factors. Yet, most of such indices rely on PMV and PPD and are unintuitive when used for building performance analysis (especially for optimization purposes), therefore, expressing discomfort in building-related terms is of great importance when using a discomfort index.

Existing temperature-based weighting factors, which can be better related to building performance, only account for differences between operative and the limiting acceptable temperatures, ignoring the effects of other personal and environmental factors, such as humidity. Furthermore, in current practice, Fanger- and adaptive model-based indices are separately utilized in mixed-mode buildings to report discomfort in air-conditioned and naturally ventilated spaces, respectively. Here, SET is used to express the severity of thermal discomfort. This index could help building engineers to easily relate thermal comfort to building performance, and the same index can be used regardless of conditioning mode.

PRELIMINARY RESULTS AND CONCLUSIONS

Preliminary comparison between existing comfort indices and the Exceedance Degree-Hours index indicates the proposed index captures more information with regards to changes in factors affecting thermal comfort. The proposed method allows the differentiation between different points with the same PPD or opposite PMV, an important aspect to consider in simulation-based building optimization. For evaluations based on the Adaptive Comfort model, environmental and personal characteristics can be taken into account when measuring discomfort severity. Furthermore, in mixed-mode buildings, the overall discomfort severity for air-conditioned and naturally-ventilated periods can be summated in a single value, expressed in terms of building-related parameters (i.e., SET differences). This applicability can be extrapolated to simulation and/or optimization analyses, which would in return have a high positive impact on the building industry.

MAIN REFERENCES

Kim, J., Tartarini, F., Parkinson, T., Cooper, P., & De Dear, R. (2019). Thermal comfort in a mixed-mode building: Are occupants more adaptive? Energy and Buildings, 203, 109436.

CEN, E. (2019). Energy performance of buildings - Ventilation for buildings - Part 1: Indoor environmental input parameters for design and assessment of energy performance of buildings addressing indoor air quality, thermal environment, lighting and acoustics. Brussels: CEN.

ASHRAE. (2017). ASHRAE Standard 55-2017: Thermal Environmental Conditions for Human Occupancy. ASHRAE.

Capozzoli, A., Piscitelli, M. S., Gorrino, A., Ballarini, I., & Corrado, V. (2017). Data analytics for occupancy pattern learning to reduce the energy consumption of HVAC systems in office buildings. Sustainable cities and society, 35, 191-208.

Hoyt, T., Schiavon, S., Piccioli, A., Cheung, T., Moon, D., & Steinfeld, K. (2017). CBE Thermal Comfort Tool. Center for the Built Environment: http://comfort.cbe.berkeley.edu

CIBSE. (2013). The limits of thermal comfort: avoiding overheating in European buildings. London: CIBSE.

Carlucci, S., & Pagliano, L. (2012). A review of indices for the long-term evaluation of the general thermal comfort conditions in buildings. Energy and Buildings, 53, 194-205.

Borgeson, S., & Brager, G. (2011). Comfort standards and variations in exceedance for mixed-mode buildings. Building Research & Information, 39(2), 118-133.

Keywords: Thermal comfort, Adaptive Comfort, Equivalent temperature sensation, Building simulation, Optimization

Paper ID 30729: Application of coupling of the human and clothing thermal system and computational fluids dynamics in the evaluation of energy and comfort in occupied spaces

Eusebio Conceição* (1), Mª Inês Conceição (2), Mª Manuela Lúcio (1), Hazim Awbi (3)

1: University of Algarve, Portugal; 2: Instituto Superior Técnico, University of Lisbon, Portugal; 3: School of Built Environment, University of Reading, United Kingdom

AIM AND APPROACH

This numerical work develops a HVAC system, based in a ceiling mounted localized air distribution systems. The study is made in winter and summer typical day conditions.

This work presents and applies a numerical model, developed by the authors in the last years, that considers the coupling of the CFD (Computational Fluids Dynamics) and HCTS (Human and Clothing Thermal System). The coupling system, itself, generates the occupation presence and transfers the inputs/output between the CFD and HCTS numerical models.

The input of the compartment, using the Computational Aid Design (CAD), the location of the occupants and the external environmental variables are introduced in the software, while the occupants' geometry is generated by empirical equations, based in height and weight.

The study is made in a virtual chamber occupied by twelve virtual manikins, six desks and twelve seats and equipped with a new ceiling mounted localized air distribution system. The inlet airflow and the exit is located above the head level.

In the present study the thermal comfort level, the air quality level and the ADI (Air Distribution Index) are calculated, in winter conditions. In accordance with the obtained results the ADI index is highest for highest inlet air velocity.

SCIENTIFIC INNOVATION AND RELEVANCE

This study considers a coupling of the Computational Fluid Dynamics, CFD, and the Human and Clothing Thermal System, HCTS numerical models.

The ADI is used to evaluate simultaneously the occupants' thermal comfort, occupant s' air quality and ventilation effectiveness for heat removal and effectiveness for contaminant removal.

This HVAC system considers an inlet (located above the head level) and an exhaust (located above the desk level).

The twelve occupants are seated in six desks. The desks are placed in two rows and three columns.

This new HVAC system promotes good thermal comfort levels, because promotes air recirculation and renovation around the occupants.

This new HVAC system promotes good indoor air quality levels, because the inlet air goes straight to the face and, afterwards breathing, goes straight to the exit.

PRELIMINARY RESULTS AND CONCLUSIONS

The thermal comfort, the indoor air quality, the Draught Risk, the effectiveness for heat removal, the effectiveness for contaminant removal and the ADI, are evaluated.

In accordance with the obtained results the airflow around the human body increases and the temperature around the occupant decreases when the airflow rate increases.

Draught Risk around the occupants' increases when the airflow increases. In summer conditions is acceptable and in winter conditions is uncomfortable, in accordance with the international standards.

The thermal comfort level increases during summer and decreases during winter when the airflow rate increases. However, the thermal comfort levels are acceptable in accordance with the international standards.

Finally, the ADI increases when the airflow rate increases.

MAIN REFERENCES

E. Conceição, M. Lúcio. Numerical simulation of the application of solar radiant systems, internal airflow and occupants' presence in the improvement of comfort in winter conditions. Buildings 38 (2016) 1-20.

E. Conceição, M. Lúcio, S. Rosa, A. Custódio, R. Andrade, M. Meira. Evaluation of comfort level in desks equipped with two personalized ventilation systems in slightly warm environments. Building and Environment 45 (2010) 601-609.

E. Z. E. Conceição and M. M. J. R. Lúcio, "Evaluation of thermal comfort conditions in a classroom equipped with radiant cooling systems and subjected to uniform convective environment," Appl. Math. Model., vol. 35, no. 3, 2011.

E. Z. E. Conceição and M. M. J. R. Lúcio, "Evaluation of thermal comfort conditions in a localized radiant system placed in front and behind two students seated nearby warmed curtains," Build. Environ., vol. 45, no. 10, 2010.

E. Z. E. Conceição, V. D. S. R. Vicente, and M. M. J. R. Lúcio, "Airflow inside school building office compartments with moderate environments," HVAC R Res., vol. 14, no. 2, 2008.

Keywords: Human Thermal Response, CFD, ADI, Heat Exchange between the Human and the Environment.

Paper ID 30980: Pre-retrofit monitoring of indoor air quality and energy use in a low-rise block of flats in the UK

Rajat Gupta*, Alastair Howard

Low Carbon Building Research Group, Oxford Brookes University, United Kingdom

AIM AND APPROACH

The UK government has set a target of net-zero greenhouse gas emissions by 2050. To meet the national carbon reduction target, deep retrofit of the existing 28 million dwellings, around half of which was built before 1960, will be necessary (1). Since the UK housing stock varies in terms of age, construction materials, dwelling form and to avoid unintended consequences, a bespoke approach to retrofit is necessary, that considers the building fabric thermal performance, energy use and indoor environment of dwellings targeted for retrofit (2).

This paper presents an empirical investigation and statistical analysis of the energy use and indoor environment (temperature, relative humidity (RH), CO2 and TVOC levels) in three flats located in a low-rise 1950s block of flats (UK) prior to undertaking deep energy retrofit. Occupant survey was used to gain deeper understanding about household characteristics such as occupancy levels and patterns, and heating habits. The results of the analysis were used to inform the retrofit strategies adopted and calibrate PHPP model predictions.

SCIENTIFIC INNOVATION AND RELEVANCE

Research into performance of retrofit projects has identified a number of trends that reinforce the need to examine buildings prior to retrofit. In a pre- and post-retrofit evaluation of fifteen homes in Ireland, improved air tightness helped to improve occupant comfort and building temperature, but post-retrofit reductions in air exchange rates were linked to increased concentrations of CO2, TVOC and PM2.5, highlighting the importance of considering indoor air quality as well as energy efficiency in the scope of retrofit projects (3).

Occupant behaviour has been found to play a significant role in the effectiveness of retrofit projects (4), given that number of occupants and occupancy patterns influence dwelling energy use (5). For energy retrofits to meet their expectations, it is vital that examination of physical characteristics of dwellings is augmented by consideration of occupancy related factors.

This study adopts a systematic socio-technical approach that deepens the understanding of the physical and social characteristics that affect the energy use and indoor environment of a block of flats to ensure that building fabric upgrade and ventilation strategies selected enhance IAQ as much energy reduction.

PRELIMINARY RESULTS AND CONCLUSIONS

For the pre-retrofit investigation, sensors were installed to record at five-minute resolution, indoor and outdoor temperature and relative humidity (Hobo UX100's), and indoor CO2 concentrations (Tinytag TGE-0011) in living rooms and main bedrooms of the three flats from January to August 2019. TVOC levels were measured using Graywolf DS11-5 device, at one-minute resolution for one hour in the living rooms of each flat on one day in February 2019. Occupant surveys gathered insight into household characteristics and occupancy patterns that influence indoor environment and energy use.

Gas consumption was found to be higher than the national median in two out of the three flats. Cross-relating energy performance with indoor environmental data and occupant surveys revealed that heating costs were the main reason for the relatively low gas use and consequent cold, damp and mouldy conditions, particularly during the heating season in the third flat. TVOCs were found to be higher in two flats with smokers. Measured TVOC levels when cross-related with concurrent temperature, RH and CO2 concentration revealed statistically significant and strong correlation. The pre-retrofit monitoring results also enabled calibrating PHPP model predictions using real occupancy data and demand temperature.

MAIN REFERENCES

1. Skidmore, C. (2019). "UK becomes first major economy to pass net zero emissions law." Retrieved 6th December, 2019

2. Baeli, M. (2019). Residential retrofit: twenty case studies, Routledge.

3. Broderick, Á., M. Byrne, S. Armstrong, J. Sheahan and A. M. Coggins (2017). "A pre and post evaluation of indoor air quality, ventilation, and thermal comfort in retrofitted co-operative social housing." Building and Environment 122: 126-133.

4. Ben, H. and K. Steemers (2017). "Tailoring domestic retrofit by incorporating occupant behaviour." Energy Procedia 122: 427-432.

5. Gupta, R., A. Howard and A. Kotopouleas "Meta-study of the energy performance gap in UK low energy housing."Proceedings of ECEEE 2019 summer study, France.

Keywords: indoor air quality, energy, housing, retrofit

Absent authors

List of abstracts
(in alphabetic order by absent author)

Paper ID 30899: Uncertainty analysis of a hardware in loop setup for testing building technology

Prasaant Balasundaram* (1,3), Benoit Delinchant (2), Stephane Ploix (3), Cristian Muresan (1)

1: ENGIE, France; 2: Grenoble INP - G2E Lab; 3: Grenoble INP - GSCOP

AIM AND APPROACH

Hardware in the loop simulations (HIL) offers to validate and evaluate the building technology. The effect of uncertainty due to the components present in a HIL test bench should be taken into account since it affects the seasonal/annual performance that is being estimated by extrapolation of measurements done on short term testing. In this paper probability boxes (p-boxes) are used to model the uncertainties related to a test bench since generally speaking one cannot specify input parameter distributions , dependencies between inputs. The uncertainty propagation is done based on Monte Carlo Simulation. The model of an entire HIL test setup is developed in Modelica and the propagation of the uncertainties is done using Modelica as well. The modelled HIL setup is compared with the real experimental setup for discrepancies and calibrated accordingly. The simulation is done for the uncertainty bounds on the performance index with a 95% confidence interval. In the future the experimental uncertainties will be taken into account regarding the HIL setup.

SCIENTIFIC INNOVATION AND RELEVANCE

The main approach in this methodology is that the entire HIL setup is modelled in Modelica. A detailed model of the Hardware in Loop setup is made. It is based on knowledge: it is quite uncommon and innovative. Secondly , for the sake of correctness, the idea of defining parameters representing uncertainties is important. Hence, a second order probability has been chosen to model uncertainties since initially one cannot know what is the type of distribution without internal detailed data. This engineering approach is relevant to the current situation since a digital twin of the HIL could help to understand everything related to the testing process of products and systems.

PRELIMINARY RESULTS AND CONCLUSIONS

The preliminary result includes the testing of the reduced simulation cycle simulation purely in a virtual platform where we were successfully able to obtain the performance index with less than one percent error. The simulation consisted of a virtual test house developed by ENGIE for field tests which is a single zone building, where the seasonal efficiency of a boiler was tested only for the space heating in the initial context. The annual performance of the boiler was around 85.9% while the extrapolated annual performance was 86.5%. The k-means algorithm was used for the selection of typical days i.e. one week of real time simulation in our case. The next step is to test it on a hardware in loop setup thereby experimentally verifying the hypothesis. Moreover the experimental setup of the HIL is modelled with Modelica (Boiler – Building – Test Bench and Interfaces). The results obtained from this simulation is compared with the simulation without test bench (Boiler - Building). The results match well with the only hurdle now: the annual performance index of the boiler between both the simulation vary by 0.49%. It's under investigation.

MAIN REFERENCES

[1]Based on Probability Bounds (P-Box) Approach in Probabilistic Safety Asessment." Risk Analysis 662-675.

[2]Philippe Mehrfeld, Markus Nurenberg, Martin Knorr, Lars Schinke, Maxmilian Beyer , Manuel Grimm, Moritz Lauster , Dirk Muller , Joachim Seifert Konstantinos

[3]Stergiaropoulous. 2020. "Dynamic evaluations of heat pump and micro combined heat and power systems using the hardware-in-the-loop approach." Journal of Building Engineering 1-14.

[4]Thomas Schutz, Markus Hans Schraven, Marcus Fuchs, Peter Remmen, Dirk Muller. 2018. "Comparison of clustering algorithms for the selection of typical demand days for energy system synthesis." Renewable Energy 570-582.

[5]Wei Tian, Yeonsook Heo , Pieter de Wilde , Zhanyong Li , Da Yan , Cheol Soo Park , Xiaohang Feng , Godfried Augenbroe. 2018. "A review of uncertainty analysis in building energy assesment ." Renewable and Sustainable Energy Reviews 285-301.

Keywords: Uncertainty Propagation, Uncertainty, Hardware in Loop, Probability box

Paper ID 30221: Exploring easy-implementable adaptation strategies to climate change scenarios of existing office buildings: a case study in the Swiss context

Sergi Aguacil Moreno (1), Verena Marie Barthelmes* (2), Cecile Anne Berquand (2), Dolaana Khovalyg (2)

1: Building2050 group, Ecole Polytechnique Fédérale de Lausanne (EPFL), Fribourg, Switzerland; 2: Thermal Engineering for the Built Environment Laboratory (TEBEL), Ecole Polytechnique Fédérale de Lausanne (EPFL), EPFL Fribourg, Switzerland

AIM AND APPROACH

Buildings are complex architectural and engineering systems that should provide throughout their lifespan a comfortable indoor environment for occupants. In the time of climate change, the ability of existing buildings to adjust to outdoor environmental changes without significant energy expenses becomes a real challenge. Exploration of the resiliency and possible adaptation strategies that can keep energy requirements at the current level is especially important for existing buildings that were designed using historical weather data without accounting for the potential future climate effect on energy demand. In this project, heating and cooling demand, and global energy balance are explored for different IPCC future climate scenarios for an office building with a net conditioned area of 4'200 m2 built in 1991 in Lausanne, Switzerland. Synthetic weather files generated with Meteonorm [3] according to the SRES scenarios A1B, B1, and A2 [1-2] for the years 2020, 2050, and 2100 were used for energy performance simulation modeling of the selected building using DesignBuilder software [4]. The goal of the work is to determine the magnitude of energy demand increase for selected future climate scenarios, outline critical issues in the design and operation for the building, and provide design solutions to tackle future climate-related challenges.

SCIENTIFIC INNOVATION AND RELEVANCE

There are many publications such as [5,6] focusing on resilience – generally defined as the ability of humans to adapt positively to adverse situations – and, in the building sector, most of the works are focused on structural resistance to catastrophic events such as earthquakes, but not on the building energy performance. There is some publication focusing on the robustness of the design decision exploring different construction scenarios, but they are applicable to new construction or deep-renovations [7-10]. This work, along with the exploration of the effect of climate change scenarios on the global energy performance, proposes low-cost and easy-implementable strategies such as acting on the setpoints, layout, and re-distributing functionality of zones to help facility managers to maintain the energy performance at the level of 2020 (or even improve) considering the future climate scenarios.

PRELIMINARY RESULTS AND CONCLUSIONS

SRES scenarios all foresee the outdoor temperature rise and more atmospheric vapor due to greenhouse gases (GHG) emissions impacting solar exposure in the future. Results of the building energy performance show that for A2 compared to the B1 scenario, the cooling demand increases, and the heating demand decreases significantly. This is due to the severity of the A2 scenario in terms of increasing temperature and GHG emissions. In this case, the cooling demand should increase by 27% in 2050 and 146% in 2100, while the most optimistic scenario B1 has a more uniform increase in cooling demand throughout the years (61% by 250 and 74% by 2100). The easiest solution to maintain energy demand is to "relax" temperature set points. A simple design solution is the re-arrangement of spaces and functionalities of different zones. More advanced solutions to minimize the effect of climate change on building performance are possible, however, this work shows that easy and cheap solutions can be applied as well. Generally, it is important to account for the future climate prediction, and a possible increase in energy demand when buildings are designed since a typical lifespan of the building nowadays can exceed 60 years [11].

MAIN REFERENCES

[1] IPCC, Emissions Scenarios, Cambridge, UK, (2000).

[2] Equivalency between SRES scenarios (A2, B1 and A1B, from AR4 IPCC report) and RCP scenarios (RCP 8.5, 6.0, 4.5 and 2.6, from latest IPCC scenarios, AR5, ...).

[3] Meteotest, Meteonorm software, (2018). https://www.meteonorm.com/.

[4] DesignBuilder, DesignBuilder software v.6, (2019). http://www.designbuilder.co.uk/ (accessed February 23, 2018).

[5] E.U. EU, Climate-ADAPT, (2020).

[6] S. Hosseini, K. Barker, J.E. Ramirez-Marquez, A review of definitions and measures of system resilience, Reliab. Eng. Syst. Saf. 145 (2016) 47–61.

[7] K. Kalvelage, U. Passe, S. Rabideau, E.S. Takle, Changing climate: The effects on energy demand and human comfort, Energy Build. 76 (2014) 373–380.

[8] N.B. Rajkovich, Y. Okour, Climate change resilience strategies for the building sector: Examining existing domains of resilience utilized by design professionals, Sustain. 11 (2019).

[9] G. Chinazzo, P. Rastogi, M. Andersen, Assessing robustness regarding weather uncertainties for energy-efficiency-driven building refurbishments, Energy Procedia. 78 (2015) 931–936.

[10] G. Chinazzo, P. Rastogi, M. Andersen, Robustness assessment methodology for the evaluation of building performance with a view to climate uncertainties, 14th Int. Conf. IBPSA - Build. Simul. 2015, BS 2015, Conf. Proc. (2015) 947–954.

[11] SIA, SIA 2032:2010 Emboided energy on buildings, (2010).

Keywords: climate change, adaptation, energy performance, easy-implementable strategies

Paper ID 30712: Coupling building simulation with virtual building management system for advanced test and validation

Patrick Béguery* (1), Thomas Murienne (1), Van Lap Ngo (2), Oskar Nilsson (3)

1: Schneider Electric, Grenoble, France; 2: RH Solutions, Grenoble, France; 3: Schneider Electric, Lund, Sweden

AIM AND APPROACH

Building Management System (BMS) control is traditionally commissioned during the first year after their implementation (or installation). During this phase of error correction and tuning, the building performance is considerably lower than expected. Simulation usage has been demonstrated to predict trends that can be compared to real one during the commissioning phase. In that case, the control itself is simulated and the validation must wait for site testing. To reduce the magnitude of the gap and duration of this period, a solution consists in co-simulating the building physics with a virtual instance of the real BMS. By providing a realistic behavior of the controlled system, the simulation will allow to pre-validate the control logic implemented in the BMS and will facilitate the pre-tuning of parameters to optimize efficiency and comfort. In the case of advanced control solution that interact with the BMS, the testbench consisting of co-simulating the building physics with the real BMS will provide a testing environment with both the power of standard simulation and the real complexity of a full BMS system.

SCIENTIFIC INNOVATION AND RELEVANCE

Several building simulation environments have co-simulation capacity, some of them using the popular FMI standard. In this paper, IES Virtual Environment and EQUA IDA-ICE software co-simulation features are compared and tested on a simple case consisting of part of a real building (one AHU and associated zones). A Python based orchestrator was developed to manage the data flow and synchronization of the two environments. It also provides to the BMS some variable that are not existing in the building physical model. Some challenges to be addressed in the production of the digital twin of the building physics is the creation of the co-simulation interface, the capacity to access model parameters from other environment and the potential issue with time delay and simulation time.

PRELIMINARY RESULTS AND CONCLUSIONS

Preliminary co-simulation proof-of-concept is planned to be developed both with VE and IDA-ICE for the building physics part, and an instance of our building BMS. Usage of the simulation-based pre-commissioning environment will be illustrated on several test cases of error detection (inverted sensor, sign error…) or parameter optimization. By providing a high-fidelity digital twin of both the physics and control, this co-simulation approach also offers interesting opportunities to be reused in operation to support fault detection analysis or evaluation of the impact of any control parameter change.

MAIN REFERENCES

R.Sunnam & al., Utilizing Sensor Trend Data to Overcome the Challenges in Calibrating a Simulation Model for Measurement and Verification in a New Construction Project, Proceedings of BS2015, Hyderabad, India, 2015.

P.Beguery & al., A Building Control Oriented Simulation Architecture, Proceedings of BS2013, Chambery, France, 2013.

Y.Lamoudi, Distributed Predictive Control for energy management in buildings, PhD report, Grenoble University, 2012.

Keywords: Co-simulation, Building Management System, Commissioning, Test and Validation, Analytics.

Paper ID 30247: The impact of glazing on thermal performance of a hotel building in Zagreb, Croatia: An early based design

Rudina Belba*, Sokol Dervishi, Ina Osmani

Epoka University

AIM AND APPROACH

It is widely known that commercial buildings contribute to a large proportion of energy consumption nationally and across Europe. Hotels are typically ranked amongst the top five energy consumers in the tertiary sector. However, energy saving potential within the hotel industry is also significant. The aim of the present study is to analyze the energy performance of a hotel design and potential retrofit measures for a Croatian Hotel Project. Energy Plus Computational Simulation is used to assess the energy performance of the building and potential retrofit scenarios. The model includes data collection on local climate, design typology, occupancy patterns, building materials and HVAC system to generate a finite-element simulation model. A set of scenarios including different glazing typologies of the fully transparent building facade design is examined to assess the respective thermal and energy performance.

SCIENTIFIC INNOVATION AND RELEVANCE

The energy optimization of the hotel is a highly complex task that involves a large number of design variables. The research can provide original contributions to the body of knowledge by addressing the energy optimization and identifying potential improvement areas of hotel design in the oceanic climate of Zagreb, Croatia with significant humid continental and humid subtropical climate influences. In addition, limited research is conducted on the impact of transparent facade typologies in the Mediterranean climate on the energy and thermal performance, addressing the need for further research in this context. The present research studies a comprehensive framework on energy performance and evaluates different output variables within the simulation such as energy loads, indoor air temperatures, and radiant temperatures, indoor and outdoor surface temperatures. In addition, evaluating results for both summer and winter conditions represents an original contribution of the study.

PRELIMINARY RESULTS AND CONCLUSIONS

The results highlight the efficiency of glazing design strategies on high WWR facades of hotel design, achieving a significant range of energy reduction from 9% up to 15% for 100% window-to-wall ratio scenarios, with 1 °C up to 3 °C indoor air and radiant temperature increase in the winter period. Additionally, the impact of improved glazing effectiveness on a fully glazed building scenario, provides a considerable reduction of 4 °C indoor air temperature and 6 °C radiant temperature in the summer period. The performed simulation results generate a fundamental framework towards early design decision-making stages.

MAIN REFERENCES

European Commission. Commission Delegated Regulation (EU) No 244/2012 of 16 January 2012 supplementing Directive 2010/31/EU on the energy performance of buildings, Official Journal 2012: L 081

Bodach, S., Lang, W., & Auer, T. (2016). Design guidelines for energy-efficient hotels in Nepal. International Journal of Sustainable Built Environment, 5(2), 411–434. doi: 10.1016/j.ijsbe.2016.05.008

Dascalaki, E., & Balaras, C. A. (2004). XENIOS - A methodology for assessing refurbishment scenarios and the potential of application of RES and RUE in hotels. Energy and Buildings, 36(11), 1091–1105. doi: 10.1016/j.enbuild.2004.03.007

Kneifel, J. (2010). Life-cycle carbon and cost analysis of energy efficiency measures in new commercial buildings. Energy and Buildings, 42(3), 333–340. doi: 10.1016/j.enbuild.2009.09.011

Jincheng, X., Peng, R., Jihong, L. (2015). Analysis of energy efficiency retrofit scheme for hotel buildings using eQuest software: A case study from Tianjin, China, Energy and Buildings, Vol. 87, Pages 14-24

Keywords: Energy Optimization, Simulation, Oceanic Climate, Hotel Design, Scenarios

Paper ID 30202: Energy and environmental assessment of urban areas: an integrated approach for urban planning

Matteo Ghellere, Lorenzo Belussi*, Benedetta Barozzi, Alice Bellazzi, Ludovico Danza, Anna Devitofrancesco, Francesco Salamone

ITC-CNR, Italy

AIM AND APPROACH

The article presents a bottom-up approach for the planning of urban areas from an energy and environmental point of view, based on the definition of specific virtual archetypes, defined analyzing the characteristics of the considered area. Moving from energy considerations, eleven geometrical archetypes are defined covering representative intended use and construction periods. The energy performance of each archetype (primary energy consumption, CO2 emissions) are simulated with the dynamic simulation tool, Energy Plus. Considering environmental topics, the main objective is to analyze, through a Grasshopper based parametric model, the performance of the outdoor areas in terms of overheating risk defined by identifying a series of archetypes characterized by urban building geometry, intended use, vegetation coverage . Using information held by Public Authorities, statistical studies and data from field surveys the model provides an estimation of the current energy performance of the building stock, in terms of primary energy consumption and CO2 emissions, and the environmental characteristics of urban fabric, in terms of outdoor potential overheating. The outcomes of energy and environmental analyses together with urban planning indexes are aggregated in a single index using Multi Criteria Analysis (MCA) allowing to provide a Municipal Reward Indicator (MRI).

SCIENTIFIC INNOVATION AND RELEVANCE

Modelling the complex flows that characterize an urban area is a hot topic for the definition of sustainable urban policies in terms of reducing consumption, environmental impacts and improving energy efficiency and safety of building sector. The development of Urban Building Energy Models (UBEM) provides response to the need to characterize the energy behavior of buildings at urban scale managing and reducing energy consumption and emissions. Most of UBEMs are focused on the residential sector. Alongside, further investigations have been carried out on the evaluation of the environmental performance of urban areas, in terms of overheating risk, runoff, landscape value and soil consumption. These analyses are mainly focused on specific neighborhoods or portion of the urban fabric, whose characteristics are well known. The attempt of the research is to extend these punctual evaluations to a wider scale, starting from the definition of specific archetypes and defining a new approach to characterize the urban area as a whole. Besides, the application of the MCA allows the integration of the energy and environmental aspects with urban planning analyses in a single index (MRI) for each Municipal area. Finally, the effect of new urban policies can be evaluated by applying MRI.

PRELIMINARY RESULTS AND CONCLUSIONS

The integrated approach was applied to a Province of Northern Italy consisting of fifty-five municipalities. The application of the energy and environmental assessment allowed to characterize the performance of the considered areas. The results of the simulation are aggregated at different scale: single buildings, census areas, single municipalities, aggregate of municipalities to provide the local authorities a growing knowledge of the performance of their territory. The results of the simulation were managed with GRASS GIS, an open source GIS tool, for the creation of thematic maps and shape file to be integrated into the local information systems. Besides, the MCA provided a ranking of the municipalities showing single strengths and weaknesses. This enables to guide future urban policies, rewarding the most deserving municipalities. The approach was also applied to check the effect of planned urban policies (new constructions, refurbishment scenarios, etc.) showing their effects on energy and environmental performance of the territories and how the ranking can change. The overall model provided a useful tool for the Public Authorities in the definition of urban policies, guiding choices with a new sustainable point of view.

MAIN REFERENCES

Reinhart, C. F., & Davila, C. C. (2016). Urban building energy modeling–A review of a nascent field. Building and Environment, 97, 196-202.

Belussi, L., Danza, L., Ghellere, M., Guazzi, G., Meroni, I., & Salamone, F. (2017). Estimation of building energy performance for local energy policy at urban scale. Energy Procedia, 122, 98-103.

Li, W., Zhou, Y., Cetin, K., Eom, J., Wang, Y., Chen, G., & Zhang, X. (2017). Modeling urban building energy use: A review of modeling approaches and procedures. Energy, 141, 2445-2457.

Salamone, F., Belussi, L., Danza, L., Di Nunzio, A., Ghellere, M., & Meroni, I. (2019). Energy and environmental analysis of urban environment: Methodology and application of an integrated approach. IOP Conference Series: Materials Science and Engineering (Vol. 609, No. 7, p. 072018). IOP Publishing.

Caputo, P., Pasetti, G., & Ferrari, S. (2019). Implementation of an urban efficiency index to comprehend post-metropolitan territories—The case of Greater Milan in Italy. Sustainable Cities and Society, 48, 101565.

Keywords: Energy and environmental policies, territorial planning, urban index

Paper ID 31013: Design approach towards Photovoltaic Thermal Solar Air Heater (PVT-SAH) assisted Solid Desiccant Cooling (SDC) system in hot and humid climates

Aditya Nibandhe, Nima Bonyadi*, Bruno Lee, Ashutosh Bagchi

Department of Building, Civil and Environmental Engineering, Concordia University, Montreal (Canada)

AIM AND APPROACH

Solar assisted Solid Desiccant Cooling (SDC) systems can operate either in ventilation mode (100% fresh air supply) or recirculation mode. Typical integrated system configurations use an "open exhaust air discharge approach", meaning the sensible heat exchanger (SW) and solar collector do not directly interact with each other; resulting in the processed air being exhausted after the SW on the regeneration side. Therefore, typical integrated system configurations offer low thermal coefficient of performance and increase the unmet hours.

A PVT-SAH offers electrical and thermal gains which can be used to assist the operation of an SDC system. This research aims to improve the performance of an existing integrated PVT-SAH assisted SDC system configuration with proposed modifications for daytime operation. Simulation results of three different configurations are analyzed and compared for their performance in cooling a retail store of a low-rise mixed-use building in a hot and humid climate:

Configuration 1 is a typical integrated system in ventilation mode (baseline). Configuration 2 is a typical integrated system in recirculation mode. And configuration 3 is an integrated system in recirculation mode with proposed modifications. Three performance indicators are employed – Thermal Coefficient of Performance (COPth), net electricity gains (kWh), and unmet hours (%).

SCIENTIFIC INNOVATION AND RELEVANCE

On the regeneration side, typical integrated PVT SAH assisted SDC system configurations use ambient air as the inlet to the solar collector (PVT SAH); whereas the air exiting the sensible heat exchanger (Sensible Wheel, or SW) is exhausted to the environment. Also, in configuration 2, a portion of the return air is exhausted to the environment. In both the above mentioned configurations, the integrated system ends up exhausting processed air, which essentially corresponds to wasted useful work.

The current research investigates an existing integrated system configuration in which the processed air is being utilized completely. First, the air exiting the sensible heat exchanger is used as inlet air to the solar collector. Second, a part of return air is recirculated on the supply side, and leftover part of return air is reused on the regeneration side.

This research showcases an example of Building Applied Photovoltaic Thermal (BAPV/T) system application in space cooling. The electrical and thermal performance improvement achievable through this research can reduce the electrical grid reliance of the integrated system, at the same time, improve the thermal comfort.

PRELIMINARY RESULTS AND CONCLUSIONS

The annual average COPth for the integrated system configurations 1, 2 and 3 comes around 0.95, 1.90 and 2.23, respectively. It is observed that the integrated system configuration 3 offers 135% and 18% improvement in COPth over the configurations 1 and 2, respectively.

Net electricity gains correspond to the difference between the PV electrical energy gain (kWh) and the total electricity consumed by the components of the integrated system (mainly auxiliary heater, fans, etc.). It is observed that the net electricity gains for all the configurations are almost identical and comes around 16,000 kWh (1.5% less for configuration 3).

Unmet hours (%) in this study are calculated as the percentage of the number of hours in the schedule during which the integrated system cannot maintain the room air temperature and relative humidity of 26oC and 65%, respectively. The unmet hours for the integrated system configurations 1, 2 and 3 are 23%, 18%, and 12%, respectively. The integrated system configuration 3 reduced the unmet hours by 48% and 33% over the configurations 1 and 2, respectively.

The integrated system configuration with proposed modifications achieved improvements in COPth, and reduced the unmet hours without sacrificing the net electricity gains.

MAIN REFERENCES

Eicker, U. Energy Efficient Buildings with Solar and Geothermal Resources; Wiley: Chichester, West Sussex, United Kingdom, 2014.

Eicker, U.; Schneider, D.; Schumacher Jürgen; Ge, T.; Dai, Y. Operational Experiences with Solar Air Collector Driven Desiccant Cooling Systems. Applied Energy 2010, 87 (12), 3735–3747 DOI: 10.1016/j.apenergy.2010.06.022.

Fan, W.; Kokogiannakis, G.; Ma, Z. Integrative Modelling and Optimisation of a Desiccant Cooling System Coupled with a Photovoltaic Thermal-Solar Air Heater. Solar Energy 2019, 193, 929–947 DOI: 10.1016/j.solener.2019.10.030.

Jani, D. B.; Mishra, M.; Sahoo, P. K. Solid Desiccant Air Conditioning - a State of the Art Review. Renewable & Sustainable Energy Reviews 2016, 60, 1451–1469.

Yang, T.; Athienitis, A. K. A Review of Research and Developments of Building-Integrated Photovoltaic/thermal (bipv/t) Systems. Renewable and Sustainable Energy Reviews 2016, 66, 886–912 DOI: 10.1016/j.rser.2016.07.011.

Keywords: PVT-SAH, Solid Desiccant Cooling, Hot and Humid Climates

Paper ID 30786: A framework to evaluate the reliability of energy performance assessment methods using buildings databases, application to the French Residential Energy Performance Certificate

Nicolas Bernard Jean Bourbon* (1,2), Thomas Recht (2), Rofaïda Lahrech (1), Mora Laurent (2)

1: CSTB - Centre scientifique et technique du bâtiment, Champs sur Marne, France; 2: University of Bordeaux, CNRS, Arts et Metiers Institute of Technology, Bordeaux INP, INRAE, I2M Bordeaux, F-33400 Talence, France

AIM AND APPROACH

The French residential building stock is responsible for 14,5% of the national greenhouse gases emissions and represents 30% of the national final energy consumption [1]. To reach national objectives and mitigate global warming, efficient retrofit is considered as a major lever and estimating correctly housing energy performance is a preliminary to do so.

The EPC (Energy Performance Certificate) is a mandatory and regulatory instrument introduced by the EPBD which estimates heating, cooling and hot water energy consumptions in a housing thanks to a static annual method (3CL –method) in France describing both the systems and envelope. Indoor and outdoor conditions are defined by the method itself allowing a comparison between housings. It makes also possible to estimate the energy gain of a refurbishment action.

However, the current assessment methodology is considered not reliable by occupants, state actors and experts. Researchers criticize both a systemic overestimation of consumption [2], a large uncertainty on final and primary building consumptions and a non-reproducible results for a few housing cases [3], [4].

We aim to assess the reliability of the 3CL-method by quantifying precision and accuracy on a stock of housing using statistical analysis, and identifying potential improvements.

SCIENTIFIC INNOVATION AND RELEVANCE

Studies on the French EPC method do not separate the different parts of the method: the data collection, the data treatment and the model. Most approaches use uncertainties propagation methods and sensibility analysis to evaluate uncertainties on final consumptions and provide global assumptions. They are also limited towards a single housing with different insulation levels depending on the year of construction / retrofit. They solely studied envelope data and do not consider uncertainty on the description of the systems [3]. A proven lack of precision is identified partly due to the input mode describing the thermal transmission coefficient of walls [4] but the model and the data treatment are not interrogated.

This paper mainly aims to evaluate the reliability of the 3CL model part on a panel of existing housing from diverse climate areas and with a diversity of inherent energy performances. We used bibliography and expert knowledge to characterize the uncertainty over every input data.

To quantify the method reliability and its dependence towards inherent characteristics, we used uncertainty propagation methods. This paper also identifies the evolution of the model sensitivity to parameters regarding intrinsic energy performances, housing types and climate areas.

PRELIMINARY RESULTS AND CONCLUSIONS

Accuracy is defined as the distance in terms of final energy consumptions in between the 3CL-method and another conventional method: Th-BCE which has been tested against the ASHRAE 140 standard [5]. We applied this comparison using a same conventional scenario of indoor and outdoor conditions. Precision is defined as the average gap between every Monte-Carlo throw and the global average consumption. Our analysis is made on a stock of existing housing and, when possible, we provide statistical results.

We applied these methods on the selected panel of buildings. A strong dependency toward inherent performance is observed. Energy effective housings are impacted by the static aspect of the model which misrepresents solar and internal gains. Consequently, the most influent parameters are changing depending on the inherent performance of housing. Finally, precision and accuracy of the method dependent on the housing type and the intrinsic energy performance, are assessed on the full panel of buildings.

These first conclusions provide leads to a later approach to improve both models and input data in order to obtain a satisfying reliability for the EPC methodology.

MAIN REFERENCES

[1] ADEME, « Climat Air et Energie Chiffres Clés - édition 2018 - Résidentiel (logement) », in Climat Air et Energie Chiffres Clés - édition 2018, 2018e éd., ADEME, 2019, p. 133-154.

[2] M. H. Laurent, B. Allibe, T. Oreszczyn, I. Hamilton, R. Galvin, et C. Tigchelaar, « Back to reality: How domestic energy efficiency policies in four European countries can be improved by using empirical data instead of normative calculation. », ECEEE Summer Study proceedings, 2013.

[3] D. Osso, T. Schnell, A. Rolland, et M. Raynaud, « The influence of uncertainties related to the inputs of the French EPC's calculation method -an analysis for individual houses », in eceee 2019 Summer Study, Presqu'île de Giens, France, juin 2019, p. 1209-1215

[4] M. Raynaud et P. Stabat, « DPE : influence des incertitudes liées aux entrées de la méthode 3CL en maison individuelle », CVC Rev. Clim., vol. juillet-octobre, no 870, p. 18-22, 2011.

[5] D. Da Silva, J. Alessandrini, J. Videau, C. Pele, et J.-R. Millet, « Évaluation et perspectives du modèle thermique de COMETh, le cœur de calcul de la réglementation thermique des bâtiments neufs », 2016.

Keywords: Energy Performance Certificate (EPC), Reliability, Housing stocks, Building energy performance simulation

Paper ID 30604: Aperture-based daylight modelling: Evaluating the airmass refinement for the sunlight beam index

John Mardaljevic (1), Eleonora Brembilla* (2)

1: Loughborough University, United Kingdom; 2: TU Delft, Netherlands

AIM AND APPROACH

Aperture-based daylight modelling (ABDM) is a new building simulation paradigm based on measures of an aperture's 'connectedness' to the sun and the external environment. ABDM comprises two distinct parts: the sunlight beam index (SBI) and the aperture view indices (AVIs). SBI is an area measure of the 'connectedness' of a building aperture to all of the possible occurring sun positions for a particular locale/scenario including any obstructing surfaces. [1] Similarly, each AVI is an area measure of the 'connectedness' of an aperture to - respectively - the three layers of the external environment: sky, ground and a middle layer of natural/man-made objects. [2] The AVI for each layer is computed in terms of the lumens received from an external environment made luminous for this purpose. [3] ABDM was conceived as a method suited for daylight/sunlight planning. A major refinement to the sunlight part is to account for the attenuation effect of airmass so that the still, essentially geometrically-based measures of SBI correspond directly to actual measures of direct sun irradiance/illuminance computed using climate data e.g. thermal/CBDM. Thus creating the potential to couple the evaluation for building planning to that for building performance.

SCIENTIFIC INNOVATION AND RELEVANCE

At the planning level, there currently does not exist anywhere an evaluative schema which is equally applicable to measures of solar energy potential (for PV performance, overheating risk, etc) and measures of sunlight/daylight amenity (for daylight, well-being, connectivity/view, etc). With ABDM, the 'aperture' is any defined planar surface which forms part of the building envelope, say, windows and/or PV panels - in any number, shape, and combination. In order to be effective, a method which forms the basis for planning must avoid the vagaries of the building simulation 'performance gap'. Otherwise, planning decisions will be contested on the basis of inconsistent/unreliable outcomes depending on the methodology/software used, and so the approval process will lose credibility. Even with the airmass refinement for SBI, the ABDM metrics remain essentially geometrical in nature. Modern surveying techniques virtually guarantee accurate geometry. And so an evaluation procedure - say, to gain planning approval - will be almost impossible to gameplay. Another key innovation is the potential to make view a material concern in evaluations to predict daylight/amenity loss due to obstructions. ABDM offers the potential for a seamless workflow from outline planning to detailed building performance evaluation, e.g. by 'climatising' ABDM.

PRELIMINARY RESULTS AND CONCLUSIONS

Confirmation that airmass-refined sunlight beam index (AR-SBI) serves as an effective proxy for direct sun irradiance/illuminance is tested by comparing annual AR-SBI with annual direct sun irradiance from standardised weather files for a variety of scenarios. These include variations in: climate locales, aperture orientation, degrees of obstruction, shading components (e.g. brise soleil) and geometrical complexity. Preliminary results show good linear correlations between AR-SBI and annual direct sun irradiance. We show how these correlations can be used to create locale-specific factors to allow for direct conversion from (essentially-geometrical) AR-SBI to annual direct sun irradiance/illuminance. Recent work on improving the basis of standardised weather files has revealed the potential for enhanced models of solar radiation. [4] Comparison is also made between AR-SBI and weather files for a locale for both standard and improved solar models. [5]

MAIN REFERENCES

[1] J. Mardaljevic and N. Roy. The sunlight beam index. Lighting Research & Technology, 48(1):55–69, 2016.

[2] British Standards Institute. BS EN 17037: 2018 – Daylight in buildings. London, BSI, 2018.

[3] J. Mardaljevic. Aperture-Based Daylight Modelling: Introducing the 'View Lumen'. In IBPSA - Building Simulation Conference, Rome, Italy, 2019.

[4] E. Brembilla, J. Mardaljevic, and A. Mylona. Improving solar data in CIBSE climate files – Survey of measuring networks and test on daylight simulation. CIBSE Technical Symposium, Sheffield, UK, 25-26 April, 2019.

[5] E. Brembilla, J. Mardaljevic, and A. Mylona. Daylight simulation and compliance: Reliability of climate data solar radiation models. In IBPSA - Building Simulation Conference, Rome, Italy, 7 2019.

Keywords: Daylight, sunlight, planning, performance

Paper ID 30665: Optimal control strategies of an ice-based thermal energy storage system in an existing office building

Seongkwon Cho*, Han Gyeong Chu, Cheol-Soo Park

Seoul National University, Korea, Republic of (South Korea)

AIM AND APPROACH

An energy-efficient and occupant responsive indoor environmental controller must be able to adapt itself to building environment and occupant's personal preferences. Recently, deep reinforcement learning (DRL) has received increasing attention because it can be used for optimal control without resorting to a simulation model (model-free). In addition, DRL has an ability to learn dynamic relations between mechanical components and/or occupants-systems.

With this in mind, the authors will introduce in the paper DRL based HVAC controller. For this purpose, the agent of deep Q-network (DQN), one of the most famous DRL algorithms, is trained to learn occupant preferences (thermal, visual, air quality) as well as to predict energy use of a single occupancy office. After being trained, the DRL controller is invoked to provide indoor environment according to the occupant preferences learned, e.g. windows opening/closing, lighting on/off, turning on/ff heating and cooling systems, changing setpoint temperature, etc. For this study, a virtual test room with a single occupancy is modeled using EnergyPlus. The DQN will be implemented with Python 3.7 and Tensorflow 1.15. In the DRL controller, occupant behavior model published by LBNL (e.g. obXML, ref.[1]) will be embedded.

SCIENTIFIC INNOVATION AND RELEVANCE

It has been long explored how the indoor environmental controller can adapt itself to occupant's personal preferences (thermal, visual, indoor air quality) but there is no viable approach reported yet. This study signifies that the DRL controller could be beneficially used for the aforementioned purpose. In other words, it will be highlighted in the paper that the DRL controller is able to learn the occupant's personal preferences, the dynamics of HVAC systems and the thermal response of the room space.

Many research efforts have been made with regard to modeling of occupant behavior as well as identifying occupant preferences. This study will propose a new way to 'view' occupant behavior because the DRL does not require any pre-defined occupant behavior model but can self-learn it from gathered data.

PRELIMINARY RESULTS AND CONCLUSIONS

Firstly, it will be investigated how well the DRL controller could learn thermal dynamics of the space (e.g. energy use, prediction of room air temperature) as well as occupant behavior (presence, actions, preferences). The performance of the DRL controller will be investigated in terms of prediction accuracy (energy, temperature), energy savings, etc.

When a decision suggested by the agent of DRL does not meet the occupant's preferences (thermal, visual, air quality), it is assumed that an overriding action is triggered. Then, the number of overridden times will be used as a performance criteria of the DRL based controller.

It will be shown in the paper that the DRL controller far outperforms rule-based or Model Predictive Control (MPC) in terms of prediction accuracy, energy use, adaptability to occupant's preferences.

MAIN REFERENCES

[1] Yan, D., Hong, T., Dong, B., Mahdavi, A., D'oca, S., Gaetani, I., Feng, X. (2017), IEA EBC Annex 66: Definition and simulation of occupant behavior in buildings, Energy and Buildings, Vol.156, September, pp. 258-270

[2] Cehn, Y., Norford, L.K., Samuelson, H.W., Malkawi, A. (2018), Optimal control of HVAC and window systems for natural ventilation through reinforcement learning, Energy and Buildings, Vol.169, March, pp.195-205

[3] Park, J.Y., Dougherty, T., Fritz, H., Nagy, Z. (2018), LightLearn: An adaptive and occupant centered controller for lighting based on reinforcement learning, Building and Environment, Vol.147, October, pp.397-414

[4] Ahn, K.U., Kim, D.W., Park, C.S. and de Wilde, P. (2017), Predictability of occupant presence and performance gap in building energy simulation, Applied Energy, Vol.208, December, pp.1639-1652

[5] Tian, W., Heo, Y., de Wilde, P., Li, Z., Yan, D., Park, C.S., Feng, X. and Augenbroe, G. (2018), A review of uncertainty analysis in building energy assessment, Renewable and Sustainable Energy Reviews Vol.93, October, pp.285-301

[6] Wang, Z., Hong, T. (2020) Reinforcement learning for building controls: The opportunities and challenges, Applied Energy, Vol.269, 115036

Keywords: ice-based thermal energy storage, optimal control, data-driven, cooling demand prediction

Paper ID 30290: Coupling of neural models for predicting indoor temperatures and heating loads in buildings

Benoit Delcroix*, Simon Sansregret, Michaël Fournier, Ahmed Daoud

Hydro-Québec Research Institute, Laboratoire des Technologies de l'Énergie, Shawinigan, QC, Canada

AIM AND APPROACH

Building energy models are critical to forecast the energy use and to improve the control of HVAC systems. However, these models are building-specific and their development is tedious, error-prone and time-consuming. Compared to traditional white-box and grey-box models, black-box models need less development time and no information about the building properties, and only rely on collected data. This black-box approach is particularly well suited for building managers whose objective is to forecast and optimize the operations of many buildings with unknown characteristics. In this work, two autoregressive neural models are coupled to simulate the building energy behaviour: the first model predicts the future indoor temperatures of each room from known conditions; the second model predicts the future heating energy use of each room from known conditions, including the outputs provided by the first model. These models are trained, validated and compared to actual experimental data obtained for 8 existing houses in Canada (province of Quebec) heated by electric baseboards controlled by connected thermostats (on average, 10 thermostats per house). Available data include room temperatures, set-point temperatures, heating energy use and weather data (dry bulb temperatures, wind speed and global horizontal irradiance).

SCIENTIFIC INNOVATION AND RELEVANCE

Grey-box modelling approach such as Resistance-Capacitance (RC) models are commonly used in multiple applications, including model predictive control. However, this approach is sometimes not readily applicable because they still need to be parameterized based on key building properties. When no information is available, RC models are then difficult to calibrate. In these situations, purely data-driven methods outperform RC models if data are adequately collected (quantity and quality). In this category of methods, neural networks are recognized as an efficient and flexible universal estimator allowing multiple inputs and outputs. Moreover, they are capable to consider nonlinearities such as radiative heat transfer and human occupancy in buildings. In this work, the main contribution is the coupling approach of 2 neural networks to mimic the techniques used in initial-value problem solvers that are commonly applied in building energy simulation tools. This modelling approach has the substantial advantage of being generalizable to all buildings without major modifications. This is particularly of interest for building managers dealing with a wide number of buildings and who need to implement in a timely manner models forecasting indoor temperatures and heating energy use for diverse applications (model predictive control, fault detection, etc.).

PRELIMINARY RESULTS AND CONCLUSIONS

The model coupling 2 neural networks (for indoor temperature and heating energy use prediction) was developed and applied to 8 houses in Canada. Data collected from connected thermostats include room temperatures, set-point temperatures and heating energy use. Weather data (dry bulb temperature, global horizontal irradiance and wind speed) were also included. These data were pre-processed to optimize the model training phase. The techniques used for pre-processing mainly include resampling, normalization and feature engineering. Both neural networks were trained and validated separately on different datasets to avoid overfitting. Finally, both neural networks were coupled to simulate all houses over a period of a week (excluded from the training and validation datasets). Results show that the model mimics accurately the thermal behaviour of all houses. For indoor temperatures, the root mean square deviations (RMSD) are always below 0.5 °C. For heating energy use, the normalized RMSD (RMSD divided by the maximum value) are between 6 and 14 %. Future work will be dedicated to verifying the evolution of the model performance according to the time of the year and assessing the relevance of using it in various applications (e.g., fault detection and predictive control).

MAIN REFERENCES

RUIZ, Luis Gonzaga Baca, CUÉLLAR, Manuel Pegalajar, CALVO-FLORES, Miguel Delgado, et al. An application of non-linear autoregressive neural networks to predict energy consumption in public buildings. Energies, 2016, vol. 9, no 9, p. 684.

KRAMER, Rick, VAN SCHIJNDEL, Jos, SCHELLEN, Henk. Simplified thermal and hygric building models: A literature review. Frontiers of architectural research, 2012, vol. 1, no 4, p. 318-325.

FRAUSTO, Hugo Uchida, PIETERS, Jan G. Modelling greenhouse temperature using system identification by means of neural networks. Neurocomputing, 2004, vol. 56, p. 423-428.

MECHAQRANE, A., ZOUAK, Mohcine. A comparison of linear and neural network ARX models applied to a prediction of the indoor temperature of a building. Neural Computing & Applications, 2004, vol. 13, no 1, p. 32-37.

MUSTAFARAJ, Giorgio, LOWRY, Gordon, CHEN, Jie. Prediction of room temperature and relative humidity by autoregressive linear and nonlinear neural network models for an open office. Energy and Buildings, 2011, vol. 43, no 6, p. 1452-1460.

DELCROIX, Benoit, LE NY, Jérôme, BERNIER, Michel, et al. Autoregressive neural networks with exogenous variables for indoor temperature prediction in buildings. Building Simulation, 2020. p. 1-14.

Keywords: Building energy simulation, Neural network, Multi-Layer Perceptron, Experimental validation

Paper ID 31080: An automated approach to Parametric Design Analysis for large-scale design projects

Jeanie Chan (1), Junru Shen (2), Oscar Valdes (3), Luisa Drope* (3), Andrea Frisque (4)

1: AECOM, Burnaby, Canada; 2: AME Group, Vancouver, Canada; 3: Stantec, Vancouver, Canada; 4: British Columbia Institute of Technology, Burnaby, Canada

AIM AND APPROACH

Parametric Design Analysis (PDA) has become popular in the building simulation industry. Large sets of design options are analyzed using parametric simulations to evaluate pathways to achieve project targets. As energy models become more complex and the number of simulations required for PDA exponentially grows, more computing power and time-savings initiatives are required to deliver results in a short amount of time. There is a need to automate the process and reduce the computing time.

This paper presents a workflow using Python, Eppy and a computer server farm to automate the PDA process. Python and Eppy are used to edit the base energy model files to generate a large set of design iterations. An integrated submitter is used to facilitate the distribution of simulation jobs to the server farm. The server farm then executes the simulation jobs in parallel on each machine. Once the simulation runs are complete, Python is further used to perform post-simulation data processing for complex modelling workarounds and calculating performance metrics. Finally, results can be collected and plotted using a Parallel Coordinates graph, which provides invaluable data to inform decision makers of the best design measures to implement.

SCIENTIFIC INNOVATION AND RELEVANCE

The proposed workflow pushes the limitations of simulation tools and gives practitioners the flexibility to customize their parametric simulation configurations and results, perform complex modelling workarounds, and visualize data based on specific project requirements. Scripting tools can significantly reduce the time needed for parametric energy model preparation and post-simulation calculations, while reducing the risk of errors. Once the scripting tools are set up, subsequent projects can be executed more efficiently as most of the process is automated and scripts can be easily applied to future models.

The use of server farm can significantly reduce the overall simulation time for the parametric runs.

PRELIMINARY RESULTS AND CONCLUSIONS

This workflow has been applied to the design of a large healthcare project in a continental Canadian climate. The project's main targets are energy use intensity (EUI), thermal energy demand intensity (TEDI), and greenhouse gas emissions (GHGI). A diverse and ambitious array of envelope, electrical, and HVAC energy efficiency measures was investigated, which resulted in 40,000 unique simulations. In the integrated design environment, following a workplan with tight deadlines, Parametric Design Analysis was used to drive the design decisions and provide valuable insight into what measures will help the project achieve all its targets.

MAIN REFERENCES

N/A

Keywords: Parametric, Energy Simulation, Design Assistance, Scripting, Automation

Paper ID 30967: Thermal bridges numerical study in building retrofitting and ground model coupling with Whole building energy software "EnergyPlus"

Ali El Saied* (1,2), Chadi Maalouf (1), Timea Bejat (3), Etienne Wurtz (4)

1: University of Reims Champagne-Ardenne, MATIM « MATériaux et Ingénierie Mécanique », Reims, France; 2: Univ Grenoble Alpes, CEA, LITEN, DTS, LELA, INES, F-38000 Grenoble, France; 3: Univ Grenoble Alpes, CEA, LITEN, DTS, LPM, INES, F-38000 Grenoble, France; 4: Univ Grenoble Alpes, CEA, LITEN, DTS, INES, F-38000 Grenoble, France

AIM AND APPROACH

Today in France, building construction represents 44% of final energy consumption, and 26% of greenhouse gas emissions [1]. One of the reasons of this energy consumption is the non-treated thermal bridges (TB). Ground heat loss has been considered negligible compared to heat loss from other surfaces of a building. In the 90's, heat loss through foundation was around 15% compared to 35% for walls, 25% roof, 15% for various air leaks and 10% for windows [2]. After the improvement of building insulation, the relative importance of heat losses through foundation increases and becomes an important factor in the energy efficiency of a dwelling. Nowadays, the heat losses by the foundation are rather of the order of 30% to 50% [3]. The purpose of this paper is to analyze the impact of wall-Slab junctions on final energy consumptions under several conditions. Actually, different models such as SLAB [4] or KIVA [5] calculate TB and heat transfer via slab, foundation and soil. These models handle the heat transmission calculation but ignore coupled heat and moisture effects. The aim of this work is compare these methods and propose a new one where heat and humidity conditions are taken into account.

SCIENTIFIC INNOVATION AND RELEVANCE

The heat flux through floors in contact with ground depends on several factors, which are difficult to determine because of their physical nature. Heat transfer at grounds level should be considered as 3D. In addition, many properties (of the soil) should be taken into consideration like conductivity, density, thermal capacity, moisture content, and ground homogeneity. Ground (soil) is considered as a porous material. Moisture has a direct influence on ground heat transfer. In fact, an excess of moisture in soil can directly affect energy performance (thermal conductivity). Most of ground heat transfer models did not consider moisture effect on thermal behavior [6] and didn't consider other characteristics such as three dimensional one or the influence of climate on ground heat transfer. A new model considering these effects will be proposed. It will reduce complexity of ground heat transfer models by defining simple correlations describing different phenomena.

PRELIMINARY RESULTS AND CONCLUSIONS

The first part of the calculation is done on THERM software (based on EN 10211 [7]) to determine a solution catalogue to overcome thermal bridges. The linear transmittance coefficient (of TB) is found based on a 2D (steady state) conduction heat transfer model. It was found that when an exterior insulation is deeper in ground, heat loss will decrease to achieve higher reduction (500 KWh/year). Also when an exterior insulation thickness is increased, energy consumption is decreased.. Then, a numerical model is written in python and coupled with EnergyPlus software to calculate slab temperature and TB. To run this code, the folowing input data is required : zone temperature, slab, foundation, insulation and soil parameters are needed (geometry, thermal and hydric properties). After the initialization of soil water content and slab temperature (for several years), the calculation begins by calculating heat transfer at soil and slab surface. Finally, interaction between soil, foundation and slab determine slab surface temperature. The model will be validated against experimental data from literature.

MAIN REFERENCES

[1] "Énergie dans les bâtiments." 2016. Ministère de la Transition écologique et solidaire. 2016. https://www.ecologique-solidaire.gouv.fr/energie-dans-batiments.

[2] Deru, M. R. 2003. "Whole-Building Energy Simulation with a Three-Dimensional Ground-Coupled Heat Transfer Model"

[3] Deru, Michael P., and Allan T. Kirkpatrick. 2002. "Ground-Coupled Heat and Moisture Transfer from Buildings Part 1–Analysis and Modeling." Journal of Solar Energy Engineering 124 (1): 10–16. https://doi.org/10.1115/1.1435652.

[4] Kruis, Nathanael John Frothingham. 2015. "Development and Application of a Numerical Framework for Improving Building Foundation Heat Transfer Calculations."

[5] Bahnfleth, William. 1989. "Three-Dimensional Modelling of Heat Transfer From Slab Floors." US Army Corps of Engineers.

[6] Janssen, Hans, Jan Carmeliet, and Hugo Hens. 2004. "The Influence of Soil Moisture Transfer on Building Heat Loss via the Ground." Building and Environment 39 (7): 825–36. https://doi.org/10.1016/j.buildenv.2004.01.004.

[7] ISO 10211. 2017. "ISO 10211:2017: Thermal Bridges in Building Construction — Heat Flows and Surface Temperatures — Detailed Calculations." https://viewer.afnor.org/Pdf/FileOpen/?token=7YaUn7Obvto1.

Keywords: Thermal bridges, Heat and moisture transfer, Energy consumption, Renovation, EnergyPlus.

Paper ID 30808: Wind turbulence impact in ventilation engineering applied on buildings

Ali Esber* (1,2,3), Xavier Faure (1), Fancois Demouge (2), Etienne Wurtz (1), Simon Rouchier (3), Aurelie Foucquier (1)

1: CEA, Commissariat à l'énergie atomique et aux énergies alternatives, France; 2: CSTB, Centre scientifique et technique du bâtiment,France; 3: USMB, Université Savoie Mont Blanc, France

AIM AND APPROACH

Residential and industrial buildings equipped with a ventilation system are complex facilities. Infiltration caused by wind turbulence is difficult to predict because this turbulence is a disorder and a non-local phenomena that depend on several parameters (terrain roughness, buildings density, buildings shapes...). The wind can cause damages or accidents into the buildings, depending on the orientation and size of the inlet/outlet openings, pressure conditions and nominal airflow values. The presented study aims to understand aerodynamic effects induced by the external wind on global performance of ventilation systems (energy, air quality, humidity). This study will help to design the ventilation systems, especially mechanical systems, in term of turbulence effects. In previous studies, many researchers showed the importance to take into account the fluctuation of the pressure inside building not only the façade pressure. For our work, a numerical approach has been chosen through the Mathis software. It deals with a French standard tool employed to analyze ventilation systems in dwellings. Depending on LEROUX 2011, the turbulence has an essential contribution on the direction of the air infiltration that it cannot be seen using mean values.

SCIENTIFIC INNOVATION AND RELEVANCE

This work can be divided into three parts; the first one is to study the wind turbulence, his characteristics, the relation between the mean wind velocity and the spectra describing the wind turbulence (Davenport, Van Der Hoven). The second part deals with the study of different types of ventilation systems and the impact of the time series generated in the first part on each ventilation systems (natural and mechanical).

The third part will be the most important, the evaluation of the effect of this turbulence on the global variables. Therefore, different indicators are studied as the pressure fluctuation inside the building, the inversion of the flux due to depression (between exterior and the buildings or between dry and humid zones), or also the variation of the air quality inside the zone by analyzing the quantities of the humidity ratio and the pollutants. Mathis is used as a nodal simulation software for which boundary conditions are given by measurement of the pressure fluctuation on the façade of a real scale building. This simulation helps to predict the situation in the ventilation engineering.

PRELIMINARY RESULTS AND CONCLUSIONS

Time series of turbulent wind velocity, used as boundary conditions, are generated using python. A simulation of an experimental house was done with changing the type of the ventilation systems and the inlet/exit ducts. A comparison is done between the different measurements time step of mean and turbulent wind velocity (1 minute,

10 minutes and 1 hour). Based on a new protocol to measure the pressure on an overall building, a new database has been implemented. Different aspects of comfort can be affected by the turbulence as the rate of humidity or pollutants concentration and this work focuses on them. Moreover, considerations about the energy required for the ventilation systems are also studied. All of this will help in resizing systems in the ventilation engineering and engineers will be able to define if it is necessary to consider the wind turbulence or it is in the safe zone.

Results showed that the effect of the turbulence cannot be neglected to evaluate the indoor environment quality and HVAC systems according to human activities. An additional work is done to find a link between the simulations results and the real experiment database to reduce the deviation in results based on hypothesis.

MAIN REFERENCES

• Ademe (2011). Site internet de l'ADEME (consulté en 2011) ; http://www2.ademe.fr/servlet/getDoc?id=11433&m=3&cid=96

• Adrien Dhalluin Etude de stratégies de ventilation pour améliorer la qualité environnementale intérieure et le confort des occupants en milieu scolaire, UNIVERSITÉ DE LA ROCHELLE,2013

• ASHRAE (1997). ASHRAE Chapter 8: Thermal Comfort. In: ASHRAE handbook of fundementals. SI Edition. Atlanta: ASHRAE, 1997, p. 8.1-8.28.

• Eurocode 1, actions sur les structures, AFNOR, 2005

• Haghighat, Fariborz, Jiwu Rao, et Paul Fazio. 1991. « The Influence of Turbulent Wind on Air Change Rates—a Modelling Approach ». Building and Environment 26 (2): 95-109. https://doi.org/10.1016/0360-1323(91)90017-6.

• Holmes J. D. (1979). Mean and fluctuating internal pressures induced by wind, Proceedings, 5th International Conference on Wind Engineering, Fort Collins, July 1979, p. IV-8-1

• Le Roux N. Etude par similitude de l'influence du vent sur les transferts de masse dans les bâtiments complexes. PhD thesis, University of La Rochelle; 2011.

• Mounajed La modélisation des transferts d'air dans les bâtiments. Application à l'étude de la ventilation, Ecole Nationale des Ponts et Chaussees, 1989

Keywords: Indoor environment quality, Ventilation system, Wind turbulence, pressure fluctuation.

Paper ID 344: BIM2SIM Workshop - Computational design as an integral part of the engineering process workflow

Manuel Frey*

Gruner Roschi AG, Switzerland

Simulations no longer have to be selective examinations, detached from the planning process. In the right workflow, they can accompany the design process integrally.

Simulations are increasingly becoming an integral part of the design and planning of buildings, not only at the client's request, but because they add value to meaningful and sustainable construction and concepts. This applies to many aspects of planning, such as daylight, thermal comfort, heating and cooling demand, energy efficiency and fire protec-tion. As more and more projects are planned using digital planning methods (BIM/VDC), the process integration of simulations into planning is becoming increasingly accepted and relevant and can also be marketed in a commercially attractive way.

At the workshop, the current status of the process integration of simulations into digital building design (BIM/VDC) will be evaluated and possible next steps, will be discussed. The workshop will start with a best practice presentation on BIM2SIM.

Key Learnings

1.) Participants know the advantages of model-based simulations as an integral part of building design.

2.) Participants will be able to clearly describe the need for information during pre-processing within the BIM method, support the ordering process and accompany the integration of simulations into building design.

3.) Participants know the possibilities of data-driven decision making using the digital twin in the context of BIM/VDC as an integral part of building design.

Target Group:

Researchers, Architects, Engineers & Consultants with interest in Building Performance Simulations & BIM.

Besides my main job as an engineer, I am the president of the IBPSA Chapter Switzerland and at buildingSMART Switzerland I am the head of the BIM2SM working group in the building room.

Keywords: BIM, VDC, Computational Design, Simulation

Paper ID 30304: Effectiveness of cloud cover on solar radiation prediction using artificial neural network algorithm

Hany Gaballa*, Soolyeon Cho

North Carolina State University, United States of America

AIM AND APPROACH

Solar energy is considered the ultimate power source of energy on our planet. For the simulation of building energy performance, it is essential to have solar radiation data in the weather file. Typical Meteorological Year files include the solar radiation data; however, these are historical data, not real-time. These days, real-time simulation is more demanding for real-time performance analysis. There is no commercial tool that can predict or generate solar radiation data. Researchers in previous papers have different disciplines in dealing with solar radiation prediction with a time interval, either hourly, daily, or monthly. Solar radiation predictions are desired in various industries, such as farming, PV or renewable energy, and building performance analysis. Solar radiation measurement requires lots of resources, which is why only a few solar stations exist all over the world. It is logical to think about finding more efficient and effective ways of predicting solar radiation. Although many papers presented different approaches, the performance ranking of such methods is complicated due to the diversity of prediction methods, type of input data, and time steps. This paper aims to find the most effective, readily available parameters that accurately predict solar radiation.

SCIENTIFIC INNOVATION AND RELEVANCE

Different methods had been used for solar radiation prediction, where Artificial Neural Network (ANN) is considered one of the most accurate methods (Kashyap et al., 2015). This paper focuses on hourly global solar radiation prediction required for real-time building simulation. According to Alam et al. (2009), ANN models predict hourly and daily solar radiation with better accuracy than empirical models.

The developed ANN model by Gaballa & Cho (2019) used in this paper aims to only use the readily available parameters as inputs such as temperature and relative humidity to predict global solar radiation. Cloud cover has a high impact on solar radiation prediction when used through statistical methods (Sarkar, 2016). However, cloud cover is not readily available, so there was a need to find if cloud cover minimally or highly affects solar radiation prediction when the ANN method is used.

Three cases with different climate conditions were chosen as case studies, which are Phoenix, AZ, Seattle, WA, and Raleigh, NC, located in climate zone 2B (hot and dry), 4C (mixed-marine), and 4A (mixed-humid), respectively. The three cases also have a different number of cloudy and sunny days.

PRELIMINARY RESULTS AND CONCLUSIONS

Two scenarios were performed using the ANN model to predict global solar radiation every hour. The first scenario is using temperature (T), relative humidity (RH), and solar zenith angle (SZA), the second scenario is using the same parameters besides cloud cover (CC).

CV(RMSE), NMBE, and R-squared values were calculated to see the difference between the two scenarios. Both scenarios show a prediction accuracy within ASHRAE Guideline 14-2014 limits (ASHRAE, 2014). Phoenix showed the best accuracy before, and after including cloud cover, however, Seattle showed the most significant difference after adding cloud cover, taking into consideration that Seattle has the lowest cloud coverage. To come up with the reason for these differences, the correlation between cloud cover and global solar radiation is presented.

Raleigh and Phoenix show no correlation between cloud cover and global solar radiation, while in Seattle, the ANN algorithm can understand some relationships. So, this clarifies why Seattle has the largest difference after including cloud cover to the input layer.

In conclusion, it was found that solar radiation prediction using ANN doesn`t mainly depend on the cloud cover percentage on a specific location but by the correlation between cloud cover and global solar radiation.

MAIN REFERENCES

• Alam, S., Kaushik, S., and Garg, S. (2009). Assessment of diffuse solar energy under general sky condition using artificial neural network. Applied Energy. Volume 86, Pages 554-564.

• Gaballa, H., & Cho, S. (2019). Prediction of Hourly Solar Radiation using Temperature and Humidity for Real-Time Building Energy Simulation. Journal of Physics, 1343, 012049.

• Guideline, A. (2014). Measurement of energy, demand, and water savings.

• Kashyap, Y., Bansal, A., and Sao, A. (2015). Solar radiation forecasting with multiple parameters neural networks. Renewable and Sustainable Energy Reviews. Volume 49, Pages 825-835.

• Sarkar, M. N. I. (2016). Estimation of solar radiation from cloud cover data of Bangladesh. Renewables: Wind, Water, and Solar. Volume 3.

Keywords: Cloud cover, Solar radiation prediction, Artificial Neural Network, Real-time building performance, Real-time weather file

Paper ID 30491: Selection of input variables for covering materials through sensitivity analysis based on greenhouse energy demands

Jabin Goo* (1), Hakjong Shin (1), Jeonga Kang (1), Sunhye Mun (2), Younghoon Kwak (1), Jungho Huh (1)

1: University Of Seoul, Korea, Republic of (South Korea); 2: Korea Institute of Civil Engineering and Building Technology, Korea, Republic of (South Korea)

AIM AND APPROACH

With recent developments in ICT convergence technology in the construction sector, the market for smart-farm production systems occupies a major portion of the overall market worldwide [1]. In greenhouses, many cooling loads caused by solar radiation are generated in the cooling unit owing to the building envelope characteristics [2], and they consume enormous amounts of energy worldwide [3]. Therefore, studying the energy demand of greenhouses is necessary. In addition, many studies have been conducted to save energy by using smart farms, away from existing greenhouses with high carbon emissions. A smart greenhouse equipped with automation and complex environmental control system is required for competitive cultivation and production of greenhouse crops, and proper design for heating and cooling of the greenhouse is required for the optimal operation of smart greenhouses. However, unlike ordinary buildings, the envelopes for greenhouses are generally fabricated of materials such as plastic or glass. The thickness of the envelope is also very less compared to that of ordinary buildings, which is greatly influenced by external weather conditions [4].

SCIENTIFIC INNOVATION AND RELEVANCE

Depending on the above-mentioned envelope properties, the thermal and energy behaviors of the greenhouse are different from those of ordinary buildings. Therefore, properly reflecting the envelope characteristics such as the light-shielding or light-transmission rates, thermal insulation by condition, reflection, and absorption in the model is crucial. However, not many studies have been conducted that consider such characteristics. In this study, we present a field data-based greenhouse modeling method that considers the envelope characteristics for the optimal operation of a smart greenhouse. Our proposed model was analyzed for its high-temperature suppression effect according to the type of greenhouse (single acting or interlocked). Our proposed modeling method can be used to establish modularity by analyzing the performance of each cooling unit technology applicable to a smart greenhouse. By expanding on this, the model in this study can be used to analyze the current state of greenhouses in South Korea and develop a database of energy consumption in terms of region and size of greenhouse. With such a database, presenting the energy consumption analysis results in terms of region, size, and type of greenhouse and suggesting an optimal operating method for a smart greenhouse is possible.

PRELIMINARY RESULTS AND CONCLUSIONS

As mentioned earlier, the energy model that considers the envelope characteristics of the greenhouse is highly useful for equipment system selection and operational control at the design stage. For greenhouses, in-depth consideration to parameters such as the material performance value of the envelope is essential. This study considered the envelope characteristics on the basis of the covering performance characteristic data of the field test bed for more accurate modeling. In this study, in order to reflect the envelope characteristics of the greenhouse, the performances related to the changes in light transmittance and temperature in accordance with the combination of covering and light-shielding materials mainly used were collected through experimental analysis in a real greenhouse. The results of these coating and/or light-shielding performances were used as the input values for the envelope in energy modeling. The performance and effectiveness of our developed model were verified by comparing these with the thermal behavior of real greenhouses using statistical techniques. The verified model was used to analyze the effect of high-temperature suppression according to the light-shielding rate or light-shielding method depending on the type of greenhouse (single acting or interlocked); subsequently, the results were presented.

MAIN REFERENCES

[1] Joan Muñoz-Liesa, Mohammad Royapoor, Elisa López-Capel, Eva Cuerva, Martí Rufí-Salís, Santiago Gassó-Domingo, Alejandro Josa, Quantifying energy symbiosis of building-integrated agriculture in a mediterranean rooftop greenhouse, Renewable Energy, 2020, Pages 696-709, ISSN 0960-1481, https://doi.org/10.1016/j.renene.2020.04.098.

[2] Ali Pakari, Saud Ghani, Evaluation of a novel greenhouse design for reduced cooling loads during the hot season in subtropical regions, Solar Energy, Volume 181, 2019, Pages 234-242, ISSN 0038-092X, https://doi.org/10.1016/j.solener.2019.02.006

[3] UN. Environment Programme, Building and Climate Change, UNEP/Earth-print, 2007.

[4] Hai-Young Jung, Seok-Hyun Lee, Boong-Joo Lee, Computer Simulation of Daylight Characteristics Considering Glass Greenhouse Materials and Structure, Journal of the Korean Institute of Illuminating and Electrical Installation Engineers 32(3), 2018.3, 20-29, http://www.dbpia.co.kr/journal/articleDetail?nodeId=NODE07406991

Keywords: Greenhouse, Envelope characteristic, Energy modeling, High-temperature suppression

Paper ID 11157: Master planning based on NZEB for Positive Energy Districts

Matthias Haase*

ZHAW, Switzerland

Bringing Energy Master Planning on a district level gives the possibility to untangle the academic discussions on the dynamics of energy. Consumers and producers are integral parts of the solution by understanding and influencing energy use/consumption both on individual and societal level. Communication issues between consumers and citizens will be eased and their role empowers them in limiting energy use/consumption and thus in creating the necessary communication and or engagement.

As a sustainable energy transition will see increased electro-mobility, its impact on the energy system needs to be understood and well integrated in planning. The Positive Energy Districts in this work consists of several buildings (new, retro-fitted or a combination of both) that actively manage their energy consumption and the energy flow between them and the wider energy system. Positive Energy Blocks/Districts make optimal use of advanced materials, local RES, local storage, smart energy grids, demand-response, cutting edge energy management (electricity, heating and cooling), user interaction/involvement and ICT. Positive Energy Districts are designed to be integral part of the district/city energy system and have a positive impact on it. Their design is intrinsically scalable and they are well embedded in the spatial, economic, technical, environmental and social context of the project site.

This workshop focuses on discussing what limiting energy consumption entails, how this is related to consumption patterns in general, how much we should or could limit energy consumption. The consumer and the producer need to act and react. In Positive Energy Districts the objectives of climate mitigation and adaptation goals, local energy, air quality and climate targets and a secure and resilient energy system were met. Furthermore, it should lead to significant increase share of i) renewable energies, ii) waste heat recovery and iii) appropriate storage solutions (including batteries) and their integration into the energy system and iv) reduce greenhouse gas emissions.

Integrated requirements for Positive Energy Districts are discussed based on NZEB and their environmental, economic and social performance indicators were identified. Related to environmental impact the ISO14031 Standard presents 3 categories of performance indicators: Management Performance Indicators (MPIs), Waste Indicators (WI) & Environmental Condition Indicators (ECI). How can these be integrated into Building Performance Simulations? Who are the stakeholders? What is the relation to EN ISO 52000? What are Base case, baseline and useful scenario developments? How does the concept of resilience fit into the Energy Master Planning?

The workshop is based on 2-3 short impact presentations. Then the participants will be asked to discuss in groups of 5 the items and summarize in a short presentation.

The target audience will be planners, designer, modellers, architects, urban planners, decision-makers, municipal representatives, etc.

The work is related to ongoing work organized within IEA Annex 83 "Positive Energy Districts".

Keywords: Energy Master Planning, database architecture, workflow, key performance indicators

Paper ID 30343: Design optimization of dense urban neighbourhood with computational simulation and genetic algorithm for improving outdoor thermal environment

Tongping Hao* (1), Jianxiang Huang (1), Phil Jones (2)

1: Department of Urban Planning and Design, The University of Hong Kong, Pokfulam, Hong Kong SAR, China; 2: Welsh School of Architecture, Cardiff University, Cardiff, Wales, UK

AIM AND APPROACH

The outdoor thermal environment is of growing concern in the urban planning and design field, especially in tropical/subtropical regions. The layout and design of urban neighbourhood may have significant environmental impacts. Although computer simulation has been increasingly used to assess outdoor thermal conditions, it is yet to be fully adopted by practitioners in design thinking and processes. This study proposes a simulation-based optimization method that combines a thermal comfort simulation with the genetic algorithms (GAs) to achieve automatic optimization of design solutions for maximizing the percentage of thermal comfortable area onsite.

SCIENTIFIC INNOVATION AND RELEVANCE

Thermal environment could be improved under appropriate urban design. The simulation-based optimization method using GA bridges the gap between computational simulation and design development. It facilitates performance evaluation in guiding design refinement based on optimization objective which shows promising results. It could be a powerful tool to enhance designers' abilities in identifying optimal design solutions for improving outdoor thermal environment.

PRELIMINARY RESULTS AND CONCLUSIONS

This paper proposes a design optimization method to reduce outdoor heat stress and increase the thermal comfortable area in urban spaces. The thermal comfort simulation software CityComfort+ is linked with genetic algorithms to optimize the thermal conditions of open space defined by a cluster of buildings. The method is tested with a hypothetical case in Hong Kong. The algorithm can identify high-performing design schemes within a reasonable period, given the site and project constraints. The method can be useful for practice and the further development of simulation-based optimization methods.

MAIN REFERENCES

Cheshmehzangi, A., 2016. Multi-spatial environmental performance evaluation towards integrated urban design: A procedural approach with computational simulations. Journal of Cleaner Production 139, 1085–1093.

Housing, Planning and Lands Bureau, 2006. Air Ventilation Assessments Technical Circular.

Huang, T., Li, J., Xie, Y., Niu, J., Mak, C.M., 2017. Simultaneous environmental parameter monitoring and human subject survey regarding outdoor thermal comfort and its modelling. Building and Environment 125, 502–514.

Ng, E., 2009. Policies and technical guidelines for urban planning of high-density cities – air ventilation assessment (AVA) of Hong Kong. Building and Environment 44, 1478–1488.

Ng, E., Yuan, C., Chen, L., Ren, C., Fung, J.C.H., 2011. Improving the wind environment in high-density cities by understanding urban morphology and surface roughness: A study in Hong Kong. Landscape and Urban Planning 101, 59–74.

Shi, X., Yang, W., 2013. Performance-driven architectural design and optimization technique from a perspective of architects. Automation in Construction 32, 125–135.

Yuan, F., Lin, Y., 2019. Research on High-Rise Building Group Morphology Generative Design Method Based on Physical Wind Tunnel and Neural Network Algorithm. Journal of Human Settlements in West China 34, 22–30.

Keywords: Outdoor Thermal Environment, Design Optimization, Computational Simulation, Genetic Algorithm, Dense Urban Neighborhood

Paper ID 30651: Annual simulation of outdoor thermal comfort with TRNSYS

Marion Hiller* (1), Ata Chokhachian (2), Christian Frenzel (1), Pablo Arango (1), Wolfgang Kessling (1), Stefan Holst (1), Prof. Thomas Auer (3)

1: Transsolar Energietechnik GmbH, Germany; 2: Technische Universität München Lehrstuhl für Architekturinformatik; 3: Technische Universität München Lehrstuhl für Gebäudetechnologie und klimagerechtes Bauen

AIM AND APPROACH

In the framework of a research and development project methodologies and tools for the dynamic evaluation of the thermal behavior of outdoor urban spaces based on the simulation software TRNSYS (TRNSYS, 2020) have been developed.

The multizone building model of TRNSYS including its GUI TRNLizard for geometric parametric modelling were originally developed for evaluation of the transient behavior of indoor spaces. In general, they aren't limited to indoor spaces but can also be used for dynamic simulation of semi-outdoor and urban outdoor spaces. However, the modelling capabilities have to be extended to consider relevant parameters for the evaluation of thermal behavior of outdoor spaces e.g. wind and vegetation effects. In addition, the radiation calculation processes have to be enhanced. Finally, the available comfort metrics have to be extended to compute for example the UTCI index for local comfort positions.

In order to simplify and rationalize the working process of the simulation tool and to improve the user-friendliness, several components of TRNLizard were revised, partly merged and new components were integrated. In addition, the templates and libraries used for the exterior comfort calculation were extended.

SCIENTIFIC INNOVATION AND RELEVANCE

The last 20 years clearly show an increasing frequency of extreme weather events with an increase in summer heat waves throughout the world. This is particularly evident in the larger cities, where the effect is further enhanced by the Urban Heat Island Effect (UHI). This has an increasing impact on the health of the inhabitants (Witt et al., 2015). At the same time, the need for cooling of buildings will increase, which in turn will increase the energy demand of the building sector. However, studies also show that a 10% increase in well-watered inner-city green areas could mitigate the urban temperature rise in 2080 to the 1970 level (Gill et al., 2007).

In this context, appropriate instruments are needed to support planning processes. In addition to statements on extreme heat waves, the frequency of the various events is also important. For example, shorter periods of heat can be buffered by the thermal mass of the buildings, whereas this is impossible for longer periods. Therefore, simulation tools must be able to model these phenomena dynamically and over longer periods of time.

PRELIMINARY RESULTS AND CONCLUSIONS

All algorithms developed within the project were implemented and tested successfully. The new tools were applied for a study of a real urban space, the Bordeauxplatz in Munich. The model of the Bordeauxplatz consists of a thermal zone with three vertical layers and two inner vegetation zones. As external shading elements the surrounding buildings as well as two vegetation volumes outside the thermal model were considered.

The results show the positive effect of the vegetation on the local comfort. In the area of the tree alleys on the south-western side of the square, acceptable comfort conditions (moderate heat stress) prevail on the path area directly under the trees at air temperatures of 31 °C and an irradiation of 740 W/m² on the horizontal. In contrast, uncomfortable conditions (very strong heat stress) occur in the unshaded area close to the façade on the walkway. In addition to point in time results and time series the frequency distribution of comfort conditions is an interesting variable for urban planning. Directly under the trees a good comfort (no heat stress) is guaranteed over 85 % of the time period considered (June til August from 9 am to 7 pm).

MAIN REFERENCES

Chokhachian, A., & Hiller, M. (2020). PANDO: Parametric Tool for Simulating Soil-Plant Atmosphere of Tree Canopies in Grasshopper. Paper presented at the SimAUD2020, Vienna.

Gill, S. E., Handley, J. F., Ennos, A. R., & Pauleit, S. (2007). Adapting Cities for Climate Change: The Role of the Green Infrastructure. Built Environment, 33(1), 115-133. doi:10.2148/benv.33.1.115

Frenzel, C., Gröger, S., Hiller, M., Kessling, W., & Müllner, K. (2011). SIMULATION OF THERMAL COMFORT IN SOCCER STADIA USING TRNSYS 17. Paper presented at the Conference of International Building Performance Simulation Association, Sydney.

Hiller, M., Aschaber, J., & Dillig, M. (2010). Integration of low-e surfaces and shortwave solar radiation into human comfort calculation in TRNSYS 17. Paper presented at the BauSIM 2010 conference, Vienna, Austria

TRNSYS (2020). Trnsys18 Retrieved from https://www.trnsys.de

Jasak, H. (2009). OpenFOAM: open source CFD in research and industry. International Journal of Naval Architecture and Ocean Engineering, 1(2), 89-94.

TRNLizard. (2019). TRNLizard Retrieved from https://www.grasshopper3d.com/group/trnlizard

Witt, C., Schubert, A. J., Jehn, M., Holzgreve, A., Liebers, U., Endlicher, W., & Scherer, D. (2015). Auswirkungen von Klimaveränderungen auf Patienten mit chronischen Lungenerkrankungen Deutsches Ärzteblatt 112. Retrieved from https://m.aerzteblatt.de/print/173347.htm?app=1

Keywords: Outdoor thermal comfort, Urban simulation, UTCI, TRNSYS

Paper ID 30467: Do we supply adequate outdoor air to buildings occupancy?

Tareq Abuimara, Brodie W. Hobson*, Burak Gunay, William O'Brien

Carleton University, Canada

AIM AND APPROACH

This paper aims at evaluating the adequacy of typical mechanical ventilation rates for varying interzonal occupant densities in commercial office buildings. The goal is to quantify the frequency of underventilation due to day-to-day interzonal occupancy diversity. To this end, occupancy and ventilation data from two academic office buildings in Ottawa, Canada were collected and analyzed. Floor and zone-level occupancy were estimated by counting the number of Wi-Fi enabled devices connected to the access points (APs) in each building at hourly timesteps for 14-months in tandem with motion detector and CO2 data. Concurrent zone-level ventilation data were obtained from the building automation system (BAS). The data were analyzed to model the ventilation rates on a per person basis compared to pre- and post-pandemic recommended values.

SCIENTIFIC INNOVATION AND RELEVANCE

Design and modelling of commercial and institutional buildings heating, ventilation, and air conditioning (HVAC) systems is typically based on standard assumptions that occupants are homogeneously distributed in the building. However, occupants are heterogeneously distributed in the building due to dynamic nature of occupants and to the interzonal diversity (e.g. different space use). Consequently, overpopulated zones are expected to experience periodic underventilation. Observing the latest studies examined the airborne transmission of the disease, underventilation can increase the spread of viruses and put occupants' health and wellbeing at risk. The recent guidelines for building operation during the COVID-19 pandemic released by the American Society of Heating, Refrigeration, and Air conditioning Engineers (ASHRAE) state that ventilation during pandemic should be maximized (higher flow rate and outdoor air fraction), run continuously (24/7) if possible, and demand-controlled ventilations should be disabled to mitigate the spread of the virus through ventilation systems. This study will model and evaluate the current ventilation rates in light of the pandemic and will put forward recommendation for addressing any issues. The findings are expected to provide insights for HVAC designers and building energy management professionals for building systems design and operation.

PRELIMINARY RESULTS AND CONCLUSIONS

The preliminary results indicated that some building zones experienced periods of underventilation due to peak occupancy that is higher than anticipated by designers while other zones experienced overventilation. These findings highlight the risk of reopening these types of facilities, as a part of reopening economy, under current building operations conditions. Recommendations will be made to address the issue of underventilated periods such as deploying occupancy-centric HVAC controls and adopting maximum permissible building occupancies to mitigate the risks of occupancy reaching a density such that ventilation requirements can no longer safely be met. The findings are also expected to provide insights HVAC design and modelling.

MAIN REFERENCES

(1) Schoen, L. J. (2020). Guidance for Building Operations During the COVID-19 Pandemic. doi:10.1056/NEJMc2004973.

(2) Bourouiba L. (2020) Turbulent Gas Clouds and Respiratory Pathogen Emissions: Potential Implications for Reducing Transmission of COVID-19. JAMA.;323(18):1837–1838. doi:10.1001/jama.2020.4756.

(3) Brodie W. Hobson, H. Burak Gunay, Araz Ashouri, Guy R. Newsham (2020) Clustering and motif identification for occupancy-centric control of an air handling unit, Energy and Buildings, Volume 223, 110179, ISSN 0378-7788, https://doi.org/10.1016/j.enbuild.2020.110179.

Keywords: Pandemic, Ventilation, Occupancy, Underventilation, Interzonal diversity

Paper ID 31075: Physiological modeling of thermal responses of the elderly under heat-stressful conditions

Lili Ji* (1), Abdelaziz Laouadi (2), Chang Shu (1), Abhishek Gaur (2), Michael Lacasse (2), Liangzhu (Leon) Wang (1)

1: Department of Building, Civil and Environmental Engineering, Concordia University, Montreal, Canada; 2: Construction Research Center, National Research Council Canada, Ottawa, Canada

AIM AND APPROACH

Overheating in buildings has been a major concern for the health and comfort of occupants during extreme heat events. Through the regulation of sweating or the increase in blood flow to the surface of the skin, occupants under conditions of heat-stress can adapt to extreme heat exposure. However, the ability of the human body to regulate its temperature declines with age. Elderly people (>60 years old) are more vulnerable during indoor overheating conditions than younger people.

The objectives of this study are to: (1) Improve a physiological model of the human body by applying more advanced sweating models, shivering models, and methods of calculation of skin blood flow, and; (2) Develop a physiological model for the elderly based on the improved human body model. Several different sweating models, shivering models, and skin blood flow calculation methods were introduced to the two-node framework based on existing studies. The physiological values and threshold parameters for thermoregulatory action were modified. The core temperature, skin temperature, water loss, and standard effective temperature (SET) were calculated based on the new physiological model for the elderly and validated with measurement data.

SCIENTIFIC INNOVATION AND RELEVANCE

In a stressful thermal environment, the elderly population (>60 years old) are more vulnerable to exposure to heat than healthy, young people due to age-related changes of their physiology and thermoregulatory functions. Given the aging population around the globe, the thermal response and comfort of elderly people within their indoor environments are important factors for their well-being. In spite of this, previous studies have mainly focused on the thermal response of healthy young adults. It is worthwhile to develop a physiological model for the more vulnerable elderly groups to re-evaluate the thermal response and overheating risks in buildings.

Previous studies have developed physiological and thermoregulatory models to mimic the human response to external thermal conditions. However, there still exists a lack of comprehensive comparison of these models and a thorough validation of the simulated results with reference to experimental data.

PRELIMINARY RESULTS AND CONCLUSIONS

The two-node physiological model will be improved for the elderly population. The core temperature, skin temperature, water loss, and standard effective temperature (SET) of elderly people will be calculated based on the new model. The results will be compared with the numerical results from the 3D multi-segment model and validated with reference to measured data. This model can be used to support the development of indoor overheating criteria for the elderly in buildings such as long-term care centers or health centers.

The validated model will, therefore, be used to demonstrate the difference in thermal comfort experienced between the elderly and younger people. The same environmental conditions from a baseline experimental scenario will be used as inputs to both models to permit calculating the thermal comfort of both the elderly and young groups. The results are intended demonstrate that the elderly would be more vulnerable to exposure to extreme heat as compared to the young group; this helps illustrate the significance of focusing on overheating risks to which the elderly may be subjected during extreme heat events.

MAIN REFERENCES

Ooka, R. et al. (2010) 'Improvement of sweating model in 2-Node Model and its application to thermal safety for hot environments', Building and Environment. Elsevier Ltd, 45(7), pp. 1565–1573.

Itani, M. et al. (2020) 'Bioheat modeling of elderly and young for prediction of physiological and thermal responses in heat-stressful conditions,' Journal of Thermal Biology. Elsevier Ltd, 88(February), p. 102533.

De Souto Barreto P, et al. (2015). How much exercise are older adults living in long-term cares doing in daily life? A cross-sectional study. J Sports Sci 33, pp.116-124.

Laouadi, A. et al. (2020) 'Development of reference summer weather years for analysis of overheating risk in buildings,' Journal of Building Performance Simulation. Taylor & Francis, 0(0), pp. 1–19.

Keywords: Elderly people, Thermal comfort model, Overheating risk, Regulatory sweating

Paper ID 30489: Pre-processing behavioral data and Model Calibration in a Passive House

Jeonga Kang*, Sun-Hye Mun, Jabin Goo, Younghoon Kwak, Jung-Ho Huh

University of Seoul, Korea, Republic of (South Korea)

AIM AND APPROACH

COVID-19 has caused a global supply chain crisis, and the instability of fossil fuels has increased the implementation of policies for energy and greenhouse gas reduction.[1] The Korean government is strengthening the passive performance requirements for buildings.[2] Simultaneously, the real-time and behavioral characteristics of residential buildings are also intensifying increasingly. Various factors in the behavior of residential buildings have high uncertainty but can be advantageous in saving the energy of the buildings.[3]

Measurement and verification (M&V) in a building is a process that measures the energy-saving effect before and after performance improvements. If the behavior is reflected in the baseline model, which is the basis of the improvement effect, the effect of pure performance improvement can be confirmed. The purpose of this paper is to propose the level of occupant behavior scenarios in a passive house for the establishment of a behavioral baseline model for reliable M&V.

In the present study, the summer cooling period data of a new single-family residential building located in Korea, collected during two weeks from July 14 to 27, 2020, were used. Occupants' energy usage behavior and the energy used by the sector are measured using the power consumption in a real-time cloud monitoring system.

SCIENTIFIC INNOVATION AND RELEVANCE

Behavior is a major factor and must be considered in setting the M&V baseline. However, the measured behavior information has many variables and is too complex to apply as is. In this study, it is possible to assess the effect of high-quality performance improvement by defining the appropriate level of the behavior scenario through simplifications. In addition, if M&V is performed based on this behavioral baseline model, it is possible to derive the result of performance improvement, in which resident comfort and energy savings are considered simultaneously.

The study is divided into an analytical stage and an empirical stage [6]. In the analytical stage, the energy requirements are checked using building design information. The behavioral characteristics are classified by applying clustering, which is a data mining technique, to the cooling period data collected after July and August. In the empirical stage, the energy consumption of the simulation, with the behavioral characteristics, is compared to the actual energy usage, which is obtained from the monitoring system, and the level of simplicity for the appropriate behavior scenario is defined.

PRELIMINARY RESULTS AND CONCLUSIONS

In the present study, we define the level of application of occupant behavior scenarios for the M&V of residential passive buildings. Through this, the M&V baseline that reflects actual behavior can be set and can come close to a pure result to confirm the effect of performance improvement. The preliminary results and conclusions are as follows.

(1) As the energy requirements were derived based on the building design information, the cooling energy requirement was 0.79kWh/㎡. The actual cooling energy consumption during the period was 1.04kWh/㎡.

(2) A monitoring system was established for the occupants' energy usage behavior(occupancy, temperature, air conditioning signals, etc.) and the energy consumption(cooling, heating, ventilation, etc.) by the sector.

(3) A behavior scenario was developed by applying the clustering technique to the collected behavior data(Python 3.8.0). The level of simplicity of the appropriate behavior scenario was defined by comparing the energy demand and the actual energy consumption derived from the information in the design.

In the future, if M&V is performed using the results of this study, performance improvement can be achieved while considering occupant comfort and energy savings simultaneously. In addition, the collected behavior scenarios can be used to optimize building operations through predictions.

MAIN REFERENCES

[1] Global energy and CO2 emissions in 2020[International Energy Agency]. (2020, July 14). https://www.iea.org/reports/global-energy-review-2020/global-energy-and-co2-emissions-in-2020

[2] Yoon, Y. S., Kim, E. J., Lee, S. J., & Sim, M. H. (2014). A Study on the Correlation of Domestic Local Climate and Annual Heating Energy Demand - Focused on passive house performance buildings -. Architectural Institute of Korea, 30(1), 209-216.

[3] Mun, S. H., Kwak, Y. H., & Huh, J. H. (2019). A case-centered behavior analysis and operation prediction of AC use in residential buildings. Energy and Buildings, 188-189, 137-148.

[4] Lee, M. J., Kim, W. S., Lee, W. J., & Lee, W. T. (2012). A Study about Reduction Rates of Building Energy Demand for a Detached House according to Building Energy Efficient Methods. Architectural Institute of Korea, 28(5), 275-282.

[5] Abreu, J. M., Pereira, F. C., & Ferrao, P. (2012). Using pattern recognition to identify habitual behavior in residential electricity consumption. Energy and Buildings, 49, 479-487.

[6] Kim, K. H., & Haberl, J. S. (2015). Development of methodology for calibrated simulation in single-family residential buildings using three-parameter change-point regression model. Energy and Buildings, 99, 140-152.

Keywords: Passive house, Residential building, Occupant behavior, Data mining, M&V(Measurement & Verification)

Paper ID 30261: Machine learning-based framework to predict single and multiple daylighting simulation outputs using neural networks

Rania Labib*

Prairie View A&M university, United States of America

AIM AND APPROACH

Building energy consumption accounts for 30% of global energy consumption [1]. In order to support the development of energy-efficient built environment and cities, architects, urban planners, and engineers have begun to utilize Building performance simulation (BPS). Supporting decision-making and steering the design towards high performance is crucial in the early design phase where decisions have the highest impact on final building's energy consumption and costs [2–4]. However, BPS tasks are usually time consuming. Therefor, there is a need for framework that would speed up the BPS process. This paper aims to develop a Machine Leaning algorithm, specifically Neural Networks (NN), that can potentially speed up the process of daylighting simulations by executing only a small subset of the simulations to predict the performance of daylighting of thousands of design configuration. Furthermore, the paper will investigate the use of evolutionary algorithm to automatically adjust the parameter of the NN without the interference of the user.

SCIENTIFIC INNOVATION AND RELEVANCE

Several researchers have examined the application of Machine Learning (ML) to predict the performance of the built environment in terms of energy consumption, daylighting harvesting, and thermal comfort. Researchers have proved that ML algorithms, specifically Artificial Neural Networks (ANN) accurately predicted the energy and other performance aspects of buildings [5–7]. For example. For example, Neto and Fiorelli used a NN models to predict the thermal performance of the Administration Building of The University of Sao Paulo. Neural networks showed an agreement between energy consumption forecasts and actual values, with an average error of about 10% [8]. Although researchers have proved that NN is a great time-efficient alternative to the existing time-consuming and computationally expensive simulations tasks, designing an NN model could be difficult for the non-technical user. The research in tis paper introduces a method that automatically deploy the NN model without any interference from users, therefor, facilitating access users without any technical knowledge.

PRELIMINARY RESULTS AND CONCLUSIONS

The author examined the daylighting performance of one small office room. The geometric model contained ten different parametric variables, such as ceiling height and the lightshelf depth, etc. Each parameter contains a set of different values, with total of 30 different values that lead to 8192 unique room configurations. For the purpose of this experiment, the average illuminance inside the office room was used as a metric to measure thee daylighting performance of the room in winter solstice.

By deploying the self-adjusting ANN model, the Mean Square Error (MSE) and Mean Absolute Error (MAE) values are equal to 0.013 and 0.085 respectively. This means that the predicted average illuminance value was only 85 points off the actual value. This is considered a highly accurate model where the actual average illuminance levels in the data set range from 6000 to 15000 Lux. The results of the preliminary work by the author confirms that the NN algorithm can be a great alternative to the existing traditional BPS tools. NN reduced the time that is required to examine over 8000 room configurations from about 15-17 hours to almost instant.

MAIN REFERENCES

[1] EIA, U . S . Energy-Related Carbon Dioxide Emissions , 2015, 2017. https://www.eia.gov/environment/emissions/carbon/pdf/2015_co2analysis.pdf.

[2] J. Kanters, M. Horvat, The Design Process known as IDP: A Discussion, 1st Int. Conf. Sol. Heat. Cool. Build. Ind. SHC 2012. 30 (2012) 1153–1162. https://doi.org/10.1016/j.egypro.2012.11.128.

[3] J.S. Hygh, J.F. DeCarolis, D.B. Hill, S. Ranji Ranjithan, Multivariate regression as an energy assessment tool in early building design, Build. Environ. 57 (2012) 165–175. https://doi.org/10.1016/j.buildenv.2012.04.021.

[4] S. Attia, E. Gratia, A. De Herde, J.L. Hensen, Simulation-based decision support tool for early stages of zero-energy building design, Energy Build. 49 (2012) 2–15.

[5] S.L. Wong, K.K.W. Wan, T.N.T. Lam, Artificial neural networks for energy analysis of office buildings with daylighting, Appl. Energy. 87 (2010) 551–557. https://doi.org/10.1016/j.apenergy.2009.06.028.

[6] S. Zhou, D. Liu, Prediction of Daylighting and Energy Performance Using Artificial Neural Network and Support Vector Machine, Am. J. Civ. Eng. Archit. Am. J. Civ. Eng. Archit. 3 (2015) 1–8. https://doi.org/10.12691/ajcea-3-3A-1.

[7] H. Zhao, F. Magoulès, A review on the prediction of building energy consumption, Renew. Sustain. Energy Rev. 16 (2012) 3586–3592.

Keywords: Machine Learning, Neural Network, Daylighting

Paper ID 30584: Agent architecture in building performance: a review of current methods

Zachary Lancaster*, Tarek Rakha

Georgia Institute of Technology, United States of America

AIM AND APPROACH

The rapidly changing landscape of working habits forces us to recognize energy usage as much a social problem as a mechanical one. There is a demonstrated lack of accuracy in current methods of energy modeling, a significant portion of which has been attributed misrepresentation of occupant behavior. This has come from the ways in which occupancy modeling methods at the building scale fail to capture diversity in our built environment and flatten any notion of diversity. This is problematic at the building scale, and when applied at the urban scale the challenge is only magnified. There is a clear need to represent the role of human occupants and their diversity within our Building Energy Models (BEMs).

This paper first reviews the current state of Agent-Based Modeling (ABM) as applied to occupancy simulation at the building and urban scales. We follow with the proposal of a conceptual framework for creating agents capable of simulating how denizens consume space; influencing local energy needs. Additionally, we propose a means of comparing off-the-shelf simulation environments for these kinds of agent simulations in the context of energy performance, namely the Unreal and Unity cross-platform gaming engines.

SCIENTIFIC INNOVATION AND RELEVANCE

Current models use schedules to determine occupancy, viewing it as the intersection between occupant density, active metabolic function and the presumed induced demand for given programmatic specifications. While such models explain the behavior of mechanical systems, they relate internal thermal loads to mechanistic models of "reasonable behavior" for a monolithically defined population.

Occupant behavior is cyclically related to the state of a space; one altering the other and vice versa. Current occupancy models that rely on the occupancy schedule fail to capture dynamic social interactions, driving a 32% variance in non-residential buildings. This paper reviews the current state of ABMs in general from both a human sciences and building energy modeling perspective in order to triangulate a view of how ABMs may be used to address common BEM use cases.

PRELIMINARY RESULTS AND CONCLUSIONS

A literature review concerning occupancy modeling demonstrated three major issues. Firstly, current tools do not adequately capture occupancy within their semantic ontology beyond basic concepts of occupancy schedules that, even with sophisticated methods of calibration, will rarely capture the social diversity driving actual energy consumption. Secondly, at the urban scale this problem compounds as frameworks represent buildings as aggregates, leading to an oversimplification of the social field. Lastly, current research in ABM for occupancy simulation focuses on reactive agents that follow rules rooted in assumptions of economic rationality and focus on calibrating existing schedules.

The results indicate that there is a sliding scale for the efficacy of ABMs when considering building energy. Generally, the more broad the scope of the work the more abstract the agent can become. For example, urban use cases can tend to only simulate economic decision making around transportation and location and may not need to consider the nuances of specific social interactions. More fine grained analysis, such as HVAC sizing though may require a deeper explanation of the interaction between physical comfort and the larger social field.

MAIN REFERENCES

J. L. M. Hensen and R. Lamberts, "Building performance simulation – challenges and opportunities," in Building Performance Simulation for Design and Operation, Routledge, 2019

Yan, D., O'Brien, W., Hong, T., Feng, X., Burak Gunay, H., Tahmasebi, F., & Mahdavi, A. (2015). Occupant behavior modeling for building performance simulation: Current state and future challenges. Energy and Buildings.

Azar, E., Menassa, C. C., Asce, S. M., Menassa, C. C., & Asce, A. M. (2012). Agent-Based Modeling of Occupants and Their Impact on Energy Use in Commercial Buildings. Journal of Computing in Civil Engineering.

C. Adam and B. Gaudou (2016). "BDI agents in social simulations: A survey," Knowledge Engineering Review.

E. Azar, C. C. Menassa, S. M. Asce, C. C. Menassa, and A. M. Asce (2012). "Agent-Based Modeling of Occupants and Their Impact on Energy Use in Commercial Buildings," J. Comput. Civ. Eng.

C. Reinhart and C. C. Davila (2019). "Urban building energy modeling," in Building Performance Simulation for Design and Operation, Routledge

Abar, S., Theodoropoulos, G. K., Lemarinier, P., & O'hare, G. M. P. (2017). Agent Based Modelling and Simulation tools: A review of the state-of-art software. Computer Science Review, 24

Keywords: Agent Based Modeling, Occupancy, Energy Performance, Semantic Web, Urban

Paper ID 30537: Indoor environmental quality survey in Research Facility: a Floor-By-Floor Analysis

Jong-Won Lee*, Deuk-Woo Kim, Seung-Eon Lee

Korea Institute of Civil Engineering and Building Technology, Korea, Republic of (South Korea)

AIM AND APPROACH

Indoor environmental quality (IEQ) helps ensure the pleasant and healthy working life of the occupants living and working in various types of buildings. However, although the IEQ has significant impacts on research efficiency and work productivity in a research office, there are very few studies involving researchers in a research facility [1]. A specific drill down analysis, such as a floor-by-floor examination, would yield further insights into the IEQ. This study used the occupant IEQ surveying approach with building-by-building and floor-by-floor analysis on research buildings in Korea using the established Korea Building Occupant Survey System (K-BOSS) system. This system was designed to analyze not only the average response of a large population for each building but also the floor-by-floor survey response based on the three-dimensional spatial information for occupant satisfaction [2]. The main aim of the study was to obtain insights on the process and measures in IEQ evaluation analysis in order to upgrade the K-BOSS system.

SCIENTIFIC INNOVATION AND RELEVANCE

In this study, the K-BOSS system was found to be useful when it is difficult to identify the reason of low IEQ evaluation in the floor level. In other words, the floor-by-floor response results can explain more detail of occupant`s dissatisfaction comparing to building`s average satisfaction that cannot explain. Thus, it provides more effective and intuitive decision making on the occupant`s IEQ satisfaction.

In addition, while previous survey-based studies were consulted in this study, the questions were amended to make them suitable for an occupant satisfaction survey in a research facility utilizing the UC Berkeley Center for the Built Environment (CBE)'s web-based occupant IEQ survey that was published in the ASHRAE guidebook [3, 4]. A total of seven IEQ criteria were established in this study including IEQ's four principal criteria including thermal comfort, air quality, lighting, and acoustic quality, with additional three important criteria including personal workspace, office layout, and the building overall.

PRELIMINARY RESULTS AND CONCLUSIONS

The occupant`s satisfaction score of overall personal workspace and overall building satisfaction for three buildings in research facility were identified to be above average at 4.6, 5.0, 4.2 and 4.7, 5.1, 4.6, respectively of 7-point scale. In other words, the three buildings were regarded as workspace or buildings where, on average, no particular abnormalities were observed. However, the result from analyzing the response data of the building A through drill-down and floor-by-floor visualization by survey question, floor, position, and gender revealed the lowest satisfaction scores for the noise-related survey questions, with the lowest scores given by the 1st and 4th floor occupants.

Future study will find the reasons of lowest scores in the 1st and 4th floor by analyzing responses and taking interview of occupants.

In addition, through the upgrade of K-BOSS, subsequent studies are planned which will investigate referring to [5]: (1) the correction of response results considering the elapse of time; (2) analysis of characteristics by floor and zoning; (3) internal consistency analysis; and (4) a photo-based analysis of workspace positions to approach personal comfort responses.

MAIN REFERENCES

[1] Kang, S., Ou, D., and Mak, C. M. 2017. The impact of indoor environmental quality on work productivity in university open-plan research offices. Building and Environment, 124, 78–89. doi:10.1016/j.buildenv.2017.07.003

[2] Lee, J., and Kim, D.-W. 2019a. Design of online platform and visualization system based on three-dimensional spatial information for occupant satisfaction with indoor environment quality. IOP Conference Series: Materials Science and Engineering, 609. doi:10.1088/1757-899x/609/4/042037

[3] ASHRAE. 2012. Performance measurement protocols for commercial buildings: Best practices guide.

[4] Huizenga, C., Laeser, K., and Arens, E. 2002. A web-based occupant satisfaction survey for benchmarking building quality. Proceedings of the Indoor Air.

[5] Li, P., Froese, T. M., and Brager, G. 2018. Post-occupancy evaluation: State-of-the-art analysis and state-of-the-practice review. Building and Environment, 133, 187–202. doi:10.1016/j.buildenv.2018.02.024

Keywords: Indoor environmental quality (IEQ), Occupant satisfaction, Web-based Survey System, Research facility, Floor-by-floor analysis

Paper ID 30144: Dynamic life cycle assessment of electricity demand of buildings with storage systems – potential for environmental impact mitigation

Jalomi Maayan Tardif* (1), Vasco Medici (1), Pierryves Padey (2)

1: Institute of Applied Sustainability to the Built Environment (ISAAC), University of Applied Sciences and Arts of Southern Switzerland (SUPSI), Canobbio CH-6952, Switzerland; 2: School of Management and Engineering Vaud (HEIG-VD), University of Applied Sciences and Arts of Western Switzerland (HES-SO), Yverdon-les-Bains CH-1400, Switzerland

AIM AND APPROACH

This study aims at evaluating the influence of building electricity self-generation and storage systems on the environmental impacts of building electricity consumption. A Dynamic Life Cycle Assessment (D-LCA) method was adopted by considering the hourly environmental impacts of the grid, and by including the impacts of Photovoltaic (PV) generation and electric storage equipment. The grid impacts are based on a database of hourly imports, exports and the national production of Switzerland and its neighboring countries. The PV and battery impacts rely on the ecoinvent Life Cycle Inventory (LCI) database. The environmental impacts are assessed in terms of the Global Warming Potential (GWP) and non-renewable primary energy indicators. The loads of a monitored Swiss multi family house are used as the reference case, for a D-LCA comparison of the performance of twelve PV and battery combinations. Additionally, alternative scenarios are developed to assess the influence of the GWP indicator time resolution, the heating technology, the control approach, and the grid supply mix on the resulting environmental impact. Considering the electricity supply sources on an hourly basis will capture the intra-day and seasonal fluctuations of building electricity consumption environmental impacts, which do not come to light with annual grid environmental impact indicators.

SCIENTIFIC INNOVATION AND RELEVANCE

Over the course of recent years has been observed in Switzerland a trend of reduced annual hydropower production, perhaps as an effect of climate change. This development, along with a decree to phase-out local nuclear sources and hydropower capacity expansion limitations, has brought the government to set directives that valorise building self-generation systems, in order to ensure reliability and resiliency under the uncertain effects of climate change. Yet, it remains to be clarified the extent of environmental benefits provided by building self-generation systems with energy storage, when considering these technologies from an LCA perspective. In literature can be found a wide range of values concerning the LCI of batteries, but mostly addressed to the automotive industry. This study offers insight into the influence of batteries on the annual environmental impact of electricity consumption in the building sector. The carbon payback time of the battery is computed in order to asses if the annual emissions savings, that it provides during its operational lifetime, are enough to compensate for its embodied emissions. Additionally will be examined the break-even points, below which the GWP of a battery complemented PV system can provide environmental benefits, given the self-consumption and self-generation capacity of contemporary technologies.

PRELIMINARY RESULTS AND CONCLUSIONS

There is seasonality to the environmental impact of the grid supply mix, due to higher portions of imports in winter and a peaking local hydropower production when snow melts. As such, the GWP for each kWh discharged from the battery is higher than the grid GWP in 31% of the hours simulated, of which 77% occur in summer. Thus, a control approach that is oblivious to GWP, could lead to a disadvantageous timing of the battery discharge in summer. The PV and battery system could reduce the GHG emissions related to grid energy consumption by up to 60%, however, due to the embedded emissions of these technologies, the actual relative change in annual GHG emissions from the reference case is between 4% and 32%, depending on the system size and the scenario. As for the non-renewable primary energy indicator, the energy from the grid has an hourly impact factor that is consistently higher than the PV energy dispensed from the battery, with the difference ranging from 0.9 to 7.1 MJp/kWh. In such case, the PV and battery combination can lead up to a 45% and 23% reduction in primary energy, compared to the grid-only and the PV-only systems, respectively.

MAIN REFERENCES

Hiremath, M., Derendorf, K., & Vogt, T. (2015). Comparative life cycle assessment of battery storage systems for stationary applications. Environmental Science and Technology, 49(8), 4825–4833. https://doi.org/10.1021/es504572q

Raugei, M., Leccisi, E., & Fthenakis, V. M. (2020). What Are the Energy and Environmental Impacts of Adding Battery Storage to Photovoltaics? A Generalized Life Cycle Assessment. Energy Technology, 1901146, 1–9. https://doi.org/10.1002/ente.201901146

Stolz, P., Frischknecht, R., Kessler, T., & Züger, Y. (2018). Life cycle assessment of PV-battery systems for a cloakroom and club building in Zurich. Progress in Photovoltaics: Research and Applications. https://doi.org/10.1002/pip.3089

Vuarnoz, D., & Jusselme, T. (2018). Neutral global warming potential target of electricity storage as threshold for greenhouse gas emission mitigation in buildings. Plea 2018.

KBOB, eco-bau, & IPB. (2016). KBOB Ökobilanzdatenbestand DQRv2:2016; Grundlage für die KBOB-Empfehlung 2009/1:2016: Ökobilanzdaten im Baubereich, Stand 2016.

Ecoinvent Centre.(2010). ecoinvent data v2.2, ecoinvent reports No. 1–25. Duebendorf, Switzerland, Swiss Centre for Life Cycle Inventories.

Keywords: Energy storage, LCA, GHG emissions, Primary Energy, Photovoltaics

Paper ID 30487: Multi-objective optimization of a simplified "Jaali" screen for daylight performance

Sanyogita Manu*, Adam Rysanek

The University of British Columbia, Canada

AIM AND APPROACH

"Jaali" is an Urdu word for traditional latticed screens, also known as "mashrabiyas", widely used in most building typologies in hot climate zones, particularly in South Asia, Middle East and parts of Europe. In vernacular architecture, Jaali served three main functions: climate control, visual privacy, and aesthetics. The objective of this study is to use a simplified version of a Jaali and demonstrate the application of evolutionary computing process for optimizing the perforation ratio for daylight, thermal and energy performance. This will be done by creating the basic room geometry in Rhino6; then adding the Jaali definition in Grasshopper3D (GH), which is a parametric modeling tool; and finally, using a multi-objective optimization engine called Octopus that runs within GH to optimize the perforation ratio for performance objectives. These performance metrics will be obtained using a combination of daylighting and energy simulation plug-ins for Rhino6 and GH. The "optimal" Jaali configuration from the Pareto front will be selected based on the multi-criteria decision making techniques of "pros and cons" and "simple prioritization".

SCIENTIFIC INNOVATION AND RELEVANCE

The parametric approach is the most commonly used method for performance optimization of fixed shading devices such as Jaali screens [1]-[6]. This study, however, illustrates the use of genetic algorithms for multi-objective optimization of a Jaali. A Pareto front approach provides the designer with several optimal solutions and enables early design decision-making in the integrated design process.

Furthermore, this study uses the example of a Jaali to focus on the need for simple and in-expensive building components that are designed to improve the performance of a standard building where energy simulation or any other kind of performance assessment is not the norm. This is especially true for countries like India where new buildings are being erected at an unprecedented rate. While there are a few examples of Jaali-inspired "smart" facades (e.g., Arab World Institute), it is not possible for a typical building in India to be designed with such high-performance components due to their complexity, cost and high operational energy. The methodology used in this study could potentially result in a generic solution (of a Jaali) that would meet the performance objectives for a wide range of applications and could be used likewise for other building components.

PRELIMINARY RESULTS AND CONCLUSIONS

The preliminary analysis was performed for daylighting performance. The optimization process resulted in six unique and recommended solutions on the Pareto front. The optimal solution was identified as the one resulting in the highest SDA (89%) and lowest ASE (8.2%), therefore complying with the LEED v4 requirements for daylighting credits. The perforation ratio of this configuration was 19.6%.

MAIN REFERENCES

[1] A. Sherif, H. Sabry, and T. Rakha, "External perforated Solar Screens for daylighting in residential desert buildings: Identification of minimum perforation percentages," Sol. Energy, vol. 86, no. 6, pp. 1929–1940, 2012, doi: https://doi.org/10.1016/j.solener.2012.02.029.

[2] A. Batool and I. M. K. Elzeyadi, "From Romance to Performance: Assessing the Impacts of Jali Screens on Energy Savings and Daylighting Quality of Office Buildings in Lahore, Pakistan," in 30th International PLEA Conference, 2014, pp. 334–343.

[3] I. M. K. Elzeyadi and A. Batool, "Learning from the Vernacular: The Impacts of Massive Perforated Screen Shades on Building Energy Savings and Thermal Comfort in Two Different Hot Climate Zones," in ASHRAE Winter Conference Proceedings, 2018.

[4] A. Sherif, A. El-Zafarany, and R. Arafa, "External perforated window Solar Screens: The effect of screen depth and perforation ratio on energy performance in extreme desert environments," Energy Build., vol. 52, pp. 1–10, 2012, doi: https://doi.org/10.1016/j.enbuild.2012.05.025.

[5] D. Gandhi, "Analysis of Indian traditional Jali (stone lattice screen) for daylighting through simulations," CEPT University, 2012.

[6] P. S. Kumar, "Traditional Perforated Screens of India," Architectural Association, 2014.

Keywords: Jaali screens, perforated screens, daylight performance, genetic computing, multi-objective optimization

Paper ID 30197: Optical and calorimetric product data in Building Information Modelling

Christoph Maurer* (1), Simon Wacker (1), Bruno Bueno (1), Jacob C Jonsson (2), Hervé Lamy (3), David Bush (3), Meiling Shi (4), Wendelin Sprenger (5), Robin Mitchell (2), Helen Rose Wilson (1), Dragan C Curcija (2), Tilmann E. Kuhn (1)

1: Fraunhofer Institute for Solar Energy Systems ISE, Freiburg, Germany; 2: Lawrence Berkeley National Laboratory, Berkeley, CA, USA; 3: European Solar-Shading Organization, Brussels, Belgium; 4: Institute of Numerical Methods and Informatics in Civil Engineering, TU Darmstadt, Germany; 5: Ed. Züblin AG, Stuttgart, Germany

AIM AND APPROACH

Many formats to exchange data about optical measurements and simulations involve a high degree of implicit information which is necessary to interpret the data correctly. Currently, data about building envelopes is often handled manually, for example as individual values retrieved from a data sheet. For buildings with high performance targets, more detailed data is necessary for example high-resolution Bidirectional Scattering Distribution Functions (BSDF) to evaluate glare. As presented in [1], the detailed data exceeds available Building Information Modelling (BIM) formats such as Industry Foundation Classes (IFC).

Currently, there are databases like the International Glazing Database (IGDB), the Complex Glazing Database (CGDB) and the European Solar Shading Database (ES-SDA). In order to facilitate the use of this data by applications with a user interface, the following approach was developed: The approach is to create a Knowledge Database (KD) which manages unique identifiers (UUID) for components and institutions. The component data is then stored at child databases (CD). KD and CD share the same specification for the Application Programming Interface (API). In this way, appplications can query KD with a component ID. KD federates the query to all CD and merges the results to one response to the user interface.

SCIENTIFIC INNOVATION AND RELEVANCE

Data has been strongly linked to specific software in the past. It would be better that different applications can use the same data provided by a back end server or cloud. At the same time, applications would benefit from the option to access easily different data sources such as the new International Glazing and Shading Database (IGSDB) and ES-SDA. The keys for this easy access is a common API framework to makes it easy to create applications which use the data. UUIDs for components make it possible to search and merge data about one component. In the future, the user management of the KD may be used to define individual access rights. The proposed system of databases will therefore increase the efficiency of using product data of glazing and shading.

PRELIMINARY RESULTS AND CONCLUSIONS

The first version of the API specification is being developed and will be presented in the full paper. A first version of the KD and CD is being developed and their interaction will be presented in the full paper.

Preliminary results indicate that the system of databases may help the Architecture, Engineering and Construction domain to make energy-efficient building more cost-effective.

MAIN REFERENCES

[1] C. Maurer, A. Wagner, J. Wengzinek, U. Rüppel, W. Sprenger, T.E. Kuhn, Rule-Based Views and linked building data for efficient planning processes, in: 16th Conference of Internation Building Performance Simulation Association (IBPSA), Rome, Italy, 2019.

Keywords: Building Information Modelling (BIM), Product Data, Optical Data, Bidirectional Scattering Density Function (BSDF), Solar Heat Gain Coefficient (SHGC)

Paper ID 31129: Enable building system simulation by deriving a Digital Twin for operation from 2D P&ID plans

Hermann Georg Mayer*

Siemens AG, Germany

AIM AND APPROACH

We propose a method for generating semantics for 2D plans of building systems, predominantly P&ID plans of HVAC systems. For most buildings, even today designing the HVAC system and other building systems is not yet an integral part of a completely centralized 3D BIM modelling process. This applies even more to legacy buildings, which are regularly renovated for better energy performance. Therefore, most of system descriptions, like for HVAC, are still 2D drawings with no semantics included and no preparation for an automated simulation pipeline. The method introduced below shows how to get a machine-readable simulation model from those 2D plans by an easy-to-use, semi-automatic process.

SCIENTIFIC INNOVATION AND RELEVANCE

Use classical pattern recognition for P&ID plans and show advantages over recent AI methods.

Demonstrate the usage of a knowledge graph as basis for simulations.

Enable the versioning of system models to reflect realistic system development processes.

PRELIMINARY RESULTS AND CONCLUSIONS

The publications presents and evaluates a method to derive a semantically enriched model out of standard 2D system drawings. The model can be used for versioning the system or for later system simulation. Additional challenges have been identified for further research (like defining a domain-specific language for pattern definition or providing context-aware merging of different system versions.

MAIN REFERENCES

Duda, R. O. and P. E. Hart (1972),. Use of the Hough Transformation to Detect Lines and Curves in Pictures. Comm. ACM, Vol. 15, 11–15.

Canny, J. (1986). A computational approach to edge detection. IEEE Transactions on Pattern Analysis and Machine Intelligence, vol 8, 679–714.

Keywords: Pattern recognition, Knowledge graphs, System model versioning

Paper ID 31070: Simulation and visualization framework for selecting appropriate urban-scale climate mitigation and adaptation strategies

Justin McCarty*, Adam Rysanek

School of Architecture and Landscape Architecture, University of British Columbia, Vancouver, Canada

AIM AND APPROACH

The importance of factoring in the embodied carbon of new construction is becoming relevant at all levels of the construction industry, particularly as many decision makers (owners, planners, developers) are faced with meeting mandated carbon targets and reaching goals of densification. A recently constructed community of 15,000 in Vancouver, BC (Wesbrook Village) offers a case study in how to recommend decarbonisation planning decisions through building simulation. Previous research on this community has shown that a planned neighbourhood energy systems is not able to compete on a carbon performance level as the regional electricity grid depending on whether operating carbon and/or embodied carbon is assessed. The same issue is foreseen on decision around construction material and community density. This paper incorporates elements of life-cycle emissions accounting in a previously-published sequential search-based building simulation process. The work assesses the Wesbrook Village site to determine, from the perspective of the municipalityy, the least-cost carbon reduction pathway that includes both operating and embodied emissions - and yet still achieves the community's growth targets.

SCIENTIFIC INNOVATION AND RELEVANCE

This work will result in the production of marginal carbon abatement cost curves for an urban community that include both operating and embodied carbon, as well as the effect of individual technologies on real estate value and revenue. This is the first known simulation study to do so, and so this work seeks to contribute to best practise building performance simulation assessment in the coming century. The work makes use of future climate weather files under varying climate scenarios, to forecast long-term trends in building performance and comfort. Likewise, variable future cost scenarios are considered in a discounted cash flow analysis. Embodied carbon is also accounted for in the simulation process, such that every design intervention, whether it be a building technology change (i.e., upgrade of double glazing to triple glazing) or a future energy system choice is attributed with a projected value of operating carbon emissions, embodied carbon emissions, and real estate value and revenue.

Visualizing and discussing the sensitivity of simulation-based building design decisions to questions of operating vs. embodied carbon are important. It is not yet established whether factoring embodied carbon (and/or energy) is necessary for robust simulation-based decision-making on low-carbon building and energy systems.

PRELIMINARY RESULTS AND CONCLUSIONS

Our initial results indicate that inclusion of embodied carbon into simulation analysis alters cost-optimal community design decision in a number of manners. For Wesbrook Village in Vancouver, BC, we see that when considering embodied carbon, the favourability of electrifying heating systems of existing buildings in a simple manner (i.e., using baseboard electric heaters) becomes preferred over upgrading existing natural gas central heating systems to electric heat pumps. Most importantly, the results indicate the importance of retaining and expanding a typology of mid-rise residential buildings, which are based on timber-frame construction and yield a significantly lower carbon footprint than any development scenario that encourages construction of high-rise, concrete buildings.

MAIN REFERENCES

Rysanek, A. M., & Choudhary, R. (2013). Optimum building energy retrofits under technical and economic uncertainty. Energy and Buildings, 57, 324-337.

Nik, V. M. (2017). Application of typical and extreme weather data sets in the hygrothermal simulation of building components for future climate – A case study for a wooden frame wall. Energy and Buildings, 154, 30–45. https://doi.org/10.1016/j.enbuild.2017.08.042

Fonseca, J. A., Nguyen, T.-A., Schlueter, A., & Marechal, F. (2016). City Energy Analyst (CEA): Integrated framework for analysis and optimization of building energy systems in neighborhoods and city districts. Energy and Buildings, 113, 202–226. https://doi.org/10.1016/j.enbuild.2015.11.055

Keywords: Urban scale simulation, carbon reduction pathways, embodied carbon

Paper ID 30735: CNN vs. DNN: comparing the two neural networks performance in predicting building operational energy with respect to the building shape

Farnaz Nazari*, Wei Yan

Texas A&M University, United States of America

AIM AND APPROACH

An often overlooked, nonetheless key contributor, to a building's operational energy use is the building shape. The relationship between a building's operational energy and it's respective orientation, aspect ratio, width to length ratio, and window-to-wall ratio have been well documented in the literature; however, the relationship between the building shape and the operational energy use has rarely been addressed [1-3]. Significantly few studies have studied the relationship between building shape and building operational energy use [4]. Generally, building shape is not approached in tandem with an energy-wise design decision. One of the reasons is that making an energy-wise decision at early stages of the design is not very straightforward since the solution pool is extremely large, and it is hard to find a trend for different shapes. Particularly, self-shaded shapes may have contradictory effects on building cooling, heating, and day-lighting performance. These complexities call for an efficient method considering time and resources limitation. Deep Learning methods, which are recently being employed in architectural design research, can be used to help designers with early stages of the design to efficiently evaluate building energy performance [5, 6].

SCIENTIFIC INNOVATION AND RELEVANCE

The intention of this study is to evaluate the applicability of Deep Learning in predicting building operational energy with respect to the building shape. In this regard, two different Deep Learning algorithms; Deep Neural Network (DNN) that employs numbers as the input and Convolutional Neural Network (CNN) that are fed with images as the input, are compared in terms of complexity, number of layers, number of required epochs, and the performance measures to evaluate the predictive power of the two models, while the benefits, limitation and potential improvements of each network are investigated. Regarding that the possibility of an energy-wise design approach emerges at the early stages of the design process, where the designer produces building design representations including sketches, schematic floor plans, elevations, sections, etc., it will be highly valuable if the Deep Learning model can employ such graphics as the input to predict the operational energy use.

PRELIMINARY RESULTS AND CONCLUSIONS

The comparison of the two models shows that the DNN can be set easier with fewer number of layers and epochs, and smaller mean square error, whereas the CNN requires a deeper architecture and a higher number of epochs and is not yet able to perform as well as the DNN model. The main reason for such a wide difference between DNN and CNN performance is that CNNs are basically coded for computer vision tasks like image classification and object detection, and are not the best alternative for regression-based problems such as predicting energy performance of a building. However, the unignorable advantage of a CNN model for employing images as inputs and thus a better visual connection between the designer and the design evaluation process cannot be overlooked, and urges us for further exploration for finding methods to improve the CNN performance for the purpose of this paper.

MAIN REFERENCES

[1] T. L. Hemsath and K. A. Bandhosseini, "Sensitivity analysis evaluating basic building geometry's effect on energy use," Renewable Energy, vol. 76, pp. 526-538, 2015.

[2] W. Tian, S. Yang, J. Zuo, Z. Li, and Y. Liu, "Relationship between built form and energy performance of office buildings in a severe cold Chinese region," Building Simulation, vol. 10, no. 1, pp. 11-24, 2016.

[3] L. Wei, W. Tian, J. Zuo, Z.-Y. Yang, Y. Liu, and S. Yang, "Effects of Building Form on Energy Use for Buildings in Cold Climate Regions," Procedia Engineering, vol. 146, pp. 182-189, 2016.

[4] A. Yeretzian, H. Partamian, M. Dabaghi, and R. Jabr, "Integrating building shape optimization into the architectural design process," Architectural Science Review, vol. 63, no. 1, pp. 63-73, 2019.

[5] S. Chen, T.-T. Ren, and Z.-C. Wu, "Research on neural network optimization algorithm for building energy consumption prediction," Journal of Computational Methods in Sciences and Engineering, vol. 18, no. 3, pp. 695-707, 2018.

[6] I. As, S. Pal, and P. Basu, "Artificial intelligence in architecture: Generating conceptual design via deep learning," International Journal of Architectural Computing, vol. 16, no. 4, pp. 306-327, 2018.

Keywords: Artificial Intelligence, Neural Network, Optimization, Building energy performance, Building shape

Paper ID 30911: Towards more efficient modeling and simulation of Large-scale Thermal Energy Storages in future Local and District Energy Systems

Michael Reisenbichler* (1,2), Keith O'Donovan (1), Carles Ribas Tugores (1), Wim van Helden (1), Franz Wotawa (2)

1: AEE – Institute for Sustainable Technologies, Gleisdorf, Austria; 2: Institute of Software Technology, University of Technology, Graz, Austria

AIM AND APPROACH

Large-scale hot-water underground tank (TTES) and pit (PTES) thermal energy storages have a high potential to massively increase the proportion of renewable energy in future local and district energy systems [1]. To fully exploit the potential of these large-scale thermal energy storage (LTES) technologies, comprehensive planning and tuning of the overall system by dynamic system simulation is necessary.

The overall objective is to enable more efficient modeling and simulation in terms of implementation time, computational performance and accuracy of LTES in multi-annual dynamic system simulations in the future. As a first step to achieve this, we are going to demonstrate in this study the development and validation of a novel Modelica LTES model for dynamic system simulations.

In accordance with the Modelica modeling approach, the new model is developed based on existing models, supplemented by various adjustments to these. In order to assess the accuracy of the model, a validation case study by comparing the simulation results against real measurement data of a Danish PTES is performed. Furthermore, a cross-comparison against other numerical models used in a similar validation case study is conducted.

SCIENTIFIC INNOVATION AND RELEVANCE

Among scientific system simulation tools for LTES, TRNSYS is most broadly applied. Several models are available for dynamic system simulations (e.g. Types: 342, 1300/1301, 1322) [2–4]. Acausal modeling with Modelica is already well suited and has a high potential for improving modeling and simulation of future local and district energy systems [5]. Despite the advantages of co-simulation, the coupling of suitable models of different simulation tools with the actual system simulation may be time-consuming and costly, both in terms of implementation and computational time [6]. Therefore, a monolithic simulation of LTES systems within one simulation tool is advantageous.

Suitable LTES models in Modelica are limited or lack essential features. For Modelica-based simulation environments mainly freestanding models without modeling of the surrounding soil are available. These models are primarily focused on ordinary hot water tanks for domestic applications and can be found, for instance, in the Modelica Buildings library [7].

This study's innovation is the development and validation of a new Modelica LTES with a coupled soil model for dynamic system simulations. It is a first step towards more efficient modeling and simulation of LTES in future local and district energy systems.

PRELIMINARY RESULTS AND CONCLUSIONS

The model is developed in the Modelica-based modeling and simulation environment Dymola. Thereby, it became already apparent that due to the object-oriented modeling approach with Modelica an efficient modeling process with a high degree of reusability, extensibility and adaptability for LTES models is possible.

The mutual interaction with other system components is for the desired application of the model of high importance. The validation case study shows that the most important factors for that, like the charged and discharged energies as well as the storage temperatures and consequently also the corresponding temperatures for the interacting system components, can be accurately represented compared to the measurement data. Somewhat higher deviations from the measured data are only seen in the simulated thermal losses, which needs to be taken into consideration for detailed design studies. The cross-comparison with other numerical models indicated that the developed Modelica model generally shows good agreement with comparable models suited for system simulations.

In summary, the results show that the developed Modelica LTES model provides sufficiently accurate results for the application in dynamic system simulations. Furthermore, the experiences and findings demonstrate the high potential of Modelica for efficient modeling and simulation of LTES systems in future.

MAIN REFERENCES

[1] Pauschinger, T., Schmidt, T., Soerensen, P.A., Snijders, A., Djebbar, R., Boulter, R., and Thornton, J. (2020). Design Aspects for Large-Scale Aquifer and Pit Thermal Energy Storage for District Heating and Cooling. Final Project Report (IEA DHC/CHP Annex XII).

[2] Ochs, F. (2010). Modelling Large-Scale Thermal Energy Stores. Shaker. Aachen (DE).

[3] Klöck, M. (2018). Validierung eines Simulationsmodells für solare Fernwärmeversorgungen mit saisonalem Erdbecken-Wärmespeicher. Master Thesis. Hochschule Reutlingen (DE).

[4] Sveinbjörnsson, D., From, N., Schmidt, T., Klöck, M., and Pauschinger, T. (2020). Pit Thermal Energy Storage for Smart District Heating and Cooling: Technical report on model validation, cost functions and results of an exemplary base case study. Final Project Report (IEA DHC/CHP Annex XII).

[5] Wetter, M., and van Treeck, C. (2017). New Generation Computing Tools for Building and Community Energy Systems. Final Project Report (IEA EBC Annex 60). Lawrence Berkeley National Laboratory, Berkeley (US) and RWTH Aachen University, Aachen (DE).

[6] Dahash, A., Ochs, F., and Tosatto, A. (2019). Co-Simulation of Dynamic Energy System Simulation and COMSOL Multiphysics®. Proceedings from COMSOL Conference 2019. Cambridge (UK), 24–26 September 2019.

[7] Wetter, M., Zuo, W., Nouidui, T.S., and Pang, X. (2014). Modelica Buildings library. Journal of Building Performance Simulation 7, 253–270.

Keywords: Large-scale Thermal Energy Storage, Dynamic System Simulation, Modelica, Model Development and Validation

Paper ID 30375: Evaluating the regional implications of income-aware building stock retrofit strategies through constrained multi-objective optimization

Vincenzo Antonio Rossi*, Bianca Howard, Jonathan Wright

Loughborough University, United Kingdom

AIM AND APPROACH

Retrofitting the existing building stock is one of the most effective solutions to reduce energy demand and carbon emissions. Interventions at a national scale, however, are not successful unless they are environmentally, economically, and socially sustainable. Therefore, it is essential to understand how sociodemographic attributes of the households might affect the implementation of Energy Efficiency Measures(EEM). This study investigates how the differences among sociodemographic attributes and physical characteristics of different regional scenarios will change the trade-offs between cost and greenhouse gas emission as well as which measures should be implemented to achieve the reductions.

The English Housing Survey was used to populate a parametric model using Grasshopper for Rhino and Energy Plus as a dynamic simulation engine. The behaviour of 72 representative building archetypes, for the East Midlands, the North East and the London region was simulated, when EEM were applied. The stocks were then optimized, minimizing capital and operational costs, and emissions. Firstly, the Pareto-optimal solutions were selected for each archetype and then sociodemographic derived constraints were applied. Solutions were then optimized at the stock level using the NSGA-II genetic algorithm. Finally, the differences between constrained and not constrained optimizations were evaluated.

SCIENTIFIC INNOVATION AND RELEVANCE

Although previous research has found that households' sociodemographic characteristics have an impact both on the uptaking rate of EEM and on dwellings' energy consumption, these aspects are hardly analytically considered in a combined approach. (BEIS, 2018)(UK CCC, 2019)

This study directly links physical and sociodemographic variables in a stock-scale modelling approach for optimization purposes. Annualized costs for EEM could not exceed a threshold of 10% of the household's annual income.

Furthermore, previous work (Rossi et al., 2020) showed that consideration of income distribution of the investigated stock has a significant impact on the optimization outcome. In this study, the emissions reduction potential of the building stock was reduced by 40%, and the uptake of solid wall insulation reduced by 50%.

By selecting regional scenarios with significant differences in climate, physical conditions and sociodemographic attributes, this study determines how the regional differences in the distribution of the building form, building construction, and social-demographic variables can significantly change the optimal policy.

This study provides policymakers and local authorities with a novel science-based approach to social-sustainability analyses when dealing with large-scale retrofit policies. Also, results can help to understand how different sociodemographic contexts react to certain policies.

PRELIMINARY RESULTS AND CONCLUSIONS

Preliminary results show that when new constraints are applied, they reshape the morphology of the Pareto front that indicates the trade-off between CO_2e emissions and cost, as well as the distribution of EEM and archetypes.

Firstly, the front of optimal solutions is farther from the original non constrained optimality, comparably to the average income differences among the investigated stocks. Also, a percentage of the most emissions-reducing solutions is lost due to the constraints. This signifies that the expected emissions reduction deriving from a certain EEM implementation can be overestimated if sociodemographic barriers are not taken into account.

Secondly, EEM distribution changes accordingly to the introduced constraints. Frequency of the uptake of more expensive measures (e.g multi-glazed windows) is reduced but is also influenced by the initial retrofit condition of the stock, as well as the climate. With a higher percentage of non-insulated solid walls and an older building stock, the London region was significantly more sensitive to the sociodemographic constraints.

Lastly, the uptaking frequency of the archetypes across the solutions is also affected. Constrained optimizations could completely exclude some of the dwellings from the optimal configurations. This phenomenon is emphasized in the decision-making process when EEM are individually considered.

MAIN REFERENCES

IET & NTU. (2018). Scaling Up Retrofit 2050.

Campbell, S. D. (2016). The Planner's Triangle Revisited: Sustainability and the Evolution of a Planning Ideal That Can't Stand Still. Journal of the American Planning Association, 82(4), 388–397.

BEIS. (2018). ECO3: 2018 - 2022 Final Stage Impact Assessment (Issue October).

UK CCC. (2019). Net Zero Technical report Committee on Climate Change (Issue May).

Ministry of Housing Communities & Local Government. (2018). English Housing Survey.

Hamilton, I. G., Shipworth, D., Summerfield, A. J., Steadman, P., Oreszczyn, T., & Lowe, R. (2014). Uptake of energy efficiency interventions in English dwellings. Building Research and Information, 42(3), 255–275.

Pelenur, M. J., & Cruickshank, H. J. (2012). Closing the Energy Efficiency Gap: A study linking demographics with barriers to adopting energy efficiency measures in the home. Energy, 47(1), 348–357.

He, M., Brownlee, A., Lee, T., Wright, J., & Taylor, S. (2015). Multi-objective optimization for a large scale retrofit program for the housing stock in the North East of England. Energy Procedia, 78(0), 854–859.

(forthcoming) Rossi, V., Howard, B., & Wright, J. A. (2020). Incorporating sociodemographic constraints in multi-objective building stock retrofit optimization. USim2020.

UKGBC. (2013). Retrofit Incentives. In UK Green building council.

Keywords: retrofit, optimization, sociodemographic, building stock, social sustainability

Paper ID 31068: Uncertainty in daylight simulations of algorithmically generated complex shading screens

Luis Santos* (1), Inês Caetano (2), António Leitão (2), Inês Pereira (2)

1: Kent State University, College of Architecture and Environmental Design, United States of America; 2: INESC-ID/Instituto Superior Tecnico, University of Lisbon, Portugal

AIM AND APPROACH

This research investigates the impact of Non-uniform Rational Basis Spline (NURBS) geometry discretization and simulation methods in daylight simulation output, particularly in climate-based daylight modeling (CBDM) of complex façade screens generated algorithmically.

During a multi-objective daylight optimization experiment of weaved shading devices, we observed that the output of conventional Radiance-based CBDM approaches was susceptible to negligible geometric variations. We hypothesize the following causes for this behavior: (i) limitations related to daylight coefficient based simulation processes, (ii) geometric errors, introduced in the translation of 3D NURBS geometry to Radiance, (iii) a combination of the previous.

To further study the advanced hypotheses, this work compares different simulation parameters to the 2-phase method and different approaches to generate geometry for Radiance. Despite its limitations, the 2-phase method is the most used daylight coefficient and it is commonly available through DAYSIM-based tools [1]-[3], and uses shares several simulation with Radiance. The methods for generating geometry for Radiance include: (i) the standard export of NURBS geometry to OBJ format and then to RAD format provided by DIVA for Rhino, which provides little control of geometry discretization, and (ii) the algorithmic generation of geometry for Radiance while enabling the control of surface discretization.

SCIENTIFIC INNOVATION AND RELEVANCE

The OBJ2RAD [4] Radiance's subroutine was introduced to facilitate the interface between Radiance [5] and CAD tools that, due to the limited geometric objects supported at the time, could rely on the OBJ format to export them. As a result, designers could hardly model and simulate Complex Fenestration Systems (CFS). With the improvements in surface modeling operations, the OBJ format no longer accurately represents these geometric objects. Hence, one might argue that although OBJ2RAD perfectly satisfied the modeling needs of its time, it might be insufficient nowadays. Thus, it is critical to assess whether the OBJ2RAD tool can accurately translate the geometry produced by current digital design tools. Moreover, although the 2-phase method has limitations in simulating CFS [6] it is still recommended by the IES LM83 standard [7], and designers often use it in scenes that have CFSs. Thus, it is equally relevant to assess the error introduced by the different simulation parameters since they can hinder feedback.

By comparing different modeling methods and by simulating Radiance scenes, this work provides relevant insights on the uncertainty of common daylight and analysis workflows, as well as useful recommendations on setting adequate parametric and optimization workflows supported by daylight simulations.

PRELIMINARY RESULTS AND CONCLUSIONS

The sensitivity analysis of a parametric architectural shading device based on complex weaving patterns showed that the daylight simulation output using the 2-phase method was susceptible to negligible geometric variations. Although the absolute difference on spatial Daylight Autonomy (sDA) and Annual Sun Exposure (ASE) varies only between 2% to 5%, this small difference has a considerable impact when architects are assessing the daylight performance of their designs for building rating processes such as LEED Version 4 [8].

In the full paper, we examine the situations where 2-phase method simulation parameters are insufficient to provide adequate feedback, as well as the shortcomings of using the current NURBS geometry translation methods for Radiance. We examine the impact of small deviations in surface normals that result from the translation of a NURBS model to a Radiance model in the 2-phase method, as these can affect the light bouncing process, particularly in the case of specular reflection of complex architectural screens. Finally, we advance alternative methods to automatically describe a Radiance scene, present its merits and limitations, and provide specific recommendations on how to describe the geometry of complex design elements for CBDM.

MAIN REFERENCES

[1] C. F. Reinhart and O. Walkenhorst, "Validation of dynamic RADIANCE-based daylight simulations for a test office with external blinds," Energy and buildings, vol. 33, no. 7, pp. 683–697, 2001.

[2] A. Jakubiec and C. Reinhart, "DIVA-FOR-RHINO 2.0: environmental parametric modeling in rhinoceros/grasshopper using RADIANCE, Daysim and EnergyPlus," 2011.

[3] Sollema LLC, "Climate Studio," 2020. https://www.solemma.com/ClimateStudio.html (accessed Jul. 27, 2020).

[4] G. Ward, "OBJ2RAD," Jun. 14, 1994. https://floyd.lbl.gov/radiance/man_html/obj2rad.1.html (accessed Jul. 31, 2020).

[5] G. J. Ward, "The RADIANCE lighting simulation and rendering system," in Proceedings of the 21st annual conference on Computer graphics and interactive techniques, 1994, pp. 459–472, Accessed: Jul. 19, 2016. [Online]. Available: http://dl.acm.org/citation.cfm?id=192286.

[6] L. Santos, A. Leitão, and L. Caldas, "A comparison of two light-redirecting fenestration systems using a modified modeling technique for Radiance 3-phase method simulations," Solar Energy, vol. 161, pp. 47–63, 2018.

[7] IESNA. " LM-83-12 IES Spatial Daylight Autonomy (sDA) and Annual Sunlight Exposure (ASE)", New York, NY, USA: IESNA Lighting Measurement, 2012.

[8] USGBC, LEED v4 for Building Design & Construction. 2013.

Keywords: Facade Shading Devices, Daylight Analysis, Uncertainty, Algorithmic Design

Paper ID 30402: Finding occupant reference profiles using self-organizing maps for thermal performance assessment

Aline Schaefer*, Enedir Ghisi

Universidade Federal de Santa Catarina (UFSC), Brazil

AIM AND APPROACH

Self-Organizing Maps (SOM) are an efficient tool for multivariate data analysis due to their capability to represent multidimensional data in 2D images. The objective of this study was to find occupant reference profiles within a sample according to their occupancy and operation schedule by means of SOM for further application on thermal performance building simulation. The method consists on data acquisition, identification of clusters and checking the suitability of the clusters division. On-site data surveys on occupancy and operation schedules were conducted over a year in low-income homes in Southern Brazil. A SOM network was applied to divide the sample into clusters to obtain occupant reference profiles. Different net configurations were performed to check which net configuration shows the best capability to identify clusters. Two reference buildings were configured with the occupant reference profiles found and submitted to computer simulation, from where thermal performance indicators were obtained. A classification network was performed to verify if it was able to classify each profile correctly according to their Target (cluster). Additionally, hypothesis tests were conducted to verify if the clusters characteristics differ from each other. Finally, each cluster was described according to their characteristics.

SCIENTIFIC INNOVATION AND RELEVANCE

We report on occupant reference profiles obtained from actual data and SOM procedures for building simulation purposes. This is a significant topic since occupants strongly influence the performance of buildings. Besides, given the variability of occupant behaviour, to identify different types of occupants can result in more reliable building performance simulation outputs. The profiles will be presented as hourly schedules, which can be useful for other researches or professionals on the development of new buildings or retrofit. As for the object of study (single-family low-income housing) there is a lack of data regarding this type of occupant behaviour in the literature, which is one of the reasons why we understand the content of this study is relevant. In addition, the systematic procedure used to obtain those profiles is new and may be replicated by other researchers in other data samples as well.

PRELIMINARY RESULTS AND CONCLUSIONS

The net configuration with hexagonal mesh and small neighbour distance provided the best results. These two parameters were the ones which had provided greater impact on occupant reference profiles, within the different configurations. The classification analysis step has confirmed the adequacy of the clustering procedure, performed using SOM network, reaching almost 100% accuracy. The descriptive analysis and the hypothesis tests showed that the clusters and occupant reference profiles obtained have important distinctive characteristics. Regarding computer simulation outputs, in general, the thermal performance indicators were much more significant for cooling than for heating. Additionally, reference profiles showed quite different results for these two indicators, some of them being more vulnerable to outdoor conditions (winter or summer).

MAIN REFERENCES

Ballabio, D.; Vasighi, M. (2012). A MATLAB toolbox for self organizing maps and supervised neural networks learning strategies. Chemometrics and Intelligent Laboratory Systems, 111: p. 24-32.

Buttitta, G. ; Turner, W.J.N.; Neu, O.; Finn, D. P. (2019) Development of occupancy-integrated archetypes: Use of data mining clustering techniques to embed occupant behaviour profiles in archetypes. Energy and Buildings 198 84–99. https://doi.org/10.1016/j.enbuild.2019.05.056

D'oca, S.; Hong, T. (2014) A data-mining approach to discover patterns of window opening and closing behaviour in offices. Building and Environment, 82 726–739. doi: 10.1016/j.buildenv.2014.10.021

Mitra, D.; Steinmetz, N.; Chu, Y.; Cetin, K. S. (2020) Typical occupancy profiles and behaviors in residential buildings in the United States. Energy and Buildings 210 109713. https://doi.org/10.1016/j.enbuild.2019.109713.

Qiu, H.; Xu, Y.; Gao, L.; Li, X.; Chi, L. (2015). Multi-stage design space reduction and metamodeling optimization method based on self-organizing maps and fuzzy clustering. Expert Systems With Aplications, p. 12-20.

Keywords: Occupant reference profile, cluster analysis, SOM Network, Building thermal performance, social housing

Paper ID 30129: Contributions and challenges of the demand response ventilation strategy in a single family house under COVID-19 occupancy scenarios

Jingchun Shen*, Xingxing Zhang, Patrick Boyle, Jonn Are Myhren

Energy and Community Buildings, Dalarna University, Falun, Sweden

AIM AND APPROACH

Occupants usually spend a substantial amount of time indoors, where the indoor climate plays a leading role in supporting occupants' health and well-being. Particularly in case of the COVID-19 pandemic outbreak, occupants have to either work or quarantine themselves at home for longer time. Such varying occupancy condition causes issue in air stagnation that may concentrate airborne viruses or dust indoor. When the virus becomes suspended as aerosols, suspending for about a half-hour. Thus, it is critical to investigate the effective means to improve healthy indoor climate.

This paper targets on the demand response ventilation strategy for a single family house in Sweden under the scenario of covid-19 outbreak. The IDA ICE tool is applied to simulate the overall performance of this building. The boundary conditions and the input parameters are set up according to the Swedish Housing Agency's building standards. Different levels of confinement for activities and occupant schedules are proposed as normal operation, 'soft' and 'strict' lock downs, which are further defined in simulation by combining the indoor mechanical ventilation modes.

SCIENTIFIC INNOVATION AND RELEVANCE

In accordance with energy saving and emission reduction mechanical ventilation implementation, the demand controlled ventilation (DCV) provides a feasible solution in assisting such a kind of demand variation adaptations. But how can it face new challenge is the new focus.

To the best knowledge from the authors, there is no research so far in investigating overall impacts on the coupled building performance under different confinement levels due to crisis as COVID-19 pandemic. This is driven by the unknown impact on different key performance indicators (KPIs) of building. So to bridge such research gap, this paper assesses the potential corresponding impacts from aspects of Indoor Air Quality (IAQ), energy consumption, and thermal satisfaction (supporting mental health). It proposes different occupancy conditions that represent different levels of confinement.

This study represents a first step in the understanding of behaviour and performance of the buildings when they confront with this kind of shock.

PRELIMINARY RESULTS AND CONCLUSIONS

Diverse mechanical ventilation strategies are studied using three occupancy schedule. Main results show the DCV generally has remarkable performance in terms of energy saving and emission reduction with the maximum energy saving in total delivered heating energy up to 4210 kWh per year, along with annual electricity saving of 1,084.64 SEK and carbon emission reduction of 1,074.29 kg CO_2e. However due to prolonged occupied hours under altered confinement level, the percentage of total occupant hours with thermal dissatisfaction (>27 °C) would deteriorate from 24% to 45%. By looking into the air age status in the biggest volume room, living space, the mean annual air age differs from original shortest of 2.12 h to the longest of 3.99 h. These infer that the traditional control algorithm would fail towards the challenge in terms of healthy indoor climate. The possible impacting reasons are thermal stratification caused by longer occupation period, fluctuating fresh air supplements resulted from different located controlled rooms, accuracy and dead band inside demand control algorithms etc. It is expected that applicable suggestions from this research can be given in terms of prevention, preparedness, resilience and recovery in relation to COVID-19 and other respiratory infections.

MAIN REFERENCES

[1] Smieszek T., Lazzari G., and Salathé M., "Assessing the Dynamics and Control of Droplet- and Aerosol-Transmitted In¬fluenza Using an Indoor Positioning System," Scientific Reports, vol. 2019/02/18 2019;9(1):2185.

[2] Mendell MJ, Mirer AG, Cheung K, and Tong M., "Respiratory and allergic health effects of dampness, mold, and dampness-related agents: a review of the epidemiologic evidence," Environmental health perspectives, vol. 2011;119(6):748–756.

[3] Bekö G, Clausen G, and Weschler CJ, "Is the use of particle air filtration justified? Costs and benefits of filtration with regard to health effects, building cleaning and occupant productivity," Building and Environment, vol. 2008/10/01/ 2008;43(10):1647–1657.

[4] Kim SW, Ramakrishnan MA, Raynor PC, and Goyal SM, "Effects of humidity and other factors on the generation and sampling of a coronavirus aerosol," Aerobiologia, vol. 2007/12/01 2007;23(4):239–248.

[5] Dietz L, Horve PF, Coil DA, Fretz M, Eisen JA, and Van Den Wymelenberg K, "2019 Novel Coronavirus (COVID-19) Pandemic: Built Environment Considerations To Reduce Transmission," mSystems, vol. 2020;5(2):e00245–20.

[6] CEN, "Indoor environmental input parameters for design and assessment of energy performance of buildings." 2007.

[7] VELUX group, Daylight, Energy and Indoor Climate Basic Book, vol. 3.0 – 2014. 2014.

Keywords: Demand response ventilation, COVID-19, Occupancy scenarios, Single family house

Paper ID 31024: Analysis of multiple building overheating assessment metrics for long-term indoor thermal patterns in 12 Canadian cities

Chang Shu* (1), Abhishek Gaur (2), Lili Ji (1), Abdelaziz Laouadi (2), Michael Lacasse (2), Liangzhu {Leon} Wang (1)

1: Department of Building, Civil and Environmental Engineering, Concordia University, Montreal, 1455 De Maisonneuve, H3G 1M8, Montreal, Quebec, Canada; 2: Construction Research Centre, National Research Council Canada 1200 Montreal Road, K1A 0R6, Ottawa, Ontario, Canada

AIM AND APPROACH

Climate change is a major contributor to extreme heat events, which have been a threat to the health of building occupants. It has been found that the rates of mortality and morbidity are much higher for urban dwellers during prolonged periods of elevated outdoor temperature. However, only limited studies have been conducted on the current overheating problems in buildings of different Canadian cities. This study considered multiple existing building overheating metrics, and the trend of evolution of indoor thermal conditions has been analyzed through linear regression. To permit analyzing the long-term variation of indoor conditions within typical buildings located in different cities across Canada, a series of building simulations were performed using weather station data over a historical period ranging from 1986 to 2016. Twelve (12) cities across Canada were selected for analysis that was located in different climate zones. This study determined the current overheating conditions under the 31-year historical climate conditions in two (2) building types: schools and offices.

SCIENTIFIC INNOVATION AND RELEVANCE

In this paper a review is first completed of existing methods for assessing overheating in buildings based on an examination of existing literature, standards, and codes. Thereafter, these methods are compared through the use of building simulation undertaken over 31 years, the results of which permit discussing the advantages and shortcomings of these methods for the different Canadian cities considered in this study. A combination of the assessment methods was selected to provide a comprehensive description of the heat events as occur in buildings. The method was then applied to evaluate the annual changes in 12 Canadian cities of overheating in four (4) different types of buildings subjected to historical climate data, and as well, climate loads as may arise for climate change. In this study, an answer is provided to the question of how changes in outdoor weather conditions will affect indoor building conditions in the future under climate change as compared to the building subjected to historical climate conditions.

PRELIMINARY RESULTS AND CONCLUSIONS

The simulations capture the effect of climate change on overheating within building spaces when subjected to long-term historical weather and the future projected weather data. The trend of increasing occurrence of overheating the buildings is captured through an assessment of building simulation results under free-running conditions. The annual cooling energy required the summer, and the heating energy consumption in winter are evaluated through separate simulations with the HVAC system activated in both these instances. The building overheating patterns can also be compared between different cities to demonstrate the necessity to conduct localized adjustments of building codes to accommodate the mitigation of overheating effects in buildings for different cities and climate zones.

MAIN REFERENCES

BS EN 15251, 2007: Indoor environmental input parameters for design and assessment of energy performance of buildings addressing indoor air quality, thermal environment, lighting, and acoustics.

Canada, N. R. C. of, 2015: National Building Code of Canada, 2015. National Research Council Canada.

CIBSE, 2013: The limits of thermal comfort: avoiding overheating in European buildings. CIBSE Tm52, 1–25, https://doi.org/10.1017/CBO9781107415324.004.

Gaur, A., Lacasse, M. and Armstrong, M., 2019. Climate Data to Undertake Hygrothermal and Whole Building Simulations Under Projected Climate Change Influences for 11 Canadian Cities. Data, 4(2), p.72.

Gagge, A. P., Fobelets, A. P. and Berglund, L. G., 1986: A standard predictive index of human response to thermal environment. Am. Soc. Heating, Refrig. Air-Conditioning Eng., 709–731.

Laouadi, A., Gaur, A., Lacasse, M.A., Bartko, M. and Armstrong, M., 2020. Development of reference summer weather years for analysis of overheating risk in buildings. Journal of Building Performance Simulation, 13(3), pp.301-319.

Lamothe, F., M. Roy, and S.-É. Racine-Hamel, 2019: Enquête épidémiologique - Vague de chaleur à l'été 2018 à Montréal.

NRCan, 2014: Survey of Commercial and Institutional Energy Use: Buildings 2009. Nat. Resour. Canada.

Keywords: Climate change, heatwave, overheating, weather files, assessment

Paper ID 30283: Impact of modelling thermal phenomena in a high-density controlled environment agriculture (CEA-HD) space

Marie-Hélène Talbot*, Danielle Monfet, Timothé Lalonde, Didier Haillot

École de technologie supérieure, Canada

AIM AND APPROACH

Several interactions need to be considered when integrating high-density controlled environment agriculture (HD-CEA) to buildings. Hydroponic crop production systems are often used in HD-CEA, where the space loads are mainly influenced by the crops and lighting heat gains/losses. During photosynthesis, the crops induce an evaporative cooling effect caused by their transpiration (heat loss), while the lamps inefficiency causes the lighting heat gain. However, three other thermal phenomena, not always modelled, might occur in a HD-CEA space with a hydroponic crop production system. First, the shading of crops decreases the lighting heat gain by converting some radiative energy to latent energy through photosynthesis. Second, there is thermal storage in the leaves that increases with crop growth. Third, thermal storage occurs in the hydroponic crop production system, where the nutritive solution is recirculating in a closed-loop in the space.

To assess the impact of those thermal phenomena on the loads of a HD-CEA space, an energetic and dynamic lettuces model coupled to the space model in TRNSYS, a building performance simulation tool (BPS) tool, is developed. Using the TRNSYS model, the space loads are assessed throughout a production cycle with and without each phenomenon to observe their magnitude.

SCIENTIFIC INNOVATION AND RELEVANCE

The integration of HD-CEA space into urban environment could improve sustainability all over the globe (Al-Kodmany, 2018). In extreme climates, outdoor conditions are not suitable to year-round agriculture and protected agriculture is essential to grow local food (McCartney & Lefsrud, 2018). However, HD-CEA spaces have high energy demands mostly due to sensible cooling and dehumidification requirements (Graamans, Baeza, van den Dobbelsteen, Tsafaras, & Stanghellini, 2018). Modelling this kind of space with the consideration of crops as internal heat gains/losses is fairly new and a consensual modelling approach that uses a common tool is still missing (Waldron, 2018).

The specific contribution of this paper is to identify which simplifications can be made regarding those phenomena when modelling a HD-CEA space. It also quantifies the impact of each phenomena and discusses if detailed modelling of those phenomena is essential. More largely, the findings of this study will improve the guidelines to model HD-CEA space, calibrate HD-CEA model and predict the space loads.

PRELIMINARY RESULTS AND CONCLUSIONS

The detailed loads obtained through this study showed that some phenomena are more influential than others. For the first phenomenon, the results showed that the impact on loads varied for different LED lamps heat gain fractions. For the second phenomenon, the results showed that the thermal storage in the leaves was higher in the second part of the production cycle and had an impact on the loads. The thermal storage in the leaves completely resorbs the need for heating during the dark period, where the lights are off. Similarly, the third phenomenon also has to be considered when estimating the loads of a HD-CEA since at any moment in the production cycle, the thermal storage that occurs in the hydroponic solution of the crop production system during photoperiod completely resorbs the need for heating during the dark period. Thus, the results showed that the two thermal storage phenomena considered increased the cooling loads during the dark period while reducing them during the photoperiod. The results of this study provide a closer examination of the magnitude of the impacts of those phenomena on loads in order to improve the modelling approach of HD-CEA spaces.

MAIN REFERENCES

Al-Kodmany, K. (2018). The vertical farm: A review of developments and implications for the vertical city. Buildings, 8(2). doi: 10.3390/buildings8020024.

Graamans, L., Baeza, E., van den Dobbelsteen, A., Tsafaras, I., & Stanghellini, C. (2018). Plant factories versus greenhouses: Comparison of resource use efficiency. Agricultural Systems, 160(Supplement C), 31-43.

McCartney, L., & Lefsrud, M. (2018). Protected Agriculture in Extreme Environments: A Review of Controlled Environment Agriculture in Tropical, Arid, Polar, and Urban Locations. Applied Engineering in Agriculture, 34(2), 455-473.

Waldron, D. (2018). EVOLUTION OF VERTICAL FARMS AND THE DEVELOPMENT OF A SIMULATION METHODOLOGY. WIT Transactions on Ecology and the Environment, 217, 975-986.

Keywords: Controlled environment agriculture, thermal loads, building modelling

Paper ID 30906: Comparison of different machine learning algorithms in predicting air conditioner operating behaviors in open-plan offices

Shuai Tian* (1), Xin Zhou (1), Xing Shi (2), Xing Jin (1)

1: School of architecture, Southeast University, China, People's Republic of; 2: College of Architecture and Urban Planning, Tongji university, Shanghai, China, People's Republic of

AIM AND APPROACH

Many researchers have realized the AC(air conditioner) operating behavior influence its energy consumption greatly and proposed algorithms and models to reveal this phenomenon. The high performance of prediction and data-based characteristic determines some studies have applied ML(machine learning) methods to the field of occupant behavior. However, the characteristics of different ML algorithms and their applicability are still unclear in this field.

In this study, five ML algorithms, KNN, SVM, logistic regression, random forest and XGBoost, were applied to predict the AC operating behavior in open-plan offices. The analysis was based on the collected data, including indoor environment data, weather data and AC operating behavior, from two open-plan offices in Nanjing, China. Both state-based and action-based modeling approaches were introduced. The main difference between them is state-based approach tries to develop relationship between the AC status and driving factors at a certain moment, while the latter aims to explore when people open or close the AC.

Finally, performance and effectiveness of the five ML models were compared using AC open rate, confusion matrixes and sensitivity of input features. The findings could shed light on the application of ML methods on the AC operating behavior modeling and other occupant behavior researches.

SCIENTIFIC INNOVATION AND RELEVANCE

To figure out the characteristics of different machine learning algorithms and their applicability in the field of the AC behavior prediction, five machine learning algorithms, KNN, SVM, logistic regression, random forest, XGBoost, were applied to research the AC operating behavior.

Two open-plan offices located in different buildings were taken as study cases, which expanded the research object from a single building to multiple buildings. It can also provide a perspective on the universality of the five algorithms.

Both state-based and action-based modeling approaches were presented in this study. Differences of the two modeling approaches in prediction results and the applicability of them were rarely discussed in previous studies using machine learning methods.

Different evaluation indicators were discussed. Sensitivity analysis of input parameters were conducted to evaluate the robustness of different algorithms, AC open rate was compared to assess the accuracy and confusion matrixes was used to estimate the time-by-time AC on-off curves.

The prediction model proposed in this study can helps researchers understand the driving factors for AC operating behavior in open-plan offices and provides reference for the application of machine learning methods on AC operating behavior modeling and other types of occupant behavior studies.

PRELIMINARY RESULTS AND CONCLUSIONS

This research compared different machine learning algorithms in predicting AC operating behaviors in open-plan office. Both state-based and action-based modeling approaches were presented in this study. Different evaluation indicators were discussed to assess the effectiveness of different machine learning methods. The preliminary conclusion are as follows:

1、In winter, the indoor relative humidity of office is the key driving factor on the prediction of the AC state, while outdoor dry-bulb temperature is the most important one in summer.

2、Four important features are proved can satisfy excellent performance in prediction, including indoor dry-bulb temperature, outdoor dry-bulb temperature, indoor relative humidity and outdoor relative humidity.

3、Compared with LR ,KNN and SVM, and XGBoost, random forest model has better performance in predicting AC operating behavior in open-plan offices.

MAIN REFERENCES

Lu Y, Meng L(2020).A simplified prediction model for energy use of air conditioner in residential buildings based on monitoring data from the cloud platform. Sustainable Cities and Society, 60(2020), 102194.

Mo H, Sun H, Liu J, et al(2019).Developing window behavior models for residential buildings using XGBoost algorithm. Energy&Buildings, 205(2019), 109564

Peng Y, Rysanek A, Nagy Z, et al(2018).Using machine learning techniques for occupancy-prediction-based cooling control in office buildings. Applied Energy, 211: 1343-1358.

Piselli C, Pisello A(2019).Occupant behavior long-term continuous monitoring integrated to prediction models: Impact on office building energy performance. Energy, 176: 667-681.

Wang W, Chen J, Hong T, et al(2018). Occupancy prediction through Markov based feedback recurrent neural network (M-FRNN) algorithm with WiFi probe technology. Building and Environment, 138: 160-170.

Keywords: air conditioner operating behavior, machine learning, state-based modeling, action-based modeling, evaluation indicators

Paper ID 11159: Use of energy simulations to support informed decision-making of key stakeholders

Tomáš Vácha*, Karolína Tomešová, Lukáš Ferkl, Kateřina Mrkvičková

Czech Technical University in Prague, Czech Republic

Energy simulations have potential to help key stakeholders to make informed decisions towards investments in energy efficient solutions in buildings, but there is a question of when, how and to whom to present the information. When is a full simulation needed and when to use a pre-calculated example or model? How to use simulations and models in the context of participatory design framework? And how to present the results of simulations in the most comprehensive way?

The workshop will engage participants in sharing their experience with the role of energy simulations in the design process and the impact they can have on decision-making of key stakeholders. Experience from research projects and social sciences will be presented to inform the discussion further. Participants will co-design recommendations for the use of energy simulations and modelling in the participatory design process. We will use one specific building as an example for the workshop.

In the first non-interactive part, the topic of building simulation and its relationship with decision making will be introduced together with examples of construction projects and key stakeholder groups. This will be followed by co-design session. The participants will be split into groups each focusing on specific typology of project and simulation methods. They will share their experience and identify benefits and barriers of introducing simulation and precalculated models into the decision making/design process in the specific scenario. Finally, the participants will suggest general principles and recommendations for use of simulations and models to inform stakeholders decision-making.

Agenda:

Introduction of the topic and project examples (30 min)

Introduction of the participative design process and introduction of participants and their experience (30 min)

First round of participative design – discussing use of simulations and models for specific case – benefits, barriers, goals (60 min)

Break (30 min)

Second round of participative design – co-design of the approach for specific case – methods to be used in individual phases of the project (60 min)

Formulation of general principles and recommendations (45 min)

Conclusions, take-aways (15 min)

The worskhop is targeted at all stakeholders involved in the decision-making process in construction projects – engineers, architects, public bodies, investors and building modellers.

Keywords: stakeholder engagement, building simulations and models, informed decision making, participatory design

Paper ID 30916: Evaluating cost-effective retrofit at the stock level; the case of Orkney Islands

Paraskevi Vatougiou*, David P. Jenkins, Peter McCallum

Heriot Watt University, United Kingdom

AIM AND APPROACH

This work focuses on evaluating the economic viability of energy efficiency measures at the regional level using a bottom-up housing stock energy model applied to Orkney Islands (Scotland), based on some 11,000 unique properties. Deriving information from Energy Performance Certificate (EPC) datasets (or equivalent) and accessing GIS (relating to geometry) data, the model automatically generates house archetypes suitable to be studied using a dynamic simulation engine, which in the case of this study is EnergyPlus. Through the consideration of multiple retrofit solutions (fabric upgrade, heating system upgrade and renewable installations), the aim of this work is to match a suitable retrofit package to the different house archetypes using both economic and environmental indicators. The payback period, in particular, is calculated for the different energy efficiency measures accounting for initial investment costs (including labour, construction materials, installations, etc.) and energy saving costs, the latter also reflecting projections for electricity price rates. One objective of the study is to prioritise interventions based on the minimization of payback period and identify whether those solutions result in carbon savings targets being met.

SCIENTIFIC INNOVATION AND RELEVANCE

The consideration of building retrofit to reduce carbon emissions and energy consumption is a well-documented research area. However, the cost-effectiveness and prioritisation of packaged retrofit solutions has yet to be investigated, this being even more crucial for off-gas grid and especially fuel poor regions found at the centre of the debate due to their limited access to alternative retrofit options. EPCs, being the official UK Government's tool to inform homeowners regarding potential retrofit solutions, provide indicative investment costs and typical energy savings for assumed usage of a home. Nevertheless, evidence exists that both the economic benefits and investment costs resulting from a particular energy efficiency improvement is subject to significant variation depending on the characteristics of the house itself, location (urban, rural), climatic conditions and occupant behaviour. The latter, in particular, is treated by the model in this paper using a dedicated calibration procedure based on analytical assessments of smart meter records, this being another novelty of the paper compared to the standard occupancy and heating patterns of EPC assessments. With the model accounting for demand-side flexibility, the paper attempts to inform investors regarding the real benefit they can get from new technologies especially when combined with suitable electricity tariffs.

PRELIMINARY RESULTS AND CONCLUSIONS

Through the study of a relatively large number of houses presenting a wide range in built forms, sizes and age of construction (with the latter resulting in different insulation levels and states of repair), the paper ranks energy efficiency measures following a cost-effective approach. The impact of retrofit packaged solutions on carbon emissions is estimated at individual house level and at the stock level; the analysis gives an indication regarding the extent to which retrofit can contribute to transforming Orkney into a zero-carbon region. Preliminary results show that there might be significant variation of both energy and carbon savings resulting from the application of very similar retrofit packages across the modeled stock. In an attempt to inform policy and support retrofit investments, the discussion then focuses on the robustness of current EPCs and highlights the fact that following a universal approach should not be the case when it comes to retrofit, thus indicating that EPCs should be only used as a preliminary indication of retrofit options and their associated profitability.

MAIN REFERENCES

Ben, H. and Steemers, K., 2020. Modelling energy retrofit using household archetypes. Energy and Buildings, 224, 110224.

BRE, 2014. SAP 2012 The Government ' s Standard Assessment Procedure for Energy Rating of Dwellings. Garston, Watford: Building Research Establishment.

Chen, Y., Hong, T., and Piette, M.A., 2017. Automatic generation and simulation of urban building energy models based on city datasets for city-scale building retrofit analysis. Applied Energy, 205 (August), 323–335.

Filippi Oberegger, U., Pernetti, R., and Lollini, R., 2020. Bottom-up building stock retrofit based on levelized cost of saved energy. Energy and Buildings, 210, 109757.

Gupta, R. and Gregg, M., 2018. Targeting and modelling urban energy retrofits using a city-scale energy mapping approach. Journal of Cleaner Production, 174, 401–412.

Shaikh, P.H., Shaikh, F., Sahito, A.A., Uqaili, M.A., and Umrani, Z., 2017. An Overview of the Challenges for Cost-Effective and Energy-Efficient Retrofits of the Existing Building Stock. Cost-Effective Energy Efficient Building Retrofitting: Materials, Technologies, Optimization and Case Studies. Elsevier Ltd.

Keywords: Building retrofit, Cost-effectiveness, Housing stock

Paper ID 30674: Measurement of hourly urban albedo of building façades with different reflective directional characteristics by a simple urban model

Jihui Yuan* (1), Craig Farnham (2), Kazuo Emura (2)

1: Toyohashi University of Technology, Japan; 2: Osaka City University, Japan

AIM AND APPROACH

In order to mitigate urban heat island (UHI) phenomenon, diffuse highly reflective (DHR) materials are usually applied to building facades. In recent years, retro-reflective (RR) material has been recommended to replace the DHR materials for application of building façades to mitigate the UHI phenomenon by scholars worldwide. However, there is still little research on evaluation of urban albedo of DHR building façade and RR building façade using model experiments.

Thus, this study used a simple urban model with different urban aspect ratios (UARs) to evaluate the hourly urban albedo of building facades with different reflective directional characteristics including DHR and RR façades.

SCIENTIFIC INNOVATION AND RELEVANCE

Compared to the previous research method that usually uses the radiation principle and form factor method for evaluating the urban albedo, the method used in this paper can measure the urban albedo of building facades with different reflective directional characteristics by changing the building façade materials of urban model more easily.

PRELIMINARY RESULTS AND CONCLUSIONS

Through the measurement of simple urban model with DHR and RR building facades, it is shown that the hourly urban albedo of RR building façade is higher than that of DHR building façade. When the sun altitude is increased, the change in hourly urban albedo of RR building façade is larger than that of DHR building façade. In addition, it is also shown that the hourly urban albedo of RR building façade is the highest at approximately 3pm on August 31st in this study.

MAIN REFERENCES

・H. Akbari, A. Gholizadeh Touchaei, 2014. Modeling and labeling heterogeneous directional reflective roofing materials, Sol Energy mater Sol Cells, 124, 192-210.

・Y. Qin, J. Liang, K. Tan, F. Li. Experimental Study the Albedo of Urban Canyon Prototype with Reflective Pavements (Streets), Advances in Geoscience, 2019, Vol.3, Issue 1, DOI: http://dx.doi.org/10.18686/ag.v3i1.1134

・J. Yuan, K. Emura, C. Farnham, H. Sakai. Application of glass beads as retro-reflective facades for urban heat island mitigation: Experimental investigation and simulation analysis, Building and Environment, 2016, Vol.105, pp.140-152

・F. Rossi, B. Castellani, A. Presciutti, E. Morini, M. Filipponi, A. Nicolini, M. Santamouris. Retroreflective façades for urban heat island mitigation: Experimental investigation and energy evaluations, Applied Energy, 2015, Vol.145, pp.8-20

・R. Levinson, S. Chen, J. Slack, H. Goudey, T. Harima, P. Berdahl. Design, characterization, and fabrication of solar-retroreflective cool-wall materials, Solar Energy Materials and Solar Cells, 2020, Vol.206, 110117

Keywords: Urban heat island, Hourly urban albedo, Urban model, Retro-reflective façade, Diffuse highly reflective façade

Acknowledgements

Thanks to our sponsors and endorsers

SPONSORS PLATINUM

DAIKIN

SWECO

VIESSMANN

SPONSORS GOLD

interalu SMART CEILINGS

daidalos peutz

boydens engineering part of Sweco

EXHIBITORS

vabi

CREATING A CLIMATE FOR GROWTH PRIVA

UPGRADE ESTATE

IES

jaga CLIMATE DESIGNERS

EQUA SIMULATION TECHNOLOGY GROUP

SUPPORTERS

NATHAN driven by knowledge

ENDORSING ORGANISATIONS

AGORIA

Atic

ehpa

EnergyVille

flux50

HPT

ode

ORI

REHVA

Techlink

THE SHIFT

Vlaamse Confederatie Bouw

WITH THE SUPPORT OF

fwo

Clean the air from
Viruses and bacteria
Fine particles of dust
Odors and allergens

Transforming society together
SWECO
boydens engineering
part of Sweco
Sweco shapes the sustainable cities and societies of tomorrow. Our work results in sustainable buildings, efficient infrastructure and innovative industrial and energy projects. As Europe's leading architecture and engineering consultancy, we are always one step ahead of evolution. Together with our customers, we take on high-quality engineering projects in both the public and private sector.
Interested in our projects and services? Discover more on www.swecobelgium.be.

Committees

Local Organising Committee

- Lieve Helsen, Conference Chair, KU Leuven / EnergyVille
- Wim Boydens, Conference Chair, boydens engineering (part of sweco) / Ghent University
- Dirk Saelens, Scientific Chair, KU Leuven / EnergyVille & President of IBPSA-NVL
- Jelle Laverge, Scientific Chair, Ghent University
- Elisa Van Kenhove, Future Generation Chair, Ghent University
- Friedl Decock, Future Generation Chair, daidalos peutz
- Marc Delghust, Technical Expert, Ghent University
- Wim Plokker, On-line Track Task Force, Vabi
- Reinier Maas, On-line Track Task Force, Actiflow

With the organisational support of the KU Leuven Conference Office and the Visit Bruges Convention Bureau

Executive Scientific Committee

- Prof. Ian Beausoleil-Morrison
- Prof. Wim Boydens
- Prof. Vincenzo Corrado
- Dru Crawley
- Prof. Pieter de Wilde
- Joseph Deringer
- Philip Haves
- Prof. Jan Hensen
- Prof. Christina Hopfe
- Erik Kolderup
- Prof. Roberto Lamberts
- Prof. Ardeshir Mahdavi
- Prof. Dirk Müller
- Prof. Yiqun Pan
- Prof. Cheol-Soo Park
- Wim Plokker
- Prof. Yoshiyuki Shimoda
- Prof. Veronica Soebarto
- Prof. Paul Strachan
- Prof. Christoph van Treeck
- Dr. Michael Wetter
- Prof. Monika Woloszyn
- Prof. Jelena Srebric
- Prof. Da Yan

WITH THE SUPPORT OF

fwo

DEPARTMENT OF ENERGY · UNITED STATES OF AMERICA

BRUGGE
VISIT BRUGES CONVENTION BUREAU

Thank you for standing tall with us and your contribution to the BS2021 International IBPSA Building Simulation Conference!

www.bs2021.org
mailto:info@bs2021.org